Molecular Mechanisms
of Protein Biosynthesis

MOLECULAR BIOLOGY

An International Series of Monographs and Textbooks

Editors: BERNARD HORECKER, NATHAN O. KAPLAN, JULIUS MARMUR, AND HAROLD A. SCHERAGA

A complete list of titles in this series appears at the end of this volume.

Molecular Mechanisms of Protein Biosynthesis

Edited by

HERBERT WEISSBACH

SIDNEY PESTKA

Roche Institute of Molecular Biology
Nutley, New Jersey

ACADEMIC PRESS   New York   San Francisco   London   1977
A Subsidiary of Harcourt Brace Jovanovich, Publishers

ACADEMIC PRESS, INC.
111 Fifth Avenue, New York, New York 10003

United Kingdom Edition published by
ACADEMIC PRESS, INC. (LONDON) LTD.
24/28 Oval Road, London NW1

Library of Congress Cataloging in Publication Data

Main entry under title:

Molecular mechanisms of protein biosynthesis.

(Molecular biology, an international series of
monographs and textbooks)
    Includes bibliographies and index.
    1.   Protein biosynthesis.   2.   Molecular biology.
I.   Weissbach, Herbert.   II.   Pestka, Sidney.   III.   Se-
ries.   [DNLM: 1.   Genetics, Biochemical–Congresses.
W3 I322   1976m / QH426 I103m 1976]
QP551.M62   574.1'9296   76-46464
ISBN 0–12–744250–2

PRINTED IN THE UNITED STATES OF AMERICA

Contents

3 Primary Structure and Three-Dimensional Arrangement of Proteins within the *Escherichia coli* Ribosome

G. STÖFFLER and H. G. WITTMANN

4 Affinity Labeling of Ribosomes

MARIA PELLEGRINI and CHARLES R. CANTOR

5 Initiation of Messenger RNA Translation into Protein and Some Aspects of Its Regulation

MICHEL REVEL

6 Factors Involved in the Transfer of Aminoacyl-tRNA to the Ribosome

DAVID L. MILLER and HERBERT WEISSBACH

7 Translocation

NATHAN BROT

8 Peptide Bond Formation

R. J. HARRIS and S. PESTKA

9 Peptide Chain Termination

C. THOMAS CASKEY

10 Inhibitors of Protein Synthesis

SIDNEY PESTKA

11 Messenger RNA and Its Translation

DAVID A. SHAFRITZ

12 DNA-Dependent Cell-Free Protein Synthesis

BENOIT de CROMBRUGGHE

13 Genetics of the Translational Apparatus

ISSAR SMITH

List of Contributors

Numbers in parentheses indicate the pages on which the authors' contributions begin.

Nathan Brot (375), Roche Institute of Molecular Biology, Nutley, New Jersey

Charles R. Cantor (203), Department of Chemistry and Biological Sciences, Columbia University, New York, New York

C. Thomas Caskey (443), Section of Medical Genetics, Departments of Medicine and Biochemistry, Kleberg Center for Human Genetics, Baylor College of Medicine, Houston, Texas

Benoit de Crombrugghe (603), Laboratory of Molecular Biology, National Cancer Institute, National Institutes of Health, Bethesda, Maryland

R. J. Harris (413), School of Pharmacy, South Australian Institute of Technology, Adelaide, South Australia

C. G. Kurland (81), The Wallenberg Laboratory, University of Uppsala, Uppsala, Sweden

Fritz Lipmann (1), The Rockefeller University, New York, New York

David L. Miller (323), Roche Institute of Molecular Biology, Nutley, New Jersey

James Ofengand (7), Roche Institute of Molecular Biology, Nutley, New Jersey

Maria Pellegrini (203), Department of Molecular Biology, University of California at Irvine, Irvine, California

Sidney Pestka (413,467), Roche Institute of Molecular Biology, Nutley, New Jersey

Michel Revel (245), Department of Biochemistry, The Weizmann Institute of Science, Rehovot, Israel

David A. Shafritz (555), Departments of Medicine and Cell Biology, and the Liver Research Center, Albert Einstein College of Medicine, Bronx, New York

Issar Smith (627), Department of Microbiology, The Public Health Research Institute of the City of New York, Inc., New York, New York

G. Stöffler (117), Max-Planck-Institut für Molekulare Genetik, Berlin-Dahlem, Germany

Herbert Weissbach (323), Roche Institute of Molecular Biology, Nutley, New Jersey

H. G. Wittmann (117), Max-Planck-Institut für Molekulare Genetik, Berlin-Dahlem, Germany

Preface

In little more than a decade since the elucidation of the genetic code, it has been possible to obtain considerable insight into the process of protein synthesis. There are, of course, many aspects of this problem still unexplained, but yet enough information is now available to warrant publication of this volume. We hope that the reader will be made aware of those areas that are still in a state of flux, and where relatively little information is yet available. As the reader can see from the Table of Contents, a wide scope of topics is covered. Transfer RNA, the initiation, elongation, and termination processes, ribosome structure and function, mRNA translation, and DNA-directed *in vitro* protein synthesis are included. A chapter on inhibitors of protein synthesis and one on the genetics of the components are also included. An attempt has been made to encompass both pro- and eukaryotic systems. Where possible, a uniform nomenclature for the various soluble factors is used throughout. This should minimize the difficulties in reading the literature in this area in which diverse designations permeate the published reports. Unfortunately, in the area of eukaryote initiation, this is not yet possible.

It is hoped that this volume will be of value to a wide audience, not only to those specializing in the field of protein synthesis, but to students and nonspecialists as well. To achieve this goal introductions to the chapters should provide an orientation and a guide to the material that follows. The subjects, however, are covered in depth so that the reader should have a basic knowledge of biochemistry and molecular biology. It should be stressed that emphasis is placed on studies at the molecular level in cell-free systems. Thus, in general, protein synthesis in intact cells, including the related aspects of hormonal action, protein turnover, and posttranslational modification, has not been included in this volume.

We would like to express our gratitude to the authors for the time and effort expended in preparing their contributions to this volume. Their cooperation in this endeavor was essential and was of great help to us. In addition, we are grateful to Janet M. Hill, Bernadette McCaskey, Diane Popowicz and Tanya Schulz for their technical help in editing and processing the individual chapters. We would also like to thank the staff of Academic Press.

Herbert Weissbach
Sidney Pestka

Twenty Years of Molecular Biology

FRITZ LIPMANN

For twenty years now we have been familiar with the double helix structure of DNA. As is well known, its follow-up led to the chemical explanation of genetic replication. The double helical structure has been interpreted as being due to a pairing by hydrogen bonding between the four side chain bases, guanosine (G) ⟵⟶ cytosine (C), and adenine (A) ⟵⟶ thymidine (T) [or uridine (U) in RNA]. In the five to ten years after its discovery, it became clear that the base-pairing mechanism represented the driving force in the whole sequence of reactions now known under the name of genetic information transfer. To summarize briefly, three main stages are distinguished: two transcriptions—(1) replication of DNA in cell division and (2) transcription of DNA into an RNA—the latter of which continues largely into the third phase, the translation of the four-base code in the "messenger" RNA into the twenty amino acid expression in the polypeptide chain of proteins. In a rather complex manner, this translation also turns out to involve a base-pairing mechanism of A→U and G→C in triplet clusters between the messenger RNA and amino acid transfer RNA.

This chemical definition of the mechanism of replication, which had remained for so long a mysterious ability of living organisms, has changed our way of thinking rather dramatically. However, so much attention has since converged on the further filling in of the details of this information transfer, that the novelty and excitement originally generated by the transition has now rather worn off. To some extent it was expressed in the preference to call what we are dealing with molecular biology rather than biochemistry. Therefore, I thought it might be worthwhile to preface this book on protein synthesis by reviewing more soberly what actually is the essential feature of this change, which one might consider, rather, a change of attitude, and how it came about.

On an earlier occasion, at a meeting at Rutgers University on Organizational Biosynthesis, I entitled my introduction "Molecular Technology." Although not quite clearly expressed in the talk, the main reason for giving it this title was that I had come to the conclusion that molecular biology is essentially a molecular technology. By this, I meant a chemical

1

technology where the working parts have the dimensions of angstroms and microns rather than of the centimeters or meters as used in man-designed technology. It was realized already then that to obtain a "miniaturization" of chemical machines to the level of the size of the cell, one has to use molecules as building blocks. And to do this, it appears one is forced to work with "soft" systems rather than with the hardware system of human technology. It would be shortsighted to attribute this change of attitude exclusively to the discovery of the double helix. It was fortunate that it went parallel with the general development of X-ray crystallography, of which it was a part, and of electron microscopy. One could now see the intracellular structures of which the biomolecular machines are composed.

I think a consequence of the biomolecular revolution is that it led biochemists to accept the essentially technological aspect of biochemical mechanisms, which has had a most salutary influence on recent developments in the biochemical domain. This is in contrast to an attitude that I remember from the early days of my studies in medicine, when we were strongly warned against using "anthropomorphic" explanations for biological reactions, an attitude which we have now long thrown away and have fared much better since doing so.

With this background, one might once more have a look at the actual events during transcription and translation where the hydrogen bonding through base pairing is the chemical means of information transmission. It is rather straightforward in the transcription sector between polynucleotides: the synthesis of a polynucleotide occurs on the template and forms an antiparallel image of the template, in that the polymerase or replicase finds first, by base pairing, a complementary preactivated nucleotide triphosphate to be added next. This is then sealed to the 3'-hydroxyl terminal of the growing polynucleotide chain by displacement of a pyrophosphate from a 5'-triphosphate annex of this particular nucleotide triphosphate. The same sequence repeats until the chain is terminated.

From this example, one may generalize that in all cases the synthesis of a polynucleotide image occurs by combination of (1) a placing mechanism by hydrogen bonding of the newly to be added part of the chain which (2) through use of energy available in the nucleotide triphosphate annex when applied by pyrophosphoryl displacement for group transfer becomes the new terminal of the extending chain.

It was, indeed, a farsighted prediction of Crick to propose that an analogous mechanism of base pairing also would have to apply in the translation step from polynucleotide sequence into the amino acid sequence of polypeptides. As is well known, he did propose that for this purpose the amino acid would be specifically sealed to an "adapter," a polynucleotide

SECOND LETTER

FIRST LETTER		U	C	A	G	THIRD LETTER
U		UUU UUC } Phe UUA UUG } Leu	UCU UCC UCA UCG } Ser	UAU UAC } Tyr UAA OCHRE UAG AMBER	UGU UGC } Cys UGA AMBER UGG Tryp	U C A G
C		CUU CUC CUA CUG } Leu	CCU CCC CCA CCG } Pro	CAU CAC } His CAA CAG } GluN	CGU CGC CGA CGG } Arg	U C A G
A		AUU AUC AUA } Ileu AUG Met	ACU ACC ACA ACG } Thr	AAU AAC } AspN AAA AAG } Lys	AGU AGC } Ser AGA AGG } Arg	U C A G
G		GUU GUC GUA GUG } Val	GCU GCC GCA GCG } Ala	GAU GAC } Asp GAA GAG } Glu	GGU GGC GGA GGG } Gly	U C A G

GENETIC CODE

Fig. 1 Code triplets for amino acids from Crick's introduction to the Cold Spring Harbor Symposium, Vol. XXI, 1966; slightly revised, ochre and amber stand for termination codons.

carrier of small chain length that contained amino acid-specific triplet sequences, and that the triplet sequence would then find its place on the polynucleotide template. This template we now know to be the messenger RNA (mRNA) translated from chromosomal DNA to guide the translation reaction, and the Crick proposition became reality when the mechanism of amino acid activation by transfer to a tRNA was revealed. The details of this translation process are one of the topics of this volume and thus omitted here.

I would like to expand, however, on one aspect of the code problem, namely, the early prediction of the redundancy, or rather degeneracy as physicists like to call it, of the code with regard to several nucleotide combinations (isotriplets) relating to the same amino acid. This is essentially the result of the presence of twenty amino acids in all complete proteins. The number twenty made a doublet code insufficient, as it would only allow for 4×4, or 16 base combinations. Therefore, the need for a triplet code was visualized early, and eventually when the triplet assignment was completed, it turned out, as described in the Cold Spring Harbor Symposium of 1966, that all the 64 triplets available were occupied by placing of amino acids or other functions. In other words, all of the possible $4 \times 4 \times 4 = 64$ base combinations were being made use of. As shown in Fig. 1, with only 20 amino acids available for this many combinations, nearly all amino acids have several different code triplets to join with, all of the ones

Amino acid	Number of codons	Codons		Amino acid	Number of codons	Codons
Leu	6	UU A,G CU U,C,A,G		Ala	4	GC U,C,A,G
Arg	6	AG A,G CG U,C,A,G		Thr	4	AC U,C,A,G
				Val	4	GU U,C,A,G
Ser	6	AG U,C UC U,C,A,G		Gly	4	GG U,C,A,G
				Pro	4	CC U,C,A,G

Fig. 2 Amino acids with triplet redundancy of 6 and 4.

specifying for one amino acid being handled within a particular cell by one specific aminoacyl-tRNA synthetase.

I found it interesting to order the amino acids in sequence of the abundance of redundancy of their code triplets; this is shown in Figs. 2 and 3. It appears that there are three amino acids with six, and five with four, different triplets. There is one amino acid that has three triplets; the largest number are those that have two. At the other end of the scale there are two amino acids that appear to have only one single triplet, namely, methionine and tryptophan. It may be worth realizing that these two amino acids are chemically the most complex compared with a general organic simplicity in the side chain structure of other amino acids; they sometimes

Amino acid	Number of codons	Codons		Amino acid	Number of codons	Codons
Ile	3	AU U,C,A		Cys	2	UG U,C
Asp	2	GA U,C		His	2	CA U,C
Asn	2	AA U,C		Lys	2	AA A,G
Glu	2	GA A,G		Met	1	AUG
Gln	2	CA A,G		Trp	1	UGG
Phe	2	UU U,C				
Tyr	2	UA U,C		Termination I, II		{ UAG { UAA { UGA

Fig. 3 Amino acids with triplet redundancy of 3 to 1.

are absent in some primitive proteins as, for example, in the apoferridoxin of *Clostridium butyricum*. Methionine is a complex case since there are, for initial and internal addition, two different methionine tRNA's both carrying the same triplet. In some ways it seems necessary, but nevertheless I find it rather remarkable, that all the 64 base permutations of the triplet code are made use of; three of these triplets are used for termination, and connect, not with tRNA's, but rather with proteins, the termination factors R_1 and R_2.

It appears, then, that protein synthesis is due to a fully automated machinery, with the 3×10^6 dalton ribosome being the reactor for mRNA combining with aminoacyl-tRNA's to start, continue, and terminate the linear production of a polypeptide chain. After completion, this chain folds, mostly spontaneously, into a rather complex structure of, for example, an enzyme. Initiation, elongation, and termination are the three phases most thoroughly explored in protein synthesis. They are, however, common to all polymerizations of mixed subunits where their sequence is the essential determinant of function, including proper start and termination, the latter being single events in contrast to the cycles many times repeated in the elongation phase.

1

tRNA and Aminoacyl-tRNA Synthetases

JAMES OFENGAND

I. Introduction

The *raison d'etre* for tRNA and aminoacyl-tRNA synthetases in the cell was first described by Francis Crick in 1955 in a privately circulated paper, and subsequently published in brief form (Crick, 1957) (for an excellent discussion, see Hoagland, 1960). In this paper, Crick pointed out the inability of a set of discrete bases in a polynucleotide to select out by physicochemical forces one amino acid from among the 19 others. At that time it was already assumed that the sequence of amino acids in protein was unique and determined by the nucleotide sequence of the template RNA, which in turn was derived from the sequence of bases in DNA. However, the *mechanism* by which a particular amino acid was selected by its code word of nucleotides was totally obscure. In order to explain this, Crick proposed the existence of an adapter oligonucleotide estimated to be at least 3 and not more than 6 bases long which would be enzymatically attached to a given amino acid, and by base pairing with the appropriate codon in the template would serve to order amino acids in their correct sequence for polymerization. The linkage between amino acid and adapter was also supposed to provide the energy for polymerization. Except for a slight error in the size of the adapter, Crick's hypothesis proposed at least two years before any hint of the existence of tRNA was available has proved to be startlingly accurate in its description of the role of tRNA and aminoacyl-tRNA synthetase in protein synthesis.

There are three essential elements to the adapter hypothesis. First, there must be at least twenty different and specific adapters, one for each amino acid. Second, there must exist a specific enzymatic mechanism, and therefore a specific enzyme, for linking each amino acid to its cognate adapter. Third, the recognition between template and adapter must be accurate. Antiparallel base pairing between RNA chains was the mechanism proposed by Crick. This aspect will be discussed below (Section VI).

By 1958, the first two of these properties had been shown by workers in the field of protein synthesis to reside in transfer RNA, or soluble RNA as it was called then, and in amino acid activating enzymes, respectively. First, the amount of amino acid linked to tRNA was shown to be characteristic and independent of other amino acids (Berg and Ofengand, 1958).

Next, these sites were found to be at the 3′-terminal adenosine and there-fore on separate molecules (Zachau *et al.*, 1958), and in the following year the site of acylation was identified as the 2′- or 3′-hydroxyl (Preiss *et al.*, 1959; Hecht *et al.*, 1959). Amino acid-activating enzymes, first described in 1955 (Hoagland, 1955), were known to be amino acid-specific and to generate enzyme-bound aminoacyladenylates (Berg, 1956; Hoagland *et al.*, 1956; DeMoss and Novelli, 1956; Davie *et al.*, 1956), but it was not until the natural acceptor, tRNA, was discovered in 1958 that it was pos-sible to show that the same enzyme involved in amino acid activation also catalyzed transfer of the amino acid to tRNA (Berg and Ofengand, 1958; Schweet *et al.*, 1958). At about the same time, it was recognized that the amino acid-tRNA bond was labile and thus had at least in part conserved the energy of activation supplied by ATP. This was subsequently con-firmed by more careful measurements (see Section II,A). Thus, all the basic requirements of Crick's adapter were met by tRNA and its cognate amino acid synthetase. The final piece of evidence, namely, that the tem-plate recognized the nature of the amino acid only through its adapter, was not to come until 1962 (Section V).

The fact that tRNA was much larger than Crick had envisioned in 1955 was already appreciated in 1959 (Preiss *et al.*, 1959), and as we shall see below this is largely due to the particular mechanism evolved by cells for the utilization of the adapter–amino acid complex. Moreover, as is often the case, the cell has taken advantage of the existence of a rather complex adapter molecule and used it for other synthetic as well as regulatory functions. These will be discussed in Section X.

In the following pages, I attempt to survey the current state of knowledge about tRNA and aminoacyl-tRNA synthetases. This would be an im-possible task in the allotted space were it not for the existence of numerous recent reviews dealing with various aspects of the subject. Therefore, this survey is mostly confined to newer topics and the reader is referred to these reviews for a more complete coverage of the earlier literature.

General aspects of aminoacyl-tRNA synthetases have been reviewed repeatedly (Mehler and Chakraburtty, 1971; Loftfield, 1971; Chapeville, 1972), most recently by Kisselev and Favorova (1974) and Söll and Schimmel (1974). In addition, the mechanism of aminoacylation has been discussed by Loftfield (1972) and interspecies acylation reactions by Ja-cobson (1971). Synthetases and tRNA in plants have been recently re-viewed (Lea and Norris, 1972).

General reviews on tRNA (Gauss *et al.*, 1971; Zachau, 1972), on tRNA tertiary structure (Cramer and Gauss, 1972), on minor bases in tRNA (Hall, 1971; Söll, 1971; Nishimura, 1972), and on the recognition between tRNA and synthetase (Chambers, 1971) are available. The primary structure of tRNA's has been reviewed more recently (Dirheimer *et al.*, 1972; Barrell and Clark, 1974), as well as specialized aspects of tRNA research (Smith, 1972; Littauer and Inouye, 1973).

To conserve space, references are confined either to the most recent or to, in my opinion, the best documented examples of the point in question in the expectation that the most recent articles will also refer to the body of work that preceded it. I apologize for any failure to refer specifically to much excellent work.

II. The Aminoacylation Reaction

A. Stoichiometry and Energetics

The aminoacylation of tRNA is conventionally represented according to Eqs. (1)–(3).*

$$AA^i + ATP + E^i \rightleftharpoons E^i \cdot AA^i\text{-}AMP + PP \qquad (1)$$
$$E^i \cdot AA\text{-}AMP + tRNA^i \rightleftharpoons AA^i\text{-}tRNA^i + AMP + E^i \qquad (2)$$
$$AA^i + ATP + tRNA^i \rightleftharpoons AA^i\text{-}tRNA^i + AMP + PP \qquad (3)$$

In the first step, amino acids are activated by formation of an enzyme-bound aminoacyladenylate (Fig. 1) with liberation of pyrophosphate. The enzyme-bound adenylate can be readily isolated by gel filtration (Allende and Allende, 1971). In the second step, the amino acid is transferred to tRNA with liberation of AMP and regeneration of the free enzyme. The stoichiometry of the overall reaction (3) is certainly correct and the reaction is definitely reversible (Berg *et al.*, 1961). Moreover, the AA-tRNA formed is at as high an energy level as ATP since the overall equilibrium constant for the reaction as written is less than 1, being 0.32 for valyl-tRNA (Berg *et al.*, 1961), 0.37 for threonyl-tRNA (Leahy *et al.*, 1960), and 0.25 for arginyl-tRNA (Papas and Peterkofsky, 1972). The high group transfer potential indicated by these measurements finds its expression in the ready reaction with nucleophiles such as the amino group of amino

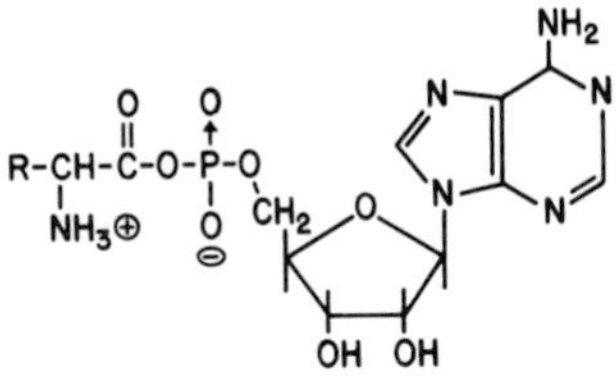

Fig. 1 Structure of aminoacyladenylate. R, amino acid side chain.

* AA = amino acid or aminoacyl; E = activating enzyme; AA-AMP = aminoacyladenylate, superscript i denotes a particular amino acid residue; · denotes a noncovalent interaction.

acids in polymerization reactions and the ease of reaction with amines such as hydroxylamine at neutral pH. The amino acid-tRNA bond is also very labile in mildly alkaline solution, although the relative lability varies widely depending on the amino acid involved (Chousterman *et al.*, 1966) and it is possible to vary the rate of hydrolysis under a given set of conditions by more than 30-fold by appropriate choice of amino acid. The stability of the bond is related both to the ionization constant for the amino group of the amino acid (Gatica *et al.*, 1966; Hentzen *et al.*, 1972) and to the presence of a vicinal hydroxyl group on the ribose, since when this group is replaced by H (Sprinzl and Cramer, 1973) or when the C-2' to C-3' bond is broken so that the hydroxyl group is no longer vicinal (Ofengand and Chen, 1972), the rate of hydrolysis is decreased. The remainder of the tRNA molecule has no influence on stability (Strickland and Jacobson, 1972). Additional discussion of properties of this ester bond can be found in Zachau and Feldmann (1965) and Schuber and Pinck (1974).

B. Structure of Aminoacyl-tRNA

The structure of the product, originally deduced by periodate inactivation and protection experiments (Preiss *et al.*, 1959; Hecht *et al.*, 1959) is shown in Fig. 2. For some years, efforts were directed toward attempts to determine which hydroxyl was enzymatically esterified and conflicting reports peppered the literature. This research effort ended in 1965–1966 when it was shown by consideration of suitable model compounds that at neutral pH the amino acid must migrate very rapidly ($t_{1/2}$ in milliseconds) between the two hydroxyls so that any attempt to determine this would only measure the equilibrium mixture, usually 70% 3'-hydroxyl and 30% 2'-hydroxyl.

Recently a new approach to this old problem became available by virtue of the ability to prepare nonisomerizable analogues of tRNA which were nevertheless capable of being enzymatically acylated. Yeast phenylalanine tRNA was modified either by substitution of 2'-deoxyadenosine

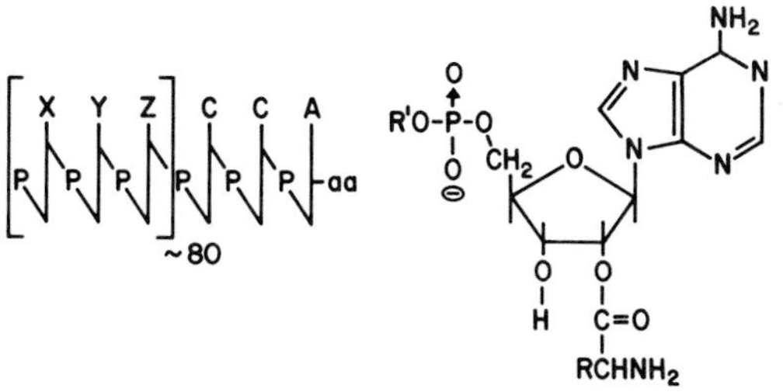

Fig. 2 Structure of aminoacyl-tRNA. R, amino acid side chain; R', nucleic acid chain.

or 3'-deoxyadenosine (cordycepin) for the normal adenosine terminus (Sprinzl and Cramer, 1973) or by opening of the terminal ribose ring by periodate oxidation followed by reductive regeneration of the hydroxyl groups (Hussain and Ofengand, 1973; Ofengand et al., 1974a). In both cases, it was possible to fully aminoacylate the modified tRNA by use of its homologous synthetase and the amino acid was bound only to the 2'-hydroxyl. The same results were obtained when E^{Phe} and tRNA from different species were tested (Ofengand et al., 1974a), including homologous reactions between yeast, E. coli, and rat liver components, heterologous cross-species reactions which in some cases required highly unusual reaction conditions in order to obtain acylation, and even one case of misacylation of E. coli valine tRNA with phenylalanine by E^{Phe} of yeast. In all these cases, the specificity of acylation for the 2'-hydroxyl was better than 90%, making it highly likely that the normal position of acylation of unmodified tRNA by E^{Phe} is the 2'-hydroxyl.

However, this is not the case for all synthetases. Recent experiments (Cramer et al., 1975; Sprinzl and Cramer, 1975; Fraser and Rich, 1975) have shown that with some other synthetases, such as E^{Ser} of yeast and E^{Lys}, E^{His}, and E^{Gly} of E. coli, acylation is specific for the 3'-hydroxyl, and with still other synthetases, such as E^{Tyr} of E. coli and of yeast, and E^{Cys} of E. coli, there is no specificity for the hydroxyl that is acylated.

At this writing, it is not possible to assess the significance, if any, of this variation in specificity of aminoacylation among different synthetases. Since the next step in protein synthesis, EF-Tu · GTP recognition (Section VII,B), probably is specific for the 2'-isomer, the rapid rate of chemical migration likely ensures that all AA-tRNA can interact regardless of the hydroxyl originally acylated. So far, 2' or 3' specificity is associated with a synthetase for a given amino acid and remains invariant for that synthetase over wide species variation. It has been suggested (Sprinzl and Cramer, 1975) that this variation in specificity may be a consequence of the mechanism particular to a given synthetase, which could explain its persistence despite evolution. There is no obvious correlation with the amino acid side chain since, for example, phenylalanine synthetases are 2' specific, yet tyrosine synthetases are not specific.

C. Assay Methods

Reaction (1) can be studied both in the forward direction by measuring E · AA-AMP formation, usually by gel filtration, and in the backward direction by measuring the exchange of ^{32}PP into ATP dependent on amino acid (Santi et al., 1974).

The assay for reaction (2) is straightforward. Since the only amino acid-containing cold acid-precipitable compound in reactions (1) to (3) is aminoacyl-tRNA, incorporation of radioactive amino acid into an acid precipitable form measures aminoacyl-tRNA formation. This assay likewise measures the overall reaction, reaction (3).

D. Requirements for Cations

The need for Mg^{2+} in reaction (1), especially when assayed by the exchange reaction, has long been recognized, although there have been more recent claims for the synthesis, at very low rates, of enzyme · AA-AMP complexes in the absence of Mg^{2+} and the presence of EDTA (Chousterman and Chapeville, 1971; Blanquet et al., 1972; Lawrence et al., 1973). In at least one case, however, this effect has turned out to be due to the presence of one mole of tightly bound Mg^{2+} per mole of enzyme (Chousterman and Chapeville, 1973). In some cases, Mg^{2+} can be substituted by Mn^{2+} (Yarus and Rashbaum, 1972) but not by spermine (Aoyama and Chaimovich, 1973), although in this particular example spermine markedly stimulated when a limiting amount of Mg^{2+} was present.

Cation requirements in reaction (2) are variable depending on the tRNA-synthetase pair involved. In some cases, Mg^{2+} is necessary (Allende et al., 1966; Charlier and Grosjean, 1972; Hirsh, 1968) and can (Bluestein et al., 1968; Chousterman and Chapeville, 1973) or cannot (Aoyama and Chaimovich, 1973) be replaced by spermine or spermidine, while in other examples no divalent cation is needed so long as sufficient monovalent cation is present (Yarus and Rashbaum, 1972; Fasiolo and Ebel, 1974) to maintain a functional tRNA structure.

Despite reports (Matsuzaki and Takeda, 1973) that reaction (3), the overall reaction, does not require Mg^{2+} if spermine is added, although Mg^{2+} cannot be replaced by spermine in reaction (1), it now appears that these findings were due to the presence of Mg^{2+} sequestered in the tRNA, or enzyme (Takeda and Ohnishi, 1975), since if the reactions are carried out in the presence of sufficient EDTA plus spermine, or if metal ions are carefully removed from the tRNA, no aminoacylation takes place (Santi and Webster, 1975; Chakraburtty et al., 1975a; Thiebe, 1975). Met-tRNA synthesis (Lawrence et al., 1973) was also reported to occur in the absence of tightly bound Mg^{2+} as long as spermine was present, but the presence of tightly bound Mg^{2+} was not definitely excluded here either.

The general conclusion to be drawn from these experiments is that Mg^{2+} is essential for the activation reaction possibly only in small amounts, but the transfer of amino acids to tRNA need not require cations at all. The function of Mg^{2+} or polyamines where needed for reaction (2) is thought to be as a structural stabilizer of the tRNA.

E. Mechanism

The order of addition of reactants has been examined in a number of cases by kinetic methods. Succinct descriptions of the methodology may be found in Midelfort and Mehler (1974), Santi *et al.* (1974), and Eigner and Loftfield (1974). In some cases, ATP, or rarely amino acid, was shown to add first. In other cases, addition was random, and in still others, a mixed type of reaction was apparent (Kisselev and Favorova, 1974). At this time there appears to be no agreement among workers in the field on a unified order of substrate binding and product release. This may be due to real differences in the mechanism of action of different synthetases or may only reflect the inability of the approaches used so far to clearly distinguish between alternatives.

In addition to the usual mechanism portrayed in reactions (1)–(3) an alternative has been advocated in which AA-AMP never occurs but reaction (3) takes place directly in a concerted manner. According to this scheme, the formation of E · AA-AMP complexes in the absence of tRNA is an aberrant reaction due to the lack of one of the normal substrates. The main argument for such a mechanism, or more precisely against the usual mechanism, reactions (1)–(3), has been summarized (Loftfield, 1972). At this time, it appears that there are no convincing arguments to support the concerted mechanism but at the same time it must be admitted that the evidence is equivocal for the definite existence of aminoacyladenylate in other than a transitory way along the main reaction path. It may well be that in the cell where the concentrations of tRNA and synthetase are such that virtually all synthetase molecules are complexed with tRNA (Yarus and Berg, 1969; Jacobson, 1969; DeLorenzo and Ames, 1970; Bonnet and Ebel, 1974) the active form of the enzyme is an E · tRNA complex, so that the *in vivo* lifetime of an aminoacyladenylate would indeed be very short. For example, although recent results by Lövgren *et al.* (1975) were interpreted by them to support a concerted mechanism in which E · Ile · ATP · tRNA forms and reacts directly to the final product, an alternative possibility was also mentioned in which the quaternary complex reacts to yield E · Ile-AMP · tRNA which then yields E · Ile-tRNA plus AMP.

One should be cautious, however, about extrapolating results from one synthetase to another. While it is attractive to suppose that a universal mechanism is used by all synthetases since the overall reactions are the same, it may well be that multiple paths differing in the details of the reaction may have evolved. For example, kinetic study of the phenylalanine synthetase of yeast was interpreted in terms of the usual AA-AMP intermediate mechanism (Berther *et al.*, 1974), as was a different kinetic analysis of the *E. coli* valyl-tRNA synthetase (Midelfort *et al.*, 1975). It has also been suggested that both pathways may coexist (Fersht and Jakes, 1975).

F. Enzymatic Deacylation

AA-tRNA synthetases can also catalyze the hydrolysis of AA-tRNA (reaction 4). This reaction is distinct from the reversal of reaction (3) which requires AMP and PP, and is 100–200 times more rapid than chemical hydrolysis which, of course, has no enzyme requirement.

$$E + AA\text{-}tRNA + H_2O \longrightarrow AA + tRNA + E \qquad (4)$$

The first careful study of this reaction, hints of which had existed in the earlier literature, was confined to a study of Ile-tRNA^{Ile} hydrolysis catalyzed by E^{Ile} (Schreier and Schimmel, 1972). Soon thereafter, these findings were extended to a study of two misacylated E. coli tRNA's, Val-tRNA^{Ile} (Eldred and Schimmel, 1972) and Ile-tRNA^{Phe} (Yarus, 1972a), and to two misacylated yeast tRNA's, Val-tRNA^{Phe} and Phe-tRNA^{Val} (Bonnet et al., 1972). In the first two tRNA's, it was observed that the hydrolysis of the misacylated tRNA was much faster than that of the correctly acylated tRNA when enzyme homologous to the tRNA was used and this led to the speculation that the physiological role of this reaction was to "verify" the correctness of the acylated tRNA (Yarus, 1972a). Indeed, this reaction explained why valine, which can readily form an E^{Ile}·Val-AMP complex, was never found attached to tRNA^{Ile} even in test tube reactions. Val-tRNA^{Ile} is transiently formed, but is hydrolyzed by E^{Ile} immediately. As the correctness of the aminoacylation reaction is essential to the fidelity of protein synthesis, this reaction appeared to provide a way to monitor aminoacylated tRNA.

However, a more extensive survey (Bonnet et al., 1972; Bonnet and Ebel, 1974) has shown that there is no relation between the relative rates of deacylation catalyzed by a given enzyme and either the tRNA or amino acid portion of the AA-tRNA. Of four yeast synthetases studied, three deacylate their completely homologous AA-tRNA's faster than either their misacylated cognate tRNA or completely noncognate AA-tRNA's (a noncognate tRNA is one normally specific for a different amino acid than is the enzyme). Approximate calculations (Bonnet and Ebel, 1974) indicate that the deacylation reaction probably contributes very little to the specificity of aminoacylation in vivo.

The broadness of specificity of the deacylation reaction, exemplified by the fact that many noncognate AA-tRNA's can be deacylated at rates comparable to cognate AA-tRNA's (Yarus, 1973; Bonnet and Ebel, 1974), does suggest, however, that the recognition between tRNA and synthetase may be much less specific than previously believed. Some recent measurements (Sourgoutchov *et al.*, 1974) support this concept. The K_m and V_{max} for deacylation of Met-tRNA^{fMet} was measured using E^{Met} and E^{Ile}. The K_m values were the same, 0.2 μM, and the V_{max} differed only by a factor of 2, but since E^{Met} has two tRNA binding sites versus one for E^{Ile} (Table I), both kinetic parameters were the same despite the fact that one reaction was with cognate AA-tRNA and the other was not. Similarly, comparison of these data with that obtained earlier (Schreier and Schimmel, 1972) for deacylation of Ile-tRNA^{Ile} catalyzed by E^{Ile} showed that here also the kinetic constants were very similar, the K_m for Ile-tRNA^{Ile} being 0.1 μM, and the turnover number being 0.8 min^{-1} versus 1.4 min^{-1} for this enzyme with Met-tRNA^{fMet}.

At this time, it is hard to conceive of a useful physiological function for this reaction. Possibly, it is merely a manifestation of labilization of the amino acid ester link to tRNA induced by particular enzyme-tRNA interactions, which can be imagined to vary widely depending on the synthetase-tRNA pair involved and is unrelated to any cellular function.

III. Structure and Specificity of Aminoacyl-tRNA Synthetases

A. Multiplicity of Enzymes for a Given Amino Acid

Although there may be multiple tRNA's (isoacceptor tRNA's) for a given amino acid (Section IV,A), so far as is known there is only one synthetase for each amino acid. Several lines of evidence show this to be the case in prokaryotic cells (Loftfield, 1971). A single synthetase, purified to homogeneity, could acylate all of its homologous isoacceptor tRNA's in the cell. Mutation to thermolability made all the synthetase activity in the cell thermolabile. Independent mutations induced in the same synthetase by different methods map at the same genetic locus. Antibody to a pure enzyme inhibited all the activity in a crude cell extract.

In eukaryotic cells, multiple enzymes exist, but their existence is correlated with the presence of cell organelles. Distinct synthetases for many amino acids are found in mitochondria, chloroplasts, and nuclei. Some, but not all, organelle synthetases are specific for their organelle tRNA, and some cytoplasmic synthetases are similarly specific. In those few cases examined, the gene for the organelle synthetase was located in the nucleus (Boguslawski *et al.*, 1974; Hecker *et al.*, 1974). The literature on organelle heterologous reactions has been reviewed in some detail (Jacobson, 1971).

Two valine synthetases appear for a time following T-even infection of *E. coli,* although the host enzyme disappears rather rapidly and the new virus-induced enzyme remains. It is now known that a virus-induced protein, of approximately 10,000 MW, is added to the host enzyme to produce the "new" enzyme (Marchin *et al.*, 1972). The purpose of this modification is unclear since substrate specificity is unchanged, and a phage mutant unable to carry out this transformation can nevertheless multiply normally in the host cell.

B. Protein Structure

The molecular weights, subunit structures, and number of binding sites for substrates is summarized in Table I for those synthetases for which subunit data are available. Although there is no apparent relation between amino acid specificity or cell type and size or subunit composition, inspection of the table does reveal certain common features. Synthetases can be grouped into 3 classes according to their subunit structure. Class I synthetases are all dimers of identical subunits, the subunit size ranging from 35,000–60,000 MW, with two substrate binding sites. There are two exceptions in this group. *E. coli* E^{Glu} is made from two dissimilar subunits and has only one substrate binding site, and *E. coli* E^{Met} (the native enzyme) is approximately twice the size of the other enzymes and has twice the number of sites for ATP than for the other substrates. Class II consists of single polypeptide chains ranging in molecular weight from 70,000–120,000 with one binding site for each substrate. The single apparent exception, yeast E^{Tyr}, may be due to the method of preparation which involved autolysis. An independent preparation from a defined yeast strain is 2–3 times larger in size (Kućan and Chambers, 1973). Only a few Class III synthetases are known, and appear to be constructed from a pair of Class I plus a pair of Class II subunits. There are either one or two binding sites in this class.

The molecular weights and classification of synthetases should, however, be accepted with some reservations in view of several recent examples which show that unsuspected protease action during purification can convert an α_1 into an α_2 (Chapeville, 1972) or an α_2 into an apparent α_4 (Koch and Bruton, 1974), and can also remove a substantial amount of protein from a synthetase without affecting its enzymatic activity (Koch and Bruton, 1974; Dimitrijevic, 1972; Chapeville, 1972; Lemaire *et al.*, 1975).

The extreme lability of certain synthetases to proteolytic splitting into halves, and the existence of two size classes of subunits, one half the molecular weight of the other, has led to the suggestion that the synthetases of Class II arose by gene duplication and fusion. Strong evidence in support of this notion has recently become available as a result of peptide mapping studies. *E. coli* E^{Leu} (Waterson and Konigsberg, 1974) and *B. stearothermophilus* synthetases for Val, Leu, and Met (the subunit) (Koch *et al.*, 1974) have been shown to be made up of 2 long repeated se-

TABLE I

Molecular Weight and Subunit Structure of AA-tRNA Synthetases[a]

Enzyme		Molecular weight			Number of binding sites			
Amino acid (source)	Type	α	β	Native	Amino acid	ATP	AA-AMP	tRNA
Glu (*E. coli*)	$\alpha\beta$	56	46	102				1
His (*E. coli*)	α_2	43		85	2	2		
His (*S. typhimurium*)	α_2	40		80				
Leu (yeast)	α_2	60		120				
Lys (*E. coli*)	α_2	52		104	2	2		
Lys (yeast)	α_2	69		138	2	2		
Met (*E. coli*)	α_2	86		172	2	4	2	2
Met (*B. stearothermophilus*)	α_2	66		135				
Pro (*E. coli*)	α_2	47		94				
Ser (*E. coli*)	α_2	48		95	2	2	2	2
Ser (yeast)	α_2	60		120		2		2
Ser (hen liver)	α_2	60		120				
Trp (*E. coli*)	α_2	37		74	2			2
Trp (yeast)	α_2	50		110				
Trp (beef pancreas)	α_2	54		108	2	2	2	2
Trp (human placenta)	α_2	58		118				
Trp (*B. stearothermophilus*	α_2	35		70				
Tyr (*E. coli*)	α_2	48		97	1–2		1–2	1–2
Tyr (*B. stearothermophilus*	α_2	44		95	1–2		1	2
Arg (*E. coli*)	α			75				
Arg (*B. stearothermophilus*	α			78	1	1		
Asp (yeast)	α			100				
Gln (*E. coli*)	α			69				1
Ile (*E. coli*)	α			112	1	1	1	1
Leu (*E. coli*)	α			105		1	1	1
Leu (*B. stearothermophilus*)	α			110				
Tyr (yeast)	α			46				
Val (*E. coli*)	α			110	1	1	1	1
Val (yeast)	α			122	1	1		1
Val (*B. stearothermophilus*	α			110				
Gly (*E. coli*)	$\alpha_2\beta_2$	80	33	225			2	
Phe (*E. coli*)	$\alpha_2\beta_2$	94	39	267	2		2	2
Phe (yeast)	$\alpha_2\beta_2$	74	63	270	2		2	2
Phe (rat liver)	$\alpha_2\beta_2$	74	69	290				

[a] The data have been summarized from Kisselev and Favorova (1974), Söll and Schimmel (1974), Berther *et al.* (1974) and Fasiolo *et al.* (1974) [yeast Phe], Bartmann *et al.* (1975) [*E. coli* Phe], Hossain and Kallenbach (1974) [yeast Trp], Penneys and Muench (1974) [human Trp], Ostrem and Berg (1974) [*E. coli* Gly], Boeker *et al.* (1973) [*E. coli* Ser], Koch *et al.* (1974) [*B. stearothermophilus* Trp, Tyr, Met, Val, Leu], Kalousek and Konigsberg (1974) [*E. coli* His], Bosshardt *et al.* (1975) [*B. stearothermophilus* Tyr], and Jakes and Fersht (1975) [*E. coli* and *B. stearothermophilus* Tyr].

quences of amino acids, and similar results have been very recently obtained for the subunit of *E. coli* E^{Met} (Bruton *et al.*, 1974) and for yeast E^{Val} (Bruton, 1975).

Thus, a unifying principle of structure can now be discerned in that all synthetases appear to be constructed from partially mutated repeats of a 35,000–55,000 molecular weight polypeptide. In some cases, the association is noncovalent (Class I), in others the association is covalent (Class II), and in still others, it is both (Class III). The discrepancy in molecular weight between the methionyl-tRNA synthetase and others of Class I can now be understood on the assumption that this subunit is constructed like the Class II polypeptides, and even the existence of 4 ATP binding sites can be rationalized if it is assumed that each primordial peptide retained its ATP binding site through the gene fusion process. The structure of E^{Glu} (*E. coli*) can also now be understood, at least hypothetically, as a pair of tandem genes that failed to fuse, one of which subsequently underwent extensive mutational alteration. It would be interesting to know if there are sequence similarities.

A recent report (Lee, 1974) further supports the above discussion. Trypsin was found to cleave a uniquely sensitive bond in E^{Ile} (*E. coli*), generating two fragments, 69,000 and 39,000 MW, respectively, which were quite resistant to further protease action. All catalytic activity for reactions (1)–(3) was lost. However, the ability to catalyze reaction (4) was retained and a complex between tRNA and trypsin-treated synthetase could be detected. No attempt was made to test activity of the separated fragments. A similar situation appears to exist for E^{Val} (*E. coli*) except that in this case, the proteolytic cleavage into a 69,000 and 46,500 MW fragment had almost no effect on activity (Paradies, 1975).

The ultimate approach to the tertiary structure of the synthetase is through X-ray crystallography and a number of synthetases have now been crystallized (Kisselev and Favorova, 1974). They include yeast lysine and leucine, *E. coli* methionine, *B. stearothermophilus* tyrosine, and pancreatic tryptophan synthetases. So far only the dimensions of the *E. coli* E^{Met} 64,000 MW enzymatically active fragment is known. It is an ellipsoid 90 × 43 × 43 Å consisting of two parts separated by a large cleft (Zelwer *et al.*, 1976).

Another aspect of interest is the location and orientation of substrate binding sites on the subunits of the synthetase. Until synthetase-tRNA complexes are crystallized and their structure determined by X-ray analysis, approaches to this question will have to come by solution chemistry. A useful method of attack is that of affinity labeling to identify that part of the synthetase which constitutes the binding site for a given substrate. Only preliminary attempts along these lines have been made so far. A chloromethylketone analogue of valine was used to alkylate the valine binding site of bovine E^{Val} (Frolova *et al.*, 1973; Silver and Laursen, 1974), but no analysis of the site of binding was described in this preliminary report. Several workers have derivatized the amino group of the amino acid of aminoacyl-tRNA with alkylating agents in attempts to form a covalent link with the synthetase in complex with AA-tRNA (Bruton and Hartley, 1970; Santi and Cunnion, 1974; Bartmann *et al.*, 1974b; Lavrik and Khutoryanskaya, 1974), and some attachment has been observed. However, in view of the fact that N-blocked amino acids are not substrates for synthetases (Section III,C,1) and the accumulated evidence that the aminoacyl group of AA-tRNA is not in the amino acid site (Bartmann *et al.*, 1974a; Bonnet and Ebel, 1975), it is doubtful whether these probes will be useful for location of the amino acid binding region.

A more useful affinity probe has recently been developed in which photoaffinity reagents are attached to rare bases in tRNA molecules (Schwartz and Ofengand, 1974). Photoaffinity probes in general are more desirable than chemical affinity probes, since activation of the probe can be delayed until the desired noncovalent complex has formed, and the reactive species are, in general, much less specific in terms of reactant partners. This feature is desirable since it avoids bias in the results due to a requirement for particular functional groups. The ability to place such probes at various locations in a tRNA should make it possible to study in some detail the effect of amino acid and ATP on the orientation of tRNA in the enzyme binding site, as well as the relation of this site to the various subunits of Class III synthetases. A preliminary report of such an approach has recently appeared (Budker *et al.*, 1974). A similar approach using photoaffinity-labeled ATP has also been described recently (Ankilova *et al.*, 1975).

C. Substrate Specificity

1. Amino Acid

As expected from the required fidelity of protein synthesis, synthetases as a class exhibit very high specificity for their cognate amino acid. Other naturally occurring amino acids are not recognized, with two exceptions. E^{Ile} of *E. coli* also forms an $E^{Ile} \cdot$ Val-AMP complex, although the K_m for valine in the exchange reaction is 50 times greater, but as pointed out in Section II,F, valine is never stably attached to tRNA, because of a rapid deacylation reaction also catalyzed by E^{Ile}. Similarly, E^{Val} of *E. coli* catalyzes an exchange with threonine, but the K_m for threonine is 100 times greater. This reaction has not been studied further. In a similar vein, naturally occurring amino acid analogues that are likely to be encountered by synthetases in the cell are inactive. This leads to the interesting situation where a particular analogue may be inactive with a synthetase from the same cell but nevertheless quite active with synthetases from other cells. The activity of unnatural analogues varies widely. Some are inactive, some are active in reaction (1) but not reaction (2), while others can be linked to tRNA [reaction (3)] and can be incorporated into protein in place of the natural amino acid. Still other amino acid analogues, natural and unnatural, are toxins that act by specifically inhibiting particular synthetases (Loftfield, 1972). In such cases, the synthetases from the organism synthesizing the toxin are insensitive to it.

In all cases but one (E^{Tyr} from *E. coli* and *B. subtilis*), synthetases are specific for the L-amino acid. In this single aberrant case, the D-isomer is also active although with a lower K_m and V_{max}.

The presence of a free α-amino group on the amino acid is essential. Many examples are known where either removal of the amino group or acylation completely blocked interaction with its synthetase, either as a substrate or as an inhibitor. Moreover, N-substituted aminoacyl-tRNA's do not undergo deacylation in the presence of AMP [reverse of reaction (3)] (Chapeville, 1972). On the other hand, the carboxyl group is not necessary, and amino acid amides or amino alcohols frequently have K_i values equivalent to the K_m for the parent amino acid. References to the original work cited above and a more detailed discussion can be found in Loftfield (1972) and Mehler and Chakraburtty (1971), as well as in recent papers (Holler *et al.*, 1973; Hecht and Hawrelak, 1974).

2. ATP

Only ATP, and with a much higher K_m and lower V_{max}, dATP, is active. Other nucleoside triphosphates are inactive. In general, ATP analogues either inhibit or are inactive except for AMP-P(CH_2)P, which is functional with some synthetases (Chapeville, 1972). Adenosine also inhibits reactions (1) and (3) with a variety of synthetases more or less competitively with ATP. Adenosine analogues are variably effective (Lawrence *et al.*, 1974).

3. AA-AMP

Under usual assay conditions some synthetases, but not all, are considerably less specific in their ability to pyrophosphorylize AA-AMP made from noncognate amino acids than they are in the reaction with free amino acid plus ATP. That is, the reverse of reaction (1) is apparently less specific than the forward reaction. This phenomenon, which appears to violate the laws of physical chemistry, can be explained by the relative K_m values for cognate and noncognate amino acid and AA-AMP and

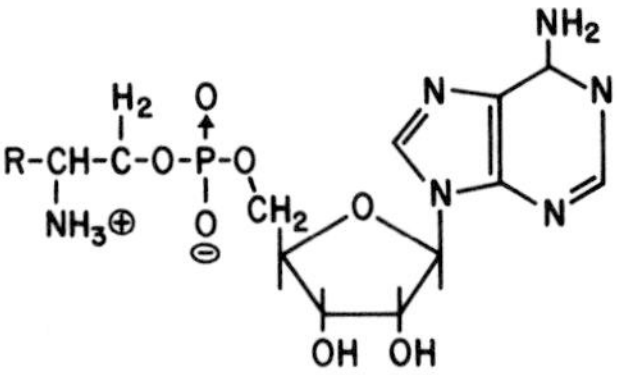

22   James Ofengand

Fig. 3 Structure of aminoalkyladenylate. R, amino acid side chain.

the concentrations used in the usual assays (Kisselev and Favorova, 1974, p. 178). The K_m for noncognate amino acids in reaction (1), ca. 0.1–1 M, would ensure lack of reaction under usual conditions while a similar 10^3–10^4 increase in K_m for AA-AMP in the reverse of reaction (1) would only decrease the cognate values of 10^{-7}–10^{-8} to 10^{-3}–10^{-4} M, a value which would still permit reaction at the concentrations usually examined. Observations on the specificity of aminoalkyladenylates (Fig. 3), structural analogues of AA-AMP, tend to substantiate this explanation. For example, the inhibition constant for isoleucinyladenylate with E^{Ile} is ca. 10^{-8} M, while the noncognate valinyladenylate value is ca. 10^{-4} M (Chapeville, 1972).

In view of the very low K_i value of aminoalkyladenylates for synthetases, they should be good inhibitors of protein synthesis, not only *in vitro* but *in vivo* since their stability and lack of charged groups should allow easy entry into cells. This possibility has been tested in a recent report (Cassio and Mathien, 1974) which shows that it is possible to inhibit the growth of *E. coli* with methioninyladenylate. Moreover, the degree of inhibition was directly correlated with the failure to acylate methionine tRNA, once the level of acylated tRNA^{Met} dropped below the minimal required level. The effect was very specific since the level of acylation of other tRNA's did not change. Very recently, similar effects were observed on chick embryo fibroblast cultures (Robert-Gero *et al.*, 1975). This class of inhibitor should prove useful for a variety of cell physiological studies.

4. tRNA

The specificity of synthetases for cognate, noncognate, and variously modified tRNA's will be the subject of Section V. However, it is appropriate to mention here the several examples known of the substitution of viral RNA for tRNA in the aminoacylation reaction. Apparently, the 3'-terminal 1% of the viral molecule (MW ca. 2×10^6) can fold into a form sufficiently like certain tRNA's so that they are specifically acylated by an appropriate synthetase. The site of attachment of amino acid is to the ribose of the 3'-terminal adenosine, just as in tRNA. This phenomenon has now been observed in the synthesis of Val-(TYMV RNA), Tyr-(BMV RNA), His-(TMV RNA), and His-(mengovirus RNA) (Salomon and Littauer, 1974; Carriquiry and Litvak, 1974; and references therein). The sequence of the 3'-terminal 71 and 74 residues of two strains of TMV RNA has recently been determined (Guilley et al., 1975; Lamy et al., 1975). The polynucleotide bears little resemblance to a tRNA-like structure and no minor bases were detected. The biological function, if any, of this reaction is not known.

D. Ligand Affinity and Allosteric Effects

Interaction between a protein and one of its substrates can be measured in a variety of ways, and most of them have been used by now in the very many reports on synthetase-substrate interactions. In addition to the usual kinetic methods, direct association has been measured by equilibrium methods (dialysis, fluorescence with and without a reporter group, optical rotation, protection from an inactivating event such as heat or protease) and by nonequilibrium methods (gel filtration, gel electrophoresis, sucrose gradient centrifugation, liquid-liquid partitioning, nitrocellulose membrane filtration). An extensive tabulation of association constants obtained by physical methods is given in Kisselev and Favorova (1974). In general, there is good agreement between the values obtained by kinetic and physical methods so long as the reaction conditions are equivalent. In view of the variable effects of tRNA on the ATP-PP exchange assay (see below) the appropriate kinetic constants for amino acid and ATP are those determined for aminoacylation, not for the exchange reaction. Overall, the K_m value for amino acid ranges from 5–150 μM, for ATP 40–2300 μM, and for tRNA 0.03–0.5 μM with most values rather close to 0.1 μM. Considerably higher affinities for tRNA have been obtained by physical methods, but these have usually been at pH 5.5–6 which is now known to result in increases of 10- to 100-fold in the affinity for synthetase. In the few cases studied, aminoacylation of the tRNA does not affect its association constant when measured under both equilibrium and nonequilibrium conditions (Bartmann et al., 1974a, and references therein). Thus, aminoacylation of tRNA does not provide the means to release tRNA from complex with its synthetase.

The maximal velocity of many of the synthetases has been determined and can be found in Loftfield (1972). The values range from 10–2200 mol/mol/min including a recent determination for E^{Gly} of E. coli (Ostrem and Berg, 1974). It does not appear worthwhile to try to relate these values to the rate of protein synthesis in the cell, since the test tube conditions are so highly artificial. For example, for some synthetases (but not all, see Fasiolo and Ebel, 1974; Bartmann et al., 1974a; Fersht and Jakes,

1975) the rate of dissociation of AA-tRNA from the synthetase is thought
to be the rate-determining step, yet in the cell it is likely that the true reaction is

$$E \cdot AA\text{-}tRNA + EF\text{-}I \cdot GTP \longrightarrow E + AA\text{-}tRNA \cdot EF\text{-}I \cdot GTP$$

(see Section II,E).

As might be expected for an enzyme with three substrates and three
products, there are numerous examples of allosteric effects of one substrate on the binding of another. Probably the most striking of these is the
effect of tRNA on reaction (1), measured by the ATP-PP exchange reaction. The allosteric effects of tRNA vary from an absolute dependence to
a stimulation, an inhibition, or no effect (references cited in Ostrem and
Berg, 1974). The absolute dependence, first noted in 1965, is the most
striking. Although it has been observed only for the arginine, glutamic
acid, and glutamine synthetases, all these synthetases show the effect
whether they are isolated from prokaryotic or eukaryotic sources. So far,
no example of a synthetase specific for one of these amino acids has been
found which is not dependent on tRNA for the exchange reaction. It is surprising that despite the potential for evolutionary divergence, the synthetases for these three amino acids have retained this peculiar feature even
as their subunit structures diverged (Table I). Since the effect appears to
be correlated with the amino acid recognized, rather than with any other
feature, one suspects that it is somehow related to the structure of the
amino acid although, since one is basic, one acidic, and one neutral, it has
not been possible to devise any plausible explanation up to now.

Stimulation or inhibition by tRNA has been observed in a number of
other cases measured both by the maximum velocity of the exchange
reaction and by changes in the K_m for amino acid and ATP (Ostrem and
Berg, 1974, and references therein). The characteristic difference
between these effects and the absolute dependence noted above, in addition to the quantitative aspect, is that tRNA with an intact 3'-terminus is
required in the cases of absolute dependence, while tRNA with an oxidized 3'-end or with some of its 3'-terminal C-C-A missing so that it is no
longer an amino acid acceptor is as good or better than intact tRNA with
synthetases of the stimulation/inhibition class. This observation appears
to indicate a fundamental difference in the mode of action of tRNA in the
two cases, but this aspect has not been studied.

Binding of amino acid can affect the binding of tRNA and of ATP. The
rates of association and dissociation of the tRNA-synthetase complex are
increased in the presence of amino acid, but the association constant is
unchanged, at least in the examples known so far (Bartmann et al.,
1974a). On the other hand, ATP can increase the affinity of the amino acid

for synthetase and vice versa. Aspects of this synergistic coupling have been discussed recently (Parfait, 1973; Kosakowski and Holler, 1973; Blanquet *et al.*, 1975; Holler *et al.*, 1975).

In one enzyme, E^{Glu} of *E. coli*, the β subunit itself acts as an allosteric effector (Lapointe and Söll, 1972). In this unusual case, the β subunit is not required for catalytic activity but modifies the K_m values for all three normal substrates. Surprisingly, while it lowers the K_m for glutamate and ATP, the K_m for tRNA^{Glu} is doubled.

E. Organization of Synthetases in Cells

Examination of crude cell lysates of prokaryotes has confirmed that the molecular weight of synthetases is as to be expected for free E or E · tRNA complexes (Nass and Stöffler, 1967). In eukaryotic cells, on the other hand, increasing evidence is accumulating that synthetases exist in the form of high molecular weight ($> 1.5 \times 10^6$) complexes containing all 20 synthetases, tRNA, and cholesterol esters (Banyopadhyay and Deutscher, 1971, 1973). These complexes are, however, very fragile and gentle homogenization is required to keep them intact which probably accounts for the failure of others (Vennegoor and Bloemendal, 1972) to find all of the synthetases in the complex. Some synthetases, especially E^{Phe}, have been found partially associated with ribosomes (Tscherne *et al.*, 1973), although in at least one case the addition of tRNA resulted in a rapid release from the ribosome and appearance in a 25 S particle (Roberts and Coleman, 1972).

It seems probable that in eukaryotic cells the existence of synthetases and tRNA in a large supramolecular complex will prove to be a general feature, whose purpose may be to increase the efficiency of the process.

IV. Structure of tRNA

A. Multiplicity of tRNA Species for a Given Amino Acid

The existence of more than one tRNA specific for a given amino acid in the same organism, called isoacceptor tRNA's, was first noted by Berg *et al.* (1961), but it was not until the development of methods for the fractionation of tRNA that isoacceptor species could be shown directly. These procedures are now highly developed (for recent reviews, see Cantoni and Davies, 1971; Gillam and Tener, 1971; Roe *et al.*, 1973a; Kelmers *et al.*, 1974; Holmes *et al.*, 1975; McCutchan *et al.*, 1975) and have led to the isolation and identification of many examples of isoacceptor tRNA families.

The reason for isoacceptor tRNA's is clear from an examination of the genetic code. There are multiple code words for each amino acid, and in view of the adapter function of tRNA, each code word should correspond to a given isoacceptor species of tRNA. By this argument, there would be 61 species of tRNA in the cell, if the extra organelle-specific tRNA's are neglected. However, as pointed out in Section VI, other base-pairing schemes in addition to the standard $A \cdot U$ and $G \cdot C$ pairs are possible in codon-anticodon matching so that one anticodon can read two or even three codons. Moreover, it appears that not all code words are utilized in all organisms, so that the number of individual tRNA's actually needed is less than 61. The actual number for a given organism is not known but is usually estimated to be in the range of 40–60.

Frequently, more isoacceptors are found than are required for codon-anticodon matching. In most cases, these are false isoacceptors which result from incomplete modification of the tRNA. There are numerous unusual nucleotides in tRNA (see below) that are made by individual modifications of a normal base after the polynucleotide chain is synthesized. Since some of these reactions are considerably slower than others, several apparent isoacceptors may be found that actually correspond to one primary sequence in various stages of completion. Another source of artifact is unsuspected chemical alteration (usually) to the unusual bases, which changes the fractionation properties of the tRNA. A good example of this is the separation of two $tRNA_f^{Met}$ species from *E. coli* (Petrissant and Favre, 1972), one of which has a covalent cross-link between 4-thiouridine and cytidine. Some tRNA's, for example, yeast $tRNA_3^{Leu}$, are also known to dimerize and to exist in multiple conformational states, which can be detected by fractionation methods (Lindahl *et al.,* 1967).

Other tRNA's are known that can be considered as *pseudo* isoacceptors. These are completed molecules in their normal physical state, but which are not true isoacceptors for several reasons. They include tRNA's that differ by only a few bases from each other, and which probably arose from a single sequence by gene duplication and subsequent mutation in one or both genes. Examples are Tyr 1:Tyr 2 and Val 2A:Val 2B of *E. coli,* and Ser 1:Ser 2 of yeast. There are also several cases of aminoacyl-tRNA's that are not used for protein synthesis, but that have been adapted by the cell for other functions such as cell-wall synthesis (Section X). These tRNA's are acylated normally by their synthetase, but cannot function in protein synthesis, at least in the one case studied in detail (glycine tRNA's of *Staphylococcus*). There is also the unique case of $tRNA_i^{Met}$, the tRNA specific for the initiation of protein synthesis (Section IV,D). Both $tRNA_m^{Met}$ and $tRNA_i^{Met}$ read the same codon, but $tRNA_i^{Met}$ only functions in initiation while $tRNA_m^{Met}$ only is active in peptide chain elongation. Although acylated by the same synthetase, their primary sequences are quite different.

In addition, many examples are known in which tRNA fractionation patterns for a given amino acid vary with growth stage or degree of differentiation of the cell (Littauer and Inouye, 1973; Chase *et al.*, 1974; Juarez *et al.*, 1975) with new peaks appearing and old one disappearing. However, it has not yet been shown that any of these represent new tRNA sequences, and, in at least one case, the changed profiles were shown to be due to variations in the degree of modification of minor nucleotides in the tRNA (White *et al.*, 1973).

So far, with the sole exception of the methionine tRNA's mentioned above, no true isoacceptor tRNA's from the same source with different nucleotide sequences have been shown to share the same anticodon (except for suppressor tRNA's), in keeping with the concept that the primary function of isoacceptor species is to utilize the degeneracy in the genetic code.

B. Primary and Secondary Structure

The methodology of tRNA sequencing is similar to protein sequencing in that various nucleases of differing specificity are used to generate collections of overlapping oligonucleotides whose individual sequences are worked out and then eventually fitted together. The theoretically serious problem of only four different bases is simplified by the existence of numerous unusual nucleotides which serve as convenient reference points for fitting together the overlapping sequences. The first sequence of a tRNA, tRNA[Ala] of yeast (Holley *et al.*, 1965), was obtained this way using conventional column methods for fractionation of oligonucleotides and chemical methods for analysis. Major technical advances, originating in Sanger's laboratory, were the use of radioactive tRNA and two-dimensional methods for oligonucleotide fractionation and analysis. These methods and subsequent modifications have been described in detail (Brownlee, 1972). The relative ease of these methods and their currently almost routine use account for the large number of tRNA sequences now known.

As of this writing the primary sequences of 62 tRNA's are known (see legend to Fig. 4 for references), not counting those tRNA's that differ by only a few residues and that probably arose by gene duplication and mutation. tRNA's from bacteria, yeast, plant, and mammalian sources and for almost all amino acids have been sequenced. They include numerous examples of isoacceptors from the same species, as well as tRNA's specific for the same amino acid from different sources, and many examples of

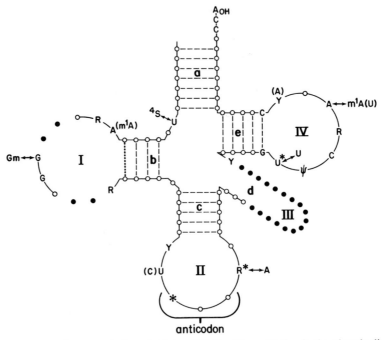

Fig. 4 Common features of all noninitiator tRNA's. The solid line is the phosphodiester backbone, base-paired nucleotides are indicated by dashed lines, and nucleotides by circles or letters. A_{OH} is the 3'-end. Nucleosides common to all structures are indicated as Y, pyrimidine; R, purine; R*, modified purine; U*, modified uridine; Ψ, pseudouridine; m^1A, 1-methyl adenosine; *, a frequently modified purine or pyrimidine; ^4S, 4-thiouridine. The ↔ means both nucleotides are frequently found, ( ) means the nucleotide in parenthesis is the single exception to the common base indicated. If there are two or more exceptions, the position is considered variable and indicated by an open circle. The loops and stem regions are numbered according to the accepted convention. The dotted line in stem b indicates that a base pair is found only in some tRNA's. Stem d plus loop III are highly variable as indicated. The solid circles indicate nucleosides which may or may not be present. Thus loop I varies from 7 to 11 residues, not counting the variably base-paired residues at the end of stem b, and stem d plus loop III varies from 3–21 bases in length. The data were compiled from Barrell and Clark (1974), Barrell *et al.* (1974) [T_4Pro,Ser], Williams *et al.* (1974) [*E. coli* Ala 1A], Holness and Atfield (1974) [yeast Cys], Roberts and Carbon (1974) [*E. coli* Gly 2], Piper and Clark (1974a) [mouse myeloma Met 4 and Val 1], Axel'rod *et al.* (1974) [yeast Val 2A], Kobayashi *et al.* (1974) [yeast Glu 3], Clarke and Carbon (1974) [*E. coli* Thr], Kimball *et al.* (1974) [*Mycoplasma* Phe], Penswick *et al.* (1975) [yeast Ala 1], Randerath *et al.* (1975) [yeast Leu (CUA)], Harada *et al.* (1975) [chicken Trp], Gruhl and Feldmann (1975) [yeast Met m], Rogg *et al.* (1975) [rat Ser 3], Yamada and Ishikura (1975) [*B. subtilis* Met f], Chakraburtty *et al.* (1975b) [*E. coli* Lys], Chakraburtty (1975a) [*E. coli* Arg 2], and Guerrier-Takada *et al.* (1975) [*B. stearothermophilus* Phe].

tRNA's for different amino acids from the same source. Thus, one has the opportunity to search for sequence homologies related to species, to amino acid specificity independent of species, and to homologies among true isoacceptor molecules. While it is not difficult to find examples of such homologies, there is no consistent pattern. Frequently, one can find more homology between two different tRNA's from the same species than between two isoacceptor tRNA's (references cited in Söll and Schimmel, 1974).

What is clear, however, is that all tRNA's can be fitted into a cloverleaf secondary structure as illustrated in Fig. 4. This figure embodies the common features of all known tRNA structures, with the exception of initiator tRNA's, *E. coli* histidine tRNA, mouse myeloma methionine tRNA, *Torula* yeast valine tRNA, *S. cerevisiae* glycine tRNA, and the glycine tRNA's from *Staphylococcus* sp. specific for cell wall synthesis. Eukaryotic initiator tRNA's differ in loop IV by replacement of the sequence U*-Ψ by A-U and Y by A. Prokaryotic initiator tRNA's differ in lacking the 3'-terminal base pair in stem a, so that there are only 6 base pairs in that stem (Ecarot and Cedergren, 1974; Walker and RajBhandary, 1975). Histidine tRNA, on the other hand, has an extra base at the 5'-end of the molecule that is base paired so that stem a has 8 pairs and only the C-C-A residues are unpaired. Mouse myeloma methionine tRNA has only 4 base pairs in stem c (Piper and Clark, 1974a). The two yeast tRNA's have only 3 bases in stem d-loop III instead of 4 or more (Barrel and Clark, 1974). The cell wall glycine tRNA's are nonfunctional in protein synthesis and differ in loops I, II, and IV from all other structures (Roberts, 1974).

Although the total number of residues ranges from 74–94, it is clear that variation is only allowed in loop I and stem d plus loop III. The number of base pairs in the other stems and the size of loops II and IV are invariant, except that stem b may have 3 or 4 base pairs. However, the size of the stem d plus loop III is not continuously variable. There appear to be three main classes of tRNA's (Levitt, 1969), those with small loops of 3–5 residues and either 3 or 4 base pairs in stem b, and those with much larger loops of 13–21 residues and only 3 base pairs in stem b. Within each of these three classes there are more common features than indicated in the figure which only shows those features common to *all* tRNA's. However, some caution should be exercised in interpretation of such data since it is not yet clear which features are truly important and which only reflect the relatively restricted range of species so far studied. The size of loop I may vary from 7 to 11 residues in addition to the two in stem b which are not always base paired. Otherwise, the tRNA structure is remarkably constant. There are 12 invariant bases if both parent and modified base are

TABLE II
Unusual Nucleosides in tRNA

Name	Symbol[a]	Location in tRNA
Pseudouridine	ψ	Loop IV always; stem C and Loop II mostly; occasionally elsewhere
Ribothymidine	T	Loop IV
5,6-Dihydrouridine	hU or D	Loop I, mostly, loop III occasionally
4-Thiouridine	⁴S or s⁴U	Residue 8 (also 9 one case)
$O^{2'}$-Methyluridine	Um	Loop II, Y; stem d
$O^{2'}$-Methylpseudouridine	ψm	Stem c
$O^{2'}$-Methylribothymidine	Tm[a]	Loop IV, U*
5-(Carboxymethoxy)uridine	cmo⁵U	Loop II, *
3-(3-Amino-3-carboxypropyl)uridine	nbt³U[b]	Loop III, 3'-side of m⁷G
5-(Methylaminomethyl)-2-thiouridine	mnm⁵s²U	Loop II, *
5-(Methoxycarbonylmethyl)-2-thiouridine	mcm⁵s²U	Loop II, *
5-Methyl-2-thiouridine	m⁵s²U[c]	Loop IV, U*
5-(Methoxycarbonylmethyl)uridine	mcm⁵U[d]	Loop II, *
$O^{2'}$-Methylcytidine	Cm	Loop II, Y, *; stems a and e
5-Methylcytidine	m⁵C	Mostly stem d, Y; occasionally elsewhere
3-Methylcytidine	m³C	Loop II, Y; loop III
N^4-Acetylcytidine	ac⁴C	Loop II, *; stem b
2-Thiocytidine	s²C	Loop II, Y
Inosine	I	Loop II, *
1-Methylinosine	m¹I	Loop II, R*
1-Methyladenosine	m¹A	Loop IV (once in loop I)
2-Methyladenosine	m²A	Loop II, R*; residue 9
N^6-Methyladenosine	m⁶A	Loop II, R*
N^6-Isopentenyladenosine	i⁶A	Loop II, R*
2-Methylthio-N^6-isopentenyladenosine	ms²i⁶A	Loop II, R*
N^6-(Threoninocarbonyl)adenosine	t⁶A	Loop II, R*
N^6-Methyl-N^6-(N-threoninocarbonyl)-adenosine	mt⁶A[e]	Loop II, R*
Complex derivatives of guanosine	W,W$_x$,W$_t$[f]	Loop II, R* (tRNA^Phe only)
1-Methylguanosine	m¹G	Loop II, R*; residue 9
N^2-Methylguanosine	m²G	Stems a and b; between stems b and c; residue 9
N^2,N^2-Dimethylguanosine	m²G	Between stems b and c
N^7-Methylguanosine	m⁷G	Central base, 5-residue loop III
O^2-Methylgunanosine	Gm	Loop I; loop II, *
7-{[(cis-4,5-Dihydroxy-2-cyclopenten-1-yl)amino]methyl}-7-deazaguanosine	Q[g]	Loop II, *

[a] Only unusual nucleosides whose locations in the tRNA molecule have been established are included. Additional unusual bases are known (Nishimura, 1972; Hall, 1971). Data were abstracted from the sequences referenced in the legend to Fig. 4 and the following: (a) Gross et al. (1974); (b) Ohashi et al. (1974), Friedman et al. (1974); (c) Watanabe et al. (1974); (d) G. Dirheimer (personal communication); (e) Clarke and Carbon (1974); (f) formerly called "Y"; Takemura et al. (1974), Keith et al. (1974), Blobstein et al. (1975); (g) Kasai et al. (1975a,b).

counted as one, and 42 residues in base pairs, leaving only 20 truly variable positions, in the case of the smallest tRNA's, to specify the uniqueness of each tRNA. Moreover, an additional restrictive feature has been noted. The A to G variation in R of loop I is correlated with the U to C variation in Y of stem d in 53 of the 55 sequences known.

tRNA is studded with unusual bases, made by sometimes quite elaborate modifications of one of the four normal nucleotides. Table II lists those whose position in the tRNA molecule is known. A number of other bases are known and can be found in the references cited in the table legend. Several favored sites for modified bases exist, especially in loops II and IV, as indicated in the figure. Those in loop II are involved in codon-anticodon recognition and are highly variable (Table II and Section VI), while those in loop IV are rather constant. Indeed, until recently, it was thought that ribothymidine was always found in the U* position, but additional data now show that uridine (Piper and Clark, 1974a; Marcu *et al.*, 1973) 2'-O-methyl ribothymidine (Gross *et al.*, 1974), 2-thioribothymidine (Watanabe *et al.*, 1974), and even pseudouridine (Harada *et al.*, 1975) may replace ribothymidine. Pseudouridine is usually localized to two regions. It is found at a specific site in loop IV, and at a variety of loci in the anticodon loop and stem. Occasionally, it is found elsewhere in the molecule. The modification of uridine to 4-thiouridine found at the eighth base from the 5'-end has so far only been observed in prokaryotes. Additional regions of the molecule where unusual bases are frequently found can be deduced by inspection of Fig. 4 and Table II.

The C-C-A sequence deserves special mention. It is found at the acceptor end of all tRNA's as well as at the 3'-end of those viral RNA's which accept amino acid. In some tRNA's, the sequence is repeated two (*E. coli* tRNA^{Ile}) or three (*E. coli* tRNA^{Tyr}) times along the stem "a" sequence. Some substitution and/or modification of these bases is possible with retention of acceptor activity (Deutscher, 1973; Sprinzl *et al.*, 1973) and a tRNA with the augmented sequence C-C-C-A is still active, but shortening the sequence to C-A is inhibitory (Rether *et al.*, 1974). A special enzyme, nucleotidyltransferase, which exists in all cells including the mitochondria of eukaryotic cells, is capable of adding or removing the C-C-A sequence of a tRNA, but despite the ubiquitous presence of this enzyme and its requirement for normal growth of *E. coli* (Deutscher *et al.*, 1974), the synthesis of the C-C-A end has not yet been shown to be the major function of this enzyme (Morse and Deutscher, 1975). Its true role remains unclear.

The experimental evidence for the cloverleaf secondary structure rests on the following facts. First, the change in optical absorption or rotatory dispersion as a function of temperature or organic solvent content established rather early that tRNA possesses some form of reversibly disruptable secondary structure. Second, as indicated above, all known primary sequences can be fitted into the cloverleaf form. Moreover, in many instances when a second mutation in a tRNA gene restored activity lost by the first mutational event, the double base change resulted in a base *pair* substitution in one of the cloverleaf stems (Smith, 1972). Third, X-ray analysis of tRNA crystals confirmed the cloverleaf structure even before sufficient resolution was attained to show how it was folded into a three-dimensional form (Section IV,C,2). Fourth, by a special application of NMR spectroscopy, the number of $G \cdot C$ and $A \cdot U$ base pairs in a tRNA were determined and distinguished from each other according to their different neighbor nucleotides. Analysis of a number of tRNA's in this way confirmed the correctness of the cloverleaf structure (Kearns and Shulman, 1974).

C. Tertiary Structure

1. Conformation in Solution

Early evidence for the existence of tertiary structure in tRNA came from a comparison of sedimentation and hyperchromism as a function of temperature which showed that an appreciable increase in viscosity and hence unfolding occurred before much disruption of secondary structure took place (Henley *et al.*, 1966). That this structure was a defined conformation and not a random collection of structures in rapid equilibrium was best demonstrated by the discovery that tRNA's can be reversibly interconverted between functionally active and inactive conformations by appropriate treatment. These so-called native and denatured forms have been most extensively studied in the case of yeast $tRNA_3^{Leu}$ and *E. coli* $tRNA^{Trp}$ because these two tRNA's are easily stabilized in each form, but the same phenomenon has been demonstrated for many other tRNA's also. A detailed discussion can be found in Kearns *et al.* (1974), Uhlenbeck *et al.* (1974), Buckingham *et al.* (1974), and Greenspan and Litt (1974), together with references to the earlier literature.

In view of the common functional requirements for all tRNA's during the course of the protein synthesis cycle, it was not likely that the defined tertiary structure described above would prove to be radically different for each tRNA, and this was borne out by small angle X-ray scattering in solution on both unfractionated and purified tRNA species. The overall dimensions of the molecule, estimated by various authors to be approximately $85-100 \times 25-40$ Å (Cramer and Gauss, 1972), are too small to accommodate an open cloverleaf and also show that folding must occur.

Many other methods, both physical and chemical in approach, were and continue to be applied to tRNA in attempts to determine its shape in solution. Summaries of the earlier work may be found in Zachau (1972), Cramer and Gauss (1972), and Gauss *et al.* (1971). Physical methods used more recently have included (a) measurement of circular dichroism (Blum *et al.*, 1972; Prinz *et al.*, 1974), laser Raman scattering (Thomas *et al.*, 1973a; Chen *et al.*, 1975), hyperchromism (Coutts *et al.*, 1974; Urbanke *et al.*, 1975) as a function of denaturing solvent and/or temperature, including T-jump experiments, intrinsic fluorescence of rare bases in tRNA (Langlois *et al.*, 1975), and (b) the use of reporter groups such as ethidium bromide which are bound noncovalently to tRNA (Urbanke *et al.*, 1973; Tritton and Mohr, 1973), or covalently attached groups either for use as spin labels (Sprinzl *et al.*, 1974) or as fluorescent probes (Yang and Söll, 1974; Maelicke *et al.*, 1974).

A number of other powerful methods that depend in one way or another on a discrimination between hidden and exposed residues have been used in order to deduce something about the conformation of tRNA in solution. Ideally, the probe should not itself perturb the system under study, and should be applicable under more or less physiological conditions. In practice, it has not always been clear that these objectives were attained. Physical methods of this type include (a) proton exchange (Englander *et al.*, 1972) or specifically purine C_8-H exchange (Gamble and Schimmel, 1974) with ^3H_2O, (b) NMR spectroscopy in water, a technique that focuses specifically on H-bonding protons shielded by structure interactions from rapid exchange with water protons (Kearns and Shulman, 1974), (c) the binding of oligonucleotides of defined sequence to a tRNA whose primary sequence is known in order to probe for complementary single-stranded sequences (Pongs *et al.*, 1973; Miller *et al.*, 1974; Uhlenbeck *et al.*, 1974; Freier and Tinoco, 1975), (d) susceptibility to nucleolytic attack by single-strand specific enzymes (Streeck and Zachau, 1972; Petrova *et al.*, 1975), and (e) chemical modification by reagents with specificity for particular nucleoside residues (Brown, 1974; Chang and Ish-Horowitz, 1974; Piper and Clark, 1974b; Singhal, 1974; Rhodes, 1975). In this last case, it is not always appreciated that "buried" or "exposed" specifically refers only to those atoms of the nucleoside actually involved in the chemical reaction. For example, U_8 (Fig. 4) forms a hydrogen-bonded structure with A_{14} according to the three-dimensional crystal structure (see Fig. 5 and related discussion) yet it can be derivatized at its 4 position (in the analogous $^4S_8$ in *E. coli* valine tRNA) by alkylating agents with full retention of functional activity in all phases of protein synthesis (Schwartz and Ofengand, 1974). The explanation in this case is that the hydrogen-bonded structure uses the oxygen at position 2 (Kim *et al.*, 1974a; Klug *et al.*, 1974) whereas it is the oxygen (or sulfur) at position 4 that is alkylated. Similar considerations apply to the observed alkylation of N-7 of purines in stem regions (Vlasov *et al.,* 1972).

A direct approach that has not yet been extensively applied in the case of tRNA is that of cross-linking. In this method two parts of the tRNA are cross-linked by the use of some suitable reagent. Depending on the size of the reagent, one can then say with confidence that the two points cross-linked cannot be separated by more than the maximum span of the cross-linker. Of course, it is necessary that the cross-linker not disturb the tertiary structure of the tRNA being measured, at least until the cross-link is formed. A limiting case of this approach, and so far the only example described, is the photochemically induced cross-linking between $^4S_8$ and C_{13} in several *E. coli* tRNA's (Favre *et al.,* 1969). This reaction only takes place efficiently in native tRNA, and since no cross-linking molecule is involved in this case, it is clear that the two pyrimidine bases must be within covalent bonding distance of each other. A more detailed discussion of this reaction and some further applications can be found in Ofengand *et al.* (1974b), Delaney *et al.* (1974), and Favre *et al.* (1975). This finding was one of the few hard facts available to model builders in the early seventies and was given considerable weight because, unlike most solution studies, it was subject to little ambiguity. Photoaffinity probes of the type described in Section III,B appear to be very useful in this regard, particularly since they can be attached at one point chemically and only photolyzed to generate the cross-link when the tRNA molecule has equilibrated to the desired structure. Although several laboratories are currently experimenting with this approach, no results have appeared so far. Recently, a chemical cross-linking study was reported (Grachev and Rivkin, 1975) which showed that the amino group of AA-tRNA was within 20 Å of the 5′-phosphate.

2. Conformation in Crystals

Despite the voluminous literature describing studies on tRNA conformation in solution, a clear understanding of this large body of data was not possible until a molecular model of tRNA was obtained independently, namely, by X-ray analysis of tRNA crystals. Two groups have now completed an analysis of monoclinic (Robertus *et al.,* 1974a; Ladner *et al.,* 1975) and orthorhombic (Kim *et al.,* 1974b; Quigley *et al.,* 1975) crystals of yeast tRNA^{Phe}, resulting in the three-dimensional models shown in Fig. 5. The two structures are virtually identical.

The estimates of the dimensions are, from top to bottom, 77 Å, from left to right, 65 Å, and 75 Å from the tip of the anticodon to the C-C-A end. The overall thickness is approximately 25 Å. The molecule consists of two helical arrays at approximately right angles to each other. One rather

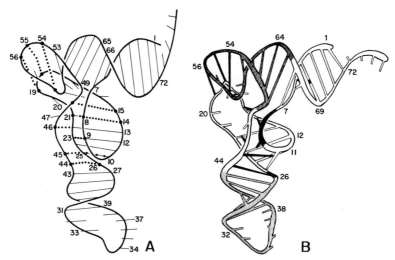

Fig. 5 Schematic diagrams of the tertiary structure of yeast tRNA^{Phe}. The molecules are oriented so that stem a points to the right, loops I and IV are in the upper left corner, and stem c and loop II points downward. The bases are numbered from the 5'-end. (A) The ribose phosphate backbone is shown by a continuous line. Double helical base pairs are indicated by long, light lines, and nonpaired bases by shorter lines. Base pairs in addition to those of Fig. 4 are indicated by dotted connecting lines. From Ladner *et al.* (1975). (B) The ribose phosphate backbone is shown as a continuous cylinder with ladder rungs to indicate hydrogen-bonded base pairs. Unpaired bases are indicated by shortened rods. Tertiary structure hydrogen bonds are shown as black rods. The stippled region corresponds to stem e and loop IV, the region marked in vertical lines is stem c and loop II, and the blackened segments of the backbone mark residues 8, 9, and 26. From Quigley *et al.* (1975).

regular helix is made from stems a and e in which the 3' chain is continuous and a much more irregular helix consists of stems b and c plus additional bases. The central part of the molecule is stabilized by a number of unusual tertiary structural features including both stacking interactions and the formation of unusual H bonds. These are, from top to bottom, the nonstandard trans base pair, $G_{15} \cdot C_{48}$, required because in this region of the molecule the two polynucleotide chains run parallel to each other instead of anti-parallel, a reverse Hoogsteen base pair between the invariant bases U_8 and A_{14} which also involves A_{21}, and the base triples $m^7G_{46} \cdot G_{22} \cdot C_{13}$, $A_9 \cdot A_{23} \cdot U_{12}$, and $C_{25} \cdot m^2G_{10} \cdot G_{45}$. In both models U_8 overlaps C_{13} such that the Favre-Yaniv light induced cross-link can be readily formed. Loops I and IV interact by forming H bonds between the invariant bases Ψ_{55}, C_{56}, and G_{18}, G_{19} with, in addition, intercalation of G_{57} between G_{18} and G_{19}. Additional stacking interactions and H bonds from bases to ribose hydroxyl and phosphate groups exist such that the only bases turned out to the surface of the molecule are D_{16}, D_{17}, and G_{20} (the variable regions of loop I), U_{47} of loop III, part of the anticodon loop (II), and the C-C-A end. The original papers, also Kim *et al.* (1974a) and Klug *et al.* (1974), should be consulted for full details.

Despite this *tour de force*, it is important to ask (a) how well the structure agrees with what has been deduced about this tRNA in solution, and (b) how generally applicable this three-dimensional model may be to other tRNA's. A detailed study of the exposed versus buried bases in this tRNA by the chemical probe method (Robertus *et al.*, 1974b; Rhodes, 1975) is in complete agreement with the prediction of the X-ray model, requiring that any alternative models must also shield the same residues, and NMR studies (Jones and Kearns, 1974) have confirmed that stems a and e are stacked, and that the internal end of stem a is close to the middle of stem b. However, other NMR studies (Kearns and Shulman, 1974; Hilbers and Shulman, 1974) failed to detect the ring NH protons involved in the various tertiary structural interactions mentioned above (a total of 6).

An extensive series of studies carried out on other tRNA species in solution are in good agreement with most aspects of the structure shown in Fig. 5, thus not only supporting this structure but also confirming that its general features are likely to be correct for all tRNA's. NMR analysis has shown stacking of stem a on stem e in *E. coli* tRNA^{Glu} (Hilbers and Shulman, 1974), an analysis of mixed (Wong and Kearns, 1974) and purified species (Wong *et al.*, 1975; Daniel and Cohn, 1975) of *E. coli* tRNA have detected the resonance due to the U_8-A_{14} base pair (in this case $^4S_8$-A_{14}), and extra ring NH protons involved in tertiary structure have recently been detected in *E. coli* valine tRNA (Reid and Robillard, 1975) and *E. coli* formylmethionine tRNA (Daniel and Cohn, 1975). Furthermore, singlet-singlet energy transfer studies using dyes attached at several sites to *E. coli* tRNA$_f^{Met}$ and tRNA^{Glu} (Yang and Söll, 1974) have yielded distance measurements in good agreement with the above model. Extensive studies on chemically modified tRNA species, including tRNA$_f^{Met}$, tRNA^{Tyr}, and tRNA$_1^{Leu}$ all from *E. coli* (Brown, 1974; Chang and Ish-Horowitz, 1974), have also confirmed that, by and large, all the residues predicted to be shielded either by H bonds or stacking interactions are indeed unreactive. In general, only the C-C-A end, loop II, and the variable regions of loop I seem to be exposed areas in those tRNA's so far examined. In tRNA^{Glu} of *E. coli,* loop I also seems to be unreactive (Singhal, 1974). A careful inspection of all the chemical data shows some variation in the reactivity of loop II, suggesting that this region may be heterogeneous in conformation. The existence of the R(loop I)-Y(stem d) base pair in other tRNA's ($G_{15} \cdot C_{48}$ in yeast tRNA^{Phe}) is supported by the finding that in *E. coli* tRNA^{Tyr} a mutation of G_{15} to A_{15} makes the corresponding C_{57} reactive to the chemical probe methoxyamine (Cashmore, 1971).

Additional support for the generality of the model has recently been obtained by the demonstration that most of the tertiary structure H bonds described for yeast tRNA^{Phe} that involve variable bases are also possible with the alternative bases found in analogous positions in other tRNA's (Kim *et al.*, 1974a; Klug *et al.*, 1974) because the base substitutions are largely correlated with each other, as illustrated by the trans base pair $G_{15} \cdot C_{48}$. In all but two out of 55 sequences, position 15 is either G or A, and correspondingly the position adjacent to the 5'-end of stem e is always either C or U, respectively, in order to allow formation of this base pair. The structures are not standard Watson-Crick base pairs, however, and in fact, the details of the model require that the purine always be at position 15, as found experimentally. The purine-pyrimidine positions are never found reversed, as would be expected to occur occasionally if standard base pairs were involved. Consequently, this feature of the model is probably invariant for all tRNA's. Similar arguments can be produced for most of the other tertiary interactions.

3. Conformational Changes during Protein Synthesis

Since tRNA functions in protein synthesis as AA-tRNA complexed with synthetases, elongation factors, and ribosomes, it is appropriate to ask if the structure of AA-tRNA is the same as tRNA, if the structure is different when complexed with the various factors involved, and if the structure is functionally dynamic, that is, if the conformation changes purposefully during the operation of the protein synthesis cycle.

The effect of aminoacylation on tRNA conformation would appear to be easy to determine, yet paradoxically there is still no complete answer to this question. Most studies by a variety of techniques have not shown any clear-cut conformational differences between tRNA and AA-tRNA (Wong *et al.*, 1973; Beres and Lucas-Lenard, 1973), yet reports do exist of ill-defined but definite differences in various physical parameters (Thomas *et al.*, 1973b, and references therein). Probably, this apparent discrepancy is due only to the use of detection methods that respond to different conformational parameters, and a difference does exist between acylated and nonacylated tRNA. However, in retrospect, such studies no longer appear relevant to processes in the cell since a consideration of the estimated concentrations of synthetases, tRNA, and elongation factors and a knowledge of the association constants involved (see Section II,E) make it highly unlikely that AA-tRNA ever exists free in the cell, uncomplexed with proteins.

Seen in this light, it is more appropriate to be concerned with the conformation and possible conformational changes of tRNA or AA-tRNA complexed with synthetases, initiation and elongation factors, or ribosome binding sites. It would appear that the solution methods described above which have been so far used only on tRNA could be profitably applied to an analysis of tRNA-protein complexes and to monitor changes in conformation during protein synthesis. This is particularly true now that many of the assumptions implicit in the methods have been validated by their good agreement with the X-ray results, but such studies are still in their infancy.

Relevant observations on tRNA-synthetase complexes are considered in Section V, and those on elongation factors and ribosome complexes and possible changes in conformation accompanying formation of these complexes are considered in Sections VII and VIII.

4. Effect of Modified Bases on Structure

One might have anticipated that the many unusual bases in tRNA would play a dominant role in controlling tertiary structural interactions, particularly those like m^7G and m^1A that carry a net positive charge at neutral pH, but this expectation has not been met to judge from the limited amount of experimental data available. The effects of modified bases at position * and R* in loop II (Fig. 4) will be treated separately (Section VI). Replacement of Ψ, T, hU, ^4S, and cmo^5U as well as all of the U residues by 5-fluorouridine did not affect the optical, electrophoretic, or nuclease susceptibility properties of *E. coli* valine tRNA (Horowitz *et al.*, 1974). On the other hand, replacement of the Ψ residues, except for the one in loop IV, by U in tRNA^{His}, tRNA^{Leu}, and tRNA^{Ile} resulted in an altered chromatographic behavior (Cortese *et al.*, 1974). No gross differences were observed in a tRNA$_f^{Met}$ of *E. coli* when its m^7G residue was substituted by an A (Egan *et al.*, 1973), although some subtle effects were detected, particularly at low ionic strength, which involved the precise positioning of residues $^4S_8$ and C_{13} (Delaney *et al.*, 1974). This effect may be related to the fact that the A_{46}-G_{22}-C_{13} base triple cannot form (Kim *et al.*, 1974a), rather than to a specific need for a positive charge, as from m^7G, at position 46.

In fact, none of the tertiary interactions described above require the participation of a modified base. Instead, just the opposite is true. The bases modified at normal H-bonding positions, such as m^1A, ^4S, and m^1G, form unusual H bonds that do not require the participation of these atoms. Moreover, no special structural role for the Wye base (W) at the R* tRNA position in yeast tRNA^{Phe} could be detected by X-ray analysis (Robertus *et al.*, 1974a) although other studies have shown that its absence perturbs the stem c helix slightly and may also have some effect on stem b (Kearns *et al.*, 1973, and references therein).

Note also that the Ψ residue, whose synthesis requires the breaking and remaking of a glycosidic bond at the polynucleotide level (Section IX) in order to generate an H bond donor site at the position of C-5 of uridine, does not appear to use this extra H bond for structure interactions (at least at the current level of resolution of the X-ray analysis). Moreover, as already noted, substitution by 5-fluorouridine has little effect structurally, or functionally in protein synthesis (Ofengand *et al.*, 1974c). At this stage of our understanding, one has the impression that minor base modifications were introduced to add increased specificity to certain functional reactions carried out by tRNA but that they have no critical structural role.

D. Initiator tRNA

1. Function in Prokaryotes and Eukaryotes

A detailed discussion of the mechanism of chain initiation can be found in Chapter 5. Here we only restate some facts necessary for a discussion of the structural differences between prokaryotic and eukaryotic initiator and internal tRNA's. The universal mechanism for chain initiation uses a special initiator methionine tRNA, distinct from the methionine tRNA that inserts methionine into internal positions in proteins, called $tRNA_i^{Met}$ in eukaryotes, and $tRNA_f^{Met}$ in prokaryotes. The f refers to formylatable species because in prokaryotes the functionally active initiator species is N-formylmethionyl-$tRNA_f^{Met}$. Unformylated Met-$tRNA_f$ is much less active in prokaryotic initiation complex formation but in eukaryotes Met-$tRNA_i$ is active both in unformylated or formylated forms. Correspondingly, prokaryotes possess a specific enzyme for catalyzing the formylation of Met-$tRNA_f$, while this enzyme is lacking in eukaryotes. The functional similarities and differences among prokaryotic and eukaryotic initiator and internal methionine tRNA's are summarized in Table III.

The following particular features should be noted. *E. coli* synthetase can acylate both *E. coli* tRNA's, as well as eukaryote $tRNA_i$ but not eukaryote $tRNA_m$. The reverse is true of eukaryotic synthetase. *E. coli* formylase recognizes common features in both types of initiator tRNA yet selects against both types of internal $tRNA^{Met}$ species. Initiation complex formation is specific for, or at least prefers, initiator tRNA (*E. coli* Ac-Phe-tRNA is also acceptable), and both $tRNA_f$ and $tRNA_i$ are active. With prokaryotic ribosomes, formylation of either tRNA is necessary (Ghosh *et al.*, 1971), but in eukaryotes it has little effect (Gupta *et al.*, 1971; Drews *et al.*, 1972). Elongation factor prefers the noninitiator tRNA. Peptidylhydrolase, a ubiquitous enzyme, is nonselective for acyl-AA-tRNA except that fMet- or acyl-Met-$tRNA_f^{Met}$ is resistant.

TABLE III
Functional Properties of Prokaryotic and Eukaryotic Initiator tRNA's

Source	Protein	Form of tRNA^{Met}	Prokaryotic		Eukaryotic	
			tRNA$_f$	tRNA$_m$	tRNA$_i$	tRNA$_m$
Prokaryote	Synthetase	tRNA	+	+	+[a]	−[a]
(*E. coli*)	Formylase	Met-tRNA	+[b]	−	+[c]	−
	Peptidyl hydrolase[d]	f Met-tRNA	−	+	+	+
	Initiation complex (70 S)	fMet-tRNA[e]	+[b]	±[f]	+[f,g]	+[f]; −[g]
	Elongation factor	Met-tRNA	−[h]	+	±[i]	+
Eukaryote	Synthetase	tRNA	+	−	+	+
	Peptidyl hydrolase[d]	Acyl-AA-tRNA	+	+	+	+
	Inititation complex (80 S)	Met-tRNA	±[l]	−	+[j]	−
	Elongation factor	Met-tRNA	−[i]	+	±[g,k]	+

[a] Stanley (1974); Samuel and Rabinowitz (1974). Yeast tRNA$_m^{Met}$ reacts under certain conditions.

[b] Misacylated tRNA$_f^{Met}$ can be formylated and will initiate (Giege *et al.*, 1973).

[c] Except wheat embryo (Leis and Keller, 1970).

[d] Menninger *et al.* (1974) and references therein.

[e] Formylation required except in T-minus tRNA (Samuel and Rabinowitz, 1974).

[f] Drews *et al.* (1973).

[g] Ghosh *et al.* (1974).

[h] Regeneration of seventh base pair in stem a activates (Schulman *et al.*, 1974).

[i] Richter and Lipmann (1970).

[j] Formylated or unformylated; Drews *et al.* (1972); Gupta *et al.* (1971).

[k] Richter *et al.* (1971); Drews *et al.* (1972).

[l] Two different initiation factors exist, only one of which can use tRNA$_f$. This situation is still unclear.

2. Structural Properties

In the hope of accounting for these functional distinctions in terms of primary structure, the sequence of *E. coli* tRNA$_f^{Met}$ and tRNA$_m^{Met}$, yeast tRNA$_i^{Met}$ and tRNA$_m^{Met}$, *B. subtilis* tRNA$_f^{Met}$, rabbit and sheep tRNA$_i^{Met}$, and mouse tRNA$_i^{Met}$ and tRNA$_m^{Met}$ were determined (refs. cited in legend to Fig. 4). Two related questions are involved. First, does a comparison of initiator and noninitiator sequences explain the functional differences between these two tRNA's, and second, does a comparison of prokaryotic and eukaryotic initiator tRNA sequences explain their behavior in formylation and initiation complex formation?

Comparison of *E. coli* tRNA$_f^{Met}$ with tRNA$_m^{Met}$ or other *E. coli* tRNA's of the same class shows only one obvious difference. The 3'-terminal base pair in stem a is missing. However, when this is repaired by a bisulfite-catalyzed conversion of the 5'-C to U, there is no effect on the ability to be formylated (Schulman *et al.*, 1974) or on the ability to bind to ribosomes in an initiation factor-dependent reaction (L. H. Schulman, personal communication). Eukaryotic tRNA$_i$ which can substitute for tRNA$_f$ in both aminoacylation and formylation has this base pair. Nevertheless, since the absence of this base pair has been conserved in tRNA$_f$ of a blue-green algae, *Anacystis nidulans,* although one of the bases is different (Ecarot and Cedergren, 1974), and in *Mycoplasma* (Walker and Raj-Bhandary, 1975) it probably serves some necessary function. The function may be to protect fMet-tRNA$_f$ in prokaryotes from hydrolysis by pepti-dylhydrolase, a ubiquitous enzyme specific for any *N*-acyl-AA-tRNA, since the bisulfite-repaired fMet-tRNA$_f$ is a good substrate for this enzyme (Schulman and Pelka, 1975). Since Met-tRNA$_i$ is not formylated in eukaryotes, it is not susceptible to attack by this enzyme, and does not need such a modification. The missing base pair also blocks elongation factor recognition (Schulman *et al.*, 1974).

Loop IV of tRNA$_f^{Met}$ is not so unusual, since the entire 7-membered loop can also be found in the serine and proline tRNA's that are made in T4-infected *E. coli,* and moreover, eukaryote tRNA$_i$, which can substitute for tRNA$_f$ has an entirely different loop IV sequence. Residue R* in loop II which is frequently modified is an unmodified A in tRNA$_f$, but this has been found in several *E. coli* tRNA's, and cannot be a distinguishing feature.

The sequences of the four eukaryote tRNA$_i$ molecules have produced a few surprises. First, all three mammalian tRNA's are identical, with only minor differences from the yeast tRNA. Second, loop IV is identical in all four sequences and completely different from *E. coli*. A-U replaced the ubiquitous U*-Ψ found in all other functional tRNA's to date, giving a loop sequence A-U-C-G-m^{1}A-A-A. The same sequence is also found in wheat embryo, except that U is replaced by an unknown U derivative (Ghosh *et al.*, 1974). Also unusual was the presence of A at the 3'-end of the loop, but since mouse valine tRNA (Piper and Clark, 1974a) also has this feature in a normal U-Ψ-C-G-m^{1}A-A-A sequence, this probably only reflects species variation. The last unusual feature, so far unique to eukaryotic initiator tRNA is the replacement of the universal U, 5' to the anticodon, by a C. This is *not* present in five other mammalian tRNA's examined so far. However, chemical probe studies of this tRNA (Piper and Clark, 1974b) showed nothing unusual about the conformation of this loop.

Since the *E. coli* methionine synthetase acylates its homologous tRNA$_f$ and tRNA$_m$ as well as all heterologous tRNA$_i$ with about the same K_m and V_{max}, it is possible to look for sequence homologies among all acylatable tRNA's in a search for the synthetase recognition site, assuming that it is so simple as a unique stretch of nucleotides. Sad to say, after subtraction of the bases universal in all tRNA sequences, very little is left (Simsek *et al.*, 1974). This point is reconsidered in Section V.

Similarly, a search for a nucleotide sequence corresponding to the formylase recognition site and to the prokaryote initiation site has not been very revealing. If one subtracts from the published common formylase recognition sequence (Piper and Clark, 1974c) those bases found also in *E. coli* tRNA$_m$ and wheat embryo tRNA$_i$ that are not recognized by the formylase, and in addition takes out the universal bases, virtually nothing is left except the set of three G · C base pairs in stem b adjacent to loop II. It is not likely that this constitutes the formylase recognition site. However, subtraction of the corresponding sections of inactive tRNA's is a risky approach since a negative result could be due to a wrong sequence elsewhere counteracting a correct sequence. Nevertheless, at this stage it is not clear that the sequence homology approach will be successful in unraveling either the prokaryote initiation-specificity or formylase-specificity features of tRNA. It is more likely that understanding will only come when the three-dimensional conformation of these molecules is known in detail.

For example, the requirement for formylation of Met-tRNA$_f$ for initiation complex formation in *S. faecalis* can be eliminated by the use of tRNA$_{Met}$ from folate-deprived cells (Samuel and Rabinowitz, 1974), in which T in loop IV is replaced by U (Delk and Rabinowitz, 1974). However, formylation could still be carried out *in vitro* and did not inhibit initiation complex formation. Possibly additional elements are involved since eukaryote tRNA$_i$, which also lacks T, is reported to require formylation before it can initiate in an *E. coli* system (Ghosh *et al.*, 1971).

The situation with regard to eukaryotic initiation may not be so bleak. The strict conservation of loop IV in all eukaryotic initiators and the marked divergence from all other eukaryotic tRNA's strongly implies a function in initiation, possibly for recognition of the initiator binding site on the ribosome. The fact that prokaryote initiator tRNA, with a completely different loop IV, can substitute for eukaryote tRNA$_i$ would contradict this proposal, but since it is not clear whether true initiation in eukaryotes can proceed with prokaryote initiator tRNA, the difference in loop IV between these two species may not be significant. Nevertheless, some unclear common feature of three-dimensional structure must exist in both tRNA$_f$ and tRNA$_i$ to enable them to be interchangeable in prokaryotic formylation and initiation complex formation.

V. Specificity of the Aminoacylation Reaction

A. The Problem

As discussed in the introduction, tRNA exists in order to adapt amino acid recognition into a form recognizable by the codon. Formal proof of the adapter role for tRNA was first supplied in 1962 by an experiment in which a misacylated tRNA was prepared and shown to place its amino acid in a polypeptide sequence according to the nature of the tRNA, not the amino acid. A brief account of this classic experiment may be found in Chapeville (1972). Consequently, the cell must maintain the correct specificity between amino acid and tRNA if errors in placement of amino acids in proteins are to be avoided. The problem then is to determine what it is about tRNA's that determine their specificity for interaction with a *particular* synthetase. Several possibilities can be considered. (1) There is a specific site in the three-dimensional structure of tRNA that is used for synthetase recognition. This site may be made up of one or several noncontiguous sequence elements but all tRNA's use analogous positions. Specificity is obtained by variation of the sequences involved. (2) There are two or more such sites which make up the recognition site, the other conditions of (1) being the same. (3) and (4). As (1) and (2) but with the proviso that different tRNA-synthetase pairs may use different regions of the tRNA to make up their site(s). (5) The tertiary structure of the molecule or part of it determines specificity. In this connection, it should be recalled that tRNA's can be denatured (Section IV,C,1) and, therefore, that some elements of tertiary structure are necessary. In the following, I will attempt to summarize the current state of knowledge. Two recent excellent reviews (Söll and Schimmel, 1974; Kisselev and Favorova, 1974) cover the earlier literature in depth, so that I will mention with specific references only newer knowledge and interpretations. References to the original older literature can be found in these two reviews.

B. Measurement of tRNA-Protein Interaction

1. Physical Methods

Direct complex formation between tRNA and synthetase measures the specificity of interaction independent of the catalysis of aminoacylation and so allows this property to be studied independently of other ligands. Many methods have been used, including nonequilibrium methods such as binding of the synthetase-tRNA complexes to membrane filters, gel chromatography, and zonal centrifugation, as well as nonperturbing methods like fluorescence spectroscopy. Recent work has used quenching of the fluorescence of tryptophan residues in the synthetase as well as naturally fluorescent nucleotides or covalently linked dyes on the tRNA to monitor complex formation between synthetase and cognate or noncognate tRNA's (Pachmann *et al.*, 1973; Blanquet *et al.*, 1973; Pingoud *et al.*, 1973; Krauss *et al.*, 1973). Noncognate interactions were found to be surprisingly strong considering the high degree of specificity found *in vivo*. Binding from 10% (Blanquet *et al.*, 1973) to 100% (Pachmann *et al.*, 1973) as strong as cognate binding has been observed, although in other examples the interactions observed were much weaker (see, for example, Krauss *et al.*, 1975). These results, as well as those discussed below, provide the basis for the current view that noncognate complex formation occurs readily. The specificity of aminoacylation comes from kinetic control over the catalytic step.

2. Reaction Catalysis

Probably the earliest method used for measuring the specificity of aminoacylation was the test of aminoacylation itself, usually without careful attention to the kinetics of the process. Early studies with purified synthetases and mixtures of tRNA's showed that in the homologous systems there was complete specificity, while heterologous systems occasionally gave evidence for misacylation reactions. They were detected initially as an increased aminoacylation in the heterologous reaction compared to the homologous case (reviewed in Jacobson, 1971). The situation has been recently clarified by a series of studies of misacylation in homologous as well as heterologous systems (Giege *et al.*, 1974, and references therein). Comparison of K_m and V_{max} for the correct and incorrect reactions has shown that the affinity of noncognate tRNA's for synthetases are in many cases as strong as cognate tRNA's, but that the maximum velocity of the aminoacylation is vastly reduced in the incorrect case. Although most of these studies have been carried out in heterologous systems and usually under nonphysiological conditions since more and more varied misacylations can be obtained in this way, a few examples of misacylation in homologous systems under standard conditions have also been studied. This is clearly necessary since ultimately one is concerned with the specificity in the cell, and not in a laboratory curiosity made by mixing extracts of different cells. In one such example studied by Ebel *et al.* (1973), pure E^{Val} of yeast was shown to misacylate tRNA^{Phe} and tRNA^{Ala} of yeast with K_m values only 360 and 63 times higher, respectively. However, since the V_{max} was only $1-2 \times 10^{-4}$ that of the correct reaction, it is clear that detectable levels of misacylation will only occur in the absence of cognate tRNA. The high degree of specificity observed in early experiments are now interpreted as resulting from the presence of both cognate and noncognate tRNA's in the unfractionated mixture which competed with each other for synthetase.

Thus, the high degree of specificity of homologous aminoacylation is now understood not in terms of specificity in complex formation, but as an interplay of (a) specificity in the catalytic step (V_{max} effect), (b) competition between $tRNA^A$ and $tRNA^B$ for $synthetase^A$ as well as between synthetase A and synthetase B for $tRNA^B$, and (c) enzyme-catalyzed deacylation of AA-tRNA (Section II,F) which can become significant when the aminoacylation V_{max} is low.

An additional feature to come out of these studies is the recognition that there exist "families" of tRNA's, members of which are more readily misacylated by synthetases of the same "family" when the barriers to misacylation are lowered, as for example by using heterologous systems and/or special acylation conditions. An interesting feature of this phenomenon is that these "families" of misacylatable homologous tRNA's seem to be the same or similar in *E. coli, B. stearothermophilus,* and yeast (Giege *et al.,* 1974) suggesting a common evolutionary origin.

C. Source of Specificity

1. Sequence Homology Approach

On the assumption that particular nucleotide sequences in tRNA determine in some unique way the recognition site for synthetase, the sequences of those tRNA's known to be acylatable by a given synthetase have been examined in the hope that some common pattern of nucleotides would emerge. Such comparisons of sets of isoacceptor tRNA's for a given synthetase (Söll and Schimmel, 1974; Piper and Clark, 1974a), sets of heterologous cognate tRNA's (Simsek *et al.,* 1974; Axel'rod *et al.,* 1974) and sets of heterologous misacylated tRNA's (Roe *et al.,* 1973b; Giege *et al.,* 1974) have been carried out. It should be mentioned that only positive cases of recognition represent useful data, since failure to acylate could be due to negative effects of other sequences not at the recognition site. The most extensive set of data has been compiled in the misacylation system. Although the findings have been criticized for the nonphysiological conditions needed to obtain misacylation, the observation that only certain misacylations are found under a given set of conditions suggests that some parameter of recognition is being explored in these studies.

According to Roe *et al.* (1973b) stem b plus the unpaired base adjacent to C-C-A are the determinants of recognition (K_m) for yeast E^{Phe}, with the size of loop I and the state of methylation of G_{10} influencing the V_{max} of the reaction. However, analogous recognition sites have not been found for *E. coli* E^{Met} (Simsek *et al.*, 1974), for yeast E^{Val} (Axel'rod *et al.*, 1974), or for *B. stearothermophilus* E^{Val} (Giege *et al.*, 1974), to cite a few examples. Moreover, in contrast to the analysis of Roe *et al.* (1973b), Giege *et al.* (1974) interpret their findings to mean that the sequences near the 3'-end of stem a and the anticodon are involved in determining the V_{max} of acylation, but not the K_m. Other examples of anticodon effects on V_{max} are also known (summarized in Roberts and Carbon, 1974). Although these differences may be explained by the possibility that different synthetases use different patterns of recognition [hypotheses (3) and (4) in Section V,A], it is still far from clear that sequence comparisons will reveal the source of specificity in aminoacylation.

2. Site Inactivation (tRNA Modifications)

Probably the earliest attempts to probe structure-function relationships in tRNA were those which analyzed tRNA modified in some specific way. This was understandable since such experiments were possible before many primary sequences were known. The very extensive literature on tRNA modifications will be discussed here according to the methods used, namely, chemical modifications, enzymatic modifications including the "dissected molecule" approach, base interconversions by mutational or chemical methods, and base analogue substitutions.

Before discussing them in turn, it should be pointed out that many of these experiments were deficient in light of our current ideas about the source of specificity in tRNA-synthetase interactions. In most cases, modification was assessed only by the loss of amino acid acceptor capacity in some standard assay, without regard for whether the K_m had become abnormally large, the V_{max} inordinately low, or both. Direct association constants were almost never measured. Only rarely was a change in specificity directly examined. Thus, it is usually not clear whether a general binding site for tRNA was blocked when inactivation occurred, or whether the specificity region was altered.

An additional difficulty in interpretation is peculiar to modification experiments. Any time a chemical group is added, phosphodiester bonds cleaved, nucleotides removed, or indeed any time *any* unnatural structures are created which are inactive, at least three interpretations are possible. First, the site modified may be (part of) the essential site. Second, the tertiary structure may be deformed, leading to inactivation according to hypothesis 5 above. Third, the tertiary structure may be disturbed by the modification so that the separate parts of the essential site, themselves unmodified, cannot come together. In view of this unsatisfactory state of affairs, the only modifications likely to be meaningful are those (a) that lead to a *retention* of activity since the modified areas are clearly unnecessary and (b) that are unlikely to affect tertiary structure because of their position. Modifications of loop II or the 3′-portion of stem a may belong in this class since these areas do not appear to interact with other regions (Fig. 5). Of course, bulky or charged modifying groups even here may interfere by disturbing the overall conformation, and destruction of base pairing in an isolated stem may even cause conformational disturbances, as described below. Unfortunately, even in those cases where activity was retained, the kinetics of acylation were not usually studied so that a quantitative assessment of the retention of activity in terms of complexing ability or acylation rate is often not available.

Chemical modification has been extensively reviewed by Chambers (1971), Gauss *et al.* (1971), and Zachau (1972) as well as in the recent reviews of Söll and Schimmel (1974) and Kisselev and Favorova (1974). Although many reagents have been used, some of them rather specific for certain minor nucleotides, the complications described above have not allowed much definitive information to be obtained, and the method is no longer much used for this purpose.

Enzymatic modifications include introduction of one or several breaks in the phosphodiester chain with or without removal of a few nucleotides, and the technique of "dissected molecules." This latter approach was pioneered by Bayev who in 1967 was the first to show that a tRNA could be enzymatically cleaved at the anticodon, separated into two pure half-molecules, and then reconstituted to an acylatable tRNA. The method has since been extended by various groups, notably in exhaustive studies of yeast tRNA^{Phe} and tRNA^{Ser} by Zachau and co-workers (summarized in Zachau *et al.*, 1974) and of yeast tRNA^{Val} by Bayev and his collaborators (reviewed in Mirzabekov and Bayev, 1974). A brief discussion of these results will illustrate the difficulties of making any generalizations about synthetase recognition sites. As discussed above, only the excision of nucleotides which lead to retention of activity are likely to be meaningful. A reasonable exception to this would be the residues in loop II since they are not supposed to interact with other parts of the tRNA. In yeast tRNA^{Ser} the 3′-half of loop II and stem c can be removed with only a 50% decrease in acylation, in yeast tRNA^{Phe} the entire anticodon can be excised with 20% activity remaining, and in *Torula* yeast tRNA^{Tyr}, removal of the anticodon leaves 25% activity (Hashimoto *et al.*, 1972). On the other hand, in yeast tRNA^{Val}, removal of the middle and 3′-nucleotide of the anticodon inactivates completely, but excision of the 5′-half of loop II does not. In a recent report, a variation of this approach was used. Half-molecules from yeast tRNA^{Ala} and *E. coli* tRNA^{Val}, chosen in order to maximize base pairing potential in stems a and c in the heterologous combination, were found to be acylatable to a small extent (6–12%), but only by the enzyme cognate to the 5'-half molecule (Wubbeler *et al.,* 1975). This result implies that loop I and/or stem b are the specificity regions. However, in contrast, the 3'-half-molecules of these same tRNA's as well as yeast tRNA^{Ser} were also found to be acylatable (up to 20–40%) under special renaturing conditions. One is forced to conclude that while it may be possible, by dint of much hard work, to learn something about the recognition site for one synthetase-tRNA pair, the information is not likely to be of much help in the analysis of any other such pairs.

Conversion of one base to another in tRNA has usually been accomplished by mutational means (reviewed in Smith, 1972; Littauer and Inouye, 1973) but C to U transitions can also be effected by *in vitro* treatment with nitrous acid (Carbon and Curry, 1968) or sodium bisulfite (reviewed in Singhal, 1974). This approach would appear to be the mildest of all since one natural base is replaced by another. However, when it is recalled that the stability of stem regions as well as tertiary structure depends on standard and nonstandard base pairing, it can be appreciated that a base change from a pairing to a nonpairing residue may have unforeseen consequences. One example was mentioned in Section IV,C,2, where a G_{15} to A_{15} mutation exposed C_{57} to reaction with methoxyamine. Another example is the UGA suppressor tRNA derived from tryptophan tRNA. This tRNA can read UGA as well as UGG, the normal codon for tryptophan, and thus suppresses the UGA termination signal. Surprisingly, the mutation which results in this capability is not in the anticodon but is a G to A change in stem b (Hirsh, 1971). This single change located some distance away is sufficient to allow base pairing with both UGA and UGG when the tRNA is bound to the ribosome.

Despite these potential ambiguities, extremely interesting results have been obtained in two studies where an actual *change* in specificity for synthetase was accomplished by a single base change in the tRNA. In the first study, mutants of *sup3* tyrosine tRNA were obtained which could insert an amino acid other than tyrosine into protein *in vivo.* Ultimately, six such mutants were isolated. All turned out to insert glutamine *in vivo* and to accept glutamine as well as tyrosine *in vitro* although some mutants were more readily acylated than others. Surprisingly, all were mutants in the first two base pairs of stem a and in the adjacent unpaired nucleotide. Apparently, a single base change in this region allows misacylation by E^{Gln}. This is all the more striking since tRNA^{Tyr} bears no simple structural relationship to tRNA^{Gln}. Most notably, tRNA^{Tyr} has a large stem d-loop III region while tRNA^{Gln} has a small loop III. Moreover, double mutants which restore base pairing by substituting a different base pair in the stem restore the original specificity for aminoacylation, at least in part. On the basis of their detailed results, the authors (Ghysen and Celis, 1974a) suggest that the unpaired fourth nucleotide from the 3'-end may have some discriminatory role in specificity (Crothers *et al.*, 1972) but that the conformation of stem a, that is whether it is base paired or not, is also important. Whatever the detailed explanation may be, these experiments for the first time show clearly that, at least in one case, the 3'-end is a *specificity* determinant.

The second clear example of a specificity change shows, on the other hand, that the anticodon is important. In this case, two suppressor tRNA's inserting glutamine in response to the codons UAG and UAA were isolated from *E. coli* and shown to correspond to mutated forms of the single tryptophan tRNA with one or two C to U changes in the anticodon which allowed it to read the above two terminator codons (Yaniv *et al.*, 1974). The single (or double) base change from C to U in the anticodon evidently also allowed acylation by E^{Gln}, and in this case blocked acylation by E^{Trp}. Recently, essentially equivalent results have been obtained by a bisulfite-induced change of C to U in the $tRNA^{Trp}$ of *E. coli* (Seno, 1975).

The significance of these two experiments cannot be overemphasized. Although one is no closer to an interpretation than before, and indeed one is tempted to throw up one's hands in the face of these diametrically opposed results and wait for the X-ray analysis of crystals of tRNA-synthetase complexes, it is clear that experiments such as these where the *specificity* of the interaction has been altered by a minimal structural change will provide the most pertinent information.

In several other cases, anticodon mutations are known which also result in a severely reduced V_{max} or complete failure to acylate. Two glycine tRNA suppressors, one of which is due to a C to U mutation in the 3'-base of the anticodon (Roberts and Carbon, 1974) and the other to a C to A mutation in the 3'-base of a different glycine isoacceptor (Carbon and Fleck, 1974), have their V_{max} reduced 600- to 10,000-fold, but with no change in specificity and little effect on K_m. Chemical mutation in yeast $tRNA^{Val}$ (Chambers *et al.*, 1973), *E. coli* $tRNA_f^{Met}$ (Schulman and Goddard, 1973), *E. coli* $tRNA^{Glu}$ (Singhal, 1974), and *E. coli* $tRNA^{Arg}$ (Chakraburtty, 1975b) by bisulfite-mediated conversion of C to U in their respective codons was also found to block aminoacylation, and at least in $tRNA_f^{Met}$ no other amino acids could be acylated. However, not all anticodon modifications have such effects. As mentioned above, it is possible to remove the anticodon entirely from some tRNA's without completely inactivating the molecule, and other examples are cited in the above references.

Base analogue substitutions have not been much studied because of the difficulties in replacing normal bases with analogues, but a few examples illustrating different approaches to analogue replacement may be of interest. In one case, all of the uridine-derived nucleotides in tRNA^{Val} of *E. coli* were replaced by 5-fluorouridine by first growing *E. coli* in the analogue and then suitably enriching the isolated tRNA for highly substituted species (Horowitz *et al.*, 1974). The final isolated tRNA^{Val} contained 14 5-fluorouridine residues, yet all functions, both acylation and the partial reactions of protein synthesis, were normal (Ofengand *et al.*, 1974c). In another approach, the terminal C-C-A residue of tRNA was chemically removed and enzymatically built back using the cytidine analogue, 2-thiocytidine (Sprinzl *et al.*, 1973). The analogue substituted tRNA^{Phe} could still be acylated with almost the same kinetics. In a third approach, chemical methods were used to excise minor bases and replace them with proflavine (Wintermeyer and Zachau, 1974). The modified tRNA's could be acylated and were also active in peptide bond formation (Odom *et al.*, 1975).

3. Synthetase Contact Sites

A different approach to the specificity of tRNA-synthetase interaction is based on the premise that whatever and wherever the specificity elements are located in tRNA, they must be in contact with the enzyme surface when the complex is formed. A potential complication of this approach is the possibility that the initial contact sites that determine specificity may not be those in the final complex if conformational rearrangements take place, but since this complication is so far not susceptible to analysis, it has been ignored. Consequently, attempts have been made simply to determine what part of the tRNA is in close contact with synthetase, with special reference to the question of whether only one area is involved, or whether the contact points are located in several discrete areas. Two approaches have been used, that of protection of the tRNA by complex formation, and direct topological probing by cross-linking studies.

Protection from oligonucleotide binding led Schimmel *et al.* (1972) to conclude that both the 3'-end and loop II of *E. coli* tRNA^{Ile} interacted with E^{Ile}, and protection from partial nuclease digestion was used by Hörz and Zachau (1973) to study shielded regions of yeast tRNA^{Phe} and tRNA^{Ser}. In this careful study, protection of tRNA^{Phe} at loops I and II was found, while in tRNA^{Ser} the 3'-end was protected but loop II was not. Other regions in the tRNA's could not be examined for technical reasons. However, when it is recalled that excision experiments on tRNA^{Phe} from the same laboratory (Section IV,C,2) showed that the anticodon was not essential for acylation, it becomes clear that while protected regions should include contact and recognition sites, they may also contain additional regions. In this case, the more meaningful results may be those where a region is *not* protected, as was the case for the anticodon region of tRNA^{Ser}, since such regions clearly cannot be contact sites.

To avoid this objection, other studies have used specific oligonucleotides to inhibit aminoacylation (Bruton and Clark, 1974; Barrett *et al.*, 1974). In the first instance, the entire loop II plus stem c of tRNA$_m^{Met}$ or tRNA$_f^{Met}$ or the codon AUG was tested as an inhibitor of aminoacylation of tRNA$_f^{Met}$ but no effect was found. This result contrasts with the inhibition reported for a C to U transition in the anticodon of this tRNA by Schulman and Goddard (1973) mentioned earlier. In the second example, some specific inhibition of acylation was observed using a tetranucleotide complementary to the anticodon region of tRNA^{Phe}. However, since inhibition might reflect conformational changes elsewhere induced by oligomer binding, probably the negative results are more meaningful in this type of analysis.

Direct measurement of changes in the conformational parameters of a tRNA in complex with its synthetases have been studied. Tritium exchange from *E. coli* tRNA^{Ile} was not affected by combination with its cognate synthetase (Yarus, 1972b) and NMR analysis of the secondary structure of *E. coli* tRNA^{Glu}·E^{Glu} complexes (Shulman *et al.*, 1974b) showed no changes either. More studies of this type are to be expected.

Cross-linking studies as direct topological probes of contact sites between tRNA and synthetase have so far only been attempted in two laboratories, although a number of other studies are in progress. In the first "brute force" approach, Schoemaker and Schimmel (1974) irradiated tRNA^{Tyr}·E^{Tyr} complexes with UV light to form nucleotide-amino acid cross-links of the type first described by Smith and Aplin (1966). Subsequent analysis of the T$_1$ RNase-generated oligonucleotide fragments bound to E^{Tyr} showed specific covalent linking of only 3 out of 14 fragments, comprising 24 out of the 85 bases in this tRNA. So far it is not known where in either the fragment or the protein the covalent links are located. One of the fragments was loop II plus part of stem c, and since the synthetase clearly must also interact with the 3'-end, this experiment, like many already mentioned, provides strong evidence that synthetases must span the distance from the anticodon to the C-C-A end. If complexes at a more physiological pH give similar results and it can be shown that the irradiation used does not inactivate the biological properties of the molecules, this method promises to be one of the more fruitful approaches to the overall problem. A recent study of cognate and noncognate complexes involving *E. coli* tRNA^{Ile} and yeast tRNA^{Phe} (Budzik *et al.*, 1975; Schoemaker *et al.*, 1975) has confirmed the notion that synthetases span the anticodon and C-C-A end, and that the covalent linking pattern at pH 7 is similar, although in lower yield, to that at pH 5.5. However, since the enzyme activity was destroyed by irradiation before most covalent linking occurred, the nature of the interaction with *native* enzyme molecules is still unclear.

Another more specific approach (Budker *et al.*, 1974) has used tRNA derivatized with a photoaffinity group at a specific base, namely, the 4-thiouridine residue in *E. coli* tRNA^{Phe}. So far only preliminary evidence for covalent attachment to synthetase has been reported, but it can be predicted that this approach will be quite useful, in no small part because of the precisely known position of the link in tRNA. (The virtues of photoaffinity labeling in general have already been cited in Section III,B.)

D. Role of Modified Nucleosides

The structural role of the modified bases in tRNA was mentioned in Section IV,C,4. Here we consider whether they play a role in the acylation reaction. Their function in codon-anticodon recognition will be considered in Section VI. Earlier work has been reviewed by Söll (1971), Nishimura (1972), and Littauer and Inouye (1973).

1. Pseudouridine

This base is found in loop IV and also elsewhere (Table II). It is not essential in loop IV since (a) substitution of Ψ by 5-fluorouridine has no effect on either the K_m or V_{max} of aminoacylation (Horowitz *et al.*, 1974); (b) the glycine tRNA of *S. epidermidis* used for cell wall synthesis has G in place of Ψ yet is acylated normally by E^{Gly} (Roberts, 1974, and references therein); (c) eukaryotic tRNA$_i$ with U instead of Ψ has the same K_m and V_{max} for acylation by *E. coli* E^{Met} as does tRNA$_f^{Met}$ which has Ψ (cited in Simsek *et al.*, 1974). Ψ in the loop II region is also unessential. tRNA^{His} with U in place of Ψ at these locations is acylated with the same kinetics (cited in Cortese *et al.*, 1974). The glycine tRNA case is particularly interesting since not only is the G-T-Ψ-C replaced by G-U-G-C, but the G-G pair in loop I is replaced by U-U. Interaction between Ψ-C and G-G is thought to hold these two loops together in the tertiary structure (Fig. 5B), but this cannot be essential for aminoacylation.

2. Ribothymidine

The nonessentiality of ribothymidine has been shown in many ways. 5-Fluorouridine replaces it (Horowitz *et al.*, 1974), it is absent from the *S. epidermidis* glycine tRNA, eukaryotic initiator tRNA, several normal tRNA's (see Section IV,B), and a number of mutant tRNA's (cited in Ofengand *et al.*, 1974c), yet in all cases examined, the tRNA's are aminoacylated normally.

3. 4-Thiouridine

The minor base 4-thiouridine is only found in prokaryotic tRNA's and has no known function. It can be replaced by fluorouridine (Horowitz *et al.*, 1974), converted to uridine (Walker and RajBhandary, 1972), alkylated, or cross-linked to cytidine without affecting aminoacylation by homologous synthetases. The latter two modifications block heterologous misacylation with phenylalanine catalyzed by yeast E^{Phe} (Kumar *et al.*, 1973).

4. Dihydrouridine

Replacement by 5-fluorouridine and other modifications (cited in Ofengand *et al.*, 1974c) have shown that dihydrouridine is unessential for aminoacylation. It is absent from $tRNA_2^{Gly}$, $tRNA^{Tyr}$, and $tRNA_2^{Glu}$ of *E. coli*.

5. 5'-Anticodon Base

This position frequently contains a wide variety of modified bases (Table II). The requirement for some of these unusual bases has been studied mostly by modification experiments since it has not been possible in most cases to obtain tRNA containing the unmodified parent base. Cyanoethylation of inosine in $tRNA^{Arg}$ has no effect (Wagner and Ofengand, 1970). Replacement of cmo^5U in $tRNA^{Val}$ by 5-fluorouridine has no effect (Horowitz *et al.*, 1974). Treatment of mnm^5s^2U in $tRNA^{Glu}$ or $tRNA^{Gln}$ with cyanogen bromide which reacts at the 2-thio position to form a thiocyanate increases the K_m tenfold but does not change the V_{max} or specificity of acylation, but replacement of the 2-thio by a 2-keto group has no effect on acylation (Seno *et al.*, 1974). Presumably in this last case, the sulfur modification per se is not important, but the presence of a thiocyanate group interferes with the fit between tRNA and synthetase.

6. R* Position

The R* position is also frequently occupied by unusual or "hypermodified" bases (Table II). The lack of any role in aminoacylation has been shown in a number of instances (Kimball and Söll, 1974, and references therein) for i^6A, ms^2i^6A, and the W (formerly called "Y") bases (Grunberger *et al.*, 1975) which are a set of related complex derivatives of guanine.

7. Methylated Bases

Earlier studies on the role of methylated bases in tRNA structure and function have been reviewed by Littauer and Inouye (1973). Roe *et al.* (1973c) have shown that conversion of G_{10} to m^2G_{10} in *E. coli* $tRNA^{Phe}$

markedly affects the V_{max} of both homologous and heterologous amino-acylation, but Spremulli *et al.* (1974) did not find any effect on the K_m or V_{max} of acylation of *E. coli* tRNA$_f^{Met}$ when G_{27} between stems b and c was converted to m^{2}G, when A_{59} in loop IV was methylated to m^{1}A, or when both methyl groups were present. The specificity of acylation was also unchanged, as was the ability to be formylated and to initiate protein synthesis. The substitution of m^{7}G in *E. coli* tRNA$_f^{Met}$ by A was already mentioned (Section IV,C,4) in connection with conformational effects. It does not appear to affect aminoacylation (Egan *et al.*, 1973). Unmethylated G in place of m^{7}G also has no effect (Nishimura *et al.*, 1974).

E. Conclusions

Some generalizations from the mass of often contradictory data can be made. First, it is clear that complexes between noncognate tRNA's and synthetases are in some cases almost as strong as the cognate ones. The specificity of aminoacylation derives mainly from the relative maximum velocity of the reaction modified by auxiliary effects. Knorre (1975) has recently proposed a mechanistic rationale for this statement. Second, many synthetases clearly interact with the anticodon region of tRNA and since of necessity they must also interact with the 3'-end, it can be concluded that the synthetase molecule must span the 80 Å distance from the anticodon to the 3'-end. The synthetases of class I have subunits corresponding to one tRNA binding site of 35,000–60,000 daltons. An average value, 47,000 daltons, would correspond to a spherical molecule only 50 Å in diameter, leading to the conclusion that either synthetases are elongated, or the structure of tRNA changes markedly when it is complexed. Third, the location of the *specific* interaction sites in tRNA-synthetase pairs must be different for different pairs since no distinct pattern has yet emerged from a study of many different combinations, yet different approaches used on the same tRNA-synthetase pair do give consistent results. Fourth, the specificity of recognition is likely to come largely from the tertiary structure of the molecule. The two clearest cases of a change in the specificity of recognition attendant on single base changes either in the anticodon or in stem a have shown that specific sequences in these regions are not likely to be responsible since the altered regions do not correspond even partially to the tRNA normally acylated by the synthetase. It seems appropriate to close with a quotation from one of the papers (Yaniv *et al.*, 1974): "Without knowing more about the structural basis for the recognition between glutamine synthetase and its cognate tRNAs, one can only guess what causes the enzyme to recognize, albeit poorly, the modified tRNA^{Trp} and the altered forms of tRNA^{Tyr}."

VI. Specificity of Codon-Anticodon Interaction

The last step in the fulfillment of the adapter role for tRNA is the recognition of a codon. Clearly, if the fidelity of this step is not maintained, all the precision of the previous steps will come to naught. This section will, therefore, be concerned with the mechanism and fidelity of codon-anticodon interaction. Experimental aspects are reviewed in Zachau (1972) and Nishimura (1972). Some theoretical considerations can be found in Ninio (1973).

A. Location and Structure

The anticodon of all tRNA molecules is found in the middle of loop II (Fig. 4). The evidence for this conclusion came initially from a comparison of the location of all possible anticodon sequences in the few tRNA's then known which could form parallel or antiparallel base pairs with the known codons. Subsequent sequence analysis and codon assignments for many more tRNAs have confirmed both the location and antiparallel nature of the pairing. More direct evidence came from an analysis of the *sup3* tyrosine tRNA, a mutant tRNA with an altered coding response. The mutation was found to be a single base change at the anticodon position predicted for antiparallel base pairing. Additional evidence was the finding that a tRNA fragment consisting of loop II plus stem c, but no other fragment of $tRNA_f^{Met}$, could be bound to the ribosome in the presence of the codon AUG.

B. Specificity

There are 61 codons, 20 amino acids, and an intermediate number of tRNA species. How does one tRNA recognize more than one codon, yet at the same time preserve the fidelity of codon-anticodon matching required by the genetic code? Separation of isoacceptor species of tRNA, determination of their codon response, and sequencing of their respective anticodons have provided the answer, at least in formal terms.

Confirming a prediction of Crick (1966), additional base pairing possibilities at the 5'-end of the anticodon, namely, $G \cdot U$ as well as $G \cdot C$ and $I \cdot A$ in addition to $I \cdot C$ and $I \cdot U$ pairs, account for the ability of one isoacceptor tRNA molecule to respond to multiple codons, as in the recognition by yeast $tRNA_1^{Ala}$ of GUU, GUC, and GUA. In this example, the anticodon is I-A-C. Another alanine tRNA species recognizes GUG exclusively. In *E. coli,* on the other hand, the four valine codons are read by two tRNA species, one with anticodon G-C-C reading GGU and GGC, and the other with anticodon cmo^5U-C-C reading GGA and GGG. This example illustrates two principles. First, $G \cdot U$ pairing allows G to pair with C or U and I can pair with U, C, or A. Second, different species have evolved different combinations of isoacceptor tRNA's to translate multiple codons. These additional base pairs were called "wobble" base pairs by Crick because the glycosidic bonds of the nucleotides involved had to be shifted somewhat in both angle and distance from those in a standard Watson-Crick double helix. For that reason, Crick chose to limit these special base pairs to the 5'-side of the anticodon. However, they are frequently found in stem regions of tRNA, especially stem a, where Ladner *et al.* (1975) have shown that the bonding scheme is exactly that predicted by Crick. Recently, the mechanism of I·A pairing has been questioned. The distance between glycosidic bonds of the wobble I·A pair is 12.8 Å, or 2 Å greater than a normal base pair, but if A pairs in the *syn* conformation, the value would be reduced to 10.8 Å. Davis *et al.* (1973) have obtained some evidence that this mechanism of pairing may exist.

The strengths of the wobble interactions have been compared with normal A·U and G·C pairing by measurement of the association constant, K_a, of various codon trinucleotides with the anticodons of tRNA (Unger and Takemura, 1973). Replacement of G·C by G·U or I·C reduced K_a 10-fold, G·U is 4 times stronger than I·U, and the dual replacement of G·C by I·U reduces K_a 50-fold. The relative strengths of association of codons ending with C, U, or A for pairing with I were 24 to 6 to < 1. Similar findings were reported by Eisinger and Gross (1974) who reported a 7- to 14-fold decrease in K_a upon substituting an A·I pair for an A·cmo⁵U pair, in an experiment which measured the association between complementary anticodons of 2 tRNA's. The latter experiments also showed that the anticodon loop was considerably more rigid than a trinucleotide, and that the conformation was well suited to formation of Watson-Crick helices (see also Eisinger and Gross, 1975).

The pairing of inosine with A, U, and C at the 5'-position of the anticodon is used widely for multiple codon recognition in bacteria, yeast, and mammals. The G·U pair is also commonly used, but with certain restrictions which are evident from a comparison of known anticodons with codons. In both prokaryotes and eukaryotes the following set of rules is obeyed.

1. Neither A nor unmodified U is ever found in the 5'-anticodon position except for U in the cell wall glycine tRNA which is inactive in protein synthesis. When the 3rd letter of the codon is A, the 5'-anticodon base is always a modified U. 2-Thiouridine derivatives, cmo⁵U, and unidentified U derivatives are found. 2-Thiouridine and derivatives base-pair strongly with A, but not G (Vormbrock *et al.*, 1974) so that in these cases a separate isoacceptor with C at the 5'-anticodon is needed. cmo⁵U pairs with both A and G, and less well with U. In the minor species of

tRNA^{Ile} reading AUA, the 5′-anticodon is an unidentified modified U which cannot read G (Harada and Nishimura, 1974). This prevents misreading of AUG (methionine) as isoleucine.

2. In most anticodons with a central U, the 5′-G is modified to Q, or a derivative. Exceptions are G-Ψ-A (yeast tRNA^{Tyr}) and G-U-C (yeast tRNA^{Asp}). Q is itself a modified G (see Table II) and pairs with both C and U.

3. When the 3′-anticodon base is A, the base at the R* position is almost always hypermodified. It is ms^{2}i^{6}A in *E. coli,* and i^{6}A in eukaryotes except that eukaryotic tRNA^{Phe} has instead the complex base W, or variants of it. All these bases are bulky and hydrophobic. There are four exceptions so far to this rule. In yeast tRNA$_3^{Leu}$, chicken tRNA^{Trp}, and *Mycoplasma* tRNA^{Phe}, R* is m^{1}G, and in yeast tRNA^{Trp}, it is A.

4. When the 3′-anticodon base is U, R* is t^{6}A or mt^{6}A. This hydrophilic base is found in both prokaryotes and eukaryotes with the sole exception of tRNA$_f^{Met}$ where it is unmodified A. Failure to modify this A suggests that there is something unique about the conformation of loop II in this tRNA since *E. coli* has been shown to modify such A residues when the 3′-anticodon base is converted to U by mutational means (Roberts and Carbon, 1974).

The mechanistic explanation for the above regularities is unknown, although one instinctively supposed that they are intended to ensure against the possibility of misrecognition. Rule 1, for example, might be related to the (almost) universal presence of U next to the 5′-anticodon position. Perhaps the anticodon U needs to be modified to distinguish it from this other U, although it is not clear why this should be so. It has been suggested that R* modifications exist to strengthen the A · U base pair of the 3′-anticodon position. The presence of hypermodified bases at R* does increase the binding constant of oligonucleotides for the anticodon of tRNA by 2- to 7-fold (Högenauer *et al.,* 1972; Miller *et al.,* 1974), but this does not explain the specificity of the modification, nor provide a rationale for the quite different electronic character of the two types of hypermodification. Since the enzyme systems for the specific synthesis of these bases have been preserved from bacteria to mammals, they appear to be necessary but it is not known why. Gefter and Russell (1969) showed that the ms^{2}i^{6}A of *E. coli* tRNA^{Tyr} was necessary for optimum *in vivo* protein synthesis, but apparently yeast tRNA^{Trp} functions perfectly well with unmodified A at that position, as does *E. coli* tRNA$_f^{Met}$ without the t^{6}A modification. The absence of t^{6}A in *E. coli* tRNA$_f^{Met}$ may be responsible for the 2-fold decrease in K_a for AUG as compared to yeast tRNA$_i^{Met}$ (Högenauer *et al.,* 1972) and might therefore explain why *E. coli* tRNA$_f^{Met}$ can read GUG as well as AUG (wobble in the first codon position) but eukaryote tRNA$_i^{Met}$ cannot do so. On the other hand, the absence of t^{6}A may be a consequence of some subtle conformational difference (see above) which is also manifested as GUG recognition.

Parthasarathy et al. (1974) have pointed out that all the modified bases at R*, including the hypermodified bases, are incapable of Watson-Crick base-pairing, and they suggest that the modifications are necessary in order to define the reading frame in the anticodon loop. The hypothesis may have merit, since it has been shown that the central four bases of an eight base loop II can be translated by the appropriate mRNA (Riddle and Carbon, 1973). Consequently there appears to be no intrinsic mechanism in the ribosome-mRNA complex which restricts reading to only three anticodon bases at a time. However, since many tRNA's are known with neither R* nor the 5'-anticodon base modified, the requirement cannot be absolute.

In Section V,C,2, the misrecognition of UGA by a mutant of the E. coli tryptophan tRNA was mentioned. The anticodon is C-C-A, normally pairing with UGG. Mutation of G to A in stem b generates an A · U base pair where there was initially a G · U pair, and this change is sufficient to allow pairing on the ribosome with UGA, corresponding to a C · A base pair at the 5'-anticodon position. Measurement of the association constant for UGA with the mutant and wild-type tRNA^{Trp} has shown that the mutation has no effect on interaction in solution (Högenauer, 1974). Consequently, the change in stem b does not alter the conformation of loop II but in some way must stabilize binding of the tRNA to the ribosome so that the UGA codon can be translated. This example serves as a warning that modifications to tRNA other than in loop II are able to modify codon-anticodon recognition.

VII. Elongation Factor Recognition of Aminoacyl-tRNA

The ability of AA-tRNA to complex with elongation factor EF-I (EF-Tu in E. coli) is essential since the ternary complex AA-tRNA · EF-I · GTP is the way AA-tRNA is carried to the ribosomal A site (Chapter 6). The specificity of formation of this intermediate ternary complex supplies the mechanism for excluding deacylated or acyl AA-tRNA, which would otherwise jam the protein-synthesizing machinery. The features of tRNA which are recognized by EF-I are interesting because EF-I must be able to pick out the common features of all AA-tRNA since it can complex with them all, and at the same time be able to overlook those features which allow synthetases to distinguish one tRNA from another. Analysis of this protein-tRNA recognition system has proceeded along two lines, tRNA modification studies and direct examination of the complex by physical methods.

A. Effect of tRNA Modifications

Since most of this work has been reviewed by Ofengand (1974), citations will only be given for the newer findings. All studies have been done with EF-Tu of *E. coli,* so far for the practical reason that it was the first to be available in pure form. It is assumed, but remains to be proved, that eukaryotic EF-I will show the same specificity. Earlier work showed that deacylated and acyl aminoacyl-tRNA did not react, and that α-hydroxyacyl-tRNA was weakly active (EF-Tu) or was inactive (EF-I of wheat embryo). Modifications to loop II by cleavage of the phosphodiester chain or by chemical modification of the bases had no effect. Modifications to stem b by cross-linking positions 8 and 13, or to loop I by converting C residues to U with bisulfite (Schulman *et al.,* 1974) had no effect. Substitution of all of the U, Ψ, T,^{4}S, hU, and cmo^{5}U by 5-fluorouridine had no effect on ternary complex formation (Ofengand *et al.,* 1974c), showing that at least these minor bases are not involved in EF-Tu recognition.

The 3'-end of the molecule is important, as might have been anticipated. Opening of the ribose ring of the 3'-terminal adenosine blocked recognition, but substitution of deoxyribose for ribose did not (Chinali *et al.,* 1974). The amino acid was located on the 2'-hydroxyl in both cases (see below). If the C-C-A end is extended to C-C-C-A, the resultant AA-tRNA makes a weak ternary complex but is transferred more rapidly than the normal tRNA to the ribosomal A site and to the same extent (Thang *et al.,* 1974). Conversion of the C-C-A end to a C-U-A by bisulfite treatment also weakens the interaction (Schulman *et al.,* 1974). Stem a is important since removal of the 5'-phosphate blocks recognition very effectively, and conversion of C · A at the 3'-end of stem a in *E. coli* tRNA$_f^{Met}$ into a proper U · A base pair by bisulfite treatment rendered a previously inactive tRNA active (see Table III; Schulman *et al.,* 1974). This very nicely explains why Met-tRNA$_f^{Met}$ is completely inactive in EF-Tu · GTP ternary complex formation, while other initiator tRNA's are weakly active (Table III).

Additional contributions to recognition must come from tertiary structure, however, since yeast Leu-tRNA$_3^{Leu}$ in the denatured conformation does not form a ternary complex, while the denatured form of *E. coli* Trp-tRNA does do so (Ofengand, 1974). The differences between these two denatured conformations should be very informative in terms of EF-Tu recognition sites. It should also be noted that initiator tRNA's do not form ternary complexes that are as strong as other tRNA's. Eukaryotic Met-tRNA$_i^{Met}$ interacts only weakly (due to its completely different loop IV?) and even the bisulfite-treated Met-tRNA$_f^{Met}$ can be easily displaced by other AA-tRNA's. The 3'-half molecule of valyl-tRNA is inactive, regaining activity after reannealing with the 5'-half-molecule, but it is not clear whether this is a reflection of a need for tertiary structure or a need for an intact double-stranded stem a. Finally it should be mentioned that the cell wall glycine tRNA which has many differences from normal tRNA also does not complex with EF-Tu (J. Ofengand and R. Roberts, unpublished results; Kawakami *et al.*, 1975).

All the above studies were done by direct measurement of AA-tRNA · EF-Tu · GTP complex formation. Recently, Ringer and Chladek (1975) have developed a new assay for interaction of tRNA fragments, that of release of GTP from the binary EF-Tu · GTP complex. They observed that 50% release could be obtained with $10^{-5} M$ C-A-Phe, C-A-Pro, or C-A-Asp, but not with C-A-(Ac-Phe), C-A, or A-Phe. These are the specificity characteristics expected for AA-tRNA interaction with EF-Tu · GTP and suggest that even small oligonucleotides may be capable of interaction when used at very high concentration. ($10^{-5} M$ is approximately 100 times that usually employed in testing tRNA substrates.) These compounds should prove useful for dissecting the recognition site into its component parts.

B. Stereochemistry of the Amino Acid

Enzymatic acylation of tRNA is specific for the 2'-hydroxyl, the 3'-hydroxyl, or is nonspecific (Section II,B). Does EF-Tu recognize both isomers, or must migration from 3' to 2' occur in some cases before formation of the ternary complex? Recent experiments from two laboratories using nonisomerizable AA-tRNA analogues have dealt with this question. Chinali *et al.* (1974) showed clearly that (2')Phe(3')deoxy-tRNA could be bound to the ribosomal A site in an EF-Tu-dependent reaction, implying very strongly that ternary complex formation occurred. More recently, Hecht *et al.* (1974) studied the binding of both isomers, the (2')Phe(3')deoxy and (3')Phe (2')deoxy-tRNA's, to the ribosome by competition methods and concluded that both were active. However, as it is not clear that these experiments actually measured EF-Tu interaction, it cannot yet be concluded that ternary complex formation is nonspecific for the position of the aminoacyl moiety. Certainly, the 2'-isomer *is* active. Recently, Ringer and Chládek (1975) showed that C-A(2'Phe)H caused the release of GTP from the EF-Tu · GTP complex but that C-A(2'H)Phe was completely inactive. Since it is believed that this reaction mimics the specificity of ternary complex formation with AA-tRNA, this result implies that ternary complex formation is 2'-specific, like aminoacylation with E^{Phe}. The reason for this selectivity for 2'-aminoacyl-tRNA may be related to a stereospecificity requirement in peptide bond formation, as discussed below (Section VIII).

C. Physical Studies

As with synthetases, studies of tRNA conformation in the ternary complex are just beginning. By NMR analysis, Shulman *et al.* (1974a) found that the secondary structure H bonds of tRNA do not change when the ternary complex is formed, and Arai *et al.* (1973) reported the crystallization of the ternary complex, although so far the crystals are too small for X-ray analysis.

VIII. tRNA-Ribosome Recognition

A. A and P Sites and 5 S RNA

Ever since it was recognized that A and P sites exist on the ribosome for the binding of all tRNA's, workers in the field have searched for the complementary common site in tRNA molecules. The existence of the sequence G-T-Ψ-C-G or G-T-Ψ-C-A in loop IV of (almost) all tRNA's implicated this region, but the first experimental evidence was not obtained until Ofengand and Henes (1969) showed that the tetranucleotide T-Ψ-C-G could inhibit binding of Phe-tRNA to both P and A sites. Subsequently, the binding of T-Ψ-C-G to 5 S RNA-ribosomal protein complexes was shown by Erdmann *et al.* (1973). The analogous tetramer U-U-C-G was bound only one-third as strongly. This finding made plausible an earlier suggestion that the common G-T-Ψ-C sequence of tRNA might interact with a complementary G-A-A-C sequence known to exist in 5 S RNA.

Although the suggestion has not yet been proved recent advances in sequence analysis of both tRNA and of 5 S RNA (Barrel and Clark, 1974) have revealed some very striking correlations that substantiate the concept of 5 S RNA-tRNA interaction (see also Dube, 1973; Nishikawa and Takemura, 1974). The three known prokaryotic 5 S RNA sequences have the pentanucleotide C-G-A-A-C corresponding to G-T-Ψ-C-G in virtually identical positions in the sequence, i.e., beginning at bases 42, 43, or 44. Two of them, *E. coli* and *P. fluorescens,* also have the tetranucleotide G-A-A-C in a second location, but the position is widely different in the two sequences; *B. stearothermophilus* does not have a second sequence. Three of the four eukaryotic 5 S RNA's have the sequences U-G-A-U-C or C-G-A-U-C corresponding to the G-A-U-C-G present in loop IV of eukaryotic initiator tRNA's at exactly the same position, which is virtually the same as the location in prokaryotes, i.e., beginning at base 40. Yeast and KB cell 5 S RNA also have an additional site approximately complementing the G-T-Ψ-C-G or G-T-Ψ-C-A sequence, but they are in different locations in the two sequences, and *Xenopus* 5 S RNA does not have any. Surprisingly, the eukaryotic algae, *Chlorella,* has two G-A-A-C sequences located almost exactly where the *E. coli* sequences are found, and has no sequence corresponding to the G-A-U-C-G of the eukaryotic initiator. It is also notable that the sequence in the region of residues 40–45 in both eukaryotes and prokaryotes includes a C residue corresponding to the G which terminates stem e in a G · C base pair. This may be further evidence for an unfolding process on binding to the ribosome (see below).

These considerations lead to the following hypothesis. In eukaryotes, the function of 5 S RNA is solely to provide a binding site on the large ribosomal subunit for initiator tRNA (the I site). The failure to find a second sequence complementary to G-T-Ψ-C-G or G-T-Ψ-C-A at the same location in the three eukaryotic 5 S RNA's implies that the A and P binding sites are elsewhere. The ubiquitous 5.8 S RNA (Rubin, 1973) may be a candidate for the A and P site loci since the yeast sequence has two G-A-A-C sequences which could pair with G-T-Ψ-C. This hypothesis thus envisages a separate binding site for the body of the initiator tRNA which is distinct from both the aminoacyl- and peptidyl-tRNA binding sites. The P and I sites would probably share a common anticodon site and peptidyl transferase donor site, however, based on available evidence in the literature.

Extension of this concept to prokaryotes by the same argument leads to the conclusion that there is only one binding site in 5 S RNA for G-T-Ψ-C, which, by analogy with eukaryotes, should correspond to the I site on the 50 S subunit. However, the ability of eukaryotic $tRNA_i^{Met}$ (which lacks T-Ψ-C-G) to bind to prokaryotic ribosomes (Section IV,D and Table III) argues against this conclusion. Since initiator and other prokaryotic tRNA's have the same loop IV G-T-Ψ-C-G or G-T-Ψ-C-A sequence, there is no compelling reason to assign 5 S RNA to the I site. In prokaryotes it may be part of the A (or P) site being adapted only subsequently by eukaryotes for a specific I site role.

The available experimental evidence supports the hypothesis at least in so far as assigning a ribosome recognition role to the common T-Ψ-C-G sequence. Richter *et al.* (1973) reexamined T-Ψ-C-G inhibition of Phe-tRNA binding to *E. coli* ribosomes under more carefully controlled conditions and were able to show a strong inhibition of EF-Tu-dependent binding to the ribosomal A site. Moreover, the effect could be localized to the 50 S subunit since T-Ψ-C-G treated 50 S subunits could be reisolated and were inactive. There was no inhibition by T-Ψ-C-G when eukaryotic Met-$tRNA_i^{Met}$ binding to rat liver ribosomes was examined (Grummt *et al.*, 1974), although binding of Met-$tRNA_m^{Met}$ was inhibited, and T-Ψ-C-G similarly fails to inhibit fMet-$tRNA_f^{Met}$ or AcPhe-tRNA binding to *E. coli* ribosomes (Erdmann *et al.*, 1974). One puzzling aspect has been the fact that loop IV is normally not exposed to solvent (Section IV,C,1) when tRNA is in solution. However, a proposal that an unfolding process occurs on the ribosome (Ofengand and Henes, 1969) has received some recent substantiation (Schwarz *et al.*, 1974). A test of the above hypothesis might be possible by examining the differential protection of exposed residues in the various ribosomal RNA species of prokaryotes and eukaryotes by binding tRNA specifically to I, P, and A sites.

Although not a study of tRNA-ribosome recognition as such, the topography of A and P sites has been explored by a variety of methods, including the direct approach of cross-linking tRNA to the ribosome. A selection of references to the recent literature can be found in Schwartz *et al.* (1975), which also reports experiments on cross-linking tRNA to both P and A sites by photoaffinity labeling techniques. Affinity labeling of ribosomes is also reviewed in Chapter 4.

B. Stringent Factor-Ribosome Interaction

In *E. coli* and other species, one mechanism for the coupling of RNA synthesis to protein synthesis involves the synthesis by stringent factor of pppGpp and ppGpp, the so-called magic spot (MS) compounds, which inhibit the synthesis of stable RNA species. This synthesis is strongly dependent on the presence of ribosomes and on the binding of codon-specific, uncharged tRNA to the ribosomal A site (Haseltine and Block, 1973; Pederson *et al.*, 1973). The dependency on unacylated tRNA makes this reaction interesting in the present context since it provides an additional type of assay for ribosome recognition of tRNA, and in fact is the only functional ribosomal recognition system for uncharged tRNA.

Although such studies are still in their infancy, it is clear that the entire panoply of modified tRNA's could be employed. It is already known that oxidation of the terminal adenosine moiety, or its removal, blocks the activity of tRNA, and m_2^2G-deficient tRNA is also inactive, although tRNA lacking T is active (cited in Block and Haseltine, 1974). tRNA whose terminal adenosine has been modified to the 2'-deoxy or 3'-deoxy derivative is active (Richter, 1975). The tRNA fragment, T-Ψ-C-G, can effectively inhibit the ribosome- and tRNA-dependent reaction (Richter *et al.*, 1973) and by itself has a weak but specific ability to stimulate MS synthesis (Richter *et al.*, 1974). Further study of this fragment reaction should yield much useful information.

C. Stereospecificity of the Peptidyl Transferase Center

A general discussion of tRNA interaction at the peptidyl transferase center can be found in Chapter 8. Here we only consider the specificity requirements of the donor and acceptor sites with respect to the stereochemistry of the aminoacyl group, that is whether it is 2' or 3'. Recent studies using nonisomerizable tRNA analogues, and fragments of these analogues have shown rather conclusively that the acceptor site is essentially 3'-specific with respect to the position of the amino acid on tRNA (cited in Ringer *et al.*, 1975). It is less clear what the specificity at the donor site may be. 2'-Aminoacyl-tRNA analogues can be bound at the P site but do not function as donors of the amino acid (Hussain and Ofengand, 1973; Chinali *et al.*, 1974). However, since the required pair of isomers has so far not been available, the lack of activity may as well be due to the modification used to create the analogue as to the position of the amino acid. A recent test of such a pair of analogues has not helped since both isomers were inactive (Hecht *et al.*, 1974).

2'-Esters do appear to interact at the acceptor site, however, since several fragment analogues are good inhibitors of peptide bond formation (Ringer *et al.*, 1975). These and other findings suggest a hypothetical model (Ringer *et al.*, 1976) in which (2')- or (3')AA-tRNA synthesized by synthetase (Section II,B) is transferred to EF-Tu·GTP as the 2'-ester (Section VII,B) and from there to the ribosomal A site without further isomerization. Upon hydrolysis of GTP and departure of EF-Tu · GDP, a shift to the 3'-position then occurs and peptide bond formation ensues. The isomerization is thus coupled to GTP hydrolysis and the attendant departure of EF-Tu · GDP from the ribosome but the reverse need not be true. In fact, GTP hydrolysis was observed when (2')Phe(3')deoxy-tRNA was enzymatically bound to the A site (Chinali *et al.*, 1974). The purpose behind this rather elaborate alternation of stereospecificity is unclear.

IX. tRNA Biosynthesis

Since the biosynthesis of tRNA has been recently reviewed (Schäfer and Söll, 1974) only a brief summary is given here with literature citations to the newer work only.

A. Synthesis and Processing of the Polynucleotide Chain

Like most stable RNA's of the cell, tRNA is transcribed from DNA as a larger precursor molecule which must be subsequently processed. RNA processing was the subject of a recent symposium (Dunn *et al.*, 1975). The genes for tRNA are not clustered in any regular way. In general, isoacceptor genes are scattered, but clustering of genes for different tRNA's is sometimes found. tRNA genes can also exist in multiple copies and sometimes one isoacceptor gene is duplicated while the other ones are present only once. tRNA genes are also found clustered in viral DNA.

Although no precursors longer than dimers have been isolated, there are indications that the original transcripts may be trimers (Grimberg and Daniel, 1974) or even larger (Kaplan and Nierlich, 1975; Daniel *et al.*, 1975; Sakano and Shimura, 1975). Dimers made up of two different tRNA precursors, as well as of a gene duplication (Ghysen and Celis, 1974b), are known as well as monomeric precursors. The monomer precursors generally range in size from 130–200 nucleotides (mature tRNA has about 80) and does or does not contain the 3′-terminal C-C-A sequence. Processing proceeds by the action of specific nucleases to give the mature length (Sakano and Shimura, 1975). Processing does not appear to be dependent on the presence of modified nucleotides, nor is it inhibited by them. Precursor tRNA's possess tertiary structure, and it is thought they assume a tRNA-like conformation, with extra bases at each end (McClain and Seidman, 1975). They can form precursor-synthetase complexes. Relatively little is known about the nature of promotor or terminator sequences for tRNA cistrons, although much progress is being made (Sekiya *et al.*, 1975; Kupper *et al.*, 1975). Regulation of tRNA synthesis is not understood but analysis of this process *in vitro* is underway (Beckman and Daniel, 1974).

B. Synthesis of the Modified Bases

This process can be divided into when and how. The sum of various studies on when modification takes place indicates that it is unrelated to tRNA processing. Depending on the genetic makeup of the cell involved, nucleolytic processing or base modification may be faster leading either to predominantly mature-sized but under modified molecules, or to modified precursor molecules. Ribose methylation of G residues may be an exception, as this modification has yet to be found in any precursor so far studied (Sakano *et al.*, 1974). There does not seem to be any sequential order to the modification reactions. In a recent study of normally growing *E. coli*, the order of methylation was apparently random, and in this case occurred mostly on mature-sized molecules (Davis and Nierlich, 1974).

The enzymatic mechanism involved in base modifications is only clear in some cases. i^6A is made by condensation of isopentenyl pyrophosphate with an adenosine residue. The enzyme has been purified. ms^2i^6A is likewise made by sequential synthesis first of i^6A, then of s^2i^6A and finally methylation to ms^2i^6A (Agris *et al.*, 1975). t^6A is made by addition of bicarbonate and threonine to an adenosine residue in tRNA in an ATP-requiring reaction. A single purified enzyme catalyzes the overall reaction (Elkins and Keller, 1974). These enzymes must be able to recognize both tRNA structure and sequence, since A residues are not derivatized at random, but only at the R* position in loop II. Moreover, they must have a unique selectivity for the 5′-adjacent base, since when the adjacent base is U, t⁶A is made, when it is A, ms²i⁶A is made, and when it is G or C, no reaction takes place (Roberts and Carbon, 1974).

The tRNA methylases have been extensively studied (for a recent review, see Littauer and Inouye, 1973). S-Adenosyl methionine is the methyl donor in almost all cases, except for recent reports of uridine methylation by a different methyl donor, probably a folate derivative (Delk and Rabinowitz, 1975; Kersten *et al.*, 1975; Schmidt *et al.*, 1975). In general, methylases are not only specific for particular residues, but also recognize the immediately adjacent nucleotide sequences which must be in the correct place in the tRNA structure (Kraus and Staehelin, 1974; Kuchino and Nishimura, 1974) and suitably unfolded (Wildenauer *et al.*, 1974).

4-Thiouridine in tRNA is made by conversion of uridine in 2 separate steps requiring 2 enzymes, ATP, and the sulfur of cysteine (Lipsett, 1972). The newly discovered nbt³U (Table II) is made by condensation of S-adenosyl methionine with a specific uridine in loop III. The enzyme has been purified (Nishimura *et al.*, 1974). The origin of the W base is now clear. It is a derivative of guanosine (Li *et al.*, 1973; Thiebe and Poralla, 1973). Part of the side chain is derived from methionine (Münch and Thiebe, 1975).

Pseudouridine is made by direct intramolecular conversion of uridine. There are no cofactor requirements. Interestingly, at least two activities exist in cells. One activity (I), coded by the *hisT* gene, produces Ψ in the loop II-stem c region while the second (II) makes the Ψ in loop IV. It is not known if additional activities exist for Ψ in other regions. Enzyme II appears to function at the precursor level, but the claim that enzyme I is restricted to modification of mature tRNA has been questioned (Sakano *et al.*, 1974).

X. Other Functions of tRNA

In addition to their role as adapters in protein synthesis, tRNA and AA-tRNA have been found to play a number of other important roles in the cell. In fact, it is possible that the function of at least some of the modified bases in tRNA is to modulate these accessory functions, rather than to be required for its protein synthetic role. The Ψ residue(s) in loop II-stem c are an example of this sort. As pointed out above, the *hisT* mutation inactivates Ψ synthetase I, the enzyme that converts U to Ψ in the loop region. Although protein synthesis proceeds normally in the mutant strain, the co-repressor activity of His-tRNA and of Leu-tRNA is lost

(Cortese *et al.*, 1974). It is not clear, however, whether the recognition by aporepressor is of this particular sequence or of a deformed tRNA structure since the mutant tRNA also chromatographs differently from the wild-type (Section IV,C,4).

Another function of AA-tRNA in the cell is as a donor of carboxyl-activated amino acid for synthetic reactions other than protein synthesis. Examples of these reactions, reviewed in Soffer (1974) and Littauer and Inouye (1973), are the donation of glycine, serine, threonine, and alanine residues for the cross-bridges in cell wall biosynthesis, the synthesis of lysyl- and alanylphosphatidylglycerol, and the terminal addition of arginine, phenylalanine, or leucine residues to certain proteins or peptides. The biological function of this latter reaction is not understood (see also Deutch and Soffer, 1975; Soffer, 1975). Cell wall synthesis has been most extensively studied, and at least in *Staphylococcus* it has been shown that while all of the Gly- or Ser-tRNA species can function in peptidoglycan synthesis, one of them is unable to function in protein synthesis. One well-studied example (Roberts, 1974) has been repeatedly referred to in earlier sections because of the insight its availability and structure has given us in understanding the relation between tRNA structure and function. While it is clearly derived from a tRNA gene, it is equally clear that not just one but several modifications have been introduced to ensure that it will not function in protein synthesis. Since the percent of glycine and serine in cell walls is at least as great as the percent in protein, it is understandable why sequestering a particular tRNA^{Gly} for exclusive use in cell wall synthesis ensures a supply of activated glycine for this purpose. It is noteworthy that this species is 40% of the total glycine tRNA of the cell.

A role for tRNA in the regulation of translation has often been proposed since by regulating the rate at which a given codon is translated, the overall rate of protein synthesis could be controlled. Despite the attractiveness of the hypothesis, no evidence for it has yet been obtained, except for two recent preliminary reports (Gupta *et al.*, 1974; Content *et al.*, 1974) which indicate that interferon-mediated inactivation of certain cellular tRNA fractions may be responsible for the failure of interferon-treated cells to effectively translate viral mRNA's while allowing host mRNA translation to continue. It is assumed that the tRNA species inactivated is one which translates a codon rarely used by the host but commonly used by the virus.

Finally, tryptophan tRNA of chick cells has been shown to be specifically required as a primer for the DNA-dependent reverse transcriptase of Rous sarcoma virus (Sawyer *et al.*, 1974; Panet *et al.*, 1975) and of avian myeloblastosis virus (Waters *et al.*, 1975). The priming activity may be species specific since preliminary evidence suggests that proline tRNA may serve this function in the AKR murine leukemia virus (Waters, 1975).

XI. Summary

In protein synthesis, tRNA is the adapter molecule that supplies the mechanism by which a given amino acid is selected by the three-nucleotide codon for correct placement in the polypeptide chain. Synthesis of aminoacyl-tRNA at the expense of ATP also provides the energy required to drive peptide bond formation. The generally accepted two-step mechanism is freely reversible with an equilibrium constant of 1. The amino acid is attached via an ester bond to the 2′- or 3′-hydroxyl of the 3′-terminal ribose of the tRNA chain, which consists of 74–94 nucleotides terminating at the 3′-end in CCA. In addition to the normal nucleotides, a large number of unusual nucleotides made by chemical modification of the normal ones are present, frequently in defined locations common to all tRNA species. Although a concerted mechanism has been proposed for aminoacyl-tRNA synthesis, the current weight of evidence still favors the two-step mechanism.

As required by the adapter hypothesis, there is at least one tRNA for each amino acid, and a specific synthetase capable of recognizing both amino acid and tRNA. There is only one synthetase for a given amino acid, but there may be more than one tRNA. The existence of multiple tRNA's, called isoacceptors, is a direct consequence of the degeneracy of the genetic code (61 code words for only 20 amino acids) and the need to read more than one codon as the same amino acid. There are fewer than 61 different tRNA molecules because some can recognize more than one codon by the process of "wobble" base-pairing. Recognition between codon and tRNA is by antiparallel base-pairing with three complementary nucleotides in the tRNA (the anticodon) which is always located in the same part of the tRNA molecule (Fig. 4). The process of codon-anticodon matching must be accurate if the adapter function of tRNA is to be fulfilled since once attachment of the amino acid to tRNA occurs, further identification of the amino acid is solely by means of the tRNA moiety.

Recognition of the proper tRNA by the synthetase is also, therefore, a critical phase in the synthesis of aminoacyl-tRNA. A given synthetase is able to recognize all isoacceptor tRNA's for its cognate amino acid but does not normally acylate noncognate tRNA's. Since all tRNA's possess the same two-dimensional cloverleaf structure (Fig. 4), recognition by synthetase leading to aminoacylation is thought to involve subtle aspects of three-dimensional structure. Although progress in understanding this structure has accelerated rapidly with the determination of the crystal structure of one tRNA (Fig. 5), the mechanism of productive recognition of a given tRNA by its cognate synthetase is not yet clear. Recognition is thought to proceed in a series of steps with a rather nonspecific interaction between tRNA and synthetase occurring at first which is then followed by a series of checking steps to ensure that only the acceptor end of the *cognate* tRNA is accommodated at the catalytic center of the enzyme.

Other reactions in protein synthesis requiring tRNA recognition include complexing with elongation factor and binding it to A and P sites on the ribosome. In contrast to synthetase recognition, these reactions are common to all aminoacyl-tRNA's and must, therefore, involve common structural elements, at least in a three-dimensional sense. Ribosomal binding is thought to involve the common G-T-Ψ-C sequence of tRNA, perhaps by interaction with 5 S RNA, and the acceptor site of the peptidyltransferase center on the 50 S subunit is known to be specific for the 3'-isomer of aminoacyl-tRNA. By contrast, elongation factor recognition is probably restricted to the 2'-isomer.

tRNA biosynthesis proceeds via precursor molecules which may contain several tRNA sequences in tandem separated by "spacer" nucleotides. Specific nucleases process the precursor. The many unusual bases in tRNA are synthesized by special enzymes either at the precursor or mature tRNA state.

tRNA and aminoacyl-tRNA are also involved in regulation of some metabolic pathways, and in some instances aminoacyl-tRNA serves as the donor of activated amino acid for other biosynthetic processes such as cell wall biosynthesis.

REFERENCES

Agris, P. F., Armstrong, D. J., Schafer, K. P., and Soll, D. (1975). *Nucleic Acids Res.* **2**, 691–698.

Allende, C. C., Allende, J. E., Gatica, M., Celis, J., Mora, G., and Matamala, M. (1966). *J. Biol. Chem.* **241**, 2245–2251.

Allende, J. E., and Allende, C. C. (1971). *In* "Methods in Enzymology" (K. Moldave and G. Grossman, eds.), Vol. 20, pp. 210–220. Academic Press, New York.

Ankilova, V. N., Knorre, D. G., Kravchenko, V. V., Lavrik, O. I., and Nevinsky, G. A. (1975). *FEBS Lett.* **60**, 172–175.

Aoyama, H., and Chaimovich, H. (1973). *Biochim. Biophys. Acta* **309**, 502–507.

Arai, K.-I., Kawakita, M., Nishimura, S., and Kaziro, Y. (1973). *Biochim. Biophys. Acta* **324**, 440–445.

Axel'rod, V. D., Kryukov, V. M., Isaenko, S. N., and Bayev, A. A. (1974). *FEBS Lett.* **45**, 333–336.

Bandyopadhyay, A. K., and Deutscher, M. P. (1971). *J. Mol. Biol.* **60**, 113–122.

Bandyopadhyay, A. K., and Deutscher, M. P. (1973). *J. Mol. Biol.* **74**, 257–261.

Barrell, B. G., and Clark, B. F. C. (1974). "Handbook of Nucleic Acid Sequences." Joynson-Bruvvers, Ltd., Eynsham, Oxford, England.

Barrell, B. G., Seidman, J. G., Guthrie, G., and McClain, W. H. (1974). *Proc. Natl. Acad. Sci. U.S.A.* **71**, 413–416.

Barrett, J. C., Miller, P. S., and Ts'o, P. O. P. (1974). *Biochemistry* **13**, 4897–4906.

Bartmann, P., Hanke, T., Hammer-Raber, B., and Holler, E. (1974a). *Biochemistry* **13**, 4171–4175.

Bartmann, P., Hanke, T., Hammer-Raber, B., and Holler, E. (1974b). *Biochem. Biophys. Res. Commun.* **60**, 743–747.

Bartmann, P., Hanke, T., and Holler, E. (1975). *J. Biol. Chem.* **250**, 7668–7674.

Beckmann, J. S., and Daniel, V. (1974). *Biochemistry* **13**, 4058–4062.

Beres, L., and Lucas-Lenard, J. (1973). *Biochemistry* **12**, 3998–4002.

Berg, P. (1956). *J. Biol. Chem.* **222**, 1025–1034.

Berg, P., and Ofengand, E. J. (1958). *Proc. Natl. Acad. Sci. U.S.A.* **44**, 78–86.

Berg, P., Bergman, F. H., Ofengand, E. J., and Dieckmann, M. (1961). *J. Biol. Chem.* **236**, 1726–1734.

Berther, J. M., Mayer, P., and Dutler, H. (1974). *Eur. J. Biochem.* **47**, 151–163.

Blanquet, S., Fayat, G., Waller, J. P., and Iwatsubo, M. (1972). *Eur. J. Biochem.* **24**, 461–469.

Blanquet, S., Petrissant, G., and Waller, J. P. (1973). *Eur. J. Biochem.* **36**, 227–233.

Blanquet, S., Fayat, G., and Waller, J. P. (1975). *J. Mol. Biol.* **94**, 1–15.

Blobstein, S. H., Gebert, R., Grunberger, D., Nakanishi, K., and Weinstein, I. B. (1975). *Arch. Biochem. Biophys.* **167**, 668–673.

Block, R., and Haseltine, W. A. (1974). *In* "Ribosomes" (M. Nomura, A. Tissières, and P. Lengyel, eds.), pp. 747–761. Cold Spring Harbor Lab., Cold Spring Harbor, New York.

Bluestein, H. G., Allende, C. C., Allende, J. E., and Cantoni, G. L. (1968). *J. Biol. Chem.* **243**, 4693–4699.

Blum, A., Uhlenbeck, O. C., and Tinoco, I., Jr. (1972). *Biochemistry* **11**, 3248–3256.

Boeker, E. A., Hays, A. P., and Cantoni, G. L. (1973). *Biochemistry* **12**, 2379–2383.

Boguslawski, G., Vodkin, M. H., Finkelstein, D. B., and Fink, G. R. (1974). *Biochemistry* **13**, 4659–4667.

Bonnet, J., and Ebel, J. P. (1974). *FEBS Lett.* **39**, 259–262.

Bonnet, J., and Ebel, J. P. (1975). *Eur. J. Biochem.* **58**, 193–201.

Bonnet, J., Giegé, R., and Ebel, J. P. (1972). *FEBS Lett.* **27**, 139–144.

Bosshardt, H. R., Koch, G. L. E., and Hartley, B. S. (1975). *Eur. J. Biochem.* **53**, 493–498.

Brown, D. M. (1974). *Basic Prin. Nucleic Acid Chem.* **2**, 1–90.

Brownlee, C. G. (1972). "Determination of Sequences in RNA," North-Holland Publ., Amsterdam.

Bruton, C. J. (1975). *Biochem. J.* **147**, 191–192.

Bruton, C. J., and Clark, B. F. C. (1974). *Nucleic Acids Res.* **1**, 217–221.

Bruton, C. J., and Hartley, B. S. (1970). *J. Mol. Biol.* **52**, 165–178.

Bruton, C. J., Jakes, R., and Koch, G. L. E. (1974). *FEBS Lett.* **45**, 26–28.

Buckingham, R. H., Danchin, A., and Grunberg-Manago, M. (1974). *Biochem. Biophys. Res. Commun.* **56**, 1–8.

Budker, V. G., Knorre, D. G., Kravchenko, V. V., Lavrik, O. I., Nevinsky, G. A., and Teplova, N. M. (1974). *FEBS Lett.* **49**, 159–163.

Budzik, G. P., Lam, S. S. M., Schoemaker, H. J. P., and Schimmel, P. R. (1975). *J. Biol. Chem.* **250**, 4433–4439.

Cantoni, G. L., and Davies, D. R., eds. (1971). "Procedures in Nucleic Acid Research," Vol. 2. Harper, New York.

Carbon, J., and Curry, J. B. (1968). *J. Mol. Biol.* **38**, 201–216.

Carbon, J., and Fleck, E. W. (1974). *J. Mol. Biol.* **85**, 371–391.

Carriquiry, E., and Litvak, S. (1974). *FEBS Lett.* **38**, 287–292.

Cashmore, T. (1971). *Nature (London), New Biol.* **230**, 236–239.

Cassio, D., and Mathien, Y. (1974). *Nucleic Acids Res.* **1**, 719–725.

Chakraburtty, K. (1975a). *Nucleic Acids Res.* **2**, 1787–1792.

Chakraburtty, K. (1975b). *Nucleic Acids Res.* **2**, 1793–1804.

Chakraburtty, K., Midelfort, C. F., Steinschneider, A., and Mehler, A. H. (1975a). *J. Biol. Chem.* **250**, 3861–3865.

Chakraburtty, K., Steinschneider, A., Case, R. V., and Mehler, A. H. (1975b). *Nucleic Acids Res.* **2**, 2069–2076.

Chambers, R. W. (1971). *Prog. Nucleic Acid Res. Mol. Biol.* **11**, 489–525.

Chambers, R. W., Aoyagi, S., Furukawa, Y., Zawadzda, H., and Bhanot, O. S. (1973). *J. Biol. Chem.* **248**, 5549–5551.

Chang, S. E., and Ish-Horowicz, D. (1974). *J. Mol. Biol.* **84**, 375–388.

Chapeville, F. (1972). *In* "The Mechanism of Protein Synthesis and Its Regulation" (L. Bosch, ed.), pp. 5–32. North-Holland Publ., Amsterdam.

Charlier, J., and Grosjean, H. (1972). *Eur. J. Biochem.* **25**, 163–174.

Chase, R., Tener, G. M., and Gillam, I. C. (1974). *Arch. Biochem. Biophys.* **163**, 303–317.

Chen, M. C., Giegè, R., Lord, R. C., and Rich, A. (1975). *Biochemistry* **14**, 4385–4391.

Chinali, G., Sprinzl, M., Parmeggiani, A., and Cramer, F. (1974). *Biochemistry* **13**, 3001–3010.

Chousterman, S., and Chapeville, F. (1971). *FEBS Lett.* **17**, 153–157.

Chousterman, S., and Chapeville, F. (1973). *Eur. J. Biochem.* **35**, 46–50.

Chousterman, S., Hervé, G., and Chapeville, F. (1966). *Bull. Soc. Chim. Biol.* **48**, 1295–1303.

Clarke, L., and Carbon, J. (1974). *J. Biol. Chem.* **249**, 6874–6885.

Content, J., Lebleu, B., Zilberstein, A., Berissi, H., and Revel, M. (1974). *FEBS Lett.* **41**, 125–130.

Cortese, R., Landsberg, R., Vonder Haar, R. A., Umbarger, H. E., and Ames, B. N. (1974). *Proc. Natl. Acad. Sci. U.S.A.* **71**, 1857–1861.

Coutts, S. M., Gangloff, J., and Dirheimer, G. (1974). *Biochemistry* **13**, 3938–3948.

Cramer, F., and Gauss, D. H. (1972). *In* "The Mechanism of Protein Synthesis and Its Regulation" (L. Bosch, ed.), pp. 219–241. North-Holland Publ., Amsterdam.

Cramer, F., Faulhammer, H., von der Haar, F., Sprinzl, M., and Sternbach, H. (1975). *FEBS Lett.* **56**, 212–214.

Crick, F. H. C. (1957). *Biochem. Soc. Symp.* **14**, 25–26.

Crick, F. H. C. (1966). *J. Mol. Biol.* **19**, 548–555.

Crothers, D. M., Seno, T., and Söll, D. G. (1972). *Proc. Natl. Acad. Sci. U.S.A.* **69**, 3063–3067.

Daniel, V., Grimberg, J. I., and Zeevi, M. (1975). *Nature (London)* **257**, 193–197.

Daniel, W. E., Jr., and Cohn, M. (1975). *Proc. Natl. Acad. Sci. U.S.A.* **72**, 2582–2586.

Davie, E. W., Koningsberger, V. V., and Lipmann, F. (1956). *Arch. Biochem. Biophys.* **65**, 21–38.

Davis, A. R., and Nierlich, D. P. (1974). *Biochim. Biophys. Acta* **374**, 23–37.

Davis, B. D., Anderson, P., and Sparling, P. F. (1973). *J. Mol. Biol.* **76**, 223–232.

Delaney, P., Bierbaum, J., and Ofengand, J. (1974). *Arch. Biochem. Biophys.* **161**, 260–267.

Delk, A. S., and Rabinowitz, J. C. (1974). *Nature (London)* **252**, 106.

Delk, A. S., and Rabinowitz, J. C. (1975). *Proc. Natl. Acad. Sci. U.S.A.* **72**, 528–530.

DeLorenzo, F., and Ames, B. N. (1970). *J. Biol. Chem.* **245**, 1710–1716.

DeMoss, J. A., and Novelli, G. D. (1956). *Biochim. Biophys. Acta* **22**, 49–61.

Deutch, C. E., and Soffer, R. L. (1975). *Proc. Natl. Acad. Sci. U.S.A.* **72**, 405–408.

Deutscher, M. P. (1973). *Prog. Nucleic Acid Res. Mol. Biol.* **13**, 51–92.

Deutscher, M. P., Foulds, J., and McClain, W. H. (1974). *J. Biol. Chem.* **249**, 6696–6699.

Dimitrijevic, L. (1972). *FEBS Lett.* **25,** 170–174.

Dirheimer, G., Ebel, J. P., Bonnet, J., Gangloff, J., Keith, G., Krebs, B., Kuntzel, B., Roy, A., Weissenbach, J., and Werner, C. (1972). *Biochimie* **54,** 127–144.

Drews, J., Grasmuk, H., and Weil, R. (1972). *Eur. J. Biochem.* **26,** 416–425.

Drews, J., Grasmuk, H., and Unger, F. M. (1973). *Biochem. Biophys. Res. Commun.* **51,** 804–812.

Dube, S. K. (1973). *FEBS Lett.* **36,** 39–42.

Dunn, J. J., Carlson, P. S., Lacks, S. A., and Studier, F. W. (1975). *Brookhaven Symp. Biol.* **26.**

Ebel, J. P., Giegé, R., Bonnet, J., Kern, D., Befort, N., Bollack, C., Fasiolo, F., Gangloff, J., and Dirheimer, G. (1973). *Biochimie* **55,** 547–557.

Ecarot, B., and Cedergren, R. J. (1974). *Biochem. Biophys. Res. Commun.* **59,** 400–405.

Egan, B. Z., Weiss, J. F., and Kelmers, A. D. (1973). *Biochem. Biophys. Res. Commun.* **55,** 320–327.

Eigner, E. A., and Loftfield, R. B. (1974). *In* "Methods in Enzymology" (L. Grossman and K. Moldave, eds.), Vol. 29, pp. 601–619. Academic Press, New York.

Eisinger, J., and Gross, N. (1974). *J. Mol. Biol.* **88,** 165–174.

Eisinger, J., and Gross, N. (1975). *Biochemistry* **14,** 4031–4041.

Eldred, E. W., and Schimmel, P. R. (1972). *J. Biol. Chem.* **247,** 2961–2964.

Elkins, B. N., and Keller, E. B. (1974). *Biochemistry* **13,** 4622–4628.

Englander, J. J., Kallenbach, N. R., and Englander, S. W. (1972). *J. Mol. Biol.* **63,** 153–169.

Erdmann, V. A., Sprinzl, M., and Pongs, O. (1973). *Biochem. Biophys. Res. Commun.* **54,** 942–948.

Erdmann, V. A., Sprinzl, M., Richter, D., and Lorenz, S. (1974). *Acta Biol. Med. Ger.* **33,** 605–608.

Fasiolo, F., and Ebel, J. P. (1974). *Eur. J. Biochem.* **49,** 257–263.

Fasiolo, F., Remy, P., Pouyet, J., and Ebel, J. P. (1974). *Eur. J. Biochem.* **50,** 227–236.

Favre, A., Yaniv, M., and Michelson, A. M. (1969). *Biochem. Biophys. Res. Commun.* **37,** 266–271.

Favre, A., Buckingham, R., and Thomas, G. (1975). *Nucleic Acids Res.* **2,** 1421–1431.

Fersht, A. R., and Jakes, R. (1976). *Biochemistry* **14,** 3350–3356.

Fraser, T. H., and Rich, A. (1975). *Proc. Natl. Acad. Sci. U.S.A.* **72,** 3044–3048.

Freier, S. M., and Tinoco, I., Jr. (1975). *Biochemistry* **14,** 3310–3314.

Friedman, S., Li, H. J., Nakanishi, K., and Van Lear, G. (1974). *Biochemistry* **13,** 2932–2937.

Frolova, L. Y., Kovaleva, G. K., Agalarova, M. B., and Kisselev, L. L. (1973). *FEBS Lett.* **34,** 213–216.

Gamble, R. C., and Schimmel, P. R. (1974). *Proc. Natl. Acad. Sci. U.S.A.* **71,** 1356–1360.

Gatica, M., Allende, C. C., Mora, G., Allende, J. E., and Medina, J. (1966). *Biochim. Biophys. Acta* **129,** 201–203.

Gauss, D. H., von der Haar, F., Maelicke, A., and Cramer, F. (1971). *Annu. Rev. Biochem.* **40,** 1045–1078.

Gefter, M. L., and Russell, R. L. (1969). *J. Mol. Biol.* **39,** 145–157.

Ghosh, K., Grishko, A., and Ghosh, H. P. (1971). *Biochem. Biophys. Res. Commun.* **42,** 462–468.

Ghosh, K., Ghosh, H. P., Simsek, M., and RajBhandary, U. L. (1974). *J. Biol. Chem.* **249,** 4720–4729.

Ghysen, A., and Celis, J. E. (1974a). *J. Mol. Biol.* **83,** 333–351.

Ghysen, A., and Celis, J. E. (1974b). *Nature (London)* **249,** 418–421.

Giege, R., Ebel, J-P., Springer, M., and Grunberg-Manago, M. (1973). *FEBS Lett.* **37,** 166–169.

Giege, R., Kern, D., Ebel, J-P., Grosjean, H., De Henau, S., and Chantrenne, H. (1974). *Eur. J. Biochem.* **45**, 351–394.

Gillam, I. C., and Tener, G. M. (1971). *In* "Methods in Enzymology" (K. Moldave and G. Grossman, eds.), Vol. 20, pp. 55–70. Academic Press, New York.

Grachev, M. A., and Rivkin, M. I. (1975). *Nucleic Acids Res.* **2**, 1237–1260.

Greenspan, C. M., and Litt, M. (1974). *FEBS Lett.* **41**, 297–302.

Grimburg, J. I., and Daniel, V. (1974). *Nature (London)* **250**, 320–322.

Gross, H. J., Simsek, M., Raba, M., Limburg, K., Heckman, J., and RajBhandary, U. L. (1974). *Nucleic Acids Res.* **1**, 35–43.

Gruhl, H., and Feldmann, H. (1975). *FEBS Lett.* **57**, 145–148.

Grummt, F., Grummt, I., Gross, H. J., Sprinzl, M., Richter, D., and Erdmann, V. A. (1974). *FEBS Lett.* **42**, 15–17.

Grunberger, D., Weinstein, I. B., and Mushinski, J. F. (1975). *Nature (London)* **253**, 66–68.

Guerrier-Takada, C., Dirheimer, G., Grosjean, H., and Keith, G. (1975). *FEBS Lett.* **60**, 286–289.

Guilley, H., Jonard, G., and Hirth, L. (1975). *Proc. Natl. Acad. Sci. U.S.A.* **72**, 864–868.

Gupta, N. K., Chatterjee, N. K., Woodley, C. L., and Bose, K. K. (1971). *J. Biol. Chem.* **246**, 7460–7469.

Gupta, S. L., Sopori, M. L., and Lengyel, P. (1974). *Biochem. Biophys. Res. Commun.* **57**, 763–770.

Hall, R. H. (1971). "The Modified Nucleosides in Nucleic Acid." Columbia Univ. Press, New York.

Harada, F., and Nishimura, S. (1974). *Biochemistry* **13**, 300–307.

Harada, F., Sawyer, R. C., and Dahlberg, J. E. (1975). *J. Biol. Chem.* **250**, 3487–3497.

Haseltine, W. A., and Block, R. (1973). *Proc. Natl. Acad. Sci. U.S.A.* **70**, 1564–1568.

Hashimoto, S., Kawata, M., and Takemura, S. (1972). *J. Biochem. (Tokyo)* **72**, 1339–1349.

Hecht, L. I., Stephenson, M. L., and Zamecnik, P. C. (1959). *Proc. Natl. Acad. Sci. U.S.A.* **45**, 505–518.

Hecht, S. M., and Hawrelak, S. D. (1974). *Biochemistry* **13**, 4967–4975.

Hecht, S. M., Kozarich, J. W., and Schmidt, F. J. (1974). *Proc. Natl. Acad. Sci. U.S.A.* **71**, 4317–4321.

Hecker, L. I., Egan, J., Reynolds, R. J., Nix, C. E., Schiff, J. A., and Barnett, W. E. (1974). *Proc. Natl. Acad. Sci. U.S.A.* **71**, 1910–1914.

Henley, D. D., Lindahl, T., and Fresco, J. R. (1966). *Proc. Natl. Acad. Sci. U.S.A.* **55**, 191–198.

Hentzen, D., Mandel, P., and Garel, J. P. (1972). *Biochim. Biophys. Acta* **281**, 228–232.

Hilbers, C. W., and Shulman, R. G. (1974). *Proc. Natl. Acad. Sci. U.S.A.* **71**, 3239–3242.

Hirsh, D. (1971). *J. Mol.. Biol.* **58**, 439–458.

Hirsh, D. I. (1968). *J. Biol. Chem.* **243**, 5731–5738.

Hoagland, M. B. (1955). *Biochim. Biophys. Acta* **16**, 288–289.

Hoagland, M. B. (1960). *In* "The Nucleic Acids" (E. Chargaff and J. N. Davidson, eds.), Vol. 3, pp. 349–408. Academic Press, New York.

Hoagland, M. B., Keller, E. B., and Zamecnik, P. C. (1956). *J. Biol. Chem.* **218**, 345–358.

Högenauer, G. (1974). *FEBS Lett.* **39**, 310–312.

Högenauer, G., Turnowsky, F., and Unger, F. M. (1972). *Biochem. Biophys. Res. Commun.* **46**, 2100–2107.

Holler, E., Rainey, P., Orme, A., Bennett, E. L., and Calvin, M. (1973). *Biochemistry* **12**, 1150–1159.

Holler, E., Hammer-Raber, B., Hanke, T., and Bartmann, P. (1975). *Biochemistry* **14**, 2496–2503.

Holley, R. W., Apgar, J., Everett, G. A., Madison, J. T., Marquisee, M., Merrill, S. H., Penswick, J. R., and Zamir, A. (1965). *Science* **147**, 1462–1465.

Holmes, W. M., Hurd, R. E., Reid, B. R., Rimerman, R. A., and Hatfield, G. W. (1975). *Proc. Natl. Acad. Sci. U.S.A.* **72**, 1068–1071.

Holness, N. J., and Atfield, G. (1974). *FEBS Lett.* **46**, 268–270.

Horowitz, J., Ou, C.-N., Ishaq, M., Ofengand, J., and Bierbaum, J. (1974). *J. Mol. Biol.* **88**, 301–312.

Hörz, W., and Zachau, H. G. (1973). *Eur. J. Biochem.* **32**, 1–14.

Hossain, A., and Kallenbach, N. R. (1974). *FEBS Lett.* **45**, 202–205.

Hussain, Z., and Ofengand, J. (1973). *Biochem. Biophys. Res. Commun.* **50**, 1143–1151.

Jacobson, K. B. (1969). *J. Cell. Physiol.* **74**, Suppl. 1, 99–101.

Jacobson, K. B. (1971). *Prog. Nucleic Acid Res. Mol. Biol.* **11**, 461–488.

Jakes, R., and Fersht, A., R. (1975). *Biochemistry* **14**, 3344–3350.

Jones, C. R., and Kearns, D. R. (1974). *Proc. Natl. Acad. Sci. U.S.A.* **71**, 4237–4240.

Juarez, H., Juarez, D., and Hedgcoth, C. (1975). *Nature (London)* **254**, 359–360.

Kalousek, F., and Konigsberg, W. H. (1974). *Biochemistry* **13**, 999–1008.

Kaplan, D. A., and Nierlich, D. P. (1975). *J. Biol. Chem.* **250**, 934–938.

Kasai, H., Kuchino, Y., Nihei, K., and Nishimura, S. (1975a). *Nucleic Acids Res.* **2**, 1931–1940.

Kasai, H., Ohashi, Z., Harada, F., Nishimura, S., Oppenheimer, N. J., Crain, P. F., Liehr, J. G., von Minden, D. L., and McCloskey, J. A. (1975b). *Biochemistry* **14**, 4198–4208.

Kawakami, M., Tanada, S., and Takemura, S. (1975). *FEBS Lett.* **51**, 321–324.

Kearns, D. R., and Shulman, R. G. (1974). *Acc. Chem. Res.* **7**, 33–39.

Kearns, D. R., Wong, K. L., and Wong, Y. P. (1973). *Proc. Natl. Acad. Sci. U.S.A.* **70**, 3843–3846.

Kearns, D. R., Wong, Y. P., Chang, S. H., and Hawkins, E. (1974). *Biochemistry* **13**, 4736–4746.

Keith, G., Ebel, J. P., and Dirheimer, G. (1974). *FEBS Lett.* **48**, 50–52.

Kelmers, A. D., Heatherly, D. E., and Egan, B. Z. (1974). *In* "Methods in Enzymology" (L. Grossman and K. Moldave, eds.), Vol. 29, pp. 483–486. Academic Press, New York.

Kersten, H., Sandig, L., and Arnold, H. H. (1975). *FEBS Lett.* **55**, 57–60.

Kim, S. H., Sussman, J. L., Suddath, F. L., Quigley, G. J., McPherson, A., Wang, W. H. J., Seeman, N. C., and Rich, A. (1974a). *Proc. Natl. Acad. Sci. U.S.A.* **71**, 4970–4974.

Kim, S. H., Suddath, F. L., Quigley, G. J., McPherson, A., Sussman, J. L., Wang, A. H. J., Seeman, N. C., and Rich, A. (1974b). *Science* **185**, 435–440.

Kimball, M. E., and Söll, D. (1974). *Nucleic Acids Res.* **1**, 1713–1720.

Kimball, M. E., Szeto, K. S., and Söll, D. (1974). *Nucleic Acids Res.* **1**, 1721–1732.

Kisselev, L. L., and Favorova, O. O. (1974). *Adv. Enzymol.* **40**, 141–238.

Klug, A., Ladner, J., and Robertus, J. D. (1974). *J. Mol. Biol.* **89**, 511–516.

Knorre, D. G. (1975). *FEBS Lett.* **58**, 50–53.

Kobayashi, T., Irie, M., Yoshida, M., Takeishi, K., and Ukita, T. (1974). *Biochim. Biophys. Acta* **366**, 168–181.

Koch, G. L. E., and Bruton, C. J. (1974). *FEBS Lett.* **40**, 180–182.

Koch, G. L. E., Boulanger, Y., and Hartley, B. S. (1974). *Nature (London)* **249**, 316–320.

Kraus, J., and Staehelin, M. (1974). *Nucleic Acids Res.* **1**, 1455–1478 and 1479–1496.

Krauss, G., Römer, R., Riesner, D., and Maass, G. (1973). *FEBS Lett.* **30**, 6–10.

Krauss, G., Pingoud, A., Boehme, D., Riesner, D., Peters, F., and Maass, G. (1975). *Eur. J. Biochem.* **55**, 517–529.

Kućan, Z., and Chambers, R. W. (1973). *J. Biochem. (Tokyo)* **73**, 811–819.

Kuchino, Y., and Nishimura, S. (1974). *Biochemistry* **13**, 3683–3688.

Kumar, S. A., Krauskopf, M., and Ofengand, J. (1973). *J. Biochem.* *(Tokyo)* **74**, 341–353.

Kupper, H., Contreras, R., Landy, A., and Khorana, H. G. (1975). *Proc. Natl. Acad. Sci. U.S.A.* **72**, 4754–4758.

Ladner, J. E., Jack, A., Robertus, J. D., Brown, R. S., Rhodes, D., Clark, B. F. C., and Klug, A. (1975). *Proc. Natl. Acad. Sci. U.S.A.* **72**, 4414–4418.

Lamy, D., Jonard, G., Guilley, H., and Hirth, L. (1975). *FEBS Lett.* **60**, 202–204.

Langlois, R., Kim, S.-H., and Cantor, C. R. (1975). *Biochemistry* **14**, 2554–2558.

Lapointe, J., and Söll, D. (1972). *J. Biol. Chem.* **247**, 4982–4985.

Lavrik, O. I., and Khutoryanskaya, L. Z. (1974). *FEBS Lett.* **39**, 287–290.

Lawrence, F., Blanquet, S., Poiret, M., Robert-Gero, M., and Waller, J. P. (1973). *Eur. J. Biochem.* **36**, 234–243.

Lawrence, F., Shire, D. J., and Waller, J. P. (1974). *Eur. J. Biochem.* **41**, 73–81.

Lea, P. J., and Norris, R. D. (1972). *Phytochemistry* **11**, 2897–2920.

Leahy, J., Glassman, E., and Schweet, R. S. (1960). *J. Biol. Chem.* **235**, 3209–3212.

Lee, M. (1974). *Biochemistry* **13**, 4747–4752.

Leis, J. P., and Keller, E. B. (1970). *Proc. Natl. Acad. Sci. U.S.A.* **67**, 1593–1599.

Lemaire, G., Gros, C., Epely S., Kaminski, M., and Labouesse, B. (1975). *Eur. J. Biochem.* **51**, 237–252.

Levitt, M. (1969). *Nature (London)* **244**, 759–763.

Li, H. J., Nakanishi, K., Grunberger, D., and Weinstein, I. B. (1973). *Biochem. Biophys. Res. Commun.* **55**, 818–823.

Lindahl, T., Adams, A., and Fresco, J. R. (1967). *J. Biol. Chem.* **242**, 3129–3134.

Lipsett, M. N. (1972). *J. Biol. Chem.* **247**, 1458–1461.

Littauer, U. Z., and Inouye, H. (1973). *Annu. Rev. Biochem.* **42**, 439–470.

Loftfield, R. B. (1971). *In* "Protein Synthesis" (E. H. McConkey, ed.), Vol. I, pp. 1–88. Dekker, New York.

Loftfield, R. B. (1972). *Prog. Nucleic Acid Res. Mol. Biol.* **12**, 87–128.

Lövgren, T., Heinonen, J., and Loftfield, R. B. (1975). *J. Biol. Chem.* **250**, 3854–3860.

McClain, W. H., and Seidman, J. G. (1975). *Nature (London)* **257**, 106–110.

McCutchan, T. F., Gilham, P. T., and Söll, D. (1975). *Nucleic Acids Res.* **2**, 853–864.

Maelicke, A., Sprinzl, M., von der Haar, F., Khwaja, T. A., and Cramer, F. (1974). *Eur. J. Biochem.* **43**, 617–625.

Marchin, G. L., Comer, M. M., and Neidhardt, F. C. (1972). *J. Biol. Chem.* **247**, 5132–5145.

Marcu, K., Mignery, R., Reszelbach, R., Roe, B., Sirover, M., and Dudock, B. (1973). *Biochem. Biophys. Res. Commun.* **55**, 477–483.

Matsuzaki, K. and Takeda, Y. (1973). *Biochim. Biophys. Acta* **308**, 339–351.

Mehler, A. H., and Chakrabutty, K. (1971). *Adv. Enzymol.* **35**, 443–501.

Menninger, J. R., Deery, S., Draper, D., and Walker, C, (1974). *Biochim. Biophys. Acta* **335**, 185–195.

Midelfort, C. F., and Mehler, A. H. (1974). *In* "Methods in Enzymology" (L. Grossman and K. Moldave, eds.), Vol. 29, pp. 627–642. Academic Press, New York.

Midelfort, C. F., Chakrabutty, K., Steinschneider, A., and Mehler, A. H. (1975). *J. Biol. Chem.* **250**, 3866–3873.

Miller, P. S., Barrett, J. C., and Ts'o, P. O. P. (1974). *Biochemistry* **13**, 4887–4896.

Mirzabekov, A. D., and Bayev, A. A. (1974). *In* "Methods in Enzymology" (L. Grossman and K. Moldave, eds.), Vol. 29, pp. 643–661. Academic Press, New York.

Morse, J. W., and Deutscher, M. P. (1975). *J. Mol. Biol.* **95**, 141–144.

Münch, H.-J., and Thiebe, R. (1975). *FEBS Lett.* **51**, 257–258.

Nass, G., and Stöffler, G. (1967). *Mol. Gen. Genet.* **100**, 378–382.

Ninio, J. (1973). *Prog. Nucleic Acid Res. Mol. Biol.* **13**, 301–337.

Nishikawa, K., and Takemura, S. (1974). . *Biochem. (Tokyo)* **76**, 935–947.

Nishimura, S. (1972). *Prog. Nucleic Acid Res. Mol. Biol.* **12,** 49–85.

Nishimura, S., Taya, Y., Kuchino, Y., and Ohashi, Z. (1974). *Biochem. Biophys. Res. Commun.* **57,** 702–708.

Odom, O. W., Hardesty, B., Wintermeyer, W., and Zachau, H. G. (1975). *Biochim. Biophys. Acta* **378,** 159–163.

Ofengand, J. (1974). *In* "Methods in Enzymology" (L. Grossman and K. Moldave, eds.), Vol. 29, pp. 661–667. Academic Press, New York.

Ofengand, J., and Chen, C.-M. (1972). *J. Biol. Chem.* **247,** 2049–2058.

Ofengand, J., and Henes, C. (1969). *J. Biol. Chem.* **244,** 6241–6253.

Ofengand, J., Chládek, S., Robilard, G., and Bierbaum, J. (1947a). *Biochemistry* **13,** 5425–5432.

Ofengand, J., Delaney, P., and Bierbaum, J. (1974b). *In* "Methods in Enzymology" (L. Grossman and K. Moldave, eds.), Vol. 29, pp. 673–684. Academic Press, New York.

Ofengand, J., Bierbaum, J., Horowitz, J., Ou, C.-N., and Ishaq, M. (1974c). *J. Mol. Biol.* **88,** 313–325.

Ohashi, Z., Maeda, M., McCloskey, J. A., and Nishimura, S. (1974). *Biochemistry* **13,** 2620–2625.

Ostrem, D. L., and Berg, P. (1974). *Biochemistry* **13,** 1338–1348.

Pachmann, U., Cronvall, E., Rigler, R., Hirsch, R., Wintermeyer, W., and Zachau, H. G. (1973). *Eur. J. Biochem.* **39,** 265–273.

Panet, A., Haseltine, W. A., Baltimore, D., Peters, G., Harada, F., and Dahlberg, J. E. (1975). *Proc. Natl. Acad. Sci. U.S.A.* **72,** 2535–2539.

Parfait, R. (1973). *FEBS Lett.* **29,** 323–325.

Paradies, H. H. (1975). *Biochem. Biophys. Res. Commun.* **64,** 1253–1262.

Papas, T. S., and Peterkofsky, A. (1972). *Biochemistry* **11,** 4602–4608.

Parthasarathy, R., Ohrt, J. M., and Chheda, G. B. (1974). *Biochem. Biophys. Res. Commun.* **60,** 211–218.

Pederson, F. S., Lund, E., and Kjeldgaard, N.O. (1973). *Nature (London), New Biol.* **243,** 13–15.

Penneys, N. S., and Muench, K. H. (1974). *Biochemistry* **13,** 560–565.

Penswick, J. R., Martin, R., and Dirheimer, G. (1975). *FEBS Lett.* **50,** 28–31.

Petrissant, G., and Favre, A. (1972). *FEBS Lett.* **23,** 191–194.

Petrova, M., Philippsen, P., and Zachau, H. G. (1975). *Biochim. Biophys. Acta* **395,** 455–467.

Pingoud, A., Riesner, D., Boehme, D., and Maass, G. (1973). *FEBS Lett.* **30,** 1–5.

Piper, P. W., and Clark, B. F. C. (1974a). *FEBS Lett.* **47,** 56–59.

Piper, P. W., and Clark, B. F. C. (1974b). *Nucleic Acids Res.* **1,** 45–51.

Piper, P. W., and Clark, B. F. C. (1974c). *Nature (London)* **247,** 516–518.

Pongs, O., Bald, R., and Reinwald, E. (1973). *Eur. J. Biochem.* **32,** 117–125.

Preiss, J., Berg, P., Ofengand, E. J., Bergman, F. H., and Dieckmann, M. (1959). *Proc. Natl. Acad. Sci. U.S.A.* **45,** 319–328.

Prinz, H., Maelicke, A., and Cramer, F. (1974). *Biochemistry* **13,** 1322–1326.

Quigley, G. J., Wang, A. H. J., Seeman, N. C., Suddath, F. L., Rich, A., Sussman, J. L., and Kim, S. H. (1975). *Proc. Natl. Acad. Sci. U.S.A.* **72,** 4866–4870.

Randerath, K., Chia, L. S. Y., Gupta, R. C., Randerath, E., Hawkins, E. R., Brum, C. K., and Chang, S. H. (1975). *Biochem. Biophys. Res. Commun.* **63,** 157–163.

Reid, B. R., and Robillard, G. T. (1975). *Nature (London)* **257,** 287–291.

Rether, B., Gangloff, J., and Ebel, J. P. (1974). *Eur. J. Biochem.* **50,** 289–295.

Rhodes, D. (1975). *J. Mol. Biol.* **94,** 449–460.

Richter, D. (1975). *In* "EMBO Workshop on tRNA," Abstract. Nof Ginossar, Israel.

Richter, D., and Lipmann, F. (1970). *Nature (London)* **227,** 1212–1214.

Richter, D., Lipmann, F., Tarrago, A., and Allende, J. E. (1971). *Proc. Natl. Acad. Sci. U.S.A.* **68**, 1805–1809.

Richter, D., Erdmann, V. A., and Sprinzl, M. (1973). *Nature (London), New Biol.* **246**, 132–135.

Richter, D., Erdmann, V. A., and Sprinzl, M. (1974). *Proc. Natl. Acad. Sci. U.S.A.* **71**, 3226–3229.

Riddle, D. L., and Carbon, J. (1973). *Nature (London), New Biol.* **242**, 230–234.

Ringer, D., and Chládek, S. (1975). *Proc. NAtl. Acad. Sci. U.S.A.* **72**, 2950–2954.

Ringer, D., Chládek, S., and Ofengand, J. (1976). *Biochemistry* **15**, 2759–2765.

Ringer, D., Quiggle, K., and Chládek, S. (1975). *Biochemistry* **14**, 514–520.

Robert-Gero, M., Lawrence, F., and Vigier, P. (1975). *Biochem. Biophys. Res. Commun.* **63**, 594–600.

Roberts, J. W., and Carbon, J. (1974). *Nature (London)* **250**, 412–414.

Roberts, R. J. (1974). *J. Biol. Chem.* **249**, 4787–4796.

Roberts, W. K., and Coleman, W. H. (1972). *Biochem. Biophys. Res. Commun.* **46**, 206–214.

Robertus, J. D., Ladner, J. E., Finch, J. T., Rhodes, D., Brown, R. S., Clark, B. F. C., and Klug, A. (1974a). *Nature (London)* **250**, 546–551.

Robertus, J. D., Ladner, J. E., Finch, J. T., Rhodes, D., Brown, R. S., Clark, B. F. C., and Klug, A. (1974b). *Nucleic Acids Res.* **1**, 927–932.

Roe, B., Marcu, K., and Dudock, B. (1973a). *Biochim. Biophys. Acta* **319**, 25–36.

Roe, B., Michael, M., and Dudock, B. D. (1973c). *Nature (London), New Biol.* **246**, 135–138.

Roe, B., Sirover, M., and Dudock, B. (1973b). *Biochemistry* **12**, 4146–4154.

Rogg, H., Muller, P., and Staehelin, M. (1975). *Eur. J. Biochem.* **53**, 115–127.

Rubin, G. M. (1973). *J. Biol. Chem.* **248**, 3860–3875.

Sakano, H., and Shimura, Y. (1975). *Proc. Natl. Acad. Sci. U.S.A.* **72**, 3369–3373.

Sakano, H., Shimura, Y., and Ozeki, H. (1974b). *FEBS Lett.* **48**, 117–121.

Salomon, R., and Littauer, U. Z. (1974). *Nature (London)* **249**, 32–34.

Samuel, C. E., and Rabinowitz, J. C. (1974). *J. Biol. Chem.* **249**, 1198–1207.

Santi, D. V., and Cunnion, S. O. (1974). *Biochemistry* **13**, 481–485.

Santi, D. V., and Webster, R. W. (1975). *J. Biol. Chem.* **250**, 3874–3877.

Santi, D. V., Webster, R. W., and Cleland, W. W. (1974). *In* "Methods in Enzymology" (L. Grossman and K. Moldave, eds.), Vol. 29, pp. 620–627. Academic Press, New York.

Sawyer, R. C., Harada, F., and Dahlberg, J. E. (1974). *J. Virol.* **13**, 1302–1311.

Schäfer, K. P., and Söll, D. (1974). *Biochimie* **56**, 795–804.

Schimmel, P. R., Uhlenbeck, O. C., Lewis, J. B., Dickson, L. A., Eldred, E. W., and Schreier, A. A. (1972). *Biochemistry* **11**, 642–646.

Schmidt, W., Arnold, H. H., and Kersten, H. (1975). *Nucleic Acids Res.* **2**, 1043–1052.

Schoemaker, H. J. P., and Schimmel, P. R. (1974). *J. Mol. Biol.* **84**, 503–513.

Schoemaker, H. J. P., Budzik, G. P., Giege, R., and Schimmel, P. R. (1975). *J. Biol. Chem.* **250**, 4440–4444.

Schreier, A. A., and Schimmel, P. R. (1972). *Biochemistry* **11**, 1582–1589.

Schuber, F., and Pinck, M. (1974). *Biochimie* **56**, 383–403.

Schulman, L. H., and Goddard, J. P. (1973). *J. Biol. Chem.* **248**, 1341–1345.

Schulman, L. H., and Pelka, H. (1975). *J. Biol. Chem.* **250**, 542–547.

Schulman, L. H., Pelka, H., and Sundari, R. M. (1974). *J. Biol. Chem.* **249**, 7102–7110.

Schwartz, I., and Ofengand, J. (1974). *Proc. Natl. Acad. Sci. U.S.A.* **71**, 3951–3955.

Schwartz, I., Gordon, E., and Ofengand, J. (1975). *Biochemistry* **14**, 2907–2914.

Schwarz, U., Luhrman, R., and Gassen, H. G. (1974). *Biochem. Biophys. Res. Commun.* **56**, 807–814.

Schweet, R. S., Bovard, F. C., Allen, E., and Glassman, E. (1958). *Proc. Natl. Acad. Sci. U.S.A.* **44**, 173–177.

Sekiya, T., van Ormondt, H., and Khorana, H. G. (1975). *J. Biol. Chem.* **250**, 1087–1098.

Seno, T. (1975). *FEBS Lett.* **51**, 325–329.

Seno, T., Agris, P. F., and Söll, D. (1974). *Biochim. Biophys. Acta* **349**, 328–338.

Shulman, R. G., Hilbers, C. W., and Miller, D. L. (1974a). *J. Mol. Biol.* **90**, 601–607.

Shulman, R. G., Hilbers, C. W., Söll, D. G., and Yang, S. K. (1974b). *J. Mol. Biol.* **90**, 609–611.

Silver, J., and Laursen, R. A. (1974). *Biochim. Biophys. Acta* **340**, 77–89.

Simsek, M., RajBhandary, U. L., Boisnard, M., and Petrissant, G. (1974). *Nature (London)* **247**, 518–520.

Singhal, R. P. (1974). *Biochemistry* **13**, 2924–2932.

Smith, J. D. (1972). *Annu. Rev. Genet.* **6**, 235–256.

Smith, K. C., and Aplin, R. T. (1966). *Biochemistry* **5**, 2125–2130.

Soffer, R. L. (1974). *Adv. Enzymol.* **40**, 91–139.

Soffer, R. L. (1975). *J. Biol. Chem.* **250**, 2626–2629.

Söll, D. (1971). *Science* **173**, 293–299.

Söll, D., and Schimmel, P. R. (1974). *In* "The Enzymes" (P. D. Boyer, ed.), 3rd ed., Vol. 10, pp. 489–538. Academic Press, New York.

Sourgoutchov, A., Blanquet, S., Fayat, G., and Waller, J.-P. (1974). *Eur. J. Biochem.* **46**, 431–438.

Spremulli, L. L., Agris, P. F., Brown, G. M., and RajBhandary, U. L. (1974). *Arch. Biochem. Biophys.* **162**, 22–37.

Sprinzl, M., and Cramer, F. (1973). *Nature (London), New Biol.* **245**, 3–5.

Sprinzl, M., and Cramer, F. (1975). *Proc. Natl. Acad. Sci. U.S.A.* **72**, 3049–3053.

Sprinzl, M., Scheit, K.-H., and Cramer, F. (1973). *Eur. J. Biochem.* **34**, 306–310.

Sprinzl, M., Kramer, E., and Stehlik, D. (1974). *Eur. J. Biochem.* **49**, 595–605.

Stanley, W. M., Jr. (1974). *In* "Methods in Enzymology" (L. Grossman and K. Moldave, eds.), Vol. 29, pp. 530–547. Academic Press, New York.

Streeck, R. E., and Zachau, H. G. (1972). *Eur. J. Biochem.* **30**, 382–391.

Strickland, J. E., and Jacobson, K. B. (1972). *Biochemistry* **11**, 2321–2323.

Takeda, Y., and Ohnishi, T. (1975). *J. Biol. Chem.* **250**, 3878–3882.

Takemura, S., Kasai, H., and Goto, M. (1974). *J. Biochem. (Tokyo)* **75**, 1169–1172.

Thang, M. N., Dondon, L., and Rether, B. (1974). *FEBS Lett.* **40**, 67–71.

Thiebe, R. (1975). *FEBS Lett.* **51**, 259–261.

Thiebe, R., and Poralla, K. (1973). *FEBS Lett.* **38**, 27–28.

Thomas, G. J., Jr., Chen, M. C., and Hartman, K. A. (1973a). *Biophys. Acta* **324**, 37–49.

Thomas, G. J., Jr., Chen, M. C., Lord, R. C., Kotsiopoulos, P. S., Tritton, T. R., and Mohr, S. C. (1973b). *Biochem. Biophys. Res. Commun.* **54**, 570–577.

Tritton, T. R., and Mohr, S. C. (1973). *Biochemistry* **12**, 905–914.

Tscherne, J. S., Weinstein, I. B., Lanks, K. W., Gersten, N. B., and Cantor, C. R. (1973). *Biochemistry* **12**, 3859–3865.

Uhlenbeck, O. C., Chirikjian, J. G., and Fresco, J. R. (1974). *J. Mol. Biol.* **89**, 495–504.

Unger, F. M., and S. Takemura, S. (1973). *Biochem. Biophys. Res. Commun.* **52**, 1141–1147.

Urbanke, C., Römer, R., and Maass, G. (1973). *Eur. J. Biochem.* **33**, 511–516.

Urbanke, C., Römer, R., and Maass, G. (1975). *Eur. J. Biochem.* **55**, 439–444.

Vennegoor, C., and Bloemendal, H. (1972). *Eur. J. Biochem.* **26**, 462–473.

Vlasov, V. V., Grineva, N. I., and Knorre, D. G. (1972). *FEBS Lett.* **20**, 66–70.

Vormbrock, R., Morawietz, R., and Gassen, H. G. (1974). *Biochim. Biophys. Acta* **340**, 348–358.

Wagner, L. P., and Ofengand, J. (1970). *Biochim. Biophys. Acta* **204**, 620–623.

Walker, R. T., and RajBhandary, U. L. (1972). *J. Biol. Chem.* **247**, 4879–4892.

Walker, R. T., and RajBhandary, U. L. (1975). *Nucleic Acids Res.* **2**, 61–78.

Watanabe, K., Oshima, T., Saneyoshi, M., and Nishimura, S. (1974). *FEBS Lett.* **43**, 59–63.

Waters, L. C. (1975). *Biochem. Biophys. Res. Commun.* **65**, 1130–1136.

Waters, L. C., Mullin, B. C., Ho, T., and Yang, W.-K. (1975). *Proc. Natl. Acad. Sci. U.S.A.* **72**, 2155–2159.

Waterson, R. M., and Konigsberg, W. H. (1974). *Proc. Natl. Acad. Sci. U.S.A.* **71**, 376–380.

White, B. N., Tener, G. M., Holden, J., and Suzuki, D. T. (1973). *J. Mol. Biol.* **74**, 635–651.

Wildenauer, D., Gross, H. S., and Riesner, D. (1974). *Nucleic Acids Res.* **1**, 1165–1182.

Williams, R. J., Nagel, W., Roe, B., and Dudock, B. (1974). *Biochem. Biophys. Res. Commun.* **60**, 1215–1221.

Wintermeyer, W., and Zachau, H. G. (1974). *In* "Methods in Enzymology" (L. Grossman and K. Moldave, eds.), Vol. 29, pp. 667–673. Academic Press, New York.

Wong, K. L., and Kearns, D. R. (1974). *Nature (London)* **252**, 738–739.

Wong, K. L., Bolton, P. H., and Kearns, D. R. (1975). *Biochim. Biophys. Acta* **383**, 446–451.

Wong, Y. P., Reid, B. R., and Kearns, D. R. (1973). *Proc. Natl. Acad. Sci. U.S.A.* **70**, 2193–2195.

Wubbeler, W., Lossow, C., Fittler, F., and Zachau, H. G. (1975). *Eur. J. Biochem.* **59**, 405–415.

Yamada, Y., and Ishikura, H. (1975). *FEBS Lett.* **54**, 155–158.

Yang, C.-H., and Söll, D. (1974). *Proc. Natl. Acad. Sci. U.S.A.* **71**, 2838–2842.

Yaniv, M., Folk, W. R., Berg, P., and Soll, L. (1974). *J. Mol. Biol.* **86**, 245–260.

Yarus, M. (1972a). *Proc. Natl. Acad. Sci. U.S.A.* **69**, 1915–1919.

Yarus, M. (1972b). *J. Biol. Chem.* **247**, 2738–2741.

Yarus, M. (1973). *J. Biol. Chem.* **248**, 6755–6758.

Yarus, M., and Berg, P. (1969). *J. Mol. Biol.* **42**, 171–189.

Yarus, M., and Rashbaum, S. (1972). *Biochemistry* **11**, 2043–2049.

Zachau, H. G. (1972). *In* "The Mechanism of Protein Synthesis and Its Regulation" (L. Bosch, ed.), pp. 173–218. North-Holland Publ., Amsterdam.

Zachau, H. G., and Feldman, H. (1965). *Prog. Nucleic Acid Res.* **4**, 217–230.

Zachau, H. G., Acs, G., and Lipmann, F. (1958). *Proc. Natl. Acad. Sci. U.S.A.* **44**, 885–889.

Zachau, H. G., Fittler, F., Harbers, K., Philippsen, P., Thiebe, R., and Wintermeyer, W. (1974). *In* "XXIVth International Congress of Pure and Applied Chemistry, Volume Two", pp. 45–56. Butterworths, London.

Zelwer, C., Risler, J. L., and Monteilhet, C. (1976). *J. Mol. Biol.* **102**, 93–101.

References Added in Proof

Brenchley, J. E., and Williams, L. S. (1975). *Annu. Rev. Microbiol.* **29**, 251–274.

Erdmann, V. A. (1976). *Prog. Nucleic Acid Res. Mol. Biol.* **18**, 45–90.

Kearns, D. R. (1976). *Prog. Nucleic Acid Res. Mol. Biol.* **18**, 91–149.

Kim, S. H. (1976). *Prog. Nucleic Acid Res. Mol. Biol.* **17**, 181–216.

Neidhardt, F. G., Parker, J., and McKeever, W. G. (1975). *Annu. Rev. Microbiol.* **29**, 215–250.

Quigley, G. J., and Rich, A. (1976). *Science* **194**, 796–806.

Rich, A., and RajBhandary, U. L. (1976). *Annu. Rev. Biochem.* **45**, 805–860.

Sigler, P. B. (1975). *Annu. Rev. Biophys. Bioeng.* **4**, 477–527.

Sussman, J. L., and Kim, S.-H. (1976). *Science* **192**, 853–858.

Aspects of Ribosome Structure and Function

C. G. KURLAND

I. Introduction

The recent solution of the structure of a transfer RNA (tRNA) molecule (Kim *et al.*, 1974; Robertus *et al.*, 1974) has two important consequences for the analysis of ribosome structure and function. First, it provides structural information of the sort needed to analyze the interactions of aminoacyl-tRNA with its ribosomal binding site during protein synthesis. Second, the crystallographic solution of this structure is a timely reminder of the kind of information that students of the ribosome have failed so far to obtain. Instead of the refined analysis afforded by crystallographic techniques we have been obliged to study the ribosome by such relatively crude techniques as affinity labeling, cross-linking, electron microscopy, immunochemistry, and neutron scattering. The consequence is that we have generated a large mass of data most of which is open to a multiplicity of structural interpretations.

We would like to be able to answer two questions: What is a ribosome? and What does it do? In spite of all the man-years that have gone into the study of these questions, it is embarrassingly difficult to provide anything but the most tentative answers. The source of our problems does not seem to be a lack of ideas. Rather, the kind of detailed information that we require to critically test our ideas seems to be just beyond the scope of the most up-to-date technology. As a consequence an outsider to this field is likely to be struck by the indirectness of the experimental approaches currently in vogue, and perhaps even more perplexed by the equivocal inferences that can be drawn from the available data. Accordingly, the present chapter cannot be anything more than a critical status report, and is therefore likely to suffer from all the defects inherent in this genre. Having made this apology, I will proceed as though everything is as clear as I can make it seem.

A. Some Ground Rules

The data discussed here concern the prokaryotic ribosome, and for the most part the ribosome of *Escherichia coli*. While this may seem to be an arbitrary restriction, it is motivated by two overriding considerations. One is the persistent dirth of nontrivial information concerning ribosomes from other sources. The other is that there is as yet no reason to believe that eukaryotic ribosomes do their job in a way fundamentally different from that of *E. coli*.

Accordingly, the term "ribosome" will be used to mean that organelle from *E. coli*, unless otherwise specified. Similarly, the other molecular components involved in protein synthesis will be referred to simply as tRNA, IF-2, etc., with the understanding that they refer to components of *E. coli* unless otherwise stated. Finally, the nomenclatures used to designate individual proteins from the ribosome will be that based on the two-dimensional gel system of Kaltschmidt and Wittmann (1971) as described for the 30 S proteins by Wittmann *et al.* (1971) and for the 50 S proteins by Kurland *et al.* (in preparation). These references contain the correlation of the common nomenclature with those intitially used by the different laboratories involved in the purification and characterization of the ribosomal proteins.

B. Elementary Functions

The adapter hypothesis requires two functions for the amino acid carriers in protein synthesis and these functions are at least formally separable (Crick, 1958). First, each adapter must be coded to accept one and only one amino acid. Second, each adapter must be coded so that it can be chosen to insert its cognate amino acid into a growing polypeptide. The experimental verification of the adapter hypothesis depended on the demonstration that the structures of the different amino acids, once joined to the tRNA adapters, play no role in determining which of the adapters are chosen by a programmed ribosome (Chappeville *et al.,* 1962). What this means from the structural point of view is that there must be at least two different centers on the programmed ribosome for the processing of a single aminoacyl-tRNA, and each of these centers must be capable of a very different kind of stereospecifc interaction with the charged adapter. One center must be involved in the selection of aminoacyl-tRNA molecules; hence, this center must search out the idiosyncratic properties of individual adapters. In contrast, the other center must be capable of recognizing common structures in the aminoacyl-tRNA's. At the very minimum, this second center must be able to recognize the structure of the ester linkage between an amino acid and its cognate tRNA, so that any amino acid can be inserted into the growing polypeptide.

As luck would have it, these two centers are so well separated in the ribosome that each can be identified exclusively with one of the two ribosomal subunits. The mRNA binding site seems to be restricted to the 30 S ribosomal subunit (Okamoto and Takanami, 1963), which means that the selection of the tRNA molecules must at the very least originate on this subunit. Furthermore, mutations that alter the error frequency of the tRNA selection process seem to be expressed exclusively as alterations in 30 S components (Gorini, 1971). Similarly, antibiotics that alter the error frequency of translation seem to be only those that interact with the 30 S subunit (Pestka, 1971).

The 50 S ribosomal subunit, on the other hand, seems on the surface to have little to do with the selection of specific tRNA molecules. All the evidence available indicates that its interactions with the tRNA are with common structural features of these molecules such as the C-C-A sequence at the 3'-end of the tRNA's or the universal T-ψ-C-G loop (Gilbert, 1963; Richter *et al.,* 1973). In addition, the peptidyl transferase activity which is responsible for peptide bond formation is located on the 50 S subunit (Monro, 1967). Therefore, the functions of this subunit seem to be restricted to the cyclic performance of identical operations on different amino acids and their cognate tRNA's.

The requirement for two sorts of aminoacyl-tRNA interaction sites hardly seems to be a sufficient condition by itself for the subdivision of the ribosome into two physically distinct subunits. However, taken together with the suggestion of Watson (1964) that the subunits might have to move relative to each other during protein synthesis, this structural feature of the ribosome becomes more understandable. Thus, there are several aspects of the translation mechanism that might be facilitated by mechanical movements of the two subunits.

First, and certainly most fundamental, is the requirement for the sequential translation of the message, one codon at a time from the N-terminal to the C-terminal amino acid of the finished protein. This means that the ribosome must be able to convert the free energy released by the hydrolysis of one or more covalent bonds into a directed, that is to say irreversible, mechanical process: mRNA movement, or more precisely a relative movement of ribosome and mRNA.

Second, a less well-defined movement of tRNA is presumed to follow the formation of each peptide bond; this is called translocation. Unlike mRNA movement, translocation is an abstraction that is usually described as a gross mechanical movement, but in reality it may not be anything more than a subtle conformational change of the tRNA (Woese, 1973). The facts are that the growing polypeptide chain seems always to be attached to a tRNA molecule (Gilbert, 1963). However, the polypeptidyl-tRNA can be bound to the ribosome in two distinct states: one that is sensitive to puromycin and the other refractory to the action of this antibiotic. These observations led Watson (1964) to a highly serviceable model in which it was assumed that the two states of reactivity with puromycin correspond to two physically distinct sites for tRNA attachment to the ribosome. According to this model, an aminoacyl-tRNA coded by the mRNA enters the ribosome at the puromycin-resistant A site. The nascent polypeptide, which was bound to another tRNA at the puromycin-sensitive D site, is then transferred to the tRNA newly arrived in the A site following the formation of the peptide bond. Then the peptidyl-tRNA is thought to be "translocated" to the D site and the system is ready for another cycle of elongation.

This two-site model is consistent with a large body of information (Lipmann, 1969), and it seems likely that a movement of tRNA corresponding to translocation will someday be observed. Nevertheless, the alternative account of the two states for peptidyl-tRNA association with the ribosome offered by Woese (1973) should not be rejected out of hand. This model as well as more conservative variants of the two-site model (Bretscher, 1968; Kurland et al., 1972; Spirin, 1969) are mentioned in the present context because they attempt to account for mRNA and tRNA movement as the outcome of cyclic rearrangements of the ribosomal subunits.

Finally, the dynamics of the ribosomal subunits may contribute to still another aspect of the translation mechanism. The view of protein synthesis developed so far has been an oversimplification of the elongation process. In reality, the genetic message is a punctuated one (Crick *et al.*, 1961) and this is reflected biochemically in the distinctive properties of the initiating, elongating, and terminating modes of the ribosome. The transitions between the different translation modes may be attended by rearrangements of the ribosomal subunits (Bretscher, 1968; Kurland *et al.*, 1972).

The foregoing account serves only to orient the subsequent discussion around some of the more obvious questions concerning the structure-function relations of the ribosome. Since several of the accompanying chapters deal with the biochemistry of the various translation factors as well as the mechanism of peptide bond synthesis, the present account will focus on the ribosomal interactions with mRNA and tRNA. However, before proceeding to these functional interactions it will be useful to briefly survey the structural features of this system.

II. Distribution of Ribosomal Components

Although it is not yet possible to describe in detail the three-dimensional arrangements of the proteins and RNA in the functional ribosome, some general features of their arrangements have become apparent. Since these features have a strong influence on the interpretation of experiments directed at the identification of "active sites," they must be discussed in some detail before we can proceed to the more functional studies. For the moment, we will concentrate on the ways in which the protein and RNA are distributed in the ribosomal subunits. In particular, evidence will be presented to substantiate the claim that there is only one kind of ribosome in *E. coli* and that it is made up of two subunits with very different structures.

A. Protein Stoichiometry

Clearly, our view of protein synthesis would be greatly complicated if it turned out that there is more than one kind of functional ribosome in a bacterium. Indeed, this was precisely the tentative conclusion drawn from an analysis of the protein compositions of the purified 30 S ribosomal subunits (Hardy *et al.*, 1969; Craven *et al.*, 1969). Here it was found that the sums of molecular weights of the individual proteins was significantly greater than the average mass of protein per particle, which implied that some of the proteins could not be present in every particle. Subsequent work provided very strong confirmation of this conclusion (Kurland *et al.*, 1969; Traut *et al.*, 1969; Voynow and Kurland, 1971; Weber, 1972). Thus, a class of proteins present in amounts significantly less than one copy per purified ribosome were identified in several laboratories. The next problem was to decide whether or not this heterogeneity is an artifact created during ribosome purification.

Experiments directed at this question showed that artifacts created by the protein extraction procedures and the salt washing of the ribosomal subunits could not account for the heterogeneity of the purified subunits (Kurland *et al.*, 1969; Voynow and Kurland, 1971). Nevertheless, there was one potential artifact that could not be controlled at the time these experiments were first done: this was the possibility that the dissociation of the ribosome in buffers containing low concentrations of magnesium could lead to a loss of proteins from the ribosomal subunits (Kurland *et al.*, 1969). Once the two-dimensional gel system had been perfected so that all the proteins from the 70 S ribosome could be fractionated at once (Kaldtschmidt and Wittmann, 1971), the effects of dissociating the ribosome on the protein compositions of the subunits could be checked.

Hardy (1975) has now completed a thorough reinvestigation of the stoichiometry of the 30 S as well as 50 S proteins and has shown that most of the proteins present in amounts less than one copy per purified subunit are, indeed, present in amounts corresponding more closely to one copy per intact 70 S ribosome. The one clear exception to this is the protein L7/L12 which is present in amounts corresponding to three or more copies per ribosome (Thammana *et al.*, 1973; Hardy, 1975).

The short period during which it was considered possible that *E. coli* ribosomes could be heterogeneous *in vivo* is probably over now. Nevertheless, this interlude had some positive effects. First, by exploiting the artifactual heterogeneity of the purified subunits it was possible to identify some functions of individual 30 S proteins (van Duin and Kurland, 1970; Randall-Hazelbauer and Kurland, 1972; van Duin *et al.*, 1972). Second, and perhaps most important, the interest generated in the stoichiometries of the proteins livened up and therefore probably hastened the completion of an otherwise dull job, namely, purifying as well as characterizing the 50 some odd proteins of the *E. coli* ribosome.

B. Repeat Structures

The stoichiometries of the ribosomal proteins indicated from the very outset that the structure of the ribosome was far more complicated than expected, and complicated in a way that created structural problems far more profound than that of the apparent heterogeneity of the purified subunits. Thus, with the notable exception of L7/L12, the proteins seem to be represented as one copy per ribosome. This conclusion dampened hopes of finding some simplifying rules to assist in the analysis of ribosome structure. Instead, each protein with the exception of L7/L12 appears to be uniquely adapted to some particular function in one particular part of the ribosome, and extensive protein repeat structures which might otherwise aid the analysis of the ribosome are not apparent.

Although the evidence discussed so far would imply that the proteins confer a radical dissymmetry on ribosome structure, this appearance may be more a reflection of the inadequate criteria used in the search for putative protein repeat structures rather than a faithful reflection of the way the proteins are ordered in the functional organelle. Thus, comparisons which depend principally on the primary sequences of the proteins (see Chapter 3) would not necessarily detect the presence of homologous tertiary structures in different proteins. Indeed, recent crystallographic studies of a variety of enzymes have revealed precisely this sort of structural homology between protein domains with apparently unrelated primary sequences (see, for example, Rossmann *et al.*, 1975).

In contrast, there are rather clear indications of extensive repeat structures among the ribosomal RNA's. First, Brownlee *et al.* (1967) have shown that approximately 60% of the 5 S RNA sequence is arranged as a rather clear structural duplication so that more than thirty nucleotides at the 5'-end of the molecule can be matched with a comparable number of nucleotides at the 3'-end of the chain. Similarly, there are several different sequences of length ten to twenty nucleotides which are each duplicated several times in different parts of the 16 S RNA molecule (Ehresmann *et al.*, 1972). A systematic search for the presence of repeated sequences in the 16 S and 23 S RNA remains to be done, and it will be a rather formidable task since nearly 5000 nucleotides are involved.

C. Subunit Structures

The first extensive definition of protein neighborhoods in the 30 S subunit was obtained from experiments in which bifunctional reagents were used to cross-link near neighboring proteins *in situ*. Here, "near" is defined by the length and reactivity of the bifunctional reagents that are employed. In most experiments the reagents used can bridge distances in the range between 6 Å and 20 Å. Since these reagents react preferentially with one sort of amino acid side chain, only those proteins that are close enough to appose reactive side chains at distances less than 20 Å can be cross-linked. Therefore, one limitation of this technique is that it does not detect all nearest neighbors, because proteins that are separated from each other by RNA or which do not have appropriate side chains within the requisite distance from each other cannot be cross-linked.

Another limitation of the cross-link technology is that it is not selective. A typical experiment consists of adding a bifunctional reagent to a solution of subunits, extracting the proteins after the reaction is over, and then trying to identify individual pairs of cross-linked proteins within the

TABLE I
Cross-Linked Pairs of 30 S Ribosomal Proteins

Pair	Smallest bridge distance (Å)	Reference
S1-S10	6	L. C. Lutter, unpublished
S2-S3	6	Lutter et al., 1974a; Sun et al., 1974
S2-S5	6	Lutter and Kurland, 1974
S2-S8	6	Lutter and Kurland, 1974
S3-S10	6	Lutter and Kurland, 1974
S4-S5	6	Lutter and Kurland, 1974
S4-S13	14	Sommer and Traut, 1974
S5-S8	6	Lutter et al., 1972; Sun et al., 1974
S6-S18	6	Lutter and Kurland, 1974
S7-S9	6	Lutter et al., 1974a
S7-S21	6	L. C. Lutter, unpublished
S13-S19	6	Lutter et al., 1974a; Sun et al., 1974
S14-S19	12	Bode et al., 1974
S18-S21	14	Lutter et al., 1972

population of extracted proteins. This latter task is extremely laborious and when it is not done with attention to detail the results are, to say the least, equivocal. For this reason I will not burden the reader with cross-link studies which appeared prematurely in the literature; these are reviewed elsewhere (Lutter et al., 1972; Kurland, 1974a).

Two sorts of reagents are now available for these studies: cleavable and noncleavable ones. Noncleavable reagents such as the diimido esters react with the amino groups of lysine while bifunctional maleimide derivatives react with the SH groups of cysteine. Such reagents have yielded valuable information about ribosomal protein neighborhoods but they have done so at a high cost of time and effort. Here, the protein extracted from ribosomal subunits reacted with any one of these reagents is fractionated either by chromatography or by electrophoresis. Purified pairs of cross-linked proteins can then be identified by one of several techniques such as isotopic transfer, Ouchterlony double diffusion tests, a modified form of affinity chromatography with specific antibodies, and reconstitution with the cross-linked proteins replacing free proteins in the reconstitution mixtures (Lutter et al., 1972; 1974a; Lutter and Kurland, 1973; Bode et al., 1974). A comprehensive description of these methods can be found in the review of Lutter and Kurland (1974). The near neighboring proteins identified by these procedures are listed in Table I.

Several reagents are now available with the attractive property that the bridge connecting the two functional groups can be disrupted under very mild conditions. Consequently, a pair of proteins cross-linked by such a reagent can be released from the complex, and then identified by gel electrophoresis. Two kinds of reagents with the common property that the cleavable center in the bridge is a disulfide bond have been developed by D. Elson's group (personal communication) and R. Traut's group (1973). Both of these reagents have two shortcomings that complicate the interpretation of the data they generate. First, oxidizing conditions must be employed when such reagents are used to cross link-proteins. Oxidizing conditions inactivate ribosomes; therefore, controls are required to show that the apparent neighborhoods generated by such reagents are present in the reduced ribosome. Second, evidence is also required to show that disulfide exchange does not create artificial cross-linked complexes during the experimental manipulation of the cross-linked proteins. Notwithstanding these technical peccadillos, such reagents show great promise. Neighborhoods detected with these reagents are listed in Table I.

Another class of cleavable reagents is available without the disadvantages of the disulfide reagents. These are bifunctional azide derivatives of tartaric acid; here, the cleavable center is provided by a pair of vicinal hydroxyl groups that are oxidizable by periodate (Lutter et al., 1974b). Unfortunately, these reagents have the disadvantage that the oxidation products are charged and, therefore, special gel systems are required for the identification of the proteins released from the cross-linked complexes (Lutter et al., 1974b). Most of the near neighboring proteins previously identified with the diimido esters have also been found with the cleavable diazo reagents. In addition a set of new neighborhoods has been discovered with the cleavable reagents (Lutter and Kurland, 1974); these are listed in Table I.

We now come to a problem that was previously alluded to: How do we know that the cross-linked proteins are near neighbors in the functional ribosome before the addition of the cross-linking reagents? This problem has been discussed in much greater detail elsewhere (Lutter and Kurland, 1973, 1974). The relevant arguments are as follows: First, most of the cross-linked pairs listed in Table I were identified with subunits that retained more than 50% of their initial activities after they had been reacted with reagent. This means that extensive disruption of the subunit structure had been avoided. Second, selected pairs of cross-linked proteins such as S5-S8 and S13-S19 could be used to reconstitute 30 S subunits, and the subunits containing the cross-linked proteins were almost as active as control subunits containing no cross-linked proteins. This shows that the cross-links themselves do not disrupt the functional arrangements of the proteins in the ribosome, at least in the case of these four proteins. Third, the pattern of the protein neighborhoods detected by the cross-link methodology is a remarkably faithful reflection of the pattern of cooperativity expressed between the same proteins during reconstitution *in vitro*. This latter finding provides the first clear structural generalization concerning the ribosome: proteins that cooperate during assembly are near neighbors in the functional ribosome. While this hardly constitutes a startling revelation, it does lend considerable weight to the conclusion that the neighborhoods detected in cross-linking experiments are real.

The physical interpretation of the protein neighborhoods is quite another matter. So far we have been discussing the arrangements of the ribosomal components as though there were only protein to contend with. In fact, there is more RNA than protein in the ribosome, and we must therefore question the simple minded inference that two proteins that can be cross-linked by a reagent which is anywhere from 6 Å to 20 Å long are necessarily two proteins that are bound to each other in the ribosome. Indeed, the best estimate that we can make now suggests that the proteins in the 30 S subunit are not, in general, arranged in RNA free domains, but are dispersed in some manner through a matrix of RNA. Accordingly, proteins that are close enough to each other to be cross-linked and that cooperate with each other during *in vitro* assembly, could be bound to neighboring sequences of the 16 S RNA rather than to each other. The arguments to support these conclusions follow.

The diimido esters have been used to cross-link oligomeric enzymes (Davies and Stark, 1970; Carpenter and Harrington, 1972). When these reagents are applied to such protein complexes, the monomeric proteins are quantitatively cross-linked to each other. In contrast, the same reagents applied to ribosomes under the same conditions produce comparatively meager yields of cross-linked complexes. This is in spite of the fact that the ribosomal proteins have an abnormally high lysine content, and most of the 30 S lysines are reactive with the diimido esters (Slobin, 1972).

Not only are the yields of cross-linked proteins limited with all of the reagents used so far, but there is a clear correlation between the size of the bifunctional reagent and the amounts of complexes created. Thus, with a series of three diimido esters ranging in size from 5 Å to 12 Å, the yield is greatest with the largest, DMS (Lutter *et al.*, 1972, 1974a). More impressive is the size dependence seen with the tartaric acid derivatives which have lengths between 6 Å and 24 Å (Lutter *et al.*, 1974b, 1975). Even a 24 Å reagent fails to give quantitative cross-linking of the 30 S proteins.

Calculations based on the volumes occupied by the 30 S proteins and 16 S RNA indicate that these results can be explained if it is assumed that the proteins are distributed within a matrix of RNA (Kurland, 1974a). The conclusions drawn from these calculations are the same whether the proteins are globular and not in general touching each other, or whether they are highly elongated and make fleeting contacts. In either case, the proteins will have well developed contacts with RNA and a minimum of direct contacts with each other if they are more or less uniformly distributed throughout the 30 S subunit.

Several lines of evidence have verified this view of 30 S subunit structure. First, a review of the relevant literature reveals that more than two-thirds of the 30 S proteins have RNA binding sites (Kurland, 1974b). Second, a neutron scattering study of deuterated subunits shows that the distribution of protein and RNA in the 30 S subunit appears to be uniform within the resolution of the measurements (Moore *et al.*, 1975). Third, energy transfer studies with fluorescence labeled 30 S subunits show that there is significant separation between most of the near neighboring proteins studied (Huang *et al.*, 1975). Finally, the proteins on the surface of the 30 S subunit can be visualized with the aid of antibodies in electron micrographs. The results of such studies by Stöffler and his colleagues show that the 30 S proteins are nearly uniformly distributed through the subunit (see Chapter 3).

The structure of the 50 S subunit is quite different. Neutron scattering studies with this subunit show that its proteins and RNA are asymmetrically distributed (Moore *et al.*, 1974). Thus, the centers of protein and RNA mass in the 50 S subunit seem to be separated by 50 Å to 60 Å. This would suggest that the proteins are preferentially packed at one end of this subunit. Indeed such an arrangement of proteins is precisely what is observed in the electron microscope with antibody-stained preparations of 50 S subunits (see Chapter 3).

It is by no means clear why the arrangements of protein and RNA in the two ribosomal subunits should be so different. However, most of the enzymatic activities involved in protein synthesis such as the peptidyl transferase and the nucleotide phosphoesterases are associated with the 50 S subunit or with factors that require interactions with this subunit to express their activities. In contrast, there is as yet no evidence that the 30 S subunit functions involve enzymatic reactions. This difference as well as the potential constraints imposed on these unequally sized subunits by the need to self-assemble may ultimately account for the observed differences in their structures. However, I must confess that I do not understand why they should be so different.

It is, however, now clear why it has been so difficult to attribute individual functions during protein synthesis to individual ribosomal components, particularly those of the 30 S subunit. When the reconstitution system of Traub and Nomura (1968) was first discovered, it was thought by many, including me, that simple experiments in which individual components are omitted from reconstitution mixtures or are chemically modified before being reconstituted would lead to a rapid cataloging of the functions of individual proteins. Nothing could have been further from the truth, and it can be fairly said that very little has been learned about individual functions of the 30 S proteins by such experiments. The reasons for this situation are explicable now in terms of the way the 30 S proteins are arranged in the ribosome.

First, the analysis of the molecular interactions during 30 S assembly *in vitro* taken together with the structure of this subunit suggest that one major function of the 30 S proteins is to help in the folding of the 16 S RNA (Kurland, 1974b). Second, the 16 S RNA apparently provides binding sites for the mRNA and tRNA (see below). Accordingly, the omission or modification of individual 30 S proteins will be expected to have a diffuse indirect effect on these RNA functions. Even in the circumstance that a protein is directly involved in some ribosome function, the fact that it is embedded in a matrix of RNA as well as other proteins means that its functions will be dependent on the proper molecular interactions with these other components. In brief, the profound cooperativity required by so complex structure as that of the 30 S subunit makes experiments designed to identify the functions of individual components quite difficult. This is well illustrated by the functional properties of one of the 30 S proteins, S4.

D. An Exemplary Protein

The 30 S protein S4 has long been known to be essential to the proper reconstitution of 30 S subunits *in vitro*. This can be explained by the finding that several other proteins require its presence for their eventual assembly into the ribosome (Nomura *et al.*, 1969). In addition, mutations that alter the structure of S4 can have the following pleitropic effects: they suppress the streptomycin dependence phenotype of the ribosome which depends on 30 S protein S12 (Birge and Kurland, 1969, 1970; Deusser *et al.*, 1970; Hasenbank *et al.*, 1973); they are responsible for increasing the error frequency of translation as expressed in the RAM phenotype (Rosset and Gorini, 1967; Zimmermann *et al.*, 1971); they cause assembly defects in the 30 S subunit both *in vivo* and *in vitro* (Birge and Kurland, 1970; Green and Kurland, 1971; Daya-Gros-Jean *et al.*, 1972); and they are responsible for a defect in the proper conversion of the precursor RNA into the 16 S RNA *in vivo* (Feunteun *et al.*, 1974; Olsson *et al.*, 1975).

This remarkable spectrum of functional defects caused by the alteration or omission of S4 is matched by the complexity of its interactions with the

16 S RNA. Thus, chemical and physical studies of the nucleotide sequences associated with this protein have revealed that it binds to dispersed sites at the 5'-end of the 16 S RNA and is able to organize the tertiary fold of more than 100,000 daltons of the RNA (Schaup et al., 1971; Schaup and Kurland, 1972; Zimmermann et al., 1972; Nanninga et al., 1972). Indeed, the structural studies of Stöffler and his colleagues have revealed that this protein is rather elongated in the ribosome and, accordingly, is likely to be in close proximity to a number of different proteins (see Chapter 3).

The pronounced effects of S4 on various aspects of the assembly of the 30 S subunit is readily explained on the basis of its complex interactions with the 16 S RNA. Thus, the assembly dependence of S16 on S4 (Green and Kurland, 1973; Held et al., 1974) probably results from the effect that S4 has on the tertiary fold of the RNA binding sites that must be organized before S16 can stably associate with them (Kurland, 1974b). Similarly, the failure to properly cleave the precursor RNA to yield a mature form of 16 S RNA is a mutant containing an altered form of S4 can be explained as a consequence of the mutant S4's failure to properly fold the cleavage site for the processing enzyme (Feunteun et al., 1974; Olsson et al., 1975).

It turns out that most mutant forms of S4 express all of the phenotypic changes discussed above (M. Olsson, personal communication). Therefore, it is tempting to account for the suppression effect on the streptomycin dependence phenotype as a similar indirect effect of this protein on the tertiary fold of the RNA which secondarily influences protein S12 (Olsson et al., 1975). This is a particularly attractive account in light of the finding that S4 mutants very often have altered affinities for their RNA binding sites (Green and Kurland, 1971; Daya-Gros-Jean et al., 1972). Nevertheless, it would be wrong to conclude that all the functional effects of this protein are necessarily indirect ones which depend on the structural interactions that it develops with the RNA.

Thus, affinity labeling experiments (Pongs and Lanka, 1975) described below suggest that this protein may be in direct contact with the codon being translated at any given time by the ribosome. This as well as the effects that altered forms of S4 can have on the error frequency of translation suggest that it may be one of the ribosomal components responsible for the presentation of the codon during translation.

The point of this minidiscourse on only one of the 54 ribosomal proteins is a simple one: Once the ribosome has been assembled, any part of it may provide a contacting domain for one of the macromolecular intermediates in protein synthesis. The corollary to this remark is that any given component of so complicated an object as the ribosome may provide an essential function by virtue of its interactions with other components. There is no way of distinguishing these effects without directly determining the nature of the molecular interactions taking place during protein synthesis. Accordingly, a rather bulky and remarkably uninformative literature can be dispensed with.

III. Programming the Ribosome

A. Complexity of the mRNA Binding Domain

The mRNA has at least four distinguishable functions, and unless the ribosome has been programmed for one of these, it can have no meaningful function in protein synthesis. Thus, the initiation of protein synthesis, the codon-dependent binding of aminoacyl-tRNA, the coordinated movements of tRNA as well as codon, and the termination of the process all represent complex functions that are guided by signals originating with the mRNA.

This multiplicity of functions would suggest that the mRNA binding sites should be complex; an expectation that is not contradicted by studies of the size of ribosome binding sites on various mRNA's. It is possible to form stable complexes between viral or synthetic mRNA and ribosomes. When such complexes are treated with nuclease, segments of the mRNA protected by the ribosome from digestion can be recovered (Takanami and Zubay, 1964; Takanami et al., 1965; Castles and Singer, 1969; Steitz, 1969). Very large nucleotide sequences resistant to nuclease digestion are observed in such experiments. However, in many such cases the interpretation of the data is complicated by the presence of double-helical structures in the protected sequences.

Double-stranded RNA is intrinsically resistant to most ribonucleases, and as a consequence, the presence of such structures could lead to an overestimation of the size of the sequences protected by the ribosome. In contrast, poly(U) will not form such helical structures, which means that the length of the poly(U) sequences protected by the ribosome would provide a reasonable index for the maximum size of the mRNA binding site during the elongation phase of protein synthesis. When the protection experiments are done with this synthetic messenger, a minimum length of 25 nucleotides can be recovered on the ribosomes (Takanami and Zubay, 1964; Castles and Singer, 1969).

A polynucleotide sequence with a chain length close to 25 would be nearly 100 Å long. This means that rather extensive interactions between the bound mRNA and elements of the ribosome are possible. In particular, it seems very unlikely that only the codon which is being translated at any one time in the A site is physically bound by the ribosome. This in turn means that the identification of one or another ribosomal component associated with the mRNA may have any number of functional interpretations. For example, a protein that provides a binding site for the mRNA may be involved in the codon-dependent binding of tRNA, or the movement of the mRNA, or alternatively it may be part of the specialized initiating and terminating machinery.

Another cautionary note must be sounded before we proceed further. Because of the paucity of information concerning the mRNA binding sites, the reader can easily form the impression that it is established that only the 30 S ribosomal subunit is involved with the messenger. This is, however, most definitely not the case; what is known is much less definitive. Thus, the isolated 30 S subunit will bind mRNA, while the 50 S subunit will not do so (Okamoto and Takanami, 1963). It still remains to be determined whether or not components of the 50 S subunit interact with the messenger in the functional 70 S ribosome. Since the 50 S subunit is so intimately involved in the movements of tRNA, as well as the functions of EF-G, it would be useful to keep an open mind with respect to this question.

B. Codon Presentation

As we shall see in a later section, the interaction of codon and anticodon is far too weak by itself to account for the precision of the tRNA selection mechanism. In their discussion of this problem Eisinger et al. (1971) argued that it is possible that the stability of interaction between synthetic trinucleotides in solution gives a deceptive impression of the stability of interaction between tRNA and mRNA. Thus, trinucleotides in solution are free to adapt whatever conformations are allowed by rotations around bonds in the backbone of the molecule. In contrast, the formation of hydrogen-bonded structures between a pair of complementary trinucleotides must be accompanied by severe restrictions upon the allowable conformations for the oligonucleotides. This means that the formation of such triplet-triplet complexes will be opposed by the decrease of the entropy of the system which accompanies complex formation.

Eisinger et al. (1971) go on to point out that by fixing the conformations of the codon and anticodon in an appropriate way, the stability of their interactions will be enhanced because this entropy effect will be nullified. Eisinger (1971) then illustrates this point by a very elegant experiment. The tRNA molecule contains its anticodon in a loop which has presumably evolved in such a way that the anticodon can be presented to the programmed ribosome in an optimum conformation. Therefore, if a pair of tRNA molecules with complementary anticodons are allowed to interact with each other, the stability of this sort of triplet-triplet interaction should be enhanced. Indeed, Eisinger has shown that instead of an association constant of 10^3 M^{-1} a pair of appropriately chosen tRNA molecules interact with an association constant of 10^5 M^{-1}.

Although, the stability of triplet interactions in restricted conformations is enhanced, this mechanism does not help with respect to the specificity of such interactions. Thus, the stability of interaction between two tRNA molecules with complementary anticodons is only ten times greater than that between two tRNA's with related noncomplementary anticodons. This level of discrimination is virtually the same as that obtained in the interaction between an oligonucleotide and a tRNA molecule. Nevertheless, there is a valuable idea here.

Eisinger's experiments suggest that at least one task for the ribosomal components which position the codon in the A site will be to fix the conformation of the nucleotides so that they are presented in an optimum configuration. We may next ask which molecules could have this function.

There are two arguments that make it likely that proteins are responsible for presenting the codon (Kurland, 1974a). One of these is the requirement to present the bases of the codon so that they are free to interact with the tRNA. The other is that this binding site must be able to accommodate any permutation of trinucleotides. This means that base interactions with the 16 S RNA are excluded unless some as yet unknown mechanism for fixing the conformation of the codon by interactions with the backbone of the 16 S RNA is discovered. It therefore seems reasonable to expect that one or more 30 S proteins make up this binding site. Indeed, it turns out that there are a number of candidates for this function.

The first ribosomal protein to be implicated in the mRNA binding function was S1 (van Duin and Kurland, 1970). When this protein is missing from the 30 S subunit, mRNA binding is weak and it increases progressively with the addition of S1. More important is the correlated proportional increase of aminoacyl-tRNA binding with increasing amounts of S1, which shows that the poly(U) or phage RNA bound under the influence of S1 is translatable (van Duin and Kurland, 1970; Randall-Hazelbauer and Kurland, 1972; Tal et al., 1972; van Duin and van Knippenberg, 1974). Finally, it has recently been shown that S1 is required for the translation of MS2-RNA as well as the other messengers (van Dieijen et al., 1975). Indeed, the only messenger which can be translated in the absence of S1 is poly(U), and here the translation requires very high Mg^{2+} concentrations.

The demonstration that a protein is required both for the binding and translation of mRNA does not by itself provide convincing evidence that this protein is providing a contacting domain for the mRNA. The functional contribution of a protein such as S1 could be a consequence of some conformational effect on other ribosomal components which in turn bind the mRNA. Clearly, what is needed is evidence that S1 and mRNA are in physical contact on the programmed ribosome.

Such evidence is plentiful. Thus, S1 is protected from oxidation by Rose Bengal as well as from trypsin degradation by a bound, synthetic mRNA (Noller and Chaires, 1972; Rummel and Noller, 1973). In addition, it is possible to covalently bind S1 to the translatable synthetic mRNA, poly-4-thiouridylic acid, if the latter is bound to the ribosome and then photoactivated (Fiser *et al.*, 1974).

Although the evidence that S1 participates in the binding of mRNA to the ribosome is quite convincing, it does not seem to be alone in this function. Fiser *et al.* (1974) noted that in addition to S1, complexes of other proteins and poly-4-thiouridylic acid could be recovered after photoactivation. These complexes are resistant to ribonuclease, but can be disrupted by base hydrolysis. More recent analysis of these complexes has revealed that they contain the proteins S7, S14, and S18 (E. Küchler, personal communication).

The evidence discussed so far certainly confirms our expectation that the mRNA binding site is complex. Furthermore, we cannot decide on the basis of the above data which if any of the four proteins S1, S7, S14, and S18 are involved in the presentation of the codon. However, another sort of affinity label brings us much closer to the solution of this problem.

Pongs and Lanka (1975) have prepared an affinity probe which is a derivative of the codon AUG. This affinity label will bind to the ribosome, and form a covalent linkage with one or more 70 S ribosomal components. After reaction the codon is able to direct the binding of fMet-tRNA (Pongs and Lanka, 1975). When the ribosomes labeled by the AUG analogue are disrupted, it can be shown that at least two proteins have been labeled. These are S4 and S18. Accordingly, there is positive evidence that only one of the proteins previously described as associated with mRNA is close enough to the codon to be involved in its presentation; this is S18.

When the same AUG analogue is bound by free 30 S ribosomal subunits, a rather different result is obtained (O. Pongs and R. Lanka, personal communication). Here, the labeled proteins are S12 and S13. Since the AUG analogue bound under these conditions will direct the binding of fMet-tRNA, this association is presumed to be a functional one. It follows from this observation that there must be at least two configurations for the binding of the codon to the 30 S subunit. The reasons for not leaping to the conclusion that these two configurations correspond to the A and P sites will become clear shortly.

In summary, there are data indicating that as many as seven different 30 S proteins can be involved in the binding of mRNA: S1, S4, S7, S12, S13, S14, and S18. Four of these (S4, S12, S13, and S14) are close enough to a codon when it is being translated to be considered as likely candidates for the presentation function. Indeed, two members of this subset are particularly provocative.

It has been known for quite some time that the proteins S4 and S12 are able to influence the error frequency of translation. In addition, their effects appear to be opposed to each other (Gorini, 1971). Thus, mutations in S4 tend to increase the error frequency, while mutations in S12 tend to lower the error frequency. Furthermore, when altered forms of S4 and S12 are together in the same ribosome, they counteract each other's effect on the tRNA selection mechanism. This would imply that they are both influencing one and the same functional site.

This naturally leads to a question that I have carefully avoided until now, namely, the problem of deciding how many codon presentation sites there are on the 30 S ribosomal subunit. It is usually assumed that there are two such sites which would correspond to the 50 S A and P sites. However, there has never been compelling evidence for this assumption, nor any theoretical reason for it. Indeed, it has been shown that the data concerning the different functional states for the association of tRNA with the ribosome are readily explained by assuming that there is only one such site at the 30 S subunit (Kurland et al., 1972).

At issue is the problem of whether or not two tRNA molecules can be simultaneously bound to two adjoining codons. Given the relatively large size of the tRNA molecule and the short distances that are involved, this seems difficult to accomplish. Thus, the tRNA molecule across its narrowest dimensions will be two to three times larger than the length of a codon (ca. 10 Å). Accordingly, it would be difficult to imagine how two such bulky objects could be fitted to the 20 Å stretch corresponding to two codons, and at the same time present their C-C-A termini very close to each other at the peptidyl transferase center. Therefore, until evidence to the contrary is forthcoming, I will assume that only one tRNA can be bound at one time at the codon presentation center.

This, of course, does not mean that an aminoacyl-tRNA · EF-Tu complex and an fMet-tRNA · IF-2 complex occupy precisely the same binding sites on the ribosome. On the contrary, it is necessary to assume that the configurations of the ribosome · mRNA · tRNA complex are different when a tRNA bound at the 30 S site is interacting with the 50 S A site or the 50 S P site (Kurland et al., 1972). Accordingly, the two kinds of labeling patterns obtained by Pongs and Lanka with the AUG analogue may reflect these two kinds of ribosome configurations.

Finally, of all the proteins considered so far for the function of codon presentation, only S4 and S12 are ones for which there is some evidence of a functional interaction with the codon. Thus, an affinity probe tells us something about physical proximity but it does not necessarily demonstrate a functional interaction, particularly in the case of large affinity probes. The evidence indicating that S4 as well as S12 can be labeled by a functional trinucleotide probe and the finding that both can influence the error frequency of translation suggest that these proteins are binding the codon during translation. Nevertheless, even in these two instances, this must be taken as a tentative conclusion.

C. Making Ends Meet

At least three formally separable functions can be identified with the initiation mode: the initial binding of the messenger, the identification of the first AUG codon, and the binding of the fMet-tRNA. Since the latter functions will be discussed in detail in an accompanying chapter, the present discussion will be restricted to the sequestering of an mRNA by the ribosome.

The first strong clue indicating a specialized ribosome contribution to the initial binding of the mRNA was the demonstration that species differences in the ability to initiate protein synthesis *in vitro* with a given coliphage RNA reside in the 30 S subunits of bacterial ribosomes (Lodish, 1970). Subsequent experiments by Goldberg and Steitz (1974) as well as Held *et al.* (1974) have implicated both the 16 S RNA as well as 30 S protein S12 as components of the ribosome that can determine the efficiency of cistron recognition with a given mRNA molecule.

More recently, comparisons of the nucleotide sequences found in the ribosome binding sites on mRNA with those at the 3'-end of the 16 S RNA have provided a most attractive explanation for the earlier experimental results (Shine and Dalgarno, 1974, 1975). Furthermore, these results will permit us to assign some additional functions to some of the proteins that have been shown to be involved in the binding of the mRNA to the 30 S ribosomal subunit.

Shine and Dalgarno (1974) have shown that the 3'-end of *E. coli* 16 S RNA contains a stretch of seven nucleotides which are complementary to a corresponding stretch of nucleotides in the ribosome binding site at the 5'-end of the A protein cistron in R17 RNA. Furthermore, a stretch of four of these nucleotides in the 16 S RNA are also complements to a corresponding sequence at the ribosomal binding site in the replicase cistron, while three of these are complements to a sequence at the corresponding region of the coat protein cistron.

The functional significance of this correlation was underscored by the observation that the degree of complementarity between the 3'-terminal sequences of the 16 S RNA and the sequences in the ribosomal binding sites on the coliphage RNA parallels the relative efficiency with which the corresponding RNA fragments can be bound to the ribosomes (Shine and Dalgarno, 1974). Subsequent analysis of the 3'-terminal sequences in 16 S RNA from three additional bacterial species revealed a similar correlation between the degree of homology with the *E. coli* sequences and the efficiency with which the coliphage RNA can be bound by ribosomes from other sources (Shine and Dalgarno, 1975).

The identification of the 3'-region of the 16 S RNA as the probable site for the sequestering of the mRNA helps us to understand why so many proteins can be shown to be in close proximity to the mRNA when it is bound by the ribosome. Thus, there is a cluster of 30 S proteins which can be cross-linked to each other, cooperate with each other during assembly *in vitro,* and are associated with the 3'-region of the 16 S RNA. This is the assembly cluster consisting of S7, S9, S13, S14, and S19 (Kurland, 1974b)

In addition, it has been suggested by Kenner (1973) that the 3'-end of the 16 S RNA may be in close apposition to the 30 S protein S1. This has now been verified by P. Czernilofsky *et al.* (1975). Accordingly, S1, S7, S13, and S14 are four proteins that are associated with the 3'-region of the 16 S RNA in the functional ribosome, and are also identified in affinity labeling experiments as ones intimately associated with a bound mRNA (see above). It may therefore be tentatively concluded that these four proteins along with the 3'-end of the 16 S RNA constitute a binding domain on the 30 S subunit for the sequestering of mRNA prior to the initiation of protein synthesis.

IV. The tRNA Selection Mechanism

A. Mountains out of Molehills

While no one would question the validity of the messenger hypothesis at this late date, the mechanism through which a codon directs the selection of its cognate tRNA is far from clear. The only aspect of this problem which does seem clear is the impossibility of mediating this function solely through triplet-triplet interactions between codon and anticodon. Such a dogmatic statement is not made gratuitously.

To begin with, the interaction between the programmed ribosome and its cognate tRNA is sufficiently stable in the temperature range 37°–60° to be characterized by an apparent association constant of 10^8 M^{-1} or greater (Kurland, 1966; McLauglin *et al.,* 1966). In contrast, the interaction of a triplet oligonucleotide and its cognate tRNA has an approximate association constant at 0° which is close to 10^3 M^{-1} (Eisinger *et al.,* 1971; Högenauer, 1970; Uhlenbeck *et al., 1970*). This means that we must account for a discrepancy of more than five orders of magnitude, and this is only part of the problem.

During the synthesis of an average size protein (chain length ca. 500) errors will accumulate at approximately 500 times the error frequency of each individual recognition step. Accordingly, the fidelity of translation of such a protein would be assured if and only if each amino acid is inserted with an error frequency less than 10^{-3}. Indeed, estimates of the error frequency for amino acid recognition during protein synthesis have been set at approximately 10^{-4} (Loftfield and Vanderjagt, 1972). Now, the stability of the interaction between a given tRNA and its cognate codon in the absence of ribosomes is only ten times greater than that with a noncognate codon at $0°$ (Uhlenbeck *et al., 1970*). This would correspond very roughly to a 0.5 error frequency for the selection of one out of twenty different aminoacyl-tRNA molecules. Clearly, the codon-anticodon interaction needs help.

There are a number of collateral phenomena which indicate potential sources for this help. On the one hand, there is the demonstration that mutations in two 30 S ribosomal proteins (S4 and S12) can influence the error frequency of translation (Gorini, 1971). On the other hand, there is the finding that the mutational alteration of a tRNA molecule at a site quite separate from the anticodon loop can change the codon specificity of that tRNA (Hirsh, 1971; Hirsh and Gold, 1971). Such observations very strongly suggest that both the structure of the ribosome as well as elements of the tRNA other than the anticodon are involved in the tRNA selection mechanism.

This conclusion will be very much reinforced in what follows. However, some qualifications are in order before proceeding. It is not my intention to argue that the ribosome is privy to the code, nor is it necessary to question in any way the prime role of the messenger RNA in determining the amino acid sequence of a protein. The view adopted here, however, will partly redirect our attention away from the codon-anticodon interaction and toward another set of nucleotide interactions.

There are three basic elements of the viewpoint to be adopted here. First, the codon-anticodon interaction is assumed to be the first weak signal which initiates a train of events that only subsequently are expressed in the stable binding of a cognate tRNA. Second, this means that the problem confronting the translational apparatus is essentially that of amplifying this weak signal without changing its sense. Finally, the amplification mechanism is recognized as a statistical process, which suggests that perturbations of this system caused by mutations of antibiotics are best viewed as nothing more than noise.

This latter point is important as a counterweight to at least one current interpretation of the effects of alterations in the ribosome which influence the error frequency of translation. It will be shown that the data concerning the tRNA selection mechanism can be explained without postulating esoteric ribosome functions such as a prior screening of classes of tRNA molecules before their interaction with the codon (Gorini, 1971).

B. Experimental Foreground

There have been a number of attempts to describe the tRNA selection mechanism as a rate process. Regardless of whatever other virtues they may have, each of these models is invalidated by one or another experimental observation. However, it is an advantage to discuss the phenomenology of this problem in the context of such models. By doing so, we can better appreciate the theoretical constraints that are imposed on the system according to what is already known.

The first of these instructive failures is that of Kurland (1970) in which three ideas are introduced. One of these is that the binding of tRNA is a stepwise process. Another is the idea that the final ternary complex is stabilized by nonspecific forces which are distinct from those that select the cognate tRNA. Finally, the selection mechanism is viewed as a rate process in which the codon as well as ribosomal proteins provide a kinetic barrier through which noncognate tRNA molecules may pass only rarely.

Thus, any tRNA is given an equal opportunity to interact with the programmed ribosome in the first step, which is viewed as a rapidly equilibrating association. If the tRNA has the correct anticodon, it may pass through the steric barrier provided by the codon. If not, it drops off the ribosome. Important to this model is the assumption that the codon does not bind the anticodon in a stable fashion. Instead, the codon blocks the passage of the noncognate tRNA's. However, any tRNA which negotiates passage through the steric barrier is then bound in the last step by the ribosome at sites that are common to all tRNA molecules.

Among the defects from which this model suffers, one is lethal. The analysis of an altered tRNA molecule which suppresses point insertion mutations shows that this suppressor tRNA has an extra base in its anticodon (Riddle and Carbon, 1973). As a consequence of this extra base, such a tRNA can read a four nucleotide codon, and, thereby, suppress the phase dislocation caused by the extra nucleotide in the mutant mRNA. Such an account would require that the codon and anticodon remain in contact after the tRNA is bound so that the movement of the mRNA can subsequently be guided by this contact. Accordingly, models such as that described above, which require at most a fleeting contact between codon and anticodon, are not consistent with the facts. However, it will be shown below that elements of this model can be salvaged in a much more realistic description of the tRNA selection mechanism.

The kinetic aspects of triplet nucleotide interactions, first discussed by Eigen (1971), have been exploited recently by Blomberg (1974) and Ninio (1974) in an attempt to explain the way in which this interaction can provide the specificity of tRNA selection. The valuable lesson derived from such models is the demonstration that differences in the rate constant for the dissociation of the codon-anticodon interaction can provide some of the discrimination between cognate and noncognate tRNA. In effect, any ribosome mechanism for checking the half-life of the codon-anticodon interaction would contribute to the selection mechanism.

The problem is that it is necessary to make some rather gratuitous assumptions in order to imagine such mechanisms. For example, Ninio (1974) suggests that the ribosome functions as an oscillator which periodically fixes a tRNA in the A site. If the periodicity of the ribosomal clock mechanism is properly adjusted, it will preferentially fix cognate tRNA molecules. This *deus ex machina* certainly works on paper, but there is no evidence that the ribosome can function this way. Furthermore, this model is silent about the relationship between the postulated clock mechanism and the forces that stably bind the tRNA once it is selected. Nevertheless, we will later have occasion to use the notion that the half-life of the codon-anticodon interaction will be longer for a cognate than for a noncognate tRNA.

The idea introduced earlier that tRNA selection is a multistep process has been exploited in rather different ways by Blomberg (1974) and Hopfield (1974). Both use a compound selection mechanism which is made up of a series of elementary discriminating processes. In Blomberg's model each elementary step is very like the single clock mechanism suggested by Ninio (1974). However, here, with only a tenfold advantage of the cognate tRNA over noncognate tRNA's at each step, a series of four such steps would yield an error frequency of the order of 10^{-4}.

Blomberg begins by assuming that all tRNA molecules have an equal chance to form the first, weak association with the programmed ribosome. This is equivalent to the statement that the first state of association is formed by a collision limited process. Second, the charged tRNA bound in this state now enters the selection sequence and at any time within this sequence it can do only one of two things: continue to the next stage or dissociate from the ribosome. In the event that a tRNA molecule drops off the ribosome, the next tRNA which binds must start the process from the very beginning. This constraint is a most important one since it means that the sequence of advancement through the screening process is fixed and energetically favored for the cognate tRNA. Finally, a terminal state is reached in which for all practical purposes the selection process is completed; this state we can associate with peptide bond formation.

Computations based on this model very convincingly demonstrate that it would work with even the most modest discrimination between cognate and noncognate tRNA's at each step of the process. Furthermore, the model makes some accurate predictions about the structure of the genetic code; in particular, it predicts some of the wobble rules. However, the model has some serious defects. In particular, a molecular mechanism is required to generate the fixed sequence of steps in the screening process at the same time that it provides the cognate tRNA with an advantage at each step. Furthermore, the model depends on an undefined potential which drives the tRNA through the postulated selection process.

Hopfield's model attempts to solve the problem of directing tRNA's through each of the "proofreading" steps by assuming that they are driven by an exogenous energy source such as the hydrolysis of GTP. However, the relationship between the postulated high energy intermediates and the sequence of physical events involving the tRNA is described by Hopfield (1974) as "an irrelevant detail." This is unfortunate for the rest of us, because it means that a realistic mechanism is still wanting. Another serious defect of this model is that it will not account for the fact that a programmed ribosome will select and stably bind a cognate tRNA in the absence of an exogenous energy source such as GTP (Kurland, 1966, McLauglin et al., 1966).

A characteristic property of the models discussed so far is their remoteness from the details of the molecular interactions of tRNA with the ribosome. It is not surprising, therefore, that we can obtain a new perspective on the tRNA selection mechanism by briefly reviewing some data concerning the structure of tRNA as well as its interactions with ribosomal components. Indeed this data very naturally leads to an explicit model for the selection mechanism.

Several lines of evidence indicate that a tRNA molecule bound at the A site interacts with ribosomal RNA as well as the mRNA. Suggestive evidence for such as interaction with the 30 S ribosomal subunit was first obtained by Noller and Chaires (1972), who studied the inactivation of 30 S particles with kethoxal, which preferentially reacts with guanine. They could show that with as few as six "hits" per 16 S RNA the 30 S particle is unable to mediate tRNA binding while mRNA binding is unaffected. Most relevant was their demonstration that the prior binding of tRNA protects the 30 S subunit from the kethoxal mediated inactivation. The simplest interpretation of this result is that the tRNA and 16 S RNA are in physical contact once the former is bound to the ribosome.

An intimate association of tRNA and 16 S RNA is also suggested by the more recent experiments of Schwartz and Ofengand (1974). They have shown that a photoaffinity probe attached to the thiouracil moiety at position 8 in Val-tRNA can be used to cross-link quantitatively that tRNA to the 16 S RNA. The binding of the tRNA to the ribosome in this experiment was done in the absence of factors; this means that the tRNA is bound in almost equal amounts to the P and A sites (V. Erdmann, personal communication). Since the probe can span a distance no greater than 9 Å and since it is capable of reacting equally well with protein and RNA, they concluded that once bound to the ribosome the tRNA is intimately associated with 16 S RNA.

A more comprehensive analysis of the interactions between aminoacyl-tRNA bound at the A site and the 5 S RNA of the 50 S subunit has been made possible by recent technical developments. This analysis began with the recognition of the potential functional relationship between the common G-T-ψ-C sequence of tRNA molecules and its complement in the eukaryotic 5 S RNA (Forget and Weissman, 1967). This was followed with the demonstration by Ofengand and Henes (1969) that the binding of aminoacyl-tRNA to the ribosome could be inhibited by the presence of excess amounts of a tRNA fragment containing the G-T-ψ-C sequence.

The subsequent analysis of this interaction with components from the bacterial system has very much strengthened the earlier interpretations. Thus, it is possible to reconstitute *in vitro* a complex of 5 S RNA as well as a few 50 S proteins, and the complex will bind the T-ψ-C-G fragment from tRNA (Erdmann *et al.*, 1973). The very same fragment will effectively inhibit the EF-Tu-dependent binding of aminoacyl-tRNA to the A site. Furthermore, once this fragment is bound in place of the tRNA, it can be recovered on the 50 S subunit (Richter *et al.*, 1973).

Clearly, the binding of the tRNA to the 5 S RNA at the A site would provide an additional stability to the ternary complex formed by the tRNA and the programmed ribosome. What has not been appreciated previously is that under certain conditions this apparently nonspecific interaction can enhance the discrimination of the cognate from noncognate tRNA's. In order to illustrate this idea we must first consider a relevant aspect of tRNA structure.

The crystal structure of yeast tRNA[Phe] (Kim *et al.*, 1974; Robertus *et al.*, 1974) as well as the properties of unfractionated tRNA in solution (Gauss *et al.*, 1971; Rhodes, 1975; Sigler, 1975) are such that the T-ψ-C-G sequence seems to be masked by interactions with the D loop. However, the evidence presented above clearly show that when tRNA is bound to the ribosomal A site, the T-ψ-C-G sequence is available for interaction with the complementary C-G-A-A sequence in the 5 S RNA. It follows, therefore, that some sort of conformational change in the tRNA must occur at some stage of the binding reaction.

A very pretty experiment performed by Schwarz *et al.* (1974) illustrates this conclusion. They asked whether the interaction of tRNA with the programmed ribosome leads to a rearrangement of the tRNA such that the T-ψ-C-G loop is now unmasked. The affirmative answer was suggested by observing the binding of radioactive C-G-A-A to the tRNA under a variety of conditions. They could show that while the EF-Tu · GTP · aminoacyl-tRNA complex does not bind the radioactive fragment, the addition of a suitably programmed 30 S subunit preparation leads to the binding of the fragment. Most important was their demonstration of the stimulatory effect of the cognate codon for this reaction both with Phe-tRNA and Pro-tRNA.

Unfortunately, there is an ambiguity in this experiment; this is the absence of direct evidence that the radioactive C-G-A-A fragment binds to the T-ψ-C-G loop of the tRNA. Nevertheless, taken together with the data discussed previously, the experiment of Schwarz *et al.* (1974) illustrates the kind of conformational changes that would be expected. Now we are in a position to see how such conformational changes could contribute to the specificity of tRNA selection.

Let us assume for the moment that the proper binding of the codon to the anticodon loop is able to increase the likelihood of a conformational change in the tRNA. Here, the T-ψ-C-G loop would have a higher probability of being available for interaction with the 5 S RNA than if the anticodon is not properly bound to its cognate codon. Were this so, the additional interaction of the T-ψ-C-G loop with the 5 S RNA would be effectively coupled to the codon-anticodon interaction so that the 5 S RNA contribution could be brought into play preferentially with cognate tRNA's.

There are at least two ways that the correct fit of the codon-anticodon interaction could influence the probability of a subsequent interaction with the 5 S RNA. One way is by affecting the half-life of the tRNA interaction with the programmed ribosome. Thus, the cognate tRNA would have a more stable interaction with the programmed ribosome than would a noncognate tRNA, and accordingly it would have a longer time during which the requisite conformational change could occur. This is equivalent to saying that the rate constant for the dissociation of the tRNA-codon interaction is smaller for the cognate than for the noncognate tRNA.

Second, a cognate codon's interaction with the anticodon loop could actually alter the preferred conformation of the tRNA in such a fashion that the ease of unmasking the T-ψ-C-G loop would be much greater than in the absence of the proper codon. Here, the codon would be functioning very much as though it were an allosteric effector. Consequently, the presence of the cognate codon would effectively raise the rate constant for the formation of the T-ψ-C-G-5 S RNA interaction.

Both of these putative effects of the codon would influence the equilibrium constant describing the tRNA-5 S RNA interaction in the same direction: they preferentially increase this equilibrium constant for the cognate tRNA. Consequently, it should be possible under some conditions to use the common binding sites on the tRNA molecules to interact with ribosomal components in such a way that both the stability and specificity of the tRNA interaction is enhanced. It now remains to rigorously describe the conditions under which such a mechanism could function.

C. Formal Description

The flow of aminoacyl-tRNA through the different steps of protein synthesis can be depicted according to the present model as follows:

$$
\begin{array}{c}
T_1 + R \rightleftharpoons T_2 + R \rightleftharpoons T_3 + R \\
\text{I} \Big\updownarrow \qquad\quad \Big\updownarrow \qquad\quad \Big\updownarrow \\
T_1R \underset{\text{II}}{\rightleftharpoons} T_2R \underset{\text{III}}{\rightleftharpoons} T_3R \underset{\text{IV}}{\longrightarrow} T_4R \underset{\text{V}}{\longrightarrow} R
\end{array}
\tag{1}
$$

Here, T_1, T_2, and T_3 represent aminoacyl-tRNA in the conformational states corresponding to unperturbed tRNA, tRNA bound at the anticodon, and tRNA bound both at the anticodon and at one or more of the common sequences. R corresponds to the programmed ribosome without an aminoacyl-tRNA molecule occupying the A site. T_1R, T_2R, and T_3R correspond to complexes of ribosomes with aminoacyl-tRNA in the indicated states. Finally, T_4R corresponds to the ribosome complexed with peptidyl-tRNA in the A site.

Calculations according to this model are greatly simplified if we assume that the free concentrations of T_2 and T_3 are negligible during protein synthesis which is merely a restatement of the data from the previous section. This assumption reduces scheme (1) to the following form:

$$
T_1 + R \underset{k_{-1}}{\overset{k_1}{\rightleftharpoons}} T_1R \underset{k_{-2}}{\overset{k_2}{\rightleftharpoons}} T_2R \underset{k_{-3}}{\overset{k_3}{\rightleftharpoons}} T_3R \overset{k_4}{\longrightarrow} T_4R \overset{k_5}{\longrightarrow} R
\tag{2}
$$
$$
\quad\;\; \text{I} \qquad\quad \text{II} \qquad\quad \text{III} \qquad\quad \text{IV} \qquad \text{V}
$$

Here, steps I, II, and III are reversible equilibria, while the enzymatic reactions of peptide bond formation (IV) and translocation (V) are taken as irreversible steps.

Steps II and III in both schemes are assumed to be the only ones in which cognate and noncognate tRNA molecules can be discriminated. This would imply that the error frequency of tRNA selection should be influenced by the fidelity of both these steps. Such a conclusion can be demonstrated after solving the rate equations for scheme (2), and then comparing the relative rates of cognate and noncognate amino acid incorporation. Unfortunately, the exact solution for scheme (2) is quite complicated. Therefore, we will discuss the steady-state solutions to the rate equation in the context of four well-defined boundary conditions which permit us to analyze the properties of this model with a minimum of arithmetical distractions.

First we consider scheme (2) when either step IV or V is rate limiting for protein synthesis. This corresponds to the situation in which k_4 or k_5 is much smaller than all the other k_i's. Under these conditions the steady-state concentrations of the intermediates in steps I, II, and III will be distributed according to the equilibria constants which characterize these steps. The rate of protein synthesis (p) will then be

$$p = k_4(T_3R) = k_5(T_4R) = k_4 K_1 K_2 K_3 (T_1)(R) \tag{3}$$

Here, the K_i's correspond to the ratios of k_i/k_{-i}.

The relative advantage (a) of a cognate tRNA over a noncognate tRNA in this system will be the ratio of the rates of protein synthesis for the two kinds of tRNA:

$$a = \frac{{}^c k_4 {}^c K_1 {}^c K_2 {}^c K_3 ({}^c T_1)({}^c R)}{{}^n k_4 {}^n K_1 {}^n K_2 {}^n K_3 ({}^n T_1)({}^n R)} = \frac{{}^c K_2 {}^c K_3 ({}^c T_1)({}^c R)}{{}^n K_2 {}^n K_3 ({}^n T_1)({}^n R)} \tag{4}$$

Here, the superscripts c and n correspond to cognate and noncognate elements. Accordingly, we obtain the simple expression (4) which shows that the fidelity of translation in this system will be a direct function of the relative discrimination at both step II and step III.

This result is not unique to the assumptions upon which equation (3) is based. Thus, we can assume instead that step III is rate limiting. This assumption reduces scheme (2) to the form:

$$T_1 + R \underset{k_{-1}}{\overset{k_1}{\rightleftharpoons}} T_1R \underset{k_{-2}}{\overset{k_2}{\rightleftharpoons}} T_2R \overset{k_3}{\longrightarrow} T_3R \overset{k_4}{\longrightarrow} T_4R \overset{k_5}{\longrightarrow} R \tag{5}$$
$$\quad\;\; \text{I} \qquad\quad \text{II} \qquad\quad \text{III} \qquad \text{IV} \qquad \text{V}$$

Here, with the assumption that k_4 is much greater than k_{-3}, step III effectively becomes an irreversible one. The steady-state rate equation for protein synthesis according to scheme (5) becomes

$$p = k_3 K_2 K_1 (T_1)(R) \tag{6}$$

and

$$a = \frac{{}^c k_3 {}^c K_2 ({}^c T_1)({}^c R)}{{}^n k_3 {}^n K_2 ({}^n T_1)({}^c R)} \tag{7}$$

Thus we end up with an analogous relation to that found in (4), except that here the forward rate constant of step III is involved instead of the corresponding equilibrium constant.

There is, however, one boundary condition that restricts the discrimination of cognate from noncognate tRNA to step II. This condition is that requiring step II to be the rate-limiting one in protein synthesis, and specifically, that k_3 is much greater than k_{-2}. Under these circomstances the flow scheme is

$$T_1 + R \underset{k_{-1}}{\overset{k_1}{\rightleftharpoons}} T_1R \xrightarrow{k_2} T_2R \underset{k_{-3}}{\overset{k_3}{\rightleftharpoons}} T_3R \xrightarrow{k_4} T_4R \xrightarrow{k_5} R \tag{8}$$

$$\ \ \ \text{I}\qquad\quad \text{II}\qquad\quad \text{III}\qquad\ \text{IV}\qquad \text{V}$$

According to scheme (8) the rate of protein synthesis is given by

$$p = k_2 K_1 (T_1)(R) \tag{9}$$

and

$$a = \frac{{}^c k_2 ({}^c T_1)({}^c R)}{{}^n k_2 ({}^n T_1)({}^n R)} \tag{10}$$

Thus, this scheme corresponds to the situation in which the only effective tRNA discrimination is at step II, and step III functions merely to stabilize the resulting tRNA interaction with the programmed ribosome.

Finally, we assume that the rate-limiting step in protein synthesis is I. Here, k_2 is much greater than k_{-1}, and we have the following scheme:

$$T_1 + R \xrightarrow{k_1} T_1R \underset{k_{-2}}{\overset{k_2}{\rightleftharpoons}} T_2R \underset{k_{-3}}{\overset{k_3}{\rightleftharpoons}} T_3R \xrightarrow{k_4} T_4R \xrightarrow{k_5} R \tag{11}$$

$$\ \ \ \text{I}\qquad\quad \text{II}\qquad\quad \text{III}\qquad\ \text{IV}\qquad \text{V}$$

Accordingly, any tRNA molecule that reaches T_1R will donate its amino acid to the nascent protein chain. Since step I is not a discriminating step, protein synthesis cannot be amino acid-specific according to this scheme.

Obviously, the virtues of calculations based on schemes in which one step at a time is assumed to be rate limiting are that the resulting rate equations are simple, and they permit the functional contributions of each step to be evaluated separately. However, one drawback to this procedure of explication is that it obscures a valuable lesson. Under these boundary conditions, the calculated discrimination advantage for cognate tRNA over noncognate tRNA is a maximum value because we have chosen conditions in which $({}^c T_1)({}^c R)$ is approximately equal to $({}^n T_1)({}^n R)$. In general, however, $({}^c T_1)({}^c R) \leq ({}^n T_1)({}^n R)$. Since a cognate tRNA would be expected to pass through the different steps faster than a noncognate tRNA, the inequality increases as the ratio of the rate constants for the irreversible steps compared to those of the equilibria increases. This suggests that an optimum system for tRNA discrimination will be one in which the equilibria are only weakly coupled to the irreversible steps in the scheme.

D. Summing Up

The results of the previous section have verified our conjecture that conformational changes of tRNA which expose common binding sites can in principle amplify the codon-anticodon interaction in such a fashion that both the stability and specificity of the resulting ternary complex are enhanced. Two conditions must be met in order for this mechanism to function as postulated. First, the conformational changes of the tRNA must be tightly coupled to the event corresponding to the codon-anticodon interaction, which here functions very much like an allosteric effector. Second, no irreversible steps may be interposed between the codon-anticodon interaction and any of the subsequent discrimination steps.

If these conditions are satisfied, it would be possible to obtain enhancement by several successive steps involving different conformational changes of the tRNA that are coupled to each other as in the simpler version of our model. From this point of view, this model accomplishes the "proofreading" function recently described, but it does so through a rather different mechanism than that postulated by Blomberg (1974) and Hopfield (1974). Furthermore, the present model will account for the codon-specific binding of tRNA in the absence of an exogenous energy source, which was not examined in the other expositions.

This naturally leads us to consider the possible functions of the GTP which is an intermediate in the formation of the aminoacyl-tRNA · EF-Tu complex that normally is bound at the ribosomal A site (Lipmann, 1969). The apparently spontaneous nature of the interaction of tRNA with a programmed ribosome suggested much earlier that energy would be required to remove a tRNA molecule from its ribosomal binding site, rather than to drive it to that site (Kurland, 1966). Accordingly, it is possible that the GTP requirement for EF-Tu-dependent binding of aminyacyl-tRNA may have more to do with the subsequent removal of the tRNA from the A site during translocation, than for the prior codon recognition step, which is our present concern.

Further experimental work is required to determine how many steps are required to obtain the high fidelity of translation that is characteristic of this system. However, it is by now clear that the stability of the aminoacyl-tRNA complex formed with the programmed ribosome is accountable in terms of but a few steps. First, by fixing the fold of the codon it is possible to obtain an enhancement of the codon-anticodon interaction, as shown by Eisinger (1971). This would yield an association constant for the triplet interaction of the order of 10^5 M^{-1}. Second, this interaction would be supplemented by the interactions of the common structural features of the tRNA with binding sites on the ribosomes. In the case of the T-ψ-C-G loop interaction with the 5 S RNA an additional stability could be obtained which corresponds to at least another 10^5 M^{-1}. Thus, with only two steps at our disposal, we can account for an association constant which is of the order of 10^{10} M^{-1}.

Assessment of the discrimination at each of these steps is quite another matter. As mentioned earlier the enhancement of the triplet-triplet interaction by Eisinger's mechanism does not help very much with respect to the specificity of the interaction. However, the existence of enzymes that can distinguish a proper double-helical region of DNA from one which has a significant dislocation (Lindahl and Nyberg, 1972; Verly and Paquette, 1972) makes an alternative conceivable. Thus, it is possible that some ribosomal elements are able to check the fit of the codon-anticodon interaction, and thereby enhance the discrimination during this stage of the binding reaction.

At first glance, the existence of mutations expressed in proteins S4 and S12 which influence the error frequency of the tRNA selection mechanism (Gorini, 1971) might seem consistent with this conjecture. Unfortunately, there are much simpler ways to account for the effects of these two proteins on the error frequency. For example, alterations of their structure may perturb the conformation of the codon, and thereby only indirectly alter the error frequency of the selection mechanism. Therefore, it is still not clear whether the ribosome has functions other than the provision of binding sites for the tRNA and mRNA during the screening process.

A particularly attractive property of the present model is that it will account for the data that show that alterations of the structure of tRNA molecules at sites separate from the anticodon loop can influence the error frequency of the translation process. This is well illustrated by the properties of one class of nonsense suppressors.

One of the codons that will cause abortive termination of translation when it appears in the middle of a mRNA is UGA, and this nonsense codon can be suppressed by mutationally altered tRNA molecules (Brenner et al., 1967; Weigert et al., 1969; Sambrook et al., 1967; Model et al., 1969). One such suppressor mutant was shown by Hirsh (1971) to contain an altered form of tRNA[Trp]. The astonishing discovery made with this suppressor tRNA was that it was altered in position 24 of the D stem rather than as expected in the anticodon loop. Nevertheless, this tRNA can insert a tryptophan at positions corresponding to the UGA nonsense codon, as well as at the normal tryptophan codon, UGG. Accordingly, Hirsh suggested that the altered suppressor tRNA could effectively read the A in the third position of the codon as though it were a G.

Here then is a very clear example of one of the necessary consequences of the present model of the tRNA discrimination mechanism. Thus, we would expect that alterations of the tRNA at sites separate from the anticodon loop could occur in which the conformational stability of the tRNA had been altered. Under these circumstances the tRNA may be able to undergo the requisite conformational change that leads to the unmasking of the common binding sites, such as the T-ψ-C-G loop, when only two out of three of the nucleotides in the anticodon are able to properly interact with the corresponding nucleotide in the codon.

Indeed, such an error would be expected to occur at very low frequencies with a normal tRNA. The only effect of the tRNA alteration in suppressor strains would be to raise the probability of such an event. It is, therefore, not surprising that Hirsh and Gold (1971) find *in vitro* that the normal tRNA^{Trp} gives a very low level of suppression of UGA codons, but that the suppressor form of the tRNA only raises the probability of this event by a factor of 30.

Strange as it may seem at first, the model developed in the previous section is in no way dependent on the factual issue of whether or not the binding of the common T-ψ-C-G loop to the 5 S RNA provides significant discrimination of the cognate from noncognate tRNA's. Rather the unmasking of the T-ψ-C-G loop during the binding reaction only provides a paradigm for the sort of conformational changes required by the model. It may be that the development of interactions between other common structural features of tRNA molecules and their binding sites on the 16 S RNA are responsible for most of the discrimination before the 5 S contribution is made. This is an experimental question.

What is important about this model is that it demonstrates the feasibility of accounting for the tRNA selection mechanism on the basis of facts already known about the mechanism of protein synthesis. No unobserved high energy intermediates or esoteric ribosome functions need to be involved. In their place, a view of of the tRNA molecule and its conjugate structures in the ribosome has been developed which may in part account for the structural complexity of the system. Here, it is worth recalling the consternation that was caused by the discovery that the amino acid adapter molecules are so large. Indeed, Crick (1958) went so far as to consider the possibility that the tRNA molecules might be precursors to the "real" adapters which could be produced by the cleavage of the more unwieldy 4 S RNA to yield a reasonable size class of adapters, e.g., trinucleotides.

In constrast, the structural complexity of the tRNA molecule is seen from the vantage point of the present model as an elegantly simple solution to a very difficult problem. Thus, the codon which directs the selection of a particular amino acid adapter cannot by itself stably bind one out of twenty triplet oligonucleotide adapters with anything like the specificity required for the reliable production of proteins. Furthermore, this problem is compounded by the need to be able to transiently trap each of the many different aminoacylated adapters at a single site, and in a fixed geometry so that each different amino acid can be inserted into the nascent protein by a single recursive operation. The solution to this formidable problem is to place the anticodon triplet in a very special molecular setting. Here, a family of nucleic acid molecules has evolved which contain a set of universal elements that can interact with their functional conjugates in the ribosome. The functionality of these universal adapters is here brought into play only on demand, being dependent on a proper sequence of conformational transitions from one metastable state to another. All of this is triggered by the weak cognate interaction between two trinucleotide sequences.

ACKNOWLEDGMENT

I am deeply indebted to R. Rigler for innumerable stimulating discussions. Thanks are also due for helpful discussions and the sharing of unpublished results to C. Blomberg, C. Cantor, M. Ehrenberg, V. Erdmann, S. J. S. Hardy, E. Küchler, O. Pongs, P. Sigler, G. Stöffler, and last, but by no means least, H. G. Wittmann.

REFERENCES

Birge, E. A., and Kurland, C. G. (1969). *Science* **166,** 1282.
Birge, E. A., and Kurland, C. G. (1970). *Mol. Gen. Genet.* **109,** 356.
Blomberg, C. (1974). Publication TRITA-TFY-74-16. Royal Institute of Technology, Stockholm.
Bode, U., Lutter, L. C., and Stöffler, G. (1974). *FEBS Lett.* **45,** 232.
Brenner, S., Barnett, L., Katz, E. R., and Crick, F. H. C. (1967). *Nature (London)* **218,** 449.
Bretscher, M. (1968). *Nature (London)* **218,** 675.
Brownlee, G. G., Sanger, F., and Barren, B. G. (1967). *Nature (London)* **215,** 735.
Carpenter, F. H., and Harrington, K. T. (1972). *J. Biol. Chem.* **247,** 5580.
Castles, B., and Singer, M. (1969) *J. Mol. Biol.* **40,** 1.
Chappeville, F., Lipmann, F., von Ehrenstein, G., Weissblum, B., Ray, W. J., Jr., and Benzer, S. (1962). *Proc. Natl. Acad. Sci. U.S.A.* **48,** 1086.
Craven, G. R., Voynow, P., Hardy, S. J. S., and Kurland, C. G. (1969). *Biochemistry* **8,** 2909.
Crick, F. H. C. (1958). *Symp. Soc. Exp. Biol.* **12,** 138.
Crick, F. H. C., Barnett, L., Brenner, S., and Watts-Tobin, R. J. (1961). *Nature (London)* **192,** 1227.
Czernilofsky, A. P., Kurland, C. G., and Stöffler, G. (1975). *FEBS Lett.* **58,** 281.

Davies, G. E., and Stark, G. R. (1970). *Proc. Natl. Acad. Sci. U.S.A.* **66,** 651.

Daya-Gros-Jean, L., Garrett, R. A., Pongs, O., Stöffler, G., and Wittmann, H. G. (1972). *Mol. Gen. Genet.* **119,** 277.

Deusser, E., Stöffler, G., Wittmann, H. G., and Apirion, D. (1970). *Mol. Gen. Genet.* **109,** 299.

Ehresmann, C., Stiegler, P., Fellner, P., and Ebel, J. P. (1972). *Biochimie* **54,** 901.

Eigen, M. (1971). *Naturwissenschaften* **58,** 465.

Eisinger, J. (1971). *Biochem. Biophys. Res. Commun.* **43,** 854.

Eisinger, J., Feuer, G., and Yamane, T. (1971). *Nature (London), New Biol.* **231,** 126.

Erdmann, V. A., Sprinzl, M., and Pongs, O. (1973). *Biochem. Biophys. Res. Commun.* **54,** 942.

Feunteun, J., Moniér, R., Vola, C., and Rosset, R. (1974). *Nucleic Acids Res.* **1,** 149.

Fiser, I., Schreit, K. H., Stöffler, G., and Küchler, E. (1974). *Biochem. Biophys. Res. Commun.* **60,** 1112.

Forget, B. G., and Weissman, S. M. (1967). *Science* **158,** 1695.

Gauss, D. H., von den Haar, F., Maelicke, A., and Cramer, F. (1971). *Annu. Rev. Biochem.* **40,** 1045.

Gilbert, W. (1963). *J. Mol. Biol.* **6,** 389.

Goldberg, M. L., and Steitz, J. A. (1974). *Biochemistry* **13,** 2123.

Gorini, L. (1971). *Nature (London), New Biol.* **234,** 261.

Green, M., and Kurland, C. G. (1971). *Nature (London), New Biol.* **234,** 273.

Green, M., and Kurland, C. G. (1973). *Mol. Biol. Rep.* **1,** 105.

Hardy, S. J. S. (1976). *Mol. Gen. Genet.* **140,** 253.

Hardy, S. J. S., Kurland, C. G., Voynow, P., and Mora, G. (1969). *Biochemistry* **8,** 2897.

Hasenbank, R., Guthrie, C., Stöffler, G., Wittmann, H. G., Rosen, H. G., and Apirion, D. (1973). *Mol. Gen. Genet.* **127,** 1.

Held, W. A., Gette, W. R., and Nomura, M. (1974). *Biochemistry* **13,** 2123.

Hirsh, D. (1971). *J. Mol. Biol.* **58,** 439.

Hirsh, D., and Gold, L. (1971). *J. Mol. Biol.* **58,** 459.

Högenauer, G. (1970). *Eur. J. Biochem.* **12,** 577.

Hopfield, J. J. (1974). *Proc. Natl. Acad. Sci. U.S.A.* **71,** 4135.

Huang, K.-H., Fairclough, R., and Cantor, C. R. (1975). *J. Mol. Biol.* **97,** 443.

Kaltschmidt, E., and Wittmann, H. G. (1971). *Proc. Natl. Acad. Sci. U.S.A.* **67,** 1296.

Kenner, R. A. (1973). *Biochem. Biophys. Res. Commun.* **51,** 932.

Kim, S. H., Suddath, F. L., Quigley, G. J., McPherson, A., Sussman, J. L., Wang, A. H. J., Seeman, W. C., and Rich, A. (1974). *Science* **185,** 435.

Kurland, C. G. (1966). *J. Mol. Biol.* **18,** 90.

Kurland, C. G. (1970). *Science* **109,** 1171.

Kurland, C. G. (1974a). *In* "Ribosomes" (M. Nomura, A. Tissières, and P. Lengyel, eds.), p. 309. Cold Spring Harbor Lab., Cold Spring Harbor, New York.

Kurland, C. G. (1974b). *J. Supramol. Struct.* **2,** 178.

Kurland, C. G., Voynow, P., Hardy, S. J. S., Randall, L., and Lutter, L. (1969). *Cold Spring Harbor Symp. Quant. Biol.* **74,** 17.

Kurland, C. G., Donner, D., van Duin, J., Green, M., Lutter, L., Randall-Hazelbauer, L., Schaup, H. W., and Zeichhardt, H. (1972). *FEBS Symp.* **27,** 225.

Lindahl, T., and Nyberg, B. (1972). *Biochemistry* **19,** 3610.

Lipmann, F. (1969). *Science* **160,** 390.

Lodish, H. F. (1970). *Nature (London)* **226,** 705.

Loftfield, R. B., and Vanderjagt, D. (1972). *Biochem J.* **128,** 1353.

Lutter, L. C., and Kurland, C. G. (1973). *Nature (London), New Biol.* **243,** 15.

Lutter, L. C., and Kurland, C. G. (1974). *Mol. Cell. Biochem.* **7,** 105.

Lutter, L. C., Zeichhardt, H., Kurland, C. G., and Stöffler, G. (1972). *Mol. Gen. Genet.* **119**, 357.

Lutter, L. C., Bode, U., Kurland, C. G., and Stöffler, G. (1974a). *Mol. Gen. Genet.* **129**, 167.

Lutter, L. C., Ortanderl, F., and Fasold, H. (1974b). *FEBS Lett.* **48**, 288.

Lutter, L. C., Kurland, C. G., and Stöffler, G. (1975). *FEBS Lett.* **54**, 1440.

McLaughlin, C. S., Dondon, J., Grunberg Manago, M., Michelson, A. M., and Saunders, G. (1966). *Cold Spring Harbor Symp. Quant. Biol.* **31**, 601.

Model, P., Webster, R. Z., and Zinder, N. (1969). *J. Mol. Biol.* **43**, 177.

Monro, R. E. (1967). *J. Mol. Biol.* **26**, 147.

Moore, P. B., Engelman, D. M., and Schoenborn, B. P. (1974). *Proc. Natl. Acad. Sci. U.S.A.* **71**, 172.

Nanninga, W., Garrett, R. A., Stöffler, G., and Klotz, G. (1972). *Mol. Gen. Genet.* **119**, 174.

Ninio, J. (1974). *J. Mol. Biol.* **84**, 297.

Noller, H. F., and Chaires, J. B. (1972). *Proc. Natl. Acad. Sci. U.S.A.* **69**, 3115.

Nomura, M., Mizushima, S., Ozaki, P., Traub, P., and Lowry, C. V. (1969). *Cold Spring Harbor Symp. Quant. Biol.* **34**, 49.

Ofengand, J., and Henes, P. (1969). *J. Biol. Chem.* **244**, 6241.

Okamoto, T., and Takanami, M. (1963). *Biochim. Biophys. Acta* **76**, 266.

Olsson, M., Isaksson, L., and Kurland, C. G. (1975). *Mol. Gen. Genet.* **135**, 191.

Pestka, S. (1971). *Annu. Rev. Biochem.* **40**, 697.

Pongs, O., and Lanka, R. (1975). *Proc. Natl. Acad. Sci. U.S.A.* **72**, 1505.

Randall-Hazelbauer, L. L., and Kurland, C. G. (1972). *Mol. Gen. Genet.* **115**, 234.

Rhodes, D. (1976). *J. Mol. Biol.* **94**, 449.

Richter, D., Erdmann, V. A., and Sprinzl, M. (1973). *Nature (London), New Biol.* **246**, 132.

Riddle, D. L., and Carbon, J. (1973). *Nature (London), New Biol.* **242**, 230.

Robertus, J. D., Ladner, J. E., Finch, J. T., Rhodes, D., Brown, R. S., Clark, B. F. C., and Klug, A. (1974). *Nature (London)* **250**, 546.

Rosset, R., and Gorini, L. (1967). *J. Mol. Biol.* **39**, 95.

Rossmann, M., Liljas, A., Brändén, C. -I., and Banaszak, L. J. (1975). *In* "The Enzymes" (P. D. Boyer, ed.), 3rd ed., Vol. 12, p. 61. Academic Press, New York.

Rummel, D. P., and Noller, H. F. (1973). *Nature (London), New Biol.* **245**, 72.

Sambrook, J. F., Fan, D. P., and Brenner, S. (1967). *Nature (London)* **214**, 452.

Schaup, H. W., and Kurland, C. G. (1972). *Mol. Gen. Genet.* **114**, 350.

Schaup, H. W., Sogin, M., Woese, C., and Kurland, C. G. (1971). *Mol. Gen. Genet.* **114**, 1.

Schwartz, I., and Offengand, J. (1974). *Proc. Natl. Acad. Sci. U.S.A.* **71**, 3951.

Schwarz, U., Reinhard, R., and Gassen, H. G. (1974). *Biochem. Biophys. Res. Commun.* **56**, 807.

Shine, J., and Dalgarno, L. (1974). *Proc. Natl. Acad. Sci. U.S.A.* **71**, 1342.

Shine, J., and Dalgarno, L. (1975). *Nature (London)* **254**, 34.

Sigler, P. (1975). *Annu. Rev. Biophys. Bioeng.* **4**, 477.

Slobin, L. (1972). *J. Mol. Biol.* **64**, 297.

Sommer, A., and Traut, R. R. (1974). *Proc. Natl. Acad. Sci. U.S.A.* **71**, 3946.

Spirin, A. (1969). *Cold Spring Harbor Symp. Quant. Biol.* **34**, 197.

Steitz, J. (1969). *Cold Spring Harbor Symp. Quant. Biol.* **34**, 621.

Sun, T. T., Bullen, A., Kahan, L., and Traut, R. R. (1974). *Biochemistry* **13**, 2334.

Takanami, M., and Zubay, G. (1964). *Proc. Natl. Acad. Sci. U.S.A.* **51**, 834.

Takanami, M., Yan, Y., and Jukes, T. H. (1965). *J. Mol. Biol.* **12**, 761.

Tal, M., Aviram, M., Kanarek, A., and Weiss, A. (1972). *Biochim. Biophys. Acta* **281**, 381.

Thammana, P., Kurland, C. G., Deusser, E., Weber, J., Maschler, R., Stöffler, G., and Wittmann, H. G. (1973). *Nature (London), New Biol.* **242**, 47.

Traub, P., and Nomura, M. (1968). *Proc. Nat. Acad. Sci. U.S.A.* **59**, 777.

Traut, R. R., Delius, H., Ahmad-Zadek, C., Bickle, T. A., Pearson, P., and Tissières, A. (1969). *Cold Spring Harbor Symp. Quant. Biol.* **34**, 25.

Traut, R. R., Bollen, A., Sun, J., Hershey, J., Sundberg, J., and Ross Pierce, L. (1973). *Biochemistry* **12**, 3266.

Uhlenbeck, O. C., Baker, J., and Doty, P. (1975). *Nature (London)* **225**, 508.

van Dieijen, G., van der Laken, C. J., van Knippenberg, P. H., and van Duin, J. (1975). *J. Mol. Biol.* **93**, 351.

van Duin, J., and Kurland, C. G. (1970). *Mol. Gen. Genet.* **109**, 169.

van Duin, J., and van Knippenberg, P. H. (1974). *J. Mol. Biol.* **54**, 185.

van Duin, J., van Knippenberg, P. H., Dieben, M., and Kurland, C. G. (1972). *Mol. Gen. Genet.* **116**, 181.

Verly, W. G., and Paquette, Y. (1972). *Can. J. Biochem.* **50**, 217.

Voynow, P., and Kurland, C. G. (1971). *Biochemistry* **10**, 517.

Watson, J. D. (1964). *Bull. Soc. Chim. Biol.* **46**, 1399.

Weber, J. (1972). *Mol. Gen. Genet.* **119**, 233.

Weigert, M. G., Lanka, E., and Barrell, B. G. (1969). *Nature (London)* **223**, 1331.

Wittmann, H. G., Stöffler, G., Hindennach, I., Kurland, C. G., Randall-Hazelbauer, L., Birge, E. A., Nomura, M., Kaldtschmidt, E., Mizushima, S., Traut, R. R., and Bickle, T. A. (1971). *Mol. Gen. Genet.* **111**, 327.

Woese, C. (1973). *J. Theor. Biol.* **38**, 203.

Zimmermann, R. A., Garvin, R. T., and Gorini, L. (1971). *Proc. Natl. Acad. Sci. U.S.A.* **68**, 2263.

Zimmermann, R. A., Muto, A., Fellner, P., Ehresmann, C., and Branlant, C. (1972). *Proc. Natl. Acad. Sci. U.S.A.* **69**, 1282.

3

Primary Structure and Three-Dimensional Arrangement of Proteins within the *Escherichia coli* Ribosome

G. STÖFFLER AND H. G. WITTMANN

Rapid progress has been made in recent years with studies of the ribosomal proteins from *Escherichia coli*. Their exact numbers have been determined by two-dimensional polyacrylamide gel electrophoresis, and all proteins from both the small and the large subunit have been isolated and characterized by chemical, physical, and immunological techniques. These studies have been recently reviewed in detail (Wittmann, 1974; Wittmann and Wittmann-Liebold, 1974; Stöffler, 1974) and will not be repeated here. Instead, we will report on the progress made within the last two years and concentrate on two subjects: a) the primary structure of ribosomal proteins from *E. coli* wild-type and mutants; b) the topographical arrangement of the ribosomal components within the particle.

I. Primary Structure of Proteins

A. *Escherichia coli* Wild-Type

There are 21 proteins in the small and 34 proteins in the large subunits of *E. coli* ribosomes. Since one protein (S20) from the small and one protein

(L26) from the large subunit are identical, the total number of proteins within the 70 S ribosome is 54, i.e., one less than the sum of the number of proteins in both subunits (see review by Wittmann, 1974). The total number of amino acids in all *E. coli* ribosomal proteins is approximately 8000. The sequences for more than 4300 of these amino acids have so far been determined in our laboratory or in collaboration with other groups. Table I lists the number of the amino acids sequenced within the individual ribosomal proteins. Those proteins whose sequences have been completely determined are shown in Figs. 1–18. It can be expected that the sequence of several additional proteins will be available within the near future.

Knowing the primary structures of proteins, their secondary structures can be estimated. Using the computer program described by Burgess *et al.* (1974) this has been done for the ribosomal proteins shown in Figs. 1–18 (B. Wittmann-Liebold and M. Dzionara, unpublished results). The predicted length and distribution of α-helical and β-structural regions within several of these proteins are depicted in Fig. 19.

The availability of the sequence for 4300 amino acids, i.e., 55% of the residues present in the *E. coli* ribosome, allows detailed information about possible homologous regions among the ribosomal proteins. From pre-

```
1                            10                            20
Ala-Arg-Tyr-Leu-Gly -Pro-Lys-Leu-Lys-Leu-Ser -Arg-Arg-Glu -Gly -Thr-Asp-Leu-Phe-Leu-

                             30                            40
Lys-Ser -Gly -Val -Arg-Ala - Ile - Asp-Thr-Lys-Cys-Lys -Ile - Glu -Gln -Ala -Pro-Gly -Gln -His -

                             50                            60
Gly -Ala -Arg-Lys-Pro-Arg-Leu-Ser -Asp-Tyr-Gly -Val -Gln -Leu-Arg-Glu - Lys-Gln - Lys-Val -

                             70                            80
Arg-Arg- Ile - Tyr-Gly -Val -Leu-Glu -Arg-Gln -Phe-Arg-Asn-Tyr-Tyr-Lys-Glu -Ala -Ala -Arg-

                             90                            100
Leu-Lys-Gly -Asn-Thr-Gly -Glu -Asn-Leu-Ala -Leu-Leu-Gln -Gly -Arg-Leu-Asp-Asn-Val -Val -

                             110                           120
Tyr-Arg-Met-Gly -Phe-Gly -Ala -Thr-Arg-Ala -Glu -Ala -Arg-Gln -Leu-Val -Ser -His- Lys-Ala -

                             130                           140
Ile - Met-Val -Asn-Gly -Arg-Val -Val -Asn -Ile - Ala -Ser -Tyr-Gln -Val -Asp-Pro-Asn-Ser -Val -

                             150                           160
Val - Ile - Arg-Glu -Lys-Ala - Lys-Lys-Glu -Ser -Arg-Val -Lys-Ala -Ala -Leu-Glu -Leu-Ala -Glu -

                             170                           180
Gln -Arg-Gln -Lys-Pro-Thr-Trp-Leu-Glu -Val -Asp-Ala -Gly -Lys-Met-Glu -Gly -Thr-Phe-Lys-

                             190                           200
Arg-Lys-Pro-Glu -Arg-Ser -Asp-Leu-Ser -Ala -Asp - Ile -Asn-Glu -His -Leu - Ile -Val -Glu -Leu-

Tyr-Ser -Lys
```

Fig. 1   Primary structure of protein S4.

```
 1                                     10                                    20
Met-Arg-His -Tyr- Glu - Ile - Val -Phe-Met-Val -His- Pro-Asp-Gln -Ser -Glu - Gln -Val -Pro- Gly -
```
```
                                      30                                    40
Met -Ile - Glu -Arg-Tyr-Thr-Ala -Ala - Ile - Thr- Gly -Ala -Glu - Gly -Lys -Ile - His -Arg-Leu- Glu -
```
```
                                      50                                    60
Asp- Trp- Gly -Arg-Arg-Gln - Leu-Ala -Tyr-Pro - Ile -Asn-Lys-Leu-His -Lys-Ala -His - Tyr- Val -
```
```
                                      70                                    80
Leu-Met-Asn-Val-Glu -Ala -Pro-Gln - Glu - Val- Ile - Asp-Glu -Leu-Glu -Thr-Thr-Phe-Arg-Phe -
```
```
                                      90                                    100
Asn-Asp-Ala -Val - Ile - Arg-Ser- Met-Val -Met-Arg-Thr-Lys-His -Ala -Val -Thr-Glu -Ala -Ser -
```
```
                                      110                                   120
Pro-Met-Val -Lys-Ala - Lys-Asp-Glu -Arg-Arg-Glu -Arg-Arg-Asp-Asp-Phe-Ala -Asn-Glu -Thr-
```
```
                                      130
Ala -Asp-Asp-Ala -Glu -Ala - Gly -Asp-Ser -Glu -Glu -Glu -Glu -Glu
```

Fig. 2 Primary structure of protein S6.

```
 1                                     10                                    20
Ser -Met- Gln -Asp-Pro - Ile -Ala -Asp-Met-Leu-Thr-Arg - Ile -Arg-Asn-Gly -Gln -Ala -Ala -Asn-
```
```
                                      30                                    40
Lys-Ala -Ala -Val -Thr-Met-Pro-Ser -Ser -Lys-Leu-Lys-Val -Ala - Ile - Ala -Asn-Val -Leu-Lys-
```
```
                                      50                                    60
Glu - Glu - Gly -Phe - Ile - Glu -Asp-Phe-Lys-Val -Glu - Gly -Asp-Thr-Lys-Pro- Glu -Leu-Glu -Leu-
```
```
                                      70                                    80
Thr-Leu-Lys-Tyr-Phe-Gln - Gly - Lys-Val -Val -Ala -Glu - Ile - Ser -Gln -Arg-Val -Ser -Arg-Pro-
```
```
                                      90                                    100
Gly -Leu-Arg -Ile - Tyr-Lys-Leu-Gln -Asp-Lys-Pro-Lys-Arg-Val -Met-Gly -Asp-Thr-Arg-Ala -
```
```
Ala -Arg-Leu-Gln - Ile - Cys-Val -Ala -Tyr
```

Fig. 3 Primary structure of protein S8.

```
 1                                     10                                    20
Ala -Glu -Asn-Gln- Tyr-Tyr-Gly -Thr- Gly -Arg-Arg-Lys-Ser -Ser -Ala -Ala -Arg-Val -Phe -Ile -
```
```
                                      30                                    40
Lys-Pro- Gly -Asn-Gly -Lys - Ile - Val - Ile - Asn-Gln -Arg-Ser -Leu-Glu -Gln -Tyr-Phe-Gly -Arg-
```
```
                                      50                                    60
Glu -Thr-Ala -Arg-Met-Val -Val -Arg-Gln -Pro-Leu-Glu -Leu-Val -Asn-Met-Val -Glu -Lys-Leu-
```
```
                                      70                                    80
Asp-Leu-Tyr - Ile -Thr-Val -Lys-Gly -Gly -Gly - Ile - Ser -Gly -Gln -Ala -Gly -Ala - Ile -Arg-His -
```
```
                                      90                                    100
Gly - Ile - Thr-Arg-Ala -Leu-Met-Glu -Tyr-Asp-Glu -Ser -Leu-Arg-Ser -Glu -Leu-Arg-Lys-Ala -
```
```
                                      110                                   120
Gly -Phe-Val -Thr-Arg-Asp-Ala -Arg-Gln -Val -Glu -Arg-Lys-Lys-Val -Gly -Leu-Arg-Lys-Ala -
```
```
Arg-Arg-Pro-Glu -Phe-Ser -Lys-Arg
```

Fig. 4 Primary structure of protein S9.

TABLE I

Compilation of Amino Acid Sequence Data of *E. coli* Ribosomal Proteins

Protein	Sequenced amino acids	Sequence completed[a]	Ref.[b]	Protein	Sequenced amino acids	Sequence completed[a]	Ref.[b]
S1	20	−	1	L7	120	+	23
S2	35	−	2	L9	46	−	1
S3	80	−	2, 3	L10	164	+	24, 25
S4	203	+	4	L11	115	−	26
S5	160	(+)	5	L12	120	+	23
S6	135	+	6	L13	60	−	21, 27
S7	80	−	7	L14	70	−	28
S8	109	+	8, 9	L15	120	−	29
S9	128	+	10	L16	136	+	30
S10	42	−	11	L17	75	−	21, 31
S11	125	(+)	2, 12	L18	117	+	32
S12	123	+	13	L19	16	−	1
S13	117	+	2, 14	L20	50	−	21
S14	36	−	2	L21	65	−	33
S15	87	+	15	L22	80	−	21, 34
S16	82	+	16	L23	50	−	21
S17	75	(+)	17	L24	75	−	21, 35
S18	74	+	18	L25	94	+	36, 37
S19	72	(+)	17	L26	43	−	1
S20	86	+	19	L27	84	+	38
S21	70	+	20	L28	77	+	39
L1	47	−	21	L29	63	+	40
L2	40	−	21	L30	58	+	41
L3	32	−	21	L32	56	+	42
L4	38	−	1	L33	54	+	43
L5	178	+	21, 22	L34	46	+	44
L6	176	+	1, 3				

[a] +, sequence completed; (+), sequence almost completed; −, sequence partially known.
[b] References: (1) B. Wittmann-Liebold, A. W. Geissler, H. Graffunder, and E. Marzinzig, unpublished; (2) Wittmann-Liebold, 1973; (3) R. Chen, unpublished; (4) Schiltz and Reinbolt, 1975; (5) B. Wittmann-Liebold, B. Greuer, and A. Lehmann, unpublished; (6) Hitz *et al.*, 1975; (7) J. Reinbolt, D. Tritsch, and B. Wittmann-Liebold, unpublished; (8) Stadler, 1974; (9) Stadler and Wittmann-Liebold, 1976; (10) Chen and Wittmann-Liebold, 1975; (11) Wittmann and Wittmann-Liebold, 1974; (12) R. Chen and U. Arfsten, unpublished; (13) G. Funatsu, M. Yaguchi, and B. Wittmann-Liebold, unpublished; (14) H. Lindemann and Wittmann-Liebold, 1976; (15) Morinaga *et al.*, 1976; (16) W. Rombauts, J. Vandekerckhove, and B. Wittmann-Liebold, unpublished; (17) M. Yaguchi and H. G. Wittmann, unpublished; (18) Yaguchi, 1975; (19) B. Wittmann-Liebold, E. Marzinzig, and A. Lehmann, 1976; (20) Vandekerckhove *et al.*, 1975; (21) Wittmann-Liebold *et al.*, 1975a; (22) R. Chen and G. Ehrke, 1976a; (23) Terhorst *et al.*, 1973; (24) Dovgas *et al.*, 1976; (25) Heiland *et al.*, 1976; (26) J. Dognin, unpublished; (27) L. Mende, unpublished; (28) G. Funatsu and H. G.

```
1                                            10                                    20
Ser -Leu-Ser -Thr-Glu -Ala -Thr-Ala - Lys - Ile -Val -Ser- Glu -Phe-Gly -Arg-Asp-Ala -Asn-Asp-

                                             30                                    40
Thr-Gly -Ser -Thr-Glu -Val -Gln -Val -Ala -Leu- Leu-Thr-Ala -Gln - Ile - Asn-His -Leu-Gln -Gly -

                                             50                                    60
His -Phe-Ala -Glu -Lys-Lys-Asp-His -His -Ser -Arg-Arg-Gly - Leu-Leu-Arg-Met-Val -Ser -Gln -

                                             70                                    80
Arg-Arg-Lys- Leu- Leu-Asp-Tyr- Leu- Lys-Arg-Lys-Asp-Val -Ala -Arg-Tyr-Thr-Gln - Leu - Ile -

                            87
Glu -Arg-Leu-Gly -Leu-Arg-Arg
```

Fig. 5   Primary structure of protein S15.

```
1                                            10                                    20
Ac · Ala -Arg-Tyr-Phe-Arg-Arg-Arg-Lys-Phe-Cys-Arg-Phe-Thr-Ala -Gln -Gly -Val -Gln -Glu - Ile -

                                             30                                    40
Asp-Tyr-Lys-Asp - Ile -Ala -Thr- Leu- Lys-Asn-Tyr - Ile - Thr-Glu -Ser -Gly - Lys - Ile - Val -Pro-

                                             50                                    60
Ser -Arg - Ile -Thr-Gly -Thr-Arg-Ala - Lys- Tyr- Gln - Arg-Gln - Leu-Ala -Arg-Ala - Ile - Lys-Arg-

                            70
Ala -Arg-Tyr- Leu-Ser - Leu- Leu-Pro-Tyr-Thr-Asp-Arg-His -Gln
```

Fig. 6   Primary structure of protein S18.

```
1                                            10                                    20
Ala -Asn- Ile - Lys-Ser -Ala - Lys- Lys-Arg-Ala - Ile - Gln -Ser -Glu - Lys-Ala -Arg-Lys-His -Asn-

                                             30                                    40
Ala -Ser -Arg-Arg-Ser -Met-Met-Arg-Thr-Phe- Ile - Lys-Lys-Val -Tyr-Ala -Ala - Ile - Glu -Ala -

                                             50                                    60
Gly -Asp-Lys-Ala -Ala -Ala -Gln - Lys-Ala -Phe-Asn-Glu -Met- Gln -Pro - Ile -Val -Asp-Arg-Gln -

                                             70                                    80
Ala -Ala - Lys-Gly - Leu - Ile -His - Lys-Asn- Lys-Ala -Ala -Arg-His - Lys-Ala -Asn- Leu-Thr-Ala -

                            86
Gln - Ile - Asn-Lys- Leu-Ala
```

Fig. 7   Primary structure of protein S20.

Wittmann, unpublished; (29) S. Giorginis, unpublished; (30) Brosius and Chen, 1976; (31) W. Rombauts, and H. G. Wittmann, unpublished; (32) Brosius *et al.*, 1975; (33) R. Chen and J. Becker, unpublished; (34) B. Wittmann-Liebold, J. Krauss, and A. Lehmann, unpublished; (35) J. F. Lontie, R. R. Crichton, and B. Wittmann-Liebold, unpublished; (36) Dovgas *et al.*, 1975; (37) Bitar and Wittmann-Liebold, 1975; (38) Chen *et al.*, 1975; (39) B. Wittmann-Liebold and E. Marzinzig, unpublished; (40) Bitar, 1975; (41) Ritter and Wittman-Liebold, 1975; (42) Wittmann-Liebold *et al.*, 1975b; (43) Wittmann-Liebold and Pannenbecker, 1976; (44) Chen and Ehrke, 1976b.

1   10   20
Pro-Val - Ile - Lys-Val- Arg-Glu -Asn-Glu -Pro-Phe-Asp-Val -Ala -Leu-Arg-Arg-Phe-Lys-Arg-

                                     30   40
Ser -Cys-Glu -Lys-Ala -Gly -Val -Leu-Ala -Glu -Val -Arg-Arg-Arg-Glu -Phe-Tyr-Glu -Lys-Pro-

                                     50   60
Thr-Thr-Glu -Arg-Lys-Arg-Ala - Lys-Ala -Ser -Ala -Val -Lys-Arg-His -Ala -Lys-Lys-Leu-Ala -

                                     70
Arg-Glu -Asn-Ala -Arg-Arg-Thr-Arg-Leu-Tyr

Fig. 8   Primary structure of protein S21.

            1   10   20
(Ac). Ser-Ile-Thr-Lys-Asp-Gln-Ile-Ile-Glu-Ala-Val-Ala-Ala-Met-Ser-Val-Met-Asp-Val-Val-

                                         30   40
       Glu-Leu-Ile-Ser-Ala-Met-Glu-Glu-Lys-Phe-Gly-Val-Ser-Ala-Ala-Ala-Ala-Val-Ala-Val-

                                         50   60
       Ala-Ala-Gly-Pro-Val-Glu-Ala-Ala-Glu-Glu-Lys-Thr-Glu-Phe-Asp-Val-Ile-Leu-Lys-Ala-

                                         70   80
       Ala-Gly-Ala-Asn-Lys-Val-Ala-Val-Ile-Lys-Ala-Val-Arg-Gly-Ala-Thr-Gly-Leu-Gly-Leu-

                                         90   100
       Lys*-Glu-Ala-Lys-Asp-Leu-Val-Glu-Ser-Ala-Pro-Ala-Ala-Leu-Lys-Glu-Gly-Val-Ser-Lys-

                                         110   120
       Asp-Asp-Ala-Glu-Ala-Leu-Lys-Lys-Ala-Leu-Glu-Glu-Ala-Gly-Ala-Glu-Val-Glu-Val-Lys-

Fig. 9   Primary structure of protein L7/L12. Lys,* ϵ-N-monomethyllysine (50%).

            1   10   20
NMM-Leu-Gln-Pro-Lys-Arg-Thr-Lys-Phe-Arg-Lys-Met-His-Lys-Gly-Arg-Asn-Arg-Gly-Leu-

                                         30   40
Ala-Gln-Gly-Thr-Asp-Val-Ser-Phe-Gly-Ser-Phe-Gly-Leu-Lys-Ala-Val-Gly-Arg-Gly-Arg-

                                         50   60
Leu-Thr-Ala-Arg-Gln-Ile-Glu-Ala-Ala-Arg-Arg-Ala-Met-Thr-Arg-Ala-Val-Lys-Arg-Gln-

                                         70   80
Gly-Lys-Ile-Trp-Ile-Arg-Val-Phe-Pro-Asp-Lys-Pro-Ile-Thr-Glu-Lys-Pro-Leu-Ala-Val-

                                         90   100
Arg*-Met-Gly-Lys-Gly-Lys-Gly-Asn-Val-Glu-Tyr-Trp-Val-Ala-Leu-Ile-Gln-Pro-Gly-Lys-

                                         110   120
Val-Leu-Tyr-Glu-Met-Asp-Gly-Val-Pro-Glu-Glu-Leu-Ala-Arg-Glu-Ala-Phe-Lys-Leu-Ala-

                                         130   136
Ala-Ala-Lys-Leu-Pro-Ile-Lys-Thr-Thr-Phe-Val-Thr-Lys-Thr-Val-Met

Fig. 10   Primary structure of protein L16. NMN, N-methylmethionine; Arg,* an unknown
amino acid related to arginine.

```
1                              10                              20
Met-Asp-Lys-Lys-Ser -Ala -Arg - Ile -Arg-Arg-Ala -Thr-Arg-Ala -Arg-Arg-Lys-Leu-Gln -Glu -

                               30                              40
Leu-Gly -Ala -Thr-Arg-Leu-Val -Val -His -Arg-Thr-Pro-Arg-His - Ile - Tyr-Ala -Gln -Val - Ile -

                               50                              60
Ala -Pro-Asn-Gly -Ser -Glu -Val -Leu-Val -Ala -Ala -Ser -Thr-Val -Glu -Lys-Ala - Ile - Ala -Glu -

                               70                              80
Gln - Leu-Lys-Tyr-Thr-Gly -Asn-Lys-Asp-Ala -Ala -Ala -Val -Gly -Lys-Ala -Val -Ala -Glu -

                               90                              100
Arg-Ala -Leu-Glu -Lys-Gly - Ile - Lys-Asp-Val -Ser -Phe-Asp-Arg-Ser -Gly -Phe-Gln -Tyr-His -

                               110
Gly -Arg-Val -Gln -Ala -Leu-Ala -Asp-Ala -Ala -Arg-Glu -Ala -Gly -Leu-Gln -Phe
```

Fig. 11   Primary structure of protein L18.

```
1                              10                              20
Met-Phe-Thr - Ile -Asn-Ala -Glu -Val -Arg-Lys-Glu -Gln -Gly -Lys-Gly -Ala -Ser -Arg-Arg-Leu-

                               30                              40
Arg-Ala -Ala -Asn-Lys-Phe-Pro-Ala - Ile - Ile - Tyr-Gly -Gly -Lys-Glu -Ala -Pro-Leu-Ala - Ile -

                               50                              60
Glu -Leu-Asp-His- Asp-Lys-Val -Met-Asn-Met-Gln -Ala -Lys-Ala -Glu -Phe-Tyr-Ser -Glu -Val -

                               70                              80
Leu-Thr - Ile -Val -Val -Asp-Gly -Lys-Glu - Ile - Lys-Val -Lys-Ala -Gln -Asp-Val -Gln -Arg-His -

                               90
Pro-Tyr-Lys-Pro-Lys-Leu-Gln -His -Ile - Asp-Phe-Val -Arg-Ala
```

Fig. 12   Primary structure of protein L25.

```
1                              10                              20
Ala -His -Lys-Lys-Ala -Gly -Gly -Ser -Thr-Arg-Asn-Gly -Arg-Asp-Ser -Glu -Ala -Lys-Arg-Leu-

                               30                              40
Gly -Val -Lys-Arg-Phe-Gly -Gly -Glu -Ser -Val -Leu-Ala -Gly - Ser -Ile - Ile -Val -Arg-Gln -Arg-

                               50                              60
Gly -Thr-Lys-Phe-His -Ala -Gly -Ala -Asn-Val -Gly -Cys-Gly -Arg-Asp-His -Thr-Leu-Phe-Ala -

                               70                              80
Lys-Ala -Asp-Gly- Lys-Val -Lys-Phe-Glu -Val -Lys-Gly -Pro-Lys-Asn-Arg-Lys-Phe - Ile - Ser-

Ile - Glu -Ala -Glu
```

Fig. 13   Primary structure of protein L27.

```
1                              10                              20
Met-Lys-Ala -Lys-Glu -Leu-Arg-Glu -Lys-Ser -Val -Glu -Glu -Leu-Asn-Thr-Glu -Leu-Leu-Asn-

                               30                              40
Leu-Leu-Arg-Glu -Gln -Phe-Asn-Leu-Arg-Met-Gln -Ala -Ala -Ser -Gly -Gln -Leu-Gln -Gln -Ser -

                               50                              60
His -Leu-Leu-Lys-Leu-Gln -Leu-Asn-Thr-Lys-Gln -Val -Arg-Arg-Asp-Val -Ala -Arg-Val -Lys-

Ala -Gly -Ala
```

Fig. 14   Primary structure of protein L29.

1   10   20
Ala -Lys-Thr - Ile - Lys - Ile - Thr-Gln - Thr-Arg-Ser - Ala - Ile - Gly -Arg-Leu-Pro-Lys-His - Lys-

                                          30   40
Ala - Thr-Leu-Leu-Gly - Leu-Gly - Leu-Arg-Arg - Ile - Gly -His -Thr-Val -Glu - Arg-Glu -Asp -Thr-

                                          50
Pro-Ala - Ile - Arg-Gly -Met - Ile -Asn-Ala -Val -Ser -Phe-Met-Val -Lys-Val -Glu - Glu

Fig. 15   Primary structure of protein L30.

1   10   20
Ala -Val -Gln -Gln -Asn-Lys-Pro-Thr-Arg-Ser - Lys-Arg-Gly -Met-Arg-Arg-Ser -His -Asp-Ala -

                                          30   40
Leu-Thr-Ala -Val -Thr-Ser - Leu-Ser - Val -Asp-Lys-Thr-Ser - Gly -Glu - Lys-His - Leu-Arg-His -

                                          50
His - Ile - Thr-Ala- Asp-Gly -Tyr-Tyr-Arg-Gly -Arg-Lys-Val - Ile - Ala -Lys

Fig. 16   Primary structure of protein L32.

1   10   20
NMA-Lys-Gly- Ile - Arg-Glu - Lys - Ile - Lys-Leu-Val -Ser -Ser -Ala - Gly -Thr-Gly -His -Phe-Tyr-

                                          30   40
Thr-Thr-Thr-Lys-Asn-Lys-Arg-Thr-Lys-Pro-Glu - Lys-Leu-Glu - Leu-Lys-Lys-Phe-Asp-Pro-

                                          50   54
Val -Val -Arg-Gln -His -Val -Tyr - Ile -Lys-Glu -Ala - Lys - Ile -Lys

Fig. 17   Primary structure of protein L33. NMA, *N*-methylalanine.

1   10   20
Met-Lys-Arg-Thr-Phe-Gln -Pro-Ser -Val -Leu-Lys-Arg-Asn-Arg-Ser -His - Gly -Phe-Arg-Ala -

                                          30   40
Arg-Met-Ala -Thr-Lys-Asn-Gly -Arg-Gln -Val -Leu-Ala -Arg-Arg-Arg-Ala - Lys-Gly -Arg-Ala -

          46
Arg-Leu-Thr-Val -Ser -Lys

Fig. 18   Primary structure of protein L34.

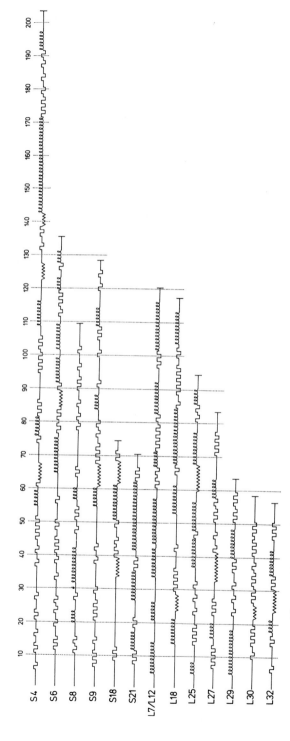

Fig. 19 Hypothetical secondary structure of ribosomal proteins. ⊔ .α-helix; w .β-structure; ⌐_ .bent.

125

TABLE II

Tetrapeptides Occurring in Three Proteins

Peptide	Protein	Positions	Protein	Positions	Protein	Positions
Lys-Ala-Ala-Val	S8	21–24	L22	42–45	L23	44–47
Gly-Lys-Val-Lys	L14	50–53	L24	22–25	L27	64–67
Ala-Lys-Phe-Val	S7	23–26	S13	60–63	L1	35–38
Ile-Arg-Glu-Lys	S4	142–145	L1	10–13	L33	4–7
Val-Glu-Lys-Ala	S17	92–95	S19	18–21	L18	54–57
Leu-Arg-Lys-Ala	S9	97–100	S9	117–120	L10	40–43

vious immunochemical studies with antisera against individual proteins isolated from both ribosomal subunits it was concluded (Stöffler and Wittmann, 1971a,b; Stöffler, 1974) that there are no *extensive* sequence similarities among ribosomal proteins with the exception of two pairs of proteins, namely, L7/L12 and S20/L26.

In a recent study, Wittmann-Liebold and Dzionara (1976a) examined the *E. coli* ribosomal proteins of known sequences for identical regions. They found identical regions consisting of three to six amino acids to be present within these proteins. However, no identical regions longer than six amino acids occur in the investigated proteins, except in the pairs L7/L12 and S20/L26.

Among the proteins used for the comparison there are more than 600 *tripeptides*, which occur in at least two proteins. In some proteins the identical tripeptides are located at the same amino acid positions of the protein chains. The search for identical *tetrapeptides* among the proteins gave the following results: Each of 97 tetrapeptides was found in at least two proteins and some of the peptides were present in three proteins (Table II).

TABLE III

Occurrence of Pentapeptides

Peptide	Protein	$p/10^{-3}$	Protein	$p/10^{-3}$
Asp-Asp-Ala-Glu-Ala	S6	1.2	L7/L12	2.2
Thr-Val-Lys-Gly-Gly	S5	1.1	S9	0.3
Leu-Gly-Leu-Arg-Arg	S15	1.3	L30	0.3
Ala-Ala-Ala-Ala-Val	L7/L12	40.0	L18	9.4
Ala-Val-Ile-Lys-Ala	L7/L12	4.8	L20	1.2
Val-Ile-Arg-Glu-Lys	S4	0.5	L1	0.1
Lys-Ser-Val-Glu-Glu	S5	0.4	L29	0.2
Glu-Leu-Arg-Lys-Ala	S9	0.6	L10	0.4
Val-Glu-Lys-Ala-Val	S17	0.7	S19	0.3
(Asp)-Gly-Lys-Val-Lys	L14	0.3	L27	0.6

p, probability to find the peptide in any position of the protein.

Each of ten *pentapeptides* was found in two proteins as listed in Table III which also gives the probabilities for the occurrence of the pentapeptides on a random basis. The largest identical region found in two proteins is the *hexapeptide* Val-Val-Ala-Asp-Ser-Arg present in proteins L10 and S16.

As mentioned above, identical regions up to a size of six amino acids were found in pairs of proteins. In some of these protein pairs identical or similar regions occur at several points. The most pronounced example is the pattern of similar regions among protein S6 and L7/L12 (Hitz *et al.*, 1975). Additional, although weaker, patterns of similarity are shown in Figs. 20 and 21. These patterns can in most cases only be arranged if one "shortens" or "stretches" the intervening nonhomologous regions.

The occurrence of numerous short identical regions within the *E. coli* ribosomal proteins poses the question whether a similar number of iden-

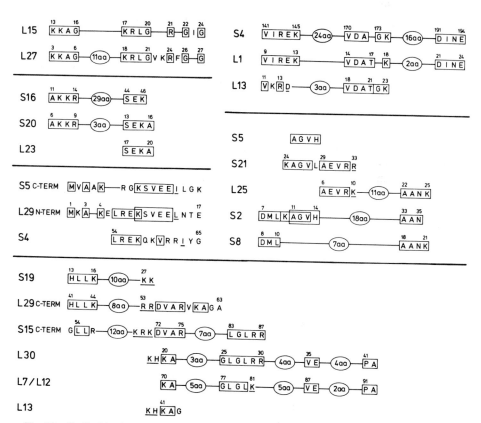

Fig. 20 Similarities in the structures of *E. coli* ribosomal proteins. The amino acids are abbreviated according to the recommendation of the IUPAC-IUB Commission on Biochemical Nomenclature. Structurally related amino acids in corresponding positions are underlined, e.g., I (Ile) and L (Leu) or K (Lys), R (Arg) and H (His) or D (Asp) and E (Glu).

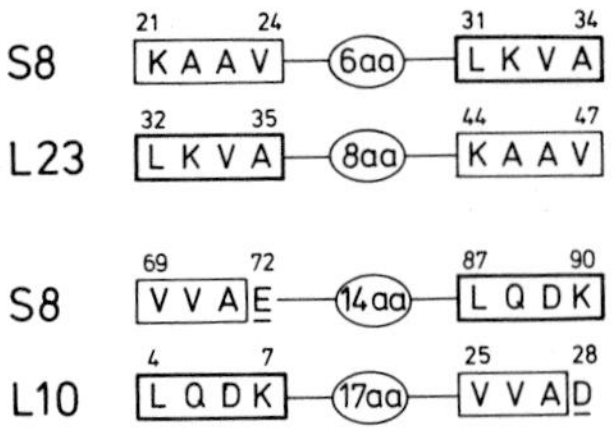

Fig. 21 Repetition of *two* identical peptides in the sequences of protein-pairs (S8-L23; S8-L10). It should be noted that the identical peptides occur in reverse order along the protein chains of each pair.

tical regions can also be found among artificial proteins which have the same amino acid compositions and lengths as the ribosomal proteins but whose amino acid sequences are randomly generated by a computer program. A comparison led to the result that the frequency pattern of the occurrence of identical peptides is similar for the native ribosomal proteins and their artificially generated isomers (Wittmann-Liebold and Dzionara, 1976b).

Considering the various factors (minimal size, amino acid composition, spatial structure, individual specificity, etc.) that determine whether a protein region is recognized as an antigenic determinant (for a review, see Arnon, 1971), it can be expected that the degree of sequence homology found by sequence analysis would lead to only low, if any, cross-reaction among most ribosomal proteins. This is in agreement with the results from immunochemical studies (Stöffler and Wittmann, 1971a,b; Stöffler, 1974).

B. *Escherichia coli* Mutants

In a recent review (Wittmann and Wittmann-Liebold, 1974) mutants with altered ribosomal proteins were described and the amino acid replacements in the altered proteins were given. Up to then, mutations in nine 30 S proteins (S4, S5, S6, S7, S8, S12, S17, S19, S20) and in two 50 S proteins (L4 and L22) were known. In a few of these proteins, e.g., S4, S5, and S12, alterations occur which are caused by mutations belonging to different types. For instance, mutations leading to spectinomycin resistance, to suppression of streptomycin dependence, or to suppression of a defect in the gene of the alanyl-tRNA synthetase all cause alterations in protein S5 although at different positions in the protein chain.

In the last two years a number of new mutants with altered proteins were isolated and analyzed which have not been described in the review by Wittmann and Wittmann-Liebold (1974). These mutations lead to alterations in the following proteins of both the small and large ribosomal subunit.*

1. Protein S2

Mutants resistant to kasugamycin are altered in protein S2 as shown by two-dimensional gel electrophoresis (Okuyama *et al.*, 1974; Yoshikawa *et al.*, 1975). This mutation (*ksgC*) is located near the *purE* locus at 12 minutes of the *E. coli* chromosomal map and its site is distinct from the locations for all other genes for ribosomal proteins so far known. In other kasugamycin-resistant mutants the 16 S RNA (and not a ribosomal protein) is altered (Helser *et al.*, 1971, 1972): two adjacent adenines in the 16 S RNA which are methylated in the wild-type are not methylated in the mutant. This is caused by a defect in a specific methylase in the mutant.

2. Protein S4

Protein S4 is affected by mutations belonging to different groups. Among the S4 mutants are those suppressing streptomycin dependence (see review by Wittmann and Wittmann-Liebold, 1974) and the so-called *ram* (ribosomal ambiguity) mutants (see review by Gorini, 1974). Although both classes were isolated by rather different isolation procedures, it can be concluded from genetic studies (Gorini, 1974) and from their protein structure (van Acken, 1975) that both types of mutants belong to the same group. In both classes amino acid replacements occur in the same region (positions 49 and 53) of the S4 protein chain (van Acken, 1975; M. Yaguchi, unpublished).

3. Protein S5

As summarized by Wittmann and Wittmann-Liebold (1974) several different types of mutants have alterations in protein S5. Since this recent review the following results were obtained:

1. Genetic and biochemical studies showed that mutations in ribosomal protein S5 can act as *ram* mutations in a very similar way to mutations in protein S4. The region of the S5 protein chain involved in control of translational fidelity is different from those regions which are altered by other mutations in protein S5, e.g., spectinomycin resistance or suppression of a defect in the alanyl-tRNA synthetase (Piepersberg *et al.*, 1975b). Similar results were recently obtained by Cabezón *et al.* (1976).

2. In mutants in which dependence on streptomycin is suppressed,

---

* *Note added in proof:* Alterations in 15 proteins from the small and 18 proteins from the large subunit have recently been observed in temperature-sensitive mutants and in revertants from a strain with a novel type of streptomycin dependence (Isono *et al.*, 1976; Dabbs and Wittmann, 1976).

alterations either in protein S4 or in protein S5 were found (Hasenbank *et al.*, 1973). The amino acid replacements in two mutants with altered S5 proteins were determined (Itoh and Wittmann, 1973; Piepersberg *et al.*, 1975b). Although both mutants were isolated from different parental strains in different years, the same amino acid replacement, namely, arginine by leucine in peptide T2, occurred at the same position of the S5 protein chain of both mutants.

3. Mutants highly resistant to the aminoglycoside neamine were found to be altered in two ribosomal proteins, namely, S5 and S12 (De Wilde *et al.*, 1975). Both mutations together, but not each of them alone, confer resistance to neamine. These mutants are structurally and functionally very similar to those in which dependence on streptomycin is suppressed (see above). In both groups of mutants the same ribosomal proteins (S5 and S12) are altered and even at identical amino acid positions, although the kinds of amino acid replacement differ in these mutants (Yaguchi *et al.*, 1976). However, not all mutants in which dependence on streptomycin is suppressed are resistant to neamine. It would appear that only particular combinations of the altered ribosomal proteins S5 and S12 can lead to the expression of the neamine-resistance phenotype.

4. Mutants in which a defect in the alanyl-tRNA synthetase is suppressed are altered in protein S5 or in protein S20 (Wittmann *et al.*, 1974). The mutant S5 protein was analyzed and it was found that it differs from the wild-type S5 protein in that it lacks five amino acids at the C-terminus probably caused by a frameshift mutation (Piepersberg *et al.*, 1975b).

5. The combination of results from genetic and protein-chemical studies on mutants with altered S5 proteins allows conclusions about the direction in which the cistron for protein S5 is transcribed. This is possible since the positions of the amino acid replacements in the S5 proteins of the various mutants are known and since the sites at which the mutations occur are localized on the chromosomal map. From these data it is clear that the cistron for protein S5 is transcribed in the counterclockwise direction on the chromosome (Piepersberg *et al.*, 1975b; Wittmann *et al.*, 1975b). This result is in good agreement with conclusions drawn from completely different genetic experiments (Jaskunas *et al.*, 1975).

4. Protein S6

In a mutant which was isolated as resistant to neo- and kanamycin, protein S6 was found to be altered. Transduction studies showed, however, that resistance to neo- and kanamycin is not caused by the altered protein S6 (H. G. Wittmann and D. Apirion, unpublished). The wild-type S6 protein has several consecutive glutamic acid residues at its C-terminus whereas there are only two of them present in the mutant S6 protein (Hitz *et al.*, 1975).

5. Protein S7

Protein S7 differs markedly in size and charge among various *E. coli* strains as reviewed by Wittmann and Wittmann-Liebold (1974). The sequences of proteins S7 from strains B and K are presently under investigation (J. Reinbolt, unpublished).

6. Protein S8

Altered S8 proteins were found in two groups of mutants in *sts* (starvation temperature sensitive) mutants (Zubke *et al.*, 1977) and in revertants of a temperature-sensitive valyl-tRNA synthetase mutant (Wittmann *et al.*, 1975a). Genetic experiments showed that in the first group of mutants the alteration in protein S8 can be separated from the *sts* phenotype whereas in the second group the mutation in protein S8 is strictly correlated with suppression of temperature sensitivity. The elucidation of the complete primary structure of the wild-type protein S8 (Stadler, 1974) was the basis for determination of the amino acid replacements in the mutants with altered S8 proteins (Zubke *et al.*, 1977).

7. Protein S12

As previously reviewed (Wittmann and Wittmann-Liebold, 1974) amino acid replacements in protein S12 lead to two groups of mutants, namely, those resistant to streptomycin and those dependent on this antibiotic. In other mutants the streptomycin requirements can be satisfied by paromomycin or by ethanol (see review by Gorini, 1974). Furthermore, mutants highly resistant to neamine are altered in two proteins, one of which is S12 (and the other S5) as described above (see protein S5).

Proteins S12 from 16 mutants belonging to these various mutant groups were isolated and analyzed (Funatsu and Wittmann, 1972; Itoh and Wittmann, 1973; van Acken, 1975; Yaguchi *et al.*, 1976; E. Dabbs and M. Yaguchi, unpublished). It is remarkable that the amino acid replacements in all mutants are located in two very small regions of the S12 protein chain, namely, around position 42 and position 87. Apparently these two regions are very important: single amino acid replacements within these regions lead to an altered response (resistance or dependence instead of sensitivity) of the ribosome toward streptomycin (or paromomycin and/or ethanol).

8. Protein S17

Alterations in protein S17 were found in the strain AT 2472 (Muto *et al.*, 1975) and in mutants with low resistance to neamine (Bollen *et al.*, 1975; Yaguchi *et al.*, 1976). In contrast to the highly resistant neamine mutants in which alterations in two ribosomal proteins, namely, S5 and S12, are necessary for the resistant phenotype (see above), a single amino acid replacement in protein S17 (histidine by proline in position 30) suffices for expression of the low level resistance to this antibiotic. The alteration in protein S17 has a similar effect in enhancing translational fidelity *in vitro* and *in vivo* as alterations in protein S12 [for reviews, see Gorini (1974), as well as Wittmann and Wittmann-Liebold (1974)]. These findings add S17 as a new member to the list of proteins that have an influence on the codon-anticodon interaction.

9. Protein S18

A mutation in the gene for protein S18 maps at 84 minutes (De Wilde *et al.*, 1974), i.e., outside of the main cluster of ribosomal genes around 64 minutes. This finding was the first evidence for the existence of genes for ribosomal proteins outside the "streptomycin region." By the determination of the complete primary structure of protein S18 (Yaguchi, 1975), it was possible to localize the amino acid replacement in a mutant S18 protein (arginine by cysteine; Kahan *et al.*, 1973) to the N-terminal region at position 11 (Yaguchi, 1975).

10. Protein S20

Alterations in protein S20 (as well as in S5 or S8; see above) are especially interesting since they can suppress mutations in nonribosomal components, namely, aminoacyl-tRNA synthetases. This was shown for alanyl-tRNA synthetase (Böck *et al.*, 1974; Wittmann *et al.*, 1975a). In several of these suppressor mutants an alteration in the S20 protein could be detected by different immunological and electrophoretic methods.

The mutations leading to altered S20 proteins are localized in the *thr-leu* region of the chromosomal map (Böck *et al.*, 1974). In order to determine whether the alteration in S20 is due to a change in the primary structure or to a secondary modification of the S20 protein by an enzyme, sequence analyses were done on the mutant S20 protein (Robinson and Wittmann-Liebold, 1977). These studies and independent genetic experiments with transducing λ phages (Friesen *et al.*, 1976) showed that the structural gene for protein S20 is located in the *thr-leu* region which is a new chromosomal location for a ribosomal protein.

11. Proteins L4 and L22

Proteins L4 and L22 of the 50 S ribosomal subunit were found to be altered in mutants resistant to erythromycin (see review by Wittmann and Wittmann-Liebold, 1974). No new results on these or other mutants with alterations in one or the other of the two proteins were published.

12. *Protein L5*

Protein L5 has been reported by Liou *et al.* (1975) to be altered in a mutant resistant to the antibiotic thiopeptin which has an effect on EF-G and EF-Tu mediated ribosomal functions similar to the more well-known antibiotics siomycin and thiostrepton. No data on the nature of the protein alteration are available.

13. *Protein L11*

Mutation from stringent to relaxed *E. coli* strains is accompanied by an alteration in protein L11 as shown by two-dimensional gel electrophoresis (Parker *et al.*, 1976). Reversion to stringency is correlated with a change to a protein L11 with normal mobility. It is very likely that the gene for protein L11 which maps in the *rif* region is identical with *relC*.

14. *Protein L18*

In an attempt to isolate mutants with alterations in ribosomal proteins, P1 phages grown in a streptomycin-resistant strain were mutagenized and used to transduce into a strain auxotrophic for *aroE*. Among the transductants with streptomycin resistance and *aroE* prototrophy, three were found in which the 50 S protein L18 was altered as shown by two-dimensional gel electrophoresis and by immunological methods (Berger *et al.*, 1975). Protein L18 is a component of the 5 S RNA protein complex possessing GTPase and ATPase activity (Gaunt-Klöpfer and Erdmann, 1975).

II. Ribosome Structure and Topography

A central step toward an understanding of the structure and function of the ribosome, a large multicomponent assembly, is the elucidation of the spatial arrangement of its proteins and rRNA. The topography of ribosomal components has been investigated by a number of elegant experimental approaches, each contributing considerable information on our present knowledge on this topic.

Among these techniques, the use of bifunctional reagents to cross-link protein pairs, has yielded the most extensive information on protein-protein neighborhoods (see Chapter 2). Another approach utilizes the study of singlet energy transfer between pairs of fluorescently labeled ribosomal proteins (Cantor *et al.*, 1974a). Chemical modification of proteins or rRNA has also been used as a probe of ribosomal topography (see Garrett and Wittmann, 1973). Several groups have attempted to identify functional sites of the ribosome by means of affinity labeling techniques

(Cantor *et al.*, 1974b; Pongs *et al.*, 1975a). Limited nuclease digestion of ribosomal subunits as well as characterization of rRNA sequences that provide binding sites for individual proteins have contributed not only very important information about the topography of proteins, but also led to some insight into the intraribosomal arrangement of rRNA (Brimacombe *et al.*, 1976). Neutron scattering of selectively deuterated ribosomes has also been exploited as a means of estimating distances separating the mass centers of proteins and RNA (Moore *et al.*, 1974; Stuhrmann *et al.*, 1976) and of ribosomal protein pairs (Hoppe *et al.*, 1975; Engelman *et al.*, 1975a,b). Each of these methods has provided data on relative protein proximity but in general without special concern for the shape of each protein.

Another technique that has recently been developed is the localization of proteins by immunoelectron microscopy. Ribosomal subunits have readily recognizable structures in the electron microscope. They also have sufficient asymmetry to provide specific structural features that can be used as landmarks to localize the binding sites of antibodies specific to individual ribosomal proteins. Thus, if ribosomal subunits are treated with a single protein-specific antibody, subunit dimers are formed by means of the bivalent IgG molecule. The Fab arms of these antibodies attach to the ribosomal subunit, and the attachment site can be taken as the point where a given ribosomal protein has an antigenic determinant exposed on the ribosomal surface.

These studies have so far enabled us to determine the location of all 21 ribosomal proteins of the 30 S subunit and 19 proteins of the large subunit. Furthermore, it was possible to investigate whether the proteins have globular or extended conformations within the ribosome, and first attempts as to the localization of particular portions of the polypeptide chain of the proteins have been successful. The data now available give for the first time an idea of the architecture of the ribosome and the "absolute" topography of ribosomal proteins.

We shall summarize the present state of these investigations and then attempt to coordinate the electron microscopy results with other data available on the topography of ribosomal proteins.

A. The Structure of the Ribosome

Two quite different approaches have been used to deduce shape and dimensions of ribosomal particles; viz., small-angle X-ray scattering and electron microscopy. The radius of gyration of particles can be determined quite unambiguously from the very low-angle region of the scattering curve. Shape and dimensions of the particle are deduced from the region at slightly higher angles. Aggregation of the particles, however, interferes with the measurements in this region. Since the deduction of the shape of a given particle from scattering curves is achieved by curve-fitting procedures, it should be remembered that similar scattering curves may fit quite different particle shapes (see van Holde and Hill, 1974).

Electron microscopy subjects hydrated particles to relatively drastic conditions during sample preparation. Ribosomal dimensions and morphology would therefore be expected to change with shrinkage. Freeze-drying is more advantageous than air-drying methods in this respect. It is conceivable that air-drying of sandwiched specimens, as used by us, might cause less damage to the highly hydrated ribosomes, since the specimen is covered by two carbon films.

1. The Structure of the 30 S Ribosomal Subunit

Escherichia coli 30 S subunits were examined by electron microscopy in several laboratories. Huxley and Zubay (1960) described flat particles with rather irregular outlines. A bipartite structure of 30 S subunits has been reported by numerous investigators (Hall and Slayter, 1959; Amelunxen, 1971; Wabl *et al.*, 1973a,b; Lake *et al.*, 1974a,b; Vasiliev, 1974; Tischendorf *et al.*, 1974b, 1975, 1976; Tischendorf and Stöffler, 1975, 1976a,b). In one of the more recent reports, 30 S subunits resembling discs which could be approximated to an oblate ellipsoid were observed (Amelunxen, 1971). The occurrence of these regular disc forms is greatly enhanced when the subunits are applied to the grids by direct ultracentrifugation (Table IV). It is conceivable that this was due to centrifugal forces causing the lobes of the body to expand (cf. below). Besides discs, rodlike structures were also found (Amelunxen, 1971). The various data obtained by electron microscopy are summarized in Table IV.

We have observed that *E. coli* 30 S subunits, when visualized by negative staining with uranyl acetate, appear in several general orientations (see Fig. 24). They are seen in two asymmetric views and two orthogonal pseudosymmetric views. 30 S subunits have an elongated, slightly bent prolate shape and are divided transversely into two parts of unequal size, a smaller "head" and a larger "body." In frontal and lateral views the lateral edges on the body protrude to different extents. The partition which resembles a hollow or a cleft is always visible in the frontal and the two lateral views. Rear views were observed in some cases with plane dorsal surfaces, others had a more or less shallow dent (cf. Fig. 27). Frontal views of 30 S subunits have more stain accumulated in the neck region than rear views. This might be explained as a particular property of the double layer–sandwich method which has been used in all our experiments (Tischendorf

TABLE IV

Shapes and Dimensions of *E. coli* 30 S Ribosomal Subunits[a]

Method	Reference	Procedures	Mg²⁺ (mM)	Heat activation	Shape	Dimensions (Å)
Electron microscopy	Hall and Slayter (1959)	AD, SC	0.2	No	PE	95 × 95 × 170
	Huxley and Zubay (1960)	AD, NS, PS	1.0, 5.0, 7.0	No	OE[b]	70 × 180 × 180
	Spirin et al. (1963)	AD, NS	1.0	No	TE	80 × 120 × 140
	Wabl et al. (1973a,b)	AD, NS	1.0	Heated[c] (37°, 10 min)	Irreg. (TE)	? × ? × 220
	Amelunxen (1971)	Susp. prep., NS, SC Sed. prep.	0.25	No	TE	75 × 100 × 200
	Vasiliev (1974)	FD, NS	0.25	No	TE OAST, T, Bip.	100 × 130 × 145 105 × 110 × 195
		FD, SC	1.0	No	OAST, T, Trip., Tetrap.	120 × 130 × 230
	Tischendorf et al. (1974b)	AD, NS	0.3	Heated[c] (37°, 10 min)	Irreg. (TE)	80 × 100 × 180
	Tischendorf et al. (1975)	AD, NS	5.0	Yes	Irreg. (TE)	80 × 100 × 180–200
	Tischendorf et al. (1976)	AD, NS	0.3, 1.0, 3.0, 5.0,10.0,20.0	No Yes	Irreg. (TE) Irreg. (TE)	80 × 100 × 180–200 80 × 100 × 180–200
	Lake et al. (1974a,b)	AD, NS	1.0, 10.0	Heated[d] (40°, 5 min)	PE (curved and flattened)	Not given[e]
Low-angle X-ray scattering	Hill et al. (1969)		1.0	No	OE	55 × 220 × 220
	Smith (1971)		1.0	No	OE	56 × 224 × 224

[a] Abbreviations: NS, negative staining; AD, air-dried; FD, freeze-dried; SC, shadow cast; OAST, obtuse-angled scalene triangle; T, triangular; Bip., bipartite; Trip., tripartite; Tetrap., tetrapartite; TE, triaxial ellipsoid; PS, positive staining. Susp. prep., suspension preparation (see Amelunxen, 1971); Sed. prep., sedimentation preparation (see Amelunxen, 1971); PS, positive staining.

[b] Also seen as triangles, trapezoids, or polygons.

[c] Subunit preparations were heated but not under ionic conditions required for activation (Zamir et al., 1971).

[d] The authors used two buffers, differing in magnesium concentration; only one of them (buffer I) fulfills the requirements for heat activation (Zamir et al., 1971).

[e] "Somewhat smaller" than rat liver 40 S subunits (115 × 140 × 230 Å).

et al., 1974b, 1975, 1976; Tischendorf and Stöffler, 1975, 1976a,b). Sometimes the subunits appear to be more bent, with the "neck" region resembling a hook; these forms were more frequently observed by Lake *et al.* (1974a,b). The dimensions of 30 S subunits were measured to be approximately 200 × 100 × 80 Å.

Recently, we investigated 30 S subunits which were freeze-dried and subsequently shadowed with tungsten (Tischendorf *et al.*, 1976). The tungsten was evaporated from a source especially designed by Tesche (1975) for minimizing the structural damage caused by tungsten on the biological specimen. This electron evaporator, in combination with charge carrier traps, is particularly suitable for the preparation of thin tungsten layers of highest purity (Tesche, 1975). Shadowing of ribosomes with ultrathin tungsten layers of less than 10 Å, prevents grooves or cavities present on ribosomes or ribosomal subunits from being filled with the evaporated material. Such hollows are no longer visible when thicker tungsten layers are applied (B. Tesche, G. W. Tischendorf, and G. Stöffler, unpublished experiments). Electron micrographs of 30 S subunits prepared by this procedure revealed similar structural features as those prepared by negative staining with uranylacetate (Tischendorf *et al.*, 1976). The various forms of shadow cast 30 S subunits observed in electron micrographs support the three-dimensional model as proposed by Tischendorf *et al.* (1974b, 1975, 1976; see also Figs. 27 and 28).

50 S subunits and 70 S ribosomes shadowed with tungsten with the same procedure (Tesche, 1975) gave very similar ribosome images as compared to the ones obtained by negative staining (Tischendorf *et al.*, 1976; B. Tesche, G. W. Tischendorf, and G. Stöffler, unpublished results; see below).

We recently proposed a three-dimensional model of the 30 S subunit which corresponds to most of the structures observed in electron micrographs. Figuratively, the structure of the 30 S subunit resembles an embryo or a telephone receiver (cf. Fig. 27). The lobes appear as distinct parts of the subunits, in particular when 30 S subparticles are shadowed with tungsten. We have observed that the two lobes protrude unequally and to different extents. Strong protrusion from one or both edges like "earlobes" has been described (Tischendorf *et al.*, 1974b). Vasiliev (1974) observed subunit images which are subdivided into three or four parts. These tri- or tetrapartite images may also be envisaged as projections of our three-dimensional model (cf. Fig. 27). All these features introduce a certain asymmetry into our model which is, however, not that extreme as the asymmetry of the three-dimensional models proposed by Lake and Kahan (1975) or by Vasiliev (1974).

The general shape of 30 S subunits as estimated from low-angle X-ray scattering has been assumed to be an ellipsoid of revolution with dimensions of 220 × 220 × 55 Å (Table IV; Hill *et al.*, 1969; Smith, 1971; van Holde and Hill, 1974). Hence, the shape and overall dimensions of the 30 S particle as estimated by electron microscopy and low-angle X-ray scattering differ significantly. The oblate ellipsoids described by Huxley and Zubay (1960) and the disc form found by Amelunxen (1971) agree somewhat with low-angle X-ray scattering data with regard to the particle shape. Freeze-dried 30 S subunits are 20–25% larger than air-dried ribosomes: their long axis measures 230 Å, a value close to 220 Å as measured by low-angle X-ray scattering (Vasiliev, 1974; Table IV). The shapes and the axial ratios of freeze-dried particles are, however, similar to those of air dried subunits (Table IV). The structures of freeze-dried 30 S subunits and of the bipartite 30 S structures described above are both incompatible with the X-ray scattering curves (Hill and Fessenden, 1974). Obviously the drying and staining procedures required for electron microscopic investigations can lead to some distortion of strongly hydrated organelles like ribosomes but drying of subunits should not lead to such enormous differences in the gross structure. On the other hand, it has not yet been excluded that the discrepancies in shape found with electron microscopy and by X-ray scattering might arise from side-to-side dimerization of ribosomal subunits during X-ray scattering measurements.

Van Holde and Hill (1974) assumed that the reason for the discrepancies in dimension, volume, and shape found between X-ray scattering and electron microscopy is due to the use of more or less unfolded 30 S particles for electron microscopic investigation. Several data reported in the literature demonstrated that ribosomes exist in different conformations (Zamir *et al.*, 1971; Spitnik-Elson *et al.*, 1972; Ginzburg *et al.*, 1973; Noll *et al.*, 1973; Spirin, 1974). Several investigators used, for electron microscopy, particles which did not fulfill the criteria for active 30 S subunits as proposed by Spitnik-Elson *et al.* (1972). We sought to determine whether "active" and "inactive" 30 S subunits show gross differences in their structures. Therefore, we examined 30 S subunits of *E. coli* at six different magnesium concentrations (from 0.3 to 20 m*M*) with and without heat activation. At high magnesium the partition between the two unequally sized globules resembles more frequently a cleft, whereas at low magnesium it resembles more a hollow or groove. Furthermore, the protrusions which extend like lobes from the "body" of the 30 S subunit are more frequently longer at low magnesium concentrations (G. W. Tischendorf and G. Stöffler, unpublished results). These slight changes in particle shape would not be sufficient to account for the discrepancy with the X-ray scattering data.

Interestingly, the shapes of small subunits from eukaryotic ribosomes seem to have very much the same features as *E. coli* 30 S subunits; bipartite particles have been observed by several electron microscopists (Shelton and Kuff, 1966; Amelunxen and Spiess, 1971; Nonomura *et al.*, 1971; Kiselev *et al.*, 1974; Lake *et al.*, 1974b; Lutsch *et al.*, 1974).

2. Structure of 50 S Subunits

A number of image types have been observed when 50 S subunits of *E. coli* were investigated by electron microscopy: rounded image types, kidney-shaped structures, circular profiles with a "nose," and more angular structures that resemble the profile of a maple leaf or a globule carrying a crown with three crests (Hall and Slayter, 1959; Huxley and Zubay, 1960; Hart, 1962; Bruskov and Kiselev, 1968; Lubin, 1968; Spirin and Gavrilova, 1969; Spiess, 1973; Wabl *et al.*, 1973a).

50 S ribosomal subunits visualized in our laboratory also by negative staining with uranyl acetate (Tischendorf *et al.*, 1974a, 1975, 1976; Wabl, 1974; Stöffler and Tischendorf, 1975; Tischendorf and Stöffler, 1976a,b; Stöffler, 1976) occur in two main forms: (1) Some subunits showed three protuberances rising from the top of a rounded base (maple leaf). The central protuberance is most prominent (Fig. 22). These images ("crown forms") have been observed in two different views, namely, from the front ("crown-front types") and from the rear ("crown-rear types"). Crown forms revealed a bilateral symmetry (Tischendorf *et al.*, 1974a). The long axis divides the subunit into two relatively similar mirror image halves. (2) Other 50 S subunits revealed crescent-shaped images with an asymmetric notch on their concave side; this notch is located closer to the more condensed pole (cf. Fig. 22). These "kidney forms" can also be subdivided depending on the orientation of the notch toward the left or right side (Tischendorf *et al.*, 1974a).

Rounded subunit structures and, infrequently, subunits with a "nose" (cf. Lubin, 1968) have been also found. Subunits observed on a typical field of the grid revealed crown forms (56%) and kidney-shaped structures (44%). Crown-front types (32%) and crown-rear types (24%) composed the crown forms. A vast majority of kidney forms bore the notch on the upper right side (Fig. 22).

The four image types of the 50 S subunit are interconvertible by successive rotations through an angle of 90° and can be interpreted as projectional forms of one and the same 50 S subunit structure. We recently proposed a three-dimensional model that resembles an "armchair" (Fig. 22). The larger protuberance in the middle corresponds to the "seat back" and confines, together with the lateral crests (the "arms" of the chair), the vaulted seat. Photographs of the model taken from various angles show that most of the images described by other investigators are more or less compatible with one of the projectional forms of the three-dimensional

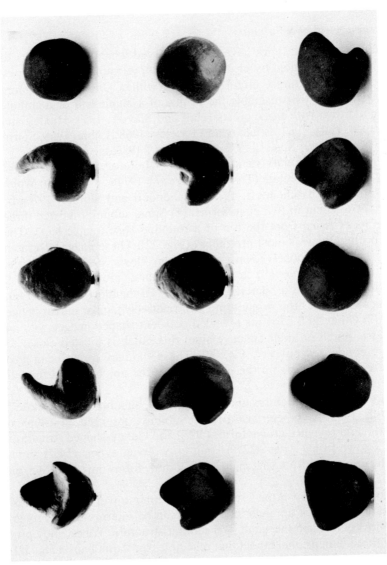

Fig. 22 Plasticine model of the 50 S subunit. The model has been photographed from various angles. The typical crown, kidney, and rounded forms are shown in the first row. Different projections, corresponding to 50 S images, as observed by other investigators (cf. Lubin's "noses") and images which were infrequently encountered are shown in rows 2 and 3. (From G. W. Tischendorf, H. Zeichhardt, and G. Stöffler, unpublished results.)

subunit structure (Fig. 22). The dimensions of 50 S subunits are $160 \times 200 \times 230$ Å.

We have examined *E. coli* 50 S subunits prepared at six magnesium concentrations (0.3 to 20 mM) and did not observe significant changes in either shape or size. Only at 0.3 mM Mg^{2+} were several subunits observed with strongly protruded lateral crests (for example, see Fig. 1a of Tischendorf *et al.*, 1974a). In addition some 50 S subunits showed a thin, rodlike projection which originated from the center of the seat region like a small stick. This projection was seen *in addition* to the three main protuberances and was scarcely observed when 50 S subunits were prepared at Mg^{2+} concentrations higher than 3.0 mM (see Tischendorf *et al.*, 1974a, 1976; G. W. Tischendorf and G. Stöffler, unpublished results). Heat activation of 50 S subunits according to Zamir *et al.* (1974) had no effect on the image forms and their observed frequency.

In contrast to the 30 S subunit data, there is gratifying agreement between X-ray scattering data and electron microscopy (see van Holde and Hill, 1974). According to Hill *et al.* (1969) the 50 S subunit is an oblate ellipsoid with dimensions of $115 \times 230 \times 230$ Å. Tolbert (1971) envisaged the 50 S particle as a triaxial ellipsoid; the scattering curves would, however, be also compatible with an oblate ellipsoid having dimensions of $113 \times 225 \times 225$ Å. The dimensions of an oblate ellipsoid are in best agreement with the 50 S structures seen by Hart (1962). In addition, Tolbert's data could be fitted to an irregular structure such as that described by Lubin (1968). The 50 S images described by Lubin also can be well explained by the three-dimensional model proposed by Tischendorf *et al.* (1974a, 1975; cf. Fig. 22). It should be mentioned that the volume of the 50 S particle calculated by Hart (1962) for both air-dried and freeze-dried specimens is significantly higher than the volume estimated from X-ray scattering (see van Holde and Hill, 1974). The best agreement between the particle volumes estimated from X-ray scattering and electron microscopy data was found by Spiess (1973) using the critical point method.

3. The Structure of 70 S Monomeric Ribosomes

For an understanding of the structure and function of the ribosome it is essential to know how the two subunits interact in 70 S monomeric ribosomes. Electron microscopy of negatively stained and tungsten-shadowed 70 S ribosomes showed that the structure and dimensions of neither subunit are grossly altered in 70 S particles as compared to isolated subunits (Hall and Slayter, 1959; Huxley and Zubay, 1960; Spirin *et al.*, 1963; Vasiliev, 1971; Lake *et al.*, 1974b; Tischendorf *et al.*, 1975, 1976). The long axis of the small subunit is within the 70 S ribosome, oriented horizontally, transverse to the central protuberance of the 50 S subunit

Fig. 23   Plasticine model of the 70 S monomeric ribosome of *E. coli.*

(Fig. 23). The hollow of the 30 S subunit faces the vaulted seat and the large protuberance of the 50 S ribosome: hollow and seat together form a tunnel. The "head" of the small subunit is in the vicinity of the right side of the large subunit. The contacting surface between the two subunits is not continuous. The hollow of the small subunit and the central protuberance of the 50 S particle separate the contacting surface into four areas (Fig. 23).

A three-dimensional model of the orientation of the two subunits within 70 S monomeric ribosomes was proposed recently (Tischendorf *et al.*, 1975; Fig. 23). The interpretation of the orientation of subunits was greatly facilitated by investigation of dimers of 70 S ribosomes which were joined by bivalent IgG molecules (Tischendorf *et al.*, 1976; see below). Our results are in conflict with a model structure suggested by Lake *et al.* (1974b). These authors interpret their electron micrographs of 70 S *E. coli* and 80 S rat liver ribosomes such that the longitudinal axis of the 30 S subunit lies horizontally across the top of the large subunit, with an orientation of 90° to that which we propose. Several electron micrographs of 70 S ribosomes published by these authors showed, at least according to our interpretation, a subunit arrangement similar to the one proposed by Tischendorf *et al.* (1975, 1976). However, some electron micrographs of 70 S ribosomes, published by Lake *et al.* (1974b) and by Tischendorf *et al.* (1975, 1976) can be explained in terms of a subunit orientation as proposed by Lake *et al.* (1974b). It could be that some subunits were dislocated during specimen preparation. 70 S ribosomes with displaced subunits have, in fact, been observed (Vasiliev, 1971; Tischendorf *et al.*, 1976; G. W. Tischendorf and G. Stöffler, unpublished data). Whether this variability of the subunit arrangement in electron micrographs of 70 S ribosomes is a corollary of interpretational differences or a consequence of preparational instability of 70 S ribosomes remains to be solved. Electron microscopic investigation of 70 S ribosomes fixed with glutaraldehyde and other bifunctional cross-linking reagents are now in progress.

Summation of the volumes of the 30 S and 50 S particle according to X-ray diffraction data gives a volume which is smaller than the one determined experimentally for the 70 S ribosome (Hill *et al.*, 1969; Smith, 1971; Tolbert, 1971). According to low-angle X-ray diffraction data, 70 S ribosomes resemble triaxial ellipsoids with dimensions of $135 \times 200 \times 400$ Å (Hill *et al.*, 1969). The length of the long axis of 70 S ribosomes was 400 Å and is probably too large; this may be explained by some end to end dimerization (van Holde and Hill, 1974). From electron microscopy it is apparent that there exists a cavity or a tunnel between the 30 S and 50 S subunit. Macromolecules possessing a central cavity can be deduced from low-angle X-ray scattering curves (see Kratky, 1971). We do not know whether such a model with a central cavity has been tested to fit the X-ray curves obtained with 70 S ribosomes.

A 70 S particle volume of 5.1×10^6 Å^3 has been calculated from electron micrographs by assuming a triaxial ellipsoid with dimensions of $170 \times 230 \times 250$ Å (Vasiliev, 1971). This volume is only slightly larger than the one derived by adding the volumes determined for the two isolated subunits. Since Vasiliev described a tunnel between the two subunits and the volume of this tunnel has not been subtracted, the agreement between the experimental and the calculated values becomes even better.

B. Protein Topography in Ribosomal Subunits

1. Each Ribosomal Protein Has Accessible
  Antigenic Determinants on the Ribosomal Surface

The availability of specific antibodies to each of the 55 ribosomal proteins from *E. coli* was a prerequisite for these studies. Apart from antisera to two protein pairs (L7 and L12 as well as S20 and L26) no immunological cross-reaction has been found between the 55 different antisera. Each antiserum was shown to react exclusively with its cognate antigen protein and with none of the other proteins. It has been demonstrated by five independent methods that each of the 21 proteins of the 30 S ribosomal subunit has antigenic determinants which are exposed on the ribosomal surface and able to interact with specific antibodies (IgG's) and with monovalent antibody fragments (Fab's). Moreover, subunit dimers joined by a bivalent antibody molecule were stable during sucrose gradient centrifugation. Thus, subunit~IgG~subunit complexes could be purified on a preparative scale and were simultaneously freed from unbound antibody molecules as well as unreacted ribosomal subunits.

Similarly, each of the antisera specific to the 34 proteins of the 50 S subunit reacted with intact large subunits from *E. coli* (Morrison *et al.*, 1977; Zeichhardt, 1976; H. Zeichhardt, G. W. Tischendorf, and G. Stöffler, unpublished results). When antisera specific to individual ribosomal proteins were assessed by immunodiffusion with intact ribosomes or ribosomal subunits, distinct precipitin lines were found with each of the 55 antisera (Stöffler *et al.*, 1973; G. Stöffler and R. Hasenbank, unpublished results). For the precipitation of an antigen the reaction of two or three determinants is required, and therefore the occurrence of precipitation as well as occurrence of aggregates larger than dimers during ultracentrifugation gave a first hint that more than one determinant should be exposed for several individual proteins. No deductions about the separation of these determinants on the ribosomal surface could, however, be drawn from these data.

2. The Topography of Ribosomal Proteins on the Surface of
  the Escherichia coli 30 S Subunit Determined by Means
  of Electron Microscopy

During the last two years we have localized all 21 ribosomal proteins on the subunit surface by immunoelectron microscopy (Tischendorf *et al.*, 1974b, 1975, 1976; Stöffler and Tischendorf, 1975; Stöffler *et al.*, 1974; Tischendorf and Stöffler, 1975, 1976a,b; Stöffler, 1976). Using the same approach proteins S4, S5, S11, S13, S14, and S19 have been localized independently (Lake *et al.*, 1974a; Lake and Kahan, 1975).

In these experiments ribosomal dimers were formed with antibodies specific to single ribosomal proteins; these subunit-antibody complexes were isolated and studied by electron microscopy. The bound antibody was visualized by negative staining and the attachment points of the Fab arms on the ribosomal surface indicate the location of the particular antigen protein. 30 S ribosomes show sufficient asymmetry to allow an accurate localization of the various antibody attachment sites. The absolute orientation of subunits observed in the pseudosymmetric front and rear views is discernible; therefore, any point on electron micrographs of 30 S subunits can be correlated to one point on the three-dimensional model. Typical examples of electron micrographs are shown in Figs. 24–26. The attachment points of the various antibodies are localized; this requires a subjective evaluation of the significance of the data. A rough localization is already possible from the few subunit dimers shown. The refined mapping of a given antibody binding site as shown in Figs. 27 and 28 required the evaluation of 50–400 enlarged dimers which have been statistically evaluated.

The results obtained with the 21 antibody preparations can be divided into two classes and are summarized in a three-dimensional model (Fig. 27) and schematically in Fig. 28: (1) Antibodies specific to ribosomal proteins S1, S8, S13, S14, S19, S20, and S21 bound to a single site on the ribosomal surface (Fig. 27). (2) Antibodies specific to the remaining 14 proteins attached at two, three, or four sites (Figs. 27 and 28). The linear distance between the various antibody binding sites of a particular protein exceeds the dimensions which would be compatible with a globular protein shape. The diameters of ribosomal proteins, if globular, should lie in the range of 25–35 Å. In consequence, detection of antibody binding sites at distances greater than 30 Å should be an indication for a nonglobular protein conformation *in situ*. In practice, however, one has to consider the size of the binding sites of the Fab arms which are 36–40 × 44–50 Å (Sarma *et al.*, 1971; Pilz *et al.*, 1973). Therefore, separation of antibody binding sites by approximately 40–50 Å could still be compatible with globular protein configuration. Among the proteins that exhibited two antibody binding sites in relative close proximity are proteins S6, S9, S16, S17; slightly more remote are the two sites of S3 and S10, respectively (Fig. 27). The separation of sites A and B of each of these six proteins could with certain assumptions still be interpreted in terms of globular or slightly ellipsoid proteins. However, the distances between the sites of eight other proteins (S2, S4, S5, S7, S11, S12, S15, and S18) are too large to be compatible with globular protein shapes (Fig. 27).

A number of conclusions can be drawn from these studies. The electron microscopic investigation confirmed previous data showing that each ri-

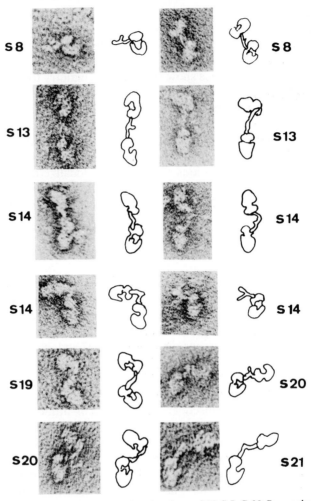

Fig. 24 Selected images and interpretive drawings of 30 S·IgG·30 S complexes obtained with anti-S8, -S13, -S14, -S19, and -S21. Antibodies bound to single subunits are shown for anti-S8 and -S14 (see also Fig. 25). ×300,000. (Taken from Tischendorf *et al.*, 1975.)

bosomal protein has antigenic sites exposed on the subunit surface (see Stöffler *et al.*, 1973; Stöffler, 1974). These antibody binding sites could now be visualized. We have recently described a total number of 37 antibody binding sites for 19 ribosomal proteins of the small subunit (Tischendorf *et al.*, 1975). Meanwhile, we have also investigated specific antibodies to the remaining proteins S1 and S17. Furthermore, we detected a third antibody binding site for protein S18 by investigating three anti-S18 sera raised in separate animals. These studies increased the total number of binding sites: the 21 proteins were found to be exposed at altogether 41 sites (Tischendorf and Stöffler, 1976a,b; Tischendorf *et al.*, 1976; Stöffler, 1976). The 41 binding sites are not uniformly distributed on the particle surface (Fig. 27). A relatively large number of sites, namely 25, have been detected on the "head" of the subunit. Only three sites were detected in the "neck" region and 13 sites on the "body."

It is noteworthy that two regions of the particle surface, the hollow and the triangle on the anterior side of the body, appear to be free of antigenic determinants and possibly of proteins. Antibody binding should be possible in either area. However, dimer formation should be possible only for antibodies attaching to the anterior surface of the body and not to the hollow. A Fab arm of an IgG molecule can easily attach to a determinant in the hollow but the other Fab arm would not jut out sufficiently in order to allow binding to the identical site on a second subunit: although antibody binding is possible, dimer formation may thus not occur. Neither subunit monomers nor Fab's bound to subunits have yet been thoroughly examined. However, in the few cases in which Fab's and/or ribosomal monomers have been investigated we have not observed antibodies that bound to the hollow region (Tischendorf and Stöffler, 1975). In view of the possibility that some areas of the subunit surface are free of proteins, the absence of antibody binding sites from the anterior surface bears more significance. Our data, however, by no means prove that RNA is exclusively located on the surface of these areas. Among other explanations it is conceivable that the exposed protein portions carry no antigenic determinant.

At first glance immunoelectron microscopy seems to allow an assessment of the protein distribution on the ribosomal surface only. Nevertheless, some conclusions can be drawn which give a rough idea of the topography of the protein chain between the antibody binding sites. Seven proteins, S2, S4, S5, S12, S15, and S18, revealed antibody binding sites on the head as well as on the body. Their protein chains must, therefore, necessarily extend through the neck, although only one of them, namely S5, has been actually found to be exposed in the neck region (Figs. 27 and 28). Indeed, impression of a more uniform protein distribution is enhanced by connecting the sites of each protein in a schematic drawing of a three-dimensional model (Stöffler and Tischendorf, 1975).

a. Are Further Proteins Elongated? It is likely that, due to the individual specificity of the immune response, additional sites will be detected for many proteins. This is illustrated by experiments with various antisera specific to proteins S4 and S18 (Tischendorf and Stöffler, 1975; G.

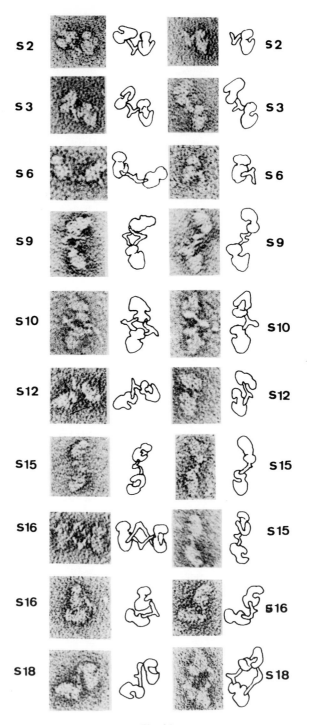

Fig. 25

Stöffler, G. W. Tischendorf, and M. Yaguchi, unpublished results). We examined the distribution of antibody binding sites as obtained with immunoglobulins and monovalent antibody fragments from six separate anti-S4 sera (Table V; Tischendorf and Stöffler, 1975). Only two of the sera contained antibodies which bound to all the four sites A, B, C, and D as shown in Fig. 28. Three other sera revealed three sites A, B, and C or D. One anti-S4 specific antiserum gave predominantly binding to sites A and B (Table V). Although binding to site C and/or D has also been observed with the latter serum, the percentage of dimers which were connected by antibodies bound to this site was only 8% (Table V), a value too low to be considered as an actual site (Tischendorf *et al.*, 1975).

Similar results were obtained with anti-S18. Antibody preparations from one serum revealed binding to only two sites (sites 18A and 18B, as described by Tischendorf *et al.*, 1975) whereas a second S18 specific serum showed that 30% of subunit dimers were connected at a third site (site 18C in Fig. 28). The systematic examination of various antisera should therefore increase the number of protein sites on the surface and eventually contribute to a more refined knowledge of the intraribosomal arrangement of individual proteins. Our data should obviously not be interpreted in such a way that a single antibody-binding site per protein implies a globular shape for that protein; it might well be that the continuation of the above studies will positively establish that more proteins have extended configuration. Several of the cross-linked protein pairs, recently characterized by Sommer and Traut (1976; see footnote *e* in Table VI) require that additional proteins with elongated shapes should be detected by immunoelectron microscopy in the near future.

b. Multiple Sites Are Not Due to Contamination or Cross-Reaction. We have been always concerned that some of the multiple sites may be due to contaminating antibodies that escaped detection during characterization of the sera by immunodiffusion. We have therefore begun to investigate specific antibodies that were purified via affinity columns to which only the cognate antigen was bound. A careful investigation of several anti-S4 sera by these techniques clearly demonstrated that the four dif-

---

Fig. 25 Electron micrographs and interpretive drawings of small subunits reacted with antibodies which bind at two separate sites. ×300,000. (i) Antibody binding sites in close vicinity to each other: S6, S9, and S16 ("double-sites"). (ii) Antibody binding sites separated by 50–60 Å: S3, S10. (iii) Antibody binding sites remotely separated (80–190 Å): S2, S12, S15, and S18. The two sites obtained with the various antibodies were encountered in ratios between 40% and 60%. Subunit pairs simultaneously connected by two antibody molecules are shown with anti-S9, -S10, -S16, and -S18 (see also Fig. 26). Antibody binding to site S18C (see Figs. 27 and 28) is not shown in this figure. (Taken from Tischendorf *et al.*, 1975.)

TABLE V

Individual-Specific Variations of the Immune Response

A. Distribution of S4-specific antibody binding sites

Source of anti-S4 immunoglobulins	Number of evaluated 30 S ~ IgG ~ 30 S complexes	Number and percentage of dimers connected at various sites						Ratio $\frac{A+B}{C+D}$
		A	B	C	D	C and/or D	Aberrant sites	
IgG R67	147	13 (9%)	16 (11%)	33 (23%)	38 (27%)	42 (30%)	5 (3%)	0.26
IgG R128	37	11 (30%)	10 (27%)			16 (43%)	0	1.3
IgG R194	48	27 (56%)	12 (25%)			9 (19%)	0	4.3
AC-S4 H4	86	29 (34%)	43 (51%)			13 (16%)	4 (5%)	5.54
AC-S4 (R478) ex anti-[S4 ~ 16 S rRNA]	120	81 (67%)	30 (25%)			9 (8%)	0	12.33
AC-S4 (H48) ex anti-30 S subunits	39	7 (23%)	4 (13%)	16 (51%)	4 (13%)		8 (20%)	0.55

B. Distribution of S18-specific antibody binding sites

Source of anti-S18 immunoglobulins	Number of evaluated 30 S ~ IgG ~ 30 S complexes	Number and percentage of dimers connected at various sites					Ratio $\frac{A}{B+C}$
		A	B	C	B + C	Aberrant sites	
IgG anti-S18 (R133)	84	55 (65%)	21 (25%)	8 (10%)	29 (34%)	0	1.90
IgG anti-S18 (R144)	80	31 (39%)	42 (52%)	7 (9%)	49 (61%)	0	0.63
AC-S18	176	82 (47%)	55 (31%)	39 (22%)	94 (53%)	0	0.87

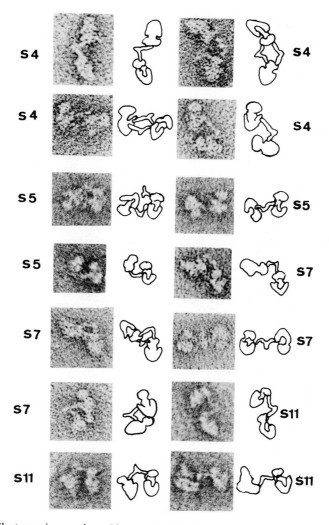

Fig. 26 Electron micrographs and interpretive drawings of 30 S subunits reacted with antibodies which attach at three or four separate sites: anti-S4, -S5, -S7, and -S11. ×300,000. The percentage of antibodies bound at each site varied. A site has only been considered when observed in at least 10% of the evaluated dimers. (Taken from Tischendorf *et al.*, 1975.)

ferent S4-specific sites can also be observed with antibodies purified by affinity chromatography (Tischendorf and Stöffler, 1975). Thus, effects of contaminating antibodies could be excluded. On the other hand, it has been shown that contaminating antibodies present in one of the antisera could be removed by this procedure (Tischendorf and Stöffler, 1975). We

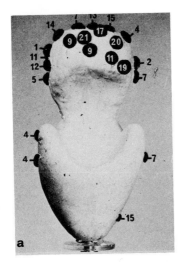

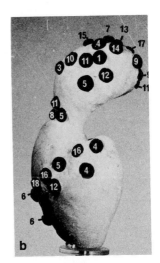

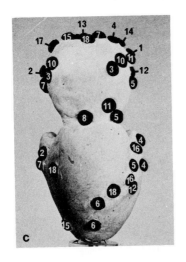

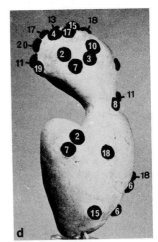

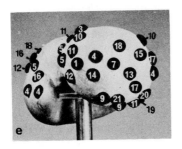

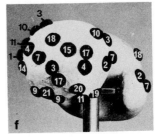

Fig. 27

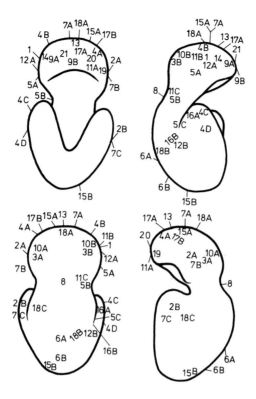

Fig. 28 Three-dimensional model of the 30 S subunit with the locations of the centers of antibody binding sites for the 21 ribosomal proteins. The four forms are related to each other by rotation through an angle of approximately 90° along the long axis. Double or multiple attachment sites of an antibody specific to one protein are designated A, B, C, D. The binding sites have to be imagined either (1) facing the observer (proteins designated within the contours of the model), (2) on the "horizon" of the subunits (proteins marked from outside). (Taken from Tischendorf *et al.,* 1975.)

more recently extended these studies to the remaining antisera; there is so far no indication that any of the 41 binding sites has originated from contamination.

These studies are also important with regard to the finding that one hexapeptide, several pentapeptides, and a rather large number of tetrapeptide

---

Fig. 27 Plasticine model of the 30 S ribosomal subunit of *E. coli*. The four forms in a,b,c and d were derived by successive rotation of the model by an angle of 90°. Panels e and f show horizontal views of the head region. The center of the antibody binding sites are indicated by numbered black circles. For better representation, the size of the pins was made smaller than the actual antibody binding sites.

sequences occur in more than one protein (Wittmann and Wittmann-Liebold, 1974; Wittmann-Liebold and Dzionara, 1976a,b). The minimal size of an antigenic determinant comprises a sequence of at least four amino acids (Sela, 1966). It is, however, obvious that the amino acid sequence determining the specificity of a determinant is in general larger in naturally occurring protein antigens (for references, see Crumpton, 1974). Nonetheless, it is conceivable that antibodies to two proteins that contain an identical pentapeptide would cross-react. This could mean that an antibody directed against this determinant would bind to either protein. This possibility is excluded by the fact that the points of attachment of a single antibody within a dimer were always identical on each subunit (Tischendorf *et al.*, 1975). Monogamous bivalent binding to one subunit has also not been observed (Tischendorf and Stöffler, 1977). In the most extreme case it might even be possible that an antibody specific to protein A may attach to the identical determinant in protein B which is exposed to the ribosomal surface and not to the inaccessible determinant in protein A. That this actually happens is already quite unlikely on a statistical basis. Sequence homologies have not only been found among the different 30 S ribosomal proteins but also among 30 S and 50 S proteins (see Section I,A and Wittmann-Liebold and Dzionara, 1976a,b). If these sequences homologies would be reflected by partial cross-reaction of the antisera, then antisera specific to individual 50 S proteins should react with 30 S subunits and vice versa. Such experiments have been made and revealed *no* cross-reaction between antisera specific to the individual proteins of one subunit with intact complementary subunits (Stöffler *et al.*, 1973; Morrison *et al.*, 1973, 1977; Zeichhardt, 1976).

    Lake and his colleagues claimed that possible cross-reactivity among the various antisera can only be eliminated by immunoelectron microscopy with hybrid ribosomes reconstituted with the antigen-homologous proteins from *B. stearothermophilus* and the remaining proteins from *E. coli*. The evanescence of this supposition is best illustrated by the fact that this control was already not applicable for two of the six antisera so far examined by Lake *et al.* (1974) and Lake and Kahan (1975). The significant cross-reaction of antisera to *E. coli* proteins S11 and S19 with their *B. stearothermophilus* counterparts did not allow these "indispensable" control experiments. Other shortcomings of this approach have already been discussed in detail (Tischendorf and Stöffler, 1975). If immunoprecipitation methods are applied, a possible cross-reactivity between antisera to two proteins which have only one (or two) determinant(s) in common might escape detection. In order to eliminate the presence of cross-reacting but not precipitating antibodies, we developed assays, utilizing antibody binding to immobilized ribosomal protein antigens (G. Stöffler, unpublished experiments). Apart from the sera to the protein pairs L7/L12 and S20/L26, the absence of any cross-reacting antibody was confirmed by this method. A careful purification and characterization of the antibodies used for immunoelectron microscopy is beyond a doubt the preferable procedure for eliminating artifacts.

Cross-reacting antibodies can also be detected and eliminated by affinity chromatography. A cross-reacting antibody would then be bound not only to a column to which the pure antigen has been linked but also to a column containing a mixture of ribosomal proteins without the cognate antigen. More evidence that this theoretical possibility does not interfere with immunoelectron microscopic data in practice is presented elsewhere (Tischendorf and Stöffler, 1977).

c. A Comparison of the Models for 30 S Ribosomal Subunits as Suggested by the Groups of Stöffler and Lake. A comparison of the data of these two groups is quite difficult. Lake and his colleagues have so far only mapped six of the 21 ribosomal proteins of the 30 S subunit, viz. S4, S5, S11, S13, S14, S19 (Lake *et al.*, 1974a; Lake and Kahan, 1975), whereas all 21 proteins have been localized by immunoelectron microscopy in Berlin (Tischendorf *et al.*, 1974b, 1975, 1976; Stöffler *et al.*, 1974; Stöffler and Tischendorf, 1975; Stöffler, 1976; Tischendorf and Stöffler, 1975, 1976a,b). Very recently a second site has been found for protein S19 (close to protein S9) and a fourth site for protein S11 was located in the proximity of site S4C (Stöffler, 1976; Tischendorf and Stöffler, 1976a,b; Tischendorf *et al.*, 1976; not incorporated in Figs. 27 and 28).

The disposition of the six proteins on the 30 S subunit which were mapped in both laboratories are in reasonably good agreement. Significant differences between the two sets of data exist, however, with regard to the shapes of these proteins: Tischendorf and colleagues found that proteins S5 and S11 both occur at multiple, well separated sites; these proteins should thus have elongated configurations (see Figs. 27 and 28). According to Lake, proteins S5 and S11 have been mapped at a single site only (Lake and Kahan, 1975). These differences are probably only methodological. Stöffler and his colleagues purified the IgG antibodies by gel filtration (Maschler, 1973) or by affinity chromatography (Stöffler, 1976; Tischendorf and Stöffler, 1975, 1976a,b; Tischendorf *et al.*, 1976) whereas Lake used elution from DEAE-Sephadex with 0.02 *M* potassium phosphate buffer. Sela and Mozes (1966) have demonstrated that the chemical nature of antibodies is dependent on the net charge of antigens: basic antigens thus stimulate the production of acidic antibodies. The antibody response to the basic ribosomal proteins follows this rule and results in a high percentage of acidic antibodies (G. Stöffler, unpublished results).

Elution of acidic IgG molecules from DEAE-Sephadex requires a higher ionic strength than used by Lake (see Sela and Mozes, 1966) and the IgGs purified by Lake *et al.* (1974a) and Lake and Kahan (1975) lack significant amounts of antibodies specific toward basic antigenic determinants. This might explain why several antibody binding sites were not found during immunoelectron microscopy by these authors.

The three-dimensional model interpretations of 30 S subunits by the groups of Lake and Stöffler differ significantly; these interpretational differences certainly have an impact on the exact three-dimensional placement of several antibody binding sites. Deduction of three-dimensional models from electron micrographs are necessarily subjective. Three-dimensional image reconstruction methods as introduced by Crowther *et al.* (1970) or by Hoppe (1973) will be required to obtain an objective three-dimensional subunit structure. Nonetheless, the necessity for knowing the exact subunit structure in order to accurately localize an antibody binding site is certainly overemphazised by Lake and Kahan (1975). Several conclusions can be drawn from the immunoelectron microscopical studies even without knowing the exact three-dimensional structure of the subunit. Such examples are illustrated in Fig. 28a: (1) It is easy to decide whether an antibody binds to the head (Fig. 28a, rows 1 and 2), to the neck (Fig. 28a, row 3), or to the body (Fig. 28a, row 4). A division of the 30 S subunit into a one-third (head) and a two-thirds portion (body) is a common feature of both models. (2) Whether a given antibody attaches on the same or on the opposite side as the two lobes (in Tischendorf's model) or as the platform (in Lake's model) can also be decided independently from the differences between the models (Fig. 28a, rows 2 and 4). (3) Proteins with elongated configurations *in situ* are detected by immunoelectron microscopy, if antibodies from an antiserum specific to *a single* protein bind at *two or more* well separated *sites* on the ribosomal surface (Fig. 28a, row 5; see also Figs. 25 and 26).

Lake and Kahan (1975) have attempted to distinguish between a symmetrical or an asymmetrical 30 S structure by correlating the location of the binding site of an antibody specific to protein S13 with structural features of 30 S subunits. This experiment is based on certain assumptions: (1) Anti-S13 labels a unique site. It has not been excluded that S13-specific antibodies do not bind at two separated regions on the head, a result which would necessarily lead to a wrong interpretation. (2) Binding of an IgG molecule to a specific site might cause the 30 S subunit to attach in a specific, non-random manner to the grid; then the deductions made with regard to the subunit structure are no longer conclusive. In fact, it has been clearly demonstrated with antibodies to individual 50 S subunit proteins that IgG molecules have the property to constrain subunits into

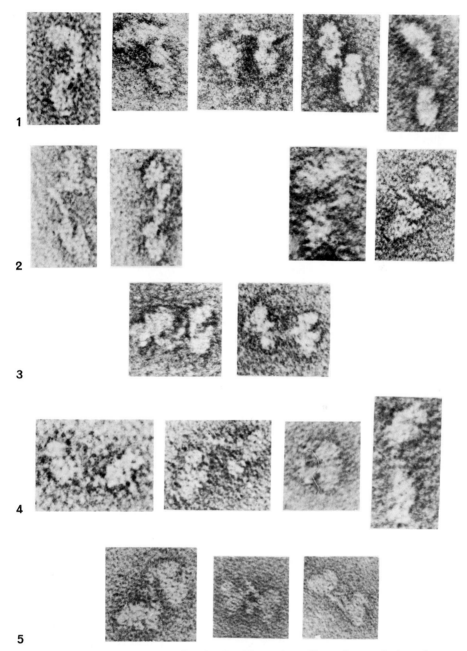

Fig. 28a Electron micrographs of 30 S·IgG·30 S complexes illustrating conclusions about the localization of antibody binding sites which can be made without knowing the exact three-dimensional subunit structure. Rows 1–5 show the following binding sites: (1) the head; (2) *left side,* front of the head; *right side,* rear of the head; (3) the neck; (4) the body; and (5) antibodies specific to a single protein (S18) bind at multiple and well-separated sites: Protein S18 is elongated.

specific orientations (see Wabl, 1974; Tischendorf *et al.*, 1974a). Therefore, more exacting controls are required in order to confirm the conclusions drawn from this experiment of Lake and Kahan (1975).

It is necessary to develop methods which make the localization of
antibody binding sites independent from an exact knowledge of the three-
dimensional subunit structure. Monovalent Fab fragments bound to the
ribosome can be visualized by negative staining (Tischendorf and Stöffler,
1975). Work is presently in progress to label 30 S subunits *simultaneously*
with an antibody against protein Sx and an Fab against protein Sy. This
approach gave promising results to map the accurate three-dimensional
position of a few ambiguous antibody binding sites (Tischendorf *et al.*,
1976; Tischendorf and Stöffler, 1976a,b, 1977).

3. *Localization of Ribosomal Proteins on the 50 S Ribosomal Subunit*

We have initiated studies of the 50 S subunits by immunoelectron microscopy. The 50 S subunit images are shown in Fig. 22 and are interpreted as projectional forms of a unique 50 S structure (armchair). This
three-dimensional interpretation of the various images of 50 S subunits
was an important prerequisite for the localization of the 19 proteins on
the 50 S subunit surface that were studied so far (cf. Figs. 22 and 29). Until
recently, the location of 13 proteins has been communicated by this laboratory (Wabl, 1973; Stöffler, 1974, 1976; Tischendorf and Stöffler, 1976a,b;
Tischendorf *et al.*, 1974b, 1975, 1976; Wabl, 1974; Stöffler and
Tischendorf, 1975; Stöffler *et al.*, 1974). An additional six proteins, L2,
L3, L5, L20, L25, L27 have now also been localized and most of the proteins previously mapped have been reinvestigated with different specific
antisera (Tischendorf and Stöffler, 1976a,b, 1977).

The distribution of the antibody binding sites for 19 of the 50 S subunit
proteins is summarized schematically in Figs. 29 and 30. Ten of the 19 proteins occur at only one site; protein L4 is exposed at two sites in close vicinity to each other and proteins L2, L6, L11, L18, L20 occur at two sites,
separated by distances which permit the assumption of elongated protein
shapes. Only protein L1 occurs at three sites and has, together with proteins L2 and L20, the most remote sites. Each of the three proteins L1,
L4, and L6 has previously been mapped only at a single site; detection of
additional sites has been achieved by the use of at least two separate antisera per protein (Tischendorf and Stöffler, 1976a,b, 1977).

Proteins L7/L12 occur together in three or four copies per 50 S subunit
(for references see reviews by Möller, 1974; Stöffler, 1974; see also Weber,
1972; Thammana *et al.*, 1973; Hardy, 1975; Subramanian, 1975). Antibodies specific to proteins L7 and L12 bound at multiple sites, all of them

50 S

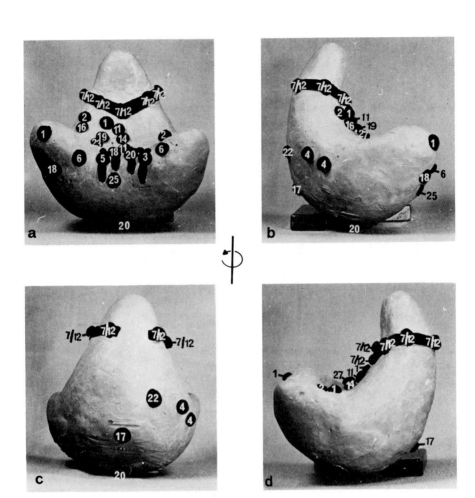

Fig. 29 Plasticine model of the 50 S ribosomal subunit of *E. coli*. The four forms were derived by successive rotation of the model by an angle of 90°. The centers of the antibody binding sites are indicated by numbered black circles. Antibody binding sites whose location was not exactly definable are indicated by black ribbons (see also legend to Fig. 27).

located on the central protuberance. These two proteins form a garland interrupted on the dorsal subunit surface only (Figs. 29 and 30). At least five separate antibody-binding sites could be discriminated for L7/L12; they are not being considered in the following calculations.

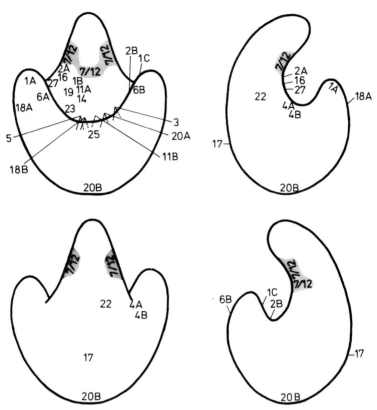

Fig. 30 Three-dimensional model of the 50 S ribosomal subunit with the location of 19 ribosomal proteins. The four views presented were derived by successive rotation through an angle of 90°. The stippled areas represent the regions of the various antibody binding sites of anti-L7/L12. (Taken from Tischendorf *et al.,* 1975, 1976.)

Twenty-five antibody-binding sites have been encountered for the seventeen proteins that are represented in one copy per ribosome. These sites are not uniformly distributed over the subunit surface. Eleven sites are within the seat area; another nine sites are on the frontal corner of the seat. Some sites of the latter group have not yet been exactly localized (indicated by the ribbons or by the dichotomous lines in Figs. 29 and 30). Antibodies attaching close to corners or edges of the subunit are very difficult to localize, since it is difficult to decide whether binding occurs in front of or behind the edges. Most difficult in this respect was the positioning in sites L18B and L20A. L11B is more likely internally, L3 and L5 externally located. Only four antibody sites have been found so far on the dorsal region of the subunit and one, L20B, is situated at the lower pole.

It is apparent that only five sites are situated on the left side of the 50 S subunit. The cluster of proteins within and around the right part of the seat region can be subdivided: One protein-rich area is on the right-hand side of the seat, the other in the center, close to the longitudinal axis at the anterior seat corner. Whether or not the 50 S proteins are arranged in clusters awaits evidence from the location of the 15 proteins not studied thus far. Systematic examination of several separate antisera per individual protein should not only provide further information in this respect but should also lead to the discovery of additional proteins with elongated shapes.

C. A Topographical Model of the 30 S Ribosomal Subunit

The location of 30 S ribosomal proteins on the subunit surface can now be estimated by comparison with results from independent research. In our construction of a ribosomal model we shall correlate the antibody mapping results with those data related to the physical structure of the ribosome. Relevant results come from the following approaches: (1) identification of cross-linked protein-protein pairs, (2) distance measurements from singlet energy transfer between pairs of fluorescently labeled ribosomal proteins, (3) neutron scattering measurements, (4) location of functional sites on the 30 S ribosomal subunit (e.g., the mRNA binding domain, the tRNA and the initiation factor binding sites), (5) characterization of ribonucleoprotein fragments and binding sites of individual proteins along the 16 S rRNA. The latter data will be used to obtain a first and rough idea of the arrangement of 16 S rRNA within the 30 S subunit.

A few models of the arrangement of 30 S ribosomal proteins have been constructed. Morgan and Brimacombe (1973) based their model on the composition of ribonucleoprotein fragments supplemented with data on cross-linked protein pairs. The model of Traut *et al.* (1974) is based on results obtained from protein-protein cross-linkage, the *in vitro* assembly map, studies of protein assembly interdependencies, and chemical modification studies (see Mizushima and Nomura, 1970; Craven *et al.*, 1974). Another model was recently suggested by Bollen *et al.* (1974). All these models assume globular shapes of the proteins. Recently, an elongated protein S4 has been considered in Traut's model (Sommer and Traut, 1975).

1. General Principles for Designing a Ribosome Model

The principle of our model construction is quite simple. The three-dimensional interpretation of electron microscopic pictures is the basis for the postulated structure of ribosomal subunits. For the 30 S subunit we propose a topographical model based upon the location of the 41 anti-

TABLE VI

Relative Distances between 30 S Ribosomal Proteins[a,b]

A. Summary of protein-protein neighborhoods (cross-linkage and singlet energy transfer)

Cross-linked proteins[c]	Very close or close by energy transfer[(a)]	Cross-linked proteins[c]	Very close or close by energy transfer[(a)]
S2-S3[(g,i,l,o)]	—	S5-S8[(c,g,i–l,o)]	—
S2-S5[(g)]	—	S5-S9[(f,h,l)]	—
S2-S8[(g)]	—	S6-S8[(n)]	—
S3-S10[(g)]	—	S6-S18[(g)]	S6-S18
S4-S3[(l)]	—	S7-S8[(l)]	—
S4-S5[(g,k,o)]	—	S7-S9[(d,g,h,k,l)]	S7-S9
S4-S6[(l)]	—	S8-S15[(m)]	S8-S15
S4-S8[(l)]	—	—	S15-S16/*S17*
S4-S9[(e)]	—	S13-S19[(d,g,i)]	S13-S19
S4-S12[(l)]	—	—	S13-S20
S4-S13[(h)]	S4-S13	S14-S19[(e)]	FAR (see Table VIB)
—	S4-*S16/S17*	—	S16/*S17*-S20
—	S4-S20	S18-S21[(b,c,h)]	—

B. Remote proteins (singlet energy transfer[(a)])

Protein pair		Assignment[d] to ribonucleoprotein fragment	
S4<	>S15	I and II	
S4≪	≫S18	I and II	
S4<	>S19	I and III	f6
S20<	>S19	I and III	f6
S16/S17≪	≫S13	I and III	f6
S15<	>S19	II and III	f6
S15<	>S18	II	
S14≪	≫S19	III	f6

[a] The data are taken from: (a) Huang *et al.*, 1975; (b) Chang and Flaks, 1972; (c) Lutter *et al.*, 1972; (d) Lutter *et al.*, 1974a; (e) Bode *et al.*, 1974; (f) Bickle *et al.*, 1972; (g) Lutter *et al.*, 1974b; (h) Sommer and Traut, 1974; (i) Sun *et al.*, 1974; (j) Clegg and Hayes, 1974; (k) Barritault *et al.*, 1975; (l) Sommer and Traut, 1975; (m) Barritault and Expert-Bezançon, 1975; (n) R. Traut, personal communication; (o) Peretz *et al.*, 1976.

[b] Several cross-linked triplets of proteins have also been identified (see Traut *et al.*, 1974; Brimacombe *et al.*, 1976); they have not been considered since the arrangements of cross-linking have not been established. Only energy transfer results with the protein mixture S16/S17 have been considered (for explanation, see text). The protein that agrees better with immune electron microscopical data is italicized; <>, far; ≪≫, very far.

body binding sites, as obtained with IgG's specific to the 21 proteins of the small subunit, together with data on physical neighborhood and functional sites.

The main obstacle in designing a model was to decide which portion of an elongated protein (i.e., which antibody binding site) is in physical proximity of a particular functional domain. If, for example, several proteins contribute to a specific ribosomal function, then the underlying principle for the location of this domain was to search for the "smallest area of common exposure." That is to say, if, for example, five proteins are involved in a certain function and they occur altogether at twelve sites, then the smallest surface area containing at least one site of each protein is considered to be the respective functional domain. We gave special emphasis to the immunochemical approach since the Fab's prepared from the 21 antisera specific for the 30 S subunit proteins and the 34 antisera specific to 50 S proteins have been studied for inhibitory effects on the various steps of the translation process (see Stöffler, 1974). Since most of these antibodies have later been investigated by immunoelectron microscopy a correlation of ribosomal structure and function through the same specific Fab probe has become possible.

2. Correlation of Antibody Mapping Data with Results on Relative Protein Proximity

Relative proximities between ribosomal proteins in the intact ribosomal subunit have been elucidated by several methods.

a. Cross-Links of Ribosomal Proteins. The most common approach has employed bifunctional reagents to covalently couple adjacent proteins. Various reagents spanning distances from 5–25 Å have been used. Identification of the cross-linked proteins can be achieved following their isolation either by cleavage of the bifunctional reagent or by identifying the proteins in the intact complex by immunological or by electrophoretic procedures. Recent and more detailed evaluations are given by Traut *et al.* (1974) and in Chapter 2.

We have summarized the pairs of 30 S subunit proteins whose characterization has been sufficiently documented (Table VI). Data on protein

---

[c] Details about the bifunctional reagents used and the methods for protein identification have been summarized by Brimacombe *et al.* (1976).

[d] I, II, III refers to the Zimmermann (1974) nucleation-sites and f6 to the ribonucleoprotein fragments of Morgan and Brimaxombe (1973).

[e] Cross-linked protein pairs characterized by Sommer and Traut (1976): S1-S18, S3-S4, S3-S5, S3-S9, S3-S12, S4-S17, S5-S13, S7-S13, S8-S11, S8-S13, S11-S13, S12-S13, S12-S20, S12-S21, and S13-S17.

neighborhoods as obtained from cross-linking studies showed a gratifying agreement with the results obtained by immunoelectron microscopy. Certain principles which should be used for such a comparison are discussed in Section II,C,2,c. Very recently, a few additional well-characterized cross-linked proteins were made available to us (Sommer and Traut, 1976). The components of these pairs are added as footnote e in Table VI. They are in good agreement with the locations of antibody binding sites described in this chapter, when the same principles as discussed above and in Section II,C,2,c are applied. Due to the fewer antibody binding sites found for proteins S5 and S11 by Lake and Kahan (1975) several cross-links listed in Table VI could hardly be explained.

   b. *Singlet Energy Transfer between Pairs of Fluorescently Labeled Ribosomal Proteins.* Cantor and his associates have introduced this approach to obtain a topological picture of the arrangement of ribosomal proteins. The rationale of this method lies in covalently attaching fluorescent labels to proteins and then, after reassembly *in vitro,* to study singlet energy transfer between various protein pairs.

   The first results obtained by singlet energy transfer studies have recently been reported (Cantor *et al.,* 1974b; Huang *et al.,* 1975). There is relatively good agreement between the data obtained by the fluorescent and immunoelectron microscopic methods (Table VI; see Section II,C,2,c). Four protein pairs used for energy transfer measurements have been made with a mixture of S16 and S17. The values obtained with this particular protein mixture should be considered with caution, since it is still uncertain to which protein the fluorescent label was attached (C. Cantor, personal communication). Proteins S16 and S17 are distinct ribosomal proteins with regard to their antigenic specificity (Stöffler and Wittmann, 1971a) and their different amino acid sequences (see Table I). Furthermore the antibody sites of these two proteins are located on remote areas of the ribosome (see Figs. 27 and 28). The component (S16 or S17) assumed to be involved in each of the pairs is printed in italics in Table VI. Good agreement exists also between cross-linkage and fluorescent studies. Five protein pairs, detected by energy transfer as being in proximity, were also cross-linked with bifunctional reagents (Table VI).

   The fluorescent approach should provide data that are additive to those of protein cross-linkage and antibody mapping. Energy transfer studies provide also distance estimates on protein-protein pairs and permit the identification of remote pairs (see Table VIB). The application of this technique is not fully dependent on protein exposure and their *in situ* reactivity, since isolated ribosomal proteins can also be labeled with fluorescent dyes and then reassembled to ribosomal subunits *in vitro.*

   If dye localization to a single amino acid on an elongated protein can be made, the fluorimetric measurement will have the potential to yield distances between specific points of a given protein pair. Though difficult, this localization should be more tractable with fluorescent dye markers than with the various cross-linking reagents. By the use of "hapten-sandwich" immunoelectron microscopy a topographical location of the attached dyes in any protein should be feasible with antibodies to the fluorescent markers.

Localization of specific regions of proteins through the use of antibodies specific for particular portions of the primary sequence has been successful for proteins S4 and S18 thus far (Tischendorf and Stöffler, 1975; G. Stöffler, G. W. Tischendorf, K. H. Rak, and M. Yaguchi, unpublished results). As more such data emerge and the amino acids to which the fluorescent dyes are bound are determined, a correlation of electron microscopic and fluorescent energy transfer results on an intramolecular level should soon be possible.

c. Possibilities and Restrictions for the Comparison of Various Data on Protein Topography. Comparison of protein distances as obtained by cross-linking and energy transfer studies on one hand and immunoelectron microscopy on the other should be done with certain precautions. Antibody mapping has two main advantages, namely to allow the determination of the "absolute" position of a given protein on the ribosomal surface and to permit the disclosure of proteins with elongated shapes. Exact distances between two ribosomal proteins can be more accurately determined by cross-linking and energy transfer studies but these methods provide no information about protein shapes or where the proteins are located on the ribosome.

Distance estimations can also be made from the immunoelectron microscopical data, however, on a different scale; this is due to the relatively large size of the antibody marker. Most of the cross-linking reagents used so far spanned distances between 10 and 20 Å. Hence the size of the Fab-arms of an antibody bound to the ribosome (approximately 40×40 Å) is larger than the distance between the cross-linked amino acids of two neighboring proteins. Similar considerations have to be made for the comparison of fluorescent studies with antibody mapping data. Separation of the centers of two neighboring antibody binding sites by ~40 Å would, due to the size of the Fab-arms, still be reflected by a slight overlap of the two binding regions in immunoelectron microscopy. On the other hand, two proteins separated by 40–50 Å in energy transfer experiments have been judged to be "far" (Huang *et al.*, 1975). A good example for such an apparent discrepancy is the protein pair S20-S19 (compare Table VIb with Fig. 27a).

Considering these limitations, we have compared the possible interprotein contacts occurring in the immunoelectron microscopic model with the data on protein-neighbors from cross-linking and energy transfer experiments by triangular contact matrices. A large number of cross-linked protein pairs and close neighbors detected by fluorescence studies (Table VIa) were found to have antibody binding sites, located in close vicinity to each other (see Figs. 27 and 28). Furthermore, a comparison of cross-linking and fluorescent data with the results obtained by immuno-electron microscopy also has to be in accordance with the notion that several 30 S subunit proteins have elongated configurations *in situ*. Hence, we have not only considered the vicinity of the antibody attachment sites but also, for elongated proteins, possible vicinities between an experimentally detected antibody site and the hypothetical intraribosomal course of a protein chain. By doing so we found an even more gratifying agreement between the three groups of results. For example, the proteins of four cross-linked protein-pairs (S2-S5, S2-S8, S4-S8, S8-S15) would not have been considered to be in close proximity from their actual antibody-binding sites. The proteins contained in these pairs might all be in contact within the narrow neck of the 30 S particle, since each of the elongated proteins S2, S4, S5, and S15 must thread through this region. We cannot, on the basis of the antibody mapping, envision a close proximity between the members of the cross-linked protein pairs S4-S6, S6-S8, and S14-S19.

In fact, energy transfer studies also suggested proteins S14 and S19 to be distant (Huang *et al.*, 1975; see Table VI). We recently have reinvestigated this apparent discrepancy by searching for other antibody-binding sites on these proteins with different antisera, and have so far discovered a second S19 attachment site closer to S14 (G. W. Tischendorf and G. Stöffler, unpublished results). By assuming an elongated shape for protein S19, one portion of this protein could be close to S14 (cross-link), another portion far from S14 (energy transfer); both measurements would hence be correct and in agreement.

  d. Neutron Scattering Measurements of the Distances of Proteins in Ribosomal Subunits. Another approach to the determination of quaternary structure is by the use of neutron scattering. This technique requires measurement of the scattering of a neutron beam by deuterated protein. So far, radius of gyration experiments have been performed mainly in order to determine the maximal possible separation between centers of mass of the protein and RNA in 30 S and 50 S ribosomal subunits (Engelman and Moore, 1972; Moore *et al.*, 1974). The value Δ for the separation between the RNA and protein centers of mass in the 50 S subunit was determined to be 57.7 Å (Moore *et al.*, 1974). These results can accommodate models for the 50 S subunit in which the protein and RNA distributions are substantially remote. More recent neutron low angle scattering studies indicated, however, that the *E. coli* 50 S subunit has a core of high scattering density (RNA) surrounded by a low density shell of proteins (Stuhrmann *et al.*, 1976).

Similar measurements on 30 S subunits of *E. coli* indicated a relatively small Δ value. Hence, the protein and RNA are more uniformly distributed in the 30 S subunit (Moore *et al.*, 1974).

Mapping of ribosomal proteins with protein specific antibodies provides direct results about the surface distribution of the individual proteins only (see above). Results obtained by the two methods can, therefore, not be compared without certain restrictions. Nonetheless, the antibody mapping data, available at the moment, indicate that the proteins of the 30 S subunits are more uniformly distributed than are the 50 S subunit proteins.

Recently, Moore and his associates have applied neutron scattering to pairwise measurements of the distances separating the centers of three protein pairs (S2-S5, S5-S8, S3-S7) in the 30 S subunit of *E. coli* (Engelman *et al.*, 1975a). The distances between centers of gravity have been determined to be 105 Å for S2-S5, 115 Å for S3-S7, but only 35 Å for S5-S8. In addition, the authors reported that the shapes of these proteins can be estimated. Accordingly, proteins S5 and S8 should be compact and protein S2 extended. With respect to protein S5, these results are in disagreement with the immunoelectron microscopic data which showed that protein S5 has an elongated configuration (Figs. 27 and 28).

Similar studies have been performed with the 50 S subunit proteins L7/L12 and L10 (Hoppe *et al.*, 1975). The distance between the mass centers of proteins L7/L12 and L10 was estimated by the label triangulation method and was found to be approximately 100 Å. Since the antibody site of L10 has so far not been exactly determined we cannot comment on these data.

Neutron scattering was expected to be among the most effective technique for protein mapping, especially when the measurement of the scattering intensities, the signal-to-noise ratio and the mathematical calculations can be further improved (Moore *et al.*, 1974; Hoppe, 1972, 1973; Engelman *et al.*, 1975a,b; Hoppe *et al.*, 1975). The fact that many ribosomal proteins have elongated configurations within the ribosome, however, diminished the expectations in this method considerably.

3. The Roles of Individual Ribosomal Proteins during Translation

With respect to the specific roles that 30 S proteins play during the translational process, numerous data are presently available. One attempt was to establish a functional classification of ribosomal components by

G. Stöffler and H. G. Wittmann

testing the activities of reconstituted subunits lacking a single entity at a time (Nomura *et al.*, 1969; Held *et al.*, 1974b). In another approach the effects of all the available anti-30 S protein antibodies at the various steps of protein synthesis have been investigated (Maschler, 1973; Lelong *et al.*, 1974; Stöffler, 1974; Jeantet *et al.*, 1977).

Similar studies with anti-50 S protein antibodies assisted in elucidating the contribution of individual 50 S proteins to the reaction of the ribosome with elongation factors EF-G and EF-Tu and with release factors (see Kischa *et al.*, 1971; Highland *et al.*, 1973, 1974a,b; Stöffler, 1974; Tate *et al.*, 1975). Partial reconstitution experiments, mainly performed by Nierhaus and collaborators, have also provided important information and have been recently reviewed (Pongs *et al.*, 1974; Werner *et al.*, 1975; Brimacombe *et al.*, 1976).

Affinity labeling techniques have contributed an enormous amount of information about the involvement of individual ribosomal proteins in certain steps of protein synthesis. The results have been reviewed in detail (Cantor *et al.*, 1974a; Pongs *et al.*, 1974, 1975a; Brimacombe *et al.*, 1976). More recently, the affinity labeling approach has also been applied to detect proteins at the codon recognition sites (Fiser *et al.*, 1974, 1975a; Pongs and Lanka, 1975a,b; Pongs *et al.*, 1975b; Lührmann *et al.*, 1976). Results obtained with this technique have been generally evaluated for deducing the specific roles of individual proteins in certain ribosomal functions. By considering affinity labels in terms of topological probes we shall attempt to localize the mRNA binding domain to a confined area on the ribosome.

a. Identification of Proteins at the Messenger RNA Binding Site. Earlier studies have indicated that protein S1 affects the binding of mRNA to the 30 S subunit (Smolarsky and Tal, 1970; van Duin and Kurland, 1970). This finding has been substantiated by several investigators (Noller *et al.*, 1971; Rummel and Noller, 1973; Tal *et al.*, 1972; Acharya and Moore, 1973; Dahlberg, 1974; Szer and Leffler, 1974; Szer *et al.*, 1975; van Duin and van Knippenberg, 1974; Fiser *et al.*, 1975a,b; van Dieijen *et al.*, 1975). Protein S1 has been shown to be identical with "interference factor" iα, and with one of the Qβ host-specific replicase subunits (Groner *et al.*, 1972; Inouye *et al.*, 1974; Wahba *et al.*, 1974).

More recently, affinity labeling techniques have been used successfully to identify proteins at the mRNA binding site of ribosomes. Poly-4-thiouridylic acid, used as an analogue of poly(U), was covalently bound to ribosomes by photo-oxidation. Most of the radioactivity was found with protein S1 (Fiser *et al.*, 1974), although S18, S21, and to a much lesser extent S11, S7, and S14 were also labeled (Fiser *et al.*, 1975a).

Pongs and co-workers studied the proteins located at the codon binding site with a chemically reactive AUG analogue, a bromoacetamidophenyl residue attached to the 5'-adenine (Pongs and Lanka, 1975a,b; Pongs *et al.*, 1975b). Pongs *et al.* (1975b) found that different proteins were labeled, depending upon the conformational state of 30 S subunits. In active 70 S ribosomes, for example, proteins S18 and S21 reacted with the codon analogue at a 9:1 ratio; reaction at low temperature led to an additional labeling of protein S4, the ram gene product. Proteins S18 (and S21) were also labeled when the AUG analogue was reacted with isolated 30 S subunits; under conditions of slow freezing and thawing 30 S subunits the proteins S4 and S12 (strA gene product) reacted with the codon. Label was also detected in the "fidelity-protein" S11 and to a very small extent, in protein S13. The results of these studies indicated that proteins S4 and S18 are part of the A site and that proteins S11 and S12 are part of the P site of the ribosomal codon binding site (Pongs *et al.*, 1975b).

Gassen and co-workers, using a reactive analogue of the tetranucleotide GUUU which carried the affinity label in the 3'-uridine, found that proteins S18, S1, and, to a very small extent, S12 were labeled (Lührmann *et al.*, 1976). Treatment of 70 S ribosomes with the reactive codon analogue, with and without the cognate Val-tRNA, resulted in quantitative differences in the distribution of the label among the three proteins. An affinity analogue with a 5-bromoacetamido uridine 5-phosphate moiety bonded to the 3' end of AUG was almost exclusively cross-linked to protein S1 (Pongs *et al.*, 1976).

Altogether nine proteins have been shown to be in proximity to the ribosomal mRNA binding domain, seven reacted with codon derivatives, viz., S1, S4, S11, S12, S13, S18, and S21, and may thus be considered as candidates for the codon presentation function (see Chapter 2). The distribution of the antibody binding sites of these proteins is shown in Fig. 31. Five of the nine proteins, found in the vicinity of the mRNA binding site, are among the most extended 30 S subunit proteins (S4, S7, S11, S12, and S18). Therefore, a topographical emplacement of the mRNA binding site becomes quite difficult. We may assume that a certain area on the surface which contains at least one binding site of all the proteins of the mRNA binding group should most likely indicate the location of the mRNA binding domain. This "minimal area of common exposure" can be localized on the anterior and right side of the "head" (Fig. 31). This particular region of the 30 S subunit is also in contact with the 50 S subunit in 70 S monomeric ribosomes. There is evidence that portions of proteins S18 and S21 occur close to the temporal area comprised of proteins S1, S4, S11, S12, S14: a monovalent Fab fragment specific to anti-S14 which reacts with the ribosome at the single S14-specific site can be cross-linked with suberimidate to proteins S18 and S21 among others (R. Maschler, S. K. Sinha, and G. Stöffler, unpublished results).

Ribosomal A and P site functions are inhibited by alkylating the sulfhy-

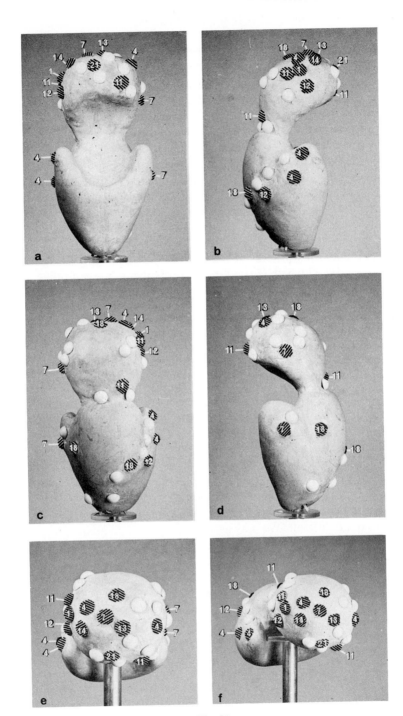

Fig. 31

dryl group of protein S18 (Moore, 1971; Ginzburg *et al.*, 1973). The AUG affinity label has recently been found to react also with the cysteine located in position 10 of the protein S18 (Yaguchi, 1975; O. Pongs, E. Lanka, and M. Yaguchi, personal communication). We have preliminary evidence from immunoelectron microscopic studies that the site S18A corresponds to a determinant which is contained in an S18 protein fragment comprising the amino acid sequence 1–34 (G. Stöffler, G. W. Tischendorf, K. H. Rak, and M. Yaguchi, unpublished results). Hence, the NH$_2$-proximal region and the reactive cysteine of protein S18 should be in the vicinity of site S18A on the head and close to the mRNA binding site (Fig. 31).

Four proteins, viz., S4, S5, S11, and S12, were shown to be involved in the streptomycin responses of the ribosome, and to affect the fidelity of translation (see Birge and Kurland, 1969; Ozaki *et al.*, 1969; Hasenbank *et al.*, 1973; Gorini, 1974; Wittmann and Wittmann-Liebold, 1974). These four proteins have been considered as the "fidelity group" of 30 S ribosomal proteins (Pongs *et al.*, 1974; Wittmann and Wittmann-Liebold, 1974; Stöffler and Tischendorf, 1975). These proteins should be located in physical proximity to the codon binding site which has already been demonstrated for three of them, namely, S4, S5, and S12. Indeed, site A of protein S5 is also located in this area and therefore close to the mRNA binding domain (see Figs. 27, 28, and 31).

b. The Location of tRNA Binding Sites on 30 S Subunits. Messenger RNA binding is a relatively simple manifestation of a partial function of the ribosome; nonetheless, at least nine proteins are involved (see above). Considering the more complex nature of tRNA binding, a large number of proteins should also contribute to this function. Attempts to elucidate the ribosomal proteins of the tRNA binding sites were made in the presence and absence of protein synthesis factors (see Pongs *et al.*, 1974; Stöffler, 1974; Traut *et al.*, 1974).

The ribosomal proteins of the 30 S P site were examined by measuring AUG-directed fMet-tRNA binding in the presence of initiation factors IF-1 and IF-2. Since a ribosomal protein required for the interaction with only one of these macromolecules can affect this complex formation, it is not surprising that pleiotropic effects are observed. At least ten proteins exert some influence on the ribosomal P site (for summaries, see Held *et al.*, 1974b; Lelong *et al.*, 1974; Stöffler, 1974; Traut *et al.*, 1974).

Fig. 31 Proteins at the ribosomal mRNA binding domain. Antibody binding sites of proteins involved in mRNA binding are shaded. Diagonal shading from right to left, proteins associated with codon-affinity probes. Diagonal shading from left to right, proteins exclusively labeled with poly-4-thiouridylic acid (see also text, pp. 168–171).

A similar large number of 30 S ribosomal proteins have been implicated for A site functions, whereas a third group of proteins is crucial for the expression of both, A and P site function. The latter results are in accordance with the concept of a 30 S "hybrid site" (Kurland, 1972).

These data have been summarized in a topographical arrangement of the 30 S proteins and are presented elsewhere (Jeantet et al., 1977). In summary, proteins associated with P site functions were located on the head of the subunit. Proteins involved in A site functions occur on the head but also on the dorsal side of the subunit around the neck region. This interpretation requires the assumption that EF-Tu is in proximity to the latter domain within 70 S ribosomes (see below).

c. *The Initiation Factor Binding Sites.* Attempts to cross-link the initiation factors IF-2 and IF-3 to 30 S subunits have been pursued during the last three years (Traut et al., 1974). Traut and his colleagues identified recently seven proteins, viz., S1, S2, S11, S12, S13, S14, and S19, to be near the binding site for initiation factor IF-2 (Bollen et al., 1975). It has not yet been established whether these many proteins were directly, or through other proteins indirectly, cross-linked to the factor. The significance of these data is further impaired by the additional presence of initiation factors IF-1 and IF-3 during the cross-linkage of IF-2 (Bollen et al., 1975). These data were, therefore, not included in our topographical model.

The attempt to localize the binding domain of the initiation factor IF-3 on the 30 S ribosomal subunit has been done with cross-linking data obtained in several laboratories: Protein S12 has been identified as near the IF-3 binding site by Hawley et al. (1974); with the use of a different reagent, a cross-link formed by IF-3 and protein S7 has been characterized (van Duin et al., 1975). These data illustrate the advantages of using different cross-linking reagents and especially the necessity to select a reagent which efficiently cross-links the factor to ribosomes but which creates very little or no cross-linkage between ribosomal proteins (see San José et al., 1976).

The recent finding that IF-3 could also be cross-linked to the 3'-end of 16 S rRNA (van Duin et al., 1976) is of particular importance since proteins S1 and S21 have also been cross-linked to the 3'-end of 16 S rRNA by the same approach (Czernilofsky et al., 1975). Thus four proteins, viz., S1, S7, S12, and S21 are located, together with the 3'-terminus of 16 S rRNA, within a domain on the 30 S ribosomal surface which is close to the binding site of initiation factor IF-3.

Most of the antibody-binding sites of the four proteins, associated with IF-3, are situated on the head of the subunit (see Figs. 27 and 28). The minimal area of common exposure comprises the unique sites of proteins

S1 and S21, as well as sites S7A and S12A (see Fig. 27e). All these sites are located at the right temporal region of the head, that is to say, within an area of the 30 S subunit which is in contact with the 50 S subunit (see Fig. 27e). This may be relevant to the mechanism of IF-3 as a "dissociation" or "antiassociation" factor. In fact, if IF-3 was cross-linked to 70 S tight couples, the factor was also found to be associated with the 3'-end of 23 S rRNA (van Duin *et al.*, 1976). The binding region of initiation factor IF-3 on the 30 S subunit as deduced from the cross-linking data can be seen best in Fig. 27e.

Additional results from Traut's IF-3–30 S subunit cross-linking experiments, which were made available to us very recently, revealed that IF-3 was recovered covalently bound to proteins S1, S11, S12, S13, S19, and S21 (Heimark *et al.*, 1976). Protein S9 has, besides S12, also been found cross-linked to IF-3 (D. A. Hawley, J. E. Sobura, and A. J. Wahba, personal communication). Indeed, these data further support our supposition that the IF-3 binding domain is located on the head of the 30 S subunit and at the subunit interface (see Figs. 27 and 28). The most straightforward approach to localize the actual ribosomal binding sites of the three initiation factors would still be a direct electron microscopic mapping, using anti-IF-1, -IF-2, and -IF-3 IgG-antibodies.

4. The Arrangement of the 16 S rRNA in the 30 S Ribosomal Subunit

Work in the elucidation of the primary sequence of 16 S rRNA has made rapid progress; more than 90% of the sequence is known (Fellner, 1974; Ehresmann *et al.*, 1975a,b). The length of the 16 S rRNA is about 1600 nucleotides (Ehresmann *et al.*, 1975b), a value lying in between that of 1500 nucleotides proposed on the basis of end-group analysis (Midgley, 1965) and the one of 1700 nucleotides, estimated by physical methods (Kurland, 1960; Stanley and Bock, 1965). A length of 1900 nucleotides as has been suggested by Ortega and Hill (1973) on the basis of low-angle X-ray diffraction studies is, however, incompatible with the nucleotide sequence determination (Ehresmann *et al.*, 1975b). Some corrections on the 16 S rRNA primary structure have recently been reported (Santer and Santer, 1973; Noller and Herr, 1974; Shine and Dalgarno, 1974; Uchida *et al.*, 1974; for a summary, see Ehresmann *et al.*, 1975b) and requires some changes to be made in the previous model. Optical rotatory dispersion studies, infrared spectroscopy, and acid-base titrations on isolated rRNA and ribosomes strongly suggest that the secondary structure of rRNA is the same within the ribosome and in isolated rRNA (for recent reviews, see Garrett and Wittmann, 1973; Brimacombe *et al.*, 1976). Approximately 30–35% of the bases occur in single-stranded and 60–70% in double-stranded "hairpin loop" regions. In an attempt to search for the most stable possible secondary structure according to the thermodynamic criteria proposed by Tinoco et al. (1971), a secondary structure model of the 16 S rRNA was recently proposed (Fellner, 1974; Ehresmann et al., 1975a,b). Very little is known about the three-dimensional and quaternary arrangement of 16 S rRNA within the 30 S ribosomal subunit (see Garrett and Wittmann, 1973; Fellner, 1974; Brimacombe et al., 1976).

Various strategies have been employed to learn the proximity of proteins and RNA in the 30 S ribosomal subunit. Combination of results from these studies together with data from immunoelectron microscopy on the distribution of ribosomal proteins should shed some light on this question. The following results have been considered in attempting to establish a model.

1. The RNA binding sites of ribosomal proteins S4, S8, S15, S20 which bind independently and directly to 16 S RNA have been characterized. Their approximate binding sites along the primary sequence of 16 S RNA are known and allow some conclusions with regard to the distribution of RNA in the 30 S subunit model (see Zimmermann et al., 1975). The binding site of protein S4, spanning the 5' one-third of 16 S rRNA, has been visualized by electron microscopy (Nanninga et al., 1972).

2. Limited nuclease treatment of 30 S subunits yields several ribonucleoprotein fragments (for a recent review, see Brimacombe et al., 1976). By means of this approach, 30 S subunits were split into several fragments of unequal size; two of them, fragments f14 and f6 of Morgan and Brimacombe (1973) were mainly considered in our model. Proteins contained in such ribonucleoprotein fragments have been interpreted as being neighbors in the intact ribosomal subunit. The larger fragment (f14) contained RNA from the 5'-proximal and the central region of 16 S RNA and the proteins S4, S5, S6, S8, S15, S16/17, S18, and S20, possibly also S13. The RNA of the smaller fragment (f6) is from the 3'-proximal region and has associated proteins S7, S9, S10, S19, and in addition proteins S14 and/or S13 (Morgan and Brimacombe, 1972, 1973; Roth and Nierhaus, 1973; Yuki and Brimacombe, 1975). A few proteins have not been found in any RNA fragment thus far (S1, S2, S3, S12, and S21).

3. Zimmermann and his collaborators located the binding sites of 12 ribosomal proteins to three specific regions of 16 S rRNA by in vitro reconstitution (Zimmermann, 1974). The proteins S4, S16, S17, and S20, bind to the 5'-proximal fragment of 16 S rRNA which spans about 500 nucleotides (nucleation site I). Four proteins, S6, S8, S15, and S18, interact with the central nucleation site II (nucleotide sequence 550–850) and another four proteins (S7, S9, S13, and S19) bind to the 3' one-third of the

16 S rRNA (nucleation site III). Similar assembly clusters were also described by Kurland and his collaborators (see Chapter 2).

4. Proteins S1 and S21, although not contained in any RNA fragment, were cross-linked to the 3′-end of the 16 S rRNA and should hence be in its proximity (Kenner, 1973; Czernilofsky *et al.*, 1975). A physical neighborhood between protein S21 and the 3′-terminal end has also been suggested by another finding: 30 S subunits reconstituted with 16 S rRNA whose 3′-terminal ends had been cleaved by colicin E3 also lack S21 (Bowman *et al.*, 1971).

The initiation factor IF-3 has also been cross-linked to the 3′ ribose of 16 S ribosomal RNA (van Duin *et al.*, 1976); in turn, proteins S7 and S12 have been cross-linked to IF-3 (see p. 172). Accordingly, proteins S7 and S12 might also be close to the 3′-terminus. In summary, there is considerable evidence that ribosomal proteins S1, S7, S12, and S21 and the initiation factor IF-3 make up a protein neighborhood organized around the 3′ region of the 16 S RNA (see also Chapter 2).

5. 30 S ribosomes from kasugamycin-resistant *E. coli* cells (ksgA) are not methylated in two adjacent adenine residues at the 3′-end of 16 S rRNA (Helser *et al.*, 1971). Although neither the free undermethylated 16 S rRNA nor intact 30 S subunits are substrates of the specific methylase, CsCl-core particles lacking proteins S3, S9, S10, and S14 were substrates (Thammana and Held, 1974). The protein S3, not found in any RNA fragment, may therefore be positioned at or near the 3′-end of the 16 S rRNA; the three other proteins S9, S10, and S14 are actually contained in the 3′-proximal fragment f6.

6. Although cross-linking of proteins to adjacent RNA with bifunctional reagents has not been fully explored, one can predict this approach will soon provide new information on protein-RNA neighborhoods and further refinement of the model (e.g., Ehresmann and Reinbolt, 1975; Möller and Brimacombe, 1975).

By correlation of these data with the antibody-binding sites of the 21 ribosomal proteins an instructive model can be derived (Fig. 32): (a) Proteins contained in nucleation site I, indicated by red pins, are found predominantly within two domains: one of them is situated on the left side of the forehead, the second area is on the right side of the body. Only a single antibody-binding site (S4B) has been found elsewhere, namely, on the right side of the head (Fig. 32a,b). (b) The eight antibody-binding sites of nucleation site II proteins S6, S8, S15, and S18 (indicated by green pins) are distributed from pole to pole over the entire dorsal subunit surface (Fig. 32c). (c) The antibody-binding sites of proteins S5 and S11 have been detected on the right lateral surface; they confine the groove (blue pins in Fig. 32b). These two proteins are associated with the larger 5'-fragment (f14) of Brimacombe but are not contained in any of the nucleation sites. The ten proteins, indicated by red, green, and blue pins, represent together these proteins which are associated with Brimacombe's fragment f14. (d) A fourth group comprises the nine proteins associated with, and hence in proximity of, the 3'-proximal region (yellow pins). This group consists of the proteins found in nucleation site III, those associated with the 3'-proximal fragment (f6) from Brimacombe, the proteins S1 and S21 (cross-linked to the 3'-terminus of 16 S rRNA), and protein S3 (from methylation studies). All the 14 antibody-binding sites of these 9 proteins, with one exception (S7C), are located on the head of the small subunit. Nine of their sites are exposed on the right side of the head, predominantly in its temporal and frontal region. A second domain (S19, S3A, S10A, S7B) is on the left temporal region. (e) Proteins S2 and S12 (white pins without numbers) have not been considered, since no data on their association with RNA are available. The indirect evidence that proteins S7 and S12 are close to the 3'-end of 16 S rRNA (see pages 172 and 175) has not been considered in the model (Fig. 32).

From the distribution of antibody-binding sites it appears that the head of the 30 S subunit contains a substantial portion of the 3'-end of 16 S rRNA . Portions of the 5'-end (nucleation site I) as well as parts of the central nucleation site II should also be contained in the head.

The body of the 30 S subunit should contain large sections from the 5'-proximal and the central region of 16 S rRNA but, according to the protein mapping data, very little RNA from the 3'-end. In fact, the only indication that a part of the 3'-proximal RNA-sequence occurs in the body comes from a single antibody site (S7C) located on the left lobe. No antibody-binding sites have been detected on the anterior surface of the body; whether RNA is exposed mainly in this area is unknown.

The model in Fig. 32 clearly shows that the antibody-binding sites of proteins associated with a given RNA fragment occur at several domains on the ribosomal surface. On the other hand, it is apparent that proteins, contained in different ribonucleoprotein fragments, have been mapped in close proximity to each other on the subunit surface (Fig. 32). This finding

Fig. 32 Three-dimensional model of the 30 S ribosomal subunit and arrangement of ribosomal proteins in specific ribonucleoprotein fragments. Proteins contained in the 5'-proximal nucleation site I are represented by red pins; nucleation site II proteins, green pins; proteins contained in the 3'-proximal RNA fragment are shown with yellow pins. Proteins not contained in any ribonucleoprotein fragment are indicated with white pins (without numbers), except proteins S1, S3, and S21 (yellow pins). The latter three proteins were considered at the 3'-end from cross-linkage to 16 S rRNA (S1 and S21) and from methylation studies (S3).

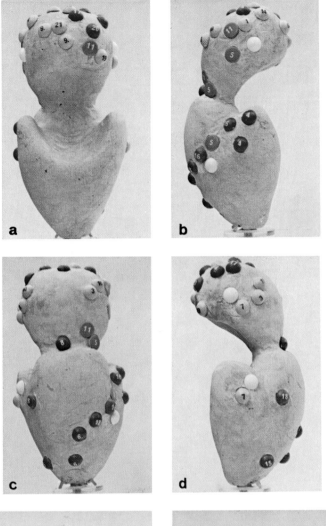

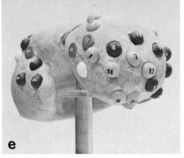
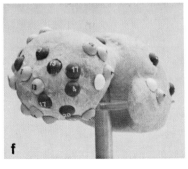

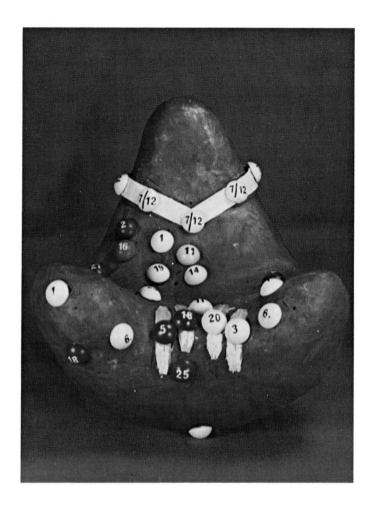

Fig. 34 Functional sites on the 50 S subunit. Frontal view of the 50 S subunit model. Yellow pins, peptidyltransferase: L11 (Nierhaus and Montejo, 1973); light green pin, L16 (Moore *et al.*, 1975); red pins, proteins associated with 5 S rRNA; green pins, proteins associated with tRNA affinity probes. (For further explanations, see text.)

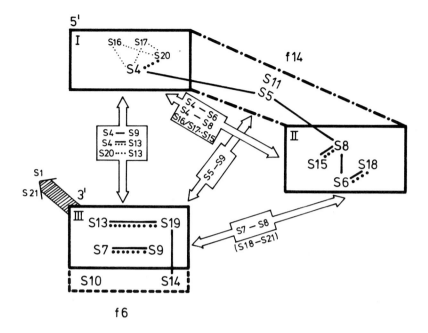

Fig. 33   Protein proximities and ribonucleoprotein fragments. The three rectangles I, II, III correspond to the Zimmermann nucleation sites; rectangles I and II, connected by —·—·—, correspond to Brimacombe's fragment f14. The proteins contained in Brimacombe fragment f6 and not in nucleation-site III are indicated by the dashed line (– – –). *Big arrows*: Neighborhood relations between proteins contained in different nucleation sites or ribonucleoprotein fragments. Sx—Sy: cross-linked protein pairs. Sx···Sy: very close or close according to energy transfer results (Sx···S16/S17, since these proteins were used as a mixture). Sx⸗Sy: proximity relations established by both methods. The cross-link between S1 and S21 and the 3'-OH end of 16 S rRNA is indicated by a shaded arrow.

is in excellent agreement with results from protein cross-linkage and singlet energy transfer (Fig. 33). The schematic summary, as shown in Fig. 33, illustrates that protein proximities have not only been found among the members of one RNA fragment but also between proteins contained in different ribonucleoprotein fragments. For example, a proximity relation between the 3'- and 5'-end of 16 S rRNA is indirectly supported by three protein pairs S4-S9, S4-S13, and S20-S13 which should be near each other according to cross-linkage and/or singlet energy transfer studies (Fig. 33). In agreement with these findings it has been shown that protein S13 occurs not only in the 3'-proximal but also in the 5'-proximal ribonucleoprotein fragment (Brimacombe *et al.*, 1976). As can be seen in Fig.

32,e,f the antibody-binding site of S13 is surrounded by proteins of all three nucleation sites. On the other hand, a few examples exist which indicate that proteins of one and the same ribonucleoprotein fragment can be quite remotely located (Fig. 33 and Table VIb).

A relatively fixed position of the 3'-terminus itself is provided by the location of proteins S1 and S21, and, although more indirectly, by proteins S7 and S12. Proteins S1 and S21 have been cross-linked to the 3'-terminal adenine (Czernilofsky *et al.*, 1975). Thus far there are no comparable data as to the location of the 5'-terminus. Nonetheless, there is circumstantial evidence suggesting that the 3'- and 5'-ends of 16 S ribosomal RNA meet within the 30 S subunit (see Fig. 33).

Extensive complementarity has been found between the 3'-end of 16 S ribosomal RNA and protein synthesis initiation sites of bacteriophage RNA (Shine and Dalgarno, 1974). They suggested that the sequence 3'-HO-A-U-U̲-C̲-C̲-U̲-C̲-C̲-A̲ . . . might be involved in the ribosome binding of bacteriophage RNA by complementary base pairing. For interaction of 16 S rRNA with the initiation site of maturation protein seven base pairs could be constructed (Shine and Dalgarno, 1974).

Hybrids with extensive base pairing can also be constructed when the 5'-terminus of 16 S rRNA and various ribosome binding sites of bacteriophages are arranged in hairpin-like structures (van Knippenberg, 1975). Several tentative models for the participation of 16 S rRNA in mRNA binding have been recently proposed by van Knippenberg (1975).

If both the 3'- and the 5'-termini of 16 S rRNA were functional in the interaction of the ribosome with mRNA they should be located in close proximity to each other in the ribosome. But what is the functional significance of these base pairs? A number of reports demonstrate that ribosomes from different bacteria show cistron specificity in the translation of natural mRNA's *in vitro*, irrespective of the source of the initiation factors used (Lodish, 1969, 1970; Steitz, 1969). While 30 S subunits from *Escherichia coli, Micrococcus cryophilus,* or *Pseudomonas* can initiate translation of the coat cistron of rRNA phage R17 with high efficiency *in vitro*, 30 S subunits from *Bacillus stearothermophilus* cannot (for a summary, see Steitz and Jakes, 1975). Studies with hybrid ribosomes show that the specificity of cistron recognition within a given mRNA molecule is determined by both the 16 S RNA and ribosomal proteins (Held *et al.*, 1974a; Goldberg and Steitz, 1974; Isono and Isono, 1975). Sequence determination of the 3'-terminal ends of 16 S RNA from *E. coli* and the three other bacterial species indicated a structural basis for these species-specific differences (Shine and Dalgarno, 1974, 1975). Whether the complementarity between mRNA and the 5'-end of the 16 S RNA plays a similar role for the sequestering of mRNA is still open.

Base complementarity between the ends of ribosomal RNA's can even be extended to the 50 S subunit. Recently, two homologous tetranucleotide sequences at the 3'-termini of 16 S RNA have been detected (C. Branlant and J. P. Ebel, personal communication; van Duin *et al.*, 1976). One of them, A-C-C-U, is completely contained in the mRNA site of 16 S rRNA, the other C-U-U-A or C-C-U-U only in part. Furthermore, the 3'-end of 23 S rRNA revealed an extensive region of base complementarity with the 3'-terminal part of 16 S rRNA located at both sides of the two N^6-dimethyladenines (C. Ehresmann, C. Branlant, J. P. Ebel, personal communication; van Duin *et al.*, 1976).

Base complementarity exists also between the two ends of 23 S rRNA and between 23 S and 5 S RNA: Recently, a heptanucleotide located at the 3'-end of 23 S rRNA has been found which is complementary to seven bases at the 5'-terminus (C. Branlant and J. P. Ebel, personal communication). In addition, a sequence complementary of residues 72–83 of 5 S rRNA with a 12-nucleotide sequence at the 3'-proximal part of 23 S rRNA has been found (Herr and Noller, 1975). These data taken together would suggest that during initiation of phage RNA translation, the 3'- and 5'-ends of both the 16 S rRNA and 23 S rRNA are together with the phage RNA initiation site, and are thus located in a small confined area of the 70 S ribosome. If all the ends of rRNA meet in 70 S ribosome and if the 3'-terminal region of 16 S rRNA is actually involved in mRNA recognition, then the initiation factor binding sites should also be located in physical proximity to the ends of RNAs on the ribosomal surface. This assumption has recently been supported by cross-links obtained between IF-3 and the modified 3'-ends of both 16 S and 23 S rRNA (van Duin *et al.*, 1976). The actual existence of these base-paired regions in the ribosome and their possible functional roles remain to be shown.

As has been discussed above, a group of seven proteins is involved in or close to the codon recognition site and two additional proteins more generally in the mRNA binding site. These are proteins S1, S4, S11, S12, S13, S18, and S21 as well as S7 and S14. All these proteins, except S12, have been found to be associated with 16 S rRNA fragments. Five of them belong to the 3'-proximal group, whereas three proteins, namely, S4, S11, S18, were found in the 5'-fragment f14. Protein S12 plays an important role in determining the specificity of mRNA sequestration (Gorini, 1974). The schematic representation (Fig. 33) shows the distribution of the various proteins within the different nucleation sites or ribonucleoprotein fragments. Among these nine proteins neighborhood relationships have been found by cross-linking and/or singlet energy transfer for four protein pairs (Tables V and VI).

We have attempted to trace the course of the RNA from antibody mapping data, characterization of protein binding sites along the 16 S rRNA, and the protein composition of ribonucleoprotein fragments. Some of the suggested tertiary structural interactions within the 5'-half of 16 S rRNA and between the 3'- and 5'-halves (Mackie and Zimmermann, 1975; Ungewickell et al., 1975; Brimacombe et al., 1976) have been considered in addition. A likely approximation of the results from protein topography with those pertinent to 16 S rRNA will be reported elsewhere (Stöffler and Tischendorf, 1977).

Apart from the physical proximity, found between the 3'-adenine and proteins S1 and S21, there are so far no means for localizing any other portions of 16 S rRNA directly. The data presently available are not sufficient to suggest a space-filling model, since we can assume a total length of approximately 6000 Å (4 Å × 1500 nucleotides) for 16 S rRNA. It is, however, quite unlikely from the data discussed above and summarized in Fig. 32 and 33 that the 16 S rRNA chain starts on one pole of the subunit, then threads through it, to end finally at the opposite pole.

D. Protein Topography of the 50 S Ribosomal Subunit

A three-dimensional model showing the locations of 19 of the 50 S ribosomal proteins has been made (Figs. 29 and 30). There are insufficient data to compile models which summarize the available structural and functional information, as has been done above with the 30 S subunit.

The reasons that limit more detailed model constructions of 50 S subunits are as follows: (1) Only 19 of the 34 ribosomal proteins of the 50 S subunit have been mapped by immunoelectron microscopy. (2) A relatively small number of protein cross-links have been identified and no distance measurements by means of singlet energy transfer have been performed thus far. (3) The characterization of the RNA binding sites of individual proteins, although making fast progress, is not as advanced as they are for 30 S ribosomal proteins. (4) Ribonucleoprotein fragments from E. coli 50 S subunits have been characterized (Allet and Spahr, 1971; Kagawa et al., 1972; Newton and Brimacombe, 1974; Roth and Nierhaus, 1975). The fragments obtained in earlier publications were isolated under conditions which lead to random exchange of ribosomal proteins (Newton et al., 1975). Specific fragments as characterized by Brimacombe's or Nierhaus' group gave either small, or rather large fragments, the latter containing most of the proteins. Although the 23 S rRNA can be cleaved into two very well-defined halves (Ehresmann and Ebel, 1970; Allet and Spahr, 1971; Saha, 1974; Hartmann and Clayton, 1974), corresponding large ribonucleoprotein fragments containing nonoverlapping groups of protein have not yet been isolated.

The main reason that makes correlation of antibody mapping data with results from other techniques difficult, at least for the moment, is the lack of overlap of the various data. That is to say, that proteins found in cross-linked pairs have so far not been mapped with antibodies and vice versa.

More data are available concerning 50 S subunit proteins involved in distinct partial steps of the translational process and in various drug binding sites. A correlation of these data with the protein topography has been reviewed recently (Stöffler and Tischendorf, 1975). We shall only summarize the data in brief.

1. The Binding Sites of the Elongation Factors EF-G and EF-Tu and of Termination Factors

Proteins L7/L12 are required for the interaction of the ribosome with initiation factor IF-2, with elongation factors EF-G and EF-Tu and with release factors RF-1 and RF-2 (for references, see Möller, 1974; Stöffler, 1974). Specific antibodies to L7/L12 inhibit the interaction of these factors with the ribosome (Kischa *et al.*, 1971; Highland *et al.*, 1973, 1974a,b; Stöffler, 1974; Tate *et al.*, 1975). As has been discussed above, L7/L12-specific antibodies bind to the central protuberance at multiple sites (Figs. 29, 30, and 34*). The domain of these antibody-binding sites resemble a garland which seems, however, to be interrupted at the dorsal side of the 50 S subunit (stippled area in Figs. 30 and 34). In an attempt to visualize bound EF-G by means of electron microscopy it has been shown that EF-G has a structure of an interrupted ring which surrounds the central protuberance in 50 S·GDP·EF-G complexes stabilized by fusidic acid (G. W. Tischendorf and G. Stöffler, unpublished results). From these experiments it can be concluded that the interaction sites of these protein synthesis factors are close to the central protuberance (Figs. 29, 30, and 34).

Additional information about further proteins in the neighborhood of the ribosomal binding sites for elongation and termination factors has been obtained from experiments in which the factors were cross-linked to the ribosome. Elongation factor EF-G has been cross-linked to proteins L7/L12 and to a few additional unidentified proteins, among them probably L2 and/or L6 (Acharya *et al.*, 1973). Eight proteins, viz., L1, L5, L7/L12, L15, L20, L30, and L33 have been individually cross-linked to EF-Tu using the reagent *p*-nitrophenylchloroformate (San José *et al.*, 1976). The formation of these cross-links was dependent on the presence of amino-acyl-tRNA and β, γ-methylene-guanosine-5'-triphosphate (GMP-PCP). With the bifunctional reagent methyl-4-mercapto-butyrimidate, proteins L1, L3, L23, L24, and L28 have been recovered cross-linked to EF-Tu; proteins L23 and L28 were significantly more reactive than the others (U. Fabian, 1976). Very recent cross-linking experiments with the termina-

---

* For Fig. 34, see color figure facing page 177.

tion factor RF-2 revealed high yields of covalent complexes of RF-2 with the 50 S subunit proteins L2, L7/L12, and L11 (R. Maschler, G. Stöffler, R. Ehrlich, and T. C. Caskey, unpublished results). Comparison of these cross-linking data with the neighborhood assignments obtained by immunoelectron microscopy provides an attractive correlation (see Figs. 29, 30, and 34). Proteins cross-linkable to EF-Tu are distributed between two 50 S domains, reaching from the central protuberance (where proteins L7/L12 are located) through the seat region down to the anterior rim, where the proteins associated with the 5 S RNA (L5, L18, L25) are positioned (see Figs. 29, 30, and 34). Finally, the cross-linking data support the accepted view that proteins L7/L12 are required for the interaction of the 50 S ribosome with each of these factors (see above). However, the data also indicate that each factor interacts in addition with distinct domains on the large subunit.

2. The Location of a 5 S rRNA Protein Complex and of Proteins in the Site Related to GTP Hydrolysis

Proteins L5, L18, and L25 form a specific complex with 5 S rRNA (Gray et al., 1972, 1973; Horne and Erdmann, 1972; see Monier, 1974). This complex has ATPase and GTPase activities in vitro which are inhibited by fusidic acid and thiostrepton (Horne and Erdmann, 1973; Roth and Nierhaus, 1975). The smallest area of common exposure of these three proteins is in the middle of the 50 S subunit, at the edge, formed by the seat and the lower anterior surface (Fig. 34). For the integration of this complex into 50 S subunits two further proteins, namely, L2 and L6, are required (Gray et al., 1972, 1973; Monier, 1974).

3. Proteins at the Peptidyl Transferase Site

Partial reconstitution experiments identified protein L11 to be involved in the peptidyl transferase activity (Nierhaus and Montejo, 1973), and showed a stimulatory effect of proteins L16 and L6 (Dietrich et al., 1974). Experiments with a peptidyl-tRNA photoaffinity analogue indicated that proteins L11 and L18 are adjacent to the peptidyl transferase center (Hsiung et al., 1974). Furthermore, anti-L11 together with anti-L16 inhibited peptidyl-tRNA hydrolysis activity during termination (Tate et al., 1975). Finally, L16 has been shown to be essential for the peptidyl transferase activity (Moore et al., 1975).

More recently, the total reconstitution of the peptidyl transferase center was achieved with highly purified proteins (Hampl and Nierhaus, to be published). These experiments revealed that the proteins L3, L16, and L20 are essential for the reconstitution of the peptidyl transferase center, i.e., if one protein exerts this activity it must be one of the three proteins. Despite a recent controversy on the participation of L11 in peptidyl transferase activity (Ballesta and Vazquez, 1974; Howard and Gordon, 1974) the total reconstitution experiments demonstrate that L11 stimulates this activity. The complex role that L11 may exert during translation is supported by the requirement of this protein for thiostrepton binding (Highland *et al.*, 1975) and by its involvement in the regulation of the balanced synthesis of RNA and proteins (stringent response, Parker *et al.*, 1976).

According to our results from immunoelectron microscopy, protein L11 is exposed at two sites; its protein chain seems to proceed through the seat region, since a line connecting the two sites follows almost exactly the longitudinal and symmetry axis of the 50 S ribosome. From the data summarized above a physical proximity of proteins L11, L16, and L18 might be deduced which reasonably agrees with the antibody-binding data of the respective proteins (Fig. 34). The finding of Maassen and Möller (1974) that L11 is involved in EF-G dependent GDP binding is, however, in better agreement with the topographical data: Site L11A is in close vicinity to proteins L7/L12 and hence to the EF-G binding site (Fig. 34).

4. Affinity Labeling Experiments with Modified tRNA

Information on the binding sites of tRNA has been obtained by the use of reactive tRNA analogues as affinity label probes (for reviews, see Cantor *et al.*, 1974a; Kuechler *et al.*, 1974). Peptidyl-tRNA derivatives that bound to the P site react with L27 and L2; furthermore, proteins L11, L13, L14, L15, and L16 reacted to a lesser degree and more variably. Thus, proteins L27 (and L2) might be considered as the major P site proteins (see Fig. 34). The site of reaction on ribosomal protein L27 with an affinity label derivative of tRNA$_f^{Met}$ has recently been determined. Tryptic fingerprints prepared from the affinity labeled protein L27 indicated that the affinity labeling reaction occurs at one (or two) distinct sites of protein L27 with the ϵ-amino groups of lysine (Collatz *et al.*, 1976).

Proteins located at the ribosomal A site were investigated with tRNA analogues and affinity probes resembling chloramphenicol. Bromacetyl-Phe-tRNA bound at the A site and iodoamphenicol showed a high specificity for protein L16. Bromoamphenicol reacted with proteins L2 and L27. Reconstitution studies clearly revealed an important role of L16 for chloramphenicol binding (Nierhaus and Nierhaus, 1973). Recent attempts to localize this particular drug binding site more directly were performed with antibodies specific to chloramphenicol. These antibodies were shown to attach to chloramphenicol~50 S complexes in the region confined by the antibody-binding sites L16, L27, L19, and L1B (Stöffler, 1976; Tischendorf *et al.*, 1976; R. Bald, G. W. Tischendorf, and G. Stöffler, unpublished results, and Fig. 34).

It can be deduced with relatively great certainty that the CCA ends of aminoacyl and peptidyl tRNA meet for peptide bond formation on the right side of the seat. This area is located in a cavity between the two ribosomal subunits (see Fig. 34 and below).

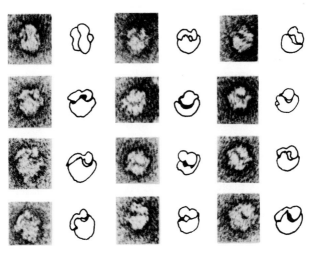

Fig. 35   Electron micrographs and interpretive drawings of 70 S monomers seen from various views. ×300,000.

E. The Structure of 70 S Monomeric Ribosomes

For an understanding of the structure and function of the ribosome it is essential to know how the two subunits interact in order to form the 70 S ribosome. It appears that the structure and the dimensions of neither subunit are grossly different in negatively stained 70 S ribosomes as compared to the isolated subunits (Fig. 35). A model of the arrangement of the two subunits has been shown in Fig. 23. The long axis of the 30 S subunit is oriented horizontally, transverse to the central protuberance of the 50 S subunit. The head of the 30 S subunit is in contact with the protein-rich side of the 50 S subunit. The body of the small subunit interacts with the left side of the seat; the contacting surfaces in this region are, in both subunits, relatively poor in antibody-binding sites. The hollow of the small subunit and the vaulted seat of the larger subunit form a tunnel. Besides this tunnel a larger cavity may exist which is formed between the two subunits in 70 S ribosomes and cannot be seen in the model (Figs. 23 and 35).

Antibody labeling studies on 70 S ribosomes are more difficult to interpret as compared to isolated subunits, since electron micrographs of 70 S ribosomes appear quite symmetric and therefore lack many of the landmarks which are required for accurate localization of antibody-binding sites. Therefore we have so far only determined the antibody attachment sites of a few proteins on 70 S antibody complexes, viz., L1, L7, L20, L27, S6, S13, and S14. These studies greatly facilitated the interpretation as to how subunits are arranged in the 70 S ribosomes; refined localization

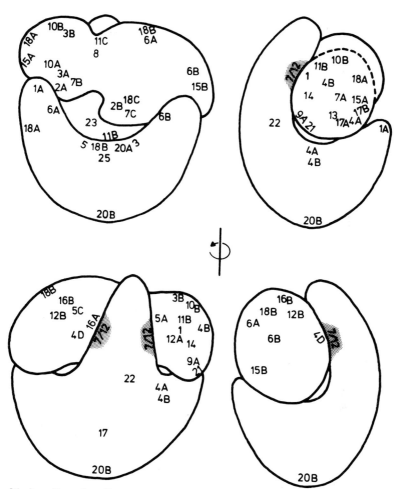

Fig. 36 Location of antibody binding sites on a three-dimensional model of the 70 S ribosome. The four projections are derived as described in legends to Figs. 28 and 29. The antibody binding sites were inferred from experiments performed with isolated subunits. Only these proteins which are fully visible in one of the projections are indicated (see Fig. 28).

of the respective antibody-binding sites is in progress. The antibody-binding sites depicted schematically in Fig. 36 have been transferred from experiments done with isolated subunits and should not be considered as real "70 S antibody-binding sites."

Due to conformational changes upon subunit reassociation additional determinants may occur, others might disappear. Only the antibody specific to protein L19 cannot bind to 70 S ribosomes although it binds to 50 S subunits. There are several binding sites found on 50 S particles to which antibody binding is difficult to conceive on 70 S ribosomes (e.g., L14 and L23). Since these antibodies also interact with 70 S particles, the detection of different antibody-binding sites is quite likely, when antibody binding to monosomes will be more thoroughly investigated.

Evidence for a conformational deformation upon subunit association has also been demonstrated by various chemical modification studies. The essence of these studies is similar: When the effects of the reagent are comparatively investigated with isolated subunits and 70 S ribosomes, some ribosomal proteins become more and others less accessible (Huang and Cantor, 1972, 1975; Chang, 1973; Miller and Sypherd, 1973; Litman and Cantor, 1974; Litman *et al.*, 1974; Michalski and Sells, 1975; Ginzburg and Zamir, 1975). Similar effects were obtained with antibodies to individual proteins: some reacted more strongly with the isolated subunit, others had stronger effects when the subunits were associated (Stöffler *et al.*, 1973; Stöffler, 1974; Zeichhardt, 1976; G. W. Tischendorf and G. Stöffler, unpublished results).

1. How the Subunits Interact

The model of 70 S ribosomes, as shown in Figs. 23, 35, and 36, demonstrates that the area of contact between the subunits is discontinuous. The region of contact can be divided into at least two but possibly into four areas. The frontal region of the 30 S head is in contact with the 50 S subunit in an area located between the central and the right lateral protuberance (area A). A second area (B) of contact is provided by the right temporal region of the 30 S head and the right side of the central protuberance. We cannot decide whether the areas A and B are continuous or interrupted. Another area of contact (C) exists between the triangular anterior region of the 30 S body on one hand and the seat region of the 50 S subunit on the other, the latter being confined by the central and the left lateral protuberance. Finally, in a fourth zone, the right lobe that protrudes from the 30 S body meets the left side of the central protuberance (area D). From the three-dimensional model there is evidence that areas C and D are disconnected. The two areas A and B are separated by the tunnel from the two other areas C and D (Figs. 23, 35, and 36).

2. 30 S Ribosomal Proteins at or near the Subunit Interface

Monovalent antibody fragments specific to five of the 30 S proteins exhibit inhibitory effects on subunit reassociation, viz., S9, S11, S12, S14, and S20 (Morrison *et al.*, 1973). Only two of them, anti-S9 and anti-S11, inhibit EF-G dependent GTP hydrolysis, a reaction that requires subunit contact (Highland *et al.*, 1974a). These data already indicate different areas of contact. Two other proteins (S11 and S12) bind to 23 S rRNA (Stöffler *et al.*, 1971). Protein S20 is identical with protein L26; they could thus be considered as one "70 S protein." The protein seems to partition with either subunit during dissociation of ribosomes (G. W. Tischendorf, G. Stöffler, and H. G. Wittmann, unpublished results; Wittmann *et al.*, 1974). From reconstitution studies proteins S5 and S9 have been implicated to be required for subunit reassociation (Marsh and Parmeggiani, 1973). Finally, a cross-link could be formed between protein S16 and one or more 50 S proteins (Sun *et al.*, 1974). In summation, at least seven of the 30 S proteins seem to be located at or near the subunit interface. A more detailed account of these data has been discussed elsewhere (Stöffler, 1974). The small inhibitory effects of antibodies to S3, S8, S10, and S21 on subunit association (Table IV in Stöffler, 1974) would also fit the topographical data discussed below.

The locations of these seven proteins and their 14 antibody binding sites, at the subunit interface are shown in Fig. 37. The agreement between the biochemical and the topographical data is quite good. All of the 14 sites were found within the contacting subunit surfaces. In each of the areas A, B, and C (defined in Section II,E,1), four sites can be encountered; two additional sites, S5B and S11C, are located in between areas B and D (Fig. 37). As mentioned above, no proteins have so far been mapped in area D. The topographical arrangement of proteins S5 and S16 also shows that antibodies to these proteins are less likely to inhibit subunit reassociation, although both proteins are at, or close to, the subunit interface (Fig. 37). One result that was always difficult to reconcile becomes quite understandable from the topographical map. Burns and Cundliffe (1973) provided evidence that the antibiotic spectinomycin has an effect on translocation. Protein S5 is altered in spectinomycin resistant ribosomes (see above). The proximity of all three sites of protein S5 to the 50 S subunit and especially to proteins L7/L12 makes these data more plausible.

3. 50 S Ribosomal Proteins at or near the Subunit Interface

The contribution of 50 S subunit proteins to subunit reassociation has been investigated by different methods (for a summary, see Table IV in Stöffler, 1974). Fourteen of the 34 proteins have been shown to be more or less involved in subunit reassociation; 11 of them have so far been mapped with antibodies. The proteins may be classified in three groups.

a. Protein L19. Anti-L19 does not react with 70 S ribosomes as opposed to 50 S subunits but is a strong inhibitor of physical reassociation (Morrison *et al.*, 1973; Stöffler, 1974; Zeichhardt, 1976; H. Zeichhardt, G. W. Tischendorf, and G. Stöffler, unpublished results). It is inhibitory for

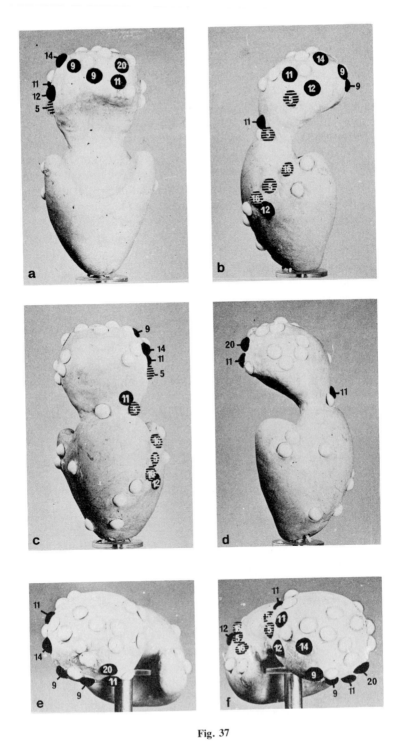

Fig. 37

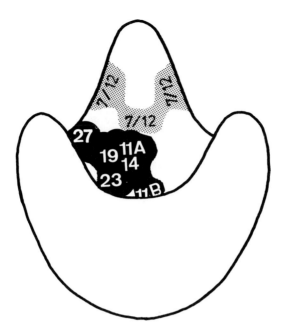

Fig. 38   Proteins at the 50 S subunit interface. Antibody binding sites of proteins located at the subunit interface by immunochemical techniques are indicated by white numbers (black background). The strippled area representing the anti-L7/L12 binding sites is only shown as a marker. (For details, see Stöffler, 1974, and text.)

functions requiring intact 70 S ribosomes, and the protein binds to 16 S rRNA.

   b. Proteins L11, L14, L23, and L27. Antibodies specific to these four proteins also strongly inhibit physical and functional reassociation. In contrast to anti-L19 they do, however, bind to 70 S ribosomes, although binding might occur to sites different from those found with 50 S subunits.

   c. Proteins L1, L2, L6, L16, L20 (L4, L10, L15, L26, S20). These antibodies exerted their effects more weakly and affected only a limited number of experimental assays (see Stöffler, 1974).

   Proteins comprised in groups a and b have their antibody sites exposed in the seat of the 50 S subunit (Fig. 38). The strongest effects on reasso-

---

Fig. 37   Proteins at the 30 S subunit interface. Proteins located at the subunit interface by immunochemical studies are indicated by numbered black circles; those found with other methods by shaded circles; the remaining proteins are represented by unnumbered white pins. (For details, see Stöffler, 1974, and text.)

ciation are correlated with the occurrence of antibody binding sites in the center of the vaulted seat. The slight effects on reassociation which were exerted by antibodies specific to the proteins of group c are also in accordance with the locations of their antibody-binding sites: most of these proteins are located within the seat, but toward the periphery. It can, therefore, be concluded that the seat region of the 50 S subunit is in contact with the 30 S particle which is in full agreement with the arrangement of subunits within 70 S monosomes, as depicted from electron micrographs. It is noteworthy that anti-L7/L12 antibodies do not interfere with subunit reassociation (Morrison et al., 1973; Highland et al., 1974a,b).

A group of monovalent Fabs, specific to proteins S9, S11, L14, L19, L23, and L27 interfere very strongly with subunit reassociation. Their antiassociation activity can be used to prepare very active ribosomal subunits (Noll et al., 1976). All these Fabs bind, at least with one site at the actual subunit interface (Figs. 37, 38). It should, however be kept in mind that an antibody which has been shown to bind at the interface by immunoelectron microscopy must, in contrast, not necessarily possess anti-association activity if its binding constant is low.

The locations of the attachment sites of those antibodies which interfere with subunit reassociation (Figs. 37, 38) support our interpretation of the subunit arrangement within the 70 S monomeric ribosome (Figs. 23, 35, 36). According to models by Lake et al. (1974b) and van Holde and Hill (1974), the interface areas B and C of the 30 S subunit would not be in contact with the 50 S subunit. In these models, the small subunit is oriented horizontally across the top of the large subunit. That is to say, the small head is on top of the large protuberance, i.e., the nose (see Figs. 1 and 2 in Lake et al., 1974b). The decision as to which of the two model structures is the correct one is not only of academic interest. Each of the models has characteristic features which have different implications with regard to ribosomal function. According to Lake's model there would be only two areas of contact between the two subunits. The tunnel formed between the two subunits is much larger in their model as compared to ours. Furthermore, in the model of Lake et al. (1974b) tRNA and mRNA could easily diffuse into the hole between the subunits, and protection of these RNA molecules from nuclease action is difficult to conceive. In contrast, a ribosome structure similar to our model would require an unlocked state for the interaction of tRNA or mRNA with ribosomal surfaces located at the subunit interface. It is a corollary of the model that mRNA as well as tRNA should be protected within 70 S ribosomes from nuclease action; similarly, the growing peptide chain should be shielded from proteases. It is hoped that three-dimensional reconstruction from tilt series of individual negatively stained ribosomes, as performed with spherical viruses (Crowther *et al.*, 1970) or the yeast fatty acid synthetase (Hoppe *et al.*, 1974) will allow an unequivocal clarification of the discrepancies between the three-dimensional models proposed by Lake and us.

In attempting to picture the mode of action of ribosomes during mRNA translation several models have been proposed (Spirin, 1969; van Holde and Hill, 1974; Rich, 1974). Spirin (1969) and van Holde and Hill (1974) suggested "locking and unlocking" or "pulsating" of the ribosomal subunits during protein synthesis. Immunochemical data also supported this assumption: anti-L19 antibodies and their monovalent Fab fragments do not bind to 70 S ribosomes. However, when the same antibodies or Fabs, specific to L19, are added to *in vitro* protein-synthesizing systems, these antibodies exerted strong inhibitory effects. It was therefore concluded that the antibody-binding site provided by protein L19 which is located at the subunit interface (Fig. 38) should become accessible during protein synthesis (Stöffler, 1974; Zeichhardt, 1976).

4. *Location of the mRNA*

As discussed in Section II,C,3,a, the mRNA binding domain is located on the head of the subunit, more precisely on its right temporal region. This area is in contact with the 50 S subunit and corresponds to the interface area B (see above). Preliminary data drawn from electron microscopical examination of *E. coli* polysomes indicated that the messenger proceed through a channel formed by the right temporal region of the small subunit and the central protuberance of the 50 S subunit (G. W. Tischendorf and G. Stöffler, unpublished results). Similar conclusions can be drawn from electron microscopic studies of rat liver polysomes (Nonomura *et al.*, 1971).

Such an entrapment of mRNA in the tunnel is in excellent agreement with biochemical studies on nuclease stability of mRNA. Castles and Singer (1969) reported that a segment of poly(U), spanning 70–120 nucleotides, is protected from nuclease attack when bound to 70 S ribosomes whereas Takanami and Zubay (1964) found this section to be smaller, namely, 25 nucleotides. A segment of natural mRNA containing 35–45 nucleotides was found to be shielded too (Steitz, 1969; Hindley and Staples, 1969). The smallest protected mRNA fragment contained information for 5 amino acids (Kuechler and Rich, 1970).

According to electron microscopy the length of the tunnel is of the order of 100 Å. The cleft formed between the central protuberance and the head adds another 30–50 Å to the entire region of mRNA protection. By assuming a length of about 4 Å per nucleotide, approximately 25–40 nucleotides should be protected, a value in agreement with the above data.

5. Where Is tRNA Located?

Transfer RNA is, like mRNA, also protected from nuclease action when bound to 70 S ribosomes (Pestka, 1968; Kuechler et al., 1972). Formation of the initiation complex is already sufficient for complete protection and tRNA was shown to endure unaffectedly even the complete digestion of both ribosomal RNA's by nucleases (Kuechler et al., 1972). Aminoacyl- and peptidyl-tRNA's are therefore likely to be entrapped in a cavity, channel, or tunnel. Rich (1974) suggested the existence of two separate channels: an aminoacyl-tRNA entrance channel and a tRNA exit channel. This suggestion can be easily accommodated by our model.

We have two fixed points in our model which aid a suggestion as to where tRNA is located in the ribosome: (1) The mRNA binding domain on the 30 S subunit should provide a topographical marker for the location of the anticodon loop; (2) Proteins L2, L16, and L27, which were associated with several tRNA affinity probes (see above), should help to localize its C-C-A end.

The structure of yeast phenylalanine tRNA was recently solved by X-ray crystallographic techniques (Kim et al., 1974; Robertus et al., 1974). The tRNA molecule has a bent, L-shaped form; one arm of the L comprises the C-C-A and the T-ψ-C stem, the other arm is composed of both the dihydrouridine and the anticodon stem. The distance from the T-ψ-C corner to the 3'-terminal adenine is ~70 Å, the distance between T-ψ-C and the anticodon is ~67 Å (see Rich, 1974). The thickness of the molecule is approximately 20 Å.

According to our model ribosomes provide sufficient space to accommodate two tRNA molecules into the cavity between the two subunits. In a frontal view of the 70 S ribosome, as shown in Fig. 23a the anticodon arm of the L would hence be oriented parallel to, and the C-C-A arm perpendicular or oblique to the paper. As a corollary of this arrangement the growing peptide chain should be protected from the action of proteases as has been suggested (Malkin and Rich, 1969). That the ribosome-bound peptide chain and hence also the C-C-A end of tRNA are located in vicinity of proteins L2, L16, and L27 is supported by preliminary findings on the location of proteins L24, L32, and L33 (see below). Their antibody-binding sites are also situated in the angle formed by the central and the right lateral protuberances. Cantor and his co-workers found that a series of peptidyl-tRNA analogues with varying chain lengths react with different proteins. According to these results, protein L2 is closest to the 3'-end of tRNA followed by L27, L32 (or L33), and last L24 (Eilat et al., 1974). The antibody-binding sites of these proteins are all in close proximity to each other and located such that protection of the growing peptide chain is very likely (G. W. Tischendorf and G. Stöffler, unpublished results).

According to the suggested model for the location of tRNA, the T-ψ-C corner should be located at the edge between the central protuberance and the seat. The T-ψ-C sequence of tRNA has been found to be a binding site for the ribosomal 5 S rRNA, which may be active during translocation (Erdmann *et al.*, 1973). There is evidence that the 5 S rRNA is required for subunit association and hence located in vicinity of the subunit interface (Yu and Wittmann, 1973). On the other hand, the 5 S rRNA binding proteins L5, L18, and L25 were mapped quite distantly from this region (Fig. 34), which does not exclude that the 5 S RNA molecule itself reaches into the seat. Supporting evidence for the occurrence of 5 S rRNA in this area comes from the location of protein L2 (Fig. 34) which is required for the formation of a complex between 5 S and 23 S rRNA (see Monier, 1974).

Several proteins have been found to react with tRNA affinity probes (see above); they should be located within a domain on the ribosomal surface. Indeed, the proteins L2, L16, and L27 were located in proximity to each other (Figs. 29, 30, and 34).

Considering the complex structure of the ribosomal particle and the completely different approaches used in studying its structure and function, the agreement between the various sets of data is surprisingly good.

F. Concluding Remarks

The elucidation of the structure of the *E. coli* ribosome has made rapid progress in recent years. This is not only true with regard to the determination of the primary structure but also for the disclosure of the spatial arrangement of the ribosomal constituents within the particle. A direct way to unravel the problem of protein topography is, at least at the present time, immunoelectron microscopy. This approach not only permitted the localization of the many ribosomal proteins on the surface of the subunits but also provided, which was at least equally important, information about the shapes of ribosomal proteins *in situ*.

In this way it has been possible to propose a model that summarizes the essential structural features of the *E. coli* ribosome, thus far known. The development of a new armada of powerful techniques, which has distinguished ribosome research during the past several years, provided various sets of independent results. It seemed propitious to summarize the presently available data on ribosome structure and/or function and to correlate them with our topographical model and in doing so, we found these other results to be in good agreement with our model. We would, however, advise caution not to overrate the conclusions drawn from the *combined* data, not only for reasons which have already been discussed in this chapter but also since the various data differ markedly in the degrees of experimental accuracy: some results are much weaker than others and a few results have in fact only been considered because no other information was available. The present discussion attempts to coordinate the current knowledge in this field of research. We have tried to sketch our view as to how further progress might be achieved and have illustrated these ideas by presenting some of our preliminary data. For example, the localization of antibiotic-binding sites with drug-specific antibodies, or, the possibility of correlating a specific region of a protein's primary structure to a particular antibody-binding site have been discussed.

We expect that the models will be a stimulus for further experiments which would lead toward an understanding of the mechanism of protein synthesis on a molecular level.

ACKNOWLEDGMENT

The authors are thankful to Richard Brimacombe, Thomas C. Caskey, Ursula Grütz-macher, Renate Hasenbank, Gilbert W. Tischendorf, and Oswald Wiener for many stimulating discussions and for their help in preparing the manuscript.

REFERENCES

Acharya, A. S., and Moore, P. B. (1973). *J. Mol. Biol.* **76**, 207–221.
Acharya, A., Moore, P. B., and Richards, F. M. (1973). *Biochemistry* **12**, 3108–3114.
Allet, B., and Spahr, P. F. (1971). *Eur. J. Biochem.* **19**, 250–255.
Amelunxen, F. (1971). *Cytobiologie* **3**, 111–126.
Amelunxen, F., and Spiess, E. (1971). *Cytobiologie* **4**, 293–306.
Arnon, R. (1971). *Curr. Top. Microbiol. Immunol.* **54**, 47–93.
Ballesta, J. P. G., and Vazquez, D. (1974). *FEBS Lett.* **48**, 226–270.
Barritault, D., Expert-Bezançon, A. E. (1975). *10th Meet. FEBS, 1975* Abstract No. 442.
Barritault, D., Expert-Bezançon, A. E., Milet, M., and Hayes, D. H. (1975). *FEBS Lett.* **50**, 114–120.
Berger, J., Geyl, D., Böck, A., Stöffler, G., and Wittmann, H. G. (1975). *Mol. Gen. Genet.* **141**, 207–211.
Bickle, T. A., Hershey, J. W. G., and Traut, R. R. (1972). *Proc. Natl. Acad. Sci. U.S.A.* **69**, 1327–1331.
Birge, E. A., and Kurland, C. G. (1969). *Science* **166**, 1282- 1284.
Bitar, K. G. (1975). *Biochim. Biophys. Acta* **386**, 99–106.
Bitar, K. G., and Wittmann-Liebold, B. (1975). *Hoppe-Seyler's Z. Physiol. Chem.* **356**, 1343–1352.
Böck, A., Ruffler, D., Piepersberg, W., and Wittmann, H. G. (1974). *Mol. Gen. Genet.* **134**, 325–332.
Bode, U., Lutter, L. C., and Stöffler, G. (1974). *FEBS Lett.* **45**, 232–236.
Bollen, A., Cedergren, R. J., Sankoff, D., and Lapalme, G. (1974). *Biochem. Biophys. Res. Commun.* **59**, 1069–1078.
Bollen, A., Heimark, R. L., Cozzone, A., Traut, R. R., Hershey, J. W. B., and Kahan, L. (1975). *J. Biol. Chem.* **250**, 4310–4314.

Bollen, A., Cabezón, T., DeWilde, M., Villarroel, R., and Herzog, A. (1975). *J. Mol. Biol.* **99**, 795–806.

Bowman, C. M., Dahlberg, J. E., Ikemura, T., Konisky, J., and Nomura, M. (1971). *Proc. Natl. Acad. Sci. U.S.A.* **68**, 964–968.

Brimacombe, R., Nierhaus, K. H., Erdmann, V. A., and Wittmann, H. G. (1976). *Prog. Nucleic Acid. Res. Mol. Biol.* **18**, 45–90.

Brosius, J., and Chen, R. (1976). *FEBS Lett.* **68**, 105–109.

Brosius, J., Schiltz, E., and Chen, R. (1975). *FEBS Lett.* **56**, 359–361.

Bruskov, V. I., and Kiselev, N. A. (1968). *J. Mol. Biol.* **37**, 367–377.

Burgess, A. W., Ponnuswamy, P. K., and Scheraga, H. A. (1974). *Isr. J. Chem.* **12**, 239–286.

Burns, D. J. W., and Cundliffe E. (1973). *Eur. J. Biochem.* **37**, 570–574.

Cabezón, T., Herzog, A., DeWilde, M., Villarroel, R., and Bollen, A. (1976). *Mol. Gen. Genet.* **144**, 59–62.

Cantor, C. R., Pellegrini, M., and Oen, H. (1974a). *In* "Ribosomes" (M. Nomura, A. Tissières, and P. Lengyel, eds.), pp. 573–585. Cold Spring Harbor Lab., Cold Spring Harbor, New York.

Cantor, C. R., Huang, K., and Fairclough, R. (1974b). *In* "Ribosomes" (M. Nomura, A. Tissières, and P. Lengyel, eds.), pp. 587–599. Cold Spring Harbor Lab., Cold Spring Harbor, New York.

Castles, B., and Singer, M. (1969). *J. Mol. Biol.* **40**, 1–17.

Chang, F. N. (1973). *J. Mol. Biol.* **78**, 563–568.

Chang, F. N., and Flaks, J. G. (1972). *J. Mol. Biol.* **68**, 177–180.

Chen, R., and Ehrke, G. (1976a). *FEBS Lett.* **69**, 240–245.

Chen, R., and Ehrke, G. (1976b). *FEBS Lett.* **63**, 215–217.

Chen, R., and Wittmann-Liebold, B. (1975). *FEBS Lett.* **52**, 139–140.

Chen, R., Mende, L., and Arfsten, U. (1975). *FEBS Lett.* **59**, 96–99.

Clegg, C., and Hayes, D. H. (1974). *Eur. J. Biochem.* **42**, 21–28.

Collatz, E., Küchler, E., Stöffler, G., and Czernilofsky, A. P. (1976). *FEBS Lett.* **63**, 283–286.

Craven, G. R., Rigby, B., and Changchien, L. M. (1974). *In* "Ribosomes" (M. Nomura, A. Tissières, and P. Lengyel, eds.), pp. 559–571. Cold Spring Harbor Lab., Cold Spring Harbor, New York.

Crowther, R. A., Amos, L. A., Finch, J. T., Rosier, D. J., and Klug, A. (1970). *Nature (London)* **226**, 421–425.

Crumpton, M. J. (1974). *In* "The Antigens" (M. Sela, ed.), Vol. 2, pp. 1–78. Academic Press New York.

Czernilofsky, A. P., Kurland, C. G., and Stöffler, G. (1975). *FEBS Lett.* **58**, 281–284.

Dabbs, E., and Wittmann, H. G. (1976). *Mol. Gen. Genet.* **149**, 303–309.

Dahlberg, J. E. (1974). *J. Biol. Chem.* **249**, 7673–7678.

DeWilde, M., Michel, F., and Broman, K. (1974). *Mol. Gen. Genet.* **133**, 329–333.

De Wilde, M., Cabezón, T., Villarroel, R., Herzog, A., and Bollen, A. (1975). *Mol. Gen. Genet.* **142**, 19–33.

Dietrich, S., Schrandt, I., and Nierhaus, K. H. (1974). *FEBS Lett.* **47**, 136–139.

Dovgas, N. V., Markova, L. F., Mednikova, T. A., Vinokurov, L. M., Alakhov, Y. B., and Ovchinnikov, Y. A. (1975). *FEBS Lett.* **53**, 351–354.

Dovgas, N. V., Vinokurov, L. M., Velmoga, I. S., Alakhov, Yu. B., and Ovchinnikov, Yu. A. (1976). *FEBS Lett.* **67**, 58–61.

Ehresmann, B., and Reinbolt, J. (1975). *10th Meet. FEBS, Paris, 1975* Abstract No. 431.

Ehresmann, C., and Ebel, J. P. (1970). *Eur. J. Biochem.* **13**, 577–582.

Ehresmann, C., Stiegler, P., Mackie, G. A. Zimmermann, R. A., Ebel, J. P., and Fellner, P. (1975a). *Nucleic Acids Res.* **2**, 265–278.

Ehresmann, C., Stiegler, P., Fellner, P., and Ebel, P. (1975b). *Biochimie* **57**, 711–748.

Eilat, D., Pellegrini, M., Oen, H., Lapidot, Y., and Cantor, C. R. (1974). *J. Mol. Biol.* **88**, 831–840.

Engelman, D. M., and Moore, P. B. (1972). *Proc. Natl. Acad. Sci. U.S.A.* **69**, 1997–1999.

Engelman, D. M., Moore, P. B., and Schoenborn, B. P. (1975a). *Proc. Natl. Acad. Sci. U.S.A.* **72**, 3888–3892.

Engelman, D. M., Moore, P. B., and Schoenborn, B. P. (1975b). *Brookhaven Symp. Biol.* **27**, IV20–IV37.

Erdmann, V. A., Sprinzl, M., and Pongs, O. (1973). *Biochem. Biophys. Res. Commun.* **54**, 942–948.

Fabian, U. (1976). *FEBS Lett.* **71**, 256–260.

Fellner, P. (1974). *In* "Ribosomes" (M. Nomura, A. Tissières, and P. Lengyel, eds.), pp. 169–191. Cold Spring Harbor Lab., Cold Spring Harbor, New York.

Fiser, I., Scheit, K. H., Stöffler, G., and Kuechler, E. (1974). *Biochem. Biophys. Res. Commun.* **60**, 1112–1118.

Fiser, I., Scheit, K. H., Stöffler, G., and Kuechler, E. (1975a). *FEBS Lett.* **56**, 226–229.

Fiser, I., Margaritella, P., and Kuechler, E. (1975b). *FEBS Lett.* **52**, 281–283.

Friesen, J. D., Parker, J., Watson, R. J., Fiil, N. P., and Pedersen, S. (1976). *Mol. Gen. Genet.* **144**, 115–118.

Funatsu, G., and Wittmann, H. G. (1972). *J. Mol. Biol.* **68**, 547–550.

Garrett, R. A., and Wittmann, H. G. (1973). *Acta Endocrinol. (Copenhagen), Suppl.* **180**, 75–94.

Gaunt-Klöpfer, M., and Erdmann, V. A. (1975). *Biochim. Biophys. Acta* **390**, 226–230.

Ginzburg, I., and Zamir, A. (1975). *J. Mol. Biol.* **93**, 465–476.

Ginzburg, I., Miskin, R., and Zamir, A. (1973). *J. Mol Biol.* **79**, 481–494.

Goldberg, M. L., and Steitz, J. A. (1974). *Biochemistry* **13**, 2123–2128.

Gorini, L. (1974). *In* "Ribosomes" (M. Nomura, A. Tissières, and P. Lengyel, eds.), pp. 791–803. Cold Spring Harbor Lab., Cold Spring Harbor, New York.

Gray, P. N., Garrett, R. A., Stöffler, G., and Monier, R. (1972). *Eur. J. Biochem.* **28**, 412–421.

Gray, P. N., Bellemare, G., Monier, R., Garrett, R. A., and Stöffler, G. (1973). *J. Mol. Biol.* **77**, 133–152.

Groner, Y., Scheps, R., Kamen, R., Kolakofsky, D., and Revel, M. (1972). *Nature (London)* **239**, 19–20.

Hall, C. E., and Slayter, H. S. (1959). *J. Mol. Biol.* **1**, 329–332.

Hardy, J. S. (1975). *Mol. Gen. Genet.* **140**, 253–274.

Hart, R. G. (1962). *Biochim. Biophys. Acta* **60**, 629–637.

Hartmann, K. A., and Clayton, N. W. (1974). *Biochim. Biophys. Acta* **335**, 201–210.

Hasenbank, R., Guthrie, G., Stöffler, G., Wittmann, H. G., Rosen, L., and Apirion, D. (1973). *Mol. Gen. Genet.* **127**, 1–18.

Hawley, D. A., Slobin, L. I., and Wahba, A. J. (1974). *Biochem. Biophys. Res. Commun.* **61**, 544–550.

Heiland, I., Brauer, D., and Wittmann-Liebold, B. (1976). *Hoppe-Seyler's Z. Physiol. Chem.* **357**, 1751–1770.

Heimark, R. L., Kahan, L., Johnston, K., Hershey, J. W. B., and Traut, R. R. (1976). *J. Mol. Biol.* **105**, 219–230.

Held, W. A., Gette, W. R., and Nomura, M. (1974a). *Biochemistry* **13**, 2115–2122.

Held, W. A., Ballou, B., Mizushima, S., and Nomura, M. (1974b). *J. Biol. Chem.* **249**, 3103–3111.

Helser, T. L., Davies, J. E., and Dahlberg, J. E. (1971). *Nature (London), New Biol.* **233**, 12–14.

Helser, T. L., Davies, J. E., and Dahlberg, J. E. (1972). *Nature (London), New Biol.* **235**, 6–9.

Herr, W., and Noller, H. F. (1975). *FEBS Lett.* **53**, 248–252.

Highland, J. H., Bodley, J. W., Gordon, J., Hasenbank, R., and Stöffler, G. (1973). *Proc. Natl. Acad. Sci. U.S.A.* **70**, 142–150.

Highland, J. H., Ochsner, E., Gordon, J., Bodley, J. W., Hasenbank, R., and Stöffler, G. (1974a). *Proc. Natl. Acad. Sci. U.S.A.* **71**, 627–630.

Highland, J. H., Ochsner, E., Gordon, J., Hasenbank R., and Stöffler, G. (1974b). *J. Mol. Biol.* **86**, 175–178.

Highland, J. H., Howard, G. A., Ochsner, E., Stöffler, G., Hasenbank, R., and Gordon, J. (1975). *J. Biol. Chem.* **250**, 1141–1145.

Hill, W. E., and Fessenden, R. J. (1974). *J. Mol. Biol.* **90**, 719–726.

Hill, W. E., Thompson, J. D., and Anderegg, J. W. (1969). *J. Mol. Biol.* **44**, 89–102.

Hindley, J., and Staples, D. H. (1969). *Nature (London)* **224**, 964–967.

Hitz, H., Schäfer, D., and Wittmann-Liebold, B. (1975). *FEBS Lett.* **56**, 259–262.

Hoppe, W. (1972). *Isr. J. Chem.* **10**, 321–333.

Hoppe, W. (1973). *J. Mol. Biol.* **78**, 581–585.

Hoppe, W., Gassmann, J., Hunsmann, N., Schramm, H. J., and Sturm, M. (1974). *Hoppe Seyler's Z. Physiol. Chem.* **355**, 1483–1487.

Hoppe, W., May, R., Stöckel, P., Lorenz, S., Erdmann, V. A., Wittmann, H. G., Crespi, H. L., Katz, J. J., and Ibel, K. (1975). *Brookhaven Symp. Neutron Scattering Biol.* **27**, IV38–IV48.

Horne, J. R., and Erdmann, V. A. (1972). *Mol. Gen. Genet.* **119**, 337–344.

Horne, J. R., and Erdmann, V. A. (1973). *Proc. Natl. Acad. Sci. U.S.A.* **70**, 2870–2873.

Howard, G. A., and Gordon, J. (1974). *FEBS Lett.* **48**, 271–274.

Hsiung, N., Reines, S. A., and Cantor, C. R. (1974). *J. Mol. Biol.* **88**, 841–855.

Huang, K., and Cantor, C. R. (1972). *J. Mol. Biol.* **67**, 265–275.

Huang, K., and Cantor, C. R. (1975). *J. Mol. Biol.* **97**, 423–441.

Huang, K., Fairclough, R. H., and Cantor, C. R. (1975). *J. Mol. Biol.* **97**, 443–470.

Huxley, H. W., and Zubay, G. (1960). *J. Mol. Biol.* **2**, 10–18.

Inouye, H., Pollack, Y., and Petre, J. (1974). *Eur. J. Biochem.* **45**, 109–117.

Isono, K., Krauss, J., and Hirota, Y. (1976). *Mol. Gen. Genet.* **149**, 297–302.

Isono, S., and Isono, K. (1975). *Eur. J. Biochem.* **56**, 15–22.

Itoh, T., and Wittmann, H. G. (1973). *Mol. Gen. Genet.* **127**, 19–32.

Jaskunas, S. R., Lindahl, L., and Nomura, M. (1975). *Nature (London)* **256**, 183–187.

Jeantet, G., Maschler, R., Tischendorf, G. W., Crépin, M., Thibault, J., Lelong, J. C., Gros, F., and Stöffler, G. (1977). To be submitted.

Kagawa, H., Jishuken, L, and Tokimatsu, H., (1972). *Nature (London)* **237**, 74–75.

Kahan, L., Zengel, J., Nomura, M., Bollen, A., and Herzog, A. (1973). *J. Mol. Biol.* **76**, 473–483.

Kenner, R. A. (1973). *Biochem. Biophys. Res. Commun.* **51**, 932–938.

Kim, S. H., Suddath, F. L., Quigley, G. J., McPherson, A., Sussman, J. L., Wang, A. H. J., Seeman, N. C., and Rich, A. (1974). *Science* **185**, 435–439.

Kischa, K., Möller, W., and Stöffler, G. (1971). *Nature (London)* **233**, 62–63.

Kiselev, N. A., Stel'mashchuk, V. Ya., Lerman, M. I., and Abakumova, O. Yu. (1974). *J. Mol. Biol.* **86**, 577–586.

Kratky, O. (1971). *Chimia* **25**, 96–97.

Kuechler, E., and Rich, A. (1970). *Nature (London)* **225**, 920–924.

Kuechler, E., Bauer, K., and Rich, A. (1972). *Biochim. Biophys. Acta* **277**, 615–627.

Kuechler, E., Hauptmann, R., Czernilofsky, A. P., Fiser, J., Barta, A., Voorma, H. O., Stöffler, G., and Scheit, K. H. (1974). *Acta Biol. Med. Ger.* **33**, 633–637.

Kurland, C. G. (1960). *J. Mol. Biol.* **2**, 83–91.

Kurland, C. G. (1972). *Annu. Rev. Biochem.* **41**, 377–408.

Lake, J. A., and Kahan, L. (1975). *J. Mol. Biol.* **99**, 631–644.

Lake, J. A., Pendergast, M., Kahan, L., and Nomura, M. (1974a). *Proc. Natl. Acad. Sci. U.S.A.* **71**, 4688–4692.

Lake, J. A., Sabatini, D. D., and Nonomura, Y. (1974b). *In* "Ribosomes" (M. Nomura, A. Tissières, and P. Lengyel, eds.), pp. 543–557. Cold Spring Harbor Lab., Cold Spring Harbor, New York.

Lelong, J. C., Gros, D., Gros, F., Bollen, A., Maschler, R., and Stöffler, G. (1974). *Proc. Natl. Acad. Sci. U.S.A.* **71**, 248–252.

Lindemann, H., and Wittmann-Liebold, B. (1976). *FEBS Lett.* **71**, 251–255.

Liou, Y. F., Yoshikawa, M., and Tanaka, N. (1975). *Biochem. Biophys. Res. Commun.* **65**, 1096–1101.

Litman, D. J., and Cantor, C. R. (1974). *Biochemistry* **13**, 512–518.

Litman, D. J., Lee, C. C., and Cantor, C. R. (1974). *FEBS Lett.* **47**, 268–271.

Lodish, H. F. (1969). *Nature (London)* **224**, 867–870.

Lodish, H. F. (1970). *Nature (London)* **226**, 705–707.

Lubin, M. (1968). *Proc. Natl. Acad. Sci. U.S.A.* **61**, 1454–1461.

Lührmann, R., Gassen, G., and Stöffler, G. (1976). *Eur. J. Biochem.* **66**, 1–9.

Lutsch, G., Noll, F., and Bielka, H. (1974). *Acta Biol. Med. Ger.* **33**, 813–816.

Lutter, L. C., Zeichhardt, H., Kurland, C. G., and Stöffler, G. (1972). *Mol. Gen. Genet.* **119**, 357–366.

Lutter, L. C., Bode, U., Kurland, C. G., and Stöffler, G. (1974a). *Mol. Gen. Genet.* **129**, 167–176.

Lutter, L. C., Ortandel, F., and Fasold, H. (1974b). *FEBS Lett.* **48**, 228–292.

Maassen, J. A., and Möller, W. (1974). *Proc. Natl. Acad. Sci. U.S.A.* **71**, 1277–1280.

Mackie, G. A., and Zimmermann, R. A. (1975). *J. Biol. Chem.* **250**, 4100–4112.

Malkin, L. I., and Rich, A. (1969). *J. Mol. Biol.* **26**, 329–346.

Marsh, R. C., and Parmeggiani, A. (1973). *Proc. Natl. Acad. Sci. U.S.A.* **70**, 151–155.

Maschler, R. (1973). *Ph.D. Thesis, Freie Universität,* Berlin (1973).

Michalski, C. J., and Sells, B. H. (1975). *Eur. J. Biochem.* **52**, 385–389.

Midgley, J. E. M. (1965). *Biochim. Biophys. Acta* **108**, 340–347.

Miller, R. V., and Sypherd, P. S. (1973). *J. Mol. Biol.* **78**, 539–550.

Mizushima, S., and Nomura, M. (1970). *Nature (London)* **226**, 1214–1218.

Möller, K., and Brimacombe, R. (1975). *Mol. Gen. Genet.* **141**, 343–355.

Möller, W. (1974). *In* "Ribosomes" (M. Nomura, A. Tissières, and P. Lengyel, eds.), pp. 771–731. Cold Spring Harbor Lab. Cold Spring Harbor, New York.

Monier, R. (1974). *In* "Ribosomes" (M. Nomura, A. Tissières, and P. Lengyel, eds.), pp. 141–168. Cold Spring Harbor Lab. Cold Spring Harbor, New York.

Moore, P. B. (1971). *J. Mol. Biol.* **60**, 169–184.

Moore, P. B., Engelman, D. M., and Schoenborn, B. P. (1974). *In* "Ribosomes" (M. Nomura, A. Tissières, and P. Lengyel, eds.), pp. 601–613. Cold Spring Harbor Lab. Cold Spring Harbor, New York.

Moore, V. G., Atchison, R. E., Thomas, G., Morgan, M., and Noller, H. F. (1975). *Proc. Natl. Acad. Sci. U.S.A.* **72**, 844–848.

Morgan, J., and Brimacombe, R. (1972). *Eur. J. Biochem.* **29**, 542–552.

Morgan, J., and Brimacombe, R. (1973). *Eur. J. Biochem.* **37**, 472–480.

Morinaga, T., Funatsu, G., Funatsu, M., and Wittmann, H. G. (1976). *FEBS Lett.* **64**, 307–309.

Morrison, C. A., Garrett, R. A., Zeichhardt, H., and Stöffler, G. (1973). *Mol. Gen. Genet.* **127**, 359–368.

Morrison, C. A., Tischendorf, G. W., Stöffler, G., and Garrett, R. A. (1977). To be submitted.

Muto, A., Otaka, E., Itoh, T., Osawa, S., and Wittmann, H. G. (1975). *Mol. Gen. Genet.* **140**, 1–5.

Nanninga, N., Garrett, R. A., Stöffler, G., and Klotz, G. (1972). *Mol. Gen. Genet.* **119**, 175–184.
Newton, I., and Brimacombe, R. (1974). *Eur. J. Biochem.* **48**, 513–518.
Newton, I., Rinke, J., and Brimacombe, R. (1975). *FEBS Lett.* **51**, 215–218.
Nierhaus, D., and Nierhaus, K. H. (1973). *Proc. Natl. Acad. Sci. U.S.A.* **70**, 2224–2228.
Nierhaus, K. H., and Montejo, V. (1973). *Proc. Natl. Acad. Sci. U.S.A.* **70**, 1931–1935.
Noll, H., Noll, M., Hapke, B., and van Dieijen, G. (1973). *In* "Regulation of Transcription and Translation in Eukaryotes" (E. Bautz, ed.), pp. 257–311. Springer-Verlag, Berlin and New York.
Noll, M., Stöffler, G., and Highland, J. H. (1976). *J. Mol. Biol.* **108**, 259–264.
Noller, H. F., and Herr, W. (1974). *Mol. Biol. Rep.* **1**, 437.
Noller, H. F., Chang, C., Thomas, G., and Aldridge, J. (1971). *J. Mol. Biol.* **61**, 669–679.
Nomura, M., Mizushima, S., Ozaki, M., Traub, P., and Lowry, C. V. (1969). *Cold Spring Harbor Symp. Quant. Biol.* **34**, 49–61.
Nonomura, Y., Blobel, G., and Sabatini, D. (1971). *J. Mol. Biol.* **60**, 303–323.
Okuyama, A., Yoshikawa, M., and Tanaka, N. (1974). *Biochem. Biophys. Res. Commun.* **60**, 1163–1169.
Ortega, J. P., and Hill, W. E. (1973). *Biochemistry* **12**, 3241–3243.
Ozaki, M., Mizushima, S., and Nomura, M. (1969). *Nature (London)* **222**, 333–339.
Parker, J., Watson, R. J., Friesen, J. D., and Fiil, N. P. (1976). *Mol. Gen. Genet.* **144**, 111–114.
Peretz, H., Towbin, H., and Elson, D. (1976). *Eur. J. Biochem.* **63**, 83–92.
Pestka, S. (1968). *J. Biol. Chem.* **243**, 4038–4044.
Piepersberg, W., Böck, A., and Wittmann, H. G. (1975a). *Mol. Gen. Genet.* **140**, 91–100.
Piepersberg, W., Böck, A., Yaguchi, M., and Wittmann, H. G. (1975b). *Mol. Gen. Genet.* **143**, 43–52.
Pilz, I., Kratky, O., Licht, A., and Sela, M. (1973). *Biochemistry* **12**, 4998–5005.
Pongs, O., and Lanka, E. (1975a). *Hoppe Seyler's Z. Physiol. Chem.* **356**, 449–458.
Pongs, O., and Lanka, E. (1975b). *Proc. Natl. Acad. Sci. U.S.A.* **72**, 1505–1509.
Pongs, O., Nierhaus, K. H., Erdmann, V. A., and Wittmann, H. G. (1974). *FEBS Lett.* **40**, S28–S37.
Pongs, O., Bald, R., Erdmann, V. A., and Reinwald, E. (1975a). *In* "Topics in Infectious Diseases. Drug Receptor Interactions in Antimicrobial Chemotherapy" (J. Drews and F. E. Hahn, eds.), Vol. 1, pp. 179–190. Springer-Verlag, Wien and New York.
Pongs, O., Stöffler, G., and Lanka, E. (1975b). *J. Mol. Biol.* **99**, 301–315.
Pongs, O., Stöffler, G., and Bald, R. W. (1976). *Nucleic Acids Res.* **3**, 1635–1646.
Rich, A. (1974). *In* "Ribosomes" (M. Nomura, A. Tissières, and P. Lengyel, eds.), pp. 871–884. Cold Spring Harbor Lab., Cold Spring Harbor, New York.
Ritter, E., and Wittmann-Liebold, B. (1975). *FEBS Lett.* **60**, 153–155.
Robertus, J. D., Ladner, J. E., Finch, J. T., Rhodes, D., Brown, R. S., Clark, B. F. C., and Klug, A. (1974). *Nature (London)* **250**, 546–551.
Robinson, S., and Wittmann-Liebold, B. (1977). In preparation.
Roth, H. E., and Nierhaus, K. H. (1973). *FEBS Lett.* **31**, 35–38.
Roth, H. E., and Nierhaus, K. H. (1975). *J. Mol. Biol.* **94**, 111–121.
Rummel, D. P., and Noller, H. F. (1973). *Nature (London)* **245**, 72–75.
Saha, B. K. (1974). *Biochim. Biophys. Acta* **353**, 292–300.
San José, C., Kurland, C. G., and Stöffler, G. (1976). *FEBS Lett.* **71**, 133–137.
Santer, M., and Santer, U. V. (1973). *J. Bacteriol.* **116**, 1304–1313.
Sarma, V. R., Silverton, E. W., Davies, D. R., and Terry, W. D. (1971). *J. Biol. Chem.* **246**, 3753–3759.
Schiltz, E., and Reinbolt, J. (1975). *Eur. J. Biochem.* **56**, 467–481.
Sela, M. (1966). *Adv. Immunol.* **5**, 29–129.

Sela, M., and Mozes, E. (1966). *Biochemistry* **55**, 445–452.
Shelton, E., and Kuff, E. L. (1966). *J. Mol. Biol.* **22**, 23–31.
Shine, J., and Dalgarno, L. (1974). *Proc. Natl. Acad. Sci. U.S.A.* **71**, 1342–1346.
Shine, J., and Dalgarno, L. (1975). *Nature (London)* **254**, 34–38.
Smith, W. S. (1971). PhD. Thesis, University of Wisconsin, Madison.
Smolarsky, M., and Tal, M. (1970). *Biochim. Biophys. Acta* **213**, 401–416.
Sommer, A., and Traut, R. R. (1974). *Proc. Natl. Acad. Sci. U.S.A.* **71**, 3946–3950.
Sommer, A., and Traut, R. R. (1975). *J. Mol. Biol.* **97**, 471–481.
Sommer, A., and Traut, R. R. (1976). *J. Mol. Biol.* **106**, 995–1015.
Spiess, E. (1973). *Cytobiologie* **7**, 28–32.
Spirin, A. S. (1969). *Cold Spring Harbor Symp. Quant. Biol.* **34**, 197–207.
Spirin, A. S. (1974). *FEBS Lett.* **40**, S28–S47.
Spirin, A. S., and Gavrilova, L. P. (1969). *In* "The Ribosome." Springer-Verlag, Berlin and New York.
Spirin, A. S., Kiselev, N. A., Shakulov, R. S., and Bogdanov, A. A. (1963). *Biokhimiya* **28**, 765.
Spitnik-Elson, P., Zamir, A., Miskin, R., Kaufmann, Y., Greenman, B., Breiman, A., and Elson, D. (1972). *FEBS Symp.* **23**, 175–195.
Stadler, H. (1974). *FEBS Lett.* **48**, 114–116.
Stadler, H., and Wittmann-Liebold, B. (1976). *Eur. J. Biochem.*, **66**, 49–56.
Stanley, W. M., Jr., and Bock, R. M. (1965). *Biochemistry* **4**, 1302–1311.
Steitz, J. A. (1969). *Nature (London)* **224**, 957–964.
Steitz, J., and Jakes, K. (1975). *Proc. Nat. Acad. Sci. U.S.A.* **72**, 4734–4738.
Stöffler, G. (1974). *In* "Ribosomes" (M. Nomura, A. Tissières, and P. Lengyel, eds.), pp. 615–667. Cold Spring Harbor Lab., Cold Spring Harbor, New York.
Stöffler, G. (1976). Abstract of the Tenth International Congress of Biochemistry, Hamburg, Germany, p. 87.
Stöffler, G., and Tischendorf, G. W. (1975). *In* "Topics in Infectious Diseases. Drug Receptor Interactions in Antimicrobial Chemotherapy" (J. Drews and F. E. Hahn, eds.), Vol. 1, pp. 117–143. Springer-Verlag, Wien and New York.
Stöffler, G., and Tischendorf, G. W. (1977). In preparation.
Stöffler, G., and Wittmann, H. G. (1971a). *Proc. Natl. Aad. Sci U.S.A.* **68**, 2283–2287.
Stöffler, G., and Wittmann, H. G. (1971b). *J. Mol. Biol.* **62**, 407–409.
Stöffler, G., Daya, L., Rak, K. H., and Garrett, R. A. (1971). *J. Mol. Biol.* **62**, 411–414.
Stöffler, G., Hasenbank, R., Lütgehaus, M., Maschler, R., Morrison, C. A., Zeichhardt, H., and Garrett, R. A. (1973). *Mol. Gen. Genet.* **127**, 89–110.
Stöffler, G., Tischendorf, G. W., and Zeichhardt, H. (1974). *Meet. Am. Chem. Soc.*
Stuhrmann, H. B., Haas, J., Ibel, K., Wolf, B. De, Koch, M. H. J., Parfait, R., and Crichton, R. R. (1976). *Proc. Nat. Acad. Sci. U.S.A.* **73**, 2379–2383.
Subramanian, R. R. (1975). *J. Mol. Biol.* **95**, 1–8.
Sun, T. T., Bollen, A., Kahan, L., and Traut, R. R. (1974). *Biochemistry* **13**, 2334–2340.
Szer, W., and Leffler, S. (1974). *Proc. Natl. Acad. Sci. U.S.A.* **71**, 3611–3615.
Szer, W., Hermoso, J. M., and Leffler, S. (1975). *Proc. Natl. Acad. Sci. U.S.A.* **72**, 2325–2329.
Takanami, M., and Zubay, G. (1964). *Proc. Natl. Acad. Sci. U.S.A.* **51**, 834–839.
Tal, M., Aviram, M., Kanarek, A., and Weiss, A. (1972). *Biochim. Biophys. Acta* **281**, 381–392.
Tate, W. P., Caskey, C. T., and Stöffler, G. (1975). *J. Mol. Biol.* **93**, 375–389.
Terhorst, C. P., Möller, W., Laursen, R., and Wittmann-Liebold, B. (1973). *Eur. J. Biochem.* **34**, 138–152.

Tesche, B. (1975). *Vakuum-Technik* **4**, 104–110.

Thammana, P., and Held, W. A. (1974). *Nature (London)* **251**, 682–686.

Thammana, P., Kurland, C. G., Deusser, E., Weber, J., Maschler, R., Stöffler, G., and Wittmann, H. G. (1973). *Nature (London)* **242**, 47–49.

Tinoco, I., Jr., Uhlenbeck, O. C., and Levine, M. D. (1971). *Nature (London)* **230**, 362–367.

Tischendorf, G. W., and Stöffler, G. (1975). *Mol. Gen. Genet.* **142**, 193–208.

Tischendorf, G. W., and Stöffler, G. (1976a). Abstract of the Tenth International Congress of Biochemistry, Hamburg, Germany, p. 113.

Tischendorf, G. W., and Stöffler, G. (1976b). *Z. Immun.-Forsch. (Immunobiology)* **152**, 120.

Tischendorf, G. W., Tesche, B., and Stöffler, G. (1976). *Proc. 6th Eur. Congr. Electron Microsc., Jerusalem* **2**, 524–526.

Tischendorf, G. W., and Stöffler, G. (1977) (in preparation).

Tischendorf, G. W., Zeichhardt, H., and Stöffler, G. (1974a). *Mol. Gen. Genet.* **134**, 187–208.

Tischendorf, G. W., Zeichhardt, H., and Stöffler, G. (1974b). *Mol. Gen. Genet.* **134**, 209–223.

Tischendorf, G. W., Zeichhardt, H., and Stöffler, G. (1975). *Proc. Natl. Acad. Sci. U.S.A.* **72**, 4820–4824.

Tolbert, W. R. (1971). Ph.D. Thesis, University of Wisconsin, Madison.

Traut, R. R., Heimark, R. L., Sun, T. T., Hershey, J. W. B., and Bollen, A. (1974). *In* "Ribosomes" (M. Nomura, A. Tissières, and P. Lengyel, eds.), pp. 271–308. Cold Spring Harbor Lab., Cold Spring Harbor, New York.

Uchida, T., Bonen, L., Schaup, H. W., Lewis, B. J., Zablen, L., and Woese, C. (1974). *J. Mol. Evol.* **3**, 63–77.

Ungewickell, E., Garrett, R. A., Ehresmann, C., Stiegler, P., and Fellner, P. (1975). *Eur. J. Biochem.* **51**, 165–180.

van Acken, U. (1975). *Mol. Gen. Genet.* **140**, 61–68.

Vandekerckhove, J., Rombauts, W., Peeters, B., and Wittmann-Liebold, B. (1975). *Hoppe Seyler's Z. Physiol. Chem.* **356**, 1955–1976.

van Dieijen, G., van der Laken, C. J., van Knippenberg, P. H., and van Duin, J. (1975). *J. Mol. Biol.* **93**, 351–364.

van Duin, J., and Kurland, C. G. (1970). *Mol. Gen. Genet.* **109**, 169–176.

van Duin, J., and van Knippenberg, P. H. (1974). *J. Mol. Biol.* **84**, 185–195.

van Duin, J., Kurland, C. G., Dondon, J., and Grunberg-Manago, M. (1975). *FEBS Lett.* **59**, 287–304.

van Duin, J., Kurland, C. G., Dondon, J., Grunberg-Manago, M., Branlant, C., and Ebel, J. P. (1976). *FEBS Lett.* **62**, 111–114.

van Holde, K. E., and Hill, W. E. (1974). *In* "Ribosomes" (M. Nomura, A. Tissières, and P. Lengyel, eds.), pp. 53–91. Cold Spring Harbor Lab., Cold Spring Harbor, New York.

van Knippenberg, P. H. (1975). *Nucleic Acids Res.* **2**, 79–85.

Vasiliev, V. D. (1971). *FEBS Lett.* **14**, 203–205.

Vasiliev, V. D. (1974) *Acta Biol. Med. Ger.* **33**, 779–793.

Wabl, M. (1973). Ph.D. Thesis, Freie Universität, Berlin.

Wabl, M. R. (1974). *J. Mol. Biol.* **84**, 241–247.

Wabl, M. R., Barends, P. J., and Nanninga, N. (1973a). *Cytobiologie* **7**, 1–9.

Wabl, M. R., Doberer, H. G., Höglund, S., and Ljung, L. (1973b). *Cytobiologie* **7**, 111–115.

Wahba, A. J., Miller, M. J., Niveleau, A., Landers, T. A., Carmichael, G. C., Weber, K., Hawley, D. A., and Slobin, L. I. (1974). *J. Biol. Chem.* **249**, 3314–3316.

Weber, H. J. (1972). *Mol. Gen. Genet.* **119**, 233–248.

Werner, R., Kollak, A., Nierhaus, D., Schreiner, G., and Nierhaus, K. H. (1975). *In* "Topics in Infectious Diseases. Drug Receptor Interactions in Antimicrobial Chemotherapy" (J. Drews and F. E. Hahn, eds.), Vol. 1, pp. 217–234. Springer-Verlag, Wien and New York.

Wittmann, H. G. (1974). *In* "Ribosomes" (M. Nomura, A. Tissières, and P. Lengyel, eds.), pp. 93–114. Cold Spring Harbor Lab. Cold Spring Harbor, New York.

Wittmann, H. G., and Wittmann-Liebold, B. (1974). *In* "Ribosomes" (M. Nomura, A. Tissières, and P. Lengyel, eds.), pp. 115–140. Cold Spring Harbor Lab., Cold Spring Harbor, New York.

Wittmann, H. G., Stöffler, G., Piepersberg, W., Buckel, P., Ruffler, D., and Böck, A. (1974). *Mol. Gen. Genet.* **134**, 225–236.

Wittmann, H. G., Stöffler, G., Geyl, D., and Böck, A. (1975a). *Mol. Gen. Genet.* **141**, 317–329.

Wittmann, H. G., Yaguchi, M., Piepersberg, W., and Böck, A. (1975b). *J. Mol. Biol.* **98**, 827–829.

Wittmann-Liebold, B. (1973). *FEBS Lett.* **36**, 247–249.

Wittmann-Liebold, B., and Dzionara, M. (1976a). *FEBS Lett.* **61**, 14–19.

Wittmann-Liebold, B., and Dzionara, M. (1976b). *FEBS Lett.* **65**, 281–283.

Wittmann-Liebold, B., and Pannenbecker, R. (1976). *FEBS Lett.* **68**, 115–118.

Wittmann-Liebold, B., Geissler, A. W., and Marzinzig, E. (1975a). *J. Supramol. Struct.* **3**, 426–447.

Wittmann-Liebold, B., Greuer, B., and Pannenbecker, R. (1975b). *Hoppe-Seyler's Z. Physiol. Chem.* **356**, 1977–1979.

Wittmann-Liebold, B., Marzinzig, E., and Lehmann, A. (1976). *FEBS Lett.* **68**, 110–114.

Yaguchi, M. (1975). *FEBS Lett.* **59**, 217–220.

Yaguchi, M., Wittmann, H. G., Cabezón, T., DeWilde, M., Villarroel, R., Herzog, A., and Bollen, A. (1976). *J. Mol. Biol.* **104**, 617–620.

Yoshikawa, M., Okuyama, A., and Tanaka, N. (1975). *J. Bacteriol.* **122**, 796–797.

Yu, R. S. T., and Wittmann, H. G. (1973). *Biochim. Biophys. Acta* **319**, 388–400.

Yuki, A., and Brimacombe, R. (1975). *Eur. J. Biochem.* **56**, 23–34.

Zamir, A., Miskin, R., and Elson, D. (1971). *J. Mol. Biol.* **60**, 347–364.

Zamir, A., Miskin, R., Vogel, Z., and Elson, D. (1974). *In* "Methods in Enzymology" (L. Grossman and K. Moldare, eds.), Vol. 30, pp. 406–601. Academic Press, New York.

Zeichhardt, H. (1976). Ph.D. Thesis, Freie Universität, Berlin.

Zimmermann, R. A. (1974). *In* "Ribosomes" (M. Nomura, A. Tissières, and P. Lengyel, eds.), pp. 225–269. Cold Spring Harbor Lab. Cold Spring Harbor, New York.

Zimmermann, R. A., Mackie, G. A., Muto, A., Garrett, R. A., Ungewickell, E., Ehresmann, C., Stiegler, P., Ebel, J. P., and Fellner, P. (1975). *Nucleic Acids Res.* **2**, 279–302.

Zubke, G., Stadler, H., Ehrlich, R., Stöffler, G., Wittmann, H. G., and Apirion, D. (1977). *Mol. Gen. Genet.* (in press).

4

Affinity Labeling of Ribosomes

MARIA PELLEGRINI AND CHARLES R. CANTOR

I. Introduction

Knowledge of the functional topography of ribosomal components and translational factors is required for the development of a molecular model of protein synthesis. The assembly map of Mizushima and Nomura (1970) combined with chemical cross-linking and electron microscopic studies of antibody-labeled ribosomes have established the general positions of certain proteins with respect to each other and the ribosomal factors (see Chapters 2 and 3). Another general approach to this problem has been to selectively remove specific proteins from the ribosome and test the remaining "core" particles for binding and catalytic sites (Nierhaus and Montejo, 1973). However, with all these methods it is very difficult to distinguish a direct from an indirect effect on the regulation of an active site. Affinity labeling offers a direct approach to identifying the components at specific functional sites. Here we shall review the current state of progress of this technique as applied to bacterial ribosomes. Much has been learned about the methodology of affinity labeling through experiments with ribosomes. These may well serve as the model for similar studies of other complex macromolecular systems, especially those that involve protein-nucleic acid interactions.

II. General Principles of Affinity Labeling

Chemical modification is one of the most informative techniques for characterizing the structure-function relationships of biologically important macromolecules. Although X-ray crystallographic studies offer much more detailed information, many large and complex systems are not readily amenable to this approach. Affinity labels are a specific class of reagents used for the chemical modification of active sites. They combine the steric complementarity of a natural substrate with a suitably small reactive group. These reagents bind to an active site and subsequently form a covalent bond to that site (for general reviews, see Singer, 1967; Shaw, 1970).

The principles for the design of affinity label reagents and the experimental criteria for specific site labeling have been well defined in efforts to characterize the active site of some proteases, nucleases, and antibodies. More complicated systems including the ribosome have recently been explored.

A. Reagent Design

A compound designed as an affinity label must contain (1) those structural elements required for binding to the functional site and (2) a group capable of covalent reaction. It should mimic as closely as possible the binding characteristics of the "natural" substrate. The reactive group should be of a size and located such that it causes minimal interference with substrate binding site interactions. Finally, the labeled products of an affinity reagent-substrate interaction must be readily identifiable. This is frequently accomplished by employing a radioactive affinity label.

The reactivity of the labeling group is an important consideration. Clearly, the labeling group should not be readily hydrolyzed or reactive toward buffers or electrolytes present in the binding system. This group must not be so reactive as to randomly label parts of the enzyme, yet it must be sufficiently reactive to form a covalent bond to an active site

α-Halocarbonyl: $X-CH_2-\overset{\overset{O}{\|}}{C}-R$ $X = Cl, Br, I$
                          $N:$

p-Nitrophenyl ester: $O_2N-\langle\bigcirc\rangle-O-\overset{\overset{O}{\|}}{C}-R$ $:N$ = nucleophile
                                              $:N$

component. Examples of this type of reagent are α-halocarbonyls and p-nitrophenyl esters. They are reactive toward nucleophiles.

An alternative to a mildly chemically reactive group is a photoactivated reagent. Examples of such reagents are diazoacetyl compounds and azides. Upon irradiation they yield carbenes and nitrenes, respectively, both high energy species which have an extremely broad spectrum of reactivity and are highly active in insertion reactions. Since diazoacetyl compounds and azides are relatively chemically inert as compared to the species they generate upon photoexcitation, it is possible to selectively control the time of covalent reaction. This is generally not possible for the chemically reactive affinity labels.

$$
\text{Diazoacetyl:} \quad R-\overset{\overset{\displaystyle O}{\|}}{C}-CH-\overset{+}{N}\equiv\overset{-}{N} \xrightarrow{h\nu} -\overset{\overset{\displaystyle O}{\|}}{C}-\overset{\overset{\displaystyle H}{|}}{C}: \; + \; N_2\uparrow
$$

$$
\text{Azide:} \quad R-N_3 \xrightarrow{h\nu} R-N: \; + \; N_2\uparrow
$$

$$
\text{Nitrene:} \quad R-N: \; + \; \overset{H}{\underset{H}{\overset{\displaystyle}{C}}}\!\!\!\!\overset{H}{\underset{|}{=}}\!\!R' \longrightarrow \overset{H}{\underset{R}{N}}\!-\!\overset{\overset{\displaystyle H}{|}}{\underset{|}{C}}-R'
$$
$$
\text{(or } R-C:) \qquad\qquad H \qquad\qquad\qquad\qquad H
$$

However, there are two major problems with photoactivated labels. Reaction yields with the macromolecules are usually low. This may be due to reaction with the solvent or internal rearrangement of the excited molecule. (For a review of photoaffinity labeling techniques, see Knowles, 1972.) In some cases the excited species may dissociate from the substrate-label complex before modification takes place and react at other random sites (Ruoho et al., 1973).

B. Site-Specific Reaction

The site specificity requirements for an affinity labeling reaction are best elucidated by asking the following questions: (1) Does the label bind to the active site? (2) Does covalent reaction occur? (3) How does this reaction relate to binding, i.e., can a competitive substrate prevent covalent reaction and/or inactivation of the active site? (4) Can proximity or concentration rate enhancement be shown for the reaction? A high local concentration of reagent at the binding site should ensure preferential reaction at that site. This situation should mimic an intramolecular reaction as compared to an intermolecular reaction between the label and random sites on the macromolecule.

C. Affinity Labeling of Ribosomes

Although affinity label studies have been performed with macromolecules involved in the protein synthesis system other than ribosomes

(Jonah and Rychlik, 1973; Bruton and Hartley, 1970; Santi and Cunnion, 1974), only those concerned directly with ribosomes will be discussed here. An affinity labeling experiment with ribosomes poses a number of potential problems which would not be as serious in a simpler system. It is not clear how well defined a given ribosomal binding site will be. Not only does the ribosome contain more than 50 components, but many of them have reactive functional groups (for illustrative data, see Hsiung and Cantor, 1973; Huang and Cantor, 1972). Thus, there is the strong possibility of nonspecific reaction, even with a carefully designed affinity analogue. Both subunits of *E. coli* ribosomes, as actually isolated, are known to be heterogeneous with respect to their protein content (Deusser *et al.*, 1974). Furthermore, in most ribosome preparations only a small fraction of the 70 S particles are fully active in protein synthesis. Therefore a low-yield chemical modification, even if highly specific, could very likely arise from interaction between the affinity analogue and nonfunctional ribosomes. An additional problem is inherent in the current procedures for analysis of ribosomal proteins. These involve long dialyses, centrifugation, and protein stripping (Traub *et al.*, 1971). There may be loss of labeled product and/or the possibility for nonspecific reaction of label not covalently bound in the initial incubation.

III. Applications to Ribosomes

The macromolecular complex responsible for protein synthesis is the ribosome. It contains several binding sites and catalytic sites. It binds mRNA, tRNA, and a number of protein factors as well as small molecules, such as GTP, ppGpp, pppGpp, and various antibiotics. The potential for affinity labeling experiments on the ribosome is great. In addition, the protein factors involved in this system which bind both small and large substrates might be profitable targets for affinity labeling.

Chemical modification studies using reagents other than affinity labels have shown that the rRNA and several different amino acid side chains of many proteins are available for reaction. Tabulations of the reactivity of 30 S and 50 S proteins are presented in Huang and Cantor (1972) and Litman and Cantor (1974), respectively. In some cases the reactivity of specific amino acid side chains can be correlated to inhibition of an active site or sites. For example, Traugh and Traut (1972) found that phosphorylation of ribosomal proteins via serine and threonine side chains destroyed protein synthesizing activity. Sulfhydryl reagents, such as *N*-ethylmaleimide can inhibit such ribosome functions as subunit association (Traut and Haenni, 1967; Moore, 1971), fMet-tRNA and aminoacyl-tRNA binding (Ginzburg *et al.*, 1973), and translation of f2 RNA (Singer and

Conway, 1973). 5,5'-Dithiobis(nitrobenzoic acid) also inhibits subunit association (Acharya and Moore, 1973), while reaction with *p*-chloromercuribenzoate stimulates translocation in the absence of EFT and EFG (Gavrilova and Spirin, 1974).

These data indicated the presence of reactive amino acid side chains at or near active centers and suggested that further insight into the nature of several ribosomal active sites could be obtained from affinity label studies.

A. Design and Synthesis of Affinity Labels for Ribosomes

Several affinity labels have been prepared for ribosomes. These include derivatives of aminoacyl and peptidyl-tRNA, mRNA, GTP, and antibiotics. Both photoactivated and chemically reactive analogues have been constructed.

1. Peptidyl Transferase—P' Site

The acceptor site (A site) and the donor site (P site) of 70 S ribosomes are responsible for the binding of aminoacyl- and peptidyl-tRNA, respectively. The A' and P' sites are defined as sections of peptidyl transferase responsible for the binding of the 3'-terminal parts of these molecules and are involved in peptide bond formation (Pestka, 1972; Harris and Symons, 1973).

Initial affinity labeling experiments with ribosomes employed a label comprised of a reactive group attached to a tRNA molecule rather than to a small molecule. *N*-bromoacetyl phenylalanyl-tRNA (BrAcPhe-tRNA was made by the condensation of the *N*-hydroxysuccinimide ester of α-bromoacetic acid and phenylalanyl-tRNA (Pellegrini *et al.*, 1972). This was designed as a peptidyl-tRNA analogue.

The Phe-tRNA is prepared enzymatically from phenylalanine, tRNA^{Phe}, and the appropriate synthetase. This procedure preserves the structural fidelity of this part of the affinity label. Subsequent acetylation of the α-amino group of aminoacyl-tRNA using succinimide esters is a mild procedure which does not disrupt the ribosome binding capacity of tRNA (de Groot *et al.*, 1966). This technique employs biological substrates and biological joining reactions or very mild procedures, thereby producing a label very similar to the natural ribosomal substrate. In contrast, for chemically prepared substrate analogues, such structural similarity to the natural substrate may not be so easily obtained.

As shown in Fig. 1, the bromoacetyl phenylalanyl moiety closely mimics peptidyl-tRNA. Although the bromine group is large, it occupies the same position as a peptide side chain on native peptidyl-tRNA.

More recently, the *N*-bromoacetyl derivative of Met-tRNA$_f^{Met}$ has been

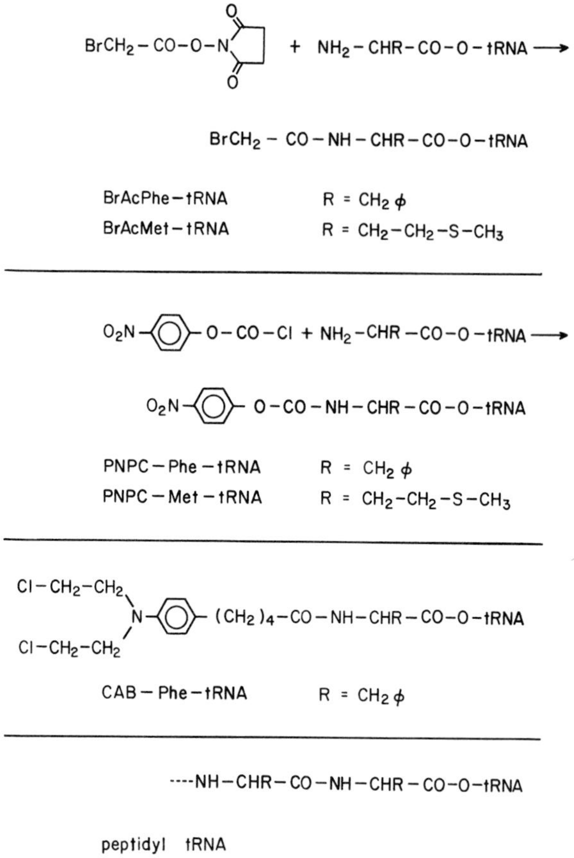

$$BrCH_2 - CO-O-N \underset{O}{\overset{O}{\diamond}} + NH_2 - CHR-CO-O-tRNA \longrightarrow$$

$$BrCH_2 - CO - NH - CHR - CO - O - tRNA$$

BrAcPhe−tRNA   R = $CH_2\,\phi$
BrAcMet−tRNA   R = $CH_2-CH_2-S-CH_3$

$$O_2N-\bigcirc-O-CO-Cl + NH_2-CHR-CO-O-tRNA \longrightarrow$$

$$O_2N-\bigcirc-O-CO-NH-CHR-CO-O-tRNA$$

PNPC−Phe−tRNA   R = $CH_2\,\phi$
PNPC−Met−tRNA   R = $CH_2-CH_2-S-CH_3$

$$\begin{matrix} Cl-CH_2-CH_2 \\ \diagdown \\ N-\bigcirc-(CH_2)_4-CO-NH-CHR-CO-O-tRNA \\ \diagup \\ Cl-CH_2-CH_2 \end{matrix}$$

CAB − Phe−tRNA   R = $CH_2\,\phi$

$$\cdots\cdots NH-CHR-CO-NH-CHR-CO-O-tRNA$$

peptidyl tRNA

Fig. 1 Synthesis and structure of some affinity analogues of peptidyl-tRNA: BrAcPhe-tRNA (Pellegrini *et al.*, 1972); BrAcMet-tRNA (Sopori *et al.*, 1974); PNPC-Phe-tRNA (Czernilofsky and Kuechler, 1972); PNPC-Met-tRNA (Hauptmann *et al.*, 1974); CAB-Phe-tRNA (Bochkareva *et al.*, 1971).

prepared (Sopori *et al.*, 1974). Its binding to ribosomes is initiation factor- and natural mRNA-dependent and it is 70% reactive with puromycin. Both reagents described above are modeled after the halocarbonyl moieties which have been successfully used to probe the active sites of the serine proteases and certain nucleases (Singer, 1967; Shaw, 1970). They are known to react with such amino acids as serine, histidine, cysteine, lysine, and methionine (Means and Feeney, 1971) and also with nucleic acid bases (Smith and Yamane, 1967; Yang and Söll, 1973). They are only mildly reactive toward water and amine buffers at the substrate concentrations needed for effective binding of tRNA to ribosomes.

Kuechler and his co-workers have prepared similar analogues of both Phe and Met-tRNA (Czernilofsky and Kuechler, 1972; Czernilofsky *et al.*, 1973; Hauptmann *et al.*, 1974) with a different chemically reactive moiety. They synthesized *p*-nitrophenylcarbamylphenylalanyl-tRNA (PNPC-Phe-tRNA) and the corresponding Met-tRNA$_f^{Met}$ analogue (see Fig. 1). *p*-Nitrophenyl chloroformate was condensed with aminoacyl-tRNA to obtain these derivatives. Again, the advantages of an enzymatically prepared substrate, Phe or Met-tRNA, were available here.

PNPC-Phe-tRNA was bound to ribosomes in the presence of poly(U), while PNPC-Met-tRNA$_f^{Met}$ was bound in the presence of a natural mRNA, R17 bacteriophage RNA, and initiation factors (Hauptmann *et al.*, 1974). *p*-Nitrophenyl esters react easily with amino or hydroxyl groups.

When either BrAcPhe-tRNA or PNPC-Phe-tRNA is used as an affinity label the resulting linkage between the label and the aminoacyl side chain is short. This suggests that reaction can only occur with a ribosomal component that is in close proximity to the aminoacyl moiety.

N-Chloroambucyl-Phe-tRNA (CAB-Phe-tRNA) was synthesized by Bochkareva *et al.*, (1971). This is a mustard analogue of peptidyl-tRNA. Although the reactive group of this compound, the mustard side chain, is further from the aminoacyl group than in the cases of PNPC-Phe-tRNA or BrAcPhe-tRNA, it extends in the same general direction as natural peptidyl-tRNA (as shown in Fig. 1).

There is a major disadvantage in using such chemically reactive labels. If, by chance, the peptidyl transferase has no available nucleophilic residue or if these reagents are incorrectly oriented for nucleophilic displacement, no reaction would occur despite the proximity of a ribosomal component and the label.

In order to overcome this difficulty photoactivated analogues of peptidyl-tRNA have been prepared as probes for investigating the peptidyl transferase center. A diazoester analogue of peptidyl-tRNA, EDM-Phe-tRNA, has been reported by Bispink and Matthaei (1973). Two azide analogues that have been studied in more detail are *p*-(2-nitro-4-azidophenoxy)phenylacetyl phenylalanyl-tRNA [NAP-Phe-tRNA (Hsiung *et al.*, 1974)] and *N*-(2-nitro-4-azidophenyl) glycylphenylalanyl-tRNA [NAG-Phe-tRNA (Hsiung and Cantor, 1974)]. The structure of these three labels is shown in Fig. 2. To synthesize NAP-Phe-tRNA or NAG-Phe-tRNA, 1-fluoro-2-nitro-4-azidobenzene was reacted with either a hydroxy acid or an amino acid. The acid was then derivatized to a succinimide ester and reacted with aminoacyl-tRNA. The azide moieties of both these reagents are relatively inert toward water, amino acids, and nucleic acids. However, once they are photoexcited, the azides generate aryl nitrenes. These species are highly reactive and are not dependent on

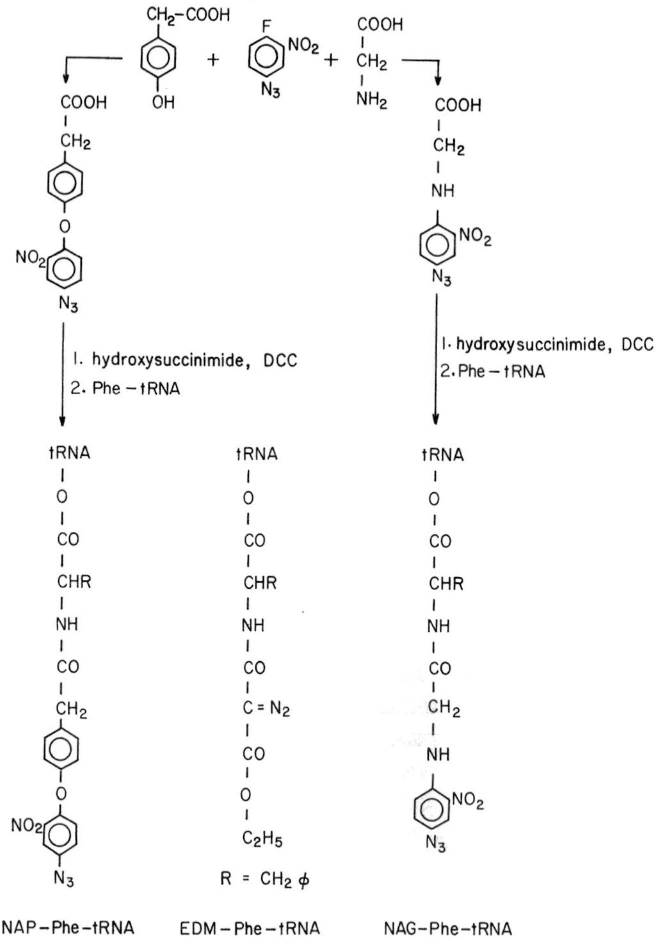

CH₂–COOH

(ring) + (ring with F, NO₂, N₃) + COOH | CH₂ | NH₂

COOH
|
CH₂
(ring)
OH

COOH
|
CH₂
(ring)
|
O
NO₂ (ring) N₃

COOH
|
CH₂
|
NH
(ring) NO₂
N₃

1. hydroxysuccinimide, DCC
2. Phe –tRNA

1. hydroxysuccinimide, DCC
2. Phe –tRNA

tRNA
|
O
|
CO
|
CHR
|
NH
|
CO
|
CH₂
(ring)
O
NO₂ (ring)
N₃

tRNA
|
O
|
CO
|
CHR
|
NH
|
CO
|
C = N₂
|
CO
|
O
|
C₂H₅

R = CH₂ φ

tRNA
|
O
|
CO
|
CHR
|
NH
|
CO
|
CH₂
|
NH
(ring) NO₂
N₃

NAP–Phe–tRNA   EDM–Phe–tRNA   NAG–Phe–tRNA

Fig. 2 Synthesis and structure of some photoaffinity analogues of peptidyl-tRNA: NAP-Phe-tRNA (Hsiung *et al.*, 1974); NAG-Phe-tRNA (Hsiung and Cantor, 1974); EDM-Phe-tRNA (Bispink and Matthaei, 1973).

electrophilic or nucleophilic displacement for reaction. In addition, the timing of the covalent reaction can be controlled with these reagents.

A structural comparison of NAG-Phe-tRNA and NAP-Phe-tRNA with the BrAc and PNPC derivatives is of interest. The reactive site of NAG-Phe-tRNA is approximately one amino acid closer to the 3′-terminus of the tRNA than NAP-Phe-tRNA, but it is one amino acid length further from the terminus than BrAcPhe-tRNA or PNPC-Phe-tRNA (see Figs. 1 and 2). The hydrophobic character of these reagents also varies. NAP has two phenyl rings, PNPC and NAG have only one, while BrAc has none. The relative hydrophobic nature of an amino acid side chain may affect both its position and peptide transfer reactivity in the fragment reaction (Nathans and Neidle, 1963), although this same center obviously accepts both hydrophobic and hydrophilic amino acids.

Affinity analogues of tRNA's altered at their 3'-termini were constructed to study the P' site binding characteristic of these molecules. Aminoacyl-tRNA$_{ox-red}$ is similar to aminoacyl-tRNA except it lacks a vicinal diol at its 3'-terminal sugar group. This tRNA analogue is formed by oxidation of the 3'-terminal diol to a dialdehyde and then reduction to a dialcohol. tRNA$_{ox-red}^{Phe}$ from yeast can accept phenylalanine and bind to ribosomes (Cramer et al., 1968; Ofengand and Chen, 1972; M. Tal et al., 1972). However, it does not function in poly(U)-directed polyphenylalanine synthesis. In order to explore the P' binding site of this tRNA, bromoacetyl-Phe tRNA$_{ox-red}$ was synthesized (M. Pellegrini, J. Ofengand and C. R. Cantor, unpublished results). It binds to 70 S ribosomes and covalently reacts with them.

A tRNA that contains the terminal sequence C-A instead of the usual C-C-A can be prepared and aminoacylated (J. Tal et al., 1972; Deutscher, 1973). BrAcPhe-tRNA (C-A) was prepared and used to affinity label ribosomes (Eilat et al., 1974b).

2. Peptidyl Transferase—A' Site

The A' site constitutes the "second half" of the peptidyl transferase center. Chemical affinity labels have also been prepared to map this site.

De Groot et al. (1971) showed that peptidyl-tRNA can be directed to the A site by blocking the P site (with deacylated tRNA) at high Mg^{2+} concentrations. Likewise, the peptidyl-tRNA analogue, BrAcPhe-tRNA, can be positioned in the A site where it covalently reacts with ribosomal proteins (Eilat et al., 1974a).

An affinity analogue of aminoacyl-tRNA (containing a free α-amino group) has not yet been synthesized. However, analogues of certain antibiotics known to bind in or overlap with the A' site have been constructed. Puromycin is an antibiotic that binds at the A' site. It acts as an acceptor for the peptidyl moiety of peptidyl-tRNA. Two analogues of this antibiotic have been synthesized as affinity labels.

Pongs and co-workers (1973b) developed the reagent α-N-iodoacetyl puromycin. It is a peptidyl puromycin (Fig. 3). This molecule mimics puromycin which has already accepted a peptide from a tRNA bound to the P site. Alternatively, Symons and co-workers (Harris et al., 1973; Greenwell et al., 1974) synthesized an analogue of puromycin 5'-O-(N-bromoacetyl-p-aminophenylphosphoryl)-3'-N-L-phenylalanyl puromycin (BAP-Pan-Phe). This compound is shown in Fig. 3. It contains an un-

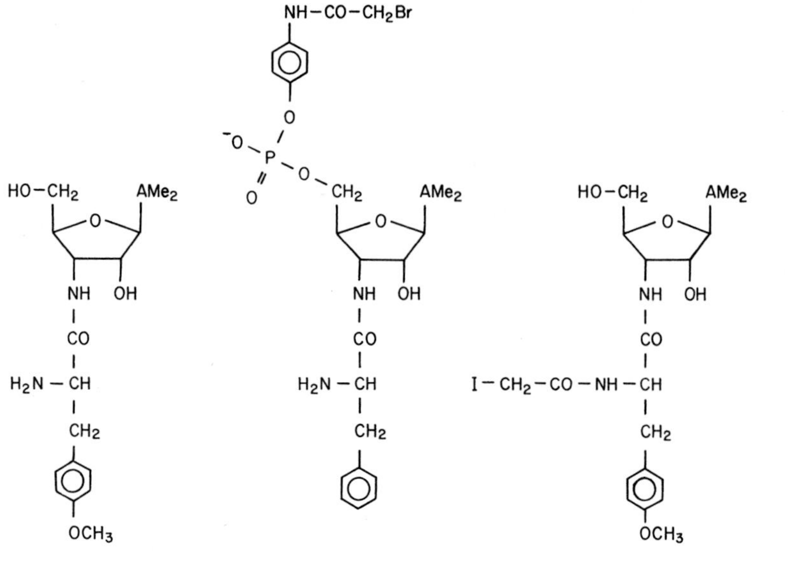

Puromycin BAP—Pan—Phe *N*—Iodoacetyl puromycin

Fig. 3 Two affinity analogues of the antibiotic puromycin: BAP-Pan-Phe (Harris *et al.*, 1973); *N*-iodoacetyl puromycin (Pongs *et al.*, 1973b).

blocked α-aminoacyl group attached to the 3′-ribose position. Thus BAP-Pan-Phe should more closely mimic the binding of puromycin to the A′ site. It should also be capable of reaction with peptidyl-tRNA, unlike *N*-iodoacetyl puromycin.

Chloramphenicol (CAP) is a strong but reversible inhibitor of protein biosynthesis. It acts at or near the peptidyl transferase center. There appear to be two ribosomal binding sites for CAP, at least one of which overlaps the binding site of puromycin and the 3′-terminus of aminoacyl-tRNA, i.e., the A′ site (Lessard and Pestka, 1972).

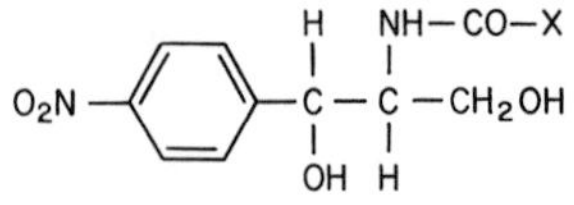

Chloramphenicol X = $CHCl_2$

Bromamphenicol X = CH_2–Br

Iodoamphenicol X = CH_2–I

Fig. 4 Chloramphenicol and its affinity analogues: bromamphenicol (Sonenberg *et al.*, 1973); iodoamphenicol (Pongs *et al.*, 1973a).

Two affinity analogues of CAP have been developed (Fig. 4). They employ either an iodoacetyl (Pongs *et al.*, 1973a) or a bromoacetyl (Sonenberg *et al.*, 1973) group to replace the dichloroacetyl group of chloramphenicol. Both were synthesized according to the method of Rebstock (1950). Neither derivatization greatly perturbs the original antibiotic structure, yet both labels covalently react with ribosomes while CAP does not. Very recently, Sonenberg *et al.* (1974) have shown direct covalent attachment of choramphenicol itself to ribosomes is induced by ultraviolet irradiation.

3. A and P Sites

Schwartz and Ofengand (1974) used tRNA affinity labels to identify the ribosomal components of the tRNA binding site which recognize structural features of the substrate other than the 3′-terminus. They designed a photoactivated tRNA label consisting of a tRNA^{Val} with *p*-azidophenacetyl attached at the sulfur atom of its 4-thiouridine residue (see Fig. 5). This should be located at least 30 Å away from the aminoacyl end. Under appropriate *in vitro* conditions this reagent can be bound to either the A or P sites and once bound photoactivated to react with any nearby ribosomal components. The derivatization of the 4-thiouridine moiety does not affect the factor-dependent binding of the label or its ability to be aminoacylated.

One possibility for further labeling the A and P sites would be to position a reactive group at the anticodon loop of tRNA, perhaps by excising the Y base and replacing it with a more active side chain. The use of a reactive 3′-terminal nucleoside or a deacylated tRNA would be other possibilities. Such reagents would place a reactive group closer to the actual site of peptide catalysis than previous labels have done.

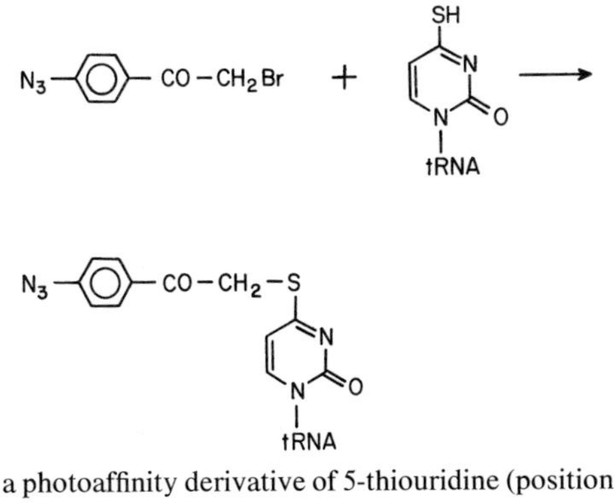

Fig. 5 Synthesis of a photoaffinity derivative of 5-thiouridine (position 8) of *E. coli* tRNA^{Val} (Schwartz and Ofengand, 1974).

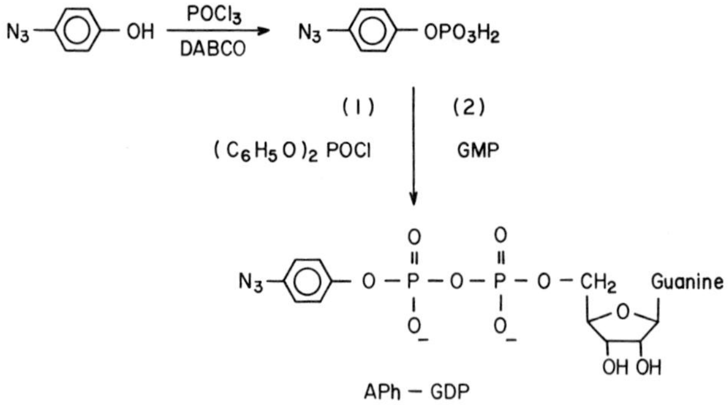

$$N_3-\!\!\bigcirc\!\!-OH \xrightarrow[\text{DABCO}]{POCl_3} N_3-\!\!\bigcirc\!\!-OPO_3H_2$$

(1) | (2)

$(C_6H_5O)_2$ POCl | GMP

$$N_3-\!\!\bigcirc\!\!-O-\overset{\overset{O}{\parallel}}{\underset{\underset{O_-}{|}}{P}}-O-\overset{\overset{O}{\parallel}}{\underset{\underset{O_-}{|}}{P}}-O-CH_2 \quad \text{Guanine}$$

OH OH

APh – GDP

Fig. 6   A photoaffinity analogue of GDP (Maasen and Möller, 1974).

4. Site of GTP Hydrolysis

Guanosine triphosphate is hydrolyzed to GDP in the presence of ribosomes and certain translational factors. To explore the ribosomal site of GTP hydrolysis, Maassen and Möller (1974) used a photolabile GDP analogue. It was tritiated in the guanine moiety and esterified with 4-azidophenol at the β-phosphate group, i.e., 1-(4-azidophenyl)-2-(5'-guanyl) pyrophosphate (APh-GDP) (see Fig. 6). The photoactive group maintains a position similar to that of the γ-phosphate of GTP. An EFT-GDP-ribosome complex can be stabilized by fusidic acid. This allows specific placement of the APh-GDP analogue on the ribosome prior to photoactivation. Upon irradiation, covalent linkage to the 50 S subunit occurs.

Bodley and Gordon (1974) have described the interaction of periodate oxidized GDP and GTP with the ribosome. These are potentially affinity labels but no evidence for covalent binding could be found.

5. Binding Site for the Growing Peptide Chain

The binding site for the growing peptide chain is located on the 50 S ribosomal subunit (Gilbert, 1963; Beller and Lubsen, 1972). Little was known about this ribosomal binding site beyond a rough estimate of its length (Malkin and Rich, 1967). In order to map this site in more detail, Eilat et al. 1974a) synthesized a series of chemically reactive peptidyl-tRNA's with the general structure BrAc(Gly)$_n$ Phe-tRNA (n = 1 to 16). They employed the method of Lapidot et al. (1967) (see Fig. 7). Procedures for the characterization of oligopeptidyl-tRNA's are well established (Lapidot et al., 1969). Although one cannot conclude that a preformed peptidyl-tRNA added to ribosomes will occupy the same binding site as the naturally growing peptide chain, there is evidence that these

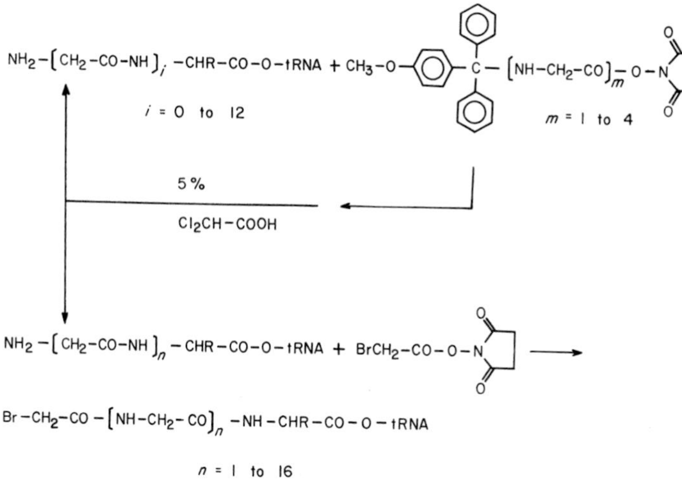

Fig. 7   Synthesis of oligopeptidyl tRNA affinity analogues (Eilat *et al.*, 1974b).

synthetic peptidyl-tRNA's do occupy a definite strong binding site on the ribosome and are active in the puromycin reaction. These data indicate the correct placement of at least the 3'-termini of these tRNA's (Hamburger *et al.*, 1972; Beller and Lubsen, 1972; Panet, 1970).

6. mRNA Binding Site

The small ribosomal subunit contains the binding site for messenger RNA. Several affinity labels have been developed to map this region of the 30 S subunit. Fiser *et al.* (1974) used the synthetic homopolymer, poly-4-thiouridylic acid, as a photoaffinity label (Fig. 8). Poly-4-thiouridine acts as a mRNA for polyphenylalanine synthesis in an *in vitro* protein-synthesizing system (Bahr *et al.*, 1973). Upon irradiation at 300 to

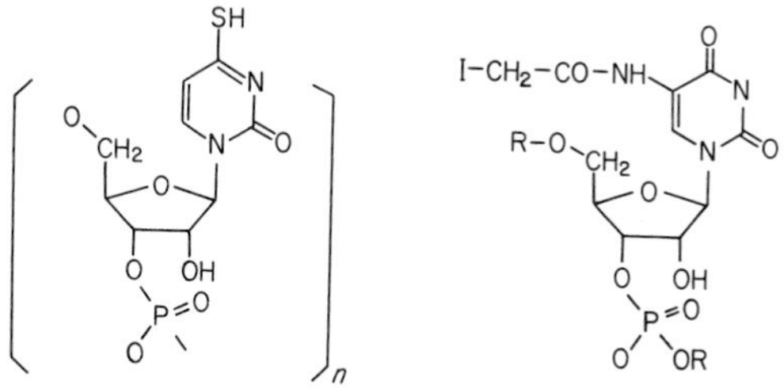

Fig. 8   Affinity analogues of mRNA: poly-4-thiouridylate (Fiser *et al.*, 1974); iodoacetyl 5-aminouridylate, containing oligonucleotides (Lurhmann *et al.*, 1973).

400 nm, the thiocarbonyl moiety becomes a highly reactive species capable of forming covalent bonds with nucleophilic groups.

Luhrmann et al. (1973) employed iodoacetyl chloride as a bifunctional reagent to try to cross-link a sulfhydryl protein chain with an oligonucleotide containing a 5-aminouridine (see Fig. 8). Schenkman et al. (1974) have used direct irradiation of poly(U) to produce covalent attachment while Pongs and his group (1974) have prepared a reactive analogue of the initiation triplet, A-U-G. Most of these studies with mRNA analogues are still in fairly preliminary stages.

7. Streptomycin Binding Site

The formyl residue in position 3 of the streptose moiety of streptomycin is known to be unimportant for the action of this antibiotic on ribosomal activity (Heding et al., 1972). Therefore, Pongs and Erdmann (1973) derivatized streptomycin at this position with 4-monoiodoacetyl aminobenzhydrazine to form the corresponding hydrazone.

B. Design of Labeling Experiments and Controls

There are a great number of potential problems that arise during affinity label experiments with a macromolecular system. The size, complexity, and heterogeneity of ribosomes as well as the large number of available reactive components make site-specific labeling very difficult. Great care must be taken to ensure that (1) reaction takes place at a functionally active site; (2) the reaction at the active site can be distinguished from random nonactive site reactions; and (3) that the reaction products can be unambiguously identified. The raw data from ribosome affinity labeling experiments are radioactive incorporations of reagent into individual ribosomal proteins. Figure 9 illustrates the labeling pattern of 50 S proteins with four different affinity labels.

Starting from such data, the problem is always to decide whether the observed specificity implies that a site-specific covalent reaction has occurred. If so, it then must be determined which of the major or minor labeled products actually represent components of the functional site. Various methods which have been used to explore these questions are discussed in detail in the following sections.

Ribosome affinity labeling studies are often plagued by low yields of reaction. For example, Czernilofsky et al. (1974) estimated that only one out of 250 to 300 ribosomes reacted with PNPC-Phe-tRNA. Schwartz and Ofengand (1974) determined that only 10–15% of the total input of 4-thiouridine derivatized tRNA^{Val} reacted with ribosomal components. Bispink and Matthaei (1973) reported a covalent yield of only 3% with their tRNA label, EDM-Phe-tRNA.

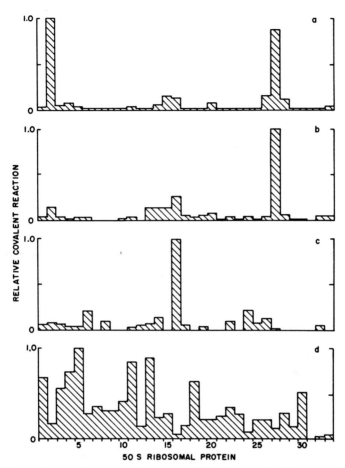

Fig. 9 Pattern of covalent attachment of various affinity labels with 50 S ribosomal proteins after incubation with 70 S ribosomes. Results have been scaled so that the protein with maximum reactivity in each case is assigned a value of 1.0. Where incorporation was originally reported as unresolved among one or more proteins it has been divided equally among them to facilitate comparison. (a) BrAcPhe-tRNA (Pellegrini *et al.*, 1974), maximum reactivity, 1650 cpm. (b) PNPC-Phe-tRNA (Czernilofsky *et al.*, 1974), maximum reactivity, 4300 cpm. (c) Iodoamphenicol (Pongs *et al.*, 1973a), maximum activity, 135 cpm. (d) APh-GDP (Maasen and Möller, 1974), maximum reactivity, 143 cpm.

1. Identification and Quantitation of Covalent Products

Since affinity labeling is presumed to take place at a specific binding site, one would expect a limited number of products. With most of the labels reported, specific reactivity appears to occur with at most three different ribosomal components out of a possible 57. The unambiguous identification of these products is not necessarily easy.

Most of the reaction products of ribosome affinity labeling studies have been proteins. The two-dimensional gel electrophoresis technique of Kaltschmidt and Wittmann (1970) offers excellent resolution of the ribosomal proteins. Stained protein spots are cut out and examined for the presence of a radioactive affinity label. However, with this method, an appreciable amount of protein and radioactive material may remain in the first dimension disc gel. Selective retention of one or more affinity labeled products may also occur. Maassen and Möller (1974) reported that the 50 S proteins that remained near the origin in the second dimension of their gels apparently showed a very high incorporation of radioactivity. However, it was then shown that this radioactivity was evenly distributed around the origin rather than being localized in spots corresponding to the proteins. Maassen and Möller speculated that these counts may have been due to polymers of their affinity label APh-GDP formed during photolysis. These polymers would be similar to those known to form during irradiation of phenyl azides.

Maassen and Möller (1974) also reported a large decrease in the stain intensity at the position of protein L16 following irradiation of the APh-GDP · EFG · ribosome · fusidic acid complex. This reduced amount of L16 may be caused by a weakening of its binding to the ribosome or, more likely, by a hydrolytic cleavage of the protein induced by nitrene insertion.

As pointed out by Czernilofsky et al. (1974), covalent attachment of an affinity label to a ribosomal protein may cause displacement of this protein relative to the unlabeled protein. This would occur as a result of the alteration in the charge caused by attachment of the label. If the extent of labeling were low, it would be impossible to detect the modified proteins in the stained pattern. Areas surrounding all stained spots may have to be analyzed. However, unless the exact protein-affinity label product is known, it is not possible to predict its electrophoretic behavior or to determine the protein from which it was produced. Czernilofsky et al. (1974) circumvented this obstacle by the use of an immunological assay. Under conditions of antibody excess, the position of radioactive proteins on a sucrose gradient can be shifted by antibody binding. Good correspondence between the results of this technique and results of two-dimensional gels were obtained.

A second convenient way to circumvent these identification problems is to analyze the same sample of labeled ribosomal proteins by separate one- and two-dimensional gel electrophoresis (Pellegrini et al., 1974). The former are sliced in their entirety. The resolution of ribosomal proteins on

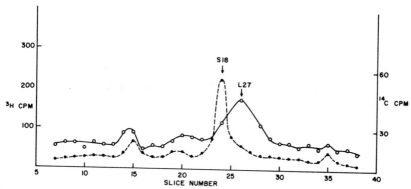

Fig. 10 One-dimensional gel analysis of a mixture of BrAc[³H]Phe-tRNA labeled 30 S proteins and BrAc[¹⁴C]Phe-tRNA labeled 50 S proteins. Standard peptidyl site labeling reactions were carried out with 70 S ribosomes and either BrAc[³H]Phe-tRNA or BrAc[¹⁴C]Phe-tRNA (Oen *et al.*, 1974). The ¹⁴C-label 50 S proteins (O—O) and the ³H-labeled 30 S proteins (●—●) were mixed together and applied to a one-dimensional acrylamide gel.

one-dimensional gels is poor; however, the location of affinity label products should not be shifted as much as on a two-dimensional gel. The two-dimensional gels are used for product identification. If results from both gels correspond quantitatively, one can be confident that all products have been detected and correctly identified.

Finally, there exists the possibility of one or more proteins from one subunit contaminating a preparation of the other. When Pellegrini *et al.* (1974) found radioactivity near 50 S protein L27, they speculated that a chemical modification had changed that protein's electrophoretic behavior. However, the 30 S protein, S18, which also reacts with the affinity label is also found in this area of the gel. Therefore, 50 S proteins labeled by BrAc[¹⁴C]Phe-tRNA and 30 S proteins labeled by BrAc[³H]Phe-tRNA were combined and coelectrophoresed. As shown in Fig. 10, neither 30 S nor 50 S labeled proteins extensively contaminated preparations of the other.

2. Controlling Nonspecific Reactions

The next step in the interpretation of labeling data is to determine which reaction products are significant. In other words, has sufficient site-specific labeling occurred to distinguish the products resulting from modification of a functional site from those of a nonspecific reaction. One good way of doing this is to lower the chances of nonspecific reaction as far as possible.

In both the BrAcPhe-tRNA and PNPC-Phe-tRNA studies, an attempt was made to limit any nonspecific reaction of these reagents with the nu-

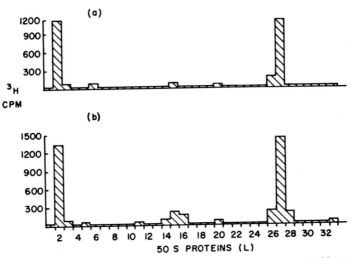

Fig. 11 Effect of treatment of a BrAcPhe-tRNA 70 S ribosome mixture with puromycin subsequent to incubation and prior to separation and analysis of covalently modified ribosomal proteins (Pellegrini *et al.*, 1974). (a) + puromycin; (b) − puromycin.

merous nucleophilic groups on the ribosome by keeping the level of non-ribosome-bound label as low as possible. In order to accomplish this, ribosomes were present in large excess (Czernilofsky *et al.*, 1974; Pellegrini *et al.*, 1974).

While excess small molecule affinity label can be easily removed from the reaction mixture by dialysis, this is not possible for these tRNA analogues. In the case of BrAcPhe-tRNA and PNPC-Phe-tRNA, the reaction mixtures were treated with puromycin following the initial incubation. This is sufficient to release most of the noncovalently attached label. The effect of this puromycin treatment on the BrAcPhe-tRNA system is shown in Fig. 11.

Pellegrini *et al.* (1974) demonstrated that breakdown products of BrAcPhe-tRNA (BrAcPhe and deacylated tRNA) do not cause labeling when added to ribosomes in the same concentrations as they are present in the intact affinity label.

Nonspecific reactions with reagents closely related to the affinity label can also be examined. Sonenberg *et al.* (1973) compared the reaction products of 50 S subunits treated with the affinity label, bromamphenicol, and three nonspecific reagents, *N*-ethylmaleimide, *N*-bromoacetyl phenylalanine, and its methyl ester. Each of these reagents was added in large excess to 50 S ribosomes. The labeling pattern obtained with bromamphenicol was significantly different and much more specific than that of the other reagents.

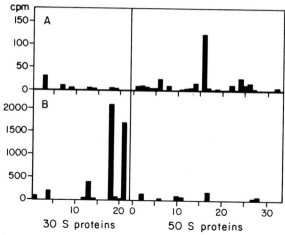

Fig. 12   Reaction pattern of (A) 70 S ribosomes incubated with a 10-fold excess of [^{14}C]iodoamphenicol and (B) 70 S ribosomes incubated with a 10-fold molar excess of [^{14}C]iodoacetamide. It is instructive to compare these patterns with the results shown in Fig. 9c (Pongs et al., 1973a).

Pongs et al. (1973a) reacted 70 S ribosomes with iodoacetamide which mimics the reactive side chain of iodoamphenicol, and contrasted the labeled proteins obtained with those labeled by iodoamphenicol. Iodoacetamide did not react with the two ribosomal proteins which were significantly labeled by iodoamphenicol. Instead, iodoacetamide reacted extensively with two other 30 S proteins, but only poorly with 50 S proteins. The labeling of ribosomal proteins with iodoamphenicol and iodoacetamide is shown in Fig. 12.

Labeling studies performed on isolated subunits rather than on the intact 70 S must be carefully examined for nonspecific reaction. Pongs and co-workers (1973a) found substantial differences in the labeling patterns obtained when iodoamphenicol treated subunits and 70 S particles were compared. These data are also presented in Fig. 12. The distribution of radioactivity in the subunit reactions resembles that obtained from labeling with the nonspecific reagent, iodoacetamide. S18, a 30 S protein, is the major reaction product in both cases. Although L16 is the major labeled 50 S component in both 50 S and 70 S reactions, it is labeled much less in the isolated subunit reaction. These data indicated that L16 and not S18 is the primary site-specific reaction product with iodoamphenicol.

Reaction of BrAcPhe-tRNA can also take place with isolated 50 S subunits but only in the presence of 20% methanol to stimulate tRNA binding (Pellegrini et al., 1974). The distribution of label among the 50 S proteins shows a much higher level of nonspecific reaction (see Fig. 13).

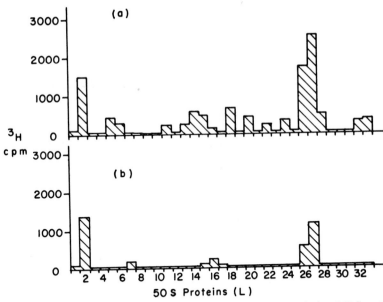

Fig. 13 Amount of [³H]Phe incorporated into 50 S proteins from (a) isolated 50 S particles reacted with BrAc[³H]Phe-tRNA in the presence of 0.5 M NaCl and 20% methanol, and (b) 70 S ribosomes reacted with BrAc[³H]Phe-tRNA (Pellegrini, 1973).

3. Competition with Nonreactive Substrates or Inhibitors

(a) One of the most important criteria for specific site labeling is the ability of substrates or inhibitors that cannot covalently label the ribosome to compete for the binding of an affinity label. This is possible when the covalent reaction is slow, under conditions where binding can still take place. With photoaffinity labels, reaction can be controlled by the presence or absence of irradiation.

Sonenberg *et al.* (1973) studied the capacity of chloramphenicol (CAP) and bromamphenicol to compete with each other for binding sites on 50 S ribosomal subunits. These experiments were carried out under conditions where the covalent attachment of the synthetic analogue was negligible. The binding of [¹⁴C]CAP in the presence of varying amounts of nonradioactive bromamphenicol was measured, and vice versa. These results are shown in Fig. 14. Although bromamphenicol seems to have a lower affinity for ribosomes than CAP, clearly each of the two compounds can prevent binding of the other. Monoiodoamphenicol and CAP or lincomycin also compete with one another. These data suggest that this affinity label binds to a CAP-binding site (Pongs *et al.*, 1973a).

Binding competition has also been examined for other labels. For ex-

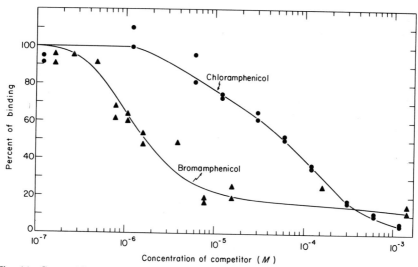

Fig. 14 Competition between chloramphenicol and bromamphenicol in binding to 50 S subunits. Unlabeled bromamphenicol was used to compete labeled chloramphenicol and vice versa (Sonenberg *et al.*, 1973).

ample, APh-GDP inhibits EFG-mediated binding of GDP to 50 S ribosomes. The inhibition was nearly maximal at an APh-GDP concentration equal to the GDP concentration used in the reaction (Maassen and Möller, 1974). As another example, at high Mg^{2+} concentrations the binding of BrAcPhe-tRNA can be successfully competed by excess deacylated tRNA which also binds to the P site (Pellegrini *et al.*, 1974).

(b) A second important criterion for site specific labeling is the inhibition of the covalent reaction by a competitive inhibitor of the active site. CAP inhibited the covalent reaction of both bromamphenicol and iodoamphenicol. Both puromycin analogues, BAP-Pan-Phe (Harris *et al.*, 1973) and *N*-iodoacetyl puromycin (Pongs *et al.*, 1973b), showed marked inhibition of covalent reaction when puromycin was present. Similarly, the covalent reaction of BrAcPhe-tRNA with ribosomes could be competed by excess deacylated tRNA. The inhibition of covalent reaction directly parallels the inhibition of binding (Pellegrini *et al.*, 1974). These results are presented in Table I.

It is important to note that data such as these are still insufficient to conclude that a site-specific reaction has occurred. Excess deacylated tRNA will inhibit the covalent reaction of BrAcPhe-tRNA with 30 S protein S18, but further experiments showed that functionally significant site-specific labeling had not occurred (Pellegrini *et al.*, 1974). There are two possible explanations. The reactive end of the molecule may not be lo-

TABLE I

Effect of Deacylated tRNA Competition on the Binding of
BrAc[³H]Phe-tRNA^Phe to *E. coli* Ribosomes[a]

tRNA additions (nmoles)	Initial binding		Covalent binding to 70 S ribosomes	
	cpm	%[b]	cpm	%[b]
None	29,000	100	1090	100
20	23,100	80	1010	93
40	20,000	69	720	67
120	14,400	50	500	46

[a] These results were adapted from the work of Pellegrini *et al.* (1974).

[b] Relative to binding in the absence of added deacylated tRNA.

cated at the functional site even though the rest of the molecule is correctly placed. Alternatively, the reactive protein, S18, may be shielded by the bound deacylated tRNA and thus protected from nonspecific reaction by BrAcPhe-tRNA in solution. These possibilities point out the major disadvantages of using a macromolecular substrate for affinity labeling studies.

4. Modulation of the Covalent Reaction

Indirect inhibition of affinity reagent binding and/or covalent reaction can also help to demonstrate correct site reaction. Modulation of the covalent reaction by other substrates or inhibitors that are not directly related to the affinity label can often be shown.

The presence of mRNA is required for correct binding of tRNA to 70 S ribosomes. Covalent attachment of BrAcPhe-tRNA^Phe (Pellegrini *et al.*, 1972), BrAcMet-tRNA_f^Met (Sopori *et al.*, 1974), and PNPC-Phe-tRNA and PNPC-Met-tRNA_f^Met (Czernilofsky *et al.*, 1974; Hauptmann *et al.*, 1974) to ribosomes are all dependent on the presence of the appropriate mRNA. Reaction of the Met-tRNA derivatives also requires initiation factors. The placement and subsequent reaction of 4-thiouridine derivatized tRNA^Val in the A or P site depended on the synthetic messenger poly(U_2G). Reaction at the A site also required the presence of the factor complex EF-Tu-GTP (Schwartz and Ofengand, 1974). NAP-Phe-tRNA and NAG-Phe-tRNA exhibit little covalent reaction in the absence of poly(U) (Hsiung *et al.*, 1974; Hsiung and Cantor, 1974).

It is interesting to note that the reaction of BrAcPhe-tRNA with S18 is actually inhibited somewhat by the presence of the mRNA, poly(U), while reaction with 50 S proteins, L2 and L27, is stimulated more than 10-fold by the presence of poly(U) (Oen et al., 1973). This is one line of evidence that an S18 reaction is not occurring from correctly bound peptidyl-tRNA. Finally, tRNA can be displaced by other tRNA's. When BrAcPhe-tRNA is positioned in the A site, its labeling pattern changes significantly. Reaction with L2 and L27 is reduced and another protein, L16, now becomes a major product.

Antibiotics can also influence covalent reaction patterns of tRNA affinity analogues. Five antibiotics, all peptidyl transferase inhibitors, show a marked effect on the labeling pattern of BrAcPhe-tRNA with ribosomes (Oen et al., 1974). In the absence of antibiotics, two proteins, L2 and L27, are the major reaction products. Although the antibiotics show only a small effect on the total binding of BrAcPhe-tRNA to ribosomes, all these antibiotics dramatically suppress the reaction of the label with a single 50 S protein, L2. They enhance the labeling of protein L27. However, no new 50 S proteins are labeled. These data call into question the specificity of BrAcPhe-tRNA reaction with L27 and/or its placement directly at the P' site.

Antibiotics also influence the binding of other affinity labels. Lincomycin which competes with chloramphenicol for a ribosome binding site also competes with the affinity label, iodoamphenicol (Pongs et al., 1973a). As cited above, fusidic acid stabilizes the EFG · GDP · 50 S ribosome complex. Maassen and Möller (1974) tested the reaction pattern of APh-GDP in the presence and absence of fusidic acid. Labeling in the absence of fusidic acid would be expected to be far less specific. The pattern shown in Fig. 15 is the difference between the radioactivity incorporated into 50 S proteins from [³H]APh-GDP ribosome · EFG · label complex both with and without fusidic acid.

Czernilofsky et al. (1974) observed a marked decrease in the covalent reaction of PNPC-Phe-tRNA and PNPC-Met-tRNA (64% and 80%, respectively) when puromycin was present at 4 mM. Puromycin would be expected to release tRNA bound at the P site, preventing covalent reaction there.

5. Direct Inhibition of Biological Function

Although irreversible inhibition of a particular ribosome function does not prove that only site-specific reaction has taken place, when taken with other data it can help to show such reaction has occurred. Incubation of 70 S ribosomes with iodoamphenicol irreversibly inhibited the poly(U)-directed peptide transfer reaction of both *Escherichia coli* and *Bacillus stearothermophilus* ribosomes to the same extent as CAP reversibly inhibited this reaction (Pongs et al., 1973a). Covalent incorporation of brom-

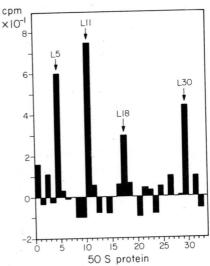

Fig. 15   Fusidic acid-dependent radioactive incorporation of [³H]APh-GDP into 50 S ribosomal proteins. This was obtained by subtracting the radioactivity incorporated in the absence of fusidic acid (see Fig. 9d) from the radioactivity incorporated in the presence of fusidic acid (Maassen and Möller, 1974).

amphenicol was accompanied by a corresponding decrease in peptidyl transferase activity. Concurrent with the decrease in this activity, the labeled ribosomes also lost their ability to bind CAP or erythromycin. The capacity of ribosomes to bind these antibiotics is related to their ability to catalyze peptide transfer (Sonenberg *et al.*, 1973).

   N-Iodoacetyl puromycin irreversibly inhibited *in vitro* polyphenylalanine synthesis by about 50%. Whereas puromycin inhibited the ribosomal activity more strongly (about 80%), extensive dialysis fully restored ribosomal activity (Pongs *et al.*, 1973b). Likewise, preincubation of ribosomes with the puromycin analogue of Symons and co-workers (Harris *et al.*, 1973), BAP-Pan-Phe, gave a 41% loss of peptidyl transferase activity (measured by the fragment reaction). The A' site-specific antibiotics puromycin and chloramphenicol gave substantial concentration-dependent protection against inactivation by the affinity label. All the inactivation, but only a small portion of the covalent labeling, could be prevented by the presence of CAP. Chloramphenicol insensitive labeling, designated nonspecific, was shown to be distributed over many ribosomal proteins. Comparison of the stoichiometry of the "specific" labeling, i.e., to the 23 S rRNA, revealed a linear correlation between peptide transfer inactivation and this covalent reaction (Greenwell *et al.*, 1974).

   Another example of inactivation by affinity labels comes from the studies of Fiser *et al.* (1974). They showed that polyphenylalanine synthesis decreased rapidly upon irradiation of the poly-4-thiouridine ribosome complex. In contrast, a parallel system containing poly(U) as the mRNA showed no such reduction in poly-Phe formation.

Although the presence of the affinity label of streptomycin did not irreversibly inhibit polyphenylalanine synthesis, Pongs and Erdmann (1973) found that this label did irreversibly distort the ribosomal tRNA binding sites. They reported irreversibly increased levels of polyphenylalanine synthesis with tRNA$_{-y}^{Phe}$. Streptomycin is known to lower ribosomal screening of tRNA's altered in their anticodon loop (Ghosh and Ghosh, 1972).

6. Direct Evidence for the Biological Function of the Affinity Label

In certain cases direct evidence for the biological functioning of covalently attached affinity labels is available.

Experiments showed that the bulk of BrAcPhe-tRNA was bound correctly in the P site. However, the low yield of covalent reaction left open the possibility that the covalent reaction occurred on ribosomes which were inactive in peptide transfer or with tRNA not bound at the P site (Oen *et al.*, 1973; Pellegrini *et al.*, 1974). To overcome these objections, 70 S ribosomes were preincubated with nonradioactive BrAcPhe-tRNA and then reacted with [³H]Phe-tRNA or [³H]puromycin. The latter were found covalently bound to 50 S ribosomal proteins. The appearance of substantial radioactivity in ribosomal components indicates that BrAcPhe-tRNA can both covalently react with ribosomes and participate in peptide bond formation with Phe-tRNA or puromycin. The ribosomal proteins labeled in this experiment are very similar to those labeled with BrAc[³H]Phe-tRNA alone. The 30 S protein, S18, is not found to be labeled by BrAcPhe-tRNA molecules which can subsequently react with Phe-tRNA or puromycin. This evidence combined with the lack of poly(U) dependence for the BrAcPhe-tRNA reaction with S18 argues that this protein is labeled from a nonfunctionally significant site, while L2 and L27 are site-specific products.

Similarly, Greenwell *et al.* (1974) showed that the 50 S · BAP-Pan-Phe complex can act as an acceptor for the transfer of acetyl-[³H]Leu from U-A-C-C-A-(acetyl-[³H]Leu) under fragment reaction conditions, and that the [³H]Leu thereby becomes bound to 23 S rRNA. It does not become attached to the already designated "nonspecific" ribosome protein reaction products with BAP-Pan-Phe. Although only 0.5% of the ribosome-affinity label complexes in each reaction mixture accepted acetyl-[³H]Leu, Greenwell *et al.* (1974) demonstrated that the amount of [³H]Leu

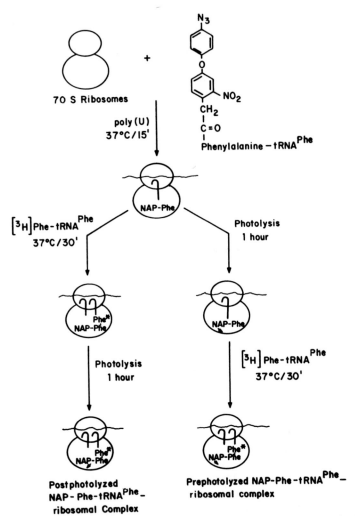

Fig. 16 Schematic diagram of two ways of combining peptidyl-tRNA photoaffinity labeling with peptide synthesis to selectively label just those products adjacent to the peptidyl transferase center (Hsiung, 1974).

covalently bound to rRNA is directly proportional to the inactivation of peptidyl transferase and therefore also directly proportional to the amount of [^{32}P]BAP-Pan-Phe attached to the rRNA.

In neither of the above cases can the sequence of the two reactions, covalent attachment and peptide transfer, be directly determined. The use of the photoactivated labels, NAP-Phe-tRNA and NAG-Phe-tRNA, in similar experiments allowed a clear reaction sequence to be defined since the timing of covalent reaction can be directly manipulated by controlling the time of photolysis (Hsiung and Cantor, 1974; Hsiung *et al.*, 1974). [^{3}H]Phe-tRNA can be added to the affinity label·ribosome complex either before or after photolysis of the sample. This is illustrated in Fig. 16. Since the affinity reagent used in these experiments was not radioactive, the covalent incorporation of [^{3}H]Phe into the ribosome could only result from peptide bond formation and any nonspecific labeling is not analyzed. Presumably the label must be functionally positioned in the P site to be able to react with [^{3}H]Phe-tRNA bound in the A site. Comparison of the results from these experiments with those of NAP-[^{3}H]Phe-tRNA reacted with ribosomes by itself indicated that a more specific pattern of labeling occurs if only the products which can participate in dipeptide formation are detected. Still, there is substantial similarity in the protein reaction pattern in all these experiments. The major products are the same two 50 S proteins, L11 and L18. This type of experiment is possible, in principle, with any enzyme-active site that catalyzes a synthetic reaction between two substrates.

7. Varying the Length of the Affinity Label

Variation of the length of an affinity label, i.e., its distance from a particular site, can provide further information about the neighboring components of that site. The best examples of this are the experiments of Eilat *et al.* (1974a). They used a series of affinity labels, BrAc(Gly)$_n$Phe-tRNA ($n = 1$ to 16) to label the ribosomal components that border the 3'-terminal binding site of peptidyl-tRNA and extend along the binding site of the growing peptide chain. As the length of the reagent increased, the reaction pattern changed. Different protein products showed maximum yields at quite different chain lengths. The results of these experiments are illustrated in Fig. 17.

IV. Survey of Affinity Labeling Results

The extensive number of affinity labeling studies already reported with ribosomes allows some useful conclusions to be drawn both about the situation of the 50 S particle and the methodology of affinity labeling experiments.

A. Similar Site-Directed Reagents Give Similar Reaction Products

In some cases several different affinity labels were designed to explore the same functional site. An important question to ask in such instances is whether such reagents give similar or at least overlapping reaction products.

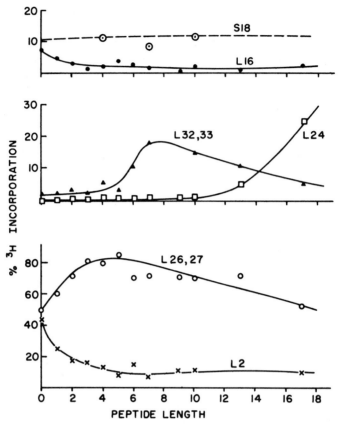

Fig. 17 Percent of [³H]Phe incorporated into various ribosomal proteins as a function of peptide chain length of the affinity label (Eilat *et al.,* 1974b). Peptide length = 0, refers to BrAcPhe-tRNA-CA; Peptide length = 1, refers to BrAcPhe-tRNA; Peptide length = 2 or greater, refers to BrAcGly$_n$Phe-tRNA, where $n + 1$ = peptide length.

1. P' Site

The P' site of peptidyl transferase has been covalently labeled by both tRNA analogues and small antibiotic analogues. Both chemically reactive and photoreactive reagents were employed. Quite a number of different covalent products have been found. The major ones implicated as functionally significant by one or more controls are summarized in Table II.

Two 50 S proteins, L2 and L27, are the major products of reaction of the peptidyl-tRNA analogue, BrAcPhe-tRNA, with the 70 S ribosome. Minor reaction products included L5, L6, L10, L11, and L14–L17. BrAcPhe-tRNA which can become attached to protein L2 and, to a lesser extent, L16 and L27 can still react with aminoacyl-tRNA or puromycin

TABLE II

Results of Affinity Labeling Studies at the Peptidyl Transferase Center

Site labeled	Label conditions	Major reaction products											
		L2	L6	L11	L14 L15	L16	L18	L27	5 S RNA	23 S RNA	U50[a]	S18	U30[b]
P'	BrAcPhe-tRNA	++											
P'	BrAcPhe-tRNA via dipeptide	++				+		++				+	
P'	BrAcMet-tRNA$_f$^Met	++						++					
P'	BrAcMet-tRNA$_f$^Met via dipeptide	++						+					
P'	PNPC-Phe-tRNA				+	+		++					
P'	PNPC-Met-tRNA$_f$^Met							++					
P'	NAG-Phe-tRNA			++			++						
P'	NAG-Phe-tRNA via dipeptide			++			++						
P'	NAP-Phe-tRNA			++			++						
P'	NAP-Phe-tRNA via dipeptide			++			++						
P'	EDM-Phe-tRNA												
A'	BrAcPhe-tRNA (excess deacylated tRNA, 30 mM Mg²⁺)	+				++	+		+	++			
A' or P'	Bromamphenicol	++											
A'	Iodoamphenicol					++		++					
A'	BAP-Pan-Phe								+	++	++		
A'	BAP-Pan-Phe via dipeptide									++	++		++
A'	N-Iodoacetyl puromycin		++										

[a] Unidentified 50 S proteins.
[b] Unidentified 30 S proteins.

(Oen *et al.*, 1973; Pellegrini *et al.*, 1974). This is the strongest evidence that these proteins are located in the P' site. Labeling studies with BrAcMet-tRNA$_f^{Met}$, directed by f2 phage mRNA, yielded the same covalent products, L2 and L27. The bound affinity label which reacts with L2 and L27 also participates efficiently in peptide transfer (Sopori *et al.*, 1974).

Kuechler and his collaborators using different chemically reactive tRNA analogues, PNPC-Phe-tRNA and PBPC-Met-tRNA$_f^{Met}$, identified L27 as the major reactive P' site component. PNPC-Met-tRNA$_f^{Met}$ in the presence of R17 phage RNA and initiation factors gave the cleanest labeling of L27. With PNPC-Phe-tRNA, proteins L2, L14, L15, and L16 were also significantly labeled (Czernilofsky *et al.*, 1974; Hauptmann *et al.*, 1974).

Photoactivated peptidyl-Phe-tRNA derivatives have shown different reaction patterns from any of the above reagents. L11 and L18 of the 50 S subunit are the major labeled products with either NAP-Phe-tRNA^{Phe} (Hsiung *et al.*, 1974) or NAG-Phe-tRNA^{Phe} (Hsiung and Cantor, 1974). Other products found with NAP-Phe-tRNA included L1, L2, L5, L22, L27, and L33 (Hsiung *et al.*, 1974). NAG-Phe-tRNA is less hydrophobic than NAP-Phe-tRNA and shows a more specific reaction pattern. It reacts almost exclusively with L11 and L18 and only to a very small extent with L2, L5, L27, and L33. Both NAP-Phe-tRNA^{Phe} and NAG-Phe-tRNA^{Phe} which have been covalently attached to L11 and L18 can subsequently participate in peptide transfer. This, once again, is quite a strong indication that these two proteins are located in the P' site, immediately adjacent to the 3'-end of the bound peptidyl-tRNA. The peptidyl-RNA photoanalogue of Bispink and Matthaei (1973) reacts with the 23 S rRNA. This suggests that the P' site is not entirely protein.

In general, the data obtained with the different chemically reactive reagents agree well with each other. The results with the two azide photoreagents are self-consistent but differ quite a bit from the other reagents. This is not surprising since the BrAc and PNPC affinity labels require nucleophilic displacement by a ribosomal component for reaction to occur. If the 3'-end of the bound tRNA is at all flexible, these reagents can search for a nearby nucleophilic side chain. The photoreagents are longer and will probably tend to react with immediate neighboring amino acid side chains. However, the minor reaction products of these labels do include L2 and L27, the major products from BrAc and PNPC labeling.

With four proteins identified at the P' site, it is possible to examine whether other reagents can bind correctly in this position. Ribosome-bound BrAcPhe-tRNA$_{ox-red}$ also reacts with proteins L2 and L27 (M. Pellegrini, J. Ofengand, and C. R. Cantor, unpublished results). Peptidyl-

$tRNA_{ox-red}$ will bind to ribosomes but it is inactive in the puromycin reaction. However, the results of parallel labeling experiments with BrAcPhe-tRNA and BrAcPhe-$tRNA_{ox-red}$ clearly shows that $tRNA_{ox-red}$ can bind correctly to the P' site.

2. The A' Site

Several different affinity analogues have been used to identify ribosomal components at the A' site. Results are summarized in Table II. BrAcPhe-tRNA, like other peptidyl-tRNA's, can be forced to bind at the A site by competition with excess deacylated tRNA at 30 mM Mg^{2+}. In the A' site this reagent reacts with 50 S protein L16 (Eilat et al., 1974a). L2 and L27 are also products but these are not labeled as efficiently by A' site-bound reagent as they were from the P' site.

Affinity analogues of antibiotics known to inhibit peptidyl transferase were also used to identify 50 S proteins in the A' site. It has been shown that the antibiotics puromycin and CAP actually compete for the same binding site as the fragment C-C-A-Phe, i.e., the A' site (Lessard and Pestka, 1972; Nierhaus and Nierhaus, 1973). Two affinity analogues of CAP gave differing results. Bromoamphenicol reacted with L2 and L27 in isolated 50 S subunits (Sonenberg et al., 1973). Iodoamphenicol reacted with L16 in the intact 70 S particle (Pongs et al., 1973a). These results perhaps suggest two different binding sites for the CAP analogues. Lessard and Pestka (1972) reported two modes of CAP binding, both of which result in the inhibition of peptidyl transferase activity (Pestka, 1972). These data may reflect the different labeling results obtained with two such similar reagents.

The puromycin analogue of Symons and co-workers bound correctly at the A' site reacts with 23 S rRNA exclusively. Once bound, this reagent can also participate in peptide transfer with acetyl-Leu bound in the P' site (Harris et al., 1973; Greenwell et al., 1974). However, the reactive moiety of this reagent is at a position somewhat removed from the amino acid binding site within the A' site. The reactive center is in a position analogous to the 5'-phosphate of the 3'-terminal adenosine of aminoacyl-tRNA.

The peptidyl puromycin analogue N-iodoacetyl puromycin, synthesized by Pongs et al. (1973b), reacts with protein L6 (O. Pongs, personal communication). L6 was seen among the minor reaction products of the P' site-directed reagents. The label also reacts to some extent with L2, a P' site protein (O. Pongs, personal communication). N-Iodoacetyl puromycin might be expected to occupy an intermediate position between what has been defined as the A' and P' sites. A similar position may be occupied by the larger tRNA analogues, NAG-Phe-tRNA and NAP-Phe-tRNA, when they were first allowed to participate in peptide bond formation with an aminoacyl-tRNA. These compounds were then photolysed to cause their covalent attachment to the ribosome. This reaction occurs mainly with proteins L11 and L18 (Hsiung and Cantor, 1974; Hsiung *et al.*, 1974).

3. P Site and A Site

Affinity label results on these and other sites are summarized in Table III.

Irradiation of the phenacyl *p*-azido derivative of the 4-thiouridine residue of tRNA^{Val} bound to the P site yielded covalent attachment of this tRNA exclusively to the 16 S rRNA. When bound to the A site of 70 S ribosomes, the overall efficiency of this label's reaction with the ribosome was reduced as compared to reaction at the P site. This tRNA reagent reacted about equally with the two ribosomal subunits from its A site position. Covalent attachment to 30 S proteins as well as 16 S rRNA was observed. 50 S reaction products included both ribosomal proteins and 5 S rRNA. The label was not found to be linked to 23 S rRNA (Ofengand and Schwartz, 1974).

4. Site of the Ribosome-Bound Peptide Chain

BrAc(Gly)$_n$Phe-tRNA's ($n = 1$ to 16) bound to 70 S ribosomes react covalently with 50 S proteins. The pattern of protein reactivity toward these P site-bound oligopeptidyl affinity reagents showed a definite dependence on the length of the peptide chain. At the very least, these results imply a hierarchy of radial distances between these proteins and the 3'-end of the P site-bound tRNA. From closest to farthest, this is L2 < L27 < L32–L33 < L24.

5. Site of GTP Hydrolysis

The GDP analogue of Maassen and Möller (1974), APh-GDP, yielded covalent reaction with the ribosome upon irradiation of the 50 S · EFG · APh-GDP · fusidic acid complex. Covalent attachment was to 50 S proteins L5, L11, L18, and L30. In addition, after irradiation the amount of L16 was greatly reduced.

6. mRNA Binding Site

Fiser *et al.* (1974) irradiated the synthetic mRNA, poly-4-thiouridine, on the ribosome in the presence of Phe-tRNA which stabilized the complex. Covalent bonds between this mRNA and 30 S protein S1 were formed.

TABLE III

Results of Other Ribosome Affinity Labeling Studies

Site labeled	Label conditions	L2	L5	L11	L18	L24	L27	L30	L32 L33	5 S RNA	U_{50}[a]	S1	S4	S18	16 S RNA	U_{30}[b]
A	p-Azidophenacyl-4-thiouridine tRNA^{Val}									++	++				++	++
P	p-Azidophenacyl-4-thiouridine tRNA^{Val}														++	
GTP hydrolysis	APh-GDP		++	++	+			++								
Growing peptide chain	BrAc(Gly)$_n$Phe-tRNA															
	n = 3	+					++		++					+		
	n = 6						++		++					+		
	n = 9						++		+					+		
	n = 12					+	++		+					+		
	n = 16					+	++							+		
mRNA	Poly-4-thiouridine											++				
Streptomycin	N-Iodoacetyl phenylhydrazone of streptomycin												++			

Major reaction products

[a] Unidentified 50 S proteins.
[b] Unidentified 30 S proteins.

235

B. Comparison of Affinity Labeling Data with Other Biochemical Data

Affinity labeling results agree well with other biochemical data on the functionality of the various ribosomal components. For a more complete review of ribosomal active sites, see Pongs *et al.* (1974).

1. Peptidyl Transferase

Studies with affinity reagents identified several 50 S proteins at or near the P′ site of peptidyl transferase. They include L2, L11, L18, and L27. It should be noted that none of these labels destroyed the peptidyl transferase activity. Therefore, none of them reacted with the active residue(s) of this enzyme. However, they were all placed close to this active center by other criteria, including their ability to function at this center.

Traugh and Traut (1972) identified L2 as the major 50 S protein phosphorylated by rabbit muscle kinase. L2 is the most rapidly phosphorylated species in the 70 S ribosome. They suggest that the destruction of the 70 S ribosome protein synthesizing activity by rabbit muscle kinase is probably due to the phosphorylation of protein L2. Protein L3 from the 50 S subunit of *B. stearothermophilus* ribosomes has been shown to correspond to protein L2 from *E. coli* (Dzionara *et al.*, 1970; Hindennach *et al.*, 1971; Stöffler *et al.*, 1971). Fahnestock *et al.* (1973) demonstrated that although L3 from *B. stearothermophilus* is not required for the *in vitro* binding of other proteins in the reconstitution of the 50 S particle, it is necessary for the assembly of active 50 S subunits. In another study, Bickle *et al.* (1973) compared the protein composition of polysomes, monosomes, single ribosomes, and free subunits. The 50 S subunit clearly showed functional heterogeneity. Proteins L2 and L20 were absent from free 50 S subunits. These proteins were found only on ribosomes actively synthesizing proteins. Partial reconstitution studies of protein-deficient 50 S core particles by the addition of individual split proteins demonstrated that L11 is a necessary protein for a functioning peptidyl transferase center (Nierhaus and Montejo, 1973). Similar experiments link L16 to the peptidyl transferase center since it is required for CAP binding (Nierhaus and Nierhaus, 1973).

2. Components Involved in Translocation

Both affinity label and other experiments suggest the proximity of the site of translocation-dependent GTPase and peptidyl transferase. It has been shown that 50 S proteins, L7 and L12, are necessary for ribosomal interaction with EFG and the subsequent hydrolysis of GTP, i.e., they are intimately involved in translocation. Protein particles deficient in L7 and L12 are inactive in factor-mediated GTP hydrolysis (Hamel *et al.*, 1972;

Sander *et al.*, 1972; Brot *et al.*, 1972; Fakunding *et al.*, 1973). A multicomponent complex can be isolated from 70 S ribosomes chemically cross-linked in the presence of EFG (Acharya *et al.*, 1973). It contains EFG, L7, L12, and probably L2 and L11 (P. B. Moore, personal communication). Yukioka and Morisawa (1970) showed that EFG and GTP counteract the CAP inhibition of polyphenylalanine synthesis. On the other hand, Nierhaus and Nierhaus (1973) and Pestka and Brot (1971) demonstrated that CAP has no significant influence on the activity of EFG-dependent GTPase. They also suggest that this GTPase site is not at the A′ site but may overlap it or be allosterically linked to it.

L6 and L10 are known to stimulate the binding of L7 and L12 to ribosomes and thus to aid EFG-dependent GTPase (Schrier *et al.*, 1973). L6 is labeled by *N*-iodoacetyl puromycin. The APh-GDP analogue labeled proteins L5, L11, L18, L30, and perhaps L16. L11, L18, and L16 are known to be at or near the peptidyl transferase center. In addition, L5, L6, L10, and L11 have all appeared as minor reaction products with BrAcPhe-tRNA or NAP-Phe-tRNA, bound to the P′ site.

5 S rRNA is known to be important for certain ribosomal functions which are linked to the A site (Erdmann *et al.*, 1973; Richter *et al.*, 1973). 5 S RNA binds 50 S proteins L25 and L18 (also a P′ site protein). Horne and Erdman (1972, 1973) observed that a similar 5 S RNA-protein complex from *B. stearothermophilus* complex contained a GTPase activity that is inhibited by some of the same antibiotics as the EFG-dependent GTPase.

The peptidyl transferase site proteins L2 and L6 along with the 23S rRNA (also identified near the P′ site by affinity labeling) are known to assist the binding of L18 and L25 to 5 S rRNA (Gray *et al.*, 1972). It is not evident whether the effects of L2 and L6 in this case are long range or neighboring interactions.

3. Subunit Interface

There is evidence that the sites of peptide transfer and GTP hydrolysis are close to the interface of the 30 S and 50 S subunits. Affinity labeling studies fit well with these data. Although BrAcPhe-tRNA reacted with S18 is no longer functionally active, ribosome binding of BrAcPhe-tRNA is clearly necessary for this reaction. Ginzburg *et al.* (1973) have suggested that S18 is essential for enzymatic binding of fMet-tRNA and for association of the 50 S subunit to form 70 S ribosomes. This is evidence for the proximity of S18 and the 30 S particle to peptidyl transferase.

Enzymatic iodination studies (Litman and Cantor, 1974) were employed to compare the reactivity of proteins in isolated 50 S particles to that of 50 S proteins in a 70 S complex. They indicate that L18 and L2, P′ site proteins, are either directly shielded from iodination by the presence of the 30 S particle or they are in some way allosterically affected by the 30 S particle. Antibodies against L14, L19, L23, and L27 inhibit subunit reassociation (Morrison *et al.*, 1974); L27 is a P′ site protein, L14 is perhaps a near neighbor.

Partial reconstitution experiments have shown that certain 30 S proteins are necessary for EFG-dependent GTP hydrolysis (Sander *et al.*, 1973; Marsh and Parmeggiani, 1973). The data of Yu and Wittmann (1973) suggest that 5 S rRNA may be involved in the association, although antibodies against L7 and L12 do not inhibit.

Ofengand and Schwartz (1974) demonstrated that a 4-thiouridine derivatized valyl-tRNA bound to the A site can react with both 30 S and 50 S subunits. They suggested that since both subunits were equally labeled, at least part of the A site for tRNA is at or near the junction of the two subunits. Other investigations have implicated certain proteins as being responsible for A site binding of tRNA (Traut *et al.*, 1974; Nomura *et al.*, 1969; Randall-Hazelbauer and Kurland, 1972; Rummel and Noller, 1973) at the 30 S–50 S subunit interface (Ginzburg *et al.*, 1973; Morrison *et al.*, 1974).

4. 30 S Sites

Numerous experiments (Van Duin and Kurland, 1970; M. Tal *et al.*, 1972; Rummel and Noller, 1973; von Knippenberg *et al.*, 1974) support the identification of 30 S protein S1 as comprising the mRNA binding site for poly(U) and poly-4-thiouridine on the ribosome (Fiser *et al.*, 1974). On irradiation poly-4-thiouridine binds irreversibly to this protein.

Several 30 S proteins have been implicated in the streptomycin interaction with ribosomes. The streptomycin analogue of Pongs and Erdmann (1973) covalently labeled S4. Mutations in the S4 gene are known to surpress streptomycin dependence (Birge and Kurland, 1970; Zimmerman *et al.*, 1971).

C. Implications for Ribosome Structure

Several general structural conclusions can be drawn from affinity labeling data in conjunction with the other biochemical and functional data presented here.

1. Peptidyl Transferase

Peptidyl transferase is composed of both the A′ and P′ sites. The distinction between these sites is an operational one, i.e., substrates bound in the P site, fMet- or peptidyl-tRNA, can transfer their amino acid to puromycin or an aminoacyl-tRNA bound in the A' site. Both substrates must be in contact with the peptidyl transferase center for the reaction to occur. This implies that the structural-functional elements of both A' and P' sites must be in close proximity if not overlapping.

The first direct proof of this comes from affinity labeling data. BrAcPhe-tRNA reacts with proteins L2 and L27 from both A' and P' sites (Pellegrini *et al.*, 1974); the same is true for the NAP-Phe-tRNA and NAG-Phe-tRNA with proteins L11 and L18 (Hsiung *et al.*, 1974; Hsiung and Cantor, 1974).

It is interesting to note that the use of two different amino acids in the construction of P' site tRNA analogues, BrAcPhe- and Met-tRNA and PNPC-Phe- and Met-tRNA did not greatly alter their reaction profiles. This indicates that the P' site has the same overall conformation for different tRNA-amino acid side chains (Oen *et al.*, 1973; Sopori *et al.*, 1974; Czernilofsky *et al.*, 1974; Hauptmann *et al.*, 1974). The fact that BrAcPhe- or Met-tRNA and PNPC-Phe- or Met-tRNA gave similar labeling patterns proved that Phe-tRNA directed by poly(U) and bound to the P site is positioned correctly in that site.

The affinity label data summarized above indicate that a number of proteins and 23 S rRNA form the structural if not necessarily the functional elements of the P' site. These data certainly make the site of protein synthesis appear crowded. In addition, the several proteins and 5 S rRNA thought to be responsible for or neighboring the site of GTP hydrolysis are located close to the peptidyl transferase. However, neutron scattering experiments by Moore *et al.* (1974) have shown that the 50 S particle is segregated into protein-rich and rRNA-rich regions. It seems reasonable to think that the protein-rich side is that in contact with the 30 S particle, and perhaps responsible for some of the enzymatic functions of the ribosome.

2. rRNA

The position of the ribosomal RNA at these sites is not so well defined. However, affinity label studies have identified the presence of 23 S rRNA at the A' site (Greenwell *et al.*, 1974). The tRNA photolabel of Ofengand and Schwartz (1974) located 16 S rRNA at the A' site.

3. Site of the Growing Peptide Chain

BrAcGly$_n$Phe-tRNA's ($n = 1$ to 16) reacted with several different ribosomal proteins. The pattern of reaction depended markedly on the length of the reagent (Eilat *et al.*, 1974b). These results suggest that the peptide chain may be bound in a fairly discrete site. The yield of reaction with various individual proteins pass through fairly sharp maxima as a function of chain length. If the peptide chain configuration were a random coil, attached to the ribosome only at one end, it would be hard to account for such a discrete reaction pattern.

A summary of the proteins that have been identified near or at the A' and P' sites of the 50 S particle is shown in Chapter 8 (Fig. 7). This figure is in qualitative agreement with the results described here. However, several limitations in the affinity labeling approach should be kept in mind when viewing such schematic summaries. A covalent derivative shows that at least one residue of a protein is located near the site of interest. If the proteins are spherical particles the location of just one point still leaves considerable ambiguity for the location of the rest of the protein. If the proteins are not compact globular shapes, as indeed is the case for at least some ribosomal proteins, then it remains possible that a protein may contact a site through one portion of its chain even though the bulk of the protein is located quite far away. What can be safely concluded at present is that a large number of different proteins can approach the A' and P' sites. This suggests the peptidyl transferase center is protein rich and involves a rather complex network of residues from many different proteins.

V. Conclusion

Affinity labeling has been successful at locating components of a number of different functional sites of the *E. coli* ribosome. The inherent advantage of this technique is that it affords direct proof of the existence of a protein or RNA at a substrate or ligand binding site. Such techniques as chemical or immunological blocking, deletion or supplementation of specific proteins, or protection of a site from chemical or enzymic inactivation all suffer from the inability to distinguish between a direct effect and an allosteric effect which alters the site. The use of active site-directed reagents is the most direct approach to defining the structure–function relationships in the ribosome.

There are two essential parts to any affinity label study. First, the chemically reactive or photoactivatible analogue must be shown to react specifically with one or more ribosomal components. Then control experiments must be performed to determine if this reaction is dependent on correct binding of the affinity label at the functional site. If so, competition with natural substrate should inhibit the covalent modification. Antibiotics or other factors which modulate the functional site may alter the pattern of covalent reaction. Sometimes the affinity analogue, after covalent attachment, is still capable of participation in a ribosome function such as peptide transfer. When this occurs, as it does with some tRNA and puromycin analogues, it is probably the most direct proof possible that true affinity labeling has occurred.

The preparation, properties, and reaction pattern of numerous affinity analogues that have been applied to ribosomes were described. These include tRNA and mRNA analogues, as well as derivatives of the antibiotics chloramphenicol and puromycin and a photoanalogue of GDP. These have identified ribosomal components at the peptidyl transferase center and others involved in the binding of tRNA and mRNA. In addition, several proteins in the GTPase center of the 50 S subunit have been mapped. Topological information about the organization of the 50 S subunit is available from affinity label studies. For example, one group of proteins, L2, L16, L27, L11, and L18, is accessible to analogues bound to the aminoacyl- or peptidyl-tRNA binding sites. This demonstrates the proximity of these two sites. Similarly, some of the same proteins, L11 and L18, react with a photoreagent bound at the GTPase center. Unless these proteins are very extended in the particle, this indicates that the site of GTPase hydrolysis is surprisingly close to the catalytic center for peptide synthesis. The use of an homologous series of reagents of increasing size can provide even more details about the ribosome structure. Reactive peptidyl-tRNA's of different chain lengths have allowed the order of four proteins along the ribosome-bound chain to be determined.

Affinity labeling offers a most promising approach to the study of binding and catalytic sites even of such large complex and heterogenous systems as the ribosome. When combined with other biochemical and physical data, it should continue to be an important tool in elucidating the functional topography of the machinery of protein synthesis.

ACKNOWLEDGMENT

M. P. is the recipient of a postdoctoral fellowship from the American Cancer Society. C. R. C. was a fellow of the John Simon Guggenheim Memorial Foundation during the preparation of this article. Support is also gratefully acknowledged for research grants from the National Science Foundation and the United States Public Health Service.

REFERENCES

Acharya, A. S., and Moore, P. B. (1973). *J. Mol. Biol.* **76**, 207–221.
Acharya, A. S., Moore, P. B., and Richards, F. M. (1973). *Biochemistry* **12**, 3108–3114.
Bahr, W., Faerber, P., and Scheit, K. H. (1973). *Eur. J. Biochem.* **33**, 535–544.
Beller, R. J., and Lubsen, N. H. (1972). *Biochemistry* **11**, 3271–3276.
Bickle, T. A., Howard, G. A., and Traut, R. R. (1973). *J. Biol. Chem.* **248**, 4862–4864.
Birge, E. A., and Kurland, C. G. (1970). *Mol. Gen. Genet.* **109**, 356–369.
Bispink, L., and Matthaei, H. (1973). *FEBS Lett.* **37**, 291–294.
Bochkareva, E. S., Budker, V. G., Girshovich, A. S., Knorre, D. G., and Teplova, N. M. (1971). *FEBS Lett.* **19**, 121–124.
Bodley, J. W., and Gordon, J. (1974). *Biochemistry* **13**, 3401–3405.

242   Maria Pellegrini and Charles R. Cantor

Brot, N., Yamasaki, E., Redfield, B., and Weissbach, H. (1972). *Arch. Biochem. Biophys.* **148**, 148–155.
Bruton, C. J., and Hartley, B. S. (1970). *J. Mol. Biol.* **52**, 165–178.
Cramer, F., von der Harr, F., and Schlimme, E. (1968). *FEBS Lett.* **2**, 136–139.
Czernilofsky, A. P., and Kuechler, E. (1972). *Biochim. Biophys. Acta* **272**, 667–671.
Czernilofsky, A. P., Collatz, E. E., Stöffler, G., and Kuechler, E. (1974). *Proc. Natl. Acad. Sci. U.S.A.* **71**, 230–234.
de Groot, N., Lapidot, Y., Panet, A., and Wolman, Y. (1966). *Biochem. Biophys. Res. Commun.* **25**, 17–22.
de Groot, N., Panet, A., and Lapidot, Y. (1971). *Eur. J. Biochem.* **23**, 523–527.
Deusser, E., Weber, H. J., and Subramanian, A. P. (1974). *J. Mol. Biol.* **84**, 249–256.
Deutscher, M. P. (1973). *Biochem. Biophys. Res. Commun.* **52**, 216–220.
Dzionara, M., Kaltschmidt, E., and Wittmann, H. G. (1970). *Proc. Natl. Acad. Sci. U.S.A.* **67**, 1909–1913.
Eilat, D., Pellegrini, M., Oen, H., de Groot, N., Lapidot, Y., and Cantor, C. R. (1974a). *Nature (London)* **250**, 514–516.
Eilat, D., Pellegrini, M., Oen, H., Lapidot, Y., and Cantor, C. R. (1974b). *J. Mol. Biol.* **88**, 831–840.
Erdmann, V. A., Spinzl, M., and Pongs, O. (1973). *Biochem. Biophys. Res. Commun.* **54**, 942–948.
Fahnestock, S., Erdmann, V., and Nomura, M. (1973). *Biochemistry* **12**, 220–224.
Fakunding, J., Traut, R. R., and Hershey, J. W. B. (1973). *J. Biol. Chem.* **248**, 8555–8559.
Fiser, I., Scheit, K. H., Stöffler, G., and Kuechler, E. (1974). *Biochem. Biophys. Res. Commun.* **60**, 1112–1118.
Gavrilova, L. P., and Spirin, A. S. (1974). *FEBS Lett.* **17**, 13–16.
Ghosh, K., and Ghosh, H. P. (1972). *J. Biol. Chem.* **247**, 3369–3374.
Gilbert, W. (1963). *J. Mol. Biol.* **6**, 374–388.
Ginzburg, I., Miskin, R., and Zamir, A. (1973). *J. Mol. Biol.* **79**, 481–495.
Gray, P. N., Garrett, R., Stöffler, G., and Monier, R. (1972). *Eur. J. Biochem.* **28**, 412–421.
Greenwell, P., Harris, R. J., and Symons, R. H. (1974). *Eur. J. Biochem.* **49**, 539–554.
Hamburger, A. D., Lapidot, Y., and de Groot, N. (1972). *Eur. J. Biochem.* **32**, 576–583.
Hamel, E., Koka, M., and Nakamoto, T. (1972). *J. Biol. Chem.* **247**, 805–814.
Harris, R. J., and Symons, R. H. (1973). *Bioorg. Chem.* **2**, 286–292.
Harris, R. J. Greenwell, P., and Symons, R. H. (1973). *Biochem. Biophys. Res. Commun.* **55**, 117–124.
Hauptmann, R., Czernilofsky, A. P., Voorma, H. D., Stöffler, G., and Kuechler, E. (1974). *Biochem. Biophys. Res. Commun.* **56**, 331–337.
Heding, H., Fredericks, G. J., and Lutzer, O. (1972). *Acta Chem. Scand.* **26**, 3251–3256.
Hindennach, I., Kaltschmidt, E., and Wittmann, H. G. (1971). *Eur. J. Biochem.* **23**, 12–16.
Horne, J. R., and Erdmann, V. A. (1972). *Mol. Gen. Genet.* **119**, 337–344.
Horne, J. R., and Erdmann, V. A. (1973). *Proc. Natl. Acad. Sci. U.S.A.* **70**, 2870–2873.
Hsiung, N. (1974). Ph.D. Thesis, Columbia University, New York.
Hsiung, N., and Cantor, C. R. (1973). *Arch. Biochem. Biophys.* **157**, 125–132.
Hsiung, N., and Cantor, C. R. (1974). *Nucleic Acids Res.* **1**, 1753–1761.
Hsiung, N., Reines, S. A., and Cantor, C. R. (1974). *J. Mol. Biol.* **88**, 841–855.
Huang, K. H., and Cantor, C. R. (1972). *J. Mol. Biol.* **67**, 265–275.
Jonak, J., and Rychlik, I. (1973). *Biochim. Biophys. Acta* **324**, 554–562.
Kaltschmidt, E., and Wittmann, H. G. (1970). *Anal. Biochem.* **36**, 401–412.
Knowles, J. R. (1972). *Act. Chem. Res.* **5**, 155–160.
Lapidot, Y., de Groot, N., Rappoport, S., and Hamburger, A. D. (1967). *Biochim. Biophys. Acta* **149**, 532–539.

Lapidot, Y., de Groot, N., Rappoport, S., and Eilat, D. (1969). *Biochim. Biophys. Acta* **190**, 304–311.

Lengyel, P., and Söll, D. (1969). *Bacteriol. Rev.* **33**, 264–301.

Lessard, J. L., and Pestka, S. (1972). *J. Biol. Chem.* **247**, 6909–6912.

Litman, D. J., and Cantor, C. R. (1974). *Biochemistry* **13**, 512–518.

Luhrmann, R., Schwarz, U., and Gassen, H. G. (1973). *FEBS Lett.* **32**, 55–58.

Maassen, J. A., and Möller, W. (1974). *Proc. Natl. Acad. Sci. U.S.A.* **71**, 1277–1280.

Malkin, I., and Rich, A. (1967). *J. Mol. Biol.* **26**, 329–346.

Marsh, R. C., and Parmeggiani, A. (1973). *Proc. Natl. Acad. Sci. U.S.A.* **70**, 151–155.

Means, G. E., and Feeney, R. E. (1971). "Chemical Modification of Proteins." Holden Day, San Francisco, California.

Mizushima, S., and Nomura, M. (1970). *Nature (London)* **226**, 1214–1218.

Moore, P. B. (1971). *J. Mol. Biol.* **60**, 169–184.

Moore, P. B., Engelman, D. M., and Schoenborn, B. P. (1974). *Proc. Natl. Acad. Sci. U.S.A.* **71**, 172–176.

Morrison, C. A., Garrett, R. A., Zeichhardt, H., and Stöffler, G. (1974). *Mol. Gen. Genet.* **127**, 359–368.

Nathans, D., and Neidle, A. (1963). *Nature (London)* **197**, 1076–1077.

Nierhaus, D., and Nierhaus, K. (1973). *Proc. Natl. Acad. Sci. U.S.A.* **70**, 2224–2228.

Nierhaus, K., and Montejo, V. (1973). *Proc. Natl. Acad. Sci. U.S.A.* **70**, 1931–1935.

Nomura, M. (1970). *Bacteriol. Rev.* **34**, 228–277.

Nomura, M., Mizushima, S., Ozaki, M., Traub, P., and Lowry, C. V. (1969). *Cold Spring Harbor Symp. Quant. Biol.* **34**, 49–61.

Oen, H., Pellegrini, M., Eilat, D., and Cantor, C. R. (1973). *Proc. Natl. Acad. Sci. U.S.A.* **70**, 2799–2803.

Oen, H., Pellegrini, M., and Cantor, C. R. (1974). *FEBS Lett.* **45**, 218–222.

Ofengand, J., and Chen, C. M. (1972). *J. Biol. Chem.* **247**, 2049–2058.

Ofengand, J., and Schwartz, I. (1974). In "Lippmann Symposium." Energy, Regulation and Biosynthesis in Molecular Biology" (D. Richter, ed.), pp. 456–469. de Gruyter, Berlin.

Panet, A. (1970). Ph.D. Thesis, Hebrew University.

Pellegrini, M. (1973). Ph.D. Thesis, Columbia University, New York.

Pellegrini, M., Oen, H., and Cantor, C. R. (1972). *Proc. Natl. Acad. Sci. U.S.A.* **69**, 837–841.

Pellegrini, M., Oen, H., Eilat, D., and Cantor, C. R. (1974). *J. Mol. Biol.* **88**, 809–829.

Pestka, S. (1972). *J. Biol. Chem.* **247**, 4669–4678.

Pestka, S., and Brot, N. (1971). *J. Biol. Chem.* **246**, 7715–7722.

Pongs, O., and Erdmann, V. (1973). *FEBS Lett.* **37**, 47–50.

Pongs, O., Bald, R., and Erdmann, V. (1973a). *Proc. Natl. Acad. Sci. U.S.A.* **70**, 2229–2233.

Pongs, O., Bald, R., Wagner, T., and Erdmann, V. (1973b). *FEBS Lett.* **35**, 137–140.

Pongs, O., Nierhaus, K. H., Erdmann, V., and Wittmann, H. G. (1974). *FEBS Lett.* **4**, S28–S37.

Randall-Hazelbauer, L. L., and Kurland, C. G. (1972). *Mol. Gen. Genet.* **115**, 234–242.

Rebstock, M. C. (1950). *J. Am. Chem. Soc.* **72**, 4800–4803.

Richter, D., Erdmann, V. A., and Sprinzl, M. (1973). *Nature (London), New Biol.* **246**, 132–135.

Rummel, D. P., and Noller, H. F. (1973). *Nature (London), New Biol.* **245**, 72–75.

Ruoho, A. E., Kiefer, H., Roeder, P. E., and Singer, S. J. (1973). *Proc. Natl. Acad. Sci. U.S.A.* **70**, 2567–2571.

Sander, G., Marsh, R. C., and Parmeggiani, A. (1972). *Biochem. Biophys. Res. Commun.* **47**, 866–873.

Sander, G., Marsh, R. C., and Parmeggiani, A. (1973). *FEBS Lett.* **33**, 132–134.

Santi, D. V., and Cunnion, S. O. (1974). *Biochemistry* **13**, 481–485.

Schenkman, M. L., Wood, D., and Moore, P. B. (1974). *Biochim. Biophys. Acta* **353**, 503–508.

Schrier, P. I., Maasen, J. A., and Möller, W. (1973). *Biochem. Biophys. Res. Commun.* **53**, 90–98.

Schwartz, I., and Ofengand, J. (1974). *Proc. Natl. Acad. Sci. U.S.A.* **71**, 3951–3955.

Shaw, E. (1970). *In* "The Enzymes" (P. D. Boyer, ed.), 3rd ed., Vol. 1, pp. 46–91. Academic Press, New York.

Singer, R. E., and Conway, T. W. (1973). *Arch. Biochem. Biophys.* **158**, 257–265.

Singer, S. J. (1967). *Adv. Protein Chem.* **22**, 1–54.

Smith, I. C., and Yamane, T. (1967). *Proc. Natl. Acad. Sci. U.S.A.* **47**, 1588–1602.

Sonenberg, N., Wilchek, M., and Zamir, A. (1973). *Proc. Natl. Acad. Sci. U.S.A.* **70**, 1423–1426.

Sonenberg, N., Zamir, A., and Wilchek, M. (1974). *Biochem. Biophys. Res. Commun.* **59**, 693–696.

Sopori, M., Pellegrini, M., Lengyel, P., and Cantor, C. R. (1974). *Biochemistry* **15**, 5432–5439.

Stöffler, G., Daya, L., Rak, K. H., and Garrett, R. A. (1971). *Mol. Gen. Genet.* **114**, 125–133.

Tal, J., Deutscher, M. P., and Littauer, U. Z. (1972). *Eur. J. Biochem.* **28**, 478–491.

Tal, M., Aviram, M., Kanarek, A., and Weiss, A. (1972). *Biochim. Biophys. Acta* **281**, 381–392.

Traub, P., Hosokawa, M. K., Craven, G., and Nomura, M. (1964). *Proc. Natl. Acad. Sci. U.S.A.* **58**, 2430–2436.

Traub, P., Mizushima, S., Lowry, C. V., and Nomura, M. (1971) *In* "Methods in Enzymology" (K. Moldave and G. Grossman, eds.), Vol. 20, pp. 391–407. Academic Press, New York.

Traugh, J. A., and Traut, R. R. (1972). *Biochemistry* **11**, 2503–2509.

Traut, R. R., and Haenni, A. L. (1967). *Eur. J. Biochem.* **2**, 64–73.

Traut, R. R., Heimark, R., Sun, T.-T., Hershey, J. W. B., and Bollen, A. (1974). *In* "Ribosomes" (M. Nomura, A. Tissières, and P. Lengyel, eds.), pp. 271–308. Cold Spring Harbor Lab., Cold Spring Harbor, New York.

Van Duin, J., and Kurland, C. G. (1970). *Mol. Gen. Genet.* **109**, 169–176.

von Knippenberg, P. H., Hooykaas, P. J. J., and Van Duin, J. (1974). *FEBS Lett.* **41**, 323–326.

Yang, C. H., and Söll, D. (1973). *J. Biochem. (Tokyo)* **73**, 1243–1247.

Yu, R. S. T., and Wittmann, H. G. (1973). *Biochim. Biophys. Acta* **324**, 375–385.

Yukioka, M., and Morisawa, S. (1970). *Biochem. Biophys. Res. Commun.* **40**, 1331–1339.

Zimmermann, R. A., Garwin, R. T., and Gorini, L. (1971). *Proc. Natl. Acad. Sci. U.S.A.* **68**, 2263–2267.

5

Initiation of Messenger RNA Translation into Protein and Some Aspects of Its Regulation

MICHEL REVEL

Introduction

Initiation of translation of a messenger RNA into protein is a series of reactions during which the separated subunits of the ribosome combine with the mRNA and with formylmethionyl transfer RNA to form the initiation complex. Interest in the mechanism of initiation stems from the fact that this is a specialized reaction, different from elongation, and a crucial step in the decoding of genetic information from nucleic acid into protein. Translation has to start on the mRNA at the precise codon corresponding to the beginning of a cistron in order to correctly assemble and terminate the amino acid sequence. A set of three initiation factors (Revel et al., 1968a; Iwasaki et al., 1968) directs these reactions on the ribosomal subunits (Fig. 1).

Many reviews have, in the past years, summarized the details of this process (the most recent being Haselkorn and Rothman-Denes, 1973; Lengyel, 1974). Recent progress has been made mainly on the structure and function of ribosomal subunits and of ribosome-associated factors. The present review deals therefore, to a great extent with this new information, assuming that the reader is familiar with the basic understanding of the mechanism of protein synthesis initiation as it was described by Rudland and Clark (1972) for fMet-tRNA$_f$ and by Revel (1972) for mRNA, ribosomal subunits, and initiation factors IF-1, IF-2, and IF-3. The literature reviewed here starts in most cases where these previous reviews end. This review was completed in December 1974 and partially revised in the spring 1976.

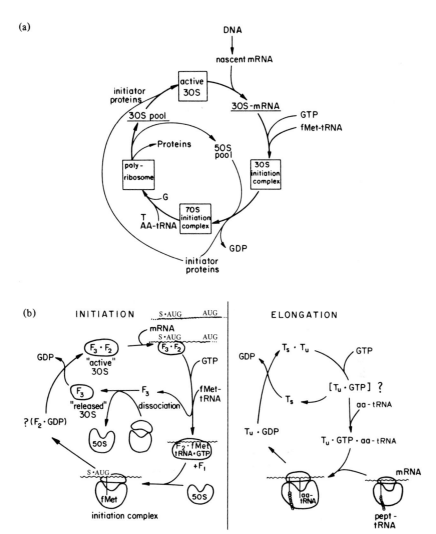

Fig. 1 The protein synthesis initiation cycle. (a) The successive steps of initiation are shown as part of the ribosome cycle in protein synthesis. (b) The functions of initiation factors of *E. coli* are schematically represented. The scheme is based on maximal analogy with the reactions of Ts and Tu in elongation, shown in the right panel. Recognition of the initiator AUG codon results from the proximity of an initiation signal S. From Revel *et al.* (1969) and Groner and Revel (1971).

PART I. Initiation of Protein Synthesis in Prokaryotes

I. Interaction of Initiation Factors with the Ribosomal Subunits

From the fact that initiation factors are found on native 30 S subunits (Parenti-Rosina et al., 1969; Miller et al., 1969) and from the dissociation activity of IF-3 (Subramanian and Davis, 1971), it was inferred that the first step in initiation is the binding of initiation factors to the ribosomal particle. This concept has been recently strengthened by direct studies on the binding of IF-1, IF-2, and IF-3 to ribosomes. In contrast to the binding of mRNA or initiator tRNA which requires initiation factors, binding of the protein factors to 30 S ribosomal subunits takes place in the complete absence of mRNA and tRNA. The binding of IF-1, IF-2, and IF-3 is, however, a highly cooperative process.

A. Binding of IF-3 to Ribosomes

Binding of radioactive IF-3 was first demonstrated by Sabol and Ochoa (1971), who found that ^{35}S-labeled IF-3 binds to 30 S ribosomal subunits, but not to 50 S subunits or to 70 S ribosomes. On the contrary, it is released upon formation of the 70 S couple which results from 30 S and 50 S subunits joining, either through an increase in magnesium concentration or during the normal process of formation of the 70 S initiation complex. Similar results were obtained by Thibault et al. (1972), Pon et al. (1972), and Vermeer et al. (1973a) and firmly established the occurrence of a cycle of binding and release, leading to a catalytic functioning of IF-3 in protein biosynthesis.

Only one molecule of IF-3 (MW 21,000) is bound to each 30 S particle with a stability sufficient to resist zonal centrifugation (Sabol et al., 1973). The average association constant was 5 μM^{-1} at 24°C although a value of $2 \times 10^7 M^{-1}$ was quoted more recently by Godefroy-Colburn et al. (1975). This relatively low affinity may explain why an excess of IF-3 is needed to dissociate ribosomes. Magnesium ions and GTP have no effect on IF-3 binding to the 30 S particle. Ribosome-bound IF-3 exchanges freely with unbound IF-3. Saturating amounts of IF-1 and IF-2 slightly stimulated IF-3 binding (40%). This is probably a stabilization of IF-3 attachment (Vermeer et al., 1973b). The slight effect of IF-2 on IF-3 binding to 30 S subunits is abolished if the complex is fixed by glutaraldehyde before analysis, or if Sephadex G-200 chromatography is used instead of a sucrose gradient to isolate the IF-3-30 S complex.

At low Mg^{2+} concentrations, IF-3 binds also to MS2 phage RNA and to 50 S ribosomal subunits, but with a much weaker affinity than to 30 S particles (Vermeer et al., 1973a).

Pon et al. (1972) found that IF-3 binding to 30 S is inhibited by streptomycin and aurintricarboxylic acid (ATA). Streptomycin, however, is a weaker inhibitor of IF-3 binding than of MS2 RNA binding to the 30 S ribosome (30% at 4 μM streptomycin instead of 90% at 0.6 μM; Sabol *et al.*, 1973). Herzog *et al.* (1971a) showed that streptomycin inhibits ribosome dissociation. The mechanism of IF-3 directed ribosome dissociation is discussed in Section I,E.

B. Binding of IF-2 to Ribosomes

Association of initiation factor IF-2 with 30 S ribosome was first demonstrated by following the biological activity of this 70,000 MW protein required for fMet-tRNA$_f$ recognition and binding, and which also promotes mRNA binding to ribosomes (Revel *et al.*, 1968a). Mixing purified IF-2 with ribosomes results in the binding of the factor mainly to 30 S particles and also to the 70 S ribosome. Lockwood *et al.* (1972) showed that IF-2 binds to 30 S subunits in the absence of mRNA and fMet-tRNA$_f$; this association was unstable unless IF-1 and IF-3 were added. The use of radioactive IF-2 has directly confirmed these early results. Fakunding and Hershey (1973) used IF-2 labeled by phosphorylation with protein kinase. By sucrose gradient analysis, IF-2 was shown to bind to 30 S particles in the absence of mRNA and tRNA; there was no interaction with 50 S particles or vacant 70 S couples. The IF-2-30 S complex is stable only at low salt concentrations. Addition of GTP (but not GDP or GMPPCP) has the same stabilizing effect as lowering the salt concentration. IF-2 dissociates from the 30 S during the sucrose gradient analysis, unless IF-1 and IF-3 are added. In the presence of *both* IF-1 and IF-3, a stable IF-2-30 S complex is obtained even at high salt concentrations. Under these conditions, one molecule of IF-2 is bound per active 30 S subunit (i.e., 30 S particles capable of binding fMet-tRNA$_f$). Fixation by glutaraldehyde leads to a stable complex, even without IF-1 and IF-3, but in this case two molecules of IF-2 are found per 30 S subunit. Using *in vivo* ^{35}S-labeled IF-2, Benne *et al.* (1973a) confirmed IF-2 binding to 30 S subunits. GTP and fMet-tRNA$_f$ had only a slight stimulatory effect, but further addition of MS2 RNA increased IF-2 binding. Although only 1% of the ribosomes were labeled, the ratio of IF-2 to fMet-tRNA$_f$ bound was 1. In further studies Benne *et al.* (1973b) showed that there is no competition between the various initiation factors for binding to 30 S subunits. IF-1 and IF-2 stabilized IF-3 binding (see Section I,A), but, in these experiments, IF-2 by itself was found to form a stable complex with 30 S subunits. The results of Chu and Mazumder (1974) also support the model that IF-2 interacts with 30 S particles, in a reversible reaction, without requirement for other elements of the initiation complex (Mazumder, 1972).

Radioactive IF-2 does not bind to 70 S ribosomes unless fMet-tRNA$_f$

and the AUG codon are added. Stable 70 S binding requires GMPPCP, while with GTP release of IF-2 occurs (see Section II,C).

C. Binding of IF-1 to Ribosomes

Interaction of the 9000 MW IF-1 with 30 S subunits was originally reported by Hershey *et al.* (1969) to require the presence of the other components of the initiation complex. More recently, however, Benne *et al.* (1973b), with [³H]IF-1, showed that IF-1 can bind by itself. This binding is weaker than that of IF-2, and was not observed by Grunberg-Manago *et al.* (1973) unless other initiation factors were also added.

D. Complexes between Initiation Factors

The cooperative effect of the other initiation factors in the binding of IF-2 (Fakunding and Hershey, 1973), and the stabilization of IF-3 binding by IF-2 and IF-1 (Vermeer *et al.*, 1973a,b) suggest that the factors may interact with each other. Existence of such interactions was proposed after the isolation of an IF-2-IF-3 complex (Groner and Revel, 1971) and of an IF-1-IF-3 complex (Hershey *et al.*, 1971) from the high salt wash of *Escherichia coli* ribosomes. These complexes represent only a small proportion of the total free factors, and it is not known if the factors can interact in solution or only on the surface of the ribosome. The IF-2-IF-3 complex could not be formed in the absence of 30 S ribosomes (Groner and Revel, 1971), and its biological activity differed from that of the free factors. An argument in favor of the formation of such complexes between initiation factors on the ribosome was recently communicated by Hershey (1974). Using dimethyl suberimidate and [³²P]IF-2, he showed that IF-1 and IF-3 could be cross-linked to IF-2 in a ribosome-dependent reaction. This suggested that the three initiation factors were located contiguously on the 30 S ribosomal subunit.

E. Ribosomal Components Involved in the Binding of Initiation Factors

Both ribosomal RNA and proteins could participate in this reaction. IF-3 has some affinity for RNA and binds well to rRNA (Sabol *et al.*, 1973). Gualerzi and Pon (1973) suggested that a segment of the 16 S rRNA provides a binding site for IF-3, and that r-proteins confer stability and specificity by restricting the number of available binding sites on the RNA. RNA-Binding dyes (acridine and pyronine) inhibit IF-3 attachment to 30 S particles. Antibodies against single 30 S r-proteins interfered only weakly with IF-3 binding, although many of these antibody fragments (most strongly anti-S19) block the formation of the initiation complex (Lelong *et al.*, 1974).

Data obtained by cross-linking experiments show that IF-3 can be cross-linked to S11, S12, S13, but also to S19, S21, and S14, S18, while IF-2 (with IF-1 and IF-3) cross-links to S13, S19, but also S2, S11, S12, S14 (Traut *et al.*, 1974; Bollen *et al.*, 1975): there is a high degree of overlap between the r-proteins found close to IF-2 and to IF-3, as well as with the mRNA binding site (Pongs *et al.*, 1975). Anti-S19 Fab fragments inhibit IF-2 binding to 30 S and so do anti-S4; but not anti-S13, S14 (Bollen *et al.*, 1975). S21 was first reported to inhibit initiation (Van Duin *et al.*, 1972) but this is now disputed: S21 may be required for initiation (Held *et al.*, 1974a; for review, see Pongs *et al.*, 1974).

An important result is that both IF-2 and IF-3 were found to cross-link with S1 (Traut *et al.*, 1974; Bollen *et al.*, 1975); in view of the identity between S1 and interference factor i-α, and its role in mRNA binding (discussed in Section III,D,1), this result gives much information on the ribosomal site for mRNA binding and recognition.

The cross-link of IF-3 to S12 was confirmed by Hawley *et al.* (1974). The role of S12, the streptomycin protein, in initiation had been suggested by Nomura (Ozaki *et al.*, 1969; Held *et al.*, 1974b). IF-1 and IF-3 inhibit the binding of dihydrostreptomycin to ribosomes (Lelong *et al.*, 1972), suggesting that binding of initiation factors to S12 can, as mutations in S12, produce a change in the 30 S subunit which prevents the attachment of streptomycin. It is not clear whether streptomycin binds to rRNA (Biswas and Gorini, 1973) or to r-proteins (Schreiner and Nierhaus, 1973; Pongs, 1973; Lelong *et al.*, 1974). It does not bind to S12.

Recently, a cross-link IF-3-S7 and IF-3-16 S RNA was reported (Van Duin *et al.*, 1975).

Further information on the binding sites of initiation factors on ribosomes is provided by studies on the location of proteins at the interface between the two ribosomal subunits. Morrison *et al.* (1973) showed that among the 30 S proteins, S9, S11, S12, S14, and S20 are located at the interface. Both S11 and S12 bind to 23 S rRNA (of the 50 S subunit) and may be involved directly in the 30 S-50 S interaction. Therefore IF-3, which cross-links to S11, S12, and S14, is bound most likely on the part of the 30 S subunit which faces the 50 S subunit and 23 S RNA. This information may be of importance for the mechanism of IF-3-induced dissociation of the ribosome, but one needs to await a more precise model of ribosome structure (see Fig. 2). The role of ribosomal proteins L7, L12 in the binding of initiation factors, is discussed below in the Section C,3.

F. Effect of Initiation Factors on Ribosome Dissociation into Subunits

There is now general agreement that dissociation of 70 S into subunits is an obligatory step for initiation of translation on natural mRNA templates, although with synthetic polynucleotides undissociated ribosomes

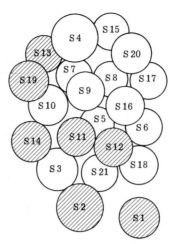

Fig. 2 Distribution, on a 30 S subunit model, of the proteins cross-linked to initiation factor IF-2 (diagonal striped circles). From Bollen *et al.* (1975).

may function (for a discussion, see H. Noll *et al.,* 1973; Kurland, 1972; Slobin, 1972). The problem of the state of vacant ribosomes in *E. coli* has been debated: Do ribosomes which leave mRNA dissociate immediately into 30 S and 50 S (Kaempfer, 1972) or do they remain as 70 S monomers (Davis, 1971)? Recently, it has been recognized that ribosomes which dissociate spontaneously may be damaged ribosomes (H. Noll *et al.,* 1973; M. Noll *et al.,* 1973) whereas tight couples (remaining 70 S at 5 mM Mg^{2+}) are the active particles. Several r-proteins may be missing from the loose "damaged" couples.

Dissociation of tight couples takes place only at initiation and is catalyzed by initiation factors. Subramanian and Davis (1971) were the first to demonstrate that IF-3 promotes dissociation, but it is still not clear whether this is an active process or that IF-3 acts as an anti-association factor, i.e., the factor binds to free 30 S subunits which dissociate spontaneously, preventing their reassociation and displacing the equilibrium toward the free subunit state as proposed by Kaempfer (1972). An important question is whether IF-3 can bind to 70 S subunits, a prerequisite for an active dissociation mechanism. Several groups (see Section I,A) have reported that they could not find binding of IF-3 to 70 S ribosomes on sucrose gradients, supporting the notion that IF-3 binds only to spontaneously dissociated 30 S particles. Vermeer *et al.* (1973a), however, succeeded in showing IF-3-70 S interaction. Since the affinity of IF-3 for 70 S ribosomes is very low, the interaction was observed only in the absence of 30 S particles. This observation may indicate that IF-3 actively dissociates 70 S ribosomes. GTP did not stimulate dissociation.

More precise information on the role of initiation factors in ribosome dissociation comes from the work of M. Noll *et al.* (1973) and Noll and Noll (1972). These authors used vacant tight couples and measured the rate of subunit exchange. IF-3 does not change the rate of dissociation but displaces the equilibrium; on the other hand (and as suggested by Miall and Tamaoki, 1972), IF-1 increases the rate of subunit exchange with tight couples and acts as a true dissociation factor. Dissociation by IF-3 alone was observed in their experiments only with loose couples. Interestingly, M. Noll *et al.* (1973) suggested that IF-2 and fMet-tRNA$_f$ may stabilize the IF-3-30 S interaction, which otherwise would be too weak even to prevent 30 S-50 S reassociation. The cooperative effect of IF-1 and IF-3 on ribosome dissociation was confirmed by Grunberg-Manago *et al.* (1973) who used light scattering to follow the changes in the state of ribosome association. IF-1 and mRNA (at least polynucleotides containing A,U,G) both stimulate IF-3-dependent dissociation. IF-2 promoted association and not dissociation (Garcia-Patrone *et al.*, 1971a). The cooperative effect of IF-1 and IF-3 in dissociating ribosomes may be explained by their cooperative binding to 30 S subunits. The IF-2-IF-3 complex (Groner and Revel, 1971) has no dissociation activity (A. Bollen, private communication).

The fact that IF-3 binds to the 30 S subunit at its interface with the 50 S subunit (see above) may be the basis for its dissociation activity. Recently, IF-3 was shown to bind to a 16 S RNA site which may base pair with 23 S RNA in the 50 S subunit (Van Duin *et al.*, 1976). Alternatively, Paradies *et al.* (1974) have demonstrated a large conformational change in the 30 S subunit upon IF-3 binding which may be related to dissociation, as suggested already by Sherman and Simpson (1969) and Chang (1973). This conformational change, which involves a transition from a sedimentation coefficient of 30 S to 37 S, may be due to IF-3 binding to the ribosomal RNA. It is not clear what effects IF-1 and IF-2 have on this transition, but it is suggested that most of the effects of initiation factors on 30 S subunit activities could be explained by such conformational changes. Strange effects of IF-3, such as inhibition of streptomycin binding (in the presence of IF-1; Lelong *et al.*, 1972) and release of Phe-tRNA (Grunberg-Manago *et al.*, 1969) may be explained in this way, as well as inhibition of dissociation by streptomycin (Herzog *et al.*, 1971a; Garcia-Patrone *et al.*, 1971b; Wallace *et al.*, 1973). A mutant of ribosome dissociation factor has been reported (Herzog *et al.*, 1971b).

G. Formation of the 30 S Initiation Complex

The first step in the formation of the 30 S initiation complex appears to be the formation of a (30 S · IF-1 · IF-2 · IF-3) particle, as a prerequisite for the specific binding of initiator tRNA and the initiator AUG codon of a mRNA. The order in which these next steps take place is still debated. Does mRNA or fMet-tRNA$_f$ come first?

Experiments have provided evidence for both fMet-tRNA$_f$ binding to 30 S in the absence of mRNA (Leder and Nau, 1967; Noll and Noll, 1972; H. Noll *et al.*, 1973) and for mRNA binding to 30 S in the absence of fMet-tRNA$_f$ (Greenshpan and Revel, 1969). In both conditions, however, only unstable complexes were formed. More recently, the use of glutaraldehyde fixation has allowed Jay and Kaempfer (1974a) to isolate a 30 S-fMet-tRNA$_f$ complex without mRNA; but at the same time, with this technique, Vermeer *et al.* (1973b) succeeded in isolating a 30 S-MS2 RNA complex without fMet-tRNA$_f$. It appears likely, therefore, that in prokaryotes both initiator tRNA and messenger RNA may be bound independently from each other, and the question of the actual order of events *in vivo* may be almost impossible to determine (unless mutants accumulating some intermediate can be isolated). This is not, however, an uninteresting situation since it indicates that ribosomes and initiation factors must have the property to select fMet-tRNA$_f$ from all other aminoacyl-tRNA's without the help of an AUG codon (which would not discriminate fMet-tRNA$_f$ from Met-tRNA$_m$). By the same token, ribosome-initiation factor complexes must also have the ability to recognize and bind mRNA. In the next sections we will review the mechanisms of these two recognition events, which together define where translation of a genetic message into proteins will begin.

II. Interaction of Initiator fMet-tRNA$_f$ with Ribosomes and Initiation Factors

A. Selective Recognition of fMet-tRNA$_f$ by Initiation Factors

Formylmethionyl-tRNA$_f$ must possess some structural features which make it differ from other bacterial tRNA's since it is recognized by the transformylase (for a review, see Rudland and Clark, 1972) but is not bound to elongation factor Tu in contrast to the other aminoacyl-tRNA's (Ono *et al.*, 1968). In addition it may wobble at the third base of the anticodon (Clark and Marcker, 1966). These particular features are likely to be also responsible for its interaction with initiation factors. Study of the primary sequence of fMet-tRNA$_f$ shows no unique feature absent from other tRNA's (Rudland and Clark, 1972) and the situation is complicated by the fact that eukaryotic initiator Met-tRNA$_f$ also reacts with *E. coli* transformylase (Caskey *et al.*, 1967; Smith and Marcker, 1970) as well as with *E. coli* ribosomes and initiation factors (Berthelot *et al.*, 1973). Comparison of the primary sequence of various initiator tRNA's does not reveal a common feature, and only information on the three-dimensional folding of the polynucleotide chain in these tRNA's will probably solve this problem (Piper and Clark, 1974).

In the formation of an initiation complex with 30 S ribosomes, initiation factors stimulate selectively the binding of fMet-tRNA$_f$ while the binding of other tRNA's may even be inhibited (Grunberg-Manago et al., 1969; Gualerzi et al., 1971). Unformylated Met-tRNA$_f$ is not stimulated, but among the other N-blocked aminoacyl-tRNA's tested none was bound with initiation factors (Rudland et al., 1969) with the notable exception of N-acetyl Phe-tRNA (Lucas-Lenard and Lipmann, 1967; Springer and Grunberg-Manago, 1972). Presence of the formyl group on Met-tRNA$_f$ is necessary, but not by itself sufficient, to explain its function. For a discussion of the role of formylation in E. coli, see Danchin (1973). In Staphylococcus, protein synthesis can be initiated without formylation of the tRNA under certain conditions (Samuel and Rabinowitz, 1974).

A direct demonstration that initiation factors recognize fMet-tRNA$_f$ even in the absence of ribosomes was provided by a series of experiments in several laboratories (Rudland et al., 1971; Groner and Revel, 1971, 1973; Lockwood et al., 1971). It was found that addition of fMet-tRNA$_f$ stimulates binding of GTP to isolated IF-2, while unformylated Met-tRNA$_f$ is less active, and other aminoacyl-tRNA's are inactive. The complex formed with free IF-2 appeared, however, very unstable: it did not contain stoichiometric amounts of GTP, and (fMet-tRNA$_f$-IF-2) could not be isolated by filtration through Sephadex G-100. A complex between fMet-tRNA$_f$ and IF-2 was isolated for the first time by glycerol gradient sedimentation (Groner and Revel, 1973). A requirement for GTP was not observed, but the free sulfhydryl group of IF-2 was needed for fMET-tRNA$_f$ binding. Other aminoacyl-tRNA's were not bound, indicating that IF-2 selectively recognizes fMet-tRNA$_f$, in line with the absolute requirement of IF-2 for fMet-tRNA$_f$ binding to 30 S ribosomes. The low affinity of IF-2 for GTP and fMet-tRNA$_f$ makes it unlikely that this free ternary complex is an intermediate in the formation of the 30 S initiation complex; also against this idea is the fact that IF-2 binds to 30 S subunits readily in the absence of fMet-tRNA$_f$ and GTP (see Section I,B).

Groner and Revel (1971) isolated a subfraction of IF-2 activity, which exhibited a much stronger affinity for GTP and fMet-tRNA$_f$ than free IF-2. This subfraction contains both IF-2 and IF-3 activities and was designated (IF-2-IF-3) complex. Stable fMet-tRNA$_f$ binding to this complex was observed on Sephadex G-200, and the specificity for initiator tRNA was very large. The (IF-2-IF-3) complex lowers by an order of magnitude the apparent K_m of 30 S subunits for fMet-tRNA$_f$ as compared to free IF-2 + IF-3. Kinetic data on the exchange of free and 30 S bound fMet-tRNA$_f$ led the authors to conclude that the ribosome-free complex is not an intermediate in initiation but most likely functions on the ribosome surface. The IF-2-IF-3 complex could be formed only on the ribosome during initiation. Existence of such a complex showed the cooperativity between IF-2 and IF-3, which is now confirmed by the effect of IF-3 on IF-2 binding to 30 S (Fakunding and Hershey, 1973) and the formation of a ribosome-dependent IF-2-IF-3 cross-link (Hershey, 1974). The question of whether other proteins (as IF-1 and other factors) are present in this complex is still an open one.

The conclusion from the data summarized in this section is that selection of fMet-tRNA$_f$ from all other aminoacyl-tRNA's, to initiate a polypeptide chain, is a function of initiation factors, IF-2 in particular. There is, however, no strong evidence to support the concept that a ternary complex of protein factors, GTP, and fMet-tRNA$_f$ is formed before binding to the ribosome occurs.

B. Role of Initiation Factors in fMet-tRNA$_f$-30 S Ribosome Interaction

1. Formation of the 30 S Initiation Complex

Binding of fMet-tRNA$_f$ has been usually studied in the presence of a template containing the initiation codon ApUpG or GpUpG. All three initiation factors have been shown to have some effect on this reaction, but IF-2 plays the most important role, its requirement being absolute. Two forms of IF-2 have been reported (Miller and Wahba, 1973), but they appear to have a similar function. Much emphasis has been placed on a search for ascribing one particular function to each of the three initiation factors IF-1, IF-2, and IF-3, but it now appears that the effects are actually cooperative. The influence of IF-1 and IF-3 on fMet-tRNA$_f$ binding could be interpreted equally well as an effect on IF-2 binding to the 30 S subunit (Fakunding and Hershey, 1973; Lockwood et al., 1972) than as a direct role in fMet-tRNA, GTP, and AUG codon interactions.

IF-1 stabilizes IF-2-dependent fMet-tRNA$_f$ binding to 30 S subunits, which otherwise would be stable only around 0° (Chae et al., 1969). IF-1 has also been shown to decrease the K_m of 30 S subunits for AUG codons (Mazumder, 1971). The effect of IF-1 on the 30 S subunit, however, is smaller than on the 70 S ribosome (Benne et al., 1973a).

The stimulation by IF-3 of 30 S-fMet-tRNA$_f$ initiation complex formation has been often attributed to the role this factor plays in 30 S subunit binding to initiation sites on natural mRNA (see Section III,C). With native phage MS2 RNA, IF-3 is required for fMet-tRNA$_f$ binding to 30 S subunits (Sabol et al., 1970; Benne et al., 1973a; Suttle and Ravel, 1974); even with unfolded MS2 RNA, which binds to 30 S subunits without IF-3, fMet-tRNA$_f$ attachment to the complex is still IF-3 dependent (Vermeer *et al.*, 1973b). With other templates, the effect of IF-3, when added to IF-2 and IF-1, is more variable. With the trinucleotide ApUpG, or poly(U,G), IF-3 even inhibits the IF-2-dependent fMet-tRNA$_f$-30 S interaction (Groner and Revel, 1971; Meier *et al.*, 1973; Suttle and Ravel, 1974). IF-3 may destabilize the 30 S complex formed with these synthethic templates (Dubnoff and Maitra, 1972). With poly(A,U,G) as template, however, a stimulation by IF-3 of 30 S-fMet-tRNA$_f$ binding was reported (Dondon *et al.*, 1974); the stimulation may depend on the concentration of ribosomes, template, and factor used, and is small as compared to the effect of IF-3 on 70 S ribosomes (Suttle and Ravel, 1974). Many authors found that IF-3 stimulates initiation by 70 S ribosomes with synthetic templates (Wahba *et al.*, 1969; Groner and Revel, 1971; Dubnoff *et al.*, 1972b; Yoshida and Rudland, 1972; Miller and Wahba, 1973; Suttle *et al.*, 1973) even with polynucleotides which do not contain AUG or GUG. With poly(U), IF-3 was found to stimulate the IF-2-dependent binding of *N*-acetyl Phe-tRNA to 30 S subunits (Bernal *et al.*, 1974). Since this particular tRNA is not a natural initiator tRNA, its interaction with IF-2 may be less strong and might have to be stabilized by IF-3 under some conditions [this may explain why Suttle and Ravel (1974) did not observe this effect]. Without any mRNA template, Jay and Kaempfer (1974a) found that fMet-tRNA-30 S interaction does not need IF-3, but this factor may have a stabilizing effect on the complex which otherwise can only be isolated after glutaraldehyde fixation.

The role of IF-3 in non-template-specific initiation could be due, as is its role in ribosome dissociation, to an effect of this protein on the 30 S subunit conformation (Paradies *et al.*, 1974). But IF-3 also appears to have more specific effects. An interesting hypothesis (Pon and Gualerzi, 1974) is that IF-3 sets the 30 S subunit in a conformation in which the initiator Met-tRNA$_f$ is most stably bound. The binding of other tRNA's, including *N*-acetyl Phe-tRNA, was shown to be destabilized by adding IF-3 to preformed 30 S-tRNA complexes. In these experiments, IF-3 appears to exhibit specificity for the initiator tRNA$_{fMet}$ molecule (formylated or not) more than for the polynucleotide used as template (Risuelo *et al.*, 1976). H. Noll *et al.* (1973) also claim that with 30 S subunits derived from tight couples, IF-3 alone is enough to bind fMet-tRNA$_f$, IF-2 being required only when 50 S are added.

In most experiments reported, however, the effect of IF-3 on fMet-tRNA$_f$ binding to 30 S subunits appears to be cooperative with IF-2. As stated above, this may represent a direct influence of IF-3 on IF-2-30 S binding (Fakunding and Hershey, 1973) and the formation of the (IF-2-IF-3) complex, which confers to the 30 S subunit a higher affinity for initiator tRNA than does IF-2 alone (Groner and Revel, 1971).

In conclusion, although the exact role of each factor cannot be defined, all three initiation factors participate in the formation of the 30 S initiation complex, in line with the evidence indicating that all three proteins are bound together to the 30 S particle.

2. Structure of the 30 S Initiation Complex

On 30 S ribosomes, IF-2 appears to be required in stoichiometric amounts to bind fMet-tRNA$_f$ (Dubnoff et al., 1972a; Benne et al., 1973a). GTP is also bound to the 30 S initiation complex in stoichiometric amounts with fMet-tRNA$_f$ (Thach and Thach, 1971a). GTP is not hydrolyzed at this stage; one of its role may be to increase the rate of binding of initiation factors to 30 S and the stability of this interaction. The nonhydrolyzable analogue of GTP, GMPPCP, may partially replace GTP, but the complex formed is less stable (Fakunding and Hershey, 1973; Benne et al., 1973a). In the presence of excess IF-2, GTP is not a prerequisite for the binding of fMet-tRNA$_f$, but only increases the rate of the reaction (Mazumder, 1972). Since fMet-tRNA$_f$ interacts with IF-2 even without GTP (Groner and Revel, 1973), it is likely that the two ligands interact independently with the factor. IF-2-GTP interaction has been well documented (Mazumder et al., 1969; Lelong et al., 1970; Rudland et al., 1971; Groner and Revel, 1971; Lockwood et al., 1971). The free sulfhydryl group of IF-2 (Mazumder et al., 1969) is essential for both GTP (Lelong et al., 1970) and fMet-tRNA$_f$ binding to this protein (Groner and Revel, 1973).

The 30 S initiation complex, therefore, contains GTP, fMet-tRNA$_f$, and an AUG (mRNA) codon. As mentioned earlier, partial complexes may be observed which lack one of these components, but these complexes are unstable, require fixation with glutaraldehyde and probably represent artifacts, which do not form in vivo.

The number of ribosomal proteins present in the 30 S initiation complex is still not well defined. As for initiation factors, IF-2 and IF-1 are part of the complex (Fakunding and Hershey, 1973; Benne et al., 1973a,b; Hershey et al., 1969). A very interesting observation made by Vermeer et al. (1973a,c) shows, on the other hand, that IF-3 is not part of the complete 30 S initiation complex.

3. Release of IF-3 upon fMet-tRNA$_f$ Binding to 30 S

Using [^{35}S]-IF-3, Vermeer et al. (1973a,c) observed that IF-3 is present in the MS2 RNA-30 S complex but is absent from the MS2 RNA-fMet-tRNA$_f$-30 S complex, suggesting that binding of initiator tRNA causes the release of IF-3. Since IF-3 is involved in RNA binding (Section III), this early release of IF-3 is strong evidence that mRNA binds to 30 S in the absence of, and prior to, the initiator tRNA.

C. Formation of the 70 S Initiation Complex and Recycling of IF-2

Upon addition of 50 S subunits to the complete 30 S initiation complex described above, a 70 S complex is formed. Addition of 50 S subunits stimulates IF-2-dependent fMet-tRNA$_f$ binding: several molecules of initiator tRNA are bound to 70 S ribosomes for each molecule of IF-2 present in the reaction. This catalytic effect of IF-2, which implies that IF-2 recycles among ribosomes, requires 50 S subunits, GTP hydrolysis, and IF-1 (Chae et al., 1969; Lockwood et al., 1972; Benne et al., 1973a).

1. Role of IF-1

The function of this small protein, about 90 amino acid long, in initiation remains ill defined. Benne et al. (1972) showed that IF-1 is needed, with GTP and 50 S, to make the 30 S-bound IF-2 recycle and act catalytically. With radioactive IF-2, Benne et al. (1973b) showed that IF-1 stimulates the release of IF-2 from 70 S initiation complexes. IF-1 stimulates the IF-2 dependent GTPase activity (Chae et al., 1969; Dubnoff and Maitra, 1972; Benne et al., 1973a; Grunberg-Manago et al., 1973) leading to the hydrolysis of more molecules of GTP than IF-2 present. IF-1 is needed to allow the bound fMet-tRNA$_f$ to react with puromycin and is absolutely required for the attachment of the (second aminoacyl tRNA-GTP-elongation factor Tu) ternary complex. To explain these results, Benne et al. (1972, 1973a,b) proposed that IF-1 acts in analogy to elongation factor Ts by catalyzing an IF-2-GDP $\rightleftharpoons$ IF-2-GTP exchange reaction. IF-1 would release IF-2 from ribosomes after GTP hydrolysis, form a new IF-2-GTP complex, and allow a new cycle of initiation to take place. Benne et al. (1972) actually showed that IF-1 stimulates a ribosome-dependent exchange of GTP and reduces the inhibition of IF-2 activity caused by GDP [the magic spot nucleotide ppGpp acts as GDP and may inhibit IF-2 in the cell (Yoshida et al., 1972)]. Several other facts, however, suggest a different model. First, IF-1 is released from the initiation complex without GTP hydrolysis, under conditions where IF-2 is still bound to the 70 S initiation complex (Hershey et al., 1969; Lockwood et al., 1972; Benne et al., 1973b). Moreover, although IF-1 stimulates more initiation when 50 S subunits are present, it does stimulate fMet-tRNA$_f$ binding to 30 S subunits even with GMPPCP (Kay and Grunberg-Manago, 1972) and must have a function on the 30 S particles, probably stabilizing IF-2 binding (Fakunding and Hershey, 1973). It is possible that IF-1 makes IF-2 bind to 30 S subunits in a configuration (Kay and Grunberg-Manago, 1972) which then allows IF-2 to trigger the GTPase site, and thereby favors the recycling of IF-2 and the accommodation of fMet-tRNA$_f$ in the donor site for puromycin reaction.

IF-3 also stimulates GTPase activity of ribosomes (Dubnoff and Maitra, 1972) and promotes the recycling of IF-2 as shown by Dondon et al. (1974)

although it is clearly not a part of the 70 S initiation complex. It seems likely that both IF-3 and IF-1 act directly on the ribosome configuration to allow the proper interactions to take place. The recycling phenomenon may be more a function of ribosomes than of IF-1. For example, *Caulobacter* seems to lack IF-1 (Leffler and Szer, 1974).

If IF-2-GDP is indeed produced during the initiation cycle, there may be no requirement for an enzymatic release of GDP from IF-2, since GDP affinity for this protein is much lower than that for elongation factor Tu (Lelong *et al.*, 1970).

2. Role of GTP Hydrolysis

When GMPPCP (the nonhydrolyzable analogue of GTP) is used, 50 S subunits join to form a 70 S initiation complex but IF-2 is not released and does not act catalytically (Lockwood *et al.*, 1972a,b; Benne *et al.*, 1973; Fakunding and Hershey, 1973) (under the same conditions, IF-3 works catalytically). In the presence of GMPPCP, sucrose gradient analyses with radioactive IF-2 show that both fMet-tRNA$_f$ and IF-2 are found on the 70 S ribosome in equimolar amounts. Binding of IF-2 to 70 S requires fMet-tRNA$_f$ and an AUG codon, which are not needed for IF-2-30 S interaction. In contrast, when GTP is present, more fMet-tRNA$_f$ is bound but IF-2 is not found anymore in the 70 S complex. It is, however, not GTP hydrolysis per se which releases IF-2, since in the absence of any guanine nucleotide, IF-2 is also released, the difference being that much less fMet-tRNA$_f$ is bound (Fakunding and Hershey, 1973).

GTP hydrolysis is required for the bound fMet-tRNA$_f$ to react with puromycin; this GTP hydrolysis-dependent step was designated "accommodation" (Thach and Thach, 1971b), but may just be the release of IF-2. When IF-2 is not released from 70 S initiation complexes because of GMPPCP, binding of the second aminoacyl-tRNA with EF-Tu is prevented as is the puromycin reaction (Benne *et al.*, 1973b; Ohta *et al.*, 1967; J. Dubnoff *et al.*, 1972a). Possibly, ribosomes are unable to accommodate IF-2 and EF-Tu (as also EF-Tu and EF-G factors) at the same time (Modollel *et al.*, 1973). It is of interest that fusidic acid, an antibiotic that freezes elongation factor G on the ribosome also inhibits initiation (Sala and Ciferri, 1970).

The fact that 30 S initiation complexes, depleted of GTP or GMPPCP, will react with puromycin, in the presence of 50 S subunits, to form fMet-puromycin (Dubnoff *et al.*, 1972a) supports the idea that GTP hydrolysis itself is not needed for accommodation. As in the case of IF-2 release discussed above, it is not the energy liberated by the hydrolysis which is needed. Rather, it appears that GTP, or its analogue, stabilizes IF-2 on the ribosome and prevents further reactions. After GTP hydrolysis the situation changes: IF-2 is released, possibly due to the low affinity of GDP for IF-2 (Lelong *et al.*, 1970; Benne *et al.*, 1973b). Nevertheless, GTP does stimulate the rate of fMet-puromycin formation (Hershey and Thach, 1967).

GTP appears therefore to have a double function: first, it increases the affinity of IF-2 for 30 S ribosomes, thereby increasing the rate at which initiation complexes are formed, and second, GTP hydrolysis permits the IF-2 to leave the ribosome, thereby allowing protein synthesis to proceed.

3. The Initiation-Coupled GTPase Activity of Ribosomes

IF-2 stimulates a ribosome-dependent GTPase activity, requiring both 30 S and 50 S subunits. In the absence of fMet-tRNA$_f$ and the AUG codon, some "uncoupled" GTP hydrolysis is observed (Kolakofsky *et al.*, 1968; Lelong *et al.*, 1970). Addition of fMet-tRNA$_f$, mRNA, IF-1, and IF-3 increases the IF-2-dependent GTPase activity by a large factor (Chae *et al.*, 1969; Dubnoff *et al.*, 1972a). IF-2 may promote the uncoupled GTPase by favoring the junction of 30 S and 50 S subunits, which in the absence of fMet-tRNA$_f$ and AUG form unstable couples, too labile to be isolated but sufficient to expose GTP to the GTPase center of the 50 S subunit (Grunberg-Manago *et al.*, 1973). Under stringent coupled conditions, one mole of GTP is hydrolyzed per mole of fMet-tRNA$_f$ bound to 70 S ribosomes (Lockwood *et al.*, 1974).

Recent studies have shed much light on the GTPase activity of ribosomes. Both elongation factors Tu- and G-dependent GTPase activities are inhibited by thiostrepton, and so is IF-2-dependent GTPase (Grunberg-Manago *et al.*, 1972; Mazumder, 1973), at least when it is coupled to fMet-tRNA$_f$ binding (Lockwood and Maitra, 1974). GTPase activity of EF-Tu, EF-G and IF-2 may therefore share a common component. The GTPase activity of 50 S ribosomes is related to proteins L7 and L12. These two proteins differ only by the presence of an N-terminal acetyl residue on L7. There are three copies of this protein per 50 S subunit, on which they form part of the "nose" of this particle seen in electron micrographs (Tischendorff *et al.*, 1975). Since L7/L12 are removed selectively by ethanol/ammonium chloride wash, their function in EF-G- and EF-Tu-dependent 50 S GTPase activity was demonstrated in reconstitution experiments (Hamel *et al.*, 1972; Brot *et al.*, 1972; Sander *et al.*, 1972). Since IF-2-dependent GTPase activity is observed only in the presence of 50 S subunits (Kolakofsky *et al.*, 1969; Dubnoff *et al.*, 1972a) several laboratories investigated the role of L7/L12 in IF-2 activity (Fakunding *et al.*, 1973; Kay *et al.*, 1973; Mazumder, 1973; Lockwood *et al.*, 1974). In the absence of L7/L12, binding of radioactive IF-2 to 70 S and IF-2-dependent fMet-tRNA$_f$ binding to 70 S is reduced by 50–75%. The IF-2-

dependent GTPase activity is reduced to even a larger extent; the coupled GTPase is reduced more than the uncoupled reaction. These results suggest that L7/L12 have a role in the binding of IF-2 as well as in the GTPase reaction. Since GTP hydrolysis is needed for IF-2 to leave the ribosome, one might have expected that with L7/L12-free ribosomes, which cannot hydrolyze GTP, IF-2 would be retained on the ribosome. Actually, Fakunding *et al.* (1973) found that IF-2 is even released faster and binds less well to these deficient ribosomes. On the other hand, Lockwood *et al.* (1974) found that IF-2 does not work catalytically and that fMet-tRNA$_f$ does not react with puromycin on L7/L12-free ribosomes. This would now indicate that the mere absence of IF-2 is not enough to allow bound fMet-tRNA$_f$ to become reactive for peptide bond formation, as was assumed (Section C,2) from the experiments in which IF-2's presence was shown to prevent EF-Tu binding (Ohta *et al.*, 1967; Benne and Voorma, 1972; Dubnoff *et al.*, 1972a). Accommodation may require an active process, involving, in addition or correlated with IF-2 release, the proteins L7/L12. Polypeptides L7/L12 may actually not be the GTPase active site but affect at a distance the GTPase activity of some other 50 S protein. A fragment of *Bacillus stearothermophilus* ribosomes with GTPase and ATPase activity was isolated (Horne and Erdman, 1973).

4. Where Is fMet-tRNA$_f$ Bound on 70 Ribosomes?

Whatever "accommodation" may be, it is clear that fMet-tRNA$_f$ is bound to ribosomes at its anticodon, in order for it to recognize the AUG codon in the piece of mRNA attached to the 30 S subunit (Steitz, 1969; Kuechler and Rich, 1970).

In the presence of streptomycin, fMet-tRNA$_f$ is released from the ribosome after GTP hydrolysis, illustrating the role the 30 S particle plays in holding the tRNA on the 70 S initiation complex (Modollel and Davis, 1970; Lelong *et al.*, 1971). In addition, fMet-tRNA$_f$ is bound to the 50 S subunit in a site corresponding to the posttranslocation stage, since no translocation, nor mRNA movement, is required to form the first dipeptide bond (Thach and Thach, 1971b; Gupta *et al.*, 1971; Kuechler, 1971). There is now some information on the position of the -CCA end of fMet-tRNA$_f$. Elegant affinity labeling experiments (Hauptmann *et al.*, 1974; Kuechler, 1974) show that the formylmethionyl residue binds to protein L27, and to a lesser extent to L15, when fMet-tRNA$_f$ is reacted with ribosomes in a GTP and IF-2-dependent reaction. When GMPPCP replaces GTP, a reaction occurs only with L27. It is difficult to ascertain if interaction with L15 reflects a change in initiator tRNA binding related to accommodation. It is, however, clear that fMet-tRNA$_f$ does not bind to the A site, characterized by protein L16. This conclusion is in line with the poor inhibition of fMet-tRNA$_f$ binding by tetracycline (Zagorska et al., 1971) and with the effects of antibody Fab fragments reported by Lelong et al. (1974). When bound to ribosomes, fMet-tRNA$_f$ may not base pair with 5 S rRNA as do the other aminoacyl-tRNA's (Erdman et al., 1973; Grummt et al., 1974). IF-2 directs the binding of initiator tRNA to a special and unique site on the ribosome: the fact that the A site is blocked as long as IF-2 is not released may still be an indication that the initiator site and the A site overlap (for review, see Pongs et al., 1974; Lelong et al., 1974).

III. Interaction of Messenger RNA with Ribosomes and Initiation Factors

Initiation of translation requires the selection by ribosomes of the proper AUG (or GUG) codon corresponding to the beginning of an mRNA cistron. Direct sequence analysis has shown for several mRNA's that the piece protected by the 70 S initiation complex against ribonuclease digestion contains the initiator AUG (Steitz, 1969; Hindley and Staples, 1969; Maizels, 1973). The mechanism of this selection will be discussed later. The first question is: How is this 70 S-mRNA complex assembled?

A. Messenger RNA Binding to 30 S Subunits without Initiation Factors

Interaction between mRNA and 30 S subunits was first reported by Takanami et al. in 1965. Indeed, 30 S particles washed free of initiation factors appear to bind even natural mRNA's, as MS2 RNA (Szer and Leffler, 1974). This binding takes place at 0° and appears quite weak. It can be demonstrated by nitrocellulose filtration but not by sucrose gradient sedimentation even with glutaraldehyde fixation (Vermeer et al., 1973b). In the absence of initiation factors, MS2 RNA is readily displaced from 30 S subunits by poly(U), and the interaction therefore appears not to involve particular nucleotide sequences on the natural mRNA.

Many studies have implicated protein S1 (factor i-α) in this function. By agarose-acrylamide gel electrophoresis, slow and fast moving 30 S particles can be separated; only the slow moving particles interact with mRNA (Szer and Leffler, 1974; Dahlberg, 1974). The slow migrating form, which is predominantly found among native 30 S subunits, contains a group of high molecular weight proteins, among which is protein S1. Addition of S1 or i-α converted fast moving 30 S particles to the slower moving species, which bind MS2 RNA (Hermoso and Szer, 1974). Protein S1 probably participates directly in the binding of mRNA to these 30 S subunits in the absence of initiation factors (Fiser et al., 1974;

Schenkman *et al.*, 1974). Aurintricarboxylic acid inhibits 30 S mRNA binding possibly as a result of the effect of the drug on S1-RNA interaction (Tal *et al.*, 1972; Inouye *et al.*, 1974), but this compound may have more than one site of action on the 30 S subunit. Edeine does not inhibit mRNA binding which occurs without initiation factors (Szer and Leffler, 1974).

The role of protein S1 appears, however, much more complicated and is discussed in Section III,D,1). It is important to note that MS2 RNA binding to 30 S subunits, at 0° and without initiation factors, is probably not the normal initiation reaction. As discussed in the next sections, binding of ribosomes to the precise initiation nucleotide sequence on MS2 RNA coat cistron strictly requires IF-3 (Iwasaki *et al.*, 1968; Berissi *et al.*, 1971; Vermeer *et al.*, 1973b). Ribosomes bound in the absence of IF-3, with S1, may be attached nonspecifically to MS2 RNA. In addition, 30 S particles alone may not even be able to bind specifically to an initiation nucleotide sequence (H. Berissi and M. Revel, unpublished data). The interaction of 30 S particles with poly(U) and MS2 RNA, without initiation factors, described by Szer and Leffler (1974) may actually be related to the process of elongation, during which mRNA-ribosome interaction should be nonspecific with regard to the nucleotide sequence. This would be consistent with the fact that poly(U) can readily displace MS2 RNA under these conditions.

B. Role of Initiation Factors in mRNA-30 S Subunit Interaction

Stable complexes between 30 S subunits and natural mRNA can be formed and detected by sucrose gradient sedimentation, in the absence of fMet-tRNA$_f$ and GTP, provided initiation factors are added (Greenshpan and Revel, 1969). Factor IF-2 (even treated with N-ethylmaleimide which destroys its fMet-tRNA$_f$ binding activity) stimulates T4 mRNA binding to 30 S subunits (Groner *et al.*, 1970). When MS2 RNA is used as template, however, little binding to 30 S subunits is seen unless IF-3 is added (Vermeer *et al.*, 1973b). In this case also, IF-2 stimulates two- to three-fold the amount of MS2 RNA bound in the absence of fMet-tRNA$_f$, and increases strongly the stability of the complex formed with IF-3. This confirms the highly cooperative action of the initiation factors. Attempts to ascribe a particular function to one particular factor may have very little physiological significance.

MS2 RNA-30 S complexes, formed with initiation factors, sediment as 40 S entities after fixation with glutaraldehyde; native 30 S subunits (still containing initiation factors) form stable complexes even without fixation (Vermeer *et al.*, 1973b). These authors used equilibrium sedimentation in cesium chloride to separate the 30 S-MS2 RNA complexes from unreacted ribosomes and MS2 RNA, which have different buoyant densities. Using [^{35}S]IF-3, they determined that 30 S subunits, IF-3, and MS2 RNA were present in equimolar amounts in the complex, when cold IF-2 and IF-1 were also present (as discussed previously, IF-3 binds to 30 S subunits without mRNA). Addition of fMet-tRNA$_f$ results in the formation of a complete 30 S initiation complex, but with the release of IF-3. Addition of fMet-tRNA$_f$ without MS2 RNA, also resulted in the partial release of IF-3. Binding of fMet-tRNA$_f$ to 30 S ribosomes can occur without mRNA (Jay and Kaempfer, 1974a; Noll and Noll, 1972) but the complex is unstable and requires fixation to be isolated. The presence of MS2 RNA stabilizes fMet-tRNA$_f$ binding, and conversely fMet-tRNA$_f$ stabilizes the mRNA-30 S interaction. As discussed in Section II, these data do not permit one to decide what is the order *in vivo*. The early release of IF-3 when fMet-tRNA is added (even without mRNA) would support the notion that mRNA binds first in the IF-3-mediated step. Kaempfer supports the opposite view (Jay and Kaempfer, 1975).

There is, however, no discussion on the requirement of fMet-tRNA$_f$ for the formation of stable 70 S-mRNA complexes. Junction of the 50 S ribosome occurs after mRNA and fMet-tRNA$_f$ have been bound to the 30 S subunit (Kondo *et al.*, 1968; Greenshpan *et al.*, 1969; Jay and Kaempfer, 1974a). With undissociated 70 S ribosomes, labile complexes of ribosomes with mRNA can be visualized by electron microscopic observation (Revel *et al.*, 1968b), but these could not be isolated on sucrose gradients.

C. Function of IF-3 in the Specific Recognition of Initiation Sites

IF-3 function in initiation appears complex. It maintains 30 S subunits from associating with 50 S subunits, and affects the 30 S subunit conformation (see Section I,F). This would be enough to explain its requirement for the initiation of translation of natural mRNA (Revel *et al.*, 1968a; Iwasaki *et al.*, 1968) and even synthetic polynucleotides (Wahba *et al.*, 1969; Schiff *et al.*, 1974; Bernal *et al.*, 1974; Dondon *et al.*, 1974). It is therefore important to consider whether it has more specific function, by using isolated 30 S ribosomes (see Section II,B,1). Binding of fMet-tRNA$_f$ to 30 S subunits is stimulated by IF-3 when natural mRNA is used as template (Sabol *et al.*, 1970; Vermeer *et al.*, 1973b), but much less when poly(U,G), poly(A,U,G), or ApUpG are used (Suttle and Ravel, 1974; Jay and Kaempfer, 1975). An inhibition by IF-3 of fMet-tRNA$_f$ binding to 30 S subunits has even been seen when the trinucleotide ApUpG is used as template (Groner and Revel, 1971; Suttle and Ravel, 1974). Although there are conflicting reports (Dondon *et al.*, 1974; Bernal, *et al.*, 1974), IF-3 shows a clear preferential effect on initiation of natural mRNA translation. Synthesis of active enzyme *in vitro*, as in the translation of the T4 lysozyme mRNA, also shows a much more stringent requirement for IF-3 than does overall protein synthesis (Brawerman *et al.*, 1969). These results, combined with the role of IF-3 in mRNA binding to 30 S subunits discussed above, suggested that IF-3 plays a role in the specific recognition of initiation signals on mRNA.

A direct demonstration of this role of IF-3 was provided by experiments in which ribosomes were added to a mixture of natural mRNA's with synthetic polynucleotides. Such template competition experiments, using for example T4 mRNA and poly(U), indicated that IF-3 increases the affinity of ribosomes for natural mRNA's. In the absence of IF-3, with IF-2 alone, 30 S subunit and 70 S ribosome binding to T4 mRNA is strongly inhibited by poly(U) or even ApUpG; addition of IF-3 to IF-2 reduces more than 10-fold the competition by ApUpG or poly(U) (Revel *et al.*, 1970). Similarly, Leffler and Szer (1974) found that, in the absence of IF-3, 30 S-MS2 RNA interaction is less stable than 30 S-poly(U) binding.

Further evidence for the role of IF-3 in the selection of initiation signals was reported by Berissi *et al.* (1971). These experiments make use of unfolded MS2 RNA. With native phage (MS2, R17, f2, etc.) RNA, the secondary structure of the molecule (Min Jou *et al.*, 1972) restricts its interaction with ribosomes; the coat protein cistron initiation site being the only one exposed on the native phage RNA (Lodish and Robertson, 1969). The phage RNA can be unfolded by mild formaldehyde treatment, and this unfolding exposes the three initiation sites of the three cistrons present on this RNA (Lodish, 1970a). Unfolded MS2 RNA makes stable complexes with 30 S subunits even in the absence of initiation factors (Vermeer *et al.*, 1973b); under these conditions more than one ribosome is bound per mRNA molecule. This contrasts with the behavior of native MS2 RNA, which forms stable complexes with 30 S subunits only in the presence of IF-3, one 30 S particle being bound per MS2 RNA molecule (Vermeer *et al.*, 1973b). The highly folded structure of the native RNA is probably responsible for the stringent requirement for IF-3; thus T4 mRNA's, which do not have a highly ordered secondary structure, bind to ribosomes in the absence of IF-3 (Revel *et al.*, 1969). Unfolded MS2 RNA allows one to examine the sites to which ribosomes bind to the mRNA in the presence and in the absence of IF-3, by studying in each case the fragments of MS2 RNA protected by ribosomes against ribonuclease digestion. The nucleotide sequence of the fragments corresponding to the three initiation sites was determined by Steitz (1969) for R17 and are similar in MS2. Berissi *et al.* (1971) found that with IF-1 and IF-2, but in the absence of IF-3, ribosomes were bound to the coat, synthetase, and maturation cistrons in a ratio of 1:2:1, respectively; ribosomes were also bound to a few other sites normally not accessible on native MS2 RNA and which may represent false initiations (Lodish, 1970a). Addition of highly purified IF-3 stimulated two- to three-fold the binding of ribosomes to the coat protein cistron initiation site, but in contrast somewhat decreased the binding to the other sites. IF-3 therefore directs ribosome binding preferentially to the coat protein cistron, while with IF-2 alone ribosomes may bind to most available AUG codons.

These experiments were carried out with 70 S initiation complexes which require fMet-tRNA$_f$. There is no proof that 30 S subunits bind specifically to the correct initiation sites on MS2 RNA.

As mentioned above, when native, intact MS2 RNA is used, the only initiation site to which ribosomes can bind is that of the coat protein cistron (S. L. Gupta et al., 1970); the preferential effect of IF-3 for the coat cistron initiation site may therefore explain why IF-3 is indispensable for ribosome attachment to native MS2 RNA (Sabol et al., 1970; Berissi et al., 1971, Vermeer et al., 1973b).

The selective effects of the IF-3 preparation used in these experiments suggested that ribosome binding to other initiation sites may be directed by other initiation factor fractions. Indeed, when assayed with two different mRNA's, IF-3 activity appears heterogeneous, and subfractions with varying activities for MS2 RNA and T4 mRNA's were isolated (Berissi et al., 1971; Groner et al., 1972a; Grunberg-Manago et al., 1971; Lee Huang and Ochoa, 1971). Differential activities of IF-3 subfractions for the various cistrons of MS2 RNA were also reported (Groner et al., 1972a; Yoshida and Rudland, 1972). In these studies, the IF-3 subfractions used were at different stages of purification. Lee-Huang and Ochoa (1973) described the purification to near homogeneity of two IF-3 subfractions IF-3 α and IF-3 β which were reported to be single polypeptide chains of 23,000 and 21,000 MW, respectively, sharing antigenic determinants. In other cases, however, the heterogeneity of IF-3 activity was shown to be due to the presence of additional protein factors accompanying IF-3 during purification, and which were designated "interference factors" (Groner et al., 1972a,b; Revel et al., 1973a,b; Revel and Groner, 1974). For example, T4 mRNA-specific subfractions of IF-3, which did not allow the translation of native MS2 RNA, yielded upon further purification a homogeneous IF-3 (MW 22,000 in SDS), active for native MS2 RNA (i.e., binding ribosomes to the coat cistron), and an interference factor which inhibits initiation on native MS2 RNA but not on T4 mRNA (Revel et al., 1973a).

The existence of interference factors has been confirmed by several laboratories (Lee-Huang and Ochoa, 1972; Miller et al., 1974; Jay and Kaempfer, 1974b; Revel et al., 1973a,b) although their number and mode of action is still unclear.

D. Function of Interference Initiation Factors in mRNA Selection by Ribosomes

Lee-Huang and Ochoa (1972) designated as interference factor i-α, an activity which inhibits translation of MS2 RNA but not of late T4 mRNA. Interference factor i-β designates an activity with opposite mRNA discrimination: it inhibits translation of T4 mRNA but not of MS2 RNA. In association with IF-1, IF-2, and IF-3 (which is not replaced by these factors), the two interference factors markedly alter the ratio of initiation on T4 mRNA over MS2 RNA.

1. Interference Factor i-α, Ribosomal Protein S1

Interference factor i-α is identical to factor i originally reported by Groner et al. 1972a,b). This large protein of 72,000 MW (in SDS) is subunit I of the Qβ replicase (Groner et al., 1972c; Kamen et al., 1972; Miller et al., 1974). When added to ribosomes in the presence of IF-3, it inhibits ribosome attachment to the coat protein initiation site, but stimulates ribosome binding and initiation at the synthetase cistron (Groner et al., 1972a). Consequently, translation of unfolded MS2 RNA is not inhibited. Interference factor i-α is not specific for MS2 or other related phage RNA's. On early T7 mRNA's, factor i-α stimulates initiation on the first cistron (starting with fMet-Ile, Arrand and Hindley, 1973) while inhibiting initiation of other T7 early mRNA cistrons (Revel et al., 1973a,b). When the in vitro translation of the E. coli lactose operon is studied, interference factor i-α also shows cistron discrimination: it inhibits strongly β-galactosidase synthesis while slightly stimulating transacetylase synthesis (Kung et al., 1975). It can therefore change the ratio at which the first and the last cistron of the lactose operon mRNA are translated into protein.

Factor i-α acts on initiation, as demonstrated by its effect on ribosome binding to specific initiation sites on mRNA (Groner et al., 1972b), or by its effects on fMet-tRNA$_f$ binding or dipeptide formation (Groner et al., 1972b; Miller et al., 1974; Jay and Kaempfer, 1974b). It acts on isolated 30 S ribosomal subunits (Revel et al., 1973a). Interference factor i-α forms a complex with ribosomes; this complex can be separated from free i-α by sedimentation, and it shows the typical inhibition of MS2 RNA initiation (Pollack, 1974; Revel et al., 1973a; Miller and Wahba, 1974). The effects of i-α are reversed by the addition of an excess of IF-3 (Revel et al., 1973a; Kung et al., 1975). The inhibitory effects of i-α can also be overcome by the addition of an excess of mRNA (Revel et al., 1973a; Jay and Kaempfer, 1974b; Miller and Wahba, 1974), although it does not degrade mRNA. The factor binds to RNA, but without clear specificity. In the absence of ribosomes, i-α binds more unfolded MS2 RNA than native MS2 RNA, whereas IF-3 has an opposite behavior (Revel et al., 1973a). It prefers, however, intact R17 RNA to degraded RNA (Jay and Kaempfer, 1974b). Recently, interference factor i-α was shown to discriminate also between synthetic polynucleotides. Miller and Wahba (1974) reported that it inhibits translation of U- and C-containing polynucleotides while no effect is seen on poly(A) or A-rich polynucleotides. Factor i-α strongly bound poly(U) and to a lesser extent poly(A).

A comparison of interference factor i-α with ribosomal proteins extracted by high salt from *E. coli* ribosomes has revealed physical and functional homology with 30 S protein S1 (Inouye *et al.*, 1974; Wahba *et al.*, 1974). What is known of the effects of interference factor i-α has, now, to be compared to the functions of S1.

S1 is a ribosomal protein and was included in Witmann's catalog (Witmann *et al.*, 1971). It can also be considered as a ribosomal associated factor, since it is easily removed from ribosomes by high salt and what remains amounts to 0.1–0.3 copies per 30 S particle (Voynow and Kurland, 1971; Weber, 1972). If ribosomes are unwashed, however, the total native 70 S ribosomes and *in vivo* polysomes contain from 0.6–0.9 copies of S1 per ribosome (Van Knippenberg *et al.*, 1974a). On polysomes made *in vitro*, there are 0.9 copies per ribosome, and S1 is neither recycled nor released at any stage of mRNA translation (Van Duin and Van Knippenberg, 1974). S1 has to be added to S1-depleted ribosomes to obtain maximal poly(U) translation (Van Duin and Kurland, 1970; Tal *et al.*, 1972; Van Duin and Van Knippenberg, 1974). At higher magnesium concentrations, ribosomes can translate poly(U) also without S1 addition (Van Dieijen *et al.*, 1975). The same is true at higher poly(U) concentrations (Inouye *et al.*, 1974). Addition of more than 0.5 copies S1 per ribosome leads to inhibition of poly(U) translation especially at high magnesium concentrations. Poly(A) translation is stimulated but not inhibited by S1 (Van Dieijen *et al.*, 1975).

S1 has been shown to participate in the binding of the template RNA on the translating ribosome. Direct affinity labeling experiments demonstrate a poly(U)–S1 interaction (Fiser *et al.*, 1974; Schenkman *et al.*, 1974). Poly(U) protects S1 on the ribosome against Rose Bengal and trypsin sensitivity (Tal *et al.*, 1972) and interacts with aurintricarboxylic acid, which inhibits mRNA binding to ribosomes in initiation and in elongation. Since S1 does not leave the ribosome during the entire translation process, this protein appears to be an obligatory component of the elongating ribosome. It should be noted that during elongation, mRNA–ribosome interaction should be nonspecific with regard to the mRNA nucleotide sequence. During initiation, on the other hand, specific recognition of an initiation site on mRNA has to take place.

Most experiments on the effects of interference factor i-α(S1) in the initiation process were performed with ribosomes washed in high salt and containing about 0.2 copies of S1 per particle. These ribosomes are capable of initiating correctly the translation of MS2 RNA or T4 mRNA when pure initiation factors IF-1, IF-2, and IF-3 are added. Native 30 S particles do not seem to contain more S1 than 30 S subunits derived at low magnesium concentrations (Van Knippenberg *et al.*, 1974a), suggesting that S1 is not always present on the free 30 S particles, in contrast to what is observed on polysomes. H. Noll *et al.* (1973) showed that 30 S subunits derived from tight ribosomal couples, which are very active in initiation, have less S1 than poorly active 30 S particles derived from loose couples. Sixty percent of the active 30 S subunits could be converted, in these experiments, into R17 RNA initiation complexes after addition of initiation factors. Bollen *et al.* (1972) also report that S1 may be absent from initiation complexes isolated on Sepharose-bound poly(A,U,G).

With ribosomes which have been depleted of S1 by dialysis at low ionic strength (Tal *et al.*, 1972), addition of S1 stimulates the formation of 30 S initiation complexes with MS2 RNA and fMet-tRNA$_f$ (Van Duin and Van Knippenberg, 1974; Van Dieijen *et al.*, 1975). Provided the procedure used to remove S1 did not alter the ribosome (it removes a polyamine as reported by Van Knippenberg *et al.*, 1974b), these data would indicate that S1 is already needed during the early steps of initiation. Active fragments of S1 antibodies can also partially inhibit the formation of initiation complexes (Lelong *et al.*, 1974). More recently, the separation by electrophoresis of slow and fast migrating 30 S ribosomes, which differ by the presence or absence of S1, has been achieved (Dahlberg, 1974; Szer and Leffler, 1974). Slow migrating 30 S ribosomes, containing S1, are much more active for MS2 RNA binding at 0° C in the absence of initiation factors. Interference factor i-α, purified from the high salt wash of ribosomes, replaces S1 in this reaction (Hermoso and Szer, 1974). The slow migrating 30 S ribosomes, containing S1, were also more active for fMet-tRNA$_f$ binding with MS2 RNA (but not ApUpG) in the presence of initiation factors than fast migrating 30 S subunits missing S1 (Szer *et al.*, 1975). On sucrose gradients, the more active 30 S subunits (containing S1) sedimented more slowly than the S1 depleted ribosome, suggesting that in addition to their difference in S1 content the two types of 30 S particles differed also in their conformation. Addition of S1 is, however, enough to convert one form into the other (Szer *et al.*, 1975).

How does the S1 requirement for the formation of an initiation complex relate to the interference effect described above, in which i-α(S1) decreases initiation at certain cistrons while increasing initiation at other cistrons? With S1 depleted ribosomes, Van Dieijen *et al.* (1975) observed that initiation on native MS2 RNA is stimulated by the addition of up to 0.6

copies of S1 per ribosome but is then inhibited by adding more S1. When undepleted ribosomes had been used, addition of factor i-α(S1) produced only inhibition of initiation on native MS2 RNA (Groner *et al.*, 1972a). Factor i-α(S1) added in excess is not, however, a general inhibitor of initiation since only the coat cistron was reduced while initiation at the synthetase cistron was increased under the same conditions (Groner *et al.*, 1972a; Revel *et al.*, 1973a). Isono and Isono (1975) have also found that addition of S1 to undepleted *E. coli* ribosomes preferentially stimulates initiation at the synthetase cistron of MS2 RNA. The same phenomenon of increased initiation at some cistrons and decreased initiation at others, upon addition of factor i-α(S1) to undepleted ribosomes, was found also with phage T7 mRNA's (Revel *et al.*, 1973b) and in the bacterial lactose operon (Kung *et al.*, 1975). The optimal amount of factor i-α(S1) appears, therefore, to differ from one initiation site to another, some cistrons being already inhibited while others are still stimulated by adding excess amounts of S1. The amount of protein needed to see such an interference effect represents the addition of about 2–3 copies of S1 per ribosome present in the system. Measures by complement fixation with anti-i-α serum (Scheps, 1973) have shown that there is, at least, a twofold excess of S1 over the number of ribosomes in intact *E. coli* cells, and over 50% of i-α(S1) is found in the soluble fraction and not ribosome bound. There is, therefore, a physiological excess of S1 per ribosome in these cells and possibly a mechanism which controls S1 binding to ribosomes since native 30 S particles are by far not saturated in this protein.

How does interference factor i-α(S1) function in the recognition of the initiator site on natural mRNA? Much has been learned from the structural analysis of the mRNA binding site on 30 S ribosomes, and it gives little doubt that S1 is a most important element in this site. As discussed in more detail below (Section III,E), the pyrimidine-rich octanucleotide sequence at the 3′ end of 16 S ribosomal RNA base pairs with a complementary purine-rich sequence preceding the AUG codon in mRNA (Shine and Dalgarno, 1974). Protein S1 can be cross-linked to the 3′ end of 16 S rRNA (Kenner, 1973) and was recently shown to bind the pyrimidine-rich 3′ end of 16 S rRNA (Dahlberg, 1974; Dahlberg and Dahlberg, 1975). In addition, an i-α(S1) complex with IF-3 was identified by Groner *et al.* (1972a,b), and, indeed, Bollen *et al.* (1974) showed that IF-3 and also IF-2 can be cross-linked to S1. Initiation factors involved in mRNA binding and interference factor i-α(S1) are therefore located around the site on rRNA to which the mRNA initiation sequence binds. These proteins, and others like S7, S18, S21 (Fisher *et al.*, 1975; Van Duin *et al.*, 1975) could influence the base pairing between 16 S rRNA and mRNA (Steitz and

Jakes, 1975), either directly or by inducing a conformational change in the site. Both IF-3 and S1 were reported to produce conformational change in the 30 S subunit (Paradies *et al.,* 1974; Szer *et al.,* 1975).

How, then, would interference factor i-α(S1) modulate initiation on different mRNA cistrons? Ribosomes which have been incubated with an excess of S1 and resedimented are inhibited for initiation on MS2 or R17 RNA (Pollack, 1974; Miller and Wahba, 1974). It seems likely, therefore, that the interference effect results from a change in the ribosome, as binding of an excess amount of S1 or a conformational change. Another possibility is that, when S1 is present in excess in the reaction, the normal order of factor binding is altered: e.g., S1 binds to the 30 S subunit before IF-3 is added. The fact that S1 stays on the ribosome even during elongation while IF-3 is released after initiation, suggests a model in which S1 promotes the mRNA site to accommodate any nucleotide sequence while IF-3 does specifically enhance binding of the initiator mRNA sequence. It is thus possible to visualize the opposite effects of IF-3 and i-α(S1) as resulting from an effect on the stability of the mRNA binding to the 3' end of 16 S rRNA. This stability, which depends on many factors, one of them being the number of base pairs stably formed between mRNA and rRNA, will be different from one mRNA cistron to another. Binding of ribosomes to the MS2 coat cistron, where the interaction is relatively weak (Shine and Dalgarno, 1974), is more dependent on IF-3 than that to other sites (Berissi *et al.,* 1971; Vermeer *et al.,* 1973b) and more rapidly inhibited by adding excess of i-α(S1) (Groner *et al.,* 1972b). Binding to the synthetase cistron initiator sequence is less dependent on IF-3, and is stimulated by adding more i-α(S1) (Groner *et al.,* 1972b; Isono and Isono, 1975). This model is supported by the fact that the interference effect of i-α(S1) is greater at low concentrations of IF-3 and is overcome by adding an excess of IF-3 (Revel *et al.,* 1973a; Kung *et al.,* 1975). An excess of mRNA also overcomes the effects of i-α(S1), suggesting that the low affinity for certain initiation sites conferred by the protein may be compensated by the higher concentration of these sites in the reaction (Groner *et al.,*1972b). On the other hand, a completely different mechanism of action has been proposed by Jay and Kaempfer (1974b) who suggest that i-α(S1) acts by binding not to ribosomes but to the mRNA as a translational repressor. Factor i-α(S1) seems, however, to bind to RNA without clear specificity for given sites, in contrast to what would be expected from a translational repressor. Even when i-α is integrated in the Qβ replicase, which inhibits initiation at the coat cistron (Kolakofsky and Weissmann, 1971), the specificity of binding to the phage RNA depends on the phage coded subunit rather than on i-α (Revel *et al.,* 1973a). Lee-Huang and Ochoa (1972) proposed that i-α "neutralizes" one of the two IF-3 activities which they isolated (Lee-Huang and Ochoa, 1971), namely, IF-3-α, while interference factor i-β would block IF-3-β activity. The effects of i-α and i-β can, however, be observed with the same homogeneous IF-3 preparation (Revel et al., 1973b).

More work is obviously needed to elucidate the mode of action of interference factor i-α and protein S1, but a final word of caution should be added. Interference factor i-α and protein S1 may still differ by some modification. Mild oxidation of the protein produces an active form which migrates as a 65,000 MW polypeptide, instead of 72,000 MW for the native protein (Inouye et al., 1974). Factor i-α contains some nucleotide-like material, which may be absent from S1. This nucleotide has not been identified, but contains guanine and is removed by phosphodiesterase treatment (Inouye and Revel, unpublished data). Phage T4 infection leads to a modification of i-α, producing in immunoelectrophoresis a more electronegative form of the protein, which is converted to the normal form by phosphodiesterase treatment (Revel et al., 1973a). Preparation of S1 extracted from ribosomes and of i-α obtained from the high-salt ribosomal wash show different migrations in immunoelectrophoresis. The multiple functions of the protein may be modulated by such modifications.

Another effect of S1 remains to be explained: it stimulates the transcription of RNA by E. coli RNA polymerase with a variety of DNA templates, including those whose translation is inhibited by the protein (Leavitt et al., 1972; Kung et al., 1975).

Isolation of a thermosensitive mutant of E. coli K12, deficient in i-α(S1) activity for Qβ replicase and in i-α antigenic activity has been reported (Scheps, 1973).

A discussion of the mode of action of interference factor i-α(S1) has to take into account two facts: (1) the mRNA discriminating properties of such factors which inhibit initiation at certain cistrons but stimulate initiation of others and (2) the existence in E. coli of several interference factors which have opposite specificities.

2. Interference Factors Other than S1

Interference factor activity i-β is defined by the inhibition of T4 mRNA translation, but not of MS2 RNA, which leads to a high MS2/T4 translation ratio. Two proteins with i-β activity have been isolated from the ribosomal wash in E. coli (Pollack, 1974; Revel and Groner, 1974). One has a MW of 26,000 (i-β_1), the other of 50,000 (i-β_2). A similar activity was described by Lee-Huang and Ochoa (1972). More recently, Wahba (1974) found that Qβ replicase host factor I (France de Fernandez et al., 1972) inhibits T4 mRNA translation but not R17 RNA. This factor also inhibits translation of poly(A) but not poly(U).

Another interference factor, i-γ (MW 13,000 in SDS) was isolated from an IF-3 preparation, called factor J before being separated in i-β_1 (MW 26,000) and i-γ (Revel *et al.*, 1973a; Revel and Groner, 1974). Interference factor i-γ added to a translation system inhibits early T4 mRNA but not late T4 mRNA translation. Electrophoretic and immunological evidence that preparation J contains ribosomal proteins S3 and S7 was obtained. It is noteworthy that S7 is a close neighbor of IF-3 on the 30 S ribosome (Van Duin *et al.*, 1975).

Stimulation of T7 lysozyme synthesis was observed by yet another interference factor i-δ, whereas overall T7 mRNA translation was inhibited. Conversely, a different factor, i-α_2 (MW 11,000 in SDS), was found to inhibit T7 lysozyme synthesis and MS2 RNA translation while not affecting overall translation of T4 and T7 mRNA's (Pollack, 1974). Factor i-α(S1), which inhibits MS2 RNA translation at the coat cistron, slightly stimulates T7 lysozyme synthesis.

When added together to the protein synthesis reaction mixture, one interference factor can reverse the effect of another, suggesting that they all act at the same step. All these observations were made in cell-free systems, and there is no evidence of whether *in vivo* these factors participate in mRNA selection. To demonstrate this, mutants in these activities would have to be isolated (see Section E). Nevertheless, these studies have shown that, *even within the same organism,* in which only one type of ribosome appears to be present, there are protein factors associated with the ribosomes which can profoundly modify the recognition and translation of different mRNA cistrons, providing a biochemical mechanism for translation control. Some of these protein factors may be ribosomal proteins, but the essential question is how do these proteins affect the affinity of ribosomes for different mRNA's, and how heterogeneous the ribosomal population is with regard to such proteins.

E. The Molecular Basis for mRNA Recognition by Ribosomes

1. Ribosomal Components Involved

Evidence for mRNA discrimination by ribosomes is very clear when ribosomes from different bacterial species are compared, as, for example, *E. coli* and *B. stearothermophilus* (Lodish, 1970b). *Escherichia coli* ribosomes initiate mainly at the coat protein cistron of phage R17 (f2, MS2, etc.) RNA, but poorly at the maturation cistron; the synthetase cistron is under strong polarity of the coat cistron and is not accessible before the RNA is unfolded by translation or formaldehyde treatment (Lodish and Robertson, 1969; Kozak and Nathans, 1972a). On the other hand, *B. stearothermophilus* ribosomes initiate very poorly at the coat cistron while initiation at the maturation cistron is comparable to that of *E. coli* ribosomes. *Bacillus stearothermophilus* ribosomes are active with T4 mRNA. The discriminating property of these ribosomes was ascribed to the 30 S particle (Lodish, 1970b). By studying the mRNA sequences protected by ribosomes against ribonuclease digestion, Steitz (1969) confirmed these observations and showed that the main fragment protected by *B. stearothermophilus* ribosomes is the maturation cistron initiation site. At 65°, a temperature at which these ribosomes are highly active and the structure of phage RNA less stable, the binding of *B. stearothermophilus* ribosomes to maturation cistron is very efficient. Instead, at lower temperature, the maturation site is shielded and one mainly observes that initiation by *B. stearothermophilus* ribosomes is less efficient. No binding of these ribosomes to the coat cistron is observed.

Some differences are found between studies on ribosome binding by mRNA protection (Steitz, 1973a) and studies on actual initiation (dipeptide or aminoterminal peptide formation; Lodish, 1970b). In the latter case only the origin of the 30 S subunit was important, while with the first method, *B. stearothermophilus* initiation factors were found to direct *E. coli* ribosomes to some noninitiation sites, absent when *E. coli* initiation factors were used. Szer and Brenowitz (1970) also observed an effect of initiation factors on differential initiation with psychrophile ribosomes. However, *Caulobacter crescenti* and *E. coli* ribosomes were shown to translate specifically their own phage RNA's (Cb5 or MS2, respectively) regardless of which initiation factors were present (Leffler and Szer, 1973). For similar studies in *Clostridium,* see Stallcup and Rabinowitz (1973).

To determine which ribosomal component is responsible for the difference between *E. coli* and *B. stearothermophilus* ribosomes, Nomura and collaborators (Held *et al.,* 1974b) prepared hybrid particles between 16 S RNA and corresponding ribosomal proteins from the two organisms. Discrimination against the coat protein cistron of R17 RNA was found to be related to the presence of both the 16 S RNA and protein S12 from *B. stearothermophilus.* When either 16 S RNA or S12 was from *B. stearothermophilus,* R17 RNA translation compared to poly(U) was decreased 50%, when both were from *Bacillus* the reduction was over 80%. Replacement of *E. coli*-S12 by *B. stearothermophilus*-S12 led to a deficient response to *E. coli* initiation factors for AUG-directed fMet-tRNA$_f$ binding and the reaction with puromycin. This is in line with the evidence that S12 is involved in the binding of IF-3 and IF-2, as shown by the cross-linking experiments (Traut *et al.,* 1974; Bollen *et al.,* 1975). Indications of the unique role of S12 in initiation had been reported from reconstitution experiments (Ozaki *et al.,* 1969), from studies with sm^d *E. coli* mutants (Lazar and Gros, 1973) or S12 *ts* mutants (Kang, 1970) and from antibody

Fab fragment inhibition studies of Lelong *et al.* (1974). Protein S12 is also involved in ribosomal restriction of certain tRNA's (Gorini, 1971). A ribosomal mutation (sm^R) in protein S12, which produces *in vivo* a lac⁻ phenotype and *in vitro* a selective change in ribosome activity for different mRNA's, was recently discovered (G. Goldberg and M. Revel, unpublished data).

Goldberg and Steitz (1974), using the assay of mRNA fragment protection by ribosomes, also found that ribosomal proteins from *B. stearothermophilus* confer the characteristic binding to the maturation cistron initiation site.

Recently, the role of protein S1 (factor i-α) in the differential activity of *E. coli* and *B. stearothermophilus,* ribosomes was also demonstrated (Isono and Isono, 1975).

2. Models for the Recognition of the Initiation Signal

The involvement of ribosomal proteins in mRNA cistron selection may be either a direct one or through interaction with initiation factors or interference factors. The nature and the configuration of the proteins which bind on the initiating 30 S around protein S12 [e.g., factor i-α(S1)] may determine the affinity of the particle for given initiation sites on the messenger RNA (see Sections III,C and D).

The involvement of 16 S ribosomal RNA, which had been suggested already by Moore (1966), may also be indirect through modification of the configuration of the various proteins in the particle. However, an attractive hypothesis was proposed by Shine and Dalgarno (1974, 1975) that a nucleotide sequence near the 3'-end of 16 S RNA base pairs directly with a common polypurine stretch found at the 5'-side of the AUG codon in the initiation region of most mRNA's.

Comparison of the initiation site sequences from various phage cistrons and bacterial operons (see Table I for details and references) shows that all contain a polypurine stretch, at the 5'-side of the initiator AUG codon. This stretch is not, however, identical in all the sites: both the nucleotide sequence and the distance from the AUG codon vary. Although other sequences (such as PuPuUUUPuPu) may be common to several initiation sites, it is tempting to consider the polypurine region of the mRNA binding site as the initiation signal for the following reasons. Template competition experiments between T4 mRNA and synthetic polynucleotides have indicated (Revel *et al.*, 1970; Okuyama and Tanaka, 1973) that poly(A,G) and poly(A,U,G) polynucleotides rich in purines are the best competitors against natural mRNA binding to ribosomes. Poly(A), poly(U), poly(C), or combinations of these nucleotides do not compete. Competition experiments with short oligonucleotides, prepared from poly(A,U,G) by partial

TABLE I

Ribosome Binding Sites and Possible Pairing with 3′-Terminus of 16 S rRNA[a]

RNA	Cistron	Possible pairing arrangement	Reference
R17, MS2	A-protein	mRNA (5′) U-A-G-G-A-G-G-U-U (3′) . . . . . . 16 S RNA HO A-U-U-C-C-U-C-C-A-C-U-A-G (5′)	Steitz, 1969 De Wachter et al., 1971
	A₂-protein	mRNA (5′) A-A-G-A-G-G-A-C-A (3′) . . . . . 16 S RNA HO A-U-U-C-C-U-C-C-A-C-U-A-G (5′) G‡ ↕	Staples et al., 1971
R17	Coat protein	mRNA (5′) C-C-G-G-A-G-U-U-U (3′) . . . . 16 S RNA HO A-U-U-C-C-U-C-C-A-C-U-A-G (5′)	Steitz, 1969
MS2	Coat protein	mRNA (5′) C-C-G-G-A-G-U-U (3′) . . . . 16 S RNA HO A-U-U-C-C-U-C-C-A-C-U-A-G (5′)	Contreras et al., 1973 Cory et al., 1970
f2	Coat protein	mRNA (5′) C-C-G-(G,A)-G-U-U (3′) . . . . 16 S RNA HO A-U-U-C-C-U-C-C-A-C-U-A-G (5′)	S. L. Gupta et al., 1970
R17, MS2	Replicase	mRNA (5′) A-U-G-A-G-G-A-U (3′) . . . . 16 S RNA HO A-U-U-C-C-U-C-C-A-C-U-A-G (5′)	Steitz, 1969 Min Jou et al., 1972
Q_B	Replicase	mRNA (5′) A-A-G-G-A-U-G-A-A (3′) . . . . 16 S RNA HO A-U-U-C-C-U-C-C-A-C-U-A-G (5′)	Staples and Hindley, 1971
"Early" T7 mRNA	—	mRNA (5′) A-U-G-A-G-G-U-A (3′) . . . . . 16 S RNA HO A-U-U-C-C-U-C-C-A-C-U-A-G (5′)	Arrand and Hindley, 1973

[a] From Shine and Dalgarno (1974).

alkali hydrolysis, showed that the sequence which competes is 6–10 nucleotides long and has a base composition very rich in A and G (Revel, 1972). The presence of the polypurine stretch in the nontranslated region of mRNA preceding the AUG codon gives a molecular basis for these competition experiments. The polypurine stretch may determine ribosome binding even without the close proximity of an AUG codon. Strikingly, Steitz (1973a) found that *B. stearothermophilus* ribosomes bind to Qβ RNA at a very specific site, containing a polypurine stretch but lacking the AUG codon altogether.

Steitz and Jakes (1975) succeeded recently in demonstrating the formation of an RNA–RNA hybrid between the complementary sequences of R17 RNA A cistron and 16 S rRNA during initiation.

3. Role of the AUG Codon

Template competition experiments (Revel *et al.*, 1970) clearly indicated that the presence of an AUG codon by itself was not the initiation signal recognized by ribosomes. ApUpG or polynucleotides $(A)_n AUG(U)_n$ did not compete against T4 mRNA when IF-3 was present. Obviously, the sequence AUG must occur often in and out of phase in a message, and ribosomes have to select the proper AUG corresponding to the beginning of a cistron. The secondary structure of mRNA which, as in phage MS2 RNA shields many AUG's in double-stranded regions (Min Jou *et al.*, 1972), is not sufficient to explain why ribosomes bind only to certain initiation sites. Although unfolding MS2 RNA does expose some additional ribosome binding and initiation sites (Lodish, 1970a; Berissi *et al.*, 1971), even quite extensive fragmentation of the RNA does not lead to random ribosome binding (Steitz, 1974), indicating that even when the secondary structure is destroyed, most AUG codons still do not serve as initiation signals.

Genetic experiments studying reinitiation after a termination event inside a cistron, however, indicate that reinitiation may occur at any AUG,GUG or even mismatched codon for fMet-tRNAf (Files *et al.*, 1974; Miller, 1974). These events do not probably occur frequently under normal conditions, but here the ribosome is already attached to the mRNA close to the potential initiator codon.

It is likely that both ribosome binding to a region lacking an AUG codon but having the polypurine stretch (Steitz, 1973a) and the reinitiation phenomenon at any AUG (Miller, 1974) are exceptional cases. The normal initiation site is probably characterized by the presence of both an initiation signal and the AUG (GUG) initiator codon, which allows fMet-tRNAf binding to the ribosomal initiation complex. The presence of two recognition elements in the initiation site is in line with the hypothesis (see Section I,G) that mRNA selection and fMet-tRNAf binding are independent events which cooperate to form a stable initiation complex, with the mRNA bound at a correct AUG codon.

4. Heterogeneity of Initiation Sites and Control of mRNA Selection in E. coli

Comparison of the dozen initiation site sequences reported to this day with the presumed complementary region of 16 S RNA (Shine and Dalgarno, 1974; Steitz and Jakes, 1975) stresses the heterogeneity of the initiation signals (Table I). Most striking is the variability in the number of nucleotides capable of forming a true base pair with the 16 S RNA: this number can vary from 7 to 3. Such variations would predict a different affinity of ribosomes for each site. An experimental approach to this question comes from the comparison of the three initiation sites of R17 RNA. Steitz (1973b, 1974) observed that when these fragments are cut out of the phage RNA molecule and studied for their ability to rebind to ribosomes, the maturation cistron initiation site is the only one to bind efficiently. Indeed, the fact that this site can form 7 base pairs would predict that it forms a more stable complex than the two other sites which can form only 3–4 base pairs. The distance of the AUG codon from the region base pairing to ribosomal RNA is also variable and may determine the affinity (the maturation site has the longest distance, 19 bases). In the intact phage RNA molecule, the maturation site is probably shielded by the secondary structure of the RNA (Lodish, 1970a; Steitz, 1973b). The same is true for Qβ RNA (Staples and Hindley, 1971). Formaldehyde treatment of the RNA exposes this site and the efficiency of initiation is increased 20-fold (Lodish, 1970a). In this case, the secondary structure of the RNA plays a *negative* role, as also in the case of the synthetase cistron initiation site (Min Jou *et al.*, 1972). The coat cistron initiation site is the only one exposed in the native RNA, which makes this relatively "weak" site the main initiation site on the native molecule. The possible *positive* contribution of secondary structure, either locally (in a hairpin loop, for example) or through interaction with a distant region in the RNA molecule, to the ribosome recognition of the coat cistron is extensively discussed by Steitz (1974).

Factors which influence the RNA structure or mask part of the RNA are likely to control translation. A good example is the inhibition of synthetase initiation by the coat protein itself, which binds to the synthetase initiation site (Sugiyama and Nakada, 1970; Bernardi and Spahr, 1972). Another example is the blocking of the coat cistron initiation site by the Qβ replicase (Kolakofsky and Weissmann, 1971; Weissman *et al.*, 1973). Processing of T7 mRNA by ribonuclease III may be another example of the importance of RNA structure (Hercules *et al.*, 1974).

Besides the structure of the mRNA, modification of the ribosome structure may also determine the affinity for different cistrons. An interesting possibility is that IF-3, which binds near the 3'-end of 16 S RNA (Van Duin *et al.*, 1976), stabilizes weak 16 S RNA-mRNA interactions. This would explain the more stringent requirement of IF-3 for the coat cistron initiation site than for other sites (Berissi *et al.*, 1971, Vermeer *et al.*, 1973b). The inhibitory effect of interference factor i-α(S1) could be explained by a specific destabilization of the rRNA-mRNA interaction, together with an increased overall binding of mRNA, expected if S1 serves for mRNA anchoring also during elongation, a process in which no base sequence recognition with 16 S rRNA, should occur. Factor i-α(S1) is bound close to the 3'-end of 16 S RNA (see Section III,D,1) and may cover part of the presumed mRNA-recognition sequence of 16 S RNA. The different initiation interference factors (discussed in Section III,D) and ribosomal proteins binding to 30 S or 50 S subunits at their interface (Morrison *et al.*, 1973) may all influence the mRNA-16 S rRNA interaction. The role of S18 and S21, and also S11, S12, S13, and S4, in mRNA and ApUpG binding has been shown recently by cross-linking studies (Fiser *et al.*, 1975; Pongs and Lanka, 1975; Pongs *et al.*, 1975) indicating an extensive overlap between mRNA and initiation factors binding sites on the ribosome. Further studies on the structure of the region of the ribosome in which the 3'-end of the 16 S RNA is found should help to understand how, in a given bacterial species, changes in mRNA selection can occur.

A group of antibiotics has been reported to affect mRNA selection. Kasugamycin, kanamycin, and gentamycin, which all act on the 30 S particle, show some cistron specificity in their inhibition of phage RNA translation (Okuyama and Tanaka, 1972; Kozak and Nathans, 1972b). It may be of interest that kasugamycin-resistant mutants have an altered methylation of the 3'-end region of 16 S RNA (Helser *et al.*, 1971), near the region which base pairs to mRNA.

Messenger RNA selection by ribosomes appears to depend on a large number of variables affecting ribosome and mRNA structure. The question of whether changes in initiation of translation take place in living bacterial cells is, however, still not clearly answered, as discussed in the Section IV.

IV. Physiological Variations in Initiation of Protein Synthesis

A. Uninfected Bacteria

Two types of regulation may be considered: (1) an overall increase or decrease in protein synthesis initiation and (2) a selective change in the relative efficiency of initiation at different mRNA cistrons.

Considering the coupling between transcription of mRNA and initiation of its translation (Revel *et al.,* 1968b; Imamoto and Kano, 1971; Crépin *et al.,* 1973), variations in translational initiation could play some role in the regulation of gene expression.

Polypeptide chain initiation is a rate-limiting step in translation, and some initiation factors are present in limiting amounts as compared to ribosomes, in contrast to the situation with elongation factors (Gordon, 1970). In exponentially growing *E. coli,* the amount of IF-3 present is 0.5 copies per native 30 S subunit, which corresponds to 0.05 copies for each 30 S ribosome in the cell (Van Knippenberg *et al.,* 1973). These measurements confirm earlier estimates, which were based on initiation factor localization (Revel *et al.,* 1969; Subramanian and Davis, 1971). While IF-3 is ribosome bound, some IF-2 is present free in the cell-soluble fraction (Scheps, 1973). The amount of initiation factors present in the cell drops to an even much lower level in stationary phase *E. coli:* the actual amount of IF-3, measured by immunoassay, is decreased about three-fold, and both IF-2 and IF-3 activities are deficient in extracts of such cells (Scheps and Revel, 1972). This is in line with the inhibition of protein synthesis in stationary phase cells; initiation resumes within a few minutes of incubation in fresh growth medium. A decrease in dissociation factor activity (IF-3) was seen also in stationary phase *B. stearothermophilus* (Bade *et al.,* 1969). The biochemical mechanism responsible for the reduction in the level of initiation factors is unclear. A prolonged chloramphenicol treatment leads to a similar effect (Young and Nakada, 1971; Legault-Demare *et al.,* 1973). On the other hand, amino acid starvation or puromycin treatment does not cause a decrease in IF-3 (Legault *et al.,* 1972a), neither does T4 infection which blocks all host protein synthesis (Scheps *et al.,* 1972). This argues against the idea that IF-3 has a rapid turnover and disappears when its synthesis is blocked. Instead, these results suggest the activation of some degradative process. Legault-Demare *et al.* (1973) propose that IF-3 becomes susceptible to biological inactivation if it cannot reassociate with 30 S subunits. An ATP-dependent inactivation of IF-3 *in vitro* was observed in their experiments. Amino acid starvation or puromycin treatment leave enough free 30 S subunits to protect IF-3.

Overall changes in initiation, such as those described above, may also produce selective effects. In particular the changes in IF-3 activity affect differentially the translation of various mRNA's. After chloramphenicol treatment, Legault-Demare *et al.* (1973) observed that translation of late T4 mRNA or Qβ RNA was much more affected than that of R17 RNA. In stationary phase *E. coli,* on the other hand, MS2 RNA translation was much more inhibited than T4 mRNA translation (Scheps and Revel, 1972). This could be explained by a change in the ratio of initiation factors to interference factors, which would inhibit more initiation at certain cistrons than at others. In stationary phase cells, the amount of interference factor i-α(S1), measured by immunoassay, does not decrease, leading to a higher i-α/IF-3 ratio (Scheps and Revel, 1972).

Amino acid starvation could also lead to inhibition of polypeptide chain initiation by another mechanism. *In vitro,* elongation factor G appears to inhibit initiation when no charged aminoacyl-tRNA is present, and this may be the basis for a feedback regulation *in vivo* (Lee-Huang *et al.,* 1973; Lee-Huang and Ochoa, 1974). A control of protein synthesis by amino acid starvation in stringent *E. coli* strains (RC$^+$) has been reported (Hall and Gallant, 1971). This may operate through the accumulation of guanosine tetraphosphate (Laffler and Gallant, 1974). This nucleotide inhibits initiation (Arai *et al.,* 1972; Yoshida *et al.,* 1972; Legault *et al.,* 1972b). Some translation control may occur also in RC$^-$ strains (Sokawa *et al.,* 1971).

Under shift-up conditions, when the *E. coli* cell growth rate doubles, initiation factor activity increases by a factor of 1.7 (Boyle and Sells, 1974).

There is as yet no hard evidence that translation control plays a role in the induction or repression of bacterial gene expression. Some indications of a translational block upon repression of the tryptophan operon were reported by Lavalle and De Hauwer (1970). In synchronized *E. coli* cells, the synthesis of β-galactosidase is discontinuous, although *lac* mRNA accumulates continuously, suggesting some translation control during the cell cycle (Goldberg and Chargaff, 1971). Some translation control of the *lac* operon expression has also been suggested during amino acid starvation (Ennis *et al.,* 1974). These results obtained *in vivo* are interesting to consider in the light of the recent evidence that interference factors affect *lac* operon expression *in vitro* (Kung *et al.,* 1975).

In developmental systems, translation control may play some role, as in spore outgrowth, in which a nucleotide (possibly a dimethyl guanosine derivative) has been implicated (Adelman and Lovett, 1974; Bacon and Sussman, 1973).

B. Phage-Infected Cells

Regulation of phage and host gene expression at the level of translation is clearer in infected bacterial cells. Three systems have been mainly studied.

1. RNA Phages

Although *in vitro,* the coat cistron is the most actively translated of the three cistrons present, the situation differs *in vivo.* There is a temporal sequence of phage protein synthesis. Immediately after infection, the synthetase is the main product translated (for review, see Kozak and Nathans, 1972a). Some translation of the coat cistron is, however, required to unfold the RNA and expose the synthetase initiation site (as shown by the study of polar coat cistron mutants; Englehardt et al., 1967). But coat synthesis remains at first very low and becomes active only after several minutes. It is tempting to propose that it is the presence of interference factor i-α(S1) which limits initiation at the coat cistron while favoring initiation at the synthetase cistron (Groner et al., 1972c). Factor i-α then combines with the accumulating synthetase polypeptide to form, with EF-Tu and EF-Ts, the active phage RNA replicase (Kamen et al., 1972). New RNA molecules are produced and inhibition at the coat cistron is decreased, the coat becoming the main product of translation. During the onset of replication of each RNA molecule, the complete phage replicase may again inhibit initiation at the coat cistron (Kolakofsky and Weissmann, 1971).

The accumulating coat protein binds to the synthetase initiation site and eventually inhibits synthetase formation, ensuring late in infection that a large excess of coat over synthetase and new RNA molecules are made (Sugiyama and Nakada, 1970; Bernardi and Spahr, 1972). Maturation protein does not seem to be made on native phage RNA, but on nascent RNA present in replicating structures (Robertson and Lodish, 1970). Although the maturation cistron may have the strongest initiation site (Steitz, 1974), the secondary structure of the phage RNA exerts a strong negative control on its activity. The role of some cistron-specific factors has also been suggested (Yoshida and Rudland, 1972). Initiation at the three cistrons of phage RNAs is therefore a strictly regulated process, controlled by the RNA structure and host ribosomal protein factors.

2. T4 Phage

Interest in this system originated from the observation by Hsu and Weiss (1969) that ribosomes from T4-infected E. coli cells did not translate MS2 or f2 RNA, but were active for late T4 mRNA translation. T4 infection in vivo causes both the cessation of host protein synthesis and the restriction of RNA phages (Hattman and Hofschneider, 1967), and both may be explained by the altered ribosomal function. Crude initiation factors from T4-infected ribosomes inhibit initiation on MS2 RNA but are active for T4 mRNA (Dube and Rudland, 1970; Schedl et al., 1970; Klem et al., 1970). The alteration is not in IF-1 and IF-2 activities but affects an IF-3-related activity for initiation on MS2 RNA and early T4 mRNA, but not for late T4 mRNA (Pollack et al., 1970). Further purification of IF-3 showed, however, that the normal E. coli IF-3 (active for MS2 RNA) is still present and in normal amounts, as shown by immunoassay (Scheps et *al.*, 1972; Revel *et al.*, 1973a). A reduction in the activity of an MS2 RNA-specific IF-3 (IF-3-α) has been suggested (Lee-Huang and Ochoa, 1971), but this conflicts with data indicating that an IF-3 active for MS2 RNA can be recovered from T4-infected crude initiation factor preparations (Revel *et al.*, 1973a; Schiff *et al.*, 1974), although these crude factors are by themselves inactive for MS2 RNA. Search for selective inhibitors of MS2 RNA translation, but not of T4 mRNA, led to the purification of several interference factors. These factors were present in both infected and noninfected cells (see Section III,D). Interference factor i-α(S1) is modified early after T4 infection; the modification, detected by immunoelectrophoresis, is blocked by chloramphenicol (Revel *et al.*, 1973a). Factor i-α inhibits initiation on MS2 RNA but not on T4 mRNA. Another interference factor, i-γ, inhibits both MS2 RNA and early T4 mRNA much more than late T4 mRNA. This could explain the decreased activity of T4-infected systems for early T4 mRNA (Pollack *et al.*, 1970; Lee-Huang and Ochoa, 1971), possibly related to the decrease of early protein synthesis late in T4 infection (Cohen, 1972).

Crude initiation factors from T4-infected *E. coli* were shown to be defective specifically in their ability to bind ribosomes to the coat initiation site of phage RNA (Steitz *et al.*, 1970), but less deficient for the maturation cistron. Cistron specificity was, however, not seen using aminoacyl-tRNA binding or dipeptide assay (Goldman and Lodish, 1972; Singer and Conway, 1973a).

The mechanism by which protein synthesis initiation is altered in T4-infected cells is unclear. It appears possible that some T4 function modifies directly or indirectly some host components of the translation apparatus (Schedl *et al.*, 1970; Revel *et al.*, 1973a) as it does for the host RNA polymerase (Schachner *et al.*, 1971). A generalized decrease in translation initiation activity has been observed with virulent T4 and T4-ghosts (Goldman and Lodish, 1972; Leder *et al.*, 1973). The selective loss of activity for certain templates may be only the first stage, followed by a more generalized inactivation as observed in stationary phase or chloramphenicol treated *E. coli* (see Section IV,A). Singer and Conway (1973a) indeed observed a general decrease, but with a much larger effect on f2 RNA translation, in particular at low template concentrations.

Hsu (1973) found that 30 S-MS2 RNA interaction is unaltered by T4 infection, while formation of the complete initiation complex with MS2 RNA is decreased. The selective alteration may be therefore apparent only under certain experimental conditions. Singer and Conway (1973b) observed that strong reducing agents present during cell breakage tend to reduce the selective alteration. Furthermore, oxidation of crude ribosomes (containing initiation factors) or blocking of —SH groups by *N*-

ethylmaleimide (NEM) led to a loss of f2 RNA translation but not of T4 mRNA translation. It is possible that the modification produced after T4 infection is directly or indirectly associated with the oxidation of ribosome-associated factors. [As an example, factor i-α(S1) may be modified in this way (Inouye et al., 1974)]. NEM also affects ribosomes themselves, but no template-specific effects were observed.

Late after T4 infection, ribosomes appear to undergo some modification which makes them resistant to streptomycin (Artman and Werthamer, 1974). In view of the relationship between the streptomycin protein (S12) and selective initiation (see Section III,E) this may be a very significant observation.

The synthesis and function of T4 coded tRNA's has been reviewed (Inouye and Littauer, 1973).

3. T7 Phage

T7 infection of E. coli inhibits the expression of the bacterial genome (Studier, 1972; Rahmsdorf et al., 1973a). The synthesis of T7 early and late proteins is, as in T4, a strictly controlled process of which only the translational aspect will be reviewed.

An early bacteriophage gene codes for a protein responsible for the block of host protein synthesis (Schweiger et al., 1972). This protein may act as a translation repressor which blocks the translation of E. coli β-galactosidase mRNA and T3 S-adenosylmethionine hydrolase mRNA (Herrlich et al., 1974). A modification of the ribosomes in T7-infected cells by phosphorylation has been reported (Rahmsdorf et al., 1973b).

A very interesting phenomenon is the abortion of T7 infection in E. coli F⁺ cells which appears to result from a lack of translation of certain classes of T7 mRNA's (Morrison and Malamy, 1971; Blumberg and Malamy, 1974). Mutants of the F episome, in RVF' E. coli which lack the restriction function, have been isolated but attempts to demonstrate that initiation factors specific for different classes of T7 mRNA are involved in this effect have remained unsuccessful. Cistron-specific effects of interference factors from uninfected cells (F⁺ or F⁻) on T7 protein synthesis in vitro have been observed, in particular on stimulation or inhibition of the lysozyme cistron (Revel et al., 1973b; see Section III,D,2).

T7 infection was shown to have an effect on initiation factors in RVF' E. coli cells. Immunoassays showed that over 80% of IF-3 disappear from F⁺ cells 1–2 min after T7 infection. This does not occur in F⁻ cells or after T4 infection (Scheps et al., 1972). Chloramphenicol or rifampicin prevents the loss of IF-3, which therefore requires protein synthesis after infection. IF-2 activity, on the other hand, decreases only by a factor of 2. The loss of IF-3 may be partially prevented by inhibitors of serine proteases with a concomitant 100-fold increase in phage yield (Revel et al., 1973a; Zeller, 1974). It is likely that in F^+ cells, T7 infection releases or activates a protease which degrades initiation factors and stops all protein synthesis, leading to cell death. Possibly IF-3 is not protected by the 30 S subunit from degradation; it is interesting that some F^+ streptomycin-resistant ribosome mutants do not restrict T7 (Chakrabarti and Gorini, 1975). Initiation factors may be most susceptible to inactivation as seen also in heat-shocked E. coli (Patterson and Gillespie, 1972). During the first two minutes of T7 infection, in F^+ cells, early mRNA may still be translated, but T7 RNA polymerase may be made in smaller amounts or IF-3 may be degraded before late T7 mRNA transcripts accumulate. Recently, a membrane permeability change has been demonstrated in T7-infected F^+ cells, at the time when IF-3 is degraded (J. A. Steitz, personal communication). There is loss of cellular ATP which may explain the arrest of mRNA and protein synthesis observed in their laboratory. Ponta et al. (1975) found no effect on initiation of protein synthesis and attribute the F^+ restriction of T7 phage to a membrane change. There was a long lag before T7 gene expression was affected in the E. coli strains used by these authors. The discrepancies could result from differences in the strains used.

In E. coli F^- cells, T7 infection (as T4 or λ) produces a generalized decrease in protein initiation activity (Leder et al., 1973) but much less pronounced than that in F^+ cells. Late T7 mRNA's are characterized by a high metabolic stability, and when isolated from the infected cell have a ribosome attached to their initiation site (Callahan and Leder, 1972).

PART II. Initiation of Protein Synthesis in the Cytoplasm of Eukaryotic Cells

V. Role of Met-tRNA$_f$

The breakthrough in understanding the initiation process in eukaryotic cells came when it was realized that unformylated Met-tRNA$_f$ species serve as initiator tRNA. Among the different Met-tRNA's, only this species can be formylated by E. coli enzymes (Caskey et al., 1967), but it is not formylated in the cytoplasm of higher cells because of the absence of a transformylase. This initiator Met-tRNA$_f$, which is coded for by AUG, has been found in protozoa (Ilan and Ilan, 1970) and plants (Leis and Keller, 1970; RajBhandary and Kumar, 1970) as well as in mammalian cells (Smith and Marcker, 1970; N. K. Gupta et al., 1970; Rudland and Clark, 1972, for review). Although there is no doubt that methionine serves to initiate peptide chains (Brown and Smith, 1970; Housman et al., 1970; Hunter and Jackson, 1971; Elson et al., 1974a), there remains some controversy on the question of whether Met-tRNA$_f$ can also donate its methionine internally in the polypeptide chain (Drews *et al.*, 1972). In contrast to prokaryotic fMet-tRNA$_f$, no formyl group prevents this reaction and AUG codes also for internal methionine. The specific utilization of Met-tRNA$_f$ in initiation results, in the eukaryotes, entirely from the selective recognition of this tRNA by the initiation factors.

Studies on the nucleotide sequence of yeast and mammalian Met-tRNA$_f$ have shown a very interesting difference with the other tRNA's. The sequence T-ψ-C-G is absent and replaced by A-U-C-G (Petrissant, 1973; Simsek and RajBhandary, 1972; Piper and Clark, 1974; Simsek *et al.*, 1973, 1974). The significance of this replacement in the function of initiator tRNA will be interesting to determine, in particular with respect to its interaction with initiation factors.

VI. Selective Recognition of Met-tRNA$_f$ by Initiation Factors

Initiation factors from mammalian cells have been obtained either from the 0.5 M KCl ribosomal wash or from the cytosol. The most specific assay uses the property of initiation factors to recognize and bind the initiator tRNA to ribosomes and to allow translation of natural messenger RNA's.

Polypeptide chain initiation factor eIF-2 (or E2, or MP; for comparison of nomenclatures see Table II) forms a complex with Met-tRNA$_f$ and GTP. This complex is more stable than the prokaryotic IF-2–fMet-tRNA$_f$ complex (see Section II,A): it is retained on Millipore filters (Chen *et al.*, 1972; Dettman and Stanley, 1972; Gupta *et al.*, 1973; Schreier and Staehelin, 1973a; Levin *et al.*, 1973a,b; Elson *et al.*, 1975; Safer *et al.*, 1975b). Formation of this ternary eIF-2–GTP–Met-tRNA$_f$ complex requires no magnesium ions. The factor may exist in two interconvertible forms, one heavy (MW 150,000) and one light (MW 40,000). Addition of magnesium, and cold, dissociate the ternary complex, the light form being less stable than the heavy form (N. K. Gupta *et al.*, 1974, 1975a). Aurintricarboxylic acid at low concentration (3×10^{-5} M) inhibits the formation of the ternary complex, and so does treatment of eIF-2 by N-ethylmaleimide.

The ternary complex contains equimolar amounts of GTP and Met-tRNA$_f$ (Dettman and Stanley, 1972). The reaction with eIF-2 is very specific for initiator tRNA; other aminoacyl-tRNA's (including Phe-tRNA) are not bound; *E. coli* fMet-tRNA$_f$ is bound although without a requirement for GTP (Gupta *et al.*, 1973). A binary complex may form (Safer *et al.*, 1975a).

The performed Met-tRNA$_f$–GTP–eIF-2 complex was found to be more active in methionine transfer reactions than free Met-tRNA$_f$. Although an exchange of bound with free Met-tRNA$_f$ was not prevented in these experiments, the results suggest that the ternary complex may be an intermediate in initiation. It has not been determined how much of this complex *in situ* exists free of ribosomes.

eIF-2 has three subunits (Table II; Staehelin *et al.*, 1975), the smallest can be phosphorylated (see Section XII, A).

VII. Binding of Met-tRNA$_f$ to 40 S Ribosomes

Ribosomes reduce the amount of Met-tRNA$_f$ retained on Millipore filters by eIF-2 (Gupta *et al.*, 1973). However, by sucrose gradient analysis, Schreier and Staehelin (1973a) showed that their factor E2 binds Met-tRNA$_f$ to 40 S subunits in a poly(A,U,G)-dependent reaction, but not with globin mRNA. No 80 S complexes are formed. More recently, N. K. Gupta *et al.* (1975a) have developed a Millipore assay for Met-tRNA$_f$ binding to ribosomes. With eIF-2 and GTP, Met-tRNA$_f$ is bound to ribosomes but the reaction is not dependent on AUG codons. Stimulation by AUG codons is observed by these authors when eIF-3 and eIF-4 (see Table II) are added to eIF-2 and 40 S subunits (Gupta *et al.*, 1975b). By sucrose gradient analysis, Cashion and Stanley (1974) and Safer *et al.* (1975a) showed that eIF-2 binds Met-tRNA$_f$ to 40 S ribosomes in a nontemplate dependent reaction. The binding of initiator tRNA in the absence of template is of interest because natural Met-tRNA$_f$–40 S ribosome complexes free of mRNA, have been found in reticulocytes (Darnbrough *et al.*, 1973; Sundkvist and Staehelin, 1975), and may be intermediates in the initiation of translation.

Initiation factor E3 (see Table II for nomenclature) plays a key role in conjunction with E2 (eIF-2) for Met-tRNA$_f$ binding to 40 S subunits (Schreier and Staehelin, 1973a). Factor E3 appears to be an aggregate of several polypeptide chains; its molecular weight is close to 300,000, and it may contain several different activities (Sundkvist and Staehelin, 1975). When both E2 and E3 are present, Met-tRNA$_f$ is bound to 40 S subunits without mRNA template (Schreier and Staehelin, 1973a,b) as well as with globin mRNA, but no 80 S ribosomes are formed. This activity of E3 is destroyed by *N*-ethylmaleimide. E3 interacts with 40 S subunits specifically, converting them to 48 S particles, and probably contains the ribosome dissociation activity (Merrick *et al.*, 1973; Kaempfer and Kaufman, 1972; Nakaya *et al.*, 1973; Lubsen and Davis, 1974a; N. K. Gupta *et al.*, 1975a; Sundkvist and Staehelin, 1975). Native 40 S subunits carry this factor E3, and also eIF-2 (E2) (Hirsch *et al.*, 1973; Ayuso-Parilla *et al.*, 1973; Lubsen and Davis, 1974b; Freienstein and Blobel, 1974; Sundkvist and Staehelin, 1975) and are the main source of these proteins in the cell extracts. Factor E3 is also involved in mRNA binding, as discussed in Section X.

Other protein factors present in the cytosol have been shown to promote Met-tRNA$_f$ binding to 40 S ribosomes. Factor M1 (MW 65,000) allows the AUG-dependent binding of eukaryotic formylated Met-tRNA$_f$ to 40 S ribosomes (Shafritz and Anderson, 1970a) and the poly(U)-dependent binding of acetyl Phe-tRNA to 40 S subunits (Shafritz and Anderson, 1970b). There is, however, no GTP requirement. With unformylated Met-tRNA$_f$ another factor named M2 (A + B) was required with GTP. The same purified M1 preparation binds both N-blocked and unformylated initiator tRNA's to 40 S subunits (Merrick *et al.*, 1974). Ochoa's laboratory reported similar data about a pure factor from *Artemia salina*, designated EIF-1 (2 subunits of MW 74,000) (Zasloff and Ochoa, 1971, 1972, 1973; Nombela *et al.*, 1974). The *A. salina* and rabbit factor have common antigenic determinants. A similar factor was isolated from rat liver cytosol (Gasior and Moldave, 1972; Leader *et al.*, 1970). This soluble factor differs clearly from eIF-2 (E2) discussed above, both in its purification and function. The major difference is the absence of specificity for Met-tRNA$_f$ (Koka and Nakamoto, 1974) and the marked template dependency. The relation between M1 (EIF-1) and the other initiation factors is not clear. It presents some analogy with a prokaryotic factor which binds tRNA to 30 S ribosomes (Kan *et al.*, 1970); it may be involved in other steps of protein synthesis besides initiation. An intriguing property of this factor is that, when 60 S ribosomes are added, 80 S ribosomes-aminoacyl-tRNA complexes are formed which react with puromycin. The ability of this single factor to form a completely active 80 S ribosomal complex contrasts with the numerous factors required when 40 S initiation complexes are first formed with eIF-2 (E2). Recently, it has become clear that M1 or EIF-1 of *A. salina* is not required as an initiation factor of protein synthesis (Filipowicz *et al.*, 1976), but eIF-2 (MP,E2) is needed. The role of M1 is unknown (Merrick and Anderson, 1975; Adams *et al.*, 1975).

VIII. Formation of the 80 S Initiation Complexes

The conversion of Met-tRNA$_f$-40 S complexes, made with eIF-2 (E2) and E3, into 80 S complexes requires 60 S subunits and several factors. The requirements were analyzed in detail by Schreier and Staehelin (1973b). They found that factor E5 (MW ~ 150,000), which is most probably identical to M2A of Anderson (Shafritz *et al.*, 1972), promotes the joining of 40 S to 60 S subunits without template. In order to have a stimulation by mRNA (in their case globin mRNA), another factor called E4 (MW ~ 50,000) was required, as well as the addition of ATP. Better dependency on mRNA for the formation of 80 S initiation complexes was observed when yet another factor, E1, was added. This latter protein inhibits joining of the subunits in the absence of mRNA and may serve to prevent fortuitous 80 S formation without mRNA. It stimulates 80 S formation with mRNA and ATP. A similar effect was reported by Nakaya and Wool (1973). Reactivity of the bound Met-tRNA$_f$ with puromycin requires all 6 initiation factors: E1,2,3,4,5,6 (recently E3 was separated into two activities E3 and E6) in addition to mRNA, GTP, and ATP (Schreier and Staehelin, 1973b; Staehelin *et al.*, 1975). E6 (M3) is an mRNA binding factor. A protein E7 (M2Bβ) is also needed to form 80 S initiation complexes (Table II).

The requirement for ATP in reticulocytes is similar to the ATP requirement for initiation in wheat embryo ribosomes originally reported by Marcus (1970; Seal and Marcus, 1973). The function of ATP is not well understood; one hypothesis proposed by Marcus (1974) was a role in the metabolism of GTP. Another hypothesis is that ATP is needed by E6 to bind mRNA (Staehelin *et al.*, 1975); an ATPase activity may be involved.

The relationships among the various initiation factors from rabbit reticulocytes reported by different laboratories are not clear. Gupta *et al.* (1973) reported that the formation of 80 S complexes with AUG templates requires in addition to eIF-2 (IF-1), two activities IF-2 and IF-3 (see Table II). Cashion and Stanley (1974) reported that only one additional factor (MW 220,000) besides eIF-2 is required with AUG-containing templates. Anderson and his co-workers (Crystal *et al.*, 1972a; Picciano *et al.*, 1973) reported that for Met-tRNA$_f$ binding with globin mRNA, M1, M2A, and M2B (α and β) (Kemper *et al.*, 1976) are required, while for the puromycin reaction and translation, M3 is also needed. As discussed above, M1 is probably not an initiation factor. Factor M3 has been recently shown

TABLE II
Tentative Nomenclature of Eukaryotic Initiation Factors

Molecular weight $\times 10^{-3a}$	Research group[b]					
	Anderson	Staehelin	Gupta	Others	Official[d]	Function
150(55 + 50 + 35)	MP	E2	IF-1(A,B)		eIF-2	Met-tRNA$_f$ binding
≥300 − 500	DF(M5)	E3	IF-3	IF-3-rr	eIF-3	Dissociation mRNA binding
50	M4	E4			eIF-4A	mRNA dependence
80	M3[c]	E6	IF-2	IF-3	eIF-4B	mRNA binding, ATP cap binding
16–19	M2B(?)	E1			eIF-1	Antijoining when no mRNA
150	M2A	E5			eIF-5	Joining subunits
17	M2Bβ	E7			eIF-4C	Joining subunits
65	M1			EIF-1		tRNA binding to 40 S

[a] Numbers in parentheses are sununit molecular weights.
[b] See Section VI to XI for details.
[c] In early papers, M3 = MP + DF + M3.
[d] Decided at Fogarty Meeting, 1976.

to contain three separable activities: DF (E3), MP (eIF-2, E2) and a real M3 (Anderson, 1974; Elson *et al.*, 1975). It is yet premature to try and compare requirements described by different laboratories. Three initiation factors from L cells have also been described (Levin *et al.*, 1973a), as well as from wheat embryos (Seal *et al.*, 1972). Table II tries to summarize these data, and presents the most recent information (Filipowicz *et al.*, 1976).

IX. Specific Assays for Initiation of Eukaryotic mRNA Translation

To measure initiation of the translation of natural mRNA, several techniques have been developed. Binding of Met-tRNA$_f$ to ribosomes, even when stimulated by exogenous mRNA, is not sufficient to ascertain that correct initiation has taken place. Formation of the initial dipeptide, containing methionine donated by Met-tRNA$_f$ and the second amino acid coded for by the cistron of mRNA studied, is a much better method.

The synthesis of the initial dipeptides of rabbit hemoglobin was described by Crystal *et al.* (1971). This reaction also requires the addition of purified elongation factor EF-1 to bind the second aminoacyl-tRNA, in addition to the different initiation factors. Both α- and β-globin chains are initiated by Met-Val. In the absence of elongation factor EF-2 and other aminoacyl-tRNA's, no chain elongation occurs. To discriminate the two chains, the synthesis of a tripeptide at least is necessary. Rabbit α-globin corresponds to Met-Val-Leu whereas the β-chain corresponds to Met-Val-His. Initiation peptides were also synthesized on reticulocyte ribosomes retaining the endogenous initiation fragment of hemoglobin mRNA after nuclease digestion (Crystal *et al.*, 1972b). The N-terminal methionine is removed from many proteins when the nascent chain is 10–20 amino acids long (Wilson and Dintzis, 1970; Chatterjee *et al.*, 1972), but with formylated mammalian Met-tRNA$_f$ the cleavage is inhibited and long methionine-labeled initiation fragments are obtained (Housman *et al.*, 1970; Elson *et al.*, 1974a). In some cases, methionine cleavage does not occur (Pemberton *et al.*, 1972; Strous *et al.*, 1972).

Another method uses the antibiotic sparsomycin to inhibit polypeptide chain elongation and avoid methionine cleavage (Jackson and Hunter, 1970; Smith and Wigle, 1973). Initiation dipeptides, tripeptides, and several longer peptides may be obtained in this system. The advantage is that sparsomycin may be added to a nonpurified cell extract, without need to purify out EF-2 or aminoacyl-tRNA's.

Such techniques are very important to ascertain that correct initiation is studied. One application has been to show that β-globin chains are initiated in systems from β-thalassemic human cells (Nienhuis *et al.*, 1971,

1973). The rate of initiation of globin synthesis in thalassemic cell extracts was also studied by following the polysome assembly and movement (Nathan et al., 1971). These experiments failed to detect any defect in translation of β-globin mRNA in β-thalassemic reticulocytes. It appears that β-globin mRNA is absent in cells from such patients (Kacian et al., 1973). Some thalassemia may still exhibit a translation defect (Conconi et al., 1972; Rowley, 1972).

Noll et al. (1972) studied the initiation of translation of rabbit globin mRNA on E. coli ribosomes, and with E. coli IF-3. The data show that, although fMet-tRNA$_f$ is bound, the second aminoacyl-tRNA bound does not correspond to valine, indicating that initiation probably occurs at a wrong site on the mRNA in this mixed system.

Studies of this kind have been performed with a variety of messenger RNA's. Initiation peptides of picornavirus RNA [encephalitis myocarditis virus (EMC) or Mengo virus] (Smith, 1973) and of plant tobacco necrosis virus (TNV) (Seal and Marcus, 1973) are some more examples of the results obtained.

X. Binding of Eukaryotic mRNA to Ribosomes

Since 40 S-Met-tRNA$_f$ complexes without mRNA are found in reticulocyte lysates (Darnbrough et al., 1973) and readily made in vitro (Schreier and Staehelin, 1973a), it has been concluded that mRNA binds to ribosomes only after the 40 S initiation complex is made. A plant viral RNA, however, binds to ribosomes without tRNA (Marcus, 1970) leaving some doubts. By studying the stimulation of 80 S initiation complex formation by rabbit globin mRNA, Schreier and Staehelin (1973b) and Staehelin et al. (1975) determined that mRNA binding requires a cooperative action of initiation factors E1, E4, E5, and E6. Factors E2 and E3 were always present in these assays.

An initiation factor first isolated by Anderson's group (Prichard et al., 1971) and called M3 was shown to be required for hemoglobin mRNA directed methionylpuromycin and methionylvaline dipeptide synthesis (Crystal et al., 1972a; Picciano et al., 1973). Some complication arose from the fact that this factor has now been separated into several activities: MP (equivalent to E2), DF (E3), and M3 itself (Anderson, 1974; Merrick et al., 1973; Elson et al., 1975). Kaempfer and Kaufman (1972) described the isolation of an activity required for recycling of ribosomes and ribosome binding to mRNA in rabbit reticulocytes. This high molecular weight factor, called IF-3-rr was considered similar to Anderson's M3; it binds by itself to mRNA. It now seems possible that this activity is a mixture of several factors, since its effect on restoration of globin synthesis in the absence of heme (see Section XII) can be obtained with purified Met-tRNA$_f$ binding factor (MP = E2 = eIF-2) (Beuzard and London, 1974; Clemens et al., 1974).

Heywood (1970a; Rourke and Heywood, 1972) isolated a factor, also called IF-3 (and considered by the author as similar to M3), from embryonic chick muscle. This activity stimulates myosin 26 S mRNA binding to ribosomes. In this case an effect on mRNA binding to 40 S subunits was also observed (Heywood, 1970b; Heywood and Thompson, 1971). As a further argument for its role in mRNA binding, this factor was shown to be rather specific for myosin mRNA and inactive for hemoglobin mRNA (see Section XI,A).

At this time, it is almost impossible to correlate the various mRNA binding activities described by different groups. Obviously, further purification and direct comparison of these factors is needed.

Studies with a homogeneous preparation of Met-tRNA$_f$ binding factor (eIF-2, MP, E2) reported by Hellerman and Shafritz (1974) have shown that this protein, in addition to its very specific recognition of initiator tRNA, also binds poly(A). Furthermore, it is found on the natural messenger ribonucleoprotein (mRNP) which can be isolated from reticulocyte or other cells polysomes (Lebleu et al., 1971; Blobel, 1972, 1973; Bryan and Hayashi, 1973). The MP factor is a minor component of the mRNP proteins, but its affinity for poly(A) suggests to the authors that it might be a protein bound to the Poly(A) end of mRNA (Kwan and Brawerman, 1972) which protects this poly(A) from degradation (Soreq et al., 1974). A role for an mRNP protein in hemoglobin mRNA binding to 40 S particles was proposed by Lebleu et al. (1971; see also Ilan and Ilan, 1973). It would be of great interest if the Met-tRNA$_f$ binding factor also participates in the attachment of mRNA to the subunit. This would strengthen the similarities with the prokaryotic system in which IF-2 (the fMet-tRNA$_f$ binding factor) also stimulates the mRNA binding to 30 S subunits (Greenshpan and Revel, 1969). This binding is, however, not specific for natural initiation sites (Revel et al., 1970). MP also binds to heterologous phage R17 RNA and double-stranded RNA (Kaempfer, 1974).

Sequence studies on the 3'-end of 18 S rRNA and on some mRNA's ribosome binding sites indicate that the model of Shine and Dalgarno (1974, 1975) (see Section III,E), for prokaryotic mRNA base pairing to rRNA, holds true also for eukaryotes (Steitz and Jakes, 1975). All species of 18 S RNA analyzed possess an identical 3'-terminal sequence -G-A-U-C-A-U-U-A$_{OH}$. The first ribosome binding site sequenced, that of brome mosaic virus (BMV) RNA, exhibits a four-base complementarity with this terminus (Dasgupta et al., 1975).

More recently, it was found that most eukaryotic cellular or viral mRNA's have at their 5'-end a capping structure of the general type: m7G(5')ppp(5')XmpY (Rottman et al., 1974). Addition of the G residue and methylation onto the first nucleotide X of the mRNA are post transcriptional events. The methylated G appears to be required as a recognition signal for ribosome binding (Both et al., 1975). Binding of capped mRNA to ribosomes is inhibited by m7pG (Hickey et al., 1976; Cnaani et al., 1976). Furthermore in the absence of ribosomes, interaction of mRNA with a preparation of initiation factor M3 (subunit MW 80,000), which seems to correspond to E6 of Staehelin et al. (1975), is inhibited by m7pG, suggesting that this protein binds the capped 5'-terminus of mRNA (Shafritz et al., 1976). In the case of BMV RNA, the cap structure is only 10 nucleotides away from the initiation codon AUG, and between the two is the sequence which base pairs to the 3'-terminus of 18 S rRNA. It may be premature to generalize, but it is possible that in monocistronic eukaryotic mRNA these features are common. Variations are, however, very likely. It should be noted that picornavirus RNA's are not capped although these are excellent mRNA's (for more detail see Both et al., 1975). Initiation factor M2 A may also bind mRNA, but not the cap (Shafritz et al., 1976).

XI. Specificity of Eukaryotic mRNA Recognition

A. Factors That Exhibit mRNA Discrimination

The variety of differentiated cell types from which cell-free protein synthesizing systems can be prepared has provided a way to investigate possible specificities in the recognition and faithful translation of natural mRNA's. In general, such studies have shown that virtually any mRNA can be faithfully translated into protein in almost any cell-free system (for review, see Bloemendal et al., 1972). More precise comparison of the efficiency at which two different mRNA's are translated in given cell-free systems indicates, nevertheless, that some mechanism of mRNA recognition exists in eukaryotic cells. As an example, mouse Krebs ascites cell extracts translate efficiently picornavirus RNA (EMC or Mengo RNA) but much less efficiently rabbit or mouse hemoglobin mRNA. Addition of initiation factors obtained from reticulocyte ribosomes strongly stimulates hemoglobin mRNA translation (Lebleu et al., 1972; Metafora et al., 1972; Hall and Arnstein, 1973). The relative efficiency at which these two types of mRNA are translated is markedly changed, therefore, by some ribosome associated component. The active factor, I_{Hb}, was purified to homogeneity, and has a molecular weight of 65,000 in SDS polyacrylamide gels (Nudel et al., 1973). This factor has a 4 times larger effect on α-globin synthesis than on β-globin; it does not stimulate Mengo RNA translation.

This effect may be related to the presence of a cap on globin mRNA, but its absence on Mengo RNA (Cnaani et al., 1976).

The fact that only one protein is sufficient to stimulate initiation of hemoglobin mRNA translation when added to the Krebs ascites cell-free system indicates that these cell extracts contain all the other initiation factors, which act without specificity on all mRNA's. It is important for these studies that the crude cell-free systems used for assaying mRNA-specific factors be very active and deficient only in the mRNA-specific activity. Many reports in the literature deal actually with nonspecific stimulation of many mRNA's, due to the addition of initiation factors to cell-free extracts which have been depleted or have lost by inactivation the endogenous initiation factors. Wheat germ cell-free extracts (Roberts and Patterson, 1973) when carefully prepared are very efficient for the translation of hemoglobin mRNA, but do not translate Mengo or EMC viral RNA. Addition of reticulocyte initiation factors to this system stimulates neither Mengo RNA translation nor that of hemoglobin mRNA; the factors, however, very strikingly increase the α-globin/β-globin ratio (U. Nudel and M. Revel, to be published). On the other hand, if a wheat germ cell-free extract is incubated at 37° instead of 25°, it rapidly loses its activity for globin synthesis, which can then be restored fully by reticulocytes initiation factors (T. Zehavi-Willner and S. Petska, 1976). This illustrates the difference between active and inactivated cell-free systems for their use in the assay of mRNA-specific activities.

The first demonstration of specificity in mRNA translation between differentiated cells was reported by Heywood (1970a,b) and Rourke and Heywood (1972), who studied the translation of myosin mRNA from embryonic chick muscle in a rabbit reticulocyte cell-free system. This system was clearly dependent on the addition of a protein factor isolated from muscle ribosomes. Translation of hemoglobin mRNA on muscle ribosomes required a reticulocyte factor. These factors were partially purified and designated IF-3 or M3. Heywood et al. (1974) further fractionated the factors from red muscle into a myosin-specific and a myoglobin-specific fraction, which stimulate selectively the translation of these two mRNA's in a reticulocyte lysate. A small molecular weight RNA, translation control RNA (tcRNA), less than half the size of tRNA, may be involved in mRNA discrimination: it inhibits initiation on heterologous mRNA (Bester et al., 1975). The tcRNA appears to act at the level of mRNA binding to 40 S subunits (Kennedy et al., 1974). Stimulatory effects of tcRNA were also reported (Bogdanovsky et al., 1973).

Using a partially fractionated Krebs ascites cell-free system, Wigle and Smith (1973; Wigle, 1973) isolated a factor from Krebs ascites cytosol which was required for initiation of EMC RNA translation but had no effect on hemoglobin mRNA translation. In addition, an independent β-globin mRNA translation activity was found in Krebs ascites cell extracts. The purified factor, I_{EMC}, had a MW of 60,000 in SDS; this relatively small factor was shown to be different from the large "M3" factor (Strycharz et al., 1974) which, although required for translation of natural mRNA, did not show tissue or mRNA specificity (Schreier and Staehelin, 1973c; Picciano et al., 1973). I_{EMC} resembles I_{Hb} discussed above, and, together with the muscle myosin and myoglobin factors, may belong to a group of proteins with functions similar to the interference factors. The step in initiation at which these factors act is not well known: I_{Hb} stimulates the conversion of 40 S to 80 S initiation complexes. It is possible that while some mRNA interaction occurs with the Met-tRNA$_f$ binding factor on 40 S subunits (Hellerman and Shafritz, 1975), the specific recognition of mRNA occurs only during the 80 S assembly. I_{EMC} has been tentatively identified as initiation factor E4 which is required in larger amounts for EMC RNA translation than for globin mRNA (Staehelin et al., 1975). A role of the cap is possible. I_{Hb} may be factor eIF-4B (E6, M3) (Golini et al., 1976).

B. mRNA Selection in Translation Control

It must be stressed that the mRNA discriminatory function of initiation factors has been demonstrated only by in vitro experiments. Such studies show that cells possess a mechanism by which translation of specific messages can be controlled, but there is no evidence as yet that these mechanisms are used for some cell regulatory purpose.

In several systems, it is clear that the amount of protein synthesized is not directly related to the amount of mRNA present, suggesting some form of translation control. A good example is the synthesis of α- and β-globin chains, which in rabbit reticulocytes is maintained at a ratio of 1, yet the amount of α-globin mRNA exceeds that of β-globin mRNA. This regulation is achieved by a lower efficiency of initiation on α-globin mRNA than on β-chain mRNA (Lodish, 1971). The rates of polypeptide chain elongation are the same for both mRNA's (Lodish and Jacobsen, 1972). Another example is the synthesis of reovirus proteins, whose mRNA's are made in equal amounts in infected cells, while up to 10-fold differences may be observed in the amounts of different protein synthesized (Both et al., 1975). Experiments in which competition between two exogenous mRNA's added to the same cell-free system is measured, also show preferential translation of one of the templates (Lebleu et al., 1972; Schreier et al., 1973a; Hall and Arnstein, 1973; Lawrence and Thach, 1974).

Recently, Lodish (1974) has proposed a mathematical model to account for the variations in α- to β-globin synthesis. The model shows that any increase in initiation will favor the "weakest" mRNA, while an increase in the rate of chain elongation will favor the "strongest" mRNA. Thus non-mRNA specific parameters may affect the ratio at which two mRNA's are translated into protein. Varying the amount of total ($\alpha + \beta$) hemoglobin mRNA added, or the ionic conditions used in the cell-free system, does change the α-globin/β-globin ratio, according to the above principle (MacKeehan, 1974). The question is: What is the mechanism that determines whether a mRNA behaves as a strong or weak template? Differences in the secondary structure of the region of mRNA to which ribosomes bind and initiate translation may exist. But in order to explain the mRNA discrimination, these differences in the initiation sites must be recognized by the ribosome and its associated factors. For example, the I_{Hb} factor (Nudel et al., 1973) increases the α to β hemoglobin translation ratio two- to threefold, over a wide range of mRNA concentrations and ionic conditions, i.e., the increase is independent of the overall efficiency of translation. Moreover, this factor increases the α/β hemoglobin ratio (from 0.4 to 1.0) in Krebs ascites cell extracts (whose efficiency for hemoglobin mRNA is low) as well as in wheat germ extracts (from 1.0 to 1.6) although the latter system is very efficient by itself for hemoglobin mRNA translation (U. Nudel and M. Revel, to be published). Such variations in the α/β globin ratio are much larger than the $\pm 20\%$ predicted by the model of Lodish (1974). This suggests that some factors actually change the affinity of the ribosomes for mRNA and may be part of the mechanism which determines what is a "strong" and a "weak" mRNA.

This situation is then similar to that in prokaryotes, where efficiency of translation at different cistrons of an RNA phage molecule is determined by the structure of the RNA, that of the ribosome, and the presence of initiation and interference factors in the system (see Section III,E).

Variations in initiation factor activities of cells have been reported in a few cases, such as reticulocyte maturation (Herzberg et al., 1969; Rowley et al., 1971), embryonic muscle development at the onset of myosin synthesis (Heywood and Kennedy, 1974), insect development (Ilan and Ilan, 1971), in which stage specific-mRNA discriminating factors have been described, and Artemia salina development (Sierra et al., 1974; Filipowicz et al., 1976). A discussion on the involvement of soluble factors in translation control may be found in the article by Pain and Clemens (1973). These authors present an extensive review of the experiments in which the control of translation in intact cells has been studied under a variety of nutritional and hormonal conditions. Since our discussion is centered on biochemical mechanisms for the regulation of mRNA translation, we shall consider a few examples of systems in which some understanding of the biochemical nature of the control has been achieved.

XII. Some Mechanisms of Control in Protein Synthesis Initiation

A. Hemin

The synthesis of proteins by reticulocytes is greatly enhanced by iron; or by hemin which is actually the active agent. Unless hemin is added, initiation of protein synthesis in reticulocyte lysates ceases abruptly after a few minutes of normal activity (Rabinowitz et al., 1969; Hunt et al., 1972). Recycling of ribosomal subunits and initiation are inhibited when the block due to the absence of hemin is established, but the effect may be overcome by the addition of an initiation factor (Kaempfer and Kaufman, 1972). The factor may be isolated from different tissues, even nonerythropoietic ones. It was first identified as IF-3-rr (Kaempfer and Kaufman, 1972; see Section X), but more recent work indicates that the Met-tRNA$_f$ binding factor (MP, E2, eIF-2) is actually responsible for this effect (Beuzard and London, 1974; Clemens et al., 1974; Kaempfer, 1974; Clemens et al., 1975).

In the absence of hemin, Met-tRNA$_f$ is not found on 40 S subunits (Balkow et al., 1973a,b; Legon et al., 1973). This results from a translational repressor which may form first by a hemin-reversible, then irreversible process (Maxwell and Rabinowitz, 1969; Adamson et al., 1972; Gross and Rabinowitz, 1972, 1973a,b). Although the repressor was partially purified (Gross and Rabinowitz, 1973c; Freedman et al., 1974) its nature is not clear (see below). The repressor decreases Met-tRNA$_f$-40 S complexes (Balkow et al., 1973a) and is overcome by initiation factors (Mizuno et al., 1972). It does not block the binding of eIF-2 (MP) to ribosomes (Clemens et al., 1974). For Balkow et al. (1973b) there is a deacylase activity which can hydrolyze specifically Met-tRNA$_f$ (Gupta and Aerni, 1973; Morrisey and Hardesty, 1972) and is blocked by the Met-tRNA$_f$ binding factor (eIF-2, MP, E2). Others (Raffel et al., 1974), however, claim that there is a direct interaction between hemin and the initiation factor IF-3-rr, with hemin stimulating its activity. A surprising set of results reported by Legon et al. (1974) indicates that the mechanism of inhibition by heme deprivation is more complex. The block can be eliminated by the addition of 3′,5′-cAMP (Mager and Giloh, 1973; Giloh and Mager, 1975) or a variety of purines, the most active being 2-aminopurine. (The same compounds overcome the effects of double-stranded RNA and oxidized glutathione on initiation, see below.) One interpretation of these results is that the repressor is an enzyme which modifies a component of the initiation machinery (factor, tRNA, etc.) and is inhibited by purines. The inhibitor produced without hemin was, indeed, recently identified as a protein kinase which phosphorylates eIF-2's small subunit (Levin et al.,

1975, 1976; T. Hunt, personal communication; Balkow *et al.*, 1975; Kramer *et al.*, 1976). Phosphorylated eIF-2 is, however, still active *in vitro*.

The effect of hemin is not restricted to hemoglobin synthesis; other proteins synthesized in reticulocyte lysates are sensitive to hemin (Beuzard *et al.*, 1973; Lodish and Desalu, 1973; Palmiter, 1973). An effect of hemin on protein synthesis in other cells was also observed (Beuzard *et al.*, 1973; Raffel *et al.*, 1974; Weber *et al.*, 1975). In *Xenopus* oocytes, hemin affects the translation of hemoglobin mRNA introduced by microinjection (Gurdon, 1973). In the absence of hemin the oocyte synthesizes mainly β-globin and hemin restores an α/β ratio close to 1. This illustrates the fact that even in intact cells, mRNA translation is a controlled process and that mRNA's are not translated at random by any ribosome present in the cell. Hemin independent translation was reported in Friend leukemia cells (Cimadevilla and Hardesty, 1975). Hemin stimulates differentiation of these cells.

B. Double-Stranded RNA

In reticulocyte lysates, double-stranded RNA (dsRNA) inhibits initiation of globin synthesis (Ehrenfeld and Hunt, 1971), with the same kinetics as hemin deprivation: initiation stops after a few minutes of normal activity. The effect also resembles hemin deprivation in that 40 S-Met-tRNA$_f$ complexes disappear (Darnbrough *et al.*, 1972; Legon *et al.*, 1973). The effect is not restricted to reticulocytes, but in these cells very small amounts of dsRNA are sufficient to block initiation. Calculations show that one base pair of dsRNA may inactivate initiation on 10^4 ribosomes; larger amounts of dsRNA inhibit less. Addition of dsRNA causes the appearance of a dominant inhibitor whose action is overcome by the addition of purines (Legon *et al.*, 1974). These observations are hard to reconcile with the suggestion by Kaempfer and Kaufman (1973) that dsRNA acts by inactivating an initiation factor. Nevertheless, it is clear that an excess of initiation factor protects the system against inactivation. Content *et al.* (1975) reported that excess Met-tRNA$_f$ also protects against dsRNA, probably by overcoming the inhibition in 40 S-Met-tRNA$_f$ assembly. Formylation also overcomes this inhibition (Cahn and Lubin, 1975).

The physiological significance of dsRNA effect on protein synthesis initiation in many animal cells is intriguing (Cordell *et al.*, 1973); particularly since dsRNA is present in virus-infected and uninfected cells (Kimball and Duesberg, 1971; Stern and Friedman, 1970, 1971; Carter and de Clercq, 1974). A dsRNA specific nuclease exists in variable amounts in mammalian cells (Robertson and Matthews, 1973). Double-stranded RNA has many cellular effects and is a universal inducer of interferon (Colby and Morgan, 1971; Carter and de Clercq, 1974) although this effect appears now to be at the cell membrane (Pitha and Pitha, 1973) and not directly on the ribosomes. Ribosomal systems from interferon treated cells are, however, more sensitive to dsRNA (Kerr *et al.*, 1974; Content *et al.*, 1975). The features of dsRNA inhibition are best explained by the involvement of a protein kinase (Levin *et al.*, 1975; T. Hunt, personal communication; Balkow *et al.*, 1975). Interferon induces a specific protein phosphorylation (Zilberstein *et al.*, 1976; Lebleu *et al.*, 1976; Roberts *et al.*, 1976).

C. Oxidized Glutathione

Kosower *et al.* (1972) showed that GSSG inhibits initiation of protein synthesis, independently of the presence of reduced glutathione. In cell-free systems, this effect appears similar to that of dsRNA or hemin deprivation (Legon *et al.*, 1974), and the precise mechanism of action which is probably the same, remains to be fully clarified.

D. Feedback Inhibition by Product

In one system it has been shown that the protein product may inhibit initiation of its synthesis. Stevens and Williamson (1973) showed that heavy chains of mouse immunoglobulin bind with high affinity to their mRNA and inhibit translation, possibly by making the initiation site unavailable to ribosomes. It is not clear whether this observation relates to control of immunoglobulin synthesis. The interaction may take place on the endoplasmic reticulum membrane. No other example of such product-mRNA interaction has been reported.

E. Inhibition of Initiation by the Lack of Aminoacyl-tRNA

In mammalian cells, both in tissue culture or in the animal, deprivation of one of the essential aminoacids leads to a block in initiation (Vaughan *et al.*, 1971; Van Venrooij *et al.*, 1972). Since this effect is observed when amino acids other than methionine are absent, it suggests some special mechanism of control. A priori, amino acid starvation, producing a decrease in aminoacylated tRNA, would be expected to result in the accuumulation of polysomes arrested in elongation: the size of the polysomes should be larger than normal. Actually, this is what is observed in reticulocytes deprived of an amino acid (Hori *et al.*, 1967). In dividing cells, on the other hand, amino acid starvation produces a breakdown in polysomes and an apparent arrest in initiation. This is not due to the loss of active mRNA (Lee *et al.*, 1971; Christman *et al.*, 1973). The initiation block is due to a defect in tRNA charging, since analogues of histidine (histidinol) or of threonine (*O*-methyl threonine), which cannot be charged onto their corresponding tRNA's, produce the same effect (Vaughan and Hansen, 1973). Although in reticulocytes, *O*-methyl threonine produces an increase in the size of β-globin polysomes (Kazazian and Freedman, 1968), in HeLa cells it produces a breakdown of polysomes and an inhibition of initiation. Mammalian cells in culture appear, therefore, to have an additional mechanism which inhibits initiation when the supply of aminoacyl-tRNA's is impaired. The supply of mRNA or ribosomal subunits is unchanged in starvation conditions.

The mechanism by which initiation is regulated by aminoacyl-tRNA is unclear. The level of ATP may decrease in starved cells and trigger the block in initiation (Freudenberg and Mager, 1971; Van Venrooij et al., 1972), but it is not affected by amino acid analogues. Guanosine tetraphosphate, which accumulates in amino acid-starved stringent E. coli strains (RC$^+$) has not been found in mammalian cells (Mamont et al., 1972), but recent reports raised the possibility of its synthesis in wheat embryos (Marcus, 1974) and mouse embryo ribosomes (Irr et al., 1974). In contrast a degradation of GTP to guanine by amino acid starvation was reported (Grummt and Speckbacher, 1975).

Studies in cell-free systems would be necessary to elucidate this regulation mechanism. It is possible that deacylated tRNA has a direct inhibitory effect on initiation (Kyner et al., 1973; Zasloff, 1973), by competing with initiator tRNA binding. The absence of inhibition of initiation in amino acid-starved reticulocytes would, however, indicate that deacylated tRNA alone cannot explain the phenomenon seen in dividing cells. An analogy could be drawn, also, with the observation that in E. coli elongation factor G inhibits initiation in the absence of aminoacyl-tRNA (Lee-Huang and Ochoa, 1974), but such an analogy would meet the same objection as above.

Studies on the effect of interferon treatment of mouse L cells on mRNA translation have shown that tRNA may in this case also regulate initiation. In cell-free extracts from interferon-treated cells, there is a block in translation of exogenous mRNA's which is overcome by the addition of mammalian tRNA (Content et al., 1974; S. L. Gupta et al., 1974). Only certain isoacceptor tRNA's have this restoration activity, initiator Met-tRNA$_f$ being completely inactive. Content et al. (1975) showed that translation is first blocked in elongation, but, after a few minutes, a block in initiation appears: met-tRNA$_f$ is no longer bound to ribosomes (40 S) and, although mRNA is bound, no active initiation complexes are formed. Formylated Met-tRNA$_f$ can still bind to initiation complexes. An inhibitory protein factor associated with ribosomes appears to mediate these effects (see Section XII,B). This inhibition is reduced by adding excess amounts of the appropriate tRNA isoacceptor species. Interferon may, thus, be an example of protein synthesis regulation by tRNA. In contrast to what happens in amino acid-starved cells, in which all protein synthesis is inhibited, the effects of interferon treatment are very selective:

host protein synthesis is unaffected while expression of foreign messages (viral or from other parasites) is inhibited. The heterogeneity in the isoacceptor tRNA population of mammalian cells may explain this discriminatory effect: tRNA species active for the translation of certain codons may limit elongation and initiation on certain mRNA's. In cells from higher organisms, tRNA is not freely diffusible, as it is in bacteria. It is found in an aggregate with amino acid-tRNA ligases, possibly other protein factors and lipids (Bandyopadhay and Deutscher, 1971; Venegoor and Bloemendal, 1972). This sequestering of tRNA may limit its availability for certain mRNA's; some tRNAs are unequally distributed on polysomes (Culp *et al.*, 1970; Katze and Mason, 1973; Smith and MacNamara, 1974; see also Anderson and Gilbert, 1969).

The possible involvement of specific tRNA's in initiation has also been suggested in other systems (Clemens and Tata, 1972).

F. Stimulation of Initiation by a Block in Protein Synthesis

Incubation of mammalian cells at elevated temperatures produces an inhibition of protein synthesis initiation (MacCormick and Penman, 1969; Vaughan *et al.*, 1971). Preincubation of the cells in low concentrations of cycloheximide restores initiation during subsequent incubation at 42°, in the absence of the drug. Moreover, cells preincubated in low levels of cycloheximide yield extracts that show a greatly enhanced rate of initiation *in vitro*. Actually little initiation occurs in HeLa cells extracts, on the preexisting cellular polysomes (as measured by Met-tRNA$_f$ incorporation or with pactamycin), unless a preincubation of the cells in cycloheximide is performed (Reichman and Penman, 1973; Goldstein and Penman, 1973). This suggested to these authors that cells react to the inhibition of protein synthesis, caused by cycloheximide or amino acid starvation, by increasing in some way the rate of initiation. Actinomycin D blocks this stimulation observed in cycloheximide, indicating that the cellular response to a decreased rate of protein synthesis may be mediated by the synthesis of some species of RNA required for initiation. The nature of this RNA species is still completely unknown. Possible candidates are the tcRNA (Bogdanovsky *et al.*, 1973) or a 7 S RNA found associated with polysomes (Walker *et al.*, 1974). Another possibility is that tRNA may be involved: cycloheximide was shown to increase tRNA synthesis over that of other RNA species (Willis *et al.*, 1974). Synthesis of mRNA and its stability were not affected (Singer and Penman, 1972) under similar conditions. One must keep in mind that other explanations are possible for these drug effects. The regulation may be more complicated and linked to the pleiotypic response described by Tomkins and his co-workers (Hersko *et al.*, 1971; Kram *et al.*, 1973), involving changes in cAMP levels as a result of cycloheximide treatment. This would be in line with the fact that addition of hemin during the preparation of cell extracts increases their initiation activity (Weber *et al.*, 1975), if hemin (see Section XII,A) acts through protein kinases.

During mitosis, inhibition of protein synthesis initiation may take place through a similar mechanism (Fan and Penman, 1970).

Translation initiation control could play some role in the arrest of protein synthesis in nondividing cells, in which mRNA continues to accumulate normally (Rudland, 1974). In a line of baby hamster kidney cells, however, Baenzinger *et al.* (1974) could find no evidence of translation control in nondividing cells. Possibly in dense cultures, cells may become starved for essential nutrients, and the situation may in some lines resemble amino acid starvation discussed in Section XII,E. Results by Englehardt (1971) may be explained in this way.

G. Virus-Infected Cells

Most viruses produce a shutoff of host protein synthesis. There are indications that the effect is on initiation and may trigger a cellular response, which tends to overcome the inhibition (Leibowitz and Penman, 1971). Since viral proteins are synthesized very efficiently, some discriminatory process must take place. Viruses as different as *Vaccinia* (Moss, 1968), vesicular stomatitis virus (VSV) (Wertz and Youngner, 1972), and small picornaviruses (Baltimore, 1969) produce the shutoff. Most biochemical studies have been concerned with the last group. Poliovirus infection produces a disaggregation of polysomes synthesizing cellular proteins, suggesting a block in initiation (Summers and Maizel, 1967), while the rate of elongation is unchanged.

It was suggested that dsRNA is responsible for this inhibition (Ehrenfeld and Hunt, 1971). The evidence is, however, not in line with this hypothesis. The shutoff does not require viral RNA replication. Cell-free extracts from EMC-infected myeloma cells do not show impaired activity for the translation of exogenously added immunoglobulin mRNA from the same cells (Lawrence and Thach, 1974). The switchover from host to virus protein synthesis is probably caused by the ability of viral RNA to outcompete cellular mRNA for a component necessary for initiation. Such a competition is actually observed in the cell-free system. The acceleration of host inhibition by actinomycin D (Leibowitz and Penman, 1971) may then be explained by the effect of the drug in preventing an increase in initiation, as in uninfected cells (Goldstein and Penman, 1973). That the initiation can be inhibited by competing RNA was shown also in reticulocyte lysates (Lodish *et al.*, 1971; Lodish and Nathan, 1972). The translation of a number of viral mRNA's is more efficiently initiated than host cell mRNA (Nuss *et al.*, 1975), as shown by hypertonic shock.

Infection by one virus may restrict the growth of another virus. Herpes-virus infection inhibits translation of poliovirus RNA in infected cells (Saxton and Stevens, 1972). Positive "helper" effects at the level of translation may also be observed. SV40 rescues adenovirus in monkey cells: translation of adenovirus mRNA is blocked in these cells unless they are superinfected with SV40 (Baum *et al.*, 1972; Hashimoto *et al.*, 1973; Fox and Baum, 1974; Nakajima *et al.*, 1974).

Virus infection may activate the translational block due to interferon treatment (Friedman *et al.*, 1972; Falcoff *et al.*, 1972). This effect may be mediated by dsRNA (Kerr *et al.*, 1974). Interestingly, *Vaccinia* virus infection releases the interferon-induced block on vesicular stomatitis virus (Thacore and Youngner, 1973). Specific phosphorylation of ribosomal proteins may occur during *Vaccinia* infection (Kaerlein and Horak, 1976).

H. Modification and Localization of Mammalian Messenger RNA

Several special features of eukaryotic mRNA may be involved in the control of its translation. The poly(A) segment at the 3'-end is not required for translation and even reinitiation (Humphries *et al.*, 1973; Soreq *et al.*, 1974) but determines the intracellular stability of the mRNA (Huez *et al.*, 1974). A model of translation control based on differences in mRNA stability was proposed by Sussmann (1970).

The 5'-end of mRNA is capped and methylated (Shatkin, 1974; Wei and Moss, 1974), a very unique structure whose function is discussed in Section X, in more detail.

The mRNA is associated with proteins (Spirin, 1972; Lebleu *et al.*, 1971; Blobel, 1972, 1973; Bryan and Hayashi, 1973), some of which are phosphorylated (Sander *et al.*, 1973). The role of the major and minor protein components of the messenger ribonucleoprotein (mRNP) is unknown (see however, Section X). Phosphorylation of ribosomal components has been documented and this reaction is mediated via cyclic nucleotides (Eil and Wool, 1973; Gressner and Wool, 1974; Cawthon *et al.*, 1974; Ventimiglia and Wool, 1974). For a review of the possible role of protein phosphorylation, see Rubin and Rosen (1975).

The presence of membrane-bound polysomes and free polysomes has for a long time suggested a regulation mechanism (Sabatini, 1971; Ganoza and Williams, 1969; Shafritz, 1974; Sunshine *et al.*, 1971). Little is known about whether mRNA goes first to the membrane (Rosbach and Penman, 1971), or a special set of initiation factors exists in the membrane (Shafritz and Isselbacher, 1972; Abraham *et al.*, 1974), or initiation first takes place on free polysomes which then attach to the membrane, due to a "signal-polypeptide" sequence (Blobel and Dobberstein, 1975). A cell-free system that may help solve these questions could be the *in vitro* synthesis of a glycoprotein of vesicular stomatitis virus which takes place only on membrane-bound polysomes in the cell-free system to which VSV mRNA is added (Grubman *et al.*, 1974). Most of the evidence supports the signal hypothesis.

XIII. Inhibitors of Protein Synthesis Initiation in Eukaryotes

A number of inhibitors that specifically interfere with eukaryote polypeptide chain initiation have been described.

Aurintricarboxylic acid (ATA) blocks the activities of the 40 S subunit: there is no mRNA binding, whereas Met-tRNA$_f$ binding is less affected (Marcus *et al.*, 1970; Lebleu *et al.*, 1970). Edeine also inhibits a very early step in initiation (Obrig *et al.*, 1971).

The compound 2-(4-methyl-2,6-dinitroanilino)-*N*-methyl propionamide (MDMP) inhibits the reassociation of the two ribosomal subunits after the binding of Met-tRNA$_f$ and mRNA (Weeks and Baxter, 1972). Pactamycin blocks initiation also by inhibiting the junction of 60 S subunits with the 40 S initiation complex, and its action is reversed by an initiation factor (may be the "joining" factor E5 of Table II) (MacDonald and Goldberg, 1970; Lodish *et al.*, 1971; Kappen *et al.*, 1973; Suzuki and Goldberg, 1974).

Sodium fluoride inhibits initiation in crude systems (cell lysates) or in whole cells. It blocks at a later stage, so that initiation complexes, which may be abnormal, are formed on the mRNA (Hoerz and MacCarthy, 1971; Lebleu *et al.*, 1967; Terada *et al.*, 1972; Geraghty *et al.*, 1973).

Phenomycin was reported to inhibit initiation of globin synthesis in a cell-free system (Yamaki *et al.*, 1974).

In intact cells in culture, dimethyl sulfoxide (DMSO) was shown to inhibit initiation of protein synthesis (Saborio and Koch, 1973), an observation which may be of importance since DMSO induces differentiation of Friend erythroid leukemia cells (Friend *et al.*, 1971). L-1-Tosylamido-2-phenylethylchloromethyl ketone (TPCK), known as a protease inhibitor, blocks initiation but in intact cells only (Summers *et al.*, 1972; Saborio and Koch, 1973; Pong *et al.*, 1975).

Exposure of cells to high salt inhibits initiation. This has been developed as a useful method to compare the efficiency of initiation on viral and cellular mRNA's *in vivo* (Nuss *et al.*, 1975).

XIV. Mixed Prokaryotic-Eukaryotic Systems

Bacteriophage mRNA's have been succesfully translated in mammalian cell-free systems from mouse Krebs ascites cells and from rabbit reticulocytes (Schreier *et al.*, 1973b; Aviv *et al.*, 1972; Morrison and Lodish,

1973). The efficiency of translation may be lower than for homologous mRNA's.

Translation of eukaryotic mRNA's in bacterial systems is much more controversial: eukaryotic cellular mRNA's (e.g., reticulocyte hemoglobin mRNA or lens crystalline mRNA) are *not* translated, while viral mRNA (e.g., RNA tumor virus RNA) may give rise to some polypeptides (Gielkens *et al.*, 1972; Noll *et al.*, 1972).

Initiator tRNA from eukaryotes, once artificially formylated, functions in *E. coli* initiation systems; on the other hand, *E. coli* fMet-tRNA$_f$, which binds to mammalian ribosomes, does not initiate the formation of polypeptides (Berthelot *et al.*, 1973; Elson *et al.*, 1975). Bacterial initiation factors function on reticulocyte 40 S ribosomes and bind even unformylated *E. coli* Met-tRNA$_f$ (on bacterial 30 S, only the formylated tRNA would be bound); eukaryotic initiation factors also give some partial reactions on *E. coli* ribosomes.

XV. Mitochondrial and Chloroplast Protein Synthesis Initiation

Formylated fMet-tRNA$_f$ is present in mammalian mitochondria (Smith and Marcker, 1968) and the initiation machinery seems to be of the bacterial type (Sala and Kuntzel, 1970; Agsteribbe and Kroon, 1973). Mitochondrial ribosomes resemble functionally, but not physically, *E. coli* ribosomes (Kroon *et al.*, 1972). Antibiotics specific for bacterial ribosomes, as chloramphenicol for example, inhibit mitochondrial protein synthesis. The same applies to chloroplasts (Leis and Keller, 1971). The reader is referred to a recent symposium on protein synthesis in these cell organelles for further information on this topic (Kroon and Saccone, 1974).

REFERENCES

Abraham, K. A., Pryme, I. F., and Eikhom, T. S. (1974). *Mol. Biol. Rep.* **1**, 371–378.
Adams, S. L., Safer, B., Anderson, W. F., and Merrick, W. C. (1975). *J. Biol. Chem.* **950**, 9083–9089.
Adamson, S. D., Yau, P. M. P., Herbert, E., and Zucker, W. V. (1972). *J. Mol. Biol.* **63**, 247–264.
Adelman, T. G., and Lovett, J. S. (1974). *Biochim. Biophys. Acta* **335**, 236–242.
Agsteribbe, E., and Kroon, A. M. (1973). *Biochem. Biophys. Res. Commun.* **51**, 8–13.
Anderson, W. F. (1974). *Abstr., EMBO Workshop Initiation Protein Synth. Prokaryotic Eukaryotic Syst., 1974.*
Anderson, W. F., and Gilbert, J. M. (1969). *Cold Spring Harbor Symp. Quant. Biol.* **34**, 585–591.
Arai, K., Arai, N., Kawakita, M., and Kaziro, Y. (1972). *Biochem. Biophys. Res. Commun.* **48**, 190–195.
Arrand, J. R., and Hindley, J. (1973). *Nature (London), New Biol.* **244**, 10–13.
Artman, M., and Werthamer, S. (1974). *Biochem. Biophys. Res. Commun.* **59**, 75–80.

Aviv, H., Boime, I., and Leder, P. (1972). *Science* **178**, 1293–1295.

Ayuso-Parilla, M., Henshaw, E. C., and Hirsch, C. (1973). *J. Biol. Chem.* **248**, 4386–4394.

Bacon, C. W., and Sussman, A. S. (1973). *J. Gen. Microbiol.* **76**, 331–344.

Bade, E. G., Gonzales, N. S., and Algranati, I. D. (1969). *Proc. Natl. Acad. Sci. U.S.A.* **64**, 654–660.

Baenzinger, N. L., Jacobi, C. H., and Thach, R. E. (1974). *J. Biol. Chem.* **249**, 3483–3488.

Balkow, K., Mizuno, S., Fisher, J. M., and Rabinowitz, M. (1973a). *Biochim. Biophys. Acta* **324**, 397–409.

Balkow, K., Mizuno, S., and Rabinowitz, M. (1973b). *Biochem. Biophys. Res. Commun.* **54**, 315–323.

Balkow, K., Hunt, T., and Jackson, R. J. (1975). *Biochem. Biophys. Res. Commun.* **67**, 366–375.

Baltimore, D. (1969). *In* "The Biochemistry of Viruses" (H. B. Levy, ed.), pp. 101–176. Dekker, New York.

Bandhyopadhay, A. K., and Deutscher, M. P. (1971). *J. Mol. Biol.* **60**, 113–120.

Baum, S., Horwitz, M., and Maizel, J. V. (1972). *J. Virol.* **10**, 211–219

Benne, R., and Voorma, H. O. (1972). *FEBS Lett.* **20**, 447–450.

Benne, R., Arentzen, A., and Voorma, H. O. (1972). *Biochim. Biophys. Acta* **269**, *304*–309.

Benne, R., Naaktgeboren, N., Gubbens, J., and Voorma, H. O. (1973a). *Eur. J. Biochem.* **32**, 372–380.

Benne, R., Ebes, F., and Voorma, H. O. (1973b). *Eur. J. Biochem.* **38**, 265–273.

Berissi, H., Groner, Y., and Revel, M. (1971). *Nature (London), New Biol.* **234**, 44–47.

Bernal, S. D., Blumberg, B. M., and Nakamoto, T. (1974). *Proc. Natl. Acad. Sci. U.S.A.* **71**, 774–778.

Bernardi, A., and Spahr, P. F. (1972). *Proc. Natl. Acad. Sci. U.S.A.* **69**, 3033–3038.

Berthelot, F., Bogdanovsky, D., Schapira, G., and Gros, F. (1973). *Mol. Cell Biochem.* **1**, 63–72.

Bester, A. J., Kennedy, D. S., and Heywood, S. M. (1975). *Proc. Natl. Acad. Sci. U.S.A.* **72**, 1523–1527.

Beuzard, Y., and London, I. M. (1974). *Proc. Natl. Acad. Sci. U.S.A.* **71**, 2863–2866.

Beuzard, Y., Rodvien, R., and London, I. M. (1973). *Proc. Natl. Acad. Sci. U.S.A.* **70**, 1022–1026.

Biswas, E., and Gorini, L. (1973). *Proc. Natl. Acad. Sci. U.S.A.* **70**, 46–51.

Blobel, G. (1972). *Biochem. Biophys. Res. Commun.* **47**, 88–93.

Blobel, G. (1973). *Proc. Natl. Acad. Sci. U.S.A.* **70**, 924–928.

Blobel, G., and Dobberstein, B. (1975). *J. Cell. Biol.* **67**, 835–851.

Bloemendal, H., Berns, A., Strous, G., Matthews, M., and Lane, C. D. (1972). "RNA Viruses/Ribosomes," FEBS Symp., pp. 237–250. North-Holland Publ., Amsterdam.

Blumberg, D. D., and Malamy, M. H. (1974). *J. Virol.* **13**, 378–385.

Bogdanovsky, D., Hermann, N., and Schapira, G. (1973). *Biochem. Biophys. Res. Commun.* **54**, 25–29.

Bollen, A., Petre, J., and Grosjean, H. (1972). *FEBS Lett.* **24**, 327–330.

Bollen, A., Heimark, R. L., Cozzone, A., Traut, R. R., Hershey, J. W. B., and Kahan, L. (1975). *J. Biol. Chem.* **250**, 4310–4314.

Both, G. W., Furuichi, Y., Mutukrishnan, S., and Shatkin, A. J. (1975). *Cell* **6**, 185–195.

Boyle, S. M., and Sells, B. H. (1974). *Biochem. Biophys. Res. Commun.* **57**, 23–30.

Brawerman, G., Revel, M., Salser, W., and Gros, F. (1969). *Nature (London)* **223**, 957–958.

Brot, N., Yamasaki, E., Redfield, B., and Weissbach, H. (1972). *Arch. Biochem. Biophys.* **148**, 148–155.

Brown, J. C., and Smith, A. E. (1970). *Nature (London)* **226**, 610–612.

Bryan, R. N., and Hayashi, M. (1973). *Nature (London) New Biol.* **244**, 271–274.

Cahn, F., and Lubin, M. (1975). *Mol. Biol. Rep.* **2**, 49–58.

Callahan, R. C., and Leder, P. (1972). *Arch. Biochem. Biophys.* **153**, 802–813.

Carter, W. A., and de Clercq, E. (1974). *Science* **186,** 1172–1178.

Cashion, L. M., and Stanley, W. M. (1974). *Proc. Natl. Acad. Sci. U.S.A.* **71,** 436–440.

Caskey, C. T., Redfield, B., and Weissbach, H. (1967). *Arch. Biochem. Biophys.* **120,** 119–123.

Cawthon, M., Bitte, L. F., Krystosek, A., and Kabat, D. (1974). *J. Biol. Chem.* **249,** 275–283.

Chae, Y. B., Mazumder, M., and Ochoa, S. (1969). *Proc. Natl. Acad. Sci. U.S.A.* **63,** 828–833.

Chakrabarti, S., and Gorini, L. (1975). *J. Bact.* **121,** 670–674.

Chang, F. N. (1973). *J. Mol. Biol.* **78,** 563–570.

Chatterjee, N. K., Kerwar, S., and Weissbach, H. (1972). *Proc. Natl. Acad. Sci. U.S.A.* **69,** 1375–1380.

Chen, Y. C., Woodley, C. L., Bose, K. K., and Gupta, N. K. (1972). *Biochem. Biophys. Res. Commun.* **48,** 1–9.

Christman, J. K., Reiss, B., Kyner, D., Levin, D. H., Klett, H., and Acs, G. (1973). *Biochim. Biophys. Acta* **294,** 138–164.

Chu, J., and Mazumder, R. (1974). *FEBS Lett.* **40,** 335–338.

Cimadevilla, J. M., and Hardesty, B. (1975). *Biochem. Biophys. Res. Commun.* **63,** 931–937.

Clark, B. F. C., and Marcker, K. A. (1966). *J. Mol. Biol.* **17,** 394–406.

Clemens, M. J., and Tata, J. R. (1972). *Biochim. Biophys. Acta* **269,** 130–138.

Clemens, M. J., Henshaw, E. C., Rahamimoff, H., and London, I. M. (1974). *Proc. Natl. Acad. Sci. U.S.A.* **71,** 2946–2950.

Clemens, M. J., Safer, B., Merrick, W. C., Anderson, W. F., and London, I. M. (1975). *Proc. Natl. Acad. Sci. U.S.A.* **72,** 1286–1290.

Cnaani, D., Revel, M., and Groner, Y. (1976). *FEBS Lett.* **64,** 326–331.

Cohen, P. S. (1972). *Virology* **47,** 780–786.

Colby, C., and Morgan, M. J. (1971). *Annu. Rev. Microbiol.* **25,** 333–350.

Conconi, F., Rowley, P. T., DelSenno, L., Pontremoli, S., and Volpato, S. (1972). *Nature (London), New Biol.* **238,** 83–87.

Content, J., Lebleu, B., Zilberstein, A., Berissi, H., and Revel, M. (1974). *FEBS Lett.* **41,** 125–129.

Content, J., Lebleu, B., Nudel, U., Zilberstein, A., Berissi, H., and Revel, M. (1975). *Eur. J. Biochem.* **54,** 1–10.

Contreras, R., Ysebaert, M., Min Jou, W., and Fiers, W. (1973). *Nature (London), New Biol.* **241,** 99–101.

Cordell-Stewart, B., Stewart, J., and Taylor, M. W. (1973). *J. Virol.* **11,** 232–240.

Cory, S., Spahr, P. F., and Adams, J. (1970). *Cold Spring Harbor Symp. Quant. Biol.* **35,** 1–12.

Crépin, M., Lelong, J. C., and Gros, F. (1973). *Acta Endocrinol. (Copenhagen), Suppl.* **180,** 34–53.

Crystal, R. G., Shafritz, D. A., Prichard, P. M., and Anderson, W. F. (1971). *Proc. Natl. Acad. Sci. U.S.A.* **68,** 1810–1814.

Crystal, R. G., Nienhuis, A. W., Prichard, P. M., Picciano, D., Elson, N. A., Merrick, W. C., Graf, H., Shafritz, D. A., Laycock, D. G., Last, J. A., and Anderson, W. F. (1972a). *FEBS Lett.* **24,** 310–314.

Crystal, R. G., Nienhuis, A. W., Elson, N. A., and Anderson, W. F. (1972b). *J. Biol. Chem.* **247,** 5357–5368.

Culp, W., Morrisey, J., and Hardesty, B. (1970). *Biochem. Biophys. Res. Commun.* **40,** 777–785.

Dahlberg, A. E. (1974). *J. Biol. Chem.* **249,** 7664–7672.

Dahlberg, A. E., and Dahlberg, J. E. (1975). *Proc. Natl. Acad. Sci. U.S.A.* **72**, 2940–2944.

Danchin, A. (1973). *FEBS Lett.* **34**, 327–332.

Darnbrough, C. H., Hunt, T., and Jackson, R. J. (1972). *Biochem. Biophys. Res. Commun.* **48**, 1556–1564.

Darnbrough, C. H., Legon, S., Hunt, T., and Jackson, R. J. (1973). *J. Mol. Biol.* **76**, 379–403.

Dasgupta, R., Shih, D. S., Saris, C., and Kaesberg, P. (1975). *Nature (London)* **256**, 624–628.

Davis, B. D. (1971). *Nature (London)* **231**, 153–155.

Dettman, G. L., and Stanley, W. M., Jr. (1972). *Biochim. Biophys. Acta* **287**, 124–133.

De Wachter, R., Merregaert, J., Vandeberghe, A., Contreras, R., and Fiers, W. (1971). *Eur. J. Biochem.* **22**, 400–414.

Dondon, J., Godefroy-Colburn, T., Graffe, M., and Grunberg-Manago, M. (1974). *FEBS Lett.* **45**, 82–87.

Drews, J., Grasmuk, H., and Weil, R. (1972). *Eur. J. Biochem.* **26**, 416–425.

Dube, S. K., and Rudland, P. S. (1970). *Nature (London)* **226**, 820–823.

Dubnoff, J. S., and Maitra, U. (1972). *J. Biol. Chem.* **247**, 2876–2883.

Dubnoff, J. S., Lockwood, A. H., and Maitra, U. (1972a). *J. Biol. Chem.* **247**, 2884–2894.

Dubnoff, J. S., Lockwood, A. H., and Maitra, U. (1972b). *Arch. Biochem. Biophys.* **149**, 528–540.

Ehrenfeld, E., and Hunt, T. (1971). *Proc. Natl. Acad. Sci. U.S.A.* **68**, 1075–1078.

Eil, C., and Wool, I. G. (1973). *J. Biol. Chem.* **248**, 513–521.

Elson, N. A., Brewer, H. B., and Anderson, W. F. (1974). *J. Biol. Chem.* **249**, 5227–5235.

Elson, N. A., Adam, S. L., Merrick, W. C., Safer, B., and Anderson, W. F. (1975). *J. Biol. Chem.* **250**, 3074–3079.

Englehardt, D. L. (1971). *J. Cell Physiol.* **78**, 333–344.

Englehardt, D. L., Webster, R. E., and Zinder, N. D. (1967). *J. Mol. Biol.* **29**, 45–58.

Ennis, H. L., Kievitt, K. D., and Artman, M. (1974). *Biochem. Biophys. Res. Commun.* **59**, 429–436.

Erdman, V. A., Sprinzl, M., and Pongs, O. (1973). *Biochem. Biophys. Res. Commun.* **54**, 942–948.

Fakunding, J. L., and Hershey, J. W. B. (1973). *J. Biol. Chem.* **248**, 4206–4212.

Fakunding, J. L., Traut, R. R., and Hershey, J. W. B. (1973). *J. Biol. Chem.* **248**, 8555–8559.

Falcoff, E., Falcoff, R., Lebleu, B., and Revel, M. (1972). *Nature (London), New Biol.* **240**, 145–147.

Fan, D., and Penman, S. (1970). *J. Mol. Biol.* **50**, 655–670.

Files, J. G., Weber, K., and Miller, J. H. (1974). *Proc. Natl. Acad. Sci. U.S.A.* **71**, 667–673.

Filipowicz, W., Sierra, J. M., Nombela, C., Ochoa, S., Merrick, W. C., and Anderson, W. F., (1976). *Proc. Natl. Acad. Sci. U.S.A.* **73**, 44–48.

Fiser, I., Scheit, K. H., Stöffler, G., and Kuechler, E. (1974). *Biochem. Biophys. Res. Commun.* **60**, 1112–1118.

Fiser, I., Scheit, K. H., Stöffler, G., and Kuechler, E. (1975). *FEBS Lett.* **56**, 226–229.

Fox, R. I., and Baum, S. G. (1974). *Virology* **60**, 45–53.

France de Fernandez, M. T., Haywood, W. S., and August, J. T. (1972). *J. Biol. Chem.* **247**, 824–834.

Freedman, M. L., Geraghty, M., and Rosman, J. (1974). *J. Biol. Chem.* **249**, 7290–7294.

Freienstein, C., and Blobel, G. (1974). *Proc. Natl. Acad. Sci. U.S.A.* **71**, 3435–3439.

Freudenberg, H., and Mager, J. (1971). *Biochim. Biophys. Acta* **232**, 537–555.

Friedman, R. M., Metz, D. H., Esteban, M., Tovell, D. R., Ball, L. A., and Kerr, I. M. (1972). *J. Virol.* **10**, 1184–1198.

Friend, C., Scher, W., Holland, J. G., and Sato, T. (1971). *Proc. Natl. Acad. Sci. U.S.A.* **68**, 378–383.

Ganoza, M. C., and Williams, C. A. (1969). *Proc. Natl. Acad. Sci. U.S.A.* **63**, 1370–1376.

Garcia-Patrone, M., Gonzales, N. S., and Algranati, I. D. (1971a). *Proc. Natl. Acad. Sci. U.S.A.* **68**, 2822–2828.

Garcia-Patrone, H., Perazzolo, C. A., Baralle, F., Gonzales, N. S., and Algranati, I. D. (1971b). *Biochim. Biophys. Acta* **246**, 291–300.

Gasior, E., and Moldave, K. (1972). *J. Mol. Biol.* **66**, 391–402.

Geraghty, M., Galler, M., Schiffman, F., and Freedman, M. (1973). *Biochem. J.* **133**, 409–417.

Gielkens, A. L. J., Salden, M. H. L., Bloemendal, H., and Konings, R. N. H. (1972). *FEBS Lett.* **28**, 348–352.

Giloh, H. and Mager, J. (1975). *Biochim. Biophys. Acta* **414**, 293–320.

Godefroy-Colburn, T., Wolfe, A., Dondon, J., Grunberg-Manago, M., Dessen, P., and Pantaloni, D. (1975). *J. Mol. Biol.* **94**, 461–478.

Goldberg, M. L., and Steitz, J. A. (1974). *Biochemistry* **13**, 2123–2130.

Goldberg, R. B., and Chargaff, E. (1971). *Proc. Natl. Acad. Sci. U.S.A.* **68**, 1702–1706.

Goldman, E., and Lodish, H. F. (1972). *J. Mol. Biol.* **67**, 37–47.

Goldstein, E. S., and Penman, S. (1973). *J. Mol. Biol.* **80**, 243–254.

Golini, F., Thach, S. S., Birge, C. H., Safer, B., Merrick, W. C., and Thach, R. E. (1976). *Proc. Natl. Acad. Sci. U.S.A.* **73**, 3040–3044.

Gordon, J. (1970). *Biochemistry* **9**, 912–917.

Gorini, L. (1971). *Nature (London), New Biol.* **234**, 261–264.

Greenshpan, H., and Revel, M. (1969). *Nature (London)* **224**, 331–335.

Gressner, A. M., and Wool, I. G. (1974). *J. Biol. Chem.* **249**, 6917–6927.

Groner, Y., and Revel, M. (1971). *Eur. J. Biochem.* **22**, 144–152.

Groner, Y., and Revel, M. (1973). *J. Mol. Biol.* **74**, 407–410.

Groner, Y., Herzberg, M., and Revel, M. (1970). *FEBS Lett.* **6**, 315–320.

Groner, Y., Pollack, Y., Berissi, H., and Revel, M. (1972a). *FEBS Lett.* **21**, 223–228.

Groner, Y., Pollack, Y., Berissi, H., and Revel, M. (1972b). *Nature (London), New Biol.* **239**, 16–19.

Groner, Y., Scheps, R., Kamen, R., Kolakofsky, D., and Revel, M. (1972c). *Nature (London), New Biol.* **239**, 19–20.

Gross, M., and Rabinowitz, M. (1972). *Biochim. Biophys. Acta* **287**, 340–352.

Gross, M., and Rabinowitz, M. (1973a). *Proc. Natl. Acad, Sci. U.S.A.* **69**, 1565–1568.

Gross, M., and Rabinowitz, M. (1973b). *Biochim. Biophys. Acta* **299**, 472–479.

Gross, M., and Rabinowitz, M. (1973c), *Biochem. Biophys. Res. Commun.* **50**, 832–838.

Grubman, M. J., Ehrenfeld, E., and Summers, D. F. (1974). *J. Virol.* **14**, 560–568.

Grummt, F., and Speckbacher, M. (1975). *Eur. J. Biochem.* **57**, 579–585.

Grummt, F., Grummt, I., Gross, H. J., Sprintzl, M., Richter, D., and Erdman, V. A. (1974). *FEBS Lett.* **42**, 15–17.

Grunberg-Manago, M., Clark, B. F. C., Revel, M., Rudland, P. S., and Dondon, J. (1969). *J. Mol. Biol.* **40**, 33–44.

Grunberg-Manago, M., Rabinowitz, J. C., Dondon, J., Lelong, J. C., and Gros, F. (1971). *FEBS Lett.* **19**, 193–200.

Grunberg-Manago, M., Dondon, J., and Graffe, M. (1972). *FEBS Lett.* **22**, 217–221.

Grunberg-Manago, M., Godefroy-Colburn, T., Wolfe, A. D., Dessen, P., Pantaloni, D., Springer, M., Graffe, M., Dondon, J., and Kay, A. (1973). *In* "Regulation of Transcription and Translation in Eukaryotes" (E. Bautz, ed.), 24th Mosbacher Colloq., pp. 213–249. Springer-Verlag, Berlin and New York.

Gualerzi, C., and Pon, C. L. (1973). *Biochem. Biophys. Res. Commun.* **52**, 792–796.

Gualerzi, C., Pon, C. L., and Kaji, A. (1971). *Biochem. Biophys. Res. Commun.* **45**, 1312–1319.

Gupta, N. K., and Aerni, R. J. (1973). *Biochem. Biophys. Res. Commun.* **51**, 907–912.

Gupta, N. K., Chatterjee, N. K., Bose, K. K., Bhaduri, S., and Chung, A. (1970). *J. Mol. Biol.* **54**, 145–154.

Gupta, N. K., Woodley, C. L., Chen, Y. C., and Bose, K. K. (1973). *J. Biol. Chem.* **248**, 4500–4511.

Gupta, N. K., Chatterjee, B., Chen, Y. C., and Majumdar, A. (1974). *Biochem. Biophys. Res. Commun.* **58**, 699–706.

Gupta, N. K., Chatterjee, B., Chen, Y. C., and Majumdar, A. (1975a). *J. Biol. Chem.* **250**, 853–862.

Gupta, N. K., Chatterjee. B., and Majumdar A. (1975b). *Biochem. Biophys. Res. Commun.* **65**, 797–805.

Gupta, S. L., Chen, J., Schaefer, L., Lengyel, P., and Weissman, S. M. (1970). *Biochem. Biophys. Res. Commun.* **39**, 883–888.

Gupta, S. L., Waterson, J., Sopori, M. L., Weissman, S. M., and Lengyel, P. (1971). *Biochemistry* **10**, 4410–4421.

Gupta, S. L., Sopori, M. L., and Lengyel, P. (1974). *Biochem. Biophys. Res. Commun.* **57**, 763–768.

Gurdon, J. (1973). *Acta Endocrinol. (Copenhagen), Suppl.* **180**, 225–243.

Hall, B. G., and Gallant, J. A. (1971). *J. Mol. Biol.* **61**, 271–273.

Hall, N. D., and Arnstein, H. R. V. (1973). *Biochem. Biophys. Res. Commun.* **54**, 1489–1494.

Hamel, E., Koka, M., and Nakamoto, T. (1972). *J. Biol. Chem.* **247**, 805–814.

Haselkorn, R., and Rothman-Denes, L. B. (1973). *Annu. Rev. Biochem.* **42**, 397–425.

Hashimoto, K., Nakajima, K., Oda, K. I., and Shimojo, H. (1973). *J. Mol. Biol.* **81**, 207–223.

Hattman, S., and Hofschneider, P. H. (1967). *J. Mol. Biol.* **29**, 173–190.

Hauptmann, R., Czernilofsky, A. P., Voorma, H. O., Stöffler, G., and Kuechler, E. (1974). *Biochem. Biophys. Res. Commun.* **56**, 331–337.

Hawley, D. A., Slobin, I. L., and Wahba, A. J. (1974). *Biochem. Biophys. Res. Commun.* **61**, 544–550.

Held, W. A., Nomura, M., and Hershey, J. W. B. (1974a). *Mol. Gen. Genet.* **128**, 11–18.

Held, W. A., Gette, W. R., and Nomura, M. (1974b). *Biochemistry* **13**, 2115–2122.

Hellerman, J. A., and Shafritz, D. A. (1975). *Proc. Natl. Acad. Sci. U.S.A.* **72**, 1021–1025.

Helser, T. L., Davies, J. E., and Dahlberg, J. E. (1971). *Nature (London), New Biol.* **233**, 12–14.

Hercules, J., Schweiger, M., and Sauerbier, W. (1974). *Proc. Natl. Acad. Sci. U.S.A.* **71**, 840–846.

Hermoso, J. M., and Szer, W. (1974). *Proc. Natl. Acad. Sci. U.S.A.* **71**, 4708–4712.

Herrlich, P., Rahmsdorf, H. J., Pai, S. H., and Schweiger, M. (1974). *Proc. Natl. Acad. Sci. U.S.A.* **71**, 1088–1092.

Hershey, J. W. B. (1974). *Abstr., EMBO Workshop Initiation Protein Synth. Prokaryotic Eukaryotic Syst., 1974.*

Hershey, J. W. B., and Thach, R. E. (1967). *Proc. Natl. Acad. Sci. U.S.A.* **57**, 759–766.

Hershey, J. W. B., Dewey, K. F., and Thach, R. E. (1969). *Nature (London)* **222**, 944–947.

Hershey, J. W. B., Remold-O'Donnell, E., Kolakofsky, D., Dewey, K. F., and Thach, R. E. (1971). *In* "Methods of Enzymology" (K. Moldave and G. Grossman, eds.) Vol. XX pp. 235–247. Academic Press, New York.

Hershko, A., Mamont, P., Shields, R., and Tomkins, G. (1971). *Nature (London), New Biol.* **232**, 206–211.

Herzberg, M., Revel, M., and Danon, D. (1969). *Eur. J. Biochem.* **11**, 148–153.

Herzog, A., Ghysen, A., and Bollen, A. (1971a). *FEBS Lett.* **15**, 291–296.

Herzog, A., Ghysen, A., and Bollen, A. (1971b). *Mol. Gen. Genet.* **110**, 211–217.

Heywood, S. M. (1970a). *Proc. Natl. Acad. Sci. U.S.A.* **67**, 1782–1788.

Heywood, S. M. (1970b). *Nature (London)* **225**, 696–698.

Heywood, S. M., and Kennedy, D. S. (1974). *Dev. Biol.* **38**, 390–393.

Heywood, S. M., and Thompson, W. C. (1971). *Biochem. Biophys. Res. Commun.* **43**, 470–475.

Heywood, S. M., Kennedy, D. S., and Bester, A. J. (1974). *Proc. Natl. Acad. Sci. U.S.A.* **71**, 2428–2431.

Hickey, E. D., Weber, L. A., and Baglioni, C. (1976). *Proc. Natl. Acad. Sci. U.S.A.* **73**, 19–23.

Hindley, J., and Staples, D. H. (1969). *Nature (London)* **224**, 964–967.

Hirsch, C. A., Cox, M. A., Van Venrooij, W. J. W., and Henshaw, E. C. (1973). *J. Biol. Chem.* **248**, 4377–4387.

Hoerz, W., and MacCarthy, K. S. (1971). *Biochim. Biophys. Acta* **228**, 526–535.

Hori, S., Fisher, J., and Rabinowitz, M. (1967). *Science* **155**, 83–84.

Horne, J. B., and Erdman, V. A. (1973). *Proc. Natl. Acad. Sci. U.S.A* **70**, 2870–2873.

Housman, D., Jacobs-Lorena, M., RajBhandary, U. L., and Lodish, H. F. (1970). *Nature (London)* **227**, 913–916.

Hsu, W. T. (1973). *Biochem. Biophys. Res. Commun.* **52**, 974–979.

Hsu, W. T., and Weiss, S. B. (1969). *Proc. Natl. Acad. Sci. U.S.A.* **64**, 345–351.

Huez, G., Marbaix, G., Hubert, E., Leclercq, M., Nudel, U., Soreq, H., Salomon, R. Lebleu, B., Revel, M., and Littauer, U. Z. (1974). *Proc. Natl. Acad. Sci. U.S.A.* **71**, 3143–3146.

Humphries, S., Doel, M., and Williamson, R. (1973). *Biochem. Biophys. Res. Commun.* **58**, 927–931.

Hunt, T., Vanderhoff, G., and London, I. (1972). *J. Mol. Biol.* **66**, 471–480.

Hunter, A. R., and Jackson, R. J. (1971). *Eur. J. Biochem.* **19**, 316–322.

Ilan, J., and Ilan, J. (1970). *Biochim. Biophys. Acta* **224**, 614–619.

Ilan, J., and Ilan, J. (1971). *Dev. Biol.* **25**, 280–292.

Ilan, J., and Ilan, J. (1973). *Nature (London)* **241**, 176–180.

Imamoto, F., and Kano, Y. (1971). *Nature (London), New Biol.* **232**, 169–173.

Inouye, H., and Littauer, U. Z. (1973). *Annu. Rev. Biochem.* **42**, 439–470.

Inouye, H., Pollack, Y., and Petre, J. (1974). *Eur. J. Biochem.* **45**, 109–117.

Irr, J. D., Kaulenas, M. S., and Unsworth, B. R. (1974). *Cell* **3**, 249–253.

Isono, S., and Isono, K. (1975). *Eur. J. Biochem.* **56**, 15–22.

Iwasaki, K., Sabol, S., Wahba, A. J., and Ochoa, S. (1968). *Arch, Biochem, Biophys.* **125**, 542–547.

Jackson, R., and Hunter, A. R. (1970). *Nature (London)* **277**, 672–674.

Jay, G., and Kaempfer, R. (1974a). *Proc. Natl. Acad. Sci. U.S.A.* **71**, 3199–3203.

Jay, G., and Kaempfer, R. (1974b). *J. Mol. Biol.* **82**, 193–212.

Jay, G., and Kaempfer, R. (1975). *J. Biol. Chem.* **250**, 5742–5750.

Kacian, D. L., Gambino, R., Dow, L. W., Grossbard, E., Natta, C., Ramirez, F., Spiegelman, S., Marks, P. A., and Bank, A. (1973). *Proc. Natl. Acad. Sci. U.S.A.* **70**, 1886–1890.

Kaempfer, R. (1972). *J. Mol. Biol.* **71**, 583–598.

Kaempfer, R. (1974). *Biochem. Biophys. Res. Commun.* **61**, 591–597.

Kaempfer, R., and Kaufman, J. (1972). *Proc. Natl. Acad. Sci. U.S.A.* **69**, 3317–3321.

Kaempfer, R., and Kaufman, J. (1973). *Proc. Natl. Acad. Sci. U.S.A.* **70**, 1222–1226.

Kaerlein, M., and Horak, I. (1976). *Nature* **259**, 150–152.

Kamen, R., Kondo, M., Romer, W., and Weissmann, C. (1972). *Eur. J. Biochem.* **31**, 44–51.

Kan, Y. W., Golini, F., and Thach, R. E. (1970). *Proc. Natl. Acad. Sci. U.S.A.* **67**, 1137–1142.

Kang, S. S. (1970). *Proc. Natl. Acad. Sci. U.S.A.* **65**, 544–550.

Kappen, L. S., Suzuki, H., and Goldberg, I. H. (1973). *Proc. Natl. Acad. Sci. U.S.A.* **70**, 22–26.

Katze, J. R., and Mason, K. H. (1973). *Biochim. Biophys. Acta* **331**, 369–381.

Kay, A. C., and Grunberg-Manago, M. (1972). *Biochim. Biophys. Acta* **277**, 225–233.

Kay, A. C., Sander, G., and Grunberg-Manago, M. (1973). *Biochem. Biophys. Res. Commun.* **51**, 979–986.

Kazazian, H. H., and Freedman, M. L. (1968). *J. Biol. Chem.* **243**, 6446–6450.

Kemper, W. M., Berry, K. W., and Merrick, W. C. (1976). *J. Biol. Chem.* **251**, 5551–5557.

Kennedy, D. S., Bester, A. J., and Heywood, S. M. (1974). *Biochem. Biophys. Res. Commun.* **61**, 365–373.

Kenner, R. A. (1973). *Biochem. Biophys. Res. Commun.* **51**, 932–938.

Kerr, I. M., Brown, R. E., and Ball, L. A. (1974). *Nature (London)* **250**, 57–59.

Kimball, P. C., and Duesberg, P. H. (1971). *J. Virol.* **7**, 697–705.

Klem, E. B., Hsu, W. T., and Weiss, S. B. (1970). *Proc. Natl. Acad. Sci. U.S.A.* **67**, 696–701.

Koka, M., and Nakamoto, T. (1974). *Biochim. Biophys. Acta* **335**, 264–274.

Kolakofsky, D., and Weissmann, C. (1971). *Nature (London), New Biol.* **231**, 42–46.

Kolakofsky, D., Dewey, K. F., Hershey, J. W. B., and Thach, R. E. (1968). *Proc. Natl. Acad. Sci. U.S.A.* **61**, 1066–1070.

Kolakofsky, D., Dewey, K., and Thach, R. E. (1969). *Nature (London)* **223**, 694–697.

Kondo, M., Eggerston, G., Eisenstadt, J., and Lengyel, P. (1968). *Nature (London)* **220**, 368–370.

Kosower, N. S., Vanderhoff, G. A., and Kosower, E. M. (1972). *Biochim. Biophys. Acta* **272**, 623–637.

Kozak, M., and Nathans, D. (1972a). *Bacteriol. Rev.* **36**, 109–134.

Kozak, M., and Nathans, D. (1972b). *J. Mol. Biol.* **70**, 41–55.

Kram, R., Mamont, P., and Tomkins, G. M. (1973). *Proc. Natl. Acad. Sci. U.S.A.* **70**, 1432–1436.

Kramer, G., Cimadevilla, J. M., and Hardesty, B. (1976). *Proc. Natl. Acad. Sci. U.S.A.* **73**, 3078–3082.

Kroon, A. M., and Saccone, C., eds. (1974). "The Biogenesis of Mitochondria: Transcriptional, Translational and Genetic Aspects." Academic Press, New York.

Kroon, A. M., Agsteribbe, E., and de Vries, H. (1972). *In* "The Mechanism of Protein Synthesis and its Regulation" (L. Bosch, ed.), p. 539. North-Holland Publ., Amsterdam.

Kuechler, E. (1971). *Nature (London), New Biol.* **234**, 216–219.

Kuechler, E. (1974). *Abstr., EMBO Workshop Initiation Protein Synth. Prokaryotic Eukaryotic Syst., 1974.*

Kuechler, E., and Rich, A. (1970). *Nature (London)* **225**, 920–924.

Kung, H. F., Morrisey, J., Revel, M., Spears, C., and Weissbach, H. (1975). *J. Biol. Chem.* **250**, 8780–8784.

Kurland, C. G. (1972). *Annu. Rev. Biochem.* **41**, 377–405.

Kwan, S. W., and Brawerman, G. (1972). *Proc. Natl. Acad. Sci. U.S.A.* **69**, 3247–3251.

Kyner, D., Zabos, P., and Levin, D. H. (1973). *Biochim. Biophys. Acta* **324**, 386–396.

Laffler, T., and Gallant, J. A. (1974). *Cell* **3**, 47–49.

Lavalle, R., and De Hauwer, G. (1970). *Cold Spring Harbor Symp. Quant. Biol.* **35**, 491–500.

Lawrence, C., and Thach, R. E. (1974). *J. Virol.* **14**, 598–610.

Lazar, D., and Gros, F. (1973). *Biochimie* **55**, 171–180.

Leader, D. P., and Wool, I. G. (1972). *Biochim. Biophys. Acta* **262**, 360–370.

Leader, D. P., Wool, I. G., and Castles, J. J. (1970). *Proc. Natl. Acad. Sci. U.S.A.* **67**, 523–528.

Leavitt, J. C., Moldave, K., and Nakada, D. (1972). *J. Mol. Biol.* **70**, 15–40.

Lebleu, B., Huez, G., Burny, A., and Marbaix, G. (1967). *Biochim. Biophys. Acta* **138**, 186–191.

Lebleu, B., Marbaix, G., Werenne, J., Burny, A., and Huez, G. (1970). *Biochem. Biophys. Res. Commun.* **40**, 731–739.

Lebleu, B., Marbaix, G., Huez, G., Temmerman, J., Burny, A., and Chantrenne, H. (1971). *Eur. J. Biochem.* **19**, 264–269.

Lebleu, B., Nudel, U., Falcoff, E., Prives, C., and Revel, M. (1972). *FEBS Lett.* **25**, 97–103.

Lebleu, B., Sen, G. C., Shaila, S., Cabrer, B., and Lengyel, P. (1976). *Proc. Natl. Acad. Sci. U.S.A.* **73**, 3107–3111.

Leder, P., and Nau, M. (1967). *Proc. Natl. Acad. Sci. U.S.A.* **58**, 774–779.

Leder, P., Skogerson, L. S., and Callahan, R. (1973). *Arch. Biochem. Biophys.* **153**, 814–822.

Lee, S. Y., Krsmanovic, V., and Brawerman, G. (1971). *Biochemistry* **10**, 895–900.

Lee-Huang, S., and Ochoa, S. (1971). *Nature (London), New Biol.* **234**, 236–239.

Lee-Huang, S., and Ochoa, S. (1972). *Biochem. Biophys. Res. Commun.* **49**, 371–376.

Lee-Huang, S., and Ochoa, S. (1973). *Arch. Biochem. Biophys.* **156**, 84–96.

Lee-Huang, S., and Ochoa, S. (1974). *Proc. Natl. Acad. Sci. U.S.A.* **71**, 2928–2931.

Lee-Huang, S., Lee, H., and Ochoa, S. (1973). *Proc. Natl. Acad. Sci. U.S.A.* **70**, 2874–2878.

Leffler, S., and Szer, W. (1973). *Proc. Natl. Acad. Sci. U.S.A.* **70**, 2364–2368.

Leffler, S., and Szer, W. (1974). *J. Biol. Chem.* **249**, 1458–1470.

Legault, L., Jeantet, C., and Gros, F. (1972a). *FEBS Symp.* **23**, 251–254.

Legault, L., Jeantet, C., and Gros, F. (1972b). *FEBS Lett.* **27**, 71–75.

Legault-Demare, L., Jeantet, C., and Gros, F. (1973). *Mol. Gen. Genet.* **125**, 301–318.

Legon, S., Jackson, R. J., and Hunt, T. (1973). *Nature (London), New Biol.* **241**, 150–152.

Legon, S., Brayley, A., Hunt, T., and Jackson, R. J. (1974). *Biochem. Biophys. Res. Commun.* **56**, 745–752.

Leibowitz, R., and Penman, S. (1971). *J. Virol.* **8**, 661–668.

Leis, J. P., and Keller, E. B. (1970). *Biochem. Biophys. Res. Commun.* **40**, 416–421.

Leis, J. P., and Keller, E. B. (1971). *Biochemistry* **10**, 889–895.

Lelong, J. C., Grunberg-Manago, M., Dondon, J., Gros, D., and Gros, F. (1970). *Nature (London)* **226**, 505–509.

Lelong, J. C., Cousin, M. A., Gros, D., Grunberg-Manago, M., and Gros, F. (1971). *Biochem. Biophys. Res. Commun.* **42**, 530–537.

Lelong, J. C., Cousin, M. A., Gros, F., Miskin, R., Vogel, Z., Groner, Y., and Revel, M. (1972). *Eur. J. Biochem.* **27**, 174–180.

Lelong, J. C., Gros D., Gros, F., Bollen, A., Maschler, R., and Stöffler, G. (1974). *Proc. Natl. Acad. Sci. U.S.A.* **71**, 248–252.

Lengyel, P. (1974). *In* "Ribosome," (M. Nomura, A. Tissieres, and P. Lengyel, eds.), Cold Spring Harbor Lab., Cold Spring Harbor, New York.

Levin, D. H., Kyner, D., and Acs, G. (1973a). *J. Biol. Chem.* **248**, 6416–6424.

Levin, D. H., Kyner, D., and Acs, G. (1973b). *Proc. Natl. Acad. Sci. U.S.A.* **70**, 41–45.

Levin, D. H., Ranu, R. S., Ernst, V., Fifer, M. A., and London, I. M. (1975). *Proc. Natl. Acad. Sci. U.S.A.* **72**, 4849–4853.

Levin, D. H., Ranu, R. S., Ernst, V., Fifer, M. A., and London, I. M. (1975). *Proc. Natl. U.S.A.* **73**, 3112–3116.

Lockwood, A. H., and Maitra, U. (1974). *J. Biol. Chem.* **249**, 346–354.

Lockwood, A. H., Chakraborty, P. R., and Maitra, U. (1971). *Proc. Natl. Acad. Sci. U.S.A.* **68**, 3122–3129.

Lockwood, A. H., Sarkar, P., and Maitra, U. (1972). *Proc. Natl. Acad. Sci. U.S.A.* **69**, 3602–3605.

Lockwood, A. H., Maitra, U., Brot, N., and Weissbach, H. (1974). *J. Biol. Chem.* **249,** 1213–1218.

Lodish, H. F. (1970a). *J. Mol. Biol.* **50,** 689–702.

Lodish, H. F. (1970b). *Nature (London)* **226,** 705–707.

Lodish, H. F. (1971). *J. Biol. Chem.* **246,** 7131–7138.

Lodish, H. F. (1974). *Nature (London)* **251,** 385–389.

Lodish, H. F., and Desalu, O. (1973). *J. Biol. Chem.* **248,** 3520–3527.

Lodish, H. F., and Jacobsen, M. (1972). *J. Biol. Chem.* **247,** 3622–3629.

Lodish, H. F., and Nathan, D. G. (1972). *J. Biol. Chem.* **247,** 7822–7830.

Lodish, H. F., and Robertson, H. D. (1969). *Cold Spring Harbor Symp. Quant. Biol.* **34,** 655–673.

Lodish, H. F., Housman, D., and Jacobsen, M. (1971). *Biochemistry* **10,** 2348–2356.

Lubsen, N. H., and Davis, B. D. (1974a). *Biochim. Biophys. Acta* **335,** 196–203.

Lubsen, N. H., and Davis, B. D. (1974b). *Proc. Natl. Acad. Sci. U.S.A.* **71,** 68–72.

Lucas-Lenard, J., and Lipmann, F. (1967). *Proc. Natl. Acad. Sci. U.S.A.* **57,** 1050–1057.

MacCormick, W., and Penman, S. (1969). *J. Mol. Biol.* **39,** 315–333.

MacDonald, J. S., and Goldberg, I. H. (1970). *Biochem. Biophys. Res. Commun.* **41,** 1–8.

MacKeehan, W. L. (1974). *J. Biol. Chem.* **249,** 6517–6526.

Mager, J., and Giloh, H. (1973). *Abstr., Int. Congr. Biochem., 9th, 1973,* p. 149.

Maizels, N. (1973). *Proc. Natl. Acad. Sci. U.S.A.* **70,** 3585–3589.

Mamont, P., Hershko, A., Kram, R., Schachter, L., Lust, J., and Tomkins, G. M. (1972). *Biochem. Biophys. Res. Commun.* **48,** 1378–1384.

Marcus, A. (1970). *J. Biol. Chem.* **245.** 962–966.

Marcus, A. (1974). *Abstr., EMBO Workshop Initiation Protein Synth. Prokaryotic Eukaryotic Syst., 1974.*

Marcus, A., Bewley, J. D., and Weeks, D. P. (1970). *Science* **167,** 1735–1736.

Maxwell, C. R., and Rabinowitz, M. (1969). *Biochem. Biophys. Res. Commun.* **35,** 79–85.

Mazumder, R. (1971). *FEBS Lett.* **18,** 64–69.

Mazumder, R. (1972). *Proc. Natl. Acad. Sci. U.S.A.* **69,** 2770–2773.

Mazumder, R. (1973). *Proc. Natl. Acad. Sci. U.S.A.* **70,** 1939–1942.

Mazumder, R., Chae, Y. B., and Ochoa, S. (1969). *Proc. Natl. Acad. Sci. U.S.A.* **63,** 98–103.

Meier, D., Lee-Huang, S., and Ochoa, S. (1973). *J. Biol. Chem.* **247,** 2884–2894.

Merrick, W. C., and Anderson, W. F. (1975). *J. Biol. Chem.* **250,** 1107–1206.

Merrick, W. C., Lubsen, N. H., and Anderson, W. F. (1973). *Proc. Natl. Acad. Sci. U.S.A.* **70,** 2220–2223.

Merrick, W. C., Safer, B., Adams, S., and Kemper, W. (1974). *Fed. Proc., Fed. Am. Soc. Exp. Biol.* **33,** 1262.

Metafora, S., Terada, M., Dow, L. W., Marks, P. A., and Bank, A. (1972). *Proc. Natl. Acad. Sci. U.S.A.* **69,** 1299–1303.

Miall, S. H., and Tamaoki, T. (1972). *Biochemistry* **11,** 4826–4832.

Miller, J. H. (1974). *Cell* **1,** 73–77.

Miller, M. J., and Wahba, A. J. (1973). *J. Biol. Chem.* **248,** 1084–1090.

Miller, M. J., and Wahba, A. J. (1974). *J. Biol. Chem.* **249,** 3808–3813.

Miller, M. J., Zasloff, M., and Ochoa, S. (1969). *FEBS Lett.* **3,** 50–53.

Miller, M. J., Niveleau, A., and Wahba, A. J. (1974). *J. Biol. Chem.* **249,** 3803–3807.

Min Jou, W., Haegeman, G., Ysebaert, M., and Fiers, W. (1972). *Nature (London)* **237,** 82–88.

Mizuno, S., Fisher, J. M., and Rabinowitz, M. (1972). *Biochim. Biophys. Acta* **272,** 638–650.

Modolell, J., and Davis, B. D. (1970). *Proc. Natl. Acad. Sci. U.S.A.* **67,** 1148–1155.

Modolell, J., Cabrer, B., and Vazquez, D. (1973). *Proc. Natl. Acad. Sci. U.S.A.* **70,** 3561–3565.

Moore, P. (1966). *J. Mol. Biol.* **22**, 145–151.

Morrisey, J., and Hardesty, B. (1972). *Arch. Biochem. Biophys.* **152**, 385–397.

Morrison, C. A., Garrett, R. A., Zeichardt, R., and Stöffler, G. (1973). *Mol. Gen. Genet.* **127**, 359–368.

Morrison, T. C., and Lodish, H. F. (1973). *Proc. Natl. Acad. Sci. U.S.A.* **70**, 315–320.

Morrison, T. C., and Malamy, M. (1971). *Nature (London), New Biol.* **231**, 37–41.

Moss, B. (1968). *J. Virol.* **2**, 1028–1037.

Nakajima, K., Ishitsuka, H., and Oda, K. I. (1974). *Nature* **252**, 649–653.

Nakaya, K., and Wool, I. G. (1973). *Biochem. Biophys. Res. Commun.* **54**, 239–245.

Nakaya, K., Randu, R. J., and Wool, I. G. (1973). *Biochem. Biophys. Res. Commun.* **54**, 246–251.

Nathan, D. G., Lodish, H. F., Kan, Y. W., and Housman, D. (1971). *Proc. Natl. Acad. Sci. U.S.A.* **68**, 2514–2518.

Nienhuis, A. W., Laycock, D. G., and Anderson, W. F. (1971). *Nature (London), New Biol.* **231**, 205–208.

Nienhuis, A. W., Canfield, P. H., and Anderson, W. F. (1973). *J. Clin. Invest.* **52**, 1735–1745.

Noll, H., Noll, M., Hapke, B., and Van Dieijen, G. (1973). *In* "Regulation of Transcription and Translation in Eukaryotes" (E. Bautz, ed.), 24th Mosbacher Colloq. pp. 257–311. Springer-Verlag, Berlin and New York.

Noll, M., and Noll, H. (1972). *Nature (London), New Biol.* **238**, 225–228.

Noll, M., Noll, H., and Lingrel, J. B. (1972). *Proc. Natl. Acad. Sci. U.S.A.* **69**, 1843–1847.

Noll, M., Hapke, B., and Noll, H. (1973). *J. Mol. Biol.* **80**, 519–530.

Nombela, C., Nombela, N. A., and Ochoa, S. (1974). *In* "Lipmann's Symposium" (W. de Gruyter, ed.), pp. 435–442. Richter, Berlin.

Nudel, U., Lebleu, B., and Revel, M. (1973). *Proc. Natl. Acad. Sci. U.S.A.* **70**, 2139–2144.

Nuss, D. L., Oppermann, H., and Koch, G. (1975). *Proc. Natl. Acad. Sci. U.S.A.* **72**, 1258–1263.

Obrig, T., Irvin, J., Culp, W., and Hardesty, B. (1971). *Eur. J. Biochem.* **21**, 31–41.

Ohta, T., Sarkar, S., and Thach, R. E. (1967). *Proc. Natl. Acad. Sci. U.S.A.* **58**, 1638–1644.

Okuyama, A., and Tanaka, N. (1972). *Biochem. Biophys. Res. Commun.* **49**, 951–957.

Okuyama, A., and Tanaka, N. (1973). *Biochem. Biophys. Res. Commun.* **52**, 1463–1469.

Ono, Y., Skoultchi, A., Klein, A., and Lengyel, P. (1968). *Nature (London)* **220**, 1304–1307.

Ozaki, M., Mizushima, S., and Nomura, M. (1969). *Nature (London)* **222**, 1333–1337.

Pain, V. M., and Clemens, M. J. (1973). *FEBS Lett.* **32**, 205–212.

Palmiter, R. (1973). *J. Biol. Chem.* **248**, 2095–2106.

Paradies, H. H., Franz, A., Pon, C. L., and Gualerzi, C. (1974). *Biochem. Biophys. Res. Commun.* **59**, 600–607.

Parenti-Rosina, R., Eisenstadt, A., and Eisenstadt, J. M. (1969). *Nature (London)* **221**, 363–365.

Patterson, D., and Gillespie, D. (1972). *J. Bacteriol.* **112**, 1177–1183.

Pemberton, R. E., Housman, D., Lodish, H. F., and Baglioni, C. (1972). *Nature (London), New Biol.* **235**, 99–102.

Petrissant, G. (1973). *Proc. Natl. Acad. Sci. U.S.A.* **70**, 1046–1049.

Picciano, D. G., Prichard, P. M., Merrick, W. C., Shafritz, D. A., Graf, H., Crystal, R. G., and Anderson, W. F. (1973). *J. Biol. Chem.* **248**, 204–214.

Piper, P. W., and Clark, B. F. C. (1974). *Nature (London)* **247**, 516–518.

Pitha, P. M., and Pitha, J. (1973). *J. Gen. Virol.* **21**, 31–37.

Pollack, Y. (1974). Ph.D. Thesis, Weizmann Institute of Science, Rehovoth, Israel.

Pollack, Y., Groner, Y., Aviv, H., and Revel, M. (1970). *FEBS Lett.* **9**, 218–221.

Pon, C. L., and Gualerzi, C. (1974). *Proc. Natl. Acad. Sci. U.S.A.* **71**, 4950–4954.

Pon, C. L., Friedman, S. M., and Gualerzi, C. (1972). *Mol. Gen. Genet.* **116**, 192–198.

Pong, S. S., Nuss, D. L., and Koch, G. B. (1975). *J. Biol. Chem.* **250**, 240–245.

Pongs, O. (1973). *FEBS Lett.* **37**, 47–55.

Pongs, O., and Lanka, E. (1975). *Proc. Natl. Acad. Sci. U.S.A.* **72**, 1505–1509.

Pongs, O., Nierhaus, K. H., Erdmann, V. A., and Wittmann, H. G. (1974). *FEBS Lett. Suppl.* **40**, S28–S37.

Pongs, O., Stöffler, G., and Lanka, E. (1975). *J. Mol. Biol.* **99**, 301–315.

Ponta, H., Pon, C. L., Herrlich, P., Gualerzi, C., Hirsch-Kaufmann, M., Pfenning-Yeh, M. L., Rahmsdorf, H. J., and Schweiger, M. (1975). *Eur. J. Biochem.* **59**, 261–270.

Prichard, P. M., Picciano, D. J., Laycock, D. G., and Anderson, W. F. (1971). *Proc. Natl. Acad. Sci. U.S.A.* **68**, 2752–2756.

Rabinowitz, M., Freedman, M., Fisher, J., and Maxwell, C. (1969). *Cold Spring Harbor Symp. Quant. Biol.* **34**, 567–575.

Raffel, C., Stein, S., and Kaempfer, R. (1974). *Proc. Natl. Acad. Sci. U.S.A.* **71**, 4020–4024.

Rahmsdorf, H. J., Herrlich, P., Tao, M., and Schweiger, M. (1973a). *Adv. Biosci.* **11**, 219–232.

Rahmsdorf H. J., Herrlich, P., Pai, S. H., Schweiger, M., and Wittmann, H. G. (1973b). *Mol. Gen. Genet.* **127**, 259–264.

RajBhandary, U. L., and Kumar, A. (1970). *J. Mol. Biol.* **50**, 707–711.

Reichman, M., and Penman, S. (1973). *Proc. Natl. Acad. Sci. U.S.A.* **70**, 2678–2682.

Revel, M. (1972). *In* "The Mechanism of Protein Synthesis and its Regulation" (L. Bosch, ed.), pp. 87–131, North-Holland Publ., Amsterdam.

Revel, M., and Groner, Y. (1974). *In* "Methods of Enzymology" (L. Grossman and K. Moldave, eds.), Vol. 30, pp. 68–78, Academic Press, New York.

Revel, M., Brawerman, G., Lelong, J. C., and Gros, F. (1968a). *Nature (London)* **219**, 1016–1021.

Revel, M., Herzberg, M., Becarevic, A., and Gros, F. (1968b). *J. Mol. Biol.* **33**, 231–249.

Revel, M., Herzberg, M., and Greenshpan, H. (1969). *Cold Spring Harbor Symp. Quant. Biol.* **34**, 261–275.

Revel, M., Greenshpan, H., and Herzberg, M. (1970). *Eur. J. Biochem.* **16**, 117–122.

Revel, M., Pollack, Y., Groner, Y., Scheps, R., Inouye, H., Berissi, H., and Zeller, H. (1973a). *Biochimie* **55**, 41–51.

Revel, M., Groner, Y., Pollack, Y., Cnaani, D., Zeller, H., and Nudel, U. (1973b). *Acta Endocrinol. (Copenhagen), Suppl.* **180**, 54–74.

Risuelo, G., Gualerzi, C., and Pon, C. (1976). *Eur. J. Biochem.* **67**, 603–613.

Roberts, B. E., and Patterson, B. M. (1973). *Proc. Natl. Acad. Sci. U.S.A.* **70**, 2330–2334.

Roberts, W. K., Hovanessian, A., Brown, R. E., Clemens, M. J., and Kerr, I. M. (1976). *Nature* **264**, 477–480.

Robertson, H. D., and Lodish, H. F. (1970). *Proc. Natl. Acad. Sci. U.S.A.* **67**, 710–716.

Robertson, H. D., and Matthews, M. (1973). *Proc. Natl. Acad. Sci. U.S.A.* **70**, 225–229.

Rosbach, M., and Penman, S. (1971). *J. Mol. Biol.* **59**, 227–241.

Rottman, F., Shatkin, A. J., and Perry, R. P. (1974). *Cell* **3**, 197–199.

Rourke, A. W., and Heywood, S. M. (1972). *Biochemistry,* **11**, 2061–2066.

Rowley, P. T. (1972). *Nature (London), New Biol.* **239**, 234–236.

Rowley, P. T., Midthum, R. A., and Adams, M. H. (1971). *Arch. Biochem. Biophys.* **145**, 6–15.

Rubin, C. S., and Rosen, O. M. (1975). *Annu. Rev. Biochem.* **44**, 831–887.

Rudland, P. S. (1974). *Proc. Natl. Acad. Sci. U.S.A.* **71**, 750–754.

Rudland, P. S., and Clark, B. F. C. (1972). *In* "The Mechanism of Protein Synthesis and its Regulation" (L. Bosch, ed.), pp. 55–86. North-Holland Publ., Amsterdam.

Rudland, P. S., Whybrow, W. A., Marcker, K. A., and Clark, B. F. C. (1969). *Nature (London)* **222**, 750–753.

Rudland, P. S., Whybrow, W. A., and Clark, B. F. C. (1971). *Nature (London), New Biol.* **231,** 76–78.

Sabatini, D. (1971). *Adv. Cytopharmacol.* **1,** 119–129.

Sabol, S., and Ochoa, S. (1971). *Nature (London), New Biol.* **234,** 233–236.

Sabol, S., Sillero, M. A., Iwasaki, K., and Ochoa, S. (1970). *Nature (London)* **228,** 1269–1275.

Sabol, S., Meier, D., and Ochoa, S. (1973). *Eur. J. Biochem.* **33,** 332–340.

Saborio, J. L., and Koch, G. (1973). *J. Biol. Chem.* **248,** 8343–8347.

Saborio, J. L., Pong, S. S., and Koch, G. (1974). *J. Mol. Biol.* **85,** 195–212.

Safer, B., Adams, S. L., Anderson, W. F., and Merrick, W. C. (1975a). *J. Biol. Chem.* **250,** 9076–9082.

Safer, B., Anderson, W. F. and Merrick, W. C. (1975b). *J. Biol. Chem.* **250,** 9067–9075.

Sala, F., and Ciferri, O. (1970). *Biochim. Biophys. Acta* **224,** 199–205.

Sala, F., and Kuntzel, H. (1970). *Eur. J. Biochem.* **15,** 280–286.

Samuel, C. E., and Rabinowitz, J. C. (1974). *J. Biol. Chem.* **249,** 1198–1206.

Sander, E. S., Stewart, A. G., Morel, C. M., and Scherrer, K. (1973). *Eur. J. Biochem.* **38,** 443–452.

Sander, G., Marsh, R. C., and Parmeggiani, A. (1972). *Biochem. Biophys. Res. Commun.* **47,** 866–873.

Saxton, R., and Stevens, J. (1972). *Virology,* **48,** 207–220.

Schachner, M., Seifert, W., and Zillig, W. (1971). *Eur. J. Biochem.* **22,** 251–259.

Schedl, P. D., Singer, R. E., and Conway, T. W. (1970). *Biochem. Biophys. Res. Commun.* **38,** 631–637.

Schenkmann, M. L., Ward, D. C., and Moore, P. B. (1974). *Biochim. Biophys. Acta* **353,** 503–508.

Scheps, R. (1973). Ph.D. Thesis, Weizmann Institute of Science, Rehovoth, Israel.

Scheps, R., and Revel, M. (1972). *Eur. J. Biochem.* **29,** 319–325.

Scheps, R., Zeller, H., and Revel, M. (1972). *FEBS Lett.* **27,** 1–4.

Schiff, N., Miller, M. J., and Wahba, A. J. (1974). *J. Biol. Chem.* **249,** 3797–3802.

Schreier, M. H., and Staehelin, T. (1973a). *Nature (London), New Biol.* **242,** 35–39.

Schreier, M. H., and Staehelin, T. (1973b). *In* "Regulation of Transcription and Translation in Eukaryotes" (E. Bautz, ed.), 24th Mosbacher Colloq., pp. 335–349. Springer-Verlag, Berlin and New York.

Schreier, M. H., and Staehelin, T. (1973c). *Proc. Natl. Acad. Sci. U.S.A.* **70,** 462–465.

Schreier, M. H., Staehelin, T., Stewart, A., Gander, E., and Scherrer, K. (1973a). *Eur. J. Biochem.* **34,** 213–218.

Schreier, M., Staehelin, T., Getseland, R. F., and Spahr, P. F. (1973b). *J. Mol. Biol.* **75,** 575–578.

Schreiner, G., and Nierhaus, K. H. (1973). *J. Mol. Biol.* **81,** 71–82.

Schweiger, M., Herrlich, P., Scherzinger, E., and Rahmsdorf, H. J. (1972). *Proc. Natl. Acad. Sci. U.S.A.* **69,** 2203–2207.

Seal, S. N., and Marcus, A. (1973). *J. Biol. Chem.* **248,** 6577–6582.

Seal, S. N., Bawley, J. D., and Marcus, A. (1972). *J. Biol. Chem.* **247,** 2592–2597.

Shafritz, D. A. (1974). *J. Biol. Chem.* **249,** 81–93.

Shafritz, D. A., and Anderson, W. F. (1970a). *Nature (London)* **227,** 918–920.

Shafritz, D. A., and Anderson, W. F. (1970b). *J. Biol. Chem.* **245,** 5553–5559.

Shafritz, D. A., and Isselbacher, K. J. (1972). *Biochem. Biophys. Res. Commun.* **46,** 1721–1727.

Shafritz, D. A., Prichard, P. M., Gilbert, J. M., Merrick, W. C., and Anderson, W. F. (1972). *Proc. Natl. Acad. Sci. U.S.A.* **69,** 983–987.

Shafritz, D. A., Weinstein, J. A., Safer, B., Merrick, W. C., Weber, L. A., Hickey, E. D., and Baglioni, C. (1976). *Nature (London)* **261,** 291–294.

Shatkin, A. J. (1974). *Proc. Natl. Acad. Sci. U.S.A.* **71,** 3204–3207.

Sherman, M. I., and Simpson, M. V. (1969). *Cold Spring Harbor Symp. Quant. Biol.* **34**, 220–229.

Shine, J., and Dalgarno, L. (1974). *Proc. Natl. Acad. Sci. U.S.A.* **71**, 1341–1346.

Shine, J., and Dalgarno, L. (1975). *Nature (London)* **254**, 34–36.

Sierra, J. M., Meier, D., and Ochoa, S. (1974). *Proc. Natl. Acad. Sci. U.S.A.* **71**, 2693–2697.

Simsek, M., and RajBhandary, U. L. (1972). *Biochem. Biophys. Res. Commun.* **49**, 508–515.

Simsek, M., Ziegenmayer, J., Heckman, J., and RajBhandary, U. L. (1973). *Proc. Natl. Acad. Sci. U.S.A.* **70**, 1041–1045.

Simsek, M., Boisnard, M., Petrissant, G., and RajBhandary, U. L. (1974). *Nature (London)* **247**, 518–520.

Singer, R., and Penman, S. (1972). *Nature (London)* **240**, 100–102.

Singer, R. E., and Conway, T. W. (1973a). *Biochim. Biophys. Acta* **331**, 102–116.

Singer, R. E., and Conway, T. W. (1973b). *Arch. Biochem. Biophys.* **158**, 257–265.

Slobin, L. I. (1972). *Proc. Natl. Acad. Sci. U.S.A.* **69**, 3769–3774.

Smith, A. E. (1973). *Eur. J. Biochem.* **33**, 301–313.

Smith, A. E., and Marcker, K. A. (1968). *J. Mol. Biol.* **38**, 241–249.

Smith, A. E., and Marcker, K. A. (1970). *Nature (London)* **226**, 607–610.

Smith, A. E., and Wigle, D. T. (1973). *Eur. J. Biochem.* **35**, 566–573.

Smith, D. W. E., and MacNamara, A. L. (1974). *J. Biol. Chem.* **249**, 1330–1334.

Sokawa, Y., Sokawa, J., and Kaziro, Y. (1971). *Nature (London), New Biol.* **234**, 7–10.

Soreq, H., Nudel, U., Salomon, R., Revel, M., and Littauer, U. Z. (1974). *J. Mol. Biol.* **88**, 233–245.

Spirin, A. (1972). *In* "The Mechanism of Protein Synthesis and its Regulation" (L. Bosch, ed.), pp. 515–537. North-Holland Publ., Amsterdam.

Springer, M., and Grunberg-Manago, M. (1972). *Biochem. Biophys. Res. Commun.* **47**, 477–484.

Staehelin, T., Trachsel, H., Erni, B., Boschetti, A., and Schreier, M. H. (1975). *In* "Proceedings 10th FEBS Meeting," pp. 309–323. North-Holland Publ., Amsterdam.

Stallcup, M. R., and Rabinowitz, J. C. (1973). *J. Biol. Chem.* **248**, 3216–3226.

Staples, D. H., and Hindley, J. (1971). *Nature (London), New Biol.* **234**, 211–212.

Staples, D. H., Hindley, J., Billeter, M. A., and Weissmann, C. (1971). *Nature (London), New Biol.* **234**, 202–204.

Steitz, J. A. (1969). *Nature (London)* **224**, 957–964.

Steitz, J. A. (1973a). *J. Mol. Biol.* **73**, 1–16.

Steitz, J. A. (1973b). *Proc. Natl. Acad. Sci. U.S.A.* **70**, 2605–2609.

Steitz, J. A. (1974). *In* "RNA Bacteriophages" (N. Zinder, ed.). Cold Spring Harbor Lab., Cold Spring Harbor, New York.

Steitz, J. A., and Jakes, K. (1975). *Proc. Natl. Acad. Sci. U.S.A.* **72**, 4734–4738.

Steitz, J. A., Dube, S. K., and Rudland, P. S. (1970). *Nature (London)* **226**, 824–827.

Stern, R., and Friedman, R. M. (1970). *Nature (London)* **226**, 612–616.

Stern, R., and Friedman, R. M. (1971). *Biochemistry* **10**, 3635–3642.

Stevens, R. H., and Williamson, A. R. (1973). *J. Mol. Biol.* **78**, 505–525.

Strous, G. J., Berns, T. J., Van Westreenin, J., and Bloemendal, H. (1972). *Eur. J. Biochem.* **30**, 48–52.

Strycharz, W. A., Ranki, M., and Dahl, H. H. M. (1974). *Eur. J. Biochem.* **48**, 303–310.

Studier, F. W. (1972). *Science* **176**, 367–376.

Subramanian, A. R., and Davis, B. D. (1971). *Nature (London)* **228**, 1273–1275.

Sugiyama, T., and Nakada, D. (1970). *J. Mol. Biol.* **48**, 349–355.

Summers, D. F., and Maizel, J. V. (1967). *Virology* **31**, 550–552.

Summers, D. F., Shaw, E. N., Stewart, M. L., and Maizel, J. V. (1972). *J. Virol.* **10**, 880–888.

Sundkvist, I. C., and Staehelin, T. (1975). *J. Mol. Biol.* **99**, 401–408.

Sunshine, G. H., Williams, D. J., and Rabin, B. R. (1971). *Nature (London), New Biol.* **230,** 133–136.

Sussmann, M. (1970). *Nature (London)* **225,** 1245–1248.

Suttle, D. P., and Ravel, J. M. (1974). *Biochem. Biophys. Res. Commun.* **57,** 386–393.

Suttle, D. P., Haralson, M. A., and Ravel, J. M. (1973). *Biochem. Biophys. Res. Commun.* **51,** 376–382.

Suzuki, H., and Goldberg, I. H. (1974). *Proc. Natl. Acad. Sci. U.S.A.* **71,** 4259–4263.

Szer, W., and Brenowitz, J. (1970). *Biochem. Biophys. Res. Commun.* **38,** 1154–1160.

Szer, W., and Leffler, S. (1974). *Proc. Natl. Acad. Sci. U.S.A.* **71,** 3611–3615.

Szer, W., Hermoso, J. M., and Leffler, S. (1975). *Proc. Natl. Acad. Sci. U.S.A.* **72,** 2325–2329.

Takanami, M., Yan, Y., and Jukes, T. H. (1965). *J. Mol. Biol.* **12,** 761–773.

Tal, M., Aviram, M., Kanarek, A., and Weiss, A. (1972). *Biochim. Biophys. Acta* **281,** 381–392.

Terada, M., Metafora, S., Dow, L. W., Bank, A., and Marks, P. A. (1972). *Biochem. Biophys. Res. Commun.* **47,** 766–770.

Thach, S., and Thach, R. E. (1971a). *Nature (London), New Biol.* **229,** 219–221.

Thach, S., and Thach, R. E. (1971b). *Proc. Natl. Acad. Sci. U.S.A.* **68,** 1791–1795.

Thacore, H. R., and Youngner, J. S. (1973). *Virology* **56,** 505–522.

Thibault, J., Chestier, A., Vidal, D., and Gros, F. (1972). *Biochimie* **54,** 829–835.

Tischendorf, G. W., Zeichardt, H., and Stöffler, G. (1975). *Proc. Natl. Acad. Sci. U.S.A.* **72,** 4820–4824.

Traut, R. R., Heimark, R. L., Sun, T. T., Bollen, A., and Hershey, J. W. B. (1974). *In* "Ribosomes" (M. Nomura, A. Tissiéres, and P. Lengyel, eds.), pp. 271–308. Cold Spring Harbor Lab., Cold Spring Harbor, New York.

Van Dieijer, G., van der Laken, C. J., Van Knippenberg, P. H., and Van Duin, J. (1975). *J. Mol. Biol.* **93,** 351–356.

Van Duin, J., and Kurland, C. G. (1970). *Mol. Gen. Genet.* **109,** 169–176.

Van Duin, J., and Van Knippenberg, P. H. (1974). *J. Mol. Biol.* **84,** 185–195.

Van Duin, J., Van Knippenberg, P. H., Dieben, M., and Kurland, C. G. (1972). *Mol. Gen. Genet.* **116,** 181–186.

Van Duin, J., Kurland, C. G., Dondon, J., and Grunberg-Manago, M. (1975). *FEBS Lett.* **59,** 287–290.

Van Duin, J., Kurland, C. G., Dondon, J., Grunberg-Manago, M., Branlant, C., and Ebel, J. P. (1976). *FEBS Lett.* **62,** 111–114.

Van Knippenberg, P. H., Van Duin, J., and Lentz, H. (1973). *FEBS Lett.* **34,** 95–98.

Van Knippenberg, P. H., Hoykaas, P. J. J., and Van Duin, J. (1974a). *FEBS Lett.* **41,** 323–326.

Van Knippenberg, P. H., Poldermans, B., and Wallaart, R. A. M. (1974b). *Abstr., EMBO Workshop Initiation Protein Synth. Prokaryotic Eukaryotic Syst., 1974.*

Van Venrooij, W. J. W., Henshaw, E. C., and Hirsch, C. A. (1972). *Biochim. Biophys. Acta* **259,** 127–137.

Vaughan, M. H., and Hansen, B. B. (1973). *J. Biol. Chem.* **248,** 7087–7096.

Vaughan, M. H., Pawlowski, P. J., and Forshammer, J. (1971). *Proc. Natl. Acad. Sci. U.S.A.* **68,** 2057–2061.

Vennegoor, C., and Bloemendal, H. (1972). *Eur. J. Biochem.* **26,** 462–473.

Ventimiglia, F. A., and Wool, I. G. (1974). *Proc. Natl. Acad. Sci. U.S.A.* **71,** 350–354.

Vermeer, C., Boon, J., Talens, A., and Bosch, L. (1973a). *Eur. J. Biochem.* **40,** 283–293.

Vermeer, C., Van Alphen, W., Van Knippenberg, P. H., and Bosch, L. (1973b). *Eur. J. Biochem.* **40,** 295–308.

Vermeer, C., De Kievit, R. J., Van Alphen, W. J., and Bosch, L. (1973c). *FEBS Lett.* **31,** 273–276.

Voynow, P., and Kurland, C. G. (1971). *Biochemistry* **10**, 517–524.

Wahba, A. J. (1974). *Abstr., EMBO Workshop Protein Synth. Prokaryotic Eukaryotic Syst. 1974.*

Wahba, A. J., Iwasaki, K., Miller, M. J., Sabol, S., Sillero, M. A. G., and Vazquez, C. (1969). *Cold Spring Harbor Symp. Quant. Biol.* **34**, 291–299.

Wahba, A. J., Miller, M. J., Niveleau, A., Landers, T. A., Carmichael, G. G., Weber, K., Hawley, D. A., and Slobin, L. I. (1974). *J. Biol. Chem.* **249**, 3314–3316.

Walker, T. A., Pace, N. R., Erikson, R. L., Erikson, E., and Behr, F. (1974). *Proc. Natl. Acad. Sci. U.S.A.* **71**, 3390–3394.

Wallace, B. J., Tai, P. C., and Davis, B. D. (1973). *J. Mol. Biol.* **75**, 391–405.

Weber, H. G. (1972). *Mol. Gen. Genet.* **119**, 233–248.

Weber, L. A., Feman, E. R., and Baglioni, C. (1975). *Biochemistry* **14**, 5315–5321.

Weeks, D. P., and Baxter, R. (1972). *Biochemistry* **11**, 3060–3064.

Wei, C. M., and Moss, B. (1974). *Proc. Natl. Acad. Sci. U.S.A.* **71**, 3014–3018.

Weissmann, C., Billeter, M. A., Goodman, H. M., Hindley, J., and Weber, H. (1973). *Annu. Rev. Biochem.* **42**, 303–330.

Wertz, G. W., and Youngner, J. S. (1972). *J. Virol.* **9**, 85–89.

Wigle, D. T. (1973). *Eur. J. Biochem.* **35**, 11–17.

Wigle, D. T., and Smith, A. E. (1973). *Nature (London), New Biol.* **242**, 136–140.

Willis, M. V., Baseman, J. B., and Amos, H. (1974). *Cell* **3**, 179–184.

Wilson, D. B., and Dintzis, H. M. (1970). *Proc. Natl. Acad. Sci. U.S.A.* **66**, 1282–1289.

Wittmann, H. G., Stöffler, G., Hindenmach, I., Kurland, C. G., Randall-Hazelbauer, L., Birge, E. A., Nomura, M., Kaltschmidt, E., Mizushima, S., Traut, R. R., and Bickle, T. A. (1971). *Mol. Gen. Genet.* **111**, 327–333.

Yamaki, H., Nishimura, T., Kubota, K., Kinoshita, T., and Tanaka, N. (1974). *Biochem. Biophys. Res. Commun.* **59**, 482–488.

Yoshida, M., and Rudland, P. S. (1972). *J. Mol. Biol.* **68**, 465–481.

Yoshida, M., Travers, A., and Clark, B. F. C. (1972). *FEBS Lett.* **23**, 163–166.

Young, R. M., and Nakada, D. (1971). *J. Mol. Biol.* **57**, 457–474.

Zagorska, L., Dondon, J., Lelong, J. C., Gros, F., and Grunberg-Manago, M. (1971). *Biochimie* **53**, 63–72.

Zasloff, M. (1973). *J. Mol. Biol.* **76**, 445–453.

Zasloff, M., and Ochoa, S. (1971). *Proc. Natl. Acad. Sci. U.S.A.* **68**, 3059–3063.

Zasloff, M., and Ochoa, S. (1972). *Proc. Natl. Acad. Sci. U.S.A.* **69**, 1796–1799.

Zasloff, M., and Ochoa, S. (1973). *J. Mol. Biol.* **73**, 65–76.

Zehavi-Willner, T., and Petska S. (1976). *Arch. Biochem. Biophys.* **172**, 706–710.

Zeller, H. (1974). M.Sc. Thesis, Bar Ilan University, Israel.

Zilberstein, A., Federman, P., Shulman, L., and Revel, M. (1976). *FEBS Lett.* **68**, 119–124.

Factors Involved in the Transfer of Aminoacyl-tRNA to the Ribosome

DAVID L. MILLER AND HERBERT WEISSBACH

The first step in the process of peptide chain elongation is the binding of the appropriate codon-specified aminoacyl-tRNA to the aminoacyl site on ribosomes containing a growing polypeptide chain on the peptidyl site. Proteins essential for this reaction have been found in both prokaryotic and eukaryotic organisms,* and although there are similarities between the binding processes in the two types of organisms, the proteins are not generally interchangeable (however, for one exception, see Krisko *et al.*, 1969) and they differ markedly in their properties. This review describes studies on the factors from *Escherichia coli* and from mammalian tissues.

I. Prokaryotic Factors—General Comments

Two factors required for polyphenylalanine synthesis by *E. coli* ribosomes with a poly(U) message were first identified by Allende *et al.* (1964) and were further characterized by Nishizuka and Lipmann (1966) who named them factor G, since this factor stimulated a ribosome-dependent GTPase, and factor T, which was later shown to promote the binding of AA-tRNA to ribosomes (Ravel, 1967; Lucas-Lenard and Haenni, 1968; Ertel *et al.*, 1968b). Analogous studies in a reticulocyte system (Arlinghaus *et al.*, 1963) indicated the presence of two factors required for peptide chain elongation.

In an effort to further characterize EF-T (using the factors from *Pseudomonas fluorescens* in the hope of avoiding a troublesome stability problem found with the *E. coli* factor), Lucas-Lenard and Lipmann (1966) were able to resolve by DEAE Sephadex chromatography two complementary factors which, taken together, exhibited the properties of EFT. These factors, now known as EF-Ts and EF-Tu, were originally distinguished by the relative stability of the former when heated at 60° at pH 7.4.

Elongation factor Tu is now known to promote the binding of AA-tRNA to ribosomes via an AA-tRNA · EFTu · GTP intermediate (Gordon, 1968; Shorey *et al.*, 1969). When this ternary complex interacts with ribosomes in the presence of mRNA, the AA-tRNA is transferred to the ribosomes and GTP is hydrolyzed with the formation of EF-Tu · GDP and P_i (Gordon, 1969; Ono *et al.*, 1969). The EFTu · GDP complex dissociates very slowly (Weissbach *et al.*, 1970a). The function of EF-Ts is to catalyze the exchange of the tightly bound GDP with free GTP to form EF-Tu · GTP, which can interact with another molecule of AA-tRNA, thus allowing EF-Tu to function catalytically in the binding cycle (Miller and Weissbach, 1970b: Weissbach *et al.*, 1970b). These reactions are summarized in the following equations:

AA-tRNA · EF-Tu · GTP + Ribosome · mRNA $\longrightarrow$
$$AA\text{-}tRNA \cdot ribosome \cdot mRNA + EF\text{-}Tu \cdot GDP + P_i \quad (1)$$

---

* For reviews of protein synthesis, the reader is directed to Lengyel and Söll (1969), Lucas-Lenard and Lipmann (1971), Leder (1973), and Lucas-Lenard and Beres (1974).

$$\text{EF-Tu} \cdot \text{GDP} + \text{EF-Ts} \rightleftharpoons \text{EF-Tu} \cdot \text{EF-Ts} + \text{GDP} \qquad (2)$$

$$\text{EF-Tu} \cdot \text{EF-Ts} + \text{GTP} \rightleftharpoons \text{EF-Tu} \cdot \text{GTP} + \text{EF-Ts} \qquad (3)$$

$$\text{AA-tRNA} + \text{EF-Tu} \cdot \text{GTP} \rightleftharpoons \text{AA-tRNA} \cdot \text{EF-Tu} \cdot \text{GTP} \qquad (4)$$

These conclusions, derived from studies performed with the factors from *E. coli,* appear to apply to all prokaryotes. The elongation factors from other bacteria which have been studied extensively, notably those from *Bacillus stearothermophilus* (Waterson *et al.,* 1970) and *Pseudomonas fluorescens* (Lucas-Lenard and Beres, 1974), resemble EF-Ts and EF-Tu in their properties and functions.

Elongation factor Tu is uniquely interesting for the number of diverse substances with which it interacts. As indicated above, EF-Tu binds EFTs, GDP, and AA-tRNA, and probably some components of the ribosome. In addition, EF-Tu and EF-Ts have been identified as two of the three host-donated components of bacteriophage Qβ polymerase (Blumenthal *et al.,* 1972; Landers *et al.,* 1974), and there is also evidence that EF-Tu may play a role in the regulation of ribosomal RNA synthesis (Travers, 1973).

II. Purification and Properties of EF-Tu and EF-Ts

A. Assay Methods for EF-Tu and EF-Ts

The original assays for EF-Tu and EF-Ts were based upon their ability to promote polypeptide synthesis on a ribosome-mRNA template in the presence of AA-tRNA and EFG. In practice, one usually measures polyphenylalanine synthesis (Nishizuka *et al.,* 1968). The polymerization of phenylalanine depends almost absolutely upon EF-Tu. Although there is not an absolute requirement for EF-Ts, one can commonly observe a 5-10-fold stimulation of the rate of polyphenylalanine formation by adding this factor when EF-Tu is present in limiting amounts (Lucas-Lenard and Lipmann, 1966; Weissbach *et al.,* 1971a).

Some of the problems encountered when using the polymerization assay as a routine method are that it requires a comparatively complex system, and that the results in units of phenylalanine polymerized cannot be directly translated into molar quantities of EF-Tu or EF-Ts. The binding of Phe-tRNA to ribosomes can also be used as an assay for EF-Tu (Ravel and Shorey, 1971), but this method is also cumbersome and subject to variability in the ribosome preparations.

We have found the most reliable and rapid assays for these factors to be the measurement of the capacity of EF-Tu to bind GDP, and the activity of EF-Ts in catalyzing the exchange of free GDP with EF-Tu · GDP (Ertel *et al.,* 1968a; Miller and Weissbach, 1974). These analyses can be performed quickly and accurately due to EF-Tu's property of forming a very strong complex with GDP, which binds quantitatively to cellulose nitrate filter discs. Using tritium-labeled GDP, picomole quantities of EF-Tu · GDP can be conveniently measured.

The details of these assays have been previously described (Miller and Weissbach, 1974). In brief, 2–40 pmoles of EF-Tu are incubated with 200 μl of a Tris-Mg^{2+} buffer containing 2 μM [^{3}H]GDP for 5 min at 37°, and the mixture passed through a cellulose nitrate filter. EF-Tu appears to be the only protein in *E. coli* which binds GDP tightly and is absorbed to the filter under these conditions.

In the assay for EF-Ts, a small aliquot containing the protein is incubated at 0° with 200 μl of a Tris-Mg^{2+} buffer containing 35 pmoles of EF-Tu · GDP and 500 pmoles of [^{3}H]GDP. The extent of exchange catalyzed by the EF-Ts extract is a measure of its activity. One unit of EF-Ts activity is defined as the amount of protein that catalyzes the exchange of 1 pmole of EF-Tu · GDP in 5 min.

B. Purification of EF-Tu

Relatively large amounts of the elongation factors Ts and Tu are found in *E. coli*. Existing in quantities roughly equivalent to the number of ribosomes, together EF-Ts and EF-Tu comprise as much as 5% of the cytoplasmic protein. The factors have readily yielded to purification, and large amounts may be obtained in a highly purified state. The current purification procedures (Miller and Weissbach, 1970a) were developed from the original method of Lucas-Lenard and Lipmann (1966), and they use the subsequent discovery that GDP stabilizes EF-Tu, and by displacing EF-Ts from EF-Tu, improves their separation (Miller and Weissbach, 1974). A brief description of the procedure follows:

Escherichia coli cells are ruptured by passage through a French press, a Manton-Gaulin mill, or by homogenization with glass beads, and the cell debris and ribosomes are removed by consecutive low and high speed centrifugations. The fraction of protein precipitating between 30–60% saturation ammonium sulfate is applied to a DEAE Sephadex column in a pH 8.0 Tris-HCl buffer containing 0.2 mM GDP. EF-Ts and a small amount of undissociated EF-Tu · EF-Ts are eluted from a column by 0.15 M KCl. After EF-Ts has been washed from the column, EF-Tu · GDP is eluted by applying a 0.15–0.35 M KCl gradient.

The EF-Tu-containing fractions are further purified by gel filtration chromatography, on Sephadex G-100. The EF-Tu-containing fractions are reprecipitated with 50% saturation ammonium sulfate, and the contaminating protein is removed by extracting the precipitate with progressively more dilute solutions of ammonium sulfate. The EF-Tu · GDP is soluble in solutions of Tris-HCl buffer containing ammonium sulfate at 30% saturation, and the protein readily crystallizes from ammonium sulfate at 40–45% saturation. The yield of EF-Tu · GDP from this procedure is routinely about 150–200 mg/pound of *E. coli.*

C. Purification of EF-Ts

Several options are available for the purification of EF-Ts. One approach is to purify the EF-Tu · EF-Ts complex and then to separate the components. Different variations of this approach have been developed by ourselves and others (Miller and Weissbach, 1969, 1970a; Hachmann *et al.,* 1971; Arai *et al.,* 1972). One such technique uses the ability of GDP to dissociate the complex. Thus, a mixture of proteins obtained by ammonium sulfate fractionation is subjected to gel filtration of Sephadex G-100, which separates the complex from many of the larger and smaller contaminants. Then the fractions containing the complex are rechromatographed on a Sephadex G-100 column which has been equilibrated with 10 μM GDP. The factors emerge together, but this time in the molecular weight range of the separate components, thus effecting a considerable purification.

Another approach used in this laboratory is to further purify EF-Ts from the DEAE Sephadex eluant obtained from the EF-Tu purification procedure (Hachmann *et al.,* 1971). By a series of steps using heat and acid precipitation and carboxymethyl Sephadex chromatography, large amounts of EF-Ts of greater than 90% purity have been prepared.

D. Properties of EF-Tu and EF-Ts

1. Crystallization of EF-Tu · GDP

The EF-Tu · GDP complex readily crystallizes in long thin needles from ammonium sulfate at 40% saturation in 50 mM Tris · HCl, pH 8.0 and 10 mM MgCl$_2$. EF-Tu · GTP and EF-Tu · GMP · PCP crystallize similarly under these conditions. Large hexagonal plates have also been obtained from this medium. In the presence of 10% isopropanol with chloride ion removed, hexagonal prisms have been obtained which are suitable for X-ray diffraction studies (Sneden *et al.,* 1973).

Chloride ion has an unusual affect upon the solubility of EF-Tu · GDP. In the presence of 0.1 M Cl$^-$, EF-Tu · GDP will begin to precipitate at an ammonium sulfate concentration of 1.3 M; however, in 1 mM Cl$^-$, the protein will crystallize at an ammonium sulfate concentration of 0.2 M or less. Thus, it is advantageous to perform the final extraction step in the preceding purification procedure in the absence of chloride, since EF-Tu remains insoluble, while the contaminants are dissolved. In this manner,

EF-Tu · GDP has been purified by extraction from precipitates containing less than 15% of the desired protein.

On the other hand, EF-Tu · GDP should not be dialyzed or chromatographed in chloride-free buffer, because under these conditions the protein readily adheres to membranes and column materials.

2. Stability of EF-Tu and EF-Ts

One of the confusing problems in the initial studies on the elongation factors was the variable stability of EF-Tu observed in the different laboratories: some workers found that the protein became inactive within a few hours at 0°, while others noted that the protein was indefinitely stable. This discrepancy now appears to be due to the necessity for having bound GDP to stabilize the protein. Those conditions which increase the rate or extent of dissociation of GDP—low pH, low Mg^{2+} concentration, elevated temperature—inactivate the protein. Adding GDP to decrease the extent of dissociation will greatly stabilize the protein. For example, in the presence of 10 μM GDP, EF-Tu · GDP may be maintained at 0° at pH's as low as 4 for several hours without a serious loss of activity, whereas EF-Tu · GDP alone would be inactivated almost immediately. It has also been found that EF-Tu denatured in 6 M urea can be reactivated if excess GDP is added before removing the urea (T. Blumenthal, unpublished observation). Free EF-Tu is denatured with a half-life of less than 10 min at 37°, or of a few days at 0°, at pH 8.0.

Despite the stabilization induced by adding GDP, EF-Tu is still more rapidly inactivated at 60° than EF-Ts; consequently, the original distinction preserved in their names is still apropos. EF-Ts is also relatively stable at low pH.

The tRNA binding protein analogous to EF-Tu found in the thermophilic organism B. stearothermophilus (S_3) is much more stable than EF-Tu at high temperature. At 60°, pH 7.5, the half-life of EF-Tu · GDP is less than 30 sec, whereas S_3 shows no loss in activity after 30 minutes under these conditions (D. L. Miller and K. M. Rassmussen, unpublished observations).

3. Molecular Weight of EF-Tu

The molecular weight of EF-Tu has been determined in several laboratories by sedimentation equilibrium and by electrophoresis to be between 42,000 and 46,000 daltons (Miller and Weissbach, 1970a; Blumenthal et al., 1972). The analogous protein from P. fluorescens may be somewhat smaller (39,000) (Lucas-Lenard and Beres, 1974), whereas the factor from B. stearothermophilus may be slightly larger (MW 51,000) (Beaud and Lengyel, 1971). An early report that the molecular weight of EF-Ts was

67,000 (Parmeggiani and Gottschalk, 1969) is likely due to the presence of EF-Tu, possibly in an inactive form (Miller and Weissbach, 1969), in a complex with EF-Ts.

4. Spectral Properties

Since EF-Tu as normally isolated contains tightly bound GDP, its ultraviolet adsorption spectrum in the 250–290 nm region is not typical of ordinary protein, the ratio of the absorbance at 280 nm to that at 260 nm being 1.16 (Miller and Weissbach, 1970a). When GDP is removed by exhaustive dialysis, the 280/260 ratio rises to 1.65, a normal ratio for proteins. The extinction coefficient of EF-Tu $\cdot$ GDP at 280 nm is 4×10^4 M^{-1} cm^{-1}.

The circular dichroic spectrum of EF-Tu $\cdot$ GDP shows negative extrema at 210 and 220 nm, with a mean residue ellipticity of 9900 at 220 nm. From this it has been calculated (using the method of Bewley et al., 1969) that EF-Tu $\cdot$ GDP contains 32% helicity (M. Boublik, unpublished observations).

5. Amino Acid Composition

The amino acid compositions of EF-Tu and EF-Ts are shown in Table I. The molar compositions are based upon molecular weights determined in this laboratory. The analyses determined by Arai et al., (1973) are in general agreement with these, if adjustments are made for the differences in molecular weights assumed. The amino terminus of EF-Tu is blocked, and its identity is at present unknown.

It is interesting that the protein as isolated contains about one zinc atom per mole of EF-Tu (Table II), which was determined by atomic absorption spectroscopy. This metal ion is not essential for the function of EF-Tu in the conventional assays. When Zn^{2+} is removed by dialysis vs. 1,2-dimercaptopropanol, EF-Tu still binds GDP, GTP, and AA-tRNA and promotes polyphenylalanine synthesis with undiminished activity. It is possible that the Zn^{2+} is picked up as a contaminant from the buffers or media. Alternatively, there may be other functions of EF-Tu for which the metal is essential.

6. Sulfhydryl Groups

The amino acid analysis of EF-Tu shows three cysteine residues (Miller and Weissbach, 1970a). Two of these —SH groups are essential for the activity of EF-Tu in promoting peptide chain elongation (Miller et al., 1971). One of these is required for the interaction with AA-tRNA, and is readily inactivated when EF-Tu $\cdot$ GDP or EF-Tu $\cdot$ GTP is allowed to react with alkylating agents or mercurials (Table III). Another —SH group is essential for the binding of GDP or GTP, and is completely protected by

TABLE I

Amino Acid Compositions of EF-Tu and EF-Ts[a]

Amino acid	Residues/mole	
	EF-Tu[b]	EF-Ts[c]
Half-cystine	2.8	0.91
Aspartic acid	33.1	24.70
Threonine	28.7	12.40
Serine	10.3	7.40
Glutamic acid	47.9	49.70
Proline	18.7	7.70
Glycine	42.0	30.00
Alanine	27.4	35.10
Valine	33.5	28.00
Methionine	8.8	8.30
Isoleucine	24.5	13.60
Leucine	30.0	14.00
Tyrosine	8.4	1.40
Phenylalanine	13.9	7.60
Tryptophan	2.9	—
Lysine	20.3	20.0
Histidine	9.9	2.7
Arginine	23.1	8.00

[a] From Miller and Weissbach (1970a) and Hachmann *et al.* (1971).

[b] Calculated for a molecular weight of 42,000.

[c] Calculated for a molecular weight of 28,000.

TABLE II

Presence of Zn^{2+} in EF-Tu[a]

Preparation	Zn^{2+} moles/mole of EF-Tu
Crystalline EF-Tu	0.90
Same dialyzed vs. BAL	0.12
Buffer blank subtracted	0.12

[a] Ten mg of EF-Tu·GDP were dialyzed for 20 hr vs. 500 μl of a buffer composed of 1 mM Tris-HCl, pH 8.0, 0.1 mM $MgCl_2$, 0.1 mM EDTA with or without 0.2 mM BAL (2,3-dimercapto-1-propanol).

TABLE III

Effects of NEM Treatment on the Activity of
Three EF-Tu Species in Phe-tRNA·EF-Tu·GTP
Complex Formation[a]

Tu Species	NEM	Phe-tRNA·EF-Tu·GTP formation (pmoles)
EF-Tu·GDP	+	3
EF-Tu·GDP	−	27
EF-Tu·GTP	+	6
EF-Tu·GTP	−	15
EF-Tu·EF-Ts	+	2
EF-Tu·EF-Ts	−	14

[a] EF-Tu·GTP or EF-Tu·GDP (1×10^{-6} M) in a buffer containing 0.05 M Tris·HCl, pH 7.4, 0.1 M NH$_4$Cl and 0.01 M MgCl$_2$ was treated with NEM (1×10^{-3} M) for 20 min at 20°. EF-Tu·EF-Ts, formed from a mixture of EF-Tu·GDP (3×10^{-7} M) and EF-Ts (4×10^{-7} M) in a buffer containing 0.05 M Tris·HCl, pH 7.4, 0.10 M NH$_4$Cl and 5×10^{-5} M EDTA, was treated with NEM (1×10^{-3} M) for 20 min at 20°. EF-Tu·GDP (25 pmoles) and EF-Tu·GTP and EF-Tu·EF-Ts (20 pmoles) were used in Phe-tRNA binding assays. The EF-Tu·GDP and EF-Tu·EF-Ts were converted to EF-Tu·GTP before forming the ternary complex. From Miller et al. (1971).

the bound nucleotide against inactivation by N-ethyl maleimide (NEM) (Fig. 1). The same —SH group essential for GDP binding is also required for the interaction of EF-Tu with EF-Ts, and is protected by EF-Ts from reaction with NEM (Fig. 2). The third cysteine which appears in the amino acid analyses has not been observed in NEM-labeling experiments performed on the native protein.

EF-Ts contains one cysteine residue, which is essential for its activity in catalyzing the exchange of free GDP with EF-Tu · GDP, and in promoting peptide chain elongation. This —SH group is also protected from reaction with NEM when EF-Ts is bound to EF-Tu.

E. Mutants of EF-Ts and EF-Tu

Temperature sensitive mutants of EF-Tu and EF-Ts have recently been reported which may assist in answering questions about the structural relationship between EF-Tu and EF-Ts (Gordon et al., 1972; Lupker et al., 1974). The observations that these proteins and EF-G exist in the cell in amounts approximately equal to the amount of ribosomes (Gordon, 1970) suggested that these factors may be synthesized by genes in the locus of,

David L. Miller and Herbert Weissbach

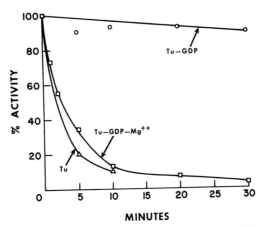

MINUTES

Fig. 1 Inactivation of EF-Tu and EF-Tu·GDP by NEM. EF-Tu·GTP ($1 \times 10^{-6} M$) was allowed to react with NEM ($1 \times 10^{-3} M$) at 20° in a buffer composed of $0.05\ M$ Tris-HCl buffer, pH 7.4, $5 \times 10^{-5} M$ EDTA, and $\pm$ MgCl$_2$. O—O, 0.01 M MgCl$_2$, □—□, no MgCl$_2$. Aliquots containing 25 pmoles of EF-Tu were withdrawn and assayed for GDP binding activity. For comparison, the rate of inactivation of GDP-free EF-Tu is included (△—△). From Miller *et al.*, (1971).

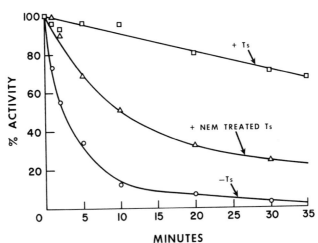

MINUTES

Fig. 2 Effect of EF-Ts on the rate of inactivation of EF-Tu. O—O, EF-Tu·GDP (150 pmoles) was allowed to react with $1 \times 10^{-3} M$ NEM in 250 μl of Tris-HCl buffer, $0.05\ M$, pH 7.4, and EDTA ($5 \times 10^{-5} M$). □—□, 150 pmoles of EF-Ts was added to the reaction mixture before adding NEM; △—△, 150 pmoles of NEM-treated EF-Ts were added before adding NEM. Aliquots containing 15 pmoles of EF-Tu were withdrawn and assayed for GDP binding activity. From Miller *et al.*, (1971).

and under the same control as, the genes for the ribosomal proteins. Indeed, the gene for EFG, identified in a fusidic acid resistant mutant, was found to lie near the locus where many ribosomal proteins map (Bernardi and Leder, 1970). However, the locus for the EF-Ts gene in temperature-sensitive mutants was not found to lie near the locus of the ribosomal proteins (Gordon et al., 1972).

It has been reported that there are two genes for EF-Tu in the *E. coli* chromsome (Jaskunas et al., 1975). One lambda transducing phage (λ fus 2) was found to carry a gene for EF-Tu as well as the genes for ribosomal proteins and EF-G. Another phage (λ rif^d 18) also carries a gene for EF-Tu as well as genes for RNA polymerase subunits β and β'. It would appear that these two genes should code for proteins of different amino acid sequences, which should be distinguishable. At present, the two gene products were identified as EF-Tu only by their migration rates in SDS gels and their reactivity with EF-Tu antiserum. It remains to be seen whether each protein possesses all of the functions of EF-Tu.

This observation is particularly pertinent to the problem of the control of protein biosynthesis. Is there a mechanism that coordinates the rate of expression of EF-Tu and EF-Ts genes with the demand for protein-synthesizing machinery? Under conditions of rapid growth, the cells contain larger amounts of EF-Tu and EF-Ts, along with EF-G and ribosomes (Gordon, 1970). There is no other evidence that the synthesis of EF-Tu and EF-Ts are controlled at present; however, as will be discussed later, it does appear that EF-Tu and EF-Ts have a wider role in protein synthesis than was originally thought.

F. Effects of Antibiotics upon the Function of EF-Tu

Until recently, no antibiotic had been found which directly inhibited any of the reactions of EF-Tu. Tetracycline derivatives, which inhibit the binding of AA-tRNA to ribosomes, appear to function by binding to a site on 30 S ribosomes essential for tRNA binding rather than EF-Tu binding, since the ribosome-catalyzed EF-Tu · GTP hydrolysis continues unabated in the presence of this antibiotic (Shorey et al., 1969).

Recently, it has been demonstrated (Wolf et al., 1974) that the antibiotic kirromycin binds to EF-Tu, stimulates the binding of GTP and AA-tRNA to EF-Tu, and also stimulates the binding of the ternary complex (now quaternary complex containing the antibiotic) to ribosomes. The authors believe that the antibiotic prevents the dissociation of EF-Tu · GDP from the ribosomes, and thus blocks peptide bond formation and translocation. This effect of the antibiotic bears some resemblance to that of fusidic acid, which prevents the dissociation of EF-G · GDP from the ribosome.

Kirromycin appears to bind at a site on EF-Tu independent from the protein's other active sites, since it does not inhibit binding of EF-Ts, GDP, GTP, AA-tRNA, or the ribosomes to EF-Tu. The antibiotic should be useful for preparing mutants of EF-Tu.

III. Interactions of EF-Tu and EF-Ts

A. The EF-Ts-EF-Tu Interaction

As described earlier, the elongation factor T activity first identified by Lipmann and co-workers (Allende *et al.,* 1964; Nishizuka and Lipmann, 1966; Krisko *et al.,* 1969) was separated into EF-Tu and EF-Ts activities by DEAE Sephadex chromatography. It now appears that the extent of separation of EF-Tu and EF-Ts obtained is related to the amount of GDP in the protein mixture applied to the column, and that the separation is achieved because under the conditions of chromatography, EF-Tu · GDP is adsorbed to the column while EF-Ts passes through.

That EF-Ts and EF-Tu interact to form a tightly bound complex was first demonstrated by the behavior of mixtures of the components during gel filtration chromatography (Fig. 3A,B) (Miller and Weissbach, 1969). It was found that separately, each component emerged in an elution volume characteristic of a protein of molecular weight about 40,000 daltons, whereas when combined, a certain proportion emerged as a larger species in a volume expected for a protein of molecular weight about 80,000 daltons. It was found that definite proportions of EF-Ts and EF-Tu combined with each other; 1 pmole of EF-Tu combines with approximately 30 units of EF-Ts (Miller and Weissbach, 1970b). If an excess of one of the factors were present in the mixture to be chromatographed, this excess would appear in the elution volume characteristic of the separate protein. The formation of the EF-Ts · EF-Tu complex has also been demonstrated using sucrose density gradient centrifugation (Lucas-Lenard *et al.,* 1969).

The molecular weight of the EF-Tu · EF-Ts complex, determined to be 65,000 daltons by equilibrium centrifugation (Hachmann *et al.,* 1971), is approximately the sum of the molecular weights of EF-Ts and EF-Tu (72,000 daltons) which indicates that the complex contains one mole of each factor. The value may be low because of some dissociation of the complex during centrifugation.

When the EF-Tu · EF-Ts complex is chromatographed on a gel filtration column which had been initially equilibrated with GDP, the proteins again emerge in the elution volume of the separate components (Fig. 3C), which suggests that GDP displaces EF-Ts when it binds to EF-Tu. This effect has also been demonstrated by a study of the inactivation of EF-Ts by NEM (Fig. 4) (Miller *et al.,* 1971). NEM rapidly reacts with free

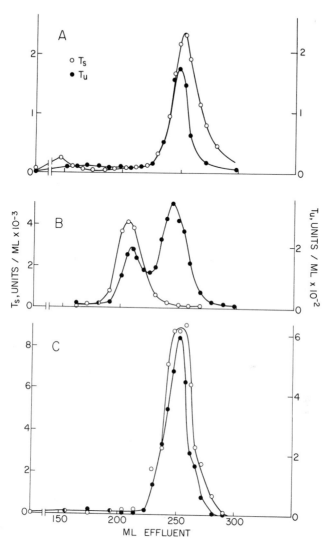

Fig. 3 Chromatography of EF-Ts and EF-Tu on Sephadex G-100. (A) 2 mg of EF-Ts (50,000 units) and 0.2 mg EF-Tu (4000 units) were chromatographed separately on a 94 × 2.5 cm column. (B) 4 mg of EF-Ts (100,000 units) and 0.5 mg of EF-Tu (10,000 units) were chromatographed under the same conditions described in (A). (C) 8 mg EF-Ts (2 × 10⁵ units) and 0.7 mg EF-Tu (1.4 × 10⁴ units) were chromatographed as described in (A). Fractions containing the EF-Ts·EF-Tu complex were concentrated to 3 ml, incubated with $5 × 10^{-5} M$ GTP for 20 min at 0°, and then rechromatographed on a 94 × 2.5 cm Sephadex G-100 column equilibrated with a buffer containing $5 × 10^5 M$ GTP. From Miller and Weissbach (1969).

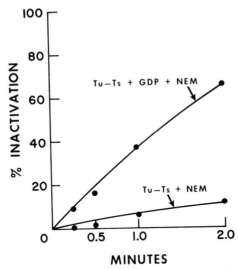

MINUTES

Fig. 4 Effect of EF-Tu and GDP on NEM inactivation of EF-Ts. The buffer used in these experiments contained 0.05 M Tris-HCl, pH 7.4, 0.1 M NH$_4$Cl and 0.01 M MgCl$_2$. The EF-Tu · EF-Ts complex (16.7 pmoles) was incubated with 4×10^{-3} M NEM, and where indicated, 4×10^{-5} M GDP was added. Five-μl aliquots were withdrawn and assayed for EF-Ts. From Miller and Weissbach (1969).

EF-Ts; however, in the EF-Tu · EF-Ts complex, EF-Tu effectively protects EF-Ts from inactivation. In contrast, when GDP is added to EF-Tu · EF-Ts, NEM rapidly inactivates EF-Ts. An effect of NEM on EF-Tu · GDP can be eliminated, since NEM does not diminish the capacity of EF-Tu · GDP to bind EF-Ts in these experiments.

It was also found that when EF-Ts interacts with EF-Tu · GDP, it displaces an equivalent amount of GDP. Figure 5 shows that this process occurs stoichiometrically rather than catalytically. Measurements of the competition between EF-Ts and GDP for EF-Tu in equilibrium mixtures of the three components using the cellulose nitrate filter procedure reveal that the affinities of EF-Ts and GDP for EF-Tu are approximately equal (see Table IV).

The rate of dissociation of EF-Tu · GDP is very slow at pH 7.4 and 0° in the presence of 10 mM Mg^{2+}, being less than 0.5% per min. This slow rate of dissociation is greatly accelerated by EF-Ts, and the ready reversal of this displacement reaction [Eq. (2)], results in EF-Ts being an efficient catalyst of the exchange of bound GDP with GDP or GTP in solution. From the specific activity of EF-Ts, about 30 units/pmole (see Section II,A for the definition of the unit), it is estimated that EF-Ts accelerates the exchange rate of EF-Tu · GDP at least 1500-fold. This value is undoubtedly

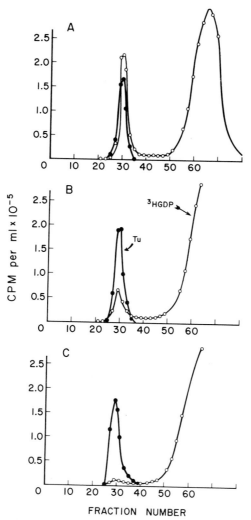

Fig. 5 Displacement of GDP from EF-Tu · [³H]GDP by EF-Ts as shown by gel filtration. EF-Tu (1000 pmoles), [³H]GDP (5000 pmoles), and three different amounts of EF-Ts were mixed and passed through a Sephadex G-25 column. (A) 500 units Ts; (B) 14,000 units Ts; (C) 25,000 units Ts. From Miller and Weissbach (1970b).

conservative, since the specific activity was not determined at saturating concentrations of EF-Tu · GDP and free GDP. Although it is routinely used to measure EF-Ts activity, a kinetic analysis of this reaction has not been performed, and nothing is known about its molecular mechanism. It is anticipated that the reaction follows a typical "ping-pong" mechanism

David L. Miller and Herbert Weissbach

TABLE IV
Dissociation Constants of Various EF-Tu Complexes

Ligand	Apparent K_{diss} (M)	K_{diss}/K_{diss} Tu·GDP
GDP	8×10^{-9} (0°), 3×10^{-9} (20°)[a]	
GDP ($-Mg^{2+}$)	1×10^{-6} (0°), 1×10^{-5} (20°)	
GTP	3×10^{-7} (20°)	
ppGpp		2.0^{b}
GMP-PCP	$2 \times 10^{-6} - 2 \times 10^{-5\,c}$	
GMP-PNP	$3 \times 10^{-7\,c}$	
EF-Ts	2×10^{-9} (20°)[a]	
GDP (ox-red)		$> 30^{d}$
dGDP		3
GDP (glucose)		600
GMP		$> 10^4$
CDP		$> 10^4$
dADP		$> 10^4$
UDP		130
IDP		300
XDP		$> 10^5$
$P_2O_7^{-4}$		

[a] From Miller and Weissbach (1970a).
[b] From Miller et al. (1973).
[c] From Arai et al. (1974a) and D. L. Miller (unpublished observation).
[d] From Bodley and Gordon (1974) and D. L. Miller (unpublished observation).

formally analogous to glutamic-oxaloacetic transaminase, in which EF-Tu would be analogous to the amino group of glutamic acid, GDP would be analogous to ketoglutarate, and EF-Ts · EF-Tu would be analogous to the pyridoxamine enzyme.

We have no evidence that EF-Ts interacts directly with GDP or GTP. Experiments to demonstrate the existence of an EF-Ts · GDP complex using the cellulose nitrate filter technique or gel filtration columns equilibrated with GDP gave negative results. We must at present conclude that EF-Ts functions by distorting or partially blocking the GDP binding site on EF-Tu, and that conversely GDP must similarly affect the EF-Ts binding site on EF-Tu.

In the preceding discussion the role of EF-Ts in catalyzing the exchange between free and bound GDP has been emphasized; however, EF-Ts also catalyzes the exchange of free GTP with bound GDP [Equations (2) and (3); Fig. 6] (Weissbach et al., 1970a). Because the dissociation constant of EF-Tu · GTP (Table IV) is one hundred times greater than that of EF-Tu · GDP (Miller and Weissbach, 1970a), the point of equilib-

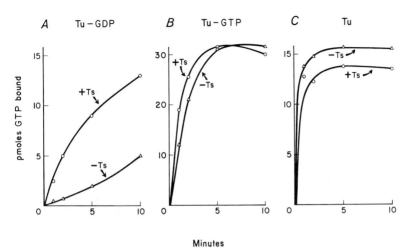

Minutes

Fig. 6   Effect of EF-Ts on the reaction of [γ-³²P]GTP with EF-Tu · GDP (A), EF-Tu · GTP (B), and EF-Tu (C). EF-Tu · GDP (30 units) or EF-Tu (15 units) were incubated at 0° in the presence of 2.5 × 10⁻⁶ M [γ-³²P]GTP. Where indicated, 60 units of EF-Ts were added. From Weissbach *et al.*, (1970a).

rium may lie toward EF-Tu · GDP; however, this is of little consequence *in vivo,* since the functional organism maintains a high GTP/GDP ratio (Gallant *et al.,* 1970), and furthermore, the strong affinity of AA-tRNA for EF-Tu · GTP displaces the equilibrium so that EF-Tu is very efficiently incorporated into the ternary complex (Weissbach *et al.,* 1970a; Miller *et al.,* 1973). There is no evidence that EF-Ts promotes any function in peptide chain elongation other than the exchange of guanosine nucleotides bound to EF-Tu (Weissbach *et al.,* 1970a). The need to reactivate the inert EF-Tu · GDP complex appears to be a sufficient reason for the existence of EF-Ts.

The rate of exchange of EF-Tu · GDP catalyzed by EF-Ts should be greater than the rate of peptide chain elongation in the intact organism. Assuming that the exchange rate at 30° is twenty times as fast as at 0°, the specific activity of 30 units/pmole at 0° can be converted to about 2 pmoles/sec exchanged per pmole of EF-Ts at 30°. This is severalfold lower than the elongation rate *in vivo,* which has been calculated to be about 10/second (Gausing, 1972). The cause of the discrepancy is most likely due to the specific activity of EF-Ts being measured at substrate concentrations far below those necessary to achieve its maximal velocity.

In preliminary experiments, we have found that the maximal velocity of EF-Ts at 0° is about 300 units/pmol and the K_m for EF-Tu · GDP is about 1 μM when the [³H]GDP concentration is 2.5 μM. Raising its [³H]GDP concentration produces an additional large increase in the exchange rate; consequently, the rate of EF-Ts catalyzed exchange *in vivo* is likely to be 10-fold greater than the rate of peptide bond formation.

In this regard, it is interesting to note that the intracellular concentrations of ribosomes, elongation factors, and tRNA's are surprisingly high. From the data of Maaløe and Kjeldgaard (1966), one calculates that the ribosome concentration is at least $2 \times 10^{-5} M$, and tRNA is about $5 \times 10^{-4} M$. Since the amount of elongation factor Tu is similar to the amount of AA-tRNA, the concentration of EF-Tu is expected to be about $5 \times 10^{-4} M$ also (Furano, 1975).

Other evidence supports the conclusion that the only function of EF-Ts is to catalyze the removal of GDP from EF-Tu in the absence of EF-Ts. Free EF-Tu, from which bound GDP has been removed by dialysis, reacts rapidly with added GTP to form EF-Tu · GTP (Fig. 6), which in turn reacts rapidly with AA-tRNA (Weissbach *et al.*, 1970a) as will be described later. In addition, the EF-Ts requirement in AA-tRNA binding and polymerization (Nishizuka *et al.*, 1968; Weissbach *et al.*, 1971a) can be substantially replaced by pyruvate kinase and phosphoenolpyruvate (PEP), which convert EF-Tu · GDP to EF-Tu · GTP (Weissbach *et al.*, 1970a). It is doubtful that pyruvate kinase acts directly upon EF-Tu · GDP. It is more likely that this phosphorylating system converts the small amount of GDP which has spontaneously dissociated from EF-Tu to GTP, and thus eliminates the unfavorable competition between GDP and GTP for EF-Tu. Pyruvate kinase is not as efficient as EF-Ts at converting EF-Tu · GDP to EF-Tu · GTP because the former enzyme may be limited by the rate of dissociation of EF-Tu · GDP.

B. Interaction of Nucleotides with EF-Tu

Elongation factor Tu binds GDP, GTP, and several other guanosine-containing polyphosphates. The protein is surprisingly selective, showing virtually no affinity for other bases or for GMP. A summary of the relative binding constants of various nucleoside phosphates is given in Table IV. One of the experimental procedures used was to allow the test nucleotide to exchange with EF-Tu · GDP in competition with [³H]GDP. The decrease in [³H]GDP bound to EF-Tu caused by the test nucleotide was used to calculate the relative affinity of the test nucleotide for the protein. Because traces of GDP may contaminate the commercial grade nucleotides, these experiments cannot reliably identify an interaction less than 1% as strong as that of GDP.

The dissociation constants of EF-Tu · GDP and EF-Tu · GTP listed in Table IV have been determined by an adaptation of the cellulose nitrate filtration method described previously, in which equilibrated solutions of EF-Tu and GDP are rapidly filtered, and the amount of EF-Tu · GDP retained on the filter is assumed to equal the amount of complex present in the solution before filtration (Miller and Weissbach, 1970b). Although this method has the apparent inadequacy of being a nonequilibrium technique, in practice it yields reliable results because of a fortunate combination of circumstances; namely, that EF-Tu · GDP and free EF-Tu are inert after being adsorbed on the filter, and the time of passage of a lamella of solution through the filter is much faster than the rate of equilibration of the components. Conventional techniques such as gel filtration and equilibrium dialysis are unsuitable because the strength of the EF-Tu · GDP association requires the use of very dilute solutions, and because the instability of free EF-Tu prohibits long equilibration times. Similar values for the dissociation constants of EF-Tu · GDP, EF-Tu · GTP, and EF-Tu · EF-Ts have been reported by Arai *et al.* (1974a). Because EF-Tu binds GDP so much more tightly than GTP and since GDP always contaminates GTP preparations, one cannot prepare EF-Tu · GTP directly from EF-Tu · GDP and GTP. One must resort to a procedure which eliminates GDP, such as treatment of the reaction mixture with pyruvate kinase and PEP (Weissbach *et al.,* 1970a), or treatment of EF-Tu · GMP-PCP with GTP (Printz and Miller, 1973).

Other more qualitative data on the interaction of guanosine derivatives with EF-Tu are also available. Bodley and Gordon (1974) found that the periodate oxidation product of GDP, presumably the 2′, 3′-dialdehyde, would not interact with EF-Tu · GDP, nor contrary to the results from this laboratory quoted in Table IV, would the dialcohol formed by the reduction of the dialdehyde. In addition, several derivatives are active in promoting protein synthesis (Uno *et al.,* 1971), and therefore, must react to some extent with EF-Tu. Among these are the N^2-methyl and N^2-dimethyl GTP derivatives.

The high specificity with which EF-Tu binds GDP suggests that there are several points of interaction between the protein and nucleotide, and accordingly, the dissociation constant is atypically low, being five orders of magnitude lower than a typical enzyme-substrate complex dissociation constant. The important requirements for the binding of guanosine nucleotides to EF-Tu are the pyrophosphate group, preferably complexed to Mg^{2+}, the exonitrogen on the guanine ring, and the stereochemical relationship between these groups imposed by the ribofuranose ring. The vicinal hydroxyls on ribose appear not to be involved in the EF-Tu-GDP interaction, since they can be substituted or eliminated with little effect.

C. Differences between EF-Tu · GTP and EF-Tu · GDP

GTP appears to bind at the same site and with the same constraints as GDP; however, there must be some distinctive differences between these complexes because only EF-Tu · GTP interacts strongly with AA-tRNA

(Gordon, 1968). Since the dissociation constant for AA-tRNA·EF-Tu·GTP is about 10^{-8} M, whereas (as will be described shortly) no interaction of EF-Tu·GDP with AA-tRNA has been detected at 10^{-3} M AA-tRNA (Shulman et $al.$, 1974), K_{diss} for the hypothetical AA-tRNA·EF-Tu·GDP complex must be at least five orders of magnitude greater than that for AA-tRNA·EF-Tu·GTP.

There is no evidence that the γ-phosphoryl group of GTP is directly responsible for the binding of AA-tRNA to EF-Tu·GTP. Such an interaction is conceivable, for example, through a magnesium bridge between GTP and tRNA; however, it is difficult to design experiments to test this hypothesis. An alternative explanation is that GTP stabilizes a conformation of EF-Tu having an AA-tRNA binding site which is not exposed in EF-Tu·GDP.

Evidence for such a conformational difference between EF-Tu·GDP and EF-Tu·GTP has been obtained by two methods. Tritium-exchange spectra show the rates at which the hydrogen atoms in a protein, principally the amide hydrogens involved in intramolecular hydrogen bonding, exchange with solvent. In these experiments EF-Tu·GDPCP was allowed to equilibrate with tritiated water for 1.6 hr, after which time the tritiated complex was converted to either EF-Tu·GTP or EF-Tu·GDP by the addition of the appropriate nucleotide, the excess tritiated water was removed, and the rate of exchange of tritium out of the complex was measured. An example of the results is shown in Fig. 7 (Printz and Miller, 1973). It was found that EF-Tu·GDP contained a set of about 14 hydrogen atoms per molecule that exchanged appreciably more slowly than those in EF-Tu·GTP. These results encourage the view that EF-Tu·GDP has a relatively tight tertiary structure in which the amide hydrogens have comparatively slight access to solvent water. The EF-Tu·GTP complex, in contrast, has a more open structure in which an appreciable number of amide hydrogens have been exposed to solvent.

Conformational differences between EF-Tu·GDP and EF-Tu·GTP have also been revealed by a study of the binding of the hydrophobic fluorescent dye 1-anilino-8-naphthalene sulfonate (ANS) (Crane and Miller, 1974). EF-Tu·GTP enhances the fluorescence of ANS to a greater extent than EF-Tu·GDP does (Fig. 8). When EF-Tu·GTP is complexed with Phe-tRNA, however, its interaction with ANS increases the fluorescence of the dye only as much as EF-Tu·GDP does (Fig. 8). Equilibrium dialysis binding measurements indicate that EF-Tu·GTP binds three molecules of ANS with an apparent K_{diss} of about 2×10^{-5} M, whereas EFTu·GDP binds two molecules with an apparent K_{diss} of 5–8 $\times$ 10^{-5} M. It appears that the differences in conformation between EF-Tu·GTP and EF-Tu·GDP revealed by ANS binding are centered chiefly in a region of

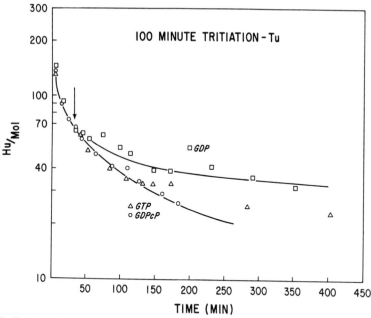

Fig. 7 Back exchange curves for EF-Tu complexes after 100 min tritiation (Gordon, 1968). In all experiments $2 \times 10^{-4}\,M$ GMP·PCP was added to the first column buffer. Sufficient GTP and GDP was added, at the time denoted by the arrow, to the pool to give a $2 \times 10^{-4}\,M$ ligand solution. The affinity of GTP and GDP for EF-Tu compared with GMP·PCP ensured that we were observing the appropriate liganded protein. From Printz and Miller (1973).

EF-Tu·GTP where AA-tRNA binds, which invites the speculation that one of the dye molecules binds at a hydrophobic site exposed when GTP interacts with EF-Tu.

The conformational difference between EF-Tu·GDP and EF-Tu·GTP does not affect the fluorescence spectrum of the protein's tryptophan residues, nor does it alter the protein's CD or ORD spectra. The reactivity toward NEM of the sulfhydryl group necessary for AA-tRNA binding is not appreciably different in EF-Tu·GTP than in EF-Tu·GDP; however, there are some small differences between the ESR spectra of EF-Tu·GTP and EF-Tu·GDP containing a paramagnetic probe bound to this —SH group (Arai et al., 1974b), which may indicate an alteration in this region of the protein.

In its interactions with GDP and GTP, EF-Tu resembles an allosteric enzyme. Indeed, nucleoside triphosphates are often allosteric effectors; for example, GTP is an essential effector for CTP synthetase when glutamine is the nitrogen donor (Levitzski and Koshland, 1972). The allosteric effect is thought in many instances to be transmitted to the enzyme's ac-

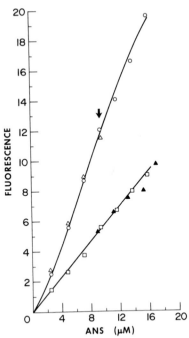

Fig. 8 Effect of Phe-tRNA upon the fluorescence of ANS-EF-Tu·GTP; comparison with ANS-EF-Tu·GDP. At the point in the titration of EF-Tu·GTP by 1-anilino-8-naphthalenesulfonate (ANS) marked by an arrow, an equimolar amount of Phe-tRNA was added to the EF-Tu · GTP solution being titrated: (O—O) control titration of EF-Tu · GTP by 1-anilino-8-naphthalenesulfonate (ANS); no Phe-tRNA added during titration; (△—△) titration of EF-Tu · GTP before addition of Phe-tRNA; (□—□) titration of EF-Tu · GTP after addition of GDP, no Phe-tRNA added during titration. From Crane and Miller (1974).

tive site by conformational changes induced by the allosteric ligand. We propose that by a similar mechanism GDP and GTP control the reactivity of EF-Tu.

Allosteric enzymes are often subject to feedback inhibition and thus represent control points in metabolic pathways. It might be supposed that the rate of protein synthesis could be controlled by the GDP/GTP ratio, which would determine the ratio of EF-Tu · GDP/EF-Tu · GTP and hence the amount of ternary complex formed.

Experiments to test this hypothesis were performed using ppGpp, the analogue of GDP which appears when protein synthesis is restricted by amino acid starvation (Miller et al., 1973). In the standard polypeptide synthesis system containing equimolar amounts of EF-Tu and EF-Ts, we have observed that ppGpp does not substantially inhibit the rate of enzymatic binding of Phe-tRNA to ribosomes, whereas in the absence of EF-Ts, ppGpp inhibits the binding strongly.

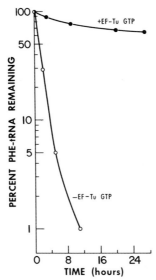

Fig. 9 Inhibition of Phe-tRNA hydrolysis by EF-Tu · GTP. The reactions were conducted at 37° in 50 mM Bis Tris · HCl, pH 7.0, 10 mM MgCl$_2$. From Jekowsky (1975).

From these results it appears that in the cell, the ratio of ppGpp/GTP (or GDP/GTP) would not effectively regulate peptide chain elongation because EF-Ts exchanges EF-Tu · GTP for EF-Tu · ppGpp (or EF-Tu · GDP) so rapidly that, even at relatively high concentrations of GDP or ppGpp, the ternary complex is formed efficiently.

D. AA-tRNA · EF-Tu · GTP Complex

The interaction of AA-tRNA with EF-Tu · GTP has been demonstrated by gel filtration chromatography, where one observes the ternary complex emerging as a new peak ahead of the separate components (Shorey *et al.*, 1969). The interaction can also be inferred from the finding that EF-Tu · GTP greatly protects AA-tRNA against hydrolysis at neutral pH (Fig. 9; Jekowsky, 1976). The most common assay technique for the ternary complex uses the discovery that unlike EF-Tu · GTP, the ternary complex is not adsorbed to cellulose nitrate filters (Gordon, 1968). In practice, about 30 pmoles of EF-Tu · GTP labeled with [³H]- and [γ-³²P]GTP are added to each of two tubes containing 200 μl of a solution composed of 0.05 M Tris buffer, pH 7.4, and 0.01 M MgCl$_2$. To one of the tubes is added 5–30 pmoles of AA-tRNA, and after one min at 0°, the mixtures are rapidly filtered. The difference in the amounts of filter bound EF-Tu · GTP between the control and the sample containing AA-tRNA represents the amount of ternary complex formed.

This method has been used for assessing the structural requirements for ternary complex formation. Ideally, the effects of structural modifications would be expressed in thermodynamic terms. This has not been done in the studies of the modified AA-tRNA's because there has been no really satisfactory method to measure dissociation constants of the ternary complex. The constant for the dissociation of AA-tRNA from EF-Tu · GTP may be estimated from the amounts of EF-Tu · GTP adsorbed when dilute solutions of the ternary complex are passed through cellulose nitrate filters. The results of these experiments indicate that the dissociation constant must be less than $1 \times 10^{-7} M$ (Ofengand and Chen, 1972); however, these results cannot be accepted without question because the forces that bind EF-Tu · GTP and not the complex to the filter are not understood. Recently, in fact, we have encountered filters which seem to strip EF-Tu · GTP from the ternary complex, thus giving erroneously low estimates of ternary complex formation.

Another technique for determining the dissociation constant of the ternary complex is to measure the extent of competition of GDP or ppGpp vs. GTP and AA-tRNA for EF-Tu (Miller *et al.*, 1973). Catalytic amounts of EF-Ts must also be present to equilibrate the components. EF-Tu · GDP and EF-Tu · ppGpp do not interact with AA-tRNA, and the decrease in the amount of EF-Tu · GDP or EF-Tu · ppGpp bound to the filter produced by GTP and AA-tRNA can be related to the dissociation constant of the ternary complex. The advantages of this procedure are that it eliminates the problem of the dissociation of EF-Tu · GTP before or during filtration and permits, by using [^{32}P]GTP, an estimation of the stripping effect of the filter. Any ^{32}P bound to the filter may be assumed to have been present in the ternary complex. The method requires, however, accurate values of the dissociation constants of EF-Tu · GTP and EF-Tu · GDP or EF-Tu · ppGpp. Using this technique, a value of 1×10^{-8} M was determined for the dissociation constant of Phe-tRNA · EF-Tu · GTP (Miller *et al.*, 1973). A similar method has been proposed by Arai *et al.*, (Arai, 1974a). Using their method, they obtain a value of $7 \times 10^{-8} M$ for the K_{diss} for Phe-tRNA. The ability of EF-Tu · GTP to protect AA-tRNA from hydrolysis could also be used as a quantitative assay for ternary complex formation.

The ternary complex is a fairly unstable species. The GTP hydrolyzes at the rate of about 25%/day at 0° and the complex dissociates. It also loses its ability to react with ribosomes when stored at 0° for a day or so, even though the filter assay indicates that the complex is substantially intact. We have found that the complex with yeast Phe-tRNA retains its activity for a week or more when stored in liquid nitrogen. It has also been reported that the ternary complex can be successfully precipitated with ethanol and stored as the frozen precipitate (Shorey *et al.*, 1969).

An interesting result of ternary complex formation is that the GTP bound in the complex becomes almost inert toward exchange with free GTP or GDP. This does not necessarily indicate that the bound GTP is sterically blocked by AA-tRNA, but is the inevitable result of the requirements introduced by the dissociation constants and the restraint that AA-tRNA binds tightly to EF-Tu · GTP but not to free EF-Tu.

The primary requirement for ternary complex formation, the EF-Tu · GTP complex, has already been described. No interactions between EF-Tu · GDP and AA-tRNA have been detected even at AA-tRNA concentrations of 1 mM. At this concentration, complex formation can be detected by NMR in the broadening of line widths of the AA-tRNA resonances in the presence of EF-Tu · GTP, but this effect was not produced by EF-Tu · GDP (Shulman et $al.$, 1974).

The only amino acid residue in EF-Tu so far identified that is essential for AA-tRNA binding is the cysteinyl residue previously described. The AA-tRNA binding protein from $B.$ $stearothermophilus$ (S_3) is also inhibited by sulfhydryl reagents; however, an —SH residue is not a general requirement for AA-tRNA binding, since EF-1 from mammalian tissues contains none. The general reactivity of an —SH group in many modification reactions complicates the search for other essential groups. It has been found that tosylphenylalanyl chloroketone (TPCK), designed as an affinity label for chymotrypsin, reacts with EF-Tu · GDP and inhibits ternary complex formation (Richman and Bodley, 1973). Whether this reagent also reacts with cysteine or whether it reacts with some other residue, such as histidine, as it does in chromotrypsin, has yet to be determined. It is not known whether any new essential residues are exposed in EF-Tu · GTP which are concealed in EF-Tu · GDP.

The most striking requirement for the interaction of AA-tRNA with EF-Tu · GTP is the aminoacyl group. There does appear to be some interaction of deacylated tRNA with EF-Tu · GTP at tRNA concentrations above 0.1 mM since at these concentrations EF-Tu · GTP broadens the NMR spectral linewidths of deacylated tRNA. Furthermore, tRNA also inhibits the adsorption of EF-Tu · GTP in the filter assay (Gordon, 1968). Nevertheless, the aminoacyl group contributes a factor of at least 10^{-4} M to the dissociation constant.

There seems to be little, if any, selectivity for the side chain aminoacyl group. Most of the bacterial and eukaryote tRNA's have been found to be about equally effective in reacting with EF-Tu · GTP (Miller and Weissbach, 1974). The amino group does not appear to be necessary since it can be replaced by an hydroxyl group via nitrous acid deamination and the resulting ester retains reactivity with EF-Tu · GTP (Fahnestock et $al.$, 1972). Although the amino group is not essential, N-acetyl-Phe-tRNA has much diminished reactivity (Ravel *et al.,* 1967) with EF-Tu · GTP, which is likely due to steric hindrance by the acetyl group.

Several pieces of information which indicate that EF-Tu · GTP interacts with additional sites on AA-tRNA besides the aminoacyl group have been described in a previous report (Ofengand, Chapter 1), and will only be summarized here. One piece of evidence is that small aminoacyl nucleotide fragments such as Phe-A-C do not bind well to EF-Tu · GTP (Ringer and Chladek, 1975). Even the aminoacylated 3'-half-molecule does not compete measurably with AA-tRNA for binding to EF-Tu · GTP (Krauskopf *et al.,* 1972). Furthermore, synthetic AA-tRNA containing a Phe-A-C-C-C-A terminus does not bind strongly to EF-Tu · GTP (Thang *et al.,* 1972).

In a recent study, it was found that the aminoacyl dinucleotide Phe-A-C binds to EF-Tu, and at a concentration of 10^{-5} M displaced GDP or GTP from the protein (Ringer and Chladek, 1975). This result was completely unexpected, and must be reconciled with other observations that Phe-tRNA does not displace GDP from EF-Tu even at concentrations above 10^{-4} M (D. Miller, unpublished observation). The loss of GDP or GTP indicates that Phe-A-C binds to free EF-Tu and weakens the interaction of the guanosine nucleotides with the protein. In contrast, as already noted, AA-tRNA stabilizes the binding of GTP to the protein. One interpretation of these data is that there is an interaction between the aminoacyl and GTP binding sites, but the latter is primarily involved in binding a region of the tRNA molecule at some distance from the C-C-A end.

As an additional interesting example of the importance of a site for EF-Tu · GTP binding in the tRNA portion of the molecule, Met-tRNA$_{fMet}$, which does not interact with EF-Tu · GTP at the concentrations where other AA-tRNA's are reactive, can be modified so that it interacts normally with EF-Tu · GTP (Schulman *et al.,* 1974). A simple bisulfite-promoted deamination procedure that converts the 5'-terminal cytosine to uracil allows the 5'-terminus to pair with an adenine in the 3'-end and produces an acceptor stem chracteristic of the other AA-tRNA's which react with EF-Tu · GTP. This result indicates that a site on the AA-tRNA at least 5 bases away from the aminoacyl group must interact with EF-Tu · GTP. The interpretation favored by Schulman *et al.,* (1974) is that the base pairing allows the 5'-terminal phosphate, which appears to be very important for ternary complex formation, to be correctly positioned for binding to a site on EF-Tu · GTP.

It is necessary to qualify this explanation of the role of the 5'-base in the function of AA-tRNA. Although most AA-tRNA's contain a paired 5'-terminal base, a mutant of the tyrosine amber suppressor su$^+$III, like tRNA$_{fMet}$, contains an unpaired 5'-terminal base (adenosine in this case) (Smith and Celis, 1973). Since the Tyr suppressor tRNA must bind to the ribosome in response to the amber mutation, it presumably interacts normally with EF-Tu.

The steric requirements for the interaction with EF-Tu · GTP are severe. In tRNA$_{fMet}$, the 5'-cytosine should be able to achieve very nearly its proper orientation without base pairing to the 3'-stem. The loss in entropy involved in placing the 5'-base into the proper orientation would not, a priori, seem large enough to cause such a large increase in the extent of dissociation of the complex. Similarly, the unreactive Phe-tRNA (ox-red) described earlier (Ofengand, Chapter 1), in which the 3'-ribose is cleaved, seems to be able to attain very nearly the configuration of the intact Phe-tRNA.

These puzzles are part of the general problem of specificity in biological interactions. Storm and Koshland (1970) have proposed that interaction energies have a very strong directional dependence, whereas Page and Jencks (1971), writing in another context, have emphasized that entropy changes in solution may be greatly underestimated. Thus, the energy needed to force either Met-tRNA$_{fMet}$ or Phe-tRNA (ox-red) into its reactive conformation may be much greater than expected.

E. Interaction of the Ternary Complex with Ribosomes

When the ternary complex interacts with ribosomes containing the appropriate codon, the complex binds to the ribosomes, GTP is hydrolyzed, EF-Tu · GDP is released, and AA-tRNA remains positioned in the aminoacyl or acceptor site of the ribosomes. If another AA-tRNA had been previously bound in the peptidyl site, the peptide bond would form immediately.

The reaction occurs very rapidly, even at 0° when either GTP hydrolysis (Fig. 10) or Phe-tRNA binding is followed. It has been reported that stable Phe-tRNA binding lags GTP hydrolysis (Weissbach et al., 1971b), but the meaning of this observation is at present unclear.

Ribosomes will also hydrolyze EF-Tu · GTP at an appreciable rate (Fig. 11) (Miller, 1972). The ribosomes can be saturated by excess ternary complex or EF-Tu · GTP, and values for the apparent dissociation constants of the complexes and maximum rates of hydrolysis can be calculated. From the data in Fig. 11, it is calculated that for the ternary complex the $K_m = 1 \times 10^{-6} M$, and $V_{max} = 12$ pmole/min/A_{260} ribosomes; whereas, for EF-Tu · GTP, $K_m = 4 \times 10^{-6} M$ and $V_{max} = 1.8$ pmole/min/A_{260} ribosomes. These data show that the ternary complex binds more tightly and hydrolyzes more rapidly than EF-Tu · GTP. However, it is apparent that EF-Tu · GTP alone has a considerable affinity for its site of activation on the ribosome.

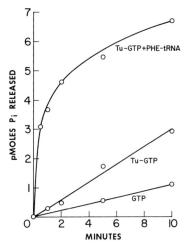

Fig. 10 Hydrolysis of GTP, EF-Tu · GTP, and Phe-tRNA · EF-Tu · GTP by ribosomes at 0°. The reaction mixtures containing (54 pmoles) [^{32}P]GTP, 7.5 pmoles of EF-Tu · GTP, 11 pmoles, where indicated, of Phe-tRNA, 2.0 A_{260} units of ribosomes, and 5 μg of poly(U) in 100 μl buffer were periodically analyzed for [^{32}P]phosphate. In the measurement of EF-Tu · GTP hydrolysis, 7.5 pmoles of EF-Tu · GTP was equilibrated 5 min with 54 pmoles of [γ-^{32}P]-GTP before addition to the reaction mixture. In the measurement of Phe-tRNA · EF-Tu · GTP hydrolysis, 11 pmoles of Phe-tRNA were added to an equilibrated EF-Tu · GTP · [γ^{32}-P]GTP mixture 5 min before the reaction was begun. The upper curves have been corrected for hydrolysis of free GTP by subtractions of the lower curve. Ribosome-independent hydrolysis was negligible. From Miller (1972).

It is also noteworthy that a primary site of interaction of the ternary complex is at the codon, because GTP hydrolysis will not occur without this complementarity. Thus, the Phe-tRNA · EF-Tu · GTP complex is not hydrolyzed appreciably by ribosomes containing bound polyadenylate.

The region on the ribosome where EF-Tu binds may overlap the EF-G binding site, since binding EF-G to the ribosome with GDP and fusidic acid prevents the binding and hydrolysis of the ternary complex (Fig. 12). It is interesting that the elongation factors from *P. fluorescens,* which are smaller than the factors from *E. coli* but still function on *E. coli* ribosomes, can apparently be bound simultaneously on *E. coli* ribosomes (Beres and Lucas-Lenard, 1973). Estimates show that the factors could cover less than 7% of the surface of the ribosome; therefore, even if the binding sites for EF-Tu and EF-G are not identical, they must be very close to each other.

The mechanism of action of EF-Tu · GTP in binding AA-tRNA to ribosomes remains conjectural. One possible function for the protein is that it might alter the structure of AA-tRNA and thereby expose a site on AA-tRNA essential for binding to the A site of ribosomes. To test this pro-

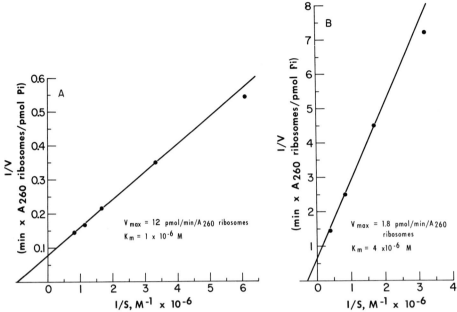

Fig. 11 Concentration dependence of hydrolysis rates of ternary complex (A) and EF-Tu · GTP (B) by ribosomes. Varying amounts of Phe-tRNA · EF-Tu · GTP or EF-Tu · GTP labeled with [γ-^{32}P]GTP were allowed to react with 2.0 A_{260} units of ribosomes and poly(U) at 0° in 0.1 ml of 0.05 M Tris acetate (pH 7.4), and 0.01 M MgCl$_2$. The rate of [^{32}P]phosphate formation was measured as described in Miller (1972).

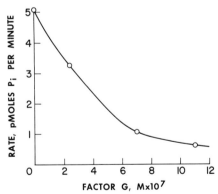

Fig. 12 Dependence of the rate of hydrolysis of the ternary complex on the amount of factor EF-G bound to the ribosomes. After mixtures containing GTP (0.1 mM), fusidic acid, (1 mM) ribosomes (2.0 A_{260} units), poly(U) (5 μg), and different amounts of factor EF-G were incubated at 0° for 45 min in 100 μl of buffer, the reaction was begun by additon of 15 pmoles of ternary complex. The extent of hydrolysis in 1 min was plotted as the "rate of hydrolysis." From Miller (1972).

posal, the effect of EF-Tu · GTP upon the double helical structure of AA-tRNA was examined by NMR. The resonances of the hydrogen atoms held in the stable H bonds of the double-helical regions are shifted downfield by the ring currents of the bases, and they occur at frequencies characteristic of the base pair. Although spectral changes caused by as few as one base pair could probably have been detected, no difference between the NMR spectra of Phe-tRNA and Phe-tRNA · EF-Tu · GTP which could be attributed to alterations in base pairing was observed (Shulman et al., 1974). Although the protein may affect the tertiary structure of AA-tRNA in a manner which would be undetectable by this technique, these results make it unlikely that EF-Tu · GTP alters the double-helical structure of AA-tRNA.

It appears likely that one of the sites on AA-tRNA that binds to ribosomes is the T-Ψ-C segment of loop III, which is thought to pair with a complementary sequence in the 5 S RNA. However, the X-ray diffraction studies (Kim et al., 1974; Robertus et al., 1974) show that this loop of tRNA is folded into the body of the molecule and is unavailable for interaction with 16 S RNA. Recently Schwartz et al. (1976) have found that the binding of poly (U) to Phe-tRNA promoted sufficient loosening of T-Ψ-C to allow some binding of the complementary tetranucleotide C-G-A-A. This process seemed to occur most extensively when the tRNA and message were bound to 30 S ribosomes. EF-Tu was not essential for the binding of C-G-A-A; however, in the absence of 30 S ribosomes, the protein seemed to stabilize the tRNA-poly(U) interaction. It has not been shown that this process occurs rapidly enough to be relevant to the mechanism of the normal binding reaction.

Another explanation for the function of EF-Tu · GTP in the binding of AA-tRNA to ribosomes is that by providing additional points of interaction with the ribosome, it directs the aminoacyl group into the active site of the peptidyl transferase. This is consistent with the observation that EF-Tu · GTP by itself reacts with the ribosome (Fig. 11). Since EF-Tu · GTP binds to ribosomes nearly as tightly as the ternary complex does, it appears that the codon-anticodon interaction does not provide much of the binding energy for the initial interaction of the ternary complex, but it does determine which ternary complexes are permitted to react with the GTPase-stimulating site. That is, a misfit at the codon may sterically inhibit binding of the wrong ternary complex.

AA-tRNA · EF-Tu · GMP-PCP, a ternary complex containing an unhydrolyzable GTP analogue, will also bind to ribosomes, producing a complex analogous to the intermediate formed with AA-tRNA · EF-Tu · GTP. When the ternary complex containing GMP-PCP is bound to ribosomes containing N-acetyl-Phe-tRNA bound in the peptidyl site, a peptide bond is not formed (Shorey *et al.*, 1971), which implies that the amin-oacyl group of the AA-tRNA is not free, but still interacts with EF-Tu · GMP-PCP. This might be expected because the aminoacyl group is an important site of interaction with EF-Tu · GTP.

This then seems to be the function of GTP hydrolysis: After EF-Tu has correctly positioned AA-tRNA in the A site of the ribosome, GTP hydrolysis allows the protein to release the aminoacyl group, and the dissociation of EF-Tu · GDP from the ribosome allows peptide bond formation to proceed. The EF-Tu · GDP is then free to be recycled in another round of AA-tRNA binding.

IV. Other Functions of the Elongation Factors

With the discovery that EF-Tu and EF-Ts form part of the RNA polymerase which replicates bacteriophage Qβ (Blumenthal *et al.*, 1972), it became apparent that the elongation factors might play a wider role in gene expression. It is of interest that the 3'-terminal region of the phage RNA closely resembles a tRNA, and that GTP is the first nucleotide incorporated. There is evidence that these factors function in the initiation of phage RNA synthesis (Landers *et al.*, 1974).

It has also been reported that EF-Tu · EF-Ts stimulates the synthesis of ribosomal RNA in *E. coli* (Travers, 1973). At 34° and 0.1 *M* KCl, EF-Tu · EF-Ts stimulates rRNA synthesis fivefold; however, the effect is not seen at 38° or in the presence of 0.01 *M* KCl. The effect is eliminated by 0.5 m*M* ppGpp. The stimulation seems to be principally due to EF-Tu, which gives a fourfold increase in the rate of rRNA synthesis. It was proposed that EF-Tu acts at the promotor region of the rRNA genes, and that the decreased requirement for EF-Tu at the higher temperature reflects a thermal opening of the promotor. These results could provide a mechanism for the coordinate control of protein synthesis by limiting the amounts of the synthetic components to that which is required by the nutritional state of the organism. It is known that such a control exists, and it has been termed the "stringent response," wherein ribosomal RNA synthesis is limited by amino acid starvation (Cashel and Gallant, 1969). Under these conditions the unusual nucleotides ppGpp and ppGppp are synthesized on the ribosome (Haseltine and Block, 1968) and can reach such levels that they are the major species of guanosine nucleotide. These substances have been shown to react with EF-Tu, and they might interfere with the function of rRNA polymerase by inhibiting the binding of EF-Tu to the promotor. A recent report (Jacobson and Rosenbush, 1976) indicates that a large fraction of EF-Tu is released by osmotic shock and is thus thought to be associated with the bacterial membrane.

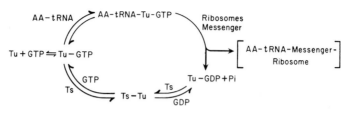

Fig. 13   The AA-tRNA binding cycle.

V. Aminoacyl-tRNA Binding Cycle

The action of EF-Ts and EF-Tu in peptide chain elongation is sum-marized in Fig. 13. In each binding cycle, EF-Tu must bind GTP in order to assume its conformation which interacts with AA-tRNA. The interaction with AA-tRNA then occurs very rapidly, and the ternary complex is ready to bind to ribosomes in response to the proper codon. When the ternary complex positions the AA-tRNA correctly on the ribosomes, GTP hydrolysis occurs, producing EF-Tu · GDP which comes off the ribosome and frees the aminoacyl group for peptide bond formation. In order to regenerate EF-Tu · GTP from the nearly inert EF-Tu · GDP at rates sufficient to sustain protein synthesis, EF-Ts catalyzes the exchange of the bound GDP for GTP via an intermediate EF-Ts · EF-Tu complex.

VI. Eukaryote Elongation Factor 1

Two factors required for peptide chain synthesis in eukaryote systems were first reported by Schweet and co-workers (Bishop and Schweet, 1961; Arlinghaus et al., 1963; Hardesty et al., 1963). Felicetti and Lipmann (1968) clearly showed that these factors corresponded to the prokaryote factors EF-T and EF-G.

The eukaryote factor corresponding to EF-T was originally called transferase 1, but is now referred to as elongation factor 1 (EF-1). It has been studied in many laboratories, although the progress with this enzyme has, until recently, been rather limited. Schneir and Moldave (1968) were first to report that EF-1 from liver was present in multiple forms, an observation that helped to explain the difficulty in obtaining a homogeneous preparation of the enzyme. A notable exception appeared to be EF-1 from reticulocytes which was obtained in a highly purified form by McKeehan and Hardesty (1969) and had a molecular weight of 186,000. However, most tissues, even including reticulocytes when examined carefully, contain EF-1 species ranging in molecular weight from as low as 50,000 to 500,000 and above. Calf brain has active EF-1 species of molecular weight well over 1×10^6 (Moon et al., 1973). At present, the

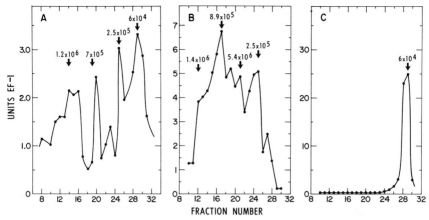

Fig. 14 Sucrose gradient profiles of calf brain EF-1 preparations. Details of centrifugation and purification of EF-1$_H$ and EF-1$_L$ are described in the papers by Moon and Weissbach (1972) and Moon *et al.*, (1972). (A) Brain postmitochondrial fraction; (B) brain EF-1$_H$; (C) brain EF-1$_L$. For results for calf liver see Liu *et al.*, (1974).

enzyme has been most thoroughly examined in liver (Gasior and Moldave, 1965; Schneir and Moldave, 1968; Collins *et al.*, 1972; Iwasaki *et al.*, 1973; Liu *et al.*, 1974), reticulocytes (McKeehan and Hardesty, 1969; Lin *et al.*, 1969), calf brain (Moon and Weissbach, 1972; Moon *et al.*, 1972, 1973; Weissbach *et al.*, 1973), wheat germ (Jerez *et al.*, 1969; Golinska and Legocki, 1973; Tarrago *et al.*, 1973; Bollini *et al.*, 1974), and Krebs ascites cells (Drews *et al.*, 1974; Nolan *et al.*, 1974). This section will concentrate primarily on studies with the calf brain and calf liver enzymes.

A. Multiple Species of EF-1

A typical example of the multiple species of EF-1 is seen in extracts of calf brain. Sucrose gradient analysis (Fig. 14, left panel) of supernatant fractions from calf brain shows a wide distribution of EF-1 activity as measured by the ability of the protein to stimulate AA-tRNA binding to ribosomes (Moon *et al.*, 1973). Similar patterns are observed in calf liver, rat liver, and mouse liver, although in the latter two tissues, the bulk of the EF-1 species are in the 300,000–500,000 range. It was apparent that in order to understand the differences among the various aspects of EF-1, it would be necessary to obtain highly purified preparations of both heavy (EF-1$_H$) and light (EF-1$_L$) forms of the enzyme so as to be able to compare their structure and properties. A procedure employing ammonium sulfate fractionation, column chromatography, and gel adsorption has been used for purification of the enzyme from calf brain and liver and has successfully separated high molecular weight species ($> 2 \times 10^5$ MW) from the lighter forms of the enzyme (Moon et al., 1973).

Figure 14 also shows the sucrose gradient analysis of the heavy species (EF-1_H) and light species (EF-1_L) from calf brain. Comparable results have been obtained with EF-1_H and EF-1_L from calf liver (Liu et al., 1974). Although EF-1_H from these sources was heterogeneous with respect to EF-1 activity, the preparations were essentially devoid of other contaminating proteins.

It was routinely observed that upon further purification or recentrifugation of EF-1_H, the sucrose gradient pattern of the enzyme often changed with a lower molecular weight species being formed. This was the first suggestion that the heavy form of the enzyme was an aggregate of a lighter species, the latter having a molecular weight of about 50,000. Although rather unstable, EF-1_L did not appear to break down to lower molecular weight active fractions. The properties of calf liver EF-1 were quite similar to the brain enzyme. However, it was possible to obtain EF-1_L from liver essentially homogeneous, whereas the brain EF-1_L very often contained about 20% contaminating proteins.

Disc gel analysis of EF-1 from brain, liver, wheat germ, and ascites cells provided the first direct evidence that EF-1_H is, in fact, an aggregate of EF-1_L (Moon et al., 1973; Golinska and Legocki, 1973; Liu et al., 1974; Drews et al., 1974). Typical results with liver EF-1 (Liu et al., 1974) are shown in Fig. 15. EF-1_H does not enter an anionic gel (5% pH 8.9), but shows a protein band at the top of the gel (Fig. 15A) which in separate experiments was shown to possess the enzymatic activity. In the presence of sodium dodecyl sulfate, EF-1_H gives one major protein band (Fig. 15B), which has a molecular weight of about 50,000 based on protein standards (Fig. 15C). EF-1_L on an SDS gel also shows a major band of molecular weight 50,000 (Fig. 15D), whereas in an anionic gel the EF-1 activity is associated with one major band (Fig. 15E). Similar results were obtained with calf brain EF-1 which was purified by a modification (Moon et al., 1973) of the same procedure used for liver. Recently, Iwasaki et al., (1974) have observed multiple forms of EF-1 in pig liver and have purified a light species which has a molecular weight of 53,000 and Nolan et al., (1974) have purified dimer, tetramer, and hexamer forms of EF-1 from Krebs ascites cells. Kotsiopoulos and Mohr (1975) have reported that EF-1 from rat liver contains 1 mole of zinc/mole of enzyme.

Further support for the postulate that EF-1_H is an aggregate of EF-1_L has come from amino acid analysis of purified EF-1 from calf liver (Liu et al., 1974). As shown in Table V, the purified preparations of EF-1_H and EF-1_L show quite similar amino acid compositions with the striking feature that they are both devoid of cysteine. The lack of cysteine was first

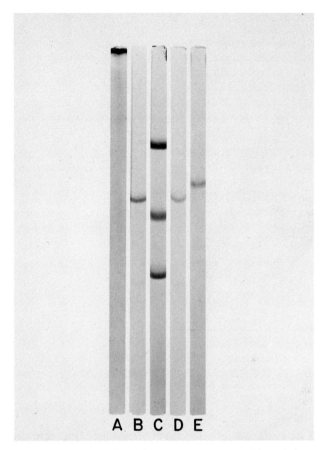

Fig. 15 Disc gel electrophoresis of calf liver EF-1$_H$ and EF-1$_L$. Disc gel electrophoresis of (A) EF-1$_H$ using a 5.0% polyacrylamide gel at pH 8.9. (B) 10% SDS polyacrylamide gel of EF-1$_H$. (C) 10% SDS polyacrylamide gel of standard proteins top-bottom; bovine serum albumin, 68,000; ovalbumin, 43,000; chymotrypsinogen, 24,000. (D) 10% SDS polyacrylamide gel of EF-1$_L$. (E) 5.0% polyacrylamide gel of EF-1$_L$ at pH 8.9. From Liu *et al.,* (1974).

observed with purified preparations of wheat germ EF-1.* In contrast, it should be noted that Bollini *et al.,* (1974) have reported that EF-1$_H$ from wheat germ is not a simple aggregate, since the enzyme consists of three subunits of differing molecular weight: 52,000, 47,000, and 17,000.

B. Nature of EF-1$_H$

The gel analysis of EF-1$_H$ from both brain and liver showed that there were essentially no other contaminating proteins in the preparation.

* A. B. Legocki (unpublished observations).

TABLE V

**Amino Acid Compositions of
EF-1$_H$ and EF-1$_L$ from Calf Liver[a]**

Amino acids	EF-1$_L$[b]	EF-1$_H$[b]
Lys	5.94	6.97
His	2.05	2.09
Arg	4.78	4.85
Asp	9.95	9.82
Thr	5.87	5.26
Ser	6.26	5.79
Glu	10.28	11.25
Gly	9.37	7.78
Ala	10.49	9.24
$\frac{1}{2}$-Cys	Trace	Trace
Val	7.49	8.11
Met	1.57	1.66
Ileu	5.69	6.34
Leu	10.42	10.34
Tyr	2.98	2.89
Phe	3.45	4.98
Pro	3.41	3.53

[a] The analysis was performed in a Joel Model 6AH analyzer. From Liu *et al.* (1974).

[b] Values are given as mole percentage of total amino acids.

Although it was conceivable that the high molecular weight aggregates of the enzyme formed spontaneously from EF-1$_L$, preliminary experiments attempting to demonstrate the conversion of the light to heavy form were unsuccessful. It was therefore decided to examine EF-1$_H$ for other non-protein components which might be necessary for aggregate formation or stability. In 1971 Hradec and co-workers reported that liver and tonsil EF-1 preparations contained a specific cholesterol ester, cholesterol-14-methylhexadecanoic acid, and that this cholesterol ester was essential for enzymatic activity. Based on this earlier report, the possible role of lipids in EF-1 from calf brain and liver was investigated. Although it was not possible to confirm the results of Hradec *et al.*, (1971) with regard to cholesterol-14-methylhexadecanoic acid, there is some evidence that lipid components are present in the very heavy aggregates of EF-1 from brain and liver. As shown in Table VI, calf brain and liver EF-1$_H$ preparations (> 500,000 MW) contain considerable amounts of total cholesterol (free plus esterified) and phospholipid (organic extractable phosphorous).

TABLE VI

Phospholipid and Cholesterol Content of Elongation Factor 1 Preparations[a]

Enzyme	Source	Specific activity	Phospholipid (pmoles/μg protein)	Total cholesterol (pmoles/μg protein)
EF-1$_H$	Calf brain	1000	70	42
EF-1$_H$	Calf brain	700	114	58
EF-1$_L$	Calf brain	800	16	4.1
EF-1$_L$	Calf brain	920	12	3.2
EF-1$_H$	Calf liver	950	487	25
EF-1$_H$	Calf liver	860	476	23
EF-1$_L$	Calf liver	1100	42	0.7
EF-1$_L$	Calf liver	1200	33	1.6

[a] For these experiments generally 50–300 units of enzyme were freeze-dried and the residue was extracted with $CHCl_3$methanol (2:1). Aliquots of the $CHCl_3$/methanol extract were assayed for phosphorus (Skipski and Barclay, 1969) and total cholesterol (Bondjers and Bjorkerud, 1971). From Legocki et al. (1974b).

However, EF-1$_L$ from both tissues contains little cholesterol and much lower amounts of phospholipid compared to EF-1$_H$, and it is not possible at this time to eliminate the possibility that the lipids present in EF-1$_L$ are due to contamination with small amounts of EF-1 aggregates or other lipid contaminating material. Figure 16 shows a thin-layer chromatogram of a lipid extract of calf brain EF-1$_H$ and EF-1$_L$. Cholesterol and cholesterol ester spots are clearly seen in the EF-1$_H$ preparation, but not in EF-1$_L$. About 60% of the total cholesterol was estimated to be unesterified. Liver EF-1$_H$ also shows a predominance of free cholesterol, although some cholesterol ester was also detected.

Fatty acid analysis by gas chromatography of the cholesterol ester fraction did not detect significant amounts of 14-methylhexadecanoic acid, although oleic, palmitic, and palmitoleic acids were present (Moon et al., 1973). Phosphatidylcholine and phosphotidylethanolamine were predominant phospholipids present in EF-1$_H$ from calf brain and liver, respectively (Legocki et al., 1974b). Although lipids are very likely present in high molecular weight aggregates of EF-1, preliminary studies with smaller EF-1 aggregates such as from reticulocytes have not shown the presence of significant amounts of lipid material.

C. Conversion of EF-1$_H$ to EF-1$_L$

The finding that the protease elastase could disaggregate reticulocyte and A. salina EF-1$_H$ first focused attention on the role of limited proteolysis in disaggregation of EF-1 (Kemper et al., 1976; Nombela et al., 1976). A typical experiment with reticulocyte EF-1 is shown in Fig. 17 where

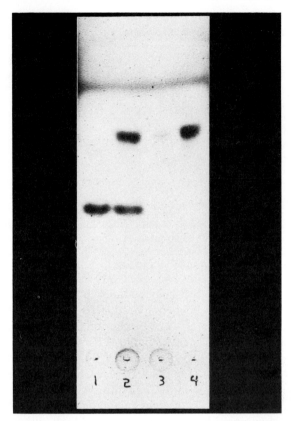

Fig. 16 Thin-layer chromatography of the lipid extract of calf brain EF-1$_H$. The plate was sprayed with $CHCl_3$ saturated with $SbCl_3$. (1) Cholesterol; (2) EF-1$_H$, 660 units; (3) EF-1$_L$, 660 units; (4) cholesterol 14-methylhexadecanoate. From Moon *et al.*, (1973).

the disaggregation of EF-1$_H$ by elastase is measured by sucrose gradient analysis. On disc gel analysis, it was shown that elastase cleaved EF-1 into two major fragments of about 30,000 and 15,000 daltons (Twardowski *et al.*, 1976). Yet, it should be noted that there was little, if any, loss of EF-1 activity associated with the disaggregation of EF-1$_H$ by elastase. The results with elastase prompted a reexamination of the earlier report that phospholipase C preparations could disaggregate EF-1 (Legocki *et al.*, 1974b). It was subsequently shown that phospholipase C preparations contained significant amounts of carboxypeptidase A activity and that this enzyme was responsible for conversion of EF-1$_H$ to EF-1$_L$ (Twardowski *et al.*, 1977). Thus both proteases and carboxypeptidases can cause disaggregation. A simple way to distinguish between protease and carboxypeptidase activity is that the former is inhibited by phenylmethenylsulfonyl fluoride whereas the latter is inhibited by o-phenanthrolene.

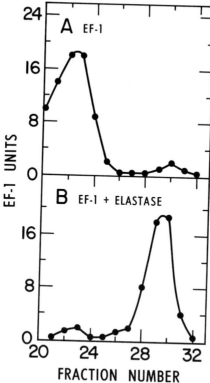

Fig. 17 The effect of elastase on the disaggregation of EF-1. About 150 units of EF-1 (one unit of EF-1 can bind 1 pmole of [^{14}C]-Phe-tRNA to ribosomes per 5 min) were incubated for 40 min at 37°C in 40 μl of a solution containing 50 mM Tris-Cl pH 7.4 and 2 mM DTT. Where indicated 2 μg of elastase were added. After incubation, 60 μl of 0.05 M Tris-HCl pH 7.4 containing 2 mM DTT were added. The mixture (100 μl) was layered on top of 4.2 ml of 5–20% sucrose gradient in a Spinco SW-56 centrifuge tube. The tubes were centrifuged for 2 hours at 50,000 rpm. After centrifugation, 32 fractions (8 drops each), beginning from the bottom, were collected. The EF-1 activity was determined in each fraction. A, EF-1 alone; B, EF-1 incubated with elastase. EF-1$_H$ appears in fractions 20–24, EF-1$_L$ in fractions 28–31 (Twardowski *et al.*, 1976).

The above findings can be summarized as follows. High molecular weight aggregates of EF-1 very often have lipids associated with the aggregates. However, although the lipids may help to stabilize the aggregate, they are very likely not required for aggregate formation or enzymatic activity. Disaggregation of EF-1$_H$ can be achieved by limited proteolysis of the polypeptide chains. Based on the effect of carboxypeptidase A, one could speculate that an essential fragment at the carboxyl-terminal end of the protein is required for aggregate formation. As will be discussed below, EF-1$_L$ is very likely the active form of the factor that interacts with the ribosome, and there is evidence that EF-1$_H$ is converted to EF-1$_L$ during AA-tRNA binding to ribosomes. This reaction does not involve limited proteolysis but may require GTP (Nombela *et al.*, 1976; Terrago *et al.*, 1973).

As also discussed below, there are two examples that have been reported in which the forms of EF-1 in tissues change during aging or development (Bolla and Brot, 1975; Slobin and Möller, 1975). Whether proteolytic cleavage of EF-1 serves as a means to regulate the amounts of EF-1$_H$ and EF-1$_L$ in tissues is not known. Further studies are obviously needed in this area since the possibility that control of total protein synthesis might occur by altering the elongation process could have many ramifications.

D. Comparison of EF-1$_H$ and EF-1$_L$ Activity

Both the heavy (aggregate) and light forms of the enzyme from calf brain and liver function in AA-tRNA binding to the ribosomes *in vitro*. Based on this assay, they have been purified to about the same specific activity, and appear to be close to homogeneity. Yet the characteristics of the EF-1 species from these sources are quite different. EF-1$_L$ is considerably more heat labile that EF-1$_H$, but forms more stable complexes with GTP and AA-tRNA than EF-1$_H$.

1. Heat Stability

Figure 18 shows a typical heat stability curve of calf brain EF-1$_H$ and EF-1$_L$ after heating at 52°. There is an initial loss of about 20% of the

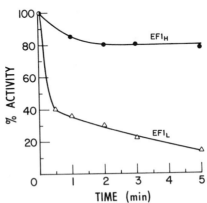

Fig. 18 Heat stability of EF-1$_H$ and EF-1$_L$ at 52°. In each case about 15 units of EF-1 in a total volume of 100 μl, containing 0.05 M Tris-Cl, pH 7.4; 0.05 M NH$_4$Cl and 0.01 M MgCl$_2$ were preincubated at 37° for 20 min before incubation at 52°. At the indicated times, 10 μl aliquots were removed and assayed for EF-1 activity.

EF-1$_H$ activity, whereafter the activity remains relatively constant. In comparison, EF-1$_L$ is quite labile with 70% of the activity destroyed within 2 min. There have been reports that the heat lability of EF-1 is increased by guanosine nucleotides and decreased by AA-tRNA (Ibuki and Moldave, 1968; Collins et al., 1972). These earlier reports did not compare the heavy and light forms of the enzyme, and although a detailed study has not been reported, these effects may be explained in part by the ability of the nucleotides to cause some dissociation of EF-1$_H$ to EF-1$_L$ and stabilization of the EF-1$_L$ by AA-tRNA.

2. Interaction with Guanosine Nucleotides

EF-1 is similar to the bacterial factor EF-Tu in that relatively stable complexes with GTP and GDP can be obtained; however, there is a striking difference between the heavy and light forms of EF-1 in their ability to form these nucleotide complexes. As seen in Table VII, EF-1$_L$ reacts with GTP to form an EF-1 · GTP complex 3–5 times more efficiently than equivalent amounts of EF-1$_H$ (Moon et al., 1973; Legocki et al., 1974a).

Figure 19 shows the kinetics of EF-1 · GTP and EF-1 · GDP formation with EF-1$_H$ and EF-1$_L$. The interaction of the nucleotides with both forms of the enzyme occurs at both 37° and 0°, although the rate and extent of the reaction is greater at the higher temperature. At all points examined using a Millipore procedure to assay for complex formation, it is seen that EF-1$_L$ and EF-1$_H$ bind more GTP than GDP per unit of activity. This is in contrast to the bacterial factor EF-Tu where the association constant for the formation of EF-Tu · GDP (3×10^8 M^{-1}) is 100 times higher than for EF-Tu · GTP. Recently, Nolan et al., (1974) have studied the interaction of dimer, tetramer, and hexamer forms of EF-1 from Krebs ascites cells with guanosine nucleotides. Association constants of 92 mM^{-1} were ob-

TABLE VII

Interaction of Different EF-1 Preparations with GTP and Aminoacyl-tRNA

Preparation	Specific activity	pmoles GTP bound unit of EF-1 activity	% of EF-1 in a ternary complex
EF-1$_H$	1400	0.30	20
EF-1$_H$	614	0.28	10
EF-1$_L$	500	0.91	58
EF-1$_L$	920	0.90	54

Formation of the ternary complex was determined by the ability of Phe-tRNA to cause EF-1 to pass through a nitrocellulose filter. From Moon et al. (1973).

EF-1$_H$

EF-1$_L$

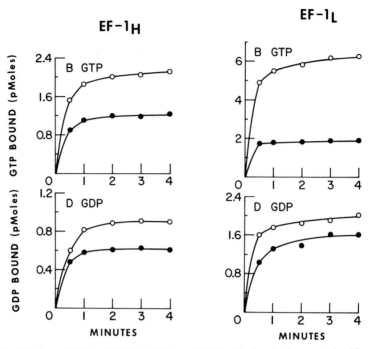

Fig. 19 Binding of GTP and GDP to EF-1$_L$ and EF-1$_H$. The incubations contained in a total volume of 50 μl: 50 mM Tris-HCl buffer (pH 7.4), 50 mM NH$_4$Cl, 10 mM MgCl$_2$, 6 units of EF-1$_L$, and where indicated, either 5×10^{-6} M [γ-^{32}P]GTP or 5×10^{-6} M [^{3}H]GDP. Incubations were done at either 0° (●—●) or 37° (○—○). For details see Legocki *et al.*, (1974a).

tained with GTP and GDP. There were no significant differences among the three forms of the enzyme. With wheat germ and *Artemia salina* EF-1, guanosine nucleotides were shown to cause disaggregation of EF-1$_H$ to EF-1$_L$ (Tarrago *et al.*, 1973; Nombela *et al.*, 1976).

3. Formation of a Ternary Complex

The EF-1 · GTP complex reacts with AA-tRNA to form a ternary complex:

$$\text{EF-1} \cdot \text{GTP} + \text{AA-tRNA} \longrightarrow \text{AA-tRNA} \cdot \text{EF-1} \cdot \text{GTP}$$

and there is also a marked difference in the behavior of EF-1$_H$ and EF-1$_L$ in this reaction. Again, EF-1$_L$ forms a much more stable complex compared to EF-1$_H$ (Table VII), as is evident by the ability of the ternary complex to pass through a Millipore filter. When the formation of ternary complex using calf brain EF-1 was examined by Sephadex chromatography (Moon *et al.*, 1972), it was apparent that any stable complex that could be detected contained EF-1$_L$. No AA-tRNA appeared in the void volume of these columns where EF-1$_H$ eluted (Moon *et al.*, 1972). Similar results have been obtained with EF-1 from wheat germ (Tarrago *et al.*, 1973; Bollini *et al.*, 1974), i.e., a stable ternary complex isolated by sucrose gradient analysis contained only the light form of the enzyme. It has also been noted with partially purified preparations of calf liver EF-1$_H$ that incubation of the enzyme with GTP and AA-tRNA results in some dissociation of the enzyme to a lighter species. Recently rat liver EF-1 has been shown to bind stoichiometric amounts of GTP and to efficiently form a ternary complex (Nagata *et al.*, 1976a).

E. Reaction of an AA-tRNA · EF-1 · GTP Complex with the Ribosomes

A ternary complex containing approximately one equivalent of [^{14}C]Phe-tRNA and [γ-^{32}P, ^{3}H]GTP bound to EF-1 has been obtained using a Millipore filter technique (Weissbach *et al.*, 1973). As indicated above, these complexes very likely contain EF-1$_L$. Upon interaction of such a complex with the ribosome in the presence of poly (U), there is evidence of transfer of the Phe-tRNA to the ribosome (^{14}C bound to the filter) and hydrolysis of GTP with the formation of EF-1 · GDP (^{3}H but not ^{32}P bound to the filter) as seen in Table VIII. In the absence of poly(U), it is of interest that the intact ternary complex appears to bind to the ribosome with no GTP hydrolysis (equivalent ^{14}C, ^{32}P, ^{3}H bound to the filter)

TABLE VIII

The Reaction of the Ternary Complex with Brain Ribosomes[a]

EF-1 in the incubations	pmoles bound + Poly U			pmoles bound − Poly (U)		
	^{14}C	^{3}H	^{32}P	^{14}C	^{3}H	^{32}P
Calf brain	1.7	1.7	0	1.5	1.4	1.4
Sheep brain	1.3	1.3	0	1.2	1.4	1.2
Calf liver	0.9	1.0	0	0.9	1.0	0.9
Reticulocytes	0.6	0.7	0	0.6	0.7	0.6
No EF-1	<0.1	<0.2	<0.1	<0.1	<0.2	<0.1

[a] The [^{14}C]Phe-tRNA·EF-1·[^{32}P-^{3}H]GTP complexes in the nitrocellulose filtrates were prepared as described by Weissbach *et al.* (1973). Aliquots of the freshly prepared filtrates containing between 1.2–5 pmoles of EF-1 activity were incubated at 0° with calf brain ribosomes in the presence or absence of poly U. The retention of ^{14}C, ^{3}H, and ^{32}P on the nitrocellulose filters was then determined. Values are corrected for any radioactivity retained by the filter in the absence of ribosomes. From Weissbach *et al.* (1973).

suggesting a high affinity of the ternary complex for the ribosome. These results indicate that poly(U) is required to bind the ternary complex at the proper ribosomal site to permit GTP hydrolysis to occur. As seen in Table VIII, similar results were obtained with EF-1 from several sources. These data suggest that EF-1 · GDP is a product of the binding reaction and, since EF-1 · GDP does not bind AA-tRNA (Moon and Weissbach, 1972), it is important, based on the similarity of the reactions seen in the prokaryote system, to consider the mechanism by which EF-1 · GTP is regenerated.

F. How Does EF-1 Function Catalytically?

It is obvious from the data above that the ratio of AA-tRNA bound to ribosomes/GTP bound to EF-1 is much greater for EF-1_H than EF-1_L. This could be interpreted to mean that EF-1_H is functioning catalytically, whereas EF-1_L functions stoichiometrically. Such a proposal is extremely attractive, for it would suggest that EF-1_H contains components required for EF-1 to recycle. Bollini *et al.*, (1974) favor this view since EF-1_H from wheat germ is composed of different subunits, only one presumably being EF-1_L. Other studies with calf brain and liver EF-1 (Moon *et al.*, 1973; Liu *et al.*, 1974) and Krebs ascites EF-1 (Drew *et al.*, 1974) do not support this view. In these tissues EF-1_H appears to be an aggregate of EF-1_L with no evidence of other proteins being present. In addition, the highly purified preparations of EF-1_H and EF-1_L from brain and liver have very close specific activities, suggesting that the factors do not differ in their ability to act catalytically. However, there is support for the presence of factors that may be involved in EF-1 recycling. Both Iwasaki *et al.*, (1973) and Prather *et al.*, (1974) have obtained protein factors from pig liver (EF-1β and reticulocyte preparations that stimulate EF-1 activity. EF-1β appears analogous (Nagata *et al.*, 1976b) to the prokaryote factor EF-ts which is believed to function in the regeneration of EF-Tu · GTP from EF-Tu · GDP by catalyzing a nucleotide exchange reaction described above (Section III, A). Attempts to show the presence of a factor that catalyzes such an exchange with calf brain enzyme EF-1_H

$$\text{EF-1} \cdot \text{GDP} + \text{GTP} \rightleftharpoons \text{EF-1} \cdot \text{GTP} + \text{GDP}$$

have not yet been successful, which might indicate that if a similar factor exists in this tissue, it has been separated from EF-1_H during the purification.

Although it is apparent that factors which stimulate EF-1 activity have now been detected (Iwasaki *et al.*, 1973; Prather *et al.*, 1974), it is not yet certain that they are similar to EF-Ts in their function. It should also be noted that in the bacterial system EF-Tu · GDP is much more tightly bound than EF-Tu · GTP (see above), so that a mechanism to regenerate EF-Tu · GTP after reaction of the ternary complex with the ribosomes is essential. With EF-1, however, the GTP complexes appear to be favored so that the need for a factor with EF-Ts- like activity may not be required if the ratio in the tissues of GTP/GDP is relatively high.

G. Summary of the Interactions of EF-1

From the available data, it is now possible to summarize some of the known reactions of EF-1 with guanosine nucleotides, AA-tRNA, and the ribosome. A major uncertainty is whether the different aggregate forms of the enzyme play specific roles. With EF-1 (MW 50,000), it appears that the interactions are similar to EF-Tu. As summarized in Fig. 20, EF-1_L reacts with GTP to form an EF-1 · GTP complex that can react with AA-tRNA to form a ternary complex.

As in the bacterial system, the ternary complex can interact with the ribosome, the AA-tRNA being transferred presumably to the A site on the ribosome. GTP hydrolysis appears to take place, and although EF-1 · GDP has not been definitively identified as a product of the reaction, the data suggest that this is so. The recycling of EF-1 · GDP to EF-1 · GTP is not fully understood. The stimulatory factors which have been noted by two laboratories (Iwasaki *et al.*, 1973; Prather *et al.*, 1974) may function at this point in a manner similar to EF-Ts. It is also not clear whether EF-1_H, which is a predominant form of EF-1 seen in most mammalian tissues, functions in the same way as EF-1_L. There is evidence that EF-1_H disaggregates to EF-1_L in the presence of guanosine nucleotides (Nombela *et al.*, 1976; Tarrago *et al.*, 1973) indicating that disaggregation occurs in this initial reaction. The results obtained with calf brain and

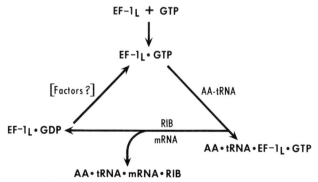

Fig. 20 Summary of interactions of EF-1_L. AA-tRNA, aminoacyl-tRNA; RIB, ribosomes; mRNA, messenger RNA.

wheat germ EF-1 support this view since the ternary complexes that have been isolated always contained the light form of the enzyme. These results can be interpreted to indicate that the light form of the enzyme is, in fact, the biologically active form, and that when EF-1$_H$ does function in AA-tRNA binding to the ribosome, it is broken down during the course of the reaction to EF-1$_L$ which participates in ternary complex formation. However, it should be stressed that it is quite possible that the aggregate forms of EF-1 do, in fact, form ternary complexes and interact with the ribosome, but these complexes have not been stable enough to be isolated by the procedures used.

H. Possible Role of the Aggregate Forms of EF-1

The occurrence of multiple species of EF-1 naturally leads one to think that the interconversion between the heavy and light forms of the enzyme could be a regulatory mechanism controlling the overall rate of protein synthesis. Preliminary experiments, both *in vivo* and *in vitro,* however, have not demonstrated any marked change in the EF-1 pattern in liver after animals have been exposed to several hormones and a variety of physiological conditions (Liu *et al.,* 1974).

Changes in the ratio of the light/heavy species have been seen in nematodes (Bolla and Brot, 1975) and Slobin and Möller (1975) have reported that EF-1$_L$ is seen in *Artemia salina* embryos 46 hr after hatching, whereas EF-1$_H$ is the predominant species before hatching. Thus there is some evidence that indicates that the multiple species of EF-1 may be in a state of flux. It is of interest that there has been one report (Lanzani *et al.,* 1974) that cyclic GMP causes a small conversion of the heavy to the light form of the wheat germ enzyme. In addition, cGMP has been reported to stimulate *in vitro* protein synthesis (Varrone *et al.,* 1973). If so, this could provide a mechanism in the cell to alter the rate of protein synthesis. However, a significant effect of cGMP was not seen with *A. salina* EF-1 (Slobin and Möller, 1975; Nombela *et al.,* 1976).

VII. Conclusions

Proteins essential for binding aminoacyl-tRNA (AA-tRNA) to ribosomes in the process of peptide chain elongation have been purified and characterized from bacteria and mammalian tissues. The prokaryotic factor, called elongation factor Tu, forms a stable complex with GTP and AA-tRNA. The protein is a single polypeptide chain of about 42,000 daltons molecular weight. The protein contains two essential sulfhydryl groups; one is necessary for binding GDP or GTP, and the other for interaction with AA-tRNA; as normally isolated EF-Tu contains one equivalent of tightly bound GDP, $K_{diss} = 3 \times 10^{-9} M$.

An auxiliary protein, EF-Ts, is required to catalyze the exchange between the slowly dissociating EF-Tu·GDP and free GDP or GTP. EF-Ts rapidly displaces GDP from EF-Tu, forming an EF-Tu·EF-Ts complex. EF-Tu·EF-Ts in turn is readily dissociated by GDP or GTP forming EF-Tu·GDP or EF-Tu·GTP. These reactions produce a rapid equilibration between GTP and EF-Tu·GDP.

Although EF-Tu binds either GDP or GTP, only EF-Tu·GTP interacts with aminoacyl-tRNA. Conformational differences between the two EF-Tu complexes have been revealed by differences in their tritium exchange spectra and their affinities for a fluorescent hydrophobic dye. The conformational change produced when GTP binds to EF-Tu may expose a site on the protein for binding AA-tRNA.

When the ternary complex binds to ribosomes, rapid hydrolysis of GTP occurs, AA-tRNA remains bound to the ribosomes, and EF-Tu·GDP is released to be reconverted to EF-Tu·GTP by reaction with EF-Ts and GTP in another round of the binding cycle. The function of GTP hydrolysis appears to be to release EF-Tu from the ribosome and thus allow AA-tRNA to participate in peptide bond formation. The function of EF-Tu may be to position the AA-tRNA correctly on the ribosome.

The analogous factor in eukaryotes is elongation factor 1 (EF-1). In most tissues multiple species of EF-1 exist with molecular weights ranging from about 5×10^4 to greater than 1×10^6. The isolation and characterization of the heavy (EF-1$_H$) and light forms of the enzyme (EF-1$_L$) have shown that EF-1$_H$ is an aggregate of EF-1$_L$, the latter having a molecular weight of about 5×10^4. In tissues such as brain and liver, the very large aggregates contain cholesterol and phospholipids. Disaggregation of EF-1$_H$ can result by treatment of the factor with elastase or carboxypeptidase A. The heavy and light forms of the enzyme appear equally active in AA-tRNA binding to ribosomes, although EF-1$_L$ forms more stable intermediate complexes with GTP (EF-1·GTP) and AA-tRNA (AA-tRNA·EF-1·GTP) than EF-1$_H$. A Phe-tRNA·EF-1·GTP complex, like the corresponding prokaryote complex, reacts with the ribosome in the presence of poly(U) with transfer of the Phe-tRNA to the ribosome, hydrolysis of GTP, and the formation of EF-1·GDP. The subsequent conversion of EF-1·GDP to EF-1·GTP, an exchange reaction catalyzed by EF-Ts in the prokaryote system, may, be catalyzed by EF-1β, a factor that stimulates EF-1 activity.

REFERENCES

Allende, J., Monro, R., and Lipmann, F. (1964). *Proc. Natl. Acad. Sci. U.S.A.* **51**, 1211–1216.
Arai, K., Kawakita, M., and Kaziro, Y. (1972). *J. Biol. Chem.* **247**, 7029–7037.

Arai, K., Kawakita, M., Kaziro, Y., Kondo, T., and Ui, N. (1973). *J. Biochem. (Tokyo)* **73**, 1095–1105.

Arai, K., Kawakita, M., and Kaziro, Y. (1974a). *J. Biochem. (Tokyo)* **76**, 293–306.

Arai, K., Kawakita, M., Kaziro, Y., Maeda, T., and Ohnishi, S. (1974b). *J. Biol. Chem.* **249**, 3311–3313.

Arlinghaus, R., Favelukes, G., and Schweet, R. (1963). *Biochem. Biophys. Res. Commun.* **11**, 92–96.

Beaud, G., and Lengyel, P. (1971). *Biochemistry* **10**, 4899–4906.

Beres, L., and Lucas-Lenard, J. (1973). *Arch. Biochem. Biophys.* **154**, 555–562.

Bernardi, A., and Leder, P. (1970). *J. Biol. Chem.* **245**, 4263–4268.

Bewley, T. A., Brovetto-Cruz, J., and Li, C. H. (1969). *Biochemistry* **8**, 4701–4708.

Bishop, J. O., and Schweet, R. S. (1961). *Biochem. Biophys. Acta* **49**, 235–236.

Blumenthal, T., Landers, T. A., and Weber, K. (1972). *Proc. Natl. Acad. Sci. U.S.A.* **69**, 1313–1317.

Bodley, J. W., and Gordon, J. (1974). *Biochemistry* **13**, 3401–3405.

Bolla, R., and Brot, N. (1975). *Arch. Biochem. Biophys.* **169**, 227–236.

Bollini, R., Soffientini, A. N., Bertani, A., Lanzani, G. A. (1974). *Biochemistry* **13**, 5421–5425.

Bondjers, G., and Bjorkerud, S. (1971). *Anal. Biochem.* **42**, 363–371.

Cashel, M., and Gallant, J. (1969). *Nature (London)* **221**, 838–841.

Chinali, G., Sprinzl, M., Parmeggiani, A., and Cramer, F. (1974). *Biochemistry* **13**, 3001–3010.

Collins, J. F., Moon H. M., and Maxwell, E. S. (1972). *Biochemistry* **11**, 4187–4194.

Crane, L. J., and Miller, D. L. (1974). *Biochemistry* **13**, 933–939.

Drews, J., Bednarik, K., and Grasmuk, H. (1974). *Eur. J. Biochem.* **41**, 217–227.

Ertel, R., Redfield, B., Brot, N., and Weissbach, H. (1968a). *Arch. Biochem. Biophys.* **128**, 331–338.

Ertel, K., Brot, N., Redfield, B., Allende, J. E., and Weissbach, H. (1968b). *Proc. Natl. Acad. Sci. U.S.A.* **59**, 861–868.

Fahnestock, S., Weissbach, H., and Rich, A. (1972). *Biochim. Biophys. Acta* **269**, 62–66.

Felicetti, L., and Lipmann, F. (1968). *Arch. Biochem. Biophys.* **125**, 548–557.

Furano, A. (1975). *Proc. Natl. Acad. Sci.* **72**, 4780–4784.

Gallant, J., Erlich, H., Hall, B., and Laffler, T. (1970). *Cold Spring Harbor Symp. Quant. Biol.* **35**, 397–405.

Gasior, E., and Moldave, K. (1965). *J. Biol. Chem.* **240**, 3346–3352.

Gausing, K. (1972). *J. Mol. Biol.* **71**, 529–545.

Golinska, B., and Legocki, A. B. (1973). *Biochim. Biophys. Acta* **324**, 156–170.

Gordon, J. (1968). *Proc. Natl. Acad. Sci. U.S.A.* **59**, 179–183.

Gordon, J. (1969). *J. Biol. Chem.* **244**, 5680–5686.

Gordon, J. (1970). *Biochemistry* **9**, 912–917.

Gordon, J., Baron, L. S., and Schweiger, M. (1972). *J. Bacteriol.* **110**, 306–312.

Hachmann, J., Miller, D. L., and Weissbach, H. (1971). *Arch. Biochem. Biophys.* **147**, 457–466.

Hardesty, B., Arlinghaus, R., Shaeffer, J., and Schweet, R. (1963). *Cold Spring Harbor Symp. Quant. Biol.* **28**, 215–222.

Haseltine, W., and Block, R. (1968). *Proc. Natl. Acad. Sci. U.S.A.* **70**, 1564–1568.

Hradec, J., Dusek, Z., Bermek, E., and Mattaeu, H. (1971). *Biochem. J.* **123**, 959–966.

Ibuki, F., and Moldave, K. (1968). *J. Biol. Chem.* **243**, 44–50.

Iwasaki, K., Mizumoto, K., Tanaka, M., and Kaziro, Y. (1973). *J. Biochem. (Tokyo)* **74**, 349–352.

Iwasaki, K., Nagata, S., Mizumoto, K., and Kaziro, Y. (1974). *J. Biol. Chem.* **249**, 5008–5010.

Jacobson, G. R., and Rosenbusch, J. P. (1976). *Nature* **261**, 23–26.

Jaskunas, S. R., Lindahl, L., Nomura, M., and Burgess, R. R. (1975). *Nature (London)* **257**, 458–462.

Jekowsky, E. J. (1976). Ph.D. Thesis, Massachusetts Institute of Technology, Cambridge.

Jerez, C., Sandoval, A., Allende, J., Henes, C., and Ofengand, J. (1969). *Biochemistry* **8**, 3006–3014.

Kemper, W. M., Merrick, W. C., Redfield, B., Liu, C. K., and Weissbach, H. (1976). *Arch. Biochem. Biophys.* (in press).

Kim, S. H., Sussman, J. L., Suddath, F. L., Quigley, G. J., McPherson, A., Wang, A.H.I., Seeman, N.C., and Rich, A. (1974). *Proc. Natl. Acad. Sci. U.S.A.* **71**, 4970–4974.

Kotsiopoulos, P. S., and Mohr, S. C. (1975). *Biochem. Biophys. Res. Commun.* **67**, 979–987.

Krauskopf, M., Chen, C. M., and Ofengand, J. (1972). *J. Biol. Chem.* **247**, 842–850.

Krisko, I., Gordon, J., and Lipmann, F. (1969). *J. Biol. Chem.* **244**, 6117–6123.

Landers, T. A., Blumenthal, T., and Weber, K. (1974). *J. Biol Chem.* **249**, 5801–5808.

Lanzani, G. A., Giannattasio, M., Manzocchi, L. A., Bollini, R., Soffientini, A. N. and Macchia, V. (1974). *Biochem. Biophys. Res. Commun.* **58**, 172–177.

Leder, P. (1973). *Adv. Protein Chem.* **27**, 213–242.

Legocki, A. B., Redfield, B., and Weissbach, H. (1974a). *Arch. Biochem. Biophys.* **161**, 709–712.

Legocki, A. B., Redfield, B., Liu, C. K., and Weissbach, H. (1974b). *Proc. Natl. Acad. Sci. U.S.A.* **71**, 2179–2182.

Lengyel, P., and Söll, D. (1969). *Bacteriol. Rev.* **33**, 264–301.

Levitski, A., and Koshland, D. E., Jr. (1972). *Biochemistry* **11**, 241–246.

Lin, S. Y., McKeehan, W. L., Culp, W., and Hardesty, B. (1969). *J. Biol. Chem.* **244**, 4340–4350.

Liu, C. K., Legocki, A. B., and Weissbach, H. (1974). *In* "Lipmann Symposium: Energy, Regulation and Biosynthesis in Molecular Biology" (D. Richter, ed.), pp. 384–398. de Gruyter, Berlin.

Lucas-Lenard, J., and Beres, L. (1974). *In* "The Enzymes" (P. D. Boyer, ed.), 3rd ed., Vol. 10, pp. 53–86.

Lucas-Lenard, J., and Haemni, A. L. (1968). *Proc. Natl. Acad. Sci. U.S.A.* **59**, 554–560.

Lucas-Lenard, J., and Lipmann, F. (1966). *Proc. Natl. Acad. Sci. U.S.A.* **55**, 1562–1566.

Lucas-Lenard, J., and Lipmann, F. (1971). *Annu. Rev. Biochem.* **40**, 409–448.

Lucas-Lenard, J., Tao, P., and Haenni, A. (1969). *Cold Spring Harbor Symp. Quant. Biol.* **34**, 455–462.

Lupker, J. H., Verschoor, G. J., DeRooij, F.W.M., Rörsch, A., and Bosch, L. (1974). *Proc. Natl. Acad. Sci. U.S.A.* **71**, 460–463.

Maaløe, O., and Kjeldgaard, N. O. (1966). *In* "Control of Macromolecular Synthesis" (O. Maaløe, and N. O. Kjeldgaard, eds.), p. 90. Benjamin, New York.

McKeehan, W. L., and Hardesty, B. (1969). *J. Biol. Chem.* **244**, 4330–4339.

Miller, D. L. (1972). *Proc. Natl. Acad. Sci. U.S.A.* **69**, 752–755.

Miller, D. L., and Weissbach, H. (1969). *Arch. Biochem. Biophys.* **132**, 146–150.

Miller, D. L., and Weissbach, H. (1970a). *Arch. Biochem. Biophys.* **141**, 26–37.

Miller, D. L., and Weissbach, H. (1970b). *Biochem. Biophys. Res. Commun.* **38**, 1016–1022.

Miller, D. L., and Weissbach, H. (1974). *In* "Methods in Enzymology" (L. Grossman and K. Moldave, eds.), Vol. 30, pp. 219–232. Academic Press, New York.

Miller, D. L., Hachmann, J., and Weissbach, H. (1971). *Arch. Biochem. Biophys.* **144**, 115–121.

Miller, D. L., Cashel, M., and Weissbach, H. (1973). *Arch. Biochem. Biophys.* **154**, 675–682.

Moon, H. M., and Weissbach, H. (1972). *Biochem. Biophys. Res. Commun.* **46**, 254–262.

Moon, H. M., Redfield, B., and Weissbach, H. (1972). *Proc. Natl. Acad. Sci. U.S.A.* **69**, 1249–1252.

Moon, H. M., Redfield, B., Millard, S., Vane, F., and Weissbach, H. (1973). *Proc. Natl. Acad. Sci. U.S.A.* **70**, 3282–3286.

Nagata, S., Iwasaki, K., and Kaziro, Y. (1976a). *Arch. Biochem. Biophys.* **172**, 168–177.

Nagata, F., Motoyoshi, K., and Iwasaki K. (1976b). *Biochem. Biophys. Res. Commun.* **71**, 933–938.

Nishizuka, Y., and Lipmann, F. (1966). *Proc. Natl. Acad. Sci. U.S.A.* **55**, 212–219.

Nishizuka, Y., Lipmann, F., and Lucas-Lenard, J. (1968). *In* "Methods in Enzymology" (L. Grossman and K. Moldave, eds.), Vol. 12. Part B, pp. 708–721. Academic Press, New York.

Nolan, R. D., Grasmuk, H., Hogenauer, G., and Drews, J. (1974). *Eur. J. Biochem.* **45**, 601–609.

Nombela, C., Redfield, B., Ochoa, S., and Weissbach, H. (1976). *Eur. J. Biochem.* **65**, 395–402.

Ofengand, J., and Chen, C. M. (1972). *J. Biol. Chem.* **247**, 2049–2058.

Ono, Y., Skoultchi, A., Waterson, J., and Lengyel, P. (1969). *Nature (London)* **222**, 645–648.

Page, M. I., and Jencks, W. P. (1971). *Proc. Natl. Acad. Sci. U.S.A.* **68**, 1678–1683.

Parmeggiani, A., and Gottschalk, E. M. (1969). *Cold Spring Harbor Symp. Quant. Biol.* **34**, 377–384.

Prather, N., Ravel, J. M., Hardesty, B., and Shive, W. (1974). *Biochem. Biophys. Res. Commun.* **57**, 578–583.

Printz, M. P., and Miller, D. L. (1973). *Biochem. Biophys. Res. Commun.* **53**, 149–156.

Ravel, J. (1967). *Proc. Natl. Acad. Sci. U.S.A.* **57**, 1811–1816.

Ravel, J. M., and Shorey, R. L. (1971). *In* "Methods in Enzymology" (K. Moldave and G. Grossman, eds.), Vol. 20, pp. 306–316. Academic Press, New York.

Ravel, J. M., Shorey, R. L., and Shive, W. (1967). *Biochem. Biophys. Res. Commun.* **29**, 68–73.

Richman, N., and Bodley, J. W. (1973). *J. Biol. Chem.* **248**, 381–383.

Ringer, D., and Chladek, S. (1975). *Proc. Natl. Acad. Sci. U.S.A.* **72**, 2950–2954.

Robertus, J. D., Ladner, J. E., Finch, J. T., Rhodes, D., Brown, R. S., Clark, B.F.C., and Klug, A. (1974). *Nature (London)* **350**, 546–551.

Schneir, M., and Moldave, K. (1968). *Biochim. Biophys. Acta* **166**, 58–67.

Schulman, L. H., Pelka, H., and Sundari, R. M. (1974). *J. Biol. Chem.* **249**, 7102–7110.

Schwartz, U., Menzel, H. M., and Gassen, H. G. (1976). *Biochemistry* **15**, 2484–2490.

Shorey, R. L. Ravel, J. M., Garner, C. W., and Shive, W. (1969). *J. Biol. Chem.* **244**, 4555–4564.

Shorey, R. L., Ravel, J. M., and Shive, W. (1971). *Arch. Biochem. Biophys.* **146**, 110–117.

Shulman, R. G., Hilbers, C. W., and Miller, D. L. (1974). *J. Mol. Biol.* **90**, 601–607.

Skipski, V. P., and Barclay, M. (1969). *In* "Methods in Enzymology" (J. M. Lowenstein, ed.), Vol. 14, pp. 530–598. Academic Press, New York.

Slobin, L. I., and Möller, W. (1975). *Nature (London)* **258**, 252–254.

Smith, J. D., and Celis, J. E. (1973). *Nature (London), New Biol.* **243**, 66–71.

Sneden, D., Miller, D. L., Kim, S. H., and Rich, A. (1973). *Nature* **241**, 530–531.

Storm, D. R., and Koshland, D. E., Jr. (1970). *Proc. Natl. Acad. Sci. U.S.A.* **66**, 445–452.

Tarrago, A., Allende, J. E., Redfield, B., and Weissbach, H. (1973). *Arch. Biochem. Biophys.* **159**, 353–361.

Thang, M. N., Dondon, L., Thang, D.C., and Rether, B. (1972). *FEBS Lett.* **26**, 145–150.

Travers, A. A. (1973). *Nature (London)* **244**, 15–18.

Twardowski, T., Redfield, B., Kemper, W. M., Merrick, W. C., and Weissbach, H. (1976). *Biochem. Biophys. Res. Commun.* **71**, 272–279.

Twardowski, T., Hill, J., and Weissbach, H. (1977). *Arch. Biochem. Biophys.* (in press).

Uno, H., Oyabu, S., Ohtsuka, E., and Ikehara, M. (1971). *Biochim. Biophys. Acta* **228**, 282–288.

Varrone, S., Di Lauro, R., and Macchia, V. (1973). *Arch. Biochem. Biophys.* **157**, 334–338.

Waterson, J., Beaud, G., and Lengyel, P. (1970). *Nature (London)* **227**, 34–38.

Weissbach, H., Miller, D. L., and Hachmann, J. (1970a). *Arch. Biochem. Biophys.* **137**, 262–269.

Weissbach, H., Redfield, B., and Hachmann, J. (1970b). *Arch. Biochem. Biophys.* **141**, 384–386.

Weissbach, H., Redfield, B., and Brot, N. (1971a). *Arch. Biochem. Biophys.* **144**, 224–229.

Weissbach, H., Redfield, B., and Brot, N. (1971b). *Arch. Biochem. Biophys.* **145**, 676–684.

Weissbach, H., Redfield, B., and Moon, H. M. (1973). *Arch. Biochem. Biophys.* **156**, 267–275.

Wolf, H., Chinali, G., and Parmeggiani, A. (1974). *Proc. Natl. Acad. Sci. U.S.A.* **71**, 4910–4914.

7

Translocation

NATHAN BROT

I. Introduction

Perhaps one of the most esoteric and complicated series of reactions in protein synthesis is the process by which the ribosome moves relative to the messenger RNA. This movement, termed translocation, is the means by which the ribosome traverses the messenger RNA a codon at a time from its initiation binding site, at or near the 5'-end of the messenger, to the termination codon located toward the 3'-end.

375

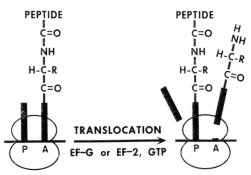

Fig. 1 Schematic representation of translocation. Solid bars represent the tRNA molecule.

The process, depicted schematically in Fig. 1, shows a ribosome with a peptidyl-tRNA on the acceptor site (A site) and a deacylated tRNA on the peptidyl site (P site). During translocation, the deacylated tRNA is ejected from the P site, the peptidyl-tRNA moves from the A to the P site, and the ribosome moves precisely one codon closer to the 3′-end of the messenger RNA. This results in a new codon appearing in the A site allowing a new aminoacyl-tRNA to bind to the vacant A site. In both prokaryotes and eukaryotes, translocation involves GTP hydrolysis and is catalyzed by supernatant proteins, referred to as elongation factor G (EF-G) and elongation factor 2 (EF-2), respectively. Although similar in many respects, there are significant differences between the two reactions, and the individual characteristics of each of the systems will be described below. For other reviews of protein synthesis, the reader is referred to recent articles by Hazelkorn and Rothman-Denes (1973), Leder (1973), and Lucas-Lenard and Beres (1974).

II. Purification and Properties of Elongation Factor G

A. Assay

Early studies had shown the presence of two complementary factors in *Escherichia coli* extracts which were required for amino acid polymerization (Allende *et al.*, 1964). It was subsequently shown that one of these fractions possessed a ribosomal dependent ability to hydrolyze GTP (Conway and Lipmann, 1964) and, because of this catalytic activity, was named factor G (Nishizuka and Lipmann, 1966b). EF-G can also be assayed by its ability to complement EF-Tu and EF-Ts in polyphenylalanine synthesis; however, the most convenient assay is based on its ability to hydrolyze GTP to GDP and P_i in the presence of ribosomes. The details of this assay have been previously described (Conway and Lipmann, 1964;

Nishizuka and Lipmann, 1966b), and involves an incubation mixture containing in a total volume of 250 μl; 12.5 μmoles Tris-Cl, pH 7.4; 40 μmoles NH$_4$Cl; 3 μmoles β-mercaptoethanol; 2.5 μmoles MgCl$_2$; 2–3 A_{260} 70 S ribosomes; and 25 nmoles [γ-^{32}P]GTP. After incubating for 10 min at 30°, the reaction is stopped by the successive addition of 0.5 ml of 0.02 M silicotungstic acid in 0.02 N H$_2$SO$_4$; 1.2 ml of 1 mM KPO$_4$, pH 6.8, and 0.5 ml of 5% ammonium molybdate in 4 N H$_2$SO$_4$. The [^{32}P]phosphomolybdate complex that is formed is then extracted into 2.5 ml of an isobutanol:benzene mixture (1:1) and an aliquot assayed for radioactivity. Since crude preparations contain various enzymes leading to the hydrolysis of GTP, it is of importance to include in the assay a control in which ribosomes are omitted. The amount of ^{32}P liberated due to EF-G is calculated as the difference obtained between an incubation performed in the presence and absence of ribosomes. One unit of elongation factor G has been defined as that amount of enzyme which in the presence of excess ribosomes will cause the hydrolysis of 1 nmole of GTP in 10 min at 30°.

This assay, as it was originally described, detected nanomole quantities of P$_i$ formed during the hydrolysis; however, because of the very high specific activity of [γ-^{32}P]GTP now available, it can be used to assay the formation of picomole quantities of ^{32}P$_i$.

B. Purification and Crystallization

Elongation factor G has been purified to apparent homogeneity in a number of laboratories using essentially standard techniques of ammonium sulfate precipitation, DEAE and hydroxyapatite chromatography, and gel filtration (Nishizuka and Lipmann, 1966c; Kaziro et al., 1969; Gordon, 1969; Leder et al., 1969a; Parmeggiani, 1968). Recently, Rohrbach et al. (1974) have shown that EF-G preparations, when purified as described in the above reports, contain a small but significant contamination by enzymes utilizing GTP or GDP as substrates, and they have described a new purification procedure that eliminates these contaminants. A summary of the purification procedure is shown in Table I This method employs a ribosome affinity purification step which takes advantage of the fact that EF-G will bind to ribosomes in the presence of GTP and fusidic acid. The ribosomes containing the bound EF-G are recovered by centrifugation after which the EF-G is eluted from the ribosome with a buffer containing 1 M NH$_4$Cl. This step yields a 25-fold purification of the enzyme with a 56% recovery. The total purification of the enzyme is about 180-fold with a yield of 23%. It does not appear, however, that the results obtained earlier utilizing somewhat less purified EF-G preparations differ from those seen with the highly purified EF-G. EF-G during logarithmic growth has been estimated to comprise about

TABLE I

Summary of Purification of *Escherichia coli* EF-G[a]

Procedure	Volume (ml)	Units	Protein (A_{260}/ml)	Purity (units/A_{260})	Yield (%)	Purification (−fold)
S-30	728	2,180,000	85	35	100	1
Affinity	148	1,220,000	9.58	861	56	25
Heat	217	945,000	2.87	1520	43	43
G-150	160	667,000	1.10	3790	31	108
DEAE-Sephadex	11	504,000	7.39	6290	23	180

[a] From Rohrbach *et al.* (1974).

2–3% of the supernatant protein (Leder *et al.*, 1969a; Parmeggiani and Gottschalk, 1969b), and since the yields from the purification procedures are in the range of 15–30%, relatively large amounts of the purified protein can be obtained. EF-G can be readily crystallized from 50–60% ammonium sulfate as either fine needles or round crystalline plates (Kaziro and Inoue, 1968; Parmeggiani, 1968; Kaziro *et al.*, 1969; Leder *et al.*, 1969a).

C. Physical Properties of EF-G

The protein consists of a single polypeptide chain with a molecular weight of 72,000–84,000 (Kaziro and Inoue, 1968; Parmeggiani, 1968; Leder *et al.*, 1969a; Kaziro *et al.*, 1972; Rohrbach *et al.*, 1974) as determined by equilibrium centrifugation both in the presence and absence of 6 *M* guanidine hydrochloride. This value is consistent with the migration of the reduced protein in SDS polyacrylamide gel electrophoresis. A previously reported value of 99,000 (Kaziro and Inoue, 1968) was high apparently due to partial aggregation of the enzyme (Kaziro *et al.*, 1969).

The amino acid composition of EF-G has been determined by Kaziro *et al.* (1969, 1972) and the protein appears to contain all of the known amino acids. It contains about 6 residues of half-cystine per mole of protein as well as significant amounts of both aspartic and glutamic acid. These results are consistent with the observations that EF-G is an acidic protein and that the presence of free thiol groups are essential for maximum activity. The purified protein has an ultraviolet absorption spectrum of a typical protein with a 280:260 ratio of between 1.75–1.87 (Parmeggiani, 1968; Kaziro *et al.*, 1972). The protein is stable at −20° for several months in a buffer containing 5 m*M* β-mercaptoethanol and either 0.25 *M* sucrose or 50% glycerol (Kaziro *et al.*, 1972; Rohrbach *et al.*, 1974). Alternatively, EF-G has been found to be stable for as long as 2 years in a Tris-Cl, pH

7.4, β-mercaptoethanol buffer when stored in liquid nitrogen (N. Brot, unpublished observations, 1975).

III. Reactions of EF-G

A. Ribosome Dependent GTP Hydrolysis

It was originally observed that EF-G had the capacity to rapidly hydrolyze GTP to GDP + P_i but only in the presence of ribosomes (Conway and Lipmann, 1964; Nishizuka and Lipmann, 1966a). It was shown by these authors that the 50 S ribosomal subunit could only partially support this reaction and that the 30 S subunit was inactive. The addition of 30 S particles to 50 S subunits stimulated the hydrolysis to that observed with 70 S ribosomes. The role of the 30 S particle in this reaction is unknown since it has been shown that under stoichiometric conditions the 50 S subunit alone can support the EF-G dependent binding and hydrolysis of GTP (Bodley and Lin, 1970; Brot et al., 1971). It is possible that the 30 S subunit facilitates the release from the ribosomes of the EF-G · GDP complex which is formed as a result of the hydrolysis. It is to be noted that this hydrolysis takes place in the absence of any translocation since there is no requirement for either mRNA or aminoacyl-tRNA, and thus the hydrolysis is "uncoupled" from any functional role of either the ribosome or EF-G.

Some characteristics of this reaction have been reported (Nishizuka and Lipmann, 1966a; Kaziro et al., 1969, 1972; Eckstein et al., 1971; Arai et al., 1972). This uncoupled hydrolysis which only requires Mg^{2+} (10 mM) in addition to EF-G and ribosomes is inhibited by several thiol inhibitors, and it was shown that the effect of these inhibitors was on EF-G and not on the ribosome. The reaction is optimal at pH 9 and the K_m for GTP was calculated to be $0.9 \times 10^{-4} M$. It is to be noted that both of these values are different from that found for polymerization where the pH maximum is about 7.5–8.0 and the affinity for GTP was much higher ($2 \times 10^{-6} M$). The reaction is inhibited by GDP, ppGpp (guanosine 5'-diphosphate, 3'-diphosphate), and the two nonhydrolyzable GTP analogues GDPCP (guanylylmethylenediphosphonate) and GDPNP (guanylylimidodiphosphate).

Recently, Rohrbach et al. (1974) have investigated the mechanism of the uncoupled hydrolysis employing EF-G prepared according to their new procedure which in the absence of ribosomes was completely devoid of GTPase activity. The position of the GTP cleavage by EF-G and ribosomes was investigated using the technique of ^{18}O incorporation from water. It was shown that solvent oxygen was incorporated only into the orthophosphate derived from the γ-phosphate of GTP. No ^{18}O was found in either the α or β phosphates of GDP which is the other product of the hydrolysis. These authors concluded that the hydrolysis of the GTP occurs between the γ phosphorus atom and the oxygen bound to the β and γ phosphorus atoms.

B. Interaction of Guanosine Nucleotides and Ribosomes with EF-G

1. Formation of Ternary Complex

Since the uncoupled GTPase reaction requires both EF-G and ribosomes, it was of interest to determine whether a ternary complex containing EF-G, ribosomes, and GTP could be observed. Figure 2 shows the results of experiments in which [³H]GTP is incubated with EF-G in the presence or absence of ribosomes and then chromatographed through Sephadex G-25. It can be seen that the formation of a complex of GTP with a high molecular weight substance is dependent upon the presence of both ribosomes and EF-G. In other experiments it was shown that [γ-³²P]GTP could not be incorporated into the complex but that [³²P]-β-γ-GTP (equally labeled in both the β and γ positions) was as efficiently bound as [³H]GTP. These results indicated that the nucleotide present in the complex isolated by Sephadex chromatography after incubating with GTP was GDP. Since the Sephadex chromatography could not separate ribosomes from EF-G, it could not be determined whether the GDP was bound to ribosomes or to EF-G or if a ternary complex had been formed.

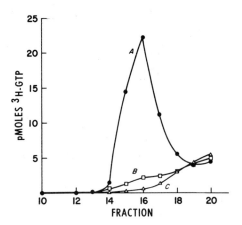

Fig. 2 The formation of a [³H]GTP high molecular weight complex by Sephadex G-25. The complete incubation mixture contained in a total volume of 350 μl: 92 μg EF-G, 25 A_{260} ribosomes; 200 pmoles of [³H]GTP; 0.01 M Tris-Cl, pH 7.0; 0.1 M magnesium acetate; and 0.001 M DDT. The incubation was carried out for 5 min at 0° and the mixture was chromatographed on a Sephadex G-25 column. (A) Complete system, (B) −G or ribosomes, (C) −G and ribosomes. From Brot *et al.* (1969).

TABLE II

The Binding of Transfer Factor G and Various Nucleotides to Ribosomes

[³H]Nucleotide incubated[a]	Nucleotide bound (pmoles per A_{260} ribosomes)	Enzymatic activity[b]
None	—	0.20
GTP	3.7	0.90
GDP	3.2	0.88
GMP	0.3	0.22
dGTP	2.9	0.68
GDPCP	6.8	2.1
ATP	0.2	0.17
UTP	0.3	0.22
GTP	0.3	0.29

[a] Incubation conditions were identical to those described in the legend to Fig. 2, except that different nucleotides were present in each incubation as noted in the table. At the end of the incubation, the reaction mixtures were centrifuged at 160,000 g and the ribosomal pellet assayed for the determination of A_{260} (recovery of ribosomes) radioactivity (nucleotide bound) and EF-G activity. The amount of EF-G present in the ribosomal pellet was assayed by its ability to hydrolyze GTP (Conway and Lipmann, 1964; Nishizuka and Lipmann, 1966b).

[b] Values are expressed as nmoles GTP hydrolyzed per A_{260} ribosomes per 10 min at 37°. From Brot et al. (1969).

To answer these questions, the ribosomes were isolated by centrifugation after the incubation and assayed for the presence of both nucleotide and EF-G. The results, shown in Table II, which were also observed by Parmeggiani and Gottschalk (1969b) and Kuriki et al. (1970), demonstrate that GTP, GDP, and dGTP bind to the ribosomes to about the same extent and that EF-G is also bound to the ribosome in the presence of these nucleotides. It can also be seen that GDPCP binds better than GTP and that this increase in binding is accompanied by an increase in the amount of EF-G bound to the ribosomes. These results showed that EF-G binding to ribosomes requires the presence of certain guanosine nucleotides, and that although GTP is hydrolyzed this binding does not necessarily require the hydrolysis of GTP. In addition, recent experiments by Rohrbach and Bodley (1976) have shown that EF-G contains a cysteine residue which is located at or near the guanosine nucleotide binding site of this protein.

2. Effect of Fusidic Acid on the Formation of the Ternary Complex

Fusidic acid is a steroid antibiotic of known structure (Godtfredsen and Vangedal, 1962) which had originally been shown to inhibit protein synthesis (Harvey *et al.*, 1966). It was later found that fusidic acid inhibited the uncoupled GTPase activity catalyzed by EF-G (Pestka, 1968; Tanaka *et al.*, 1968). In an attempt to understand which component of the system was being affected by the antibiotic, Kinoshita *et al.* (1968) isolated a fusidic acid-resistant mutant. They showed that EF-G from this mutant was resistant to the antibiotic while the EF-G isolated from the parent strain was sensitive to fusidic acid and that the source of the ribosomes had no effect on these results. This clearly showed that EF-G was the target of the fusidic acid inhibition of the GTP hydrolysis. This was further verified in other studies (Okura *et al.*, 1970) which showed by equilibrium dialysis that [³H]fusidic acid could bind to EF-G and not to ribosomes. In addition the EF-G from the fusidic acid-resistant organism showed much less binding of the antibiotic.

The mechanism of action of fusidic acid became clearer during studies which investigated the effect of that antibiotic on the formation of an EF-G · GDP · ribosome ternary complex (Bodley *et al.*, 1969, 1970b,d; Bodley and Lin, 1970; Kuriki *et al.*, 1970; Brot *et al.*, 1971; Okura *et al.*, 1971). It was found that when fusidic acid was added to an incubation containing ribosomes, EF-G and [³H]GTP, there was a large increase in both the amount of nucleotide and EF-G present in the ternary complex when isolated by ultracentrifugation (Table III). Thus, it appears that fusidic

TABLE III

The Effect of Fusidic Acid on the Binding of [³H]GTP and Factor G to Ribosomes[a,b]

Exp.	Addition	Nucleotide bound (pmoles per A_{260} ribosomes)	Factor G activity (nmoles GTP hydrolyzed)
1	None	4.2	1.3
	Fusidic acid	10.3	6.6
2	None	4.4	3.7
	Fusidic acid	15.7	22.7
3	None	4.0	—
	Fusidic acid	12.8	—
4	None[c]	0.02	2.9
	Fusidic acid	0.03	19.3

[a] From Brot *et al.* (1971).

[b] The incubation conditions and assays are described in the legends to Figure 2 and Table II except that 1.4 m*M* fusidic acid was added where indicated.

[c] [γ-³²P]GTP.

acid increases the amount of both guanosine nucleotide and EF-G bound to the ribosomes. The above results could be explained if fusidic acid inhibited GTP hydrolysis and caused a greater stability of a ribosome complex containing [^{3}H]GTP. It was observed however, that no radioactivity from [γ-^{32}P]GTP could be found associated with the ternary complex (Table III, Experiment 4) and direct analysis of the guanosine nucleotide bound to the complex showed that it was GDP. It was apparent that fusidic acid did not prevent the hydrolysis of the GTP bound to the ribosomes but stabilized an EF-G · GDP · ribosome complex. It has been further shown (Okura et al., 1971) that under these conditions there is a stoichiometric amount of EF-G, GDP, ribosomes, and fusidic acid in the complex.

Although nucleotide binding to ribosomes could be assayed by centrifugation or Sephadex chromatography, both of the procedures are long and preclude obtaining kinetic data. This is especially the case since the formation of the complex can take place at 0° and there is no convenient way to stop the reaction. Preliminary experiments revealed that it was possible to show that an EF-G · GDP complex could be retained by a nitrocellulose filter although the amount of complex retained was always less than that observed by either centrifugation or Sephadex G-25 chromatography. However, when fusidic acid was added to the incubation mixture the values obtained by the filter assay were very close to a duplicate incubation that had been assayed by either column chromatography or centrifugation. The filter method, because of its ease and rapidity, became the assay of choice. It briefly involves diluting the reaction mixture at the end of the incubation with a buffer containing 10 mM each of Tris-Cl, pH 7.4, $MgCl_2$, and NH_4Cl, and then filtering this mixture through a nitrocellulose filter. The filter is then washed three more times with 2 ml of this buffer. It is important to note that care must be taken not to wash the filter exhaustively since the complex tends to dissociate under these conditions. Using this assay, the requirements for the retention of [^{3}H]GTP by the nitrocellulose filter were studied in the presence and absence of fusidic acid. It was found that similar to the results obtained by centrifugation and chromatography the binding of [^{3}H]GTP was completely dependent upon the presence of both EF-G and ribosomes and that this binding was stimulated about three- to fourfold in the presence of fusidic acid (Table IV). Mg^{2+} but not NH_4^+ is required for the binding, and neither chloramphenicol nor sparsomycin has any effect on the reaction. In addition, the 50 S subunit can completely substitute for the 70 S ribosome while the 30 S particle by itself was inactive and had no stimulatory effect when added to an incubation containing the 50 S subunit. With the filter assay it was again observed that when [γ-^{32}P]GTP was substituted for [^{3}H]GTP with

TABLE IV

Requirements for the Binding of [³H]GTP to a Nitrocellulose Membrane in the Presence and Absence of Fusidic Acid[a,b]

Omission or addition	− Fusidic acid (pmoles)	+ Fusidic acid (pmoles)
Complete	6.2	23.0
−Ribosomes	0.0	0.0
−Factor G	0.1	0.0
−Mg^{2+}	0.2	5.3
−NH_4^+	7.3	24.4
+Chloramphenicol	5.8	24.0
+Sparsomycin	7.0	23.1
−Ribosomes + 30 S ribosomal subunits	0.3	0.1
−Ribosomes + 50 S ribosomal subunits	6.1	25.0
30 S + 50 S ribosomal subunits	6.0	24.6
Complete[c]	0.1	0.0
−Ribosomes + 50 S ribosomal subunits	0.1	0.1

[a] From Brot et al. (1971).

[b] The incubations were carried out for 1 min at 37° and the complete system contained in a total volume of 0.1 ml: 10 mM Tris-Cl buffer, pH 7.4; 10 mM ammonium acetate; 10 mM magnesium acetate; 1 mM DTT; 13 μg of EF-G; 97 pmoles of ribosomes; 6 μM [³H]GTP, and where indicated 1 mM fusidic acid. The following additions were added to each incubation where indicated: 1 mM chloramphenicol; 100 μM sparsomycin; 91 pmoles 30 S subunits, and 93 pmoles of 50 S subunits. The assay employed is based on the retention of an [³H]GTP-EF-G·ribosome complex to a nitrocellulose filter and is described in detail in the text (Section III,B,2).

[c] [³²P]GTP substituted for [³H]GTP.

either the 70 S or 50 S subunit, no radioactivity could be found on the filter even when the incubation was carried out in the presence of fusidic acid. Since the filter assay can be completed in less than 15 sec, these results ruled out the possibility that the hydrolysis of GTP occurred during the long time required for both the centrifugation and chromatography assay. Thus, these experiments verify that fusidic acid does not inhibit the hydrolysis of GTP which is associated with the binding of EF-G to ribosomes. It was concluded that the stable ribosomal complex formed contained GDP, not GTP. Some characteristics of the nucleotide binding reaction were studied (Figs. 3 and 4) and it is seen that there is a linear response with increasing concentrations of EF-G and ribosomes until about 25–30 pmoles of nucleotide are bound. The reaction is very rapid at both 37° and 0° and is essentially complete at 30 sec and 2 min, respectively. In this system GTP saturation occurs at about 5×10^{-6} M both in the presence or absence of fudisic acid while fusidic acid at a concentra-

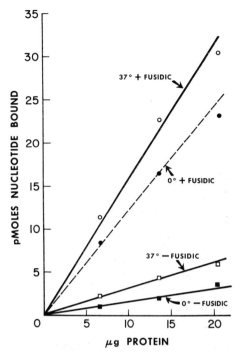

Fig. 3   The effect of protein concentration on the extent of nucleotide binding as measured by the filter assay. The incubations were carried out for either 1 min at 37° or 5 min at 0° and contained in a total volume of 0.1 ml: 10 mM Tris-Cl buffer, pH 7.4; 10 mM magnesium acetate; 10 mM ammonium acetate; 1 mM DTT, factor G; 97 pmoles ribosomes; 1.2 × 10^{-5} M [^{3}H]GTP; and, where indicated, 1 mM fusidic acid. The amount of nucleotide binding was assayed by the nitrocellulose filter technique. From Brot *et al.* (1971).

tion of 5 × 10^{-5} M yields maximum binding of nucleotide. The specificity of the binding of various nucleotides to the nitrocellulose filter in the presence of EF-G and ribosomes was found to be identical to the results obtained by the centrifugation technique (see Table II). Thus [^{3}H]GTP, [^{3}H]GDP, and [^{3}H]dGTP all bound about equally well and the binding in each case was stimulated about threefold in the presence of fusidic acid. One notable exception is the binding of GDPCP (Table V). The binding of this nucleotide is not affected by the presence of fusidic acid although the requirements for its binding were identical to those seen when GTP was used. Thus, this nucleotide in the presence of EF-G can bind as well to the 50 S ribosomal subunit as to the 70 S ribosome and it binds about twofold better than GTP in the absence of fusidic acid.

Experiments were carried out to investigate whether a stoichiometry existed between the amounts of [^{3}H]GDP bound (from [^{3}H]GTP) and

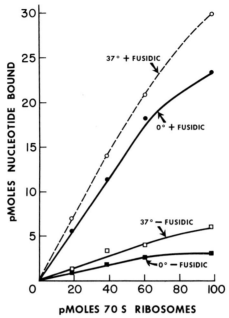

Fig. 4 The effect of ribosome concentration on the extent of nucleotide binding. The incubation conditions are described in the legend to Fig. 3 and each incubation contained 13.4 μg of EF-G. From Brot *et al.* (1971).

TABLE V

The Reaction of [³H]GTP and [³H]GDPCP with 70 S Ribosomes and 50 S Ribosomal Subunits[a]

System	−Fusidic acid (pmoles bound)	+ Fusidic acid (pmoles bound)
GTP + 70 S	9.5	22.7
GTP + 50 S subunit	8.5	23.8
GDPCP + 70 S	17.3	16.4
GDPCP + 50 S subunit	14.2	15.8

[a] The incubation conditions are identical to those described in Table IV. Both nucleotides were employed at a concentration of 6 μM, and 93 pmoles of 50 S subunits were used where indicated. From Brot *et al.* (1971).

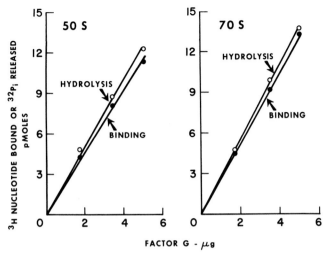

Fig. 5 The effect of protein concentration on the binding and hydrolysis of GTP. The incubations were carried out at 0° for 5 min and contained the buffer mixture described in the legend to Fig. 3. At the end of the incubation aliquots were removed and the amount of nucleotide bound and GTP hydrolyzed were determined as described in the text.

[γ-^{32}P]GTP hydrolysis. As shown in Fig. 5, stoichiometric amounts of tritium bound to ribosomes and ^{32}P$_i$ released could be obtained. This stoichiometry is linear with increasing amounts of EF-G and was observed with the 50 S subunit as well as the 70 S ribosome.

All the above results in addition to those obtained by other investigators show that fusidic acid per se does not inhibit the hydrolysis of GTP but that its main action appears to be the stabilization of an EF-G · GDP complex on the ribosome. This stabilization would trap EF-G on the ribosome and essentially inhibit its catalytic activity. Thus, fusidic acid allows one round of GTP hydrolysis along with one round of translocation (Modolell *et al.*, 1973). The reactions leading to the formation of a ternary complex and its subsequent reaction with fusidic acid are shown in Fig. 6. An EF-G · GTP complex has not been shown to occur, presumably due to the lability of this complex. However, this complex has been demonstrated with the eukaryote factor EF-2 (see Section V). It is also to be

1. EF-G + GTP(GDPCP) ⇌ EF-G · GTP(GDPCP)

2. EF-G · GTP(GDPCP) ⇌ EF-G · 50 S · GTP or EF-G · 50 S · GDPCP

3. EF-G · 50 S · GTP ⟶ EF-G · 50 S · GDP + P$_i$

4. EF-G · 50 S · GDP $\xrightarrow{\text{Fusidic acid}}$ EF-G · 50 S · GDP · Fusidic

Fig. 6 Interaction of EF-G with GTP and the 50 S ribosomal subunit.

noted that an EF-G · 50 S · GTP complex has not been isolated because of its rapid hydrolysis to an EF-G · 50 S · GDP complex. However, with GDPCP, a stable EF-G · 50 S · GDPCP complex is formed and can be isolated.

One consequence of the action of fusidic acid would be the prediction that cells grown in the presence of fusidic acid would accumulate peptidyl-tRNA in the P site of the ribosome. This was confirmed by studies (Celma *et al.*, 1972; Cundliffe, 1972) in which it was found that when protoplasts of *Bacillus megaterium* were treated with fusidic acid, protein synthesis stopped promptly and that 80% of the nascent chains could be released by puromycin indicating that protein synthesis stopped after translocation had occurred. Essentially, the same results were found *in vitro* using polysomes in the protein-synthesizing system. The results of both the *in vivo* and *in vitro* studies suggested that in the presence of fusidic acid the nascent chains accumulated in the P site of the ribosome which allowed them to react with puromycin. It was, therefore, concluded that fusidic acid is not a direct inhibitor of translocation but inhibits protein synthesis by preventing the functional binding of aminoacyl-tRNA to the ribosomes. These results were further verified by a series of reports (Cabrer *et al.*, 1972; Miller, 1972; Richman and Bodley, 1972; Richter, 1972) in which it was shown that the formation of an EF-G · GDP · ribosome · fusidic acid complex inhibits the EF-Tu dependent binding of aminoacyl-tRNA to the ribosome. These results suggest that EF-G and EF-Tu bind to a common or overlapping site on the ribosome and that the stable binding of one of the factors inhibits the reactions carried out by the other by preventing its interaction with the ribosomes.

3. Effect of Thiostrepton

Thiostrepton is a peptide antibiotic that primarily inhibits the growth of gram-positive organisms. Its chemical structure has been determined by X-ray crystallographic techniques (Anderson *et al.*, 1970) and it is very similar to siomycin A and thiopeptin, two other sulfur-containing antibiotics. These antibiotics have been shown to bind irreversibly to the 50 S ribosomal subunit (Tanaka *et al.*, 1970; Weisblum and Demohn, 1970a; Kinoshita *et al.*, 1971; Modolell *et al.*, 1971b) and are strong inhibitors of the EF-G ribosome-dependent hydrolysis of GTP as well as translocation (Pestka, 1970, Pestka and Brot, 1971; Watanabe and Tanaka, 1971). The mechanism of action of thiostrepton has been studied and it has been shown that the binding of this antibiotic to the 50 S ribosomal subunit prevents the formation of an EF-G · GDP · ribosome complex (Bodley *et al.*, 1970c; Weisblum and Demohn, 1970b; Modolell *et al.*, 1971a,b; Pestka and Brot, 1971). In view of the fact that both EF-G and thiostrepton ap-

pear to bind to the 50 S ribosomal subunit and that the binding of the antibiotic prevented the interaction of EF-G and guanosine nucleotide with
the ribosome, it was possible that these were mutually exclusive events.
That this was the case was demonstrated by experiments in which it was
shown (Highland *et al.*, 1971) that ribosomes to which an EF-G · GDP
complex was prebound were protected against inactivation by thiostrepton. These results suggest that the binding of the complex as well as
thiostrepton occur at or near the same physical region of the 50 S subunit
and that the binding of one interferes with the binding of the other. It is of
interest that although fusidic acid and thiostrepton both inhibit the EF-G
ribosomal dependent hydrolysis of GTP their mechanisms of action are
quite different. In effect, thiostrepton is a true inhibitor of translocation
while fusidic acid is not. It is to be noted, however, that thiostrepton also
possesses the ability to inhibit the EF-Tu · GTP-dependent binding of
aminoacyl-tRNA to the ribosomes (Kinoshita *et al.*, 1971; Modolell *et al.*,
1971b; Weissbach *et al.*, 1972), a not too surprising finding in view of the
fact that EF-G and EF-Tu appear to have common binding sites on the ribosome. In the intact organism, it appears, however, that the inhibition of
aminoacyl-tRNA binding by thiostrepton is predominant over the inhibition of translocation (Cannon and Burns, 1971; Cundliffe, 1971). The
reason for this is not clear.

C. Mutants of EF-G

Two different kinds of EF-G mutants have been described. One involves the isolation of an organism that is resistant to fusidic acid, due to a
specific alteration in EF-G (Kinoshita *et al.*, 1968; Bernardi and Leder,
1970; Kuwano *et al.*, 1971; Tanaka *et al.*, 1971), while the other mutant
contains a temperature-sensitive EF-G (Tocchini-Valentini and Mattoccia, 1968; Felicetti *et al.*, 1969). Bernardi and Leder (1970) have purified EF-G from a fusidic acid resistant strain of *E. coli* and have compared
various characteristics of the protein with EF-G isolated from the wild-
type. They found that although the resistance was due to a mutation affecting this protein, the resistant EF-G was indistinguishable from the
sensitive parent EF-G by electrophoretic chromatography and immunological techniques. They did observe, however, that the specific activity
of the resistant factor was less than that of the sensitive factor when assayed by phenylalanine polymerization and that the K_m for GTP in the ribosomal dependent GTPase reaction for the resistant factor was about
twice that of the sensitive EF-G. Thus, the mutation that confers fusidic
acid resistance to the organism appears to specifically affect EF-G and
may alter a binding or catalytic site on the factor.

Both the fusidic acid and temperature-sensitive mutants have been

mapped and have been shown to be 97% cotransducible with streptomycin resistance (Kuwano *et al.*, 1971; Tanaka *et al.*, 1971). In addition, other
genetic evidence indicates that the EF-G gene is transcribed simultaneously with other ribosomal protein genes. It is located at minute 72 and
appears to be at the distal end of a cistron coding for resistance to erythromycin, spectinomycin, streptomycin, and fusidic acid (Nomura and Engback, 1972). It is to be noted that the erythromycin locus determines a 50
S ribosomal protein while the spectinomycin and streptomycin loci involve the transcription of 30 S ribosomal proteins. The finding of the gene
coding for EF-G in this operon is compatible with the observation that
there is a stoichiometric relationship between the amount of EF-G and ribosomes synthesized during the growth of *E. coli*. Thus, Gordon (1970)
has shown that when *E. coli* is grown under conditions where the generation time is altered by a factor of about 3 there always exists a 1:1 stoichiometry between the amount of EF-G and ribosomes in the cell. This
relationship has led to the suggestion that perhaps EF-G might be considered a ribosomal protein especially since it is so difficult to remove contaminating EF-G from ribosomes. The ribosomes, however, contain only
a small amount of EF-G when compared to the ribosome-free supernatant
and no protein corresponding to EF-G has been found corresponding to
any of the purified ribosomal proteins. Thus, although EF-G and the ribosomal proteins appear to be synthesized in a coordinate fashion, EF-G is
probably not a bona fide ribosomal protein.

D. EF-G Interaction with Ribosomal Proteins

1. L7L12

Since it was shown that all the activity associated with EF-G involved
the participation of the ribosome it seemed reasonable to investigate
which ribosomal protein(s) was involved in these reactions. A number of
studies have now shown that two 50 S ribosomal proteins, L7 and L12,
are involved in reactions which concern EF-G (Kischa *et al.*, 1971; Hamel
et al., 1972; Brot *et al.*, 1972; Sander *et al.*, 1972; Ballesta and Vazquez,
1972; Sopori and Lengyel, 1972). These two acidic ribosomal proteins
have been shown to have identical amino acid sequences and only differ
in that L7 contains an acetyl group on the N-terminal serine (Terhorst *et
al.*, 1972a,b). It has been shown that these proteins can be removed from
the 50 S ribosomal subunit by treatment with 50% ethanol containing 1 *M*
NH₄Cl (Hamel *et al.*, 1972). Table VI shows that when L7 and L12 are removed from the 50 S ribosomal subunit with ethanol and NH₄Cl, the ribosome loses its ability to bind GTP in the presence of EF-G. When a mixture of L7 and L12 is added to the treated 50 S subunits, there is about

TABLE VI

Effect of Ribosomal Proteins
L7L12 on the Binding of [^{3}H]GTP
to the 50 S Ribosomal Subunit[a]

Omission	[^{3}H]GTP bound (pmoles)
None	6.4
L7L12	0.8
50 S Subunit	0.2
EF-G	0.1

[a] 50 S ribosomal subunits were extracted with 50% ethanol containing 1 M NH$_4$Cl. The subunits were then recovered by centrifugation and proteins L7L12 purified from the supernatant. The complete incubation mixture contained 10 mM each of Tris-Cl buffer, pH 7.4, ammonium acetate, and magnesium acetate; 1 mM DTT, 1 mM fusidic acid; 12 pmoles of ethanol-NH$_4$Cl treated 50 S subunits; 60 pmoles L7L12, 9 μg EF-G and 5×10^{-6} M [^{3}H]GTP. The amount of [^{3}H]GTP bound was assayed by the Millipore filtration technique. From Brot et al. (1972).

an eightfold stimulation in the binding which is dependent upon both ribosomes and EF-G. Other experiments have shown that either L7 or L12 is equally active in restoring this activity to the depleted subunits. These results suggest that L7 or L12 depleted ribosomes are unable to react with EF-G and GTP. This suggestion was further substantiated when it was shown that the L7 and L12 depleted ribosomes could also not support the EF-G dependent hydrolysis of GTP and that the addition of either of these two proteins to the incubation mixtures restored the activity (Fig. 7). It has been shown that these proteins rebind efficiently to the 50 S subunit and that the reconstituted particle contains between 2.4–2.8 L7L12 molecules/50 S subunit (Brot et al., 1973). This value is similar to that found by Thammana et al. (1973) who showed that the 50 S subunit contains about three copies of L7L12, but somewhat lower than the four copies per ribosome recently reported by Subramanian (1975).

   The above results show that L7 and L12 must be present on the ribosome in order for the EF-G-dependent activities to be expressed and

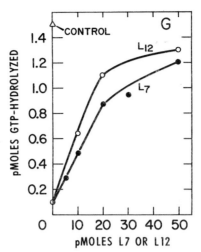

Fig. 7 The effect of L7 or L12 on GTP hydrolysis dependent on EF-G. Each incubation contained in a final volume of 50 μl: 50 mM Tris-Cl buffer, pH 7.4; 10 mM MgCl$_2$; 80 mM NH$_4$Cl; 17 pmoles of 30 S subunits; 4 pmoles of 50 S subunits depleted of L7L12; 1.6 μg EF-G and 13 pmoles of [γ-^{32}P]GTP. The reactions were carried out for 2 min at 37° and the ^{32}P$_i$ released was assayed by the method of Conway and Lipmann (1964). From Brot *et al.* (1973).

suggest that L7L12 might constitute the EF-G binding site on the ribosome. Alternatively, the presence of L7L12 on the ribosome might cause a conformation change which then permits EF-G to bind to the ribosome. That the former suggestion is probably the correct one was shown by experiments (Acharya *et al.*, 1973) in which an EF-G · GDPCP · 50 S complex was isolated and then treated with a bifunctional cross-linking reagent. About 120,000 daltons of ribosomal proteins became cross-linked to EF-G of which about 30,000 daltons were made up of proteins L7 and L12. These authors concluded that the complex contained about 2.0 to 2.5 equivalents of L7L12 per EF-G and that L7 and L12 are at or very near the EF-G binding site. It is to be noted that L7L12 have also been shown to be required for initiation (Kay *et al.*, 1973; Lockwood *et al.*, 1973; Mazumder, 1973), EF-Tu-dependent, aminoacyl-tRNA binding (Hamel *et al.*, 1972; Sander *et al.*, 1972; Weissbach *et al.*, 1972), and termination (Brot *et al.*, 1974) of protein synthesis. If further direct experiments show that these factors also bind to L7L12 it will be most intriguing to speculate how proteins of such diverse nature can recognize the same ribosomal protein.

2. Other Ribosomal Proteins

Although an extensive study (Highland *et al.*, 1973) using antibodies produced against 50 of the 55 ribosomal proteins showed that only antibodies against either L7 or L12 inhibited the formation of an EF-G · GDP · ribosome complex, other studies have implicated a variety of other ribosomal proteins in EF-G dependent reactions. Thus, Marsh and Parmeggiani (1973) showed that a 30 S ribosomal subunit depleted of proteins S1, S2, S3, S5, S9, S10, and S14 is unable to stimulate GTP hydrolysis dependent upon the presence of EF-G and 50 S subunit. The activity could only be partially restored by adding back either S5 or S9; however, when both of these proteins were added together they acted synergistically to produce a particle even more active than the original 30 S subunit. The data suggest that in the absence of these proteins the 30 S subunit is unable to associate with the 50 S particle to form a 70 S ribosome and therefore is unable to stimulate the hydrolysis of GTP.

In another approach to investigate the interaction of an EF-G · GDP complex with the ribosome Maassen and Möller (1974) prepared [³H]GDP containing a photolabile 4-azidophenol at the β-phosphate group. They showed that this analogue could be bound to the ribosome in the presence of EF-G and that the characteristics of its binding were comparable to that of GDP. They further demonstrated that UV irradiation of the ribosome containing EF-G and the azidophenyl-GDP analogue resulted in the covalent binding of the photoaffinity label to proteins L5, L11, L13, L18, and L30. These authors suggest that these proteins are involved in the EF-G-dependent GTP hydrolysis.

IV. Translocation

It was first suggested by Nishizuka and Lipmann (1966a) that EF-G might be involved in the process of translocation during protein synthesis and that GTP was required for this reaction. Direct evidence for the participation of EF-G and GTP in this reaction have come from several sources. As noted earlier, the translocation process appears to involve at least three reactions: (1) movement of acylated tRNA from one site on the ribosome to another, (2) the movement of mRNA, and (3) the ejection of deacylated tRNA from the ribosome. All these reactions have been used as criteria for translocation and will be described below.

A. Movement of Acylated tRNA

The fact that puromycin can only react with an acylated tRNA which is bound to the P site on the ribosome and not the A site yielded a convenient assay to locate the position of an acylated tRNA bound to the ribosome. Thus, as seen in Table VII, when Phe-tRNA is bound to ribosomes under conditions favoring A site binding, and then EF-G, GTP, and puromycin are added in subsequent incubations, there is good formation of a puromycin peptide. However, if EF-G and GTP are omitted from

TABLE VII

Requirements for the Formation of a Puromycin Product[a]

Omissions from first incubation for 5 min at 23°	Omissions from second incubation for 5 min at 37°	Puromycin product (pmoles)
None	None	2.1
None	−EF-G	0.18
None	−GTP	0.15
None	−EF-G, −GTP	0.12
None	−GTP + GMP	0.15
None	−GTP + GDP	0.25
None	−GTP + GDPCP	0.20
None	−GTP + ATP	0.18
None	−EF-G + EF-Tu, EF-Ts	0.15
None	+EF-Tu, EF-Ts	2.4
−Poly (U)	None	0.09
−Ribosome	None	0

[a] Three separate incubations were used to study the formation of a puromycin product. The first incubation (binding reaction) was carried out for 5 min at 23° and contained in a final volume of 85 μl: to mM Tris-Cl, pH 7.4; 12 mM β-mercaptoethanol; 160 mM NH$_4$Cl; 20 mM mgcL$_2$; 5 μg poly(U); 1.1 A_{260} ribosomes; and 10 pmoles [^{14}C]phe-tRNA. This binding was carried out to 85% completion and then EF-G (0.2 μg) and nucleotide (10 nmoles) added and the second incubation (translocation reaction) carried out for 5 min at 37°. The reaction mixtures were then placed into an ice-water mixture and 10 μl of an 0.01 M neutralized solution of puromycin were added. The third incubation was done at 0° for 30 min and the reaction was stopped by the addition of 0.3 ml potassium phosphate buffer, pH 8.0. The puromycin peptide formed was extracted with 3 ml of ethyl acetate and an aliquot removed and assayed for radioactivity. From Brot et al. (1968).

the incubation after the binding of the Phe-tRNA, there is little reaction of the ribosomal bound Phe-tRNA with puromycin. Other guanosine nucleotides including GDPCP and ATP could not substitute for GTP. In addition, elongation factors Tu and Ts could not substitute for EF-G nor did they stimulate the reaction when added to an incubation mixture containing EF-G. The reaction was completely-dependent upon the presence of both poly(U) and ribosomes. Tanaka et al. (1968) have also shown that in the presence of EF-G, GTP, and puromycin there is a decrease in the amount of Phe-tRNA bound to ribosomes. Similarly, Haenni and Lucas-Lenard (1968) found that EF-G and GTP stimulated the conversion of ribosomal bound N-acetyldiphenylalanyl-tRNA from a puromycin unreactive form to one which could react with puromycin to yield N-acetyldiphenylalanyl puromycin. They also found that GDPCP could not substitute for GTP and that the reaction was inhibited by fusidic acid, an antibiotic that inhibits the catalytic functioning of EF-G. All these results are consistent with the idea that EF-G and GTP are required to effect the movement of an acylated tRNA (or peptidyl-tRNA) from the A site on the ribosome to the P site.

B. Movement of Messenger RNA

During the translation of mRNA into a polypeptide the ribosome moves along the messenger RNA from the 5'- to 3'-end, three nucleotides at a time. This process allows the translation of successive codons in the message. This movement of the mRNA has been assayed by two different techniques and both have been shown to involve EF-G and GTP.

1. Appearance of a New Codon

The first method assays for the appearance of a new codon which appears in the A site of the ribosome after translocation has occurred. These studies have involved the use of both synthetic messenger RNA (Erbe and Leder, 1968; Haenni and Lucas-Lenard, 1968; Pestka, 1968; Erbe *et al.*, 1969; Leder *et al.*, 1969b) as well as natural mRNA from the virus Qβ Leder *et al.*, 1969c; Roufa *et al.*, 1970). Thus, for example, it was shown that if a synthetic polymer such as A-U-G-U-U-U-U-U was used as the template the translation of the first codon required only initiation factors, fMet-tRNA and GTP. The formation of fMetPhe was seen when only EF-T, Phe-tRNA, and GTP were also added to the incubation mixture. The synthesis of the tripeptide fMetPhePhe, an indication that the third codon had been translated, was dependent upon the presence of EF-G in addition to initiation factors, the acylated tRNA's, EF-T, and GTP. All the studies came to the conclusion that a dipeptide could be formed in the absence of EF-G but that this protein was required for making the third codon available for translation. This apparently occurs by the simultaneous movement of both peptidyl-tRNA and mRNA. Similar results were obtained using mRNA coding for the coat cistron of Qβ.

2. Protection of Messenger RNA

This assay is based on the identification of a fragment of mRNA which is protected by the ribosome from ribonuclease digestion (Gupta *et al.*, 1971; Thach and Thach, 1971). In these studies, it was shown that when a pre-translocation complex containing either fMetAla-tRNA or fMetVal-tRNA was incubated in the presence of EF-G and GTP, an additional three nucleotides toward the 3'-end of the messenger RNA were protected against ribonuclease digestion. The ribosomal protection of the additional triplet paralleled both the formation of a puromycin-reactive product as well as the introduction of a new codon into the A site of the ribosome. These results provided further evidence that the movement of the messenger

RNA along the ribosome was coupled to the translocation of the nascent peptidyl-tRNA on the ribosome. It was further shown that there was no movement of the messenger RNA associated with (1) the formation of the initiation complex, (2) the EF-Tu-dependent binding of aminoacyl-tRNA, and (3) the formation of the peptide bond.

C. Release of Deacylated tRNA

During protein synthesis, each time a peptide bond is formed by the transfer of an amino acid from aminoacyl-tRNA to the growing polypeptide chain, a deacylated tRNA is produced on the ribosomes. This deacylated tRNA, which is located at the P site of the ribosome, must in some fashion be released from the ribosome prior to the next round of translocation. It has been found that EF-G and GTP are required for the removal of this deacylated tRNA (Kurika and Kaji, 1968; Lucas-Lenard and Haenni, 1969; Ishitsuka et al., 1970; Roufa et al., 1970). In addition, the data show that the release of the deacylated tRNA is dependent upon the presence of either acylated or deacylated tRNA bound to the A site. Thus EF-G and GTP were unable to release deacylated tRNA from the ribosome if the A site was not occupied. It was also found that the rate of release of the deacylated tRNA was about the same as the translocation of peptidyl-tRNA from the A site to the P site. These results suggest that there is a coupling between these two events.

D. GTP Hydrolysis Associated with Translocation

The role of GTP hydrolysis during translocation is still not clear. It is known that GTP is hydrolyzed during the translocation process and that the release of deacylated tRNA, movement of messenger RNA, movement of the peptidyl-tRNA to a puromycin-sensitive site, and introduction of the next codon do not occur in the presence of GDP. However, recent studies have indicated that GTP hydrolysis is not absolutely required for translocation. Thus, it has been shown (Inoue-Yokosawa et al., 1973; Belitsina et al., 1975) that GDPCP, when incubated with stoichiometric amounts of EF-G, could substitute for GTP in the translocation of an aminoacyl-tRNA from the A site to the P site of the ribosome. These authors suggest that translocation occurs as a result of the correct binding of EF-G to the ribosome and that the hydrolysis of GTP is required to release EF-G from the ribosome and thus allow it to recycle. This mechanism would make the role of GTP in this reaction quite analogous to that already proposed for both EF-Tu (Shorey et al., 1971; Yokosawa et al., 1973, 1975) and IF-2 (Benne and Voorma, 1972; Dubnoff et al., 1972).

The question of how many GTP molecules are hydrolyzed for each peptide bond formed is still unanswered. The major difficulty with these studies is the presence of a very active uncoupled ribosome-dependent

GTPase activity. An early report that only one GTP is hydrolyzed per peptide bond formed (Nishizuka and Lipmann, 1966b) has not been confirmed, and at present there is no evidence that one GTP molecule can function in both the binding of aminoacyl-tRNA to the ribosome and translocation (Ono *et al.*, 1969; Ravel *et al.*, 1969). The best estimate is that at least two GTP molecules are hydrolyzed for each peptide bond formed during elongation (Cabrer *et al.*, 1976).

E. Nonenzymatic Translocation

Although, as discussed previously, maximum rates of translocation are dependent on the presence of EF-G and GTP, this process can occur nonenzymatically, i.e., in the absence of both EF-G and GTP (Pestka, 1968, 1969b,c; Gavrilova and Spirin, 1971). The assay employed for this reaction involves the determination of oligophenylalanine formation in an incubation containing only ribosomes, poly(U), Phe-tRNA, and Mg^{2+} and K^+ ions. It was shown that the nonenzymatic translocation could not be accounted for by small amounts of contaminating EF-G since: (1) fusidic acid did not inhibit the reaction under conditions where enzymatic translocation would be inhibited, (2) p-chloromercuribenzoate (PCMB) which inactivates EF-G not only did not inhibit the reaction but stimulated it (Gavrilova and Spirin, 1972), and (3) GDPCP had no effect on the nonenzymatic reaction but completely inhibited the enzymatic reaction. In addition the activation energy for translocation in the presence of EF-G and GTP was found to be 6.6 kcal/mole as contrasted to 23 kcal/mole when the reaction was carried out in the absence of EF-G and GTP. Finally, with the use of a mutant which contained a temperature sensitive EF-G, it was shown that ribosomes from this strain could carry out nonenzymatic translocation at both permissive and nonpermissive temperatures (Pestka, 1974). Recently, it has been shown (Gavrilova *et al.*, 1974) that ribosomal protein S12 is directly involved in nonenzymatic translocation, since both its alkylation by PCMB and its complete depletion from the 30 S subunit stimulate translocation in the absence of EF-G. Conversely, if the 30 S subunit is saturated with S12 nonenzymatic translocation is inhibited.

It is not known whether the mechanism of the nonenzymatic translocation is similar to that catalyzed by EF-G and GTP, but it is to be noted that this reaction (nonenzymatic translocation) has only been demonstrated using poly(U) as template. Indeed, when poly(A) was used no oligolysine formation was observed in the absence of EF-G and GTP (Pestka, 1969b).

F. Mechanism of Translocation

In spite of the fact that a great deal of information has been obtained at the molecular level, dealing with the interaction of EF-G, GTP, and ribosomes, the exact mechanism by which these components affect translocation is unknown. However, in the past number of years, a few theories of how translocation occurs have been proposed.

1. Locking-Unlocking Ribosome Model

The locking-unlocking ribosome model (Spirin, 1969) is based on the idea that the 70 S ribosome can exist in two different states, locked (with the subparticles closely associated) and unlocked (with the 30 S and 50 S subunits drawn somewhat apart). The two subunits are envisioned as being hinged to each other and that the locking and unlocking of the subunits of the ribosome provides the driving mechanism for translocation. In this model the A and P sites of the ribosome are pictured as being located on the contacting surfaces of the 30 S and 50 S subunits. In addition, the mRNA binding site is also on the contacting surface of the 30 S particle and the mRNA chain can slide along it between the 30 S and 50 S subunits. The ribosome in the pretranslocation state (peptidyl-tRNA in the A site and deacylated tRNA in the P site) is viewed as being very closed or "locked." During translocation EF-G is bound to the ribosome and GTP is hydrolyzed. This process causes the ribosome to become "unlocked." During this unlocking, the subparticles are pictured as drawing slightly apart from each other, by turning about a hinge axis. This process induces the movement of the peptidyl-tRNA from the A site to the P site. In addition, the peptidyl-tRNA residue drags along the mRNA triplet associated with its anticodon, thus in effect moving the mRNA chain by one codon. This theory would predict that different conformational states of the ribosome would exist during the various stages of elongation. This will be discussed below.

2. Inchworm Theory

The inchworm theory (Hardesty et al., 1969) suggests that during the enzymatic binding of aminoacyl-tRNA to the A site of the ribosome a kink is created in the mRNA. This kink may be formed by a rotation of the tRNA about its long axis and by doing so it brings its amino acid into close proximity of the peptide in the P site so that a peptide bond can now be formed. The effect of EF-G and GTP is then envisioned as straightening out the kink in the mRNA, and by doing so the peptidyl-tRNA is translocated from the A site to the P site of the ribosome. This movement of the mRNA along the ribosome has been compared to an inchworm moving across a surface.

3. Conformational Changes

It might be expected that the process of translocation might induce the ribosome to undergo some change in conformation. However, although there have been a number of reports suggesting that the ribosome undergoes conformational changes during translocation (Chuang and Simpson, 1971; Schreier and Noll, 1971; Waterson *et al.,* 1972), all these results were obtained by observing differences in the sedimentation values of pre- and posttranslocation ribosomal complexes after high-speed centrifugation. The interpretation of these results have been questioned (Infante and Baierlein, 1971; Infante and Krauss, 1971) since it has been shown that ribosomes have a tendency to undergo partial dissociation when subjected to high centrifugal forces.

V. Elongation Factor 2

The eukaryote factor corresponding to EF-G is elongation factor 2 (EF-2). Initially, Schweet and co-workers (Bishop and Schweet, 1961; Arlinghaus *et al.,* 1963; Hardesty *et al.,* 1963) separated two factors required for polypeptide synthesis from rabbit reticulocytes which were referred to as T_1 and T_2. The former fraction corresponded to EF-T from prokaryotes whereas the latter factor complimented T_1 and possessed similar characteristics to EF-G (Felicetti and Lipmann, 1968). The early studies on EF-2 were performed mainly with enzyme preparations from rabbit reticulocytes (Bishop and Schweet, 1961; Arlinghaus *et al.,* 1963; Hardesty *et al.,* 1963) and rat liver (Fessenden and Moldave, 1963; Gasior and Moldave, 1965; Schneir and Moldave, 1968; Skogerson and Moldave, 1969a,b; Raeburn *et al.,* 1971).

A. Purification

The initial purification of EF-2 from rat liver was achieved by Galasinski and Moldave (1969). Based on this procedure Raeburn *et al.* (1971) were able to obtain a preparation that appeared to be close to 95% pure. The purification involved chromatography on DEAE cellulose, Sephadex G-150, and phosphocellulose followed by isoelectric focusing. The overall purification was slightly greater than 100-fold, and the material showed one band on disc gel electrophoresis and one symmetrical peak in the analytical ultracentrifuge. However, more recently, Henriksen and co-workers (1975a) showed that this highly purified preparation of EF-2 catalyzes a GTP-GDP transphosphorylation due to a contaminating protein. A second isoelectric focusing step was employed to remove the trace of the transphosphorylase activity.

B. Properties

The molecular weight of EF-2 has been estimated at 96,500 by sedimentation analysis (Raeburn *et al.,* 1971) and 110,000 by amino acid analysis (Robinson and Maxwell, 1972). An amino acid analysis is shown in Table

TABLE VIII

Amino Acid Composition of Elongation Factor 2[a]

Amino acid	(A) [³H]ADPR EF-2 (residues/mole protein)	(B) EF-2 (residues/moles protein)	(C) [¹⁴C]ADPR EF-2 (residues/moles ADPR)
Lysine	69.5	75.5	72.4
Histidine	17.2	18.4	18.0
Aminoethylated cysteine	21.5		
Arginine	53.7	51.8	49.6
Aspartic acid	93.5	91.8	95.3
Threonine	47.4	48.9	47.1
Serine	54.0	58.6	54.6
Glutamic acid	107.8	107.0	103.4
Proline	49.3	52.3	50.6
Glycine	75.5	84.2	81.9
Alanine	74.4	77.1	77.0
Valine	76.9	78.5	81.0
Methionine	27.8	30.0	29.9
Isoleucine	58.8	61.6	62.3
Leucine	94.3	86.3	85.8
Tyrosine	25.3	25.1	24.0
Phenylalanine	44.4	40.1	40.3
Tryptophan	8.1		

[a] The values for A and C represent the average or extrapolated values for six determinations (two each after 20, 46, and 70 hr hydrolysis), while those for B are based on only one determination after 46 hr hydrolysis; 1.3–1.7 nmoles of EF-2 were hydrolyzed for each determination. From Robinson and Maxwell (1972).

VIII. Valine is the N-terminal amino acid and the protein contains 18 sulf-hydryl groups and 2 disulfide bonds per mole of enzyme. EF-2 can be ADP ribosylated as first shown by Honjo *et al.* (1968) in the presence of diphtheria toxin and NAD (to be described below), but the reaction is prevented if EF-2 is first treated with *p*-hydroxymercuribenzoate.

C. Assay

There are four convenient ways of measuring EF-2 activity, which include (1) polypeptide synthesis dependent on EF-2, (2) puromycin peptide formation using polysomes containing nascent protein chains (Schneider *et al.,* 1968) or ribosomes containing endogenous peptides (Skogerson and Moldave, 1968a,b), (3) GTP hydrolysis dependent on EF-2 and ribosomes, and (4) ADP ribosylation of EF-2 in the presence of diphtheria toxin. The assays are described in detail by Raeburn *et al.* (1971). The most unique and very specific assay for EF-2 is the ADP ribosylation which is carried out as follows.

[³H]ADP ribosylation: Fifty μl of a mixture containing 38,000 dpm (405 pmoles) of [³H]NAD labeled in the adenosine moiety, 2.9 μg of diphtheria toxin, 7.5 μmoles of dithiothreitol, and 2.5 μmoles of EDTA are added to varying amounts of EF-2 in 0.15 ml of a buffer containing 5 mM Tris-Cl pH 7.2, 1 mM DTT, 0.1 mM EDTA, and 0.25 to 0.5 mg of albumin to stabilize the enzyme. The reaction mixtures are incubated for 1 hr at 30°. The reactions are stopped by the addition of 3 ml of 5% tricholoracetic acid. The protein is precipitated by heating at 90°, then washed onto glass fiber filters, and counted. The bound radioactivity is a measure of the amount of EF-2.

D. Interaction of EF-2 with Ribosomes and Guanosine Nucleotides

Skogerson and Moldave (1967) first reported that rat liver EF-2 could bind to ribosomes in the presence of GTP and a sulfhydryl compound. The bound EF-2 remained with the ribosomes during centrifugation and ribosomes prepared in this manner were active in amino acid incorporation in the absence of added EF-2. In subsequent studies (Skogerson and Moldave, 1968a,b), these authors showed that the binding of EF-2 to ribosomes could be obtained with the nonhydrolyzable GTP analogue GDPCP, as well as GDP. However, GTP was required to yield a ribosome EF-2 complex that was active in stimulating AA-tRNA transfer to the ribosomes. These results indicated that a ribosome EF-2 complex prepared in the presence of GTP made available an AA-tRNA binding site on the ribosome due to translocation of any peptidyl-tRNA on the ribosome from the A site to the P site. The binding of EF-2 to the ribosome requires sulfhydryl groups on both the ribosome and EF-2, and it has been reported that the binding site can be occluded by the presence of peptidyl-tRNA at the acceptor site (Baliga and Munro, 1971). In addition the association of EF-2 with ribosomes has been demonstrated *in vivo*. Smulson and Rideau (1970) found that EF-2 was associated with HeLa cell monoribosomes, but not heavy or light polyribosomes. The amount of EF-2 on the monoribosome increased under growth conditions which increased the amount of monoribosomes, such as amino deprivation or when cells were exposed to puromycin.

It is now apparent that the interaction of EF-2 with guanosine nucleotides and ribosomes is quite similar to that observed with the prokaryote factor EF-G (see above). The basic difference, however, between the two systems is that EF-G does not form a stable complex with either GTP or GDP, whereas EF-2 does. An EF-2 · GTP complex has been

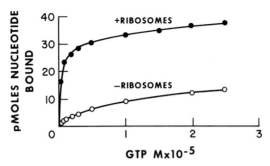

Fig. 8 The effect of GTP concentration on the extent of tritium binding from [³H]GTP. The incubations containing 7.2 μg of EF-2 were performed at 37° for 3 min in the absence or the presence of 2.0 A_{260} units ribosomes and the nucleotide complex measured by a nitrocellulose filter procedure. From Chuang and Weissbach (1972).

shown with EF-2 preparations from liver (Baliga and Munro, 1972; Henriksen *et al.*, 1975a), reticulocytes (Bodley and Lin, 1970), human tonsil (Bermek and Matthaei, 1971), and calf brain (Chuang and Weissbach, 1972). The characteristics of the reaction of EF-2 with guanosine nucleotides and the ribosomes from all sources appear quite similar and some of the properties of the reactions will be presented. One can conveniently assay for nucleotide binding to EF-2 using a nitrocellulose filter (Millipore) procedure, since an EF-2 nucleotide complex is retained on the filter whereas the free nucleotide is not. Figure 8 shows the effect of GTP concentration on complex formation when EF-2 is incubated with and without ribosomes. Radioactive nucleotide is bound to the filter in the presence of ribosomes although a significant amount is bound to the filter with EF-2 alone. However, the nature of the bound nucleotide depends on the presence of ribosomes. Using [³H,³²P-γ]GTP as shown in Table IX, GTP was present on the filter without ribosomes (³H/³²P = ~1) but GDP (³H³²P = ~7) was in the complex when ribosomes were present in the incubation. These results suggested the following reaction sequence.

$$\text{EF-2} + \text{GTP} \rightleftharpoons \text{EF-2} \cdot \text{GTP}$$
$$\text{EF-2} \cdot \text{GTP} + \text{Rib} \longrightarrow \text{Rib} \cdot \text{EF-2} \cdot \text{GDP} + \text{P}_i$$

It should be noted that GTP also binds to EF-2 and that the EF-2 · GDP can also interact with ribosomes.

The interaction of guanosine nucleotides with EF-2 in the absence of ribosomes could also be shown by chromatography on Sephadex G-100. The results with calf brain EF-2 are seen in Fig. 9. Radioactive GTP or GDP could be detected eluting with EF-2 near the void volume of the column. However, the EF-2 · GDP complex appeared more labile as seen by the elution pattern. It was estimated that the K_d values for EF-2 · GDP and

TABLE IX

The Binding of [³H]GTP and [γ-³²P]GTP to EF-2 and Ribosomes[a]

Addition	[³H]GTP bound (pmoles)	[³²P]GTP bound (pmoles)
None	0.1	0.1
EF-2	4.8	4.6
Ribosomes	2.2	2.0
EF-2 + ribosomes	20.3	3.1

[a] The reaction mixtures contained in a total volume of 100 μl 10 mM Tris-Cl, pH 7.4; 10 mM MgCl$_2$; 10 mM ammonium acetate; 1 mM DTT; 5 × 10^{-6} M of [³H, ³²P − γ]GTP; and, where indicated, 7.2 μg EF-2 and 2.0 A_{260} units ribosomes. The mixture was incubated for 3 min at 37° and the binding of the nucleotide was assayed by a filter procedure. From Chuang and Weissbach (1972).

EF-2 · GTP complexes from calf brain were 3.2 × 10^{-7} M and 1.2 × 10^{-7} M, respectively. The latter value is similar to that found by Montanaro *et al.* (1971). More recently, however, Henriksen *et al.* (1975a) have done a careful study on the binding of GTP and GDP to liver EF-2 using equilibrium dialysis and have shown that the enzyme has one guanosine nucleotide binding site with K_d values for the EF-2 · GDP and EF-2 · GTP complexes being 4 × 10^{-7} M and 2 × 10^{-6} M, respectively. This higher value

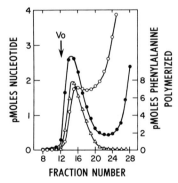

Fig. 9 Sephadex G-100 chromatography of EF-2 after incubating with [³H]GTP or [³H]GDP. Each incubation contained in a total volume of 0.5 ml: 140 μg EF-2; 0.8 × 10^{-5} M labeled GTP or GDP; 10 mM Tris-Cl, pH 7.4; 10 mM MgCl$_2$; 10 mM ammonium acetate; and 5 mM DDT. The mixtures were incubated for 3 min at 37° and then placed on a Sephadex G-100 column. Aliquots were assayed for radioactivity and for the presence of EF-2 by polyphenylalanine polymerization. ●—●, [³H]GTP; ○—○, [³H]GDP; △—△, EF-2 activity. From Chuang and Weissbach (1972).

than what has been reported previously for the EF-2 · GTP complex may be due to their correction for the amount of GDP formed in the incubations. The nonhydrolyzable analogue of GTP, GDPCP, was found to have a K_d value of $2 \times 10^{-5} M$, and no binding of GMP to the enzyme was observed.

In a further study on the binding of guanosine nucleotides to EF-2, Henriksen *et al.* (1975b) demonstrated that both ribosomes and Mg^{2+} inhibit the binding of GDP, but not GTP, to the enzyme. It was suggested that the GTP/GDP ratio may play an important role in the regulation of translocation due to different conformational states of EF-2 in the presence of either GTP or GDP.

In the prokaryote system, fusidic acid is known to greatly stabilize the ribosome · EF-G · GDP complex (Table III). Although fusidic acid also affects the corresponding complex with EF-2, the effect is much less pronounced. This has been confirmed in several eukaryote systems (Bodley *et al.*, 1970a; Bermek and Matthaei, 1971; Chuang and Weissbach, 1972; Montanaro *et al.*, 1971). Thus two basic differences between the prokaryote and eukaryote translocation factors are that (1) the eukaryote factor forms a stable complex with GDP or GTP in the absence of ribosomes and (2) the ribosome · EF-2 · GDP complex is more stable than the prokaryote complex and therefore is less affected by fusidic acid. It should be noted that the EF-2 · GTP(GDP) complex is much less stable than the ribosome · EF-2 · GDP complex (Chuang and Weissbach, 1972).

The results suggest the following sequence of reactions quite similar to what has been proposed for EF-G (see Fig. 6).

$$
\text{EF-2} + \text{GTP} \rightleftarrows \text{EF-2} \cdot \text{GTP}
$$

$$
\downarrow \text{ribosome}
$$

$$
\text{Fusidic} \cdot \text{Rib} \cdot \text{EF-2} \cdot \text{GDP} \xleftarrow{\text{fusidic acid}} \text{Rib} \cdot \text{EF-2} \cdot \text{GDP} + \text{P}_i
$$

Presumably during the course of polypeptide synthesis EF-2 comes off the ribosome, interacts with another mole of GTP and the ribosome, and in this way is able to catalyze both translocation and GTP hydrolysis. Fusidic acid, which traps EF-2 on the ribosome, prevents recycling of EF-2 and by so doing inhibits AA-tRNA binding since EF-1 cannot interact with ribosomes that contain bound EF-2 (Nombela and Ochoa, 1973; Richter, 1973). The role of the ribosomal subunits in GTP hydrolysis in eukaryotes appear similar to the prokaryote system. McKeehan (1972) has shown that the 60 S subunit is required for GTP hydrolysis with EF-2. The 40 S subunit, however, did stimulate the hydrolysis in the presence of the larger subunit. It is also of interest that EF-2 can bind to RNA, such as ribosomal RNA as well as DNA (Traugh and Collier, 1970, 1971). This interaction is inhibited by either GDP or GTP, but the physiological significance of this reaction is not known.

E. Effect of Diphtheria Toxin on EF-2

One of the most interesting features of EF-2 is its inhibition by diphtheria toxin. The events leading to our current knowledge of this exciting phenomenon started with the observation of Strauss and Hendee (1959) who showed that protein synthesis in HeLa cells was inhibited within 2 hr after exposure to toxin, whereas other metabolic processes did not seem to be affected. Collier and Pappenheimer (1964) were then able to demonstrate an effect of diphtheria toxin on protein synthesis in cell-free extracts of HeLa cells and reticulocytes. Of special importance was the finding that the toxin effect *in vitro* required the cofactor NAD. Collier (1967) and Goor and Pappenheimer (1967a,b) then showed that the effect of the toxin and NAD was on EF-2. Subsequently, Honjo *et al.* (1968) helped to elucidate the mechanism of this inhibition by showing that the toxin caused the inactivation of EF-2 by catalyzing the transfer of the adenosine diphosphate ribose portion of NAD to the protein.

$$\text{Nic}^+\text{-ADPR} + \text{EF-2} \xrightarrow{\text{toxin}} \text{ADPR-EF-2} + \text{Nic} + \text{H}^+$$
$$\text{(NAD)} \qquad\qquad\qquad \text{(Inactive)}$$

There has been some question as to whether the GTPase activity in the EF-2 preparations was associated with the enzyme. This was clearly shown by Raeburn *et al.* (1968) who demonstrated that the GTPase activity of EF-2 is also inhibited by toxin and NAD. However, the binding of GTP to EF-2 was not affected by the toxin in the absence of ribosomes, and both GTP and ribosomes partially protect EF-2 from toxin inactivation. It was also noted in these studies that GTP binding to EF-2 in the presence of ribosomes was inhibited by toxin. This is in contrast to the results with brain EF-2, where it was reported that the toxin inhibited polypeptide synthesis but did not appear to affect nucleotide binding to EF-2 in the presence or absence of ribosomes (Chuang and Weissbach, 1972). A typical experiment described in Fig. 10 shows the effect of toxin on the various activities of calf brain EF-2.

The site of the ADP ribosylation of EF-2 is under careful study. Recently, Robinson *et al.* (1974) have purified a single peptide obtained by tryptic digestion of [^{14}C]ADP-ribosylated EF-2. The peptide consists of 15 amino acids, Phe-Asp-Val-His-Asp-Val-Thr-Leu-His-Ala-Asp-Ala-Ile-X-Arg. The ADP ribose is linked to the, as yet, unknown weakly basic residue X.

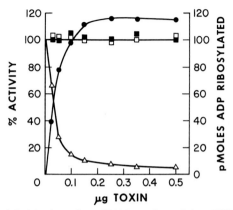

Fig. 10 The effect of diphtheria toxin and NAD on the activity of EF-2. ADP ribosylation was performed in a total volume of 500 μl containing: 10 mM Tris-Cl, pH 7.4; 10 mM MgCl$_2$; 10 mM ammonium acetate; 1 mM DTT; 30 μg EF-2; 2 μM [^{3}H]NAD; and various amounts of diphtheria toxin. The mixture was incubated at 37° for 20 min. Aliquots of 200 μl were drawn and assayed for ADP ribosylation. Aliquots of 100 μl were used to assay either polymeriza-tion, the binding of [γ-^{32}P]GTP to EF-2 in the absence of ribosomes, or the binding of [^{14}C]GTP to ribosomes in the presence of EF-2. ●—●, [^{3}H]ADP ribosylation of EF-2; △—△, polymerization activity of EF-2; □—□, [γ-^{32}P]GTP binding activity in the absence of ribosomes; ■—■ [^{14}C]GTP binding activity in the presence of ribosomes. From Chuang and Weissbach (1972).

VI. Conclusions

During the biosynthesis of a protein, the ribosome moves relative to the messenger RNA from the 5′- to the 3′-end. This process, which is called translocation, appears to involve three coordinated reactions: (1) ejection of a deacylated tRNA from the ribosome, (2) movement of peptidyl-tRNA from one site on the ribosome to another and, (3) movement of the mRNA so that a new codon can be translated.

All the above events of translocation have been shown to involve GTP and are catalyzed by a single 100,000 g supernatant protein in both prokaryotes and eukaryotes.

The prokaryote factor, called elongation factor G has been puri-fied to homogeneity and has a molecular weight of about 80,000. This pro-tein binds to the 50 S ribosomal subunit and forms a ternary complex in the presence of a number of guanosine nucleotides including GTP, GDP, and the GTP analogue, guanylylmethylenediphosphonate. When GTP is employed, it is rapidly hydrolyzed during the binding reaction to GDP, and the ternary complex then contains EF-G, GDP, and the 50 S ribo-somal subunit. Fusidic acid, an antibiotic that inhibits the enzymatic activity of EF-G, has been shown to greatly increase the stability of the ternary complex by forming stoichiometric amounts of a very stable EF-G · GDP · ribosome · fusidic acid complex. Thus, fusidic acid allows one round of GTP hydrolysis and one round of translocation. The mechanism of fusidic acid inhibition of protein synthesis involves the trapping of EF-G on the ribosome and, in this way, prevents it from acting catalytically.

All of the reactions concerning EF-G involve the participation of two 50 S ribosomal proteins, L7L12. If these two proteins are removed from the ribosome, the ribosome loses its ability to bind EF-G. Studies using bifunctional cross-linking reagents have shown that L7L12 are at or very near the EF-G binding site.

The eukaryote protein analogous to EF-G is elongation factor 2. This protein, which is about 97,000 daltons, forms a complex with the 80 S ribosome and guanosine nucleotides very similar to that described for EF-G. The essential difference between the two systems is that EF-2 forms a stable complex with either GTP or GDP, whereas EF-G does not.

Diphtheria toxin in the presence of NAD inhibits the enzymatic activity of EF-2. This inhibition appears to be caused by the transfer of the adenosine diphosphate ribose moiety of NAD to the protein. A peptide of 15 amino acids has been isolated which contains the ADP ribose.

REFERENCES

Acharya, A. S., Moore, P. B., and Richards, F. M. (1973). *Biochemistry* **12**, 3108–3114.
Allende, J. E., Monro, R., and Lipmann, F. (1964). *Proc. Natl. Acad. Sci. U.S.A.* **51**, 1211–1216.
Anderson, B., Hodgkin, D. C., and Viswamitra, M. A. (1970). *Nature (London)* **225**, 233–235.
Arai, K., Arai, N., Kawakita, M., and Kaziro, Y. (1972). *Biochem. Biophys. Res. Commun.* **48**, 190–196.
Arlinghaus, R., Favelukes, G., and Schweet, R. (1963). *Biochem. Biophys. Res. Commun.* **11**, 92–96.
Baliga, B. S., and Munro, H. N. (1971). *Nature (London), New Biol.* **233**, 257–258.
Baliga, B. S., and Munro, H. N. (1972). *Biochem. Biophys. Acta* **277**, 368–383.
Ballesta, J. P. G., and Vazquez, D. (1972). *FEBS Lett.* **28**, 337–342.
Belitsina, N. V., Glukhova, M. A., and Spirin, A. S. (1975). *FEBS Lett.* **54**, 35–38.
Benne, R., and Voorma, H. O. (1972). *FEBS Lett.* **20**, 347–351.
Bermek, E., and Matthaei, H. (1971). *Biochemistry* **10**, 4906–4912.
Bernardi, A., and Leder, P. (1970). *J. Biol. Chem.* **245**, 4263–4268.
Bishop, J. O., and Schweet, R. S. (1961). *Biochim. Biophys. Acta* **54**, 617–619.
Bodley, J. W., and Lin, L. (1970). *Nature (London)* **227**, 60–62.
Bodley, J. W., Zieve, F. J., Lin, L., and Zieve, S. T. (1969). *Biochem. Biophys. Res. Commun.* **37**, 437–443.
Bodley, J. W., Lin, L., Salas, M. L., and Tao, M. (1970a). *FEBS Lett.* **11**, 153–156.
Bodley, J. W., Zieve, F. J., Lin, L., and Zieve, S. T. (1970b). *J. Biol. Chem.* **245**, 5656–5661.

Bodley, J. W., Lin, L., and Highland, J. H. (1970c). *Biochem. Biophys. Res. Commun.* **41,** 1406–1411.

Bodley, J. W., Zieve, F. J., and Lin, L. (1970d). *J. Biol. Chem.* **245,** 5662–5667.

Brot, N., Ertel, R., and Weissbach, H. (1968). *Biochem. Biophys. Res. Commun.* **31,** 563–570.

Brot, N., Spears, C., and Weissbach, H. (1969). *Biochem. Biophys. Res. Commun.* **34,** 843–848.

Brot, N., Spears, C., and Weissbach, H. (1971). *Arch. Biochem. Biophys.* **143,** 286–296.

Brot, N., Yamasaki, E., Redfield, B., and Weissbach, H. (1972). *Arch. Biochem. Biophys.* **148,** 148–155.

Brot, N., Marcel, R., Yamasaki, E., and Weissbach, H. (1973). *J. Biol. Chem.* **248,** 6952–6959.

Brot, N., Tate, W. P., Caskey, C. T., and Weissbach, H. (1974). *Proc. Natl. Acad. Sci. U.S.A.* **71,** 89–92.

Cabrer, B., Vazquez, D., and Modolell, J. (1972). *Proc. Natl. Acad. Sci. U.S.A.* **69,** 733–736.

Cabrer, B., San-Millan, M. J., Vazquez, D., and Modolell, J. (1976). *J. Biol. Chem.* **251,** 1718–1722.

Cannon, M., and Burns, K. (1971). *FEBS Lett.* **18,** 1–5.

Celma, M. L., Vazquez, D., and Modolell, J. (1972). *Biochem. Biophys. Res. Commun.* **48,** 1240–1246.

Chuang, D. M., and Simpson, M. V. (1971). *Proc. Natl. Acad. Sci. U.S.A.* **68,** 1474–1478.

Chuang, D. M., and Weissbach, H. (1972). *Arch. Biochem. Biophys.* **152,** 114–124.

Collier, R. J. (1967). *J. Mol. Biol.* **25,** 83–98.

Collier, R. J., and Pappenheimer, A. M., Jr. (1964). *J. Exp. Med.* **120,** 1019–1039.

Conway, T., and Lipmann, F. (1964). *Proc. Natl. Acad. Sci. U.S.A.* **52,** 1462–1469.

Cundliffe, E. (1971). *Biochem. Biophys. Res. Commun.* **44,** 912–917.

Cundliffe, E. (1972). *Biochem. Biophys. Res. Commun.* **46,** 1794–1801.

Dubnoff, J. S., Lockwood, A. H., and Maitra, U. (1972). *J. Biol. Chem.* **247,** 2884–2894.

Eckstein, F., Kettler, M., and Parmeggiani, A. (1971). *Biochem. Biophys. Res. Commun.* **45,** 1151–1158.

Erbe, R. W., and Leder, P. (1968). *Biochem. Biophys. Res. Commun.* **31,** 798–803.

Erbe, R. W., Nau, M. M., and Leder, P. (1969). *J. Mol. Biol.* **39,** 441–460.

Felicetti, L., and Lipmann, F. (1968). *Arch. Biochem. Biophys.* **125,** 548–557.

Felicetti, L., Tocchini-Valentini, G. P., and Matteo, F. (1969). *Biochemistry* **8,** 3428–3442.

Fessenden, J. M., and Moldave, K. (1963). *J. Biol. Chem.* **238,** 1479–1484.

Galasinski, W., and Moldave, K. (1969). *J. Biol. Chem.* **244,** 6527–6532.

Gasior, E., and Moldave, K. (1965). *J. Biol. Chem.* **240,** 3346–3352.

Gavrilova, L. I., and Spirin, A. S. (1971). *FEBS Lett.* **17,** 324–326.

Gavrilova, L. I., and Spirin, A. S. (1972). *FEBS Lett.* **22,** 91–92.

Gavrilova, L. I., Koteliansky, V. E., and Spirin, A. S. (1974). *FEBS Lett.* **45,** 324–328.

Godtfredsen, W. O., and Vangedal, S. (1962). *Tetrahedron* **18,** 1029–1048.

Goor, R. S., and Pappenheimer, A. M., Jr. (1967a). *J. Exp. Med.* **126,** 899–912.

Goor, R. S., and Pappenheimer, A. M., Jr. (1967b). *J. Exp. Med.* **126,** 913–921.

Gordon, J. (1969). *J. Biol. Chem.* **244,** 5680–5686.

Gordon, S. (1970). *Biochemistry* **9,** 912–917.

Gupta, S. L., Waterson, J., Sopori, M. L., Weissman, S. M., and Lengyel, P. (1971). *Biochemistry* **10,** 4410–4421.

Haenni, A. L., and Lucas-Lenard, J. (1968). *Proc. Natl. Acad. Sci. U.S.A.* **61,** 1363–1369.

Hamel, E., Koka, M., and Nakamoto, T. (1972). *J. Biol. Chem.* **247,** 807–814.

Hardesty, B., Arlinghaus, R., Schaeffer, J., and Schweet, R. (1963). *Cold Spring Harbor Symp. Quant. Biol.* **28**, 215–222.

Hardesty, N., Culp, W., and McKeehan, W. (1969). *Cold Spring Harbor Symp. Quant. Biol.* **39**, 331–344.

Harvey, C. L., Knight, S. G., and Sih, C. J. (1966). *Biochemistry* **5**, 3320–3327.

Hazelkorn, R., and Rothman-Denes, L. B. (1973). *Annu. Rev. Biochem.* **42**, 397–438.

Henriksen, O., Robinson, E. A., and Maxwell, E. S. (1975a). *J. Biol. Chem.* **250**, 720–724.

Henriksen, O., Robinson, E. A., and Maxwell, E. S. (1975b). *J. Biol. Chem.* **250**, 725–730.

Highland, J. H., Lin, L., and Bodley, J. W. (1971). *Biochemistry* **10**, 4404–4409.

Highland, J. H., Bodley, J. W., Gordon, J., Hasenbank, R., and Stöffler, G. (1973). *Proc. Natl. Acad. Sci. U.S.A.* **70**, 142–150.

Honjo, T., Nishizuka, Y., Hayaishi, O., and Kato, I. (1968). *J. Biol. Chem.* **243**, 3553–3555.

Infante, A. A., and Baierlein, R. (1971). *Proc. Natl. Acad. Sci. U.S.A.* **68**, 1780–1785.

Infante, A. A., and Krauss, M. (1971). *Biochim. Biophys. Acta* **246**, 81–99.

Inoue-Yokosawa, N., Ishikawa, C., and Kaziro, Y. (1973). *J. Biol. Chem.* **249**, 4321–4323.

Ishitsuka, H., Kuriki, Y., and Kaji, A. (1970). *J. Biol. Chem.* **245**, 3346–3351.

Kay, A., Sander, G., and Grunberg-Manago, M. (1973). *Biochem. Biophys. Res. Commun.* **51**, 979–986.

Kaziro, Y., and Inoue, N. (1968). *J. Biochem. (Tokyo)* **64**, 423–425.

Kaziro, Y., Inoue, N., Kuriki, Y., Mizumoto, K., Tanaka, M., and Kawakita, M. (1969). *Cold Spring Harbor Symp. Quant. Biol.* **34**, 385–393.

Kaziro, Y., Inoue-Yokosawa, N., and Kawakita, M. (1972). *J. Biochem. (Tokyo)* **72**, 853–863.

Kinoshita, T., Kuwano, G., and Tanaka, M. (1968). *Biochem. Biophys. Res. Commun.* **33**, 769–773.

Kinoshita, T., Liou, Y. F., and Tanaka, N. (1971). *Biochem. Biophys. Res. Commun.* **44**, 859–863.

Kischa, K., Möller, W., and Stöffler, G. (1971). *Nature (London), New Biol.* **233**, 62–63.

Kuriki, Y., and Kaji, A. (1968). *Proc. Natl. Acad. Sci. U.S.A.* **61**, 1399–1405.

Kuriki, Y., Inoue, N., and Kaziro, Y. (1970). *Biochim. Biophys. Acta* **224**, 487–497.

Kuwano, M., Schlessinger, D., Rinaldi, G., Felicetti, I., and Tocchini-Valentini, G. P. (1971). *Biochem. Biophys. Res. Commun.* **42**, 441–444.

Leder, P. (1973). *Adv. Protein Chem.* **27**, 213–242.

Leder, P., Skogerson, L. E., and Nau, M. M. (1969a). *Proc. Natl. Acad. Sci. U.S.A.* **62**, 454–460.

Leder, P., Skogerson, L. E., and Roufa, D. J. (1969b). *Proc. Natl. Acad. Sci. U.S.A.* **62**, 928–933.

Leder, P., Bernardi, A., Livingston, D., Lloyd, B., Roufa, D. J., and Skogerson, L. (1969c). *Cold Spring Harbor Symp. Quant. Biol.* **34**, 411–417.

Lockwood, A. H., Maitra, U., Brot, N., and Weissbach, H. (1973). *J. Biol. Chem.* **249**, 1213–1218.

Lucas-Lenard, J., and Beres, L. (1974). *In* "The Enzymes" (P. D. Boyer, ed.), 3rd ed., Vol. 10, pp. 53–86. Academic Press, New York.

Lucas-Lenard, J., and Haenni, A. L. (1969). *Proc. Natl. Acad. Sci. U.S.A.* **63**, 93–97.

Maassen, J. A., and Möller, W. (1974). *Proc. Natl. Acad. Sci. U.S.A.* **71**, 1277–1280.

McKeehan, W. (1972). *Biochem. Biophys. Res. Commun.* **48**, 1117–1122.

Marsh, R. C., and Parmeggiani, A. (1973). *Proc. Natl. Acad. Sci. U.S.A.* **70**, 151–155.

Mazumder, R. (1973). *Proc. Natl. Acad. Sci. U.S.A.* **70**, 1939–1942.

Miller, D. L. (1972). *Proc. Natl. Acad. Sci. U.S.A.* **69**, 752–755.

Modolell, J., Cabrer, B., and Vazquez, D. (1973). *J. Biol. Chem.* **248**, 8356–8360.

Modolell, J., Vazquez, D., and Monro, R. E. (1971a). *Nature (London), New Biol.* **230,** 109–112.

Modolell, J., Cabrer, B., Parmeggiani, A., and Vazquez, D. (1971b). *Proc. Natl. Acad. Sci. U.S.A.* **68,** 1796–1800.

Montanaro, L., Sperti, S., and Mattioli, A. (1971). *Biochim. Biophys. Acta* **238,** 493–497.

Nishizuka, Y., and Lipmann, F. (1966a). *Arch. Biochem. Biophys.* **116,** 344–351.

Nishizuka, Y., and Lipmann, F. (1966b). *Proc. Natl. Acad. Sci. U.S.A.* **55,** 212–219.

Nombela, C., and Ochoa, S. (1973). *Proc. Natl. Acad. Sci. U.S.A.* **70,** 3556–3560.

Nomura, M., and Engback, F. (1972). *Proc. Natl. Acad. Sci. U.S.A.* **69,** 1526–1530.

Okura, A., Kinoshita, T., and Tanaka, N. (1970). *Biochem. Biophys. Res. Commun.* **41,** 1545–1550.

Okura, A., Kinoshita, T., and Tanaka, N. (1971). *J. Antibiot.* **24,** 655–661.

Ono, Y., Skoultchi, A., Waterson, J., and Lengyel, P. (1969). *Nature (London)* **223,** 697–701.

Parmeggiani, A. (1968). *Biochem. Biophys. Res. Commun.* **30,** 613–619.

Parmeggiani, A., and Gottschalk, E. M. (1969a). *Biochem. Biophys. Res. Commun.* **35,** 861–867.

Parmeggiani, A., and Gottschalk, E. M. (1969b). *Cold Spring Harbor Symp. Quant. Biol.* **34,** 377–384.

Pestka, S. (1968). *Proc. Natl. Acad. Sci. U.S.A.* **61,** 726–733.

Pestka, S. (1969a). *J. Biol. Chem.* **243,** 2810–2820.

Pestka, S. (1969b). *J. Biol. Chem.* **244,** 1533–1539.

Pestka, S. (1969). *Cold Spring Harbor Symp. Quant. Biol.* **34,** 395–410.

Pestka, S. (1970). *Biochem. Biophys. Res. Commun.* **40,** 667–674.

Pestka, S. (1974). *In* "Methods in Enzymology" (L. Grossman and K. Moldave, eds.), Vol. 30, pp. 462–470. Academic Press, New York.

Pestka, S., and Brot, N. (1971). *J. Biol. Chem.* **246,** 7715–7722.

Raeburn, S., Goor, R. S., Schneider, J. A., and Maxwell, E. S. (1968). *Proc. Natl. Acad. Sci. U.S.A.* **61,** 1428–1434.

Raeburn, S., Collins, J. F., Moon, H. M., and Maxwell, E. S. (1971). *J. Biol. Chem.* **246,** 1041–1048.

Ravel, J. M., Shorey, R. L., Garner, C. W., Dawkins, R. C., and Shive, W. (1969). *Cold Spring Harbor Symp. Quant. Biol.* **34,** 321–330.

Richman, N., and Bodley, J. W. (1972). *Proc. Natl. Acad. Sci. U.S.A.* **69,** 686–689.

Richter, D. (1972). *Biochem. Biophys. Res. Commun.* **46,** 1850–1856.

Richter, D. (1973). *J. Biol. Chem.* **248,** 2853–2857.

Robinson, E. A., and Maxwell, E. S. (1972). *J. Biol. Chem.* **247,** 7023–7028.

Robinson, E. A., Henriksen, O., and Maxwell, E. S. (1974). *J. Biol. Chem.* **249,** 5088–5093.

Rohrbach, M. S., and Bodley, J. W. (1976). *J. Biol. Chem.* **251,** 930–933.

Rohrbach, M. S., Dempsey, M. E., and Bodley, J. W. (1974). *J. Biol. Chem.* **249,** 5094–5101.

Roufa, D. J., Skogerson, L. E., Leder, P. (1970). *Nature (London)* **227,** 567–570.

Sander, G., Marsh, R. C., and Parmeggiani, A. (1972). *Biochem. Biophys. Res. Commun.* **47,** 866–873.

Schneider, J. A., Raeburn, S., and Maxwell, E. S. (1968). *Biochem. Biophys. Res. Commun.* **33,** 177–181.

Schneir, M., and Moldave, K. (1968). *Biochim. Biophys. Acta* **166,** 58–67.

Schreier, M. H., and Noll, H. (1971). *Proc. Natl. Acad. Sci. U.S.A.* **68,** 805–809.

Shorey, R. L., Ravel, J. M., and Shive, W. (1971). *Arch. Biochem. Biophys.* **146,** 110–117.

Skogerson, L., and Moldave, K. (1967). *Biochem. Biophys. Res. Commun.* **27,** 568–572.

Skogerson, L., and Moldave, K. (1968a). *J. Biol. Chem.* **243,** 5354–5360.

Skogerson, L., and Moldave, K. (1968b). *J. Biol. Chem.* **243,** 5361–5367.

Smulson, M. E., and Rideau, C. (1970). *J. Biol. Chem.* **245**, 5350–5353.

Sopori, M. L., and Lengyel, P. (1972). *Biochem. Biophys. Res. Commun.* **46**, 238–244.

Spirin, A. S. (1969). *Cold Spring Harbor Symp. Quant. Biol.* **39**, 197–207.

Strauss, N., and Hendee, D. (1959). *J. Exp. Med.* **109**, 145–163.

Subramanian, A. R. (1975). *J. Mol. Biol.* **95**, 1–8.

Tanaka, K., Watanabe, S., Teraoka, H., and Tamaki, M. (1970). *Biochem. Biophys. Res. Commun.* **39**, 1189–1193.

Tanaka, K., Kuwano, G., and Kinoshita, T. (1971). *Biochem. Biophys. Res. Commun.* **42**, 564–567.

Tanaka, N., Kinoshita, T., and Masukawa, H. (1968). *Biochem. Biophys. Res. Commun.* **30**, 278–283.

Terhorst, C., Wittmann-Liebold, B., and Möller, W. (1972a). *Eur. J. Biochem.* **25**, 13–19.

Terhorst, C., Möller, W., Laursen, R., and Wittmann-Liebold, B. (1972b). *FEBS Lett.* **28**, 324–328.

Thach, S. S., and Thach, R. E. (1971). *Proc. Natl. Acad. Sci. U.S.A.* **68**, 1791–1795.

Thammana, P., Kurland, C. G., Deusser, E., Weber, J., Maschler, R., Stöffler, G., and Wittmann, H. G. (1973). *Nature (London), New Biol.* **242**, 47–49.

Tocchini-Valentini, G. P., and Mattoccia, E. (1968). *Proc. Natl. Acad. Sci. U.S.A.* **61**, 146–151.

Traugh, J. A., and Collier, R. J. (1970). *Biochem. Biophys. Res. Commun.* **40**, 1437–1444.

Traugh, J. A., and Collier, R. J. (1971). *Biochemistry* **10**, 2357–2366.

Watanabe, S., and Tanaka, K. (1971). *Biochem. Biophys. Res. Commun.* **45**, 728–734.

Waterson, J., Sopori, M. L., Gupta, S. L., and Lengyel, P. (1972). *Biochemistry* **11**, 1377–1382.

Weisblum, B., and Demohn, V. (1970a). *J. Bacteriol.* **101**, 1073–1075.

Weisblum, B., and Demohn, V. (1970b). *FEBS Lett.* **11**, 149–152.

Weissbach, H., Redfield, B., Yamasaki, E., Davis, R. C., Jr., Pestka, S., and Brot, N. (1972). *Arch. Biochem. Biophys.* **149**, 110–117.

Yokosawa, H., Inoue-Yokosawa, N., Arai, K., Kawakita, M., and Kaziro, Y. (1973). *J. Biol. Chem.* **248**, 375–377.

Yokosawa, H., Kawakita, M., Arai, K.-I., Inoue-Yokosawa, N., and Kaziro, Y. (1975). *J. Biochem. (Tokyo)* **77**, 719–728.

8

Peptide Bond Formation

R. J. HARRIS AND S. PESTKA

I. Introduction

Proteins in the form of enzymes and structural components perform many of the basic functions for all living things. In the last decade there has been an elucidation of the steps of protein biosynthesis. Some of the processes that occur during protein biosynthesis include formation of an mRNA · ribosome · initiator-tRNA complex, binding of aminoacyl-tRNA, peptide bond formation, translocation, and termination. These processes are carried out on the ribosome, a multiple component particle. This

413

chapter is concerned with one facet of this process, peptide bond formation or transpeptidation.

Peptide bond formation is catalyzed by the peptidyl transferase complex of the large ribosomal subunit. No external energy source or soluble factors are required in addition to ribosomes containing appropriate substrates. The process involves the transfer of the peptide chain from peptidyl-tRNA to aminoacyl-tRNA. In the process the carboxyl ester linkage of peptidyl-tRNA is broken, and a new peptide bond is formed with the amino acid of aminoacyl-tRNA on the carboxy terminal end. In addition to native substrates a large number of analogues of peptidyl- and aminoacyl-tRNA serve as appropriate donors and acceptors. The peptidyl transferase consists of two substrate binding sites (A', acceptor site, and P', donor site) and a catalytic center. The major structural components of the ribosomal peptidyl transferase sites have been outlined. In large part the stereochemical requirements for acceptor and donor substrates have been defined. In addition, as will be described in this chapter, physiological, structural, and inhibitor studies have contributed to our understanding of the biochemistry of peptide bond formation.

II. The Elongation Epicycle

A. Peptide Bond Formation by Peptidyl Transferase

Peptide bond formation takes place on the ribosome by the transfer of the nascent peptide chain from the 3'-end of peptidyl-tRNA to the α-

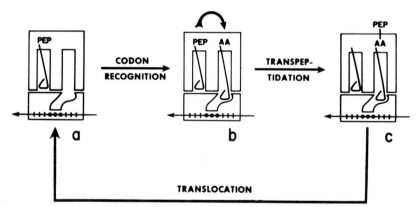

Fig. 1 The ribosome epicycle. The ribosome epicycle represents those steps through which the ribosome passes upon the addition of each amino acid during elongation. It encompasses aminoacyl-tRNA binding to ribosomes, transpeptidation, and translocation. The ribosome epicycle is presented in terms of the original donor-acceptor site model where P represents the donor site for peptidyl-tRNA and A the acceptor site for aminoacyl-tRNA.

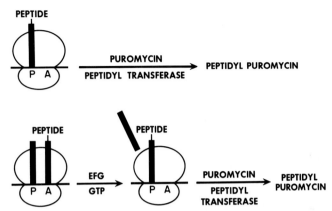

Fig. 2 Schematic illustration of the puromycin reaction with native peptidyl-tRNA. Puromycin being an analogue of aminoacyl-tRNA reacts with peptidyl-tRNA in the P site (top equation). If peptidyl-tRNA is in the A site, it must first be translocated into the P site for reaction with puromycin (bottom equation).

amino group of aminoacyl-tRNA. In this process the carboxyl ester link of peptidyl-tRNA is broken and a new peptide bond is formed (Figs. 1–4). No external energy source or soluble factors are required in addition to ribosomes containing appropriate substrates (Maden *et al.,* 1968; Monro, 1967; Pulkrábek and Rychlík, 1968; Pestka, 1968a). The process involves a group transfer and is, therefore, termed transpeptidation. The catalytic entity of the ribosome responsible for this process is termed the peptidyl transferase (Monro *et al.,* 1967). The enzymatic activity is an integral part of the large ribosomal subunit (Monro, 1967; Maden *et al.,* 1968; Vazquez *et al.,* 1969; Traut and Monro, 1964).

B. The Puromycin Reaction

Puromycin, a nucleoside antibiotic, is an analogue of the 3'-*O*-aminoacyladenosine portion of aminoacyl-tRNA (Yarmolinsky and de la Haba, 1959). It can react with peptidyl-tRNA in the presence of ribosomes (Figs. 2–4) to form peptidyl puromycin containing a newly formed peptide bond (Morris and Schweet, 1961; Rabinovitz and Fisher, 1962; Allen and Zamecnik, 1962; Nathans, 1964). Various peptidyl donors can be transferred to the puromycin acceptor. The puromycin reaction has provided a model system for studying transpeptidation.

C. Acceptor and Donor Two-Site Model of the Ribosome

Figure 1 shows the usual conceptional model of the ribosome epicycle, those events occurring during the stepwise addition of each amino acid. Peptidyl-tRNA is shown bound to the peptidyl donor site (P site), and

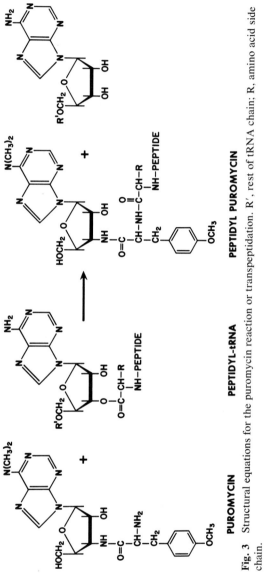

Fig. 3 Structural equations for the puromycin reaction or transpeptidation. R', rest of tRNA chain; R, amino acid side chain.

PUROMYCIN REACTION

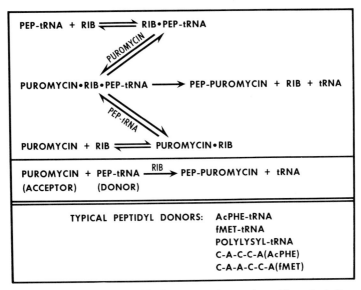

PEP-tRNA + RIB ⇌ RIB•PEP-tRNA

PUROMYCIN•RIB•PEP-tRNA ⟶ PEP-PUROMYCIN + RIB + tRNA

PUROMYCIN + RIB ⇌ PUROMYCIN•RIB

PUROMYCIN + PEP-tRNA —RIB→ PEP-PUROMYCIN + tRNA
(ACCEPTOR)   (DONOR)

TYPICAL PEPTIDYL DONORS: AcPHE-tRNA
                          fMET-tRNA
                          POLYLYSYL-tRNA
                          C-A-C-C-A(AcPHE)
                          C-A-A-C-C-A(fMET)

Fig. 4 Outline of the steps involved in the puromycin reaction with washed ribosomes. A donor (peptidyl-tRNA) must bind to the P site and an acceptor to the A site before catalysis of peptide bond formation can proceed. Some typical peptidyl donors are shown in the figure. Normally during protein biosynthesis peptidyl-tRNA is bound to ribosomes throughout the elongation cycle.

aminoacyl-tRNA to the acceptor site (A site; Watson, 1964). The peptidyl transferase is depicted as the region on the large ribosomal subunit which binds the 3′-termini of the C-C-A(amino acid) (A′ site) and C-C-A(peptide) (P′ site) of the bound esterified tRNA molecules (Rychlík, 1968; Monro *et al.*, 1968; Hishizawa *et al.*, 1970; Harris and Symons, 1973).

Studies of reactivity of peptidyl- and aminoacyl-tRNA with puromycin initially led to the formulation of the two-site model of ribosome function for peptidyl- and aminoacyl-tRNA attachment to ribosomes and have provided an operational definition for the donor (P) and acceptor (A) sites: aminoacyl-tRNA or α-N-blocked derivatives which react with puromycin in the absence of supernatant factors and GTP are considered to be in the P site while those unable to react with puromycin (unless EF-G and GTP are present, Fig.2) are considered to be in the A site (Traut and Monro, 1964). This operational definition although useful must be employed cautiously and has significant limitations. For example, it was reported that on initiation fMet-tRNA$_f$ enters the A site because it was found to be unreactive with puromycin (Ohta *et al.*, 1967). Subsequent studies, however, have demonstrated that the unreactivity of bound fMet-tRNA$_f$ with puromycin was due to the presence of initiation factor IF-2 on the fMet-tRNA$_f$·ribosome complex (Benne and Voorma, 1972; Dubnoff et al., 1972). Upon removal of IF-2, fMet-tRNA$_f$ could react with puromycin.

III. Peptidyl Transferase Assays

Normally during protein biosynthesis peptidyl transferase activity is expressed in the presence of the large and small ribosome subunits, mRNA, aminoacyl-tRNA, various small cations, the elongation factors, and GTP. Many of these components are not required in numerous assays devised to study transpeptidation. As described below, the puromycin reaction has been used extensively in these assays.

A. Peptide Bond Formation with Whole Cells

Various cell-free assays of peptide bond formation have been extensively used to examine the effects of various antibiotics and compounds on peptidyl transferase activity. However, it is important that the effects of antibiotics could be examined in intact bacteria in order to complement the cell-free studies. Cundliffe and McQuillen (1967) devised an assay that involves the puromycin-dependent release of labeled nascent polypeptides from polyribosomes of spheroplasts. Such an assay has the advantage of being free of certain artifacts of the cell-free systems. However, the assay suffers from problems due to the impermeability of the cell membrane to antibiotics, and, thus, necessitates the use of rather high concentrations of antibiotics. In addition the site of action of an antibiotic can only be determined in general terms (Cundliffe, 1972).

B. Peptide Bond Formation in Cell-Free Extracts

1. Synthetic Peptidyl-tRNA as Donor Substrate

Assays based on the use of aminoacyl-tRNA and α-N-blocked aminoacyl-tRNA molecules as donor substrates have been devised and include Phe-, fMet-, and AcPhe-tRNA, and various other synthetic peptidyl derivatives of tRNA (Fig. 4). These have either been used in conjunction with ribosomes, mRNA, and puromycin or bound to form a peptidyl-tRNA·mRNA·ribosome complex (Mao, 1973; Tanaka et al., 1971; Bretscher and Marcker, 1966; Weissbach et al., 1968; Pestka, 1970a). The reactions of the preformed complexes with puromycin have the advantage that the system is more resolved: the transfer reaction can be studied in the absence of reactions leading up to and including binding of the donor substrate. One such assay is the reaction of puromycin with the AcPhe-tRNA·poly(U)·ribosome complex isolated on cellulose nitrate filters. Formation of AcPhe-puromycin followed simple kinetics giving linear Lineweaver-Burk plots (Fico and Coutsogeorgopoulos, 1972).

Another assay that has been extensively studied is the polymerization of Phe-tRNA under the direction of poly(U) in the presence of ribosomes, supernatant factors, GTP, K^+, and Mg^{2+}. At low Mg^{2+}, elongation factor EF-Tu stimulates the binding of aminoacyl-tRNA to the ribosome, but at high Mg^{2+} (0.02 M) the requirement for EF-Tu is circumvented. In the presence of 0.02 M Mg^{2+} and 0.05 M K^+ binding of Phe-tRNA to ribosomes and peptide bond formation is rapid. Factor EF-G and GTP are needed to translocate the newly formed diphenylalanyl-tRNA from the A to the P site although nonenzymatic oligophenylalanine synthesis can occur (Pestka, 1968a,b, 1969c). Studies of the requirements and antibiotic sensitivity of the synthesis of diphenylalanine, triphenylalanine, and higher oligophenylalanines have been carried out (Pestka, 1969c, 1970b). Diphenylalanine synthesis occurred in the absence of supernatant factors and GTP. Synthesis of diphenylalanine required only Phe-tRNA, ribosomes, K^+, and Mg^{2+} and, thus, represented peptide bond synthesis in the absence of translocation (Fig. 5).

In addition a polyphenylalanyl-tRNA · poly(U) · ribosome complex was

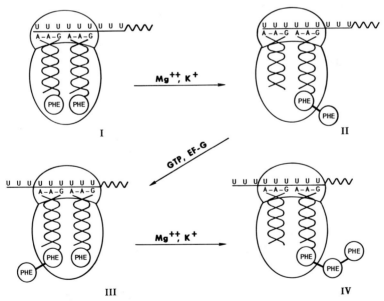

Fig. 5 Schematic illustration of di- and triphenylalanine synthesis. At 0.02 M Mg^{2+} the requirement for EF-Tu is obviated so that the formation of diphenylalanine-tRNA from Phe-tRNA and ribosomes can be used as an assay for peptide bond synthesis.

isolated and used in a model system with puromycin (Maden *et al.*, 1968). The heterogeneous kinetics of polyphenylalanyl puromycin formation were most likely due to reactivation of peptidyl transferase during the assay (Miskin *et al.*, 1970). The polyphenylalanyl-tRNA · ribosome complex was prepared in the absence of monovalent cations, a procedure that inactivates peptidyl transferase (Miskin *et al.*, 1970). This emphasizes the necessity of maintaining an appropriate ionic environment during ribosome preparation. The reaction with puromycin was rapid even at 0° and proceeded under simple ionic conditions. Successive washes of the ribosomes with attached polyphenylalanyl-tRNA under conditions known to remove soluble factors enhanced reactivity toward puromycin, further demonstrating that the puromycin reaction does not require GTP or soluble factors. Polyphenylalanyl-tRNA · 50 S ribosomal subunit complexes were also shown to react with puromycin, thus demonstrating that the peptidyl transferase was an integral part of the 50 S subunit. Similarly, the reactivity of oligolysyl-tRNA with puromycin has also been found to be independent of supernatant factors and GTP (Rychlík, 1966a,b; Gottesman, 1967).

2. Natural Peptidyl-tRNA as Donor Substrate

A large amount of data on the effects of antibiotics on the above model systems utilizing synthetic donors and high salt washed ribosomes has appeared in the literature (Pestka, 1971a, 1974a; Vazquez, 1974; also see Chapter 10). However, the peptidyl puromycin reaction has been reevaluated using native (not high salt washed) *Escherichia coli* polyribosomes carrying nascent polypeptides (Pestka, 1972a). The K_m for puromycin $(2.4 \times 10^{-6} M)$ in this system was shown to be two orders of magnitude lower than in the model systems discussed above. As summarized in Fig. 6 several antibiotics (lincomycin, various macrolides, and streptogramins A) found to be excellent inhibitors of model systems (Teraoka *et al.*, 1969; Černa *et al.*, 1969; Monro and Vazquez, 1967; Pestka, 1970a,b, 1971a; Vogel *et al.*, 1969; Pestka and Brot, 1971; Mao and Robishaw, 1971) had no effect on peptidyl puromycin synthesis on polyribosomes (Pestka and Hintikka, 1971; Pestka, 1972b). The observations that lincomycin, streptogramin A, and spiramycin were unable to prevent rapid polysome breakdown in intact cells and cell-free extracts (Cundliffe, 1968; Ennis, 1970) is consistent with their inability to inhibit peptidyl puromycin synthesis on native polyribosomes. It appears that these macrolides and streptogramins A cannot bind to polyribosomes containing peptidyl-tRNA (Oleinick and Corcoran, 1970; Pestka, 1974b). Thus, although results with native polyribosomes reflected effects on intact cells, effects on model systems for transpeptidation do not necessarily represent their mode of action in whole cells (Pestka, 1972b).

3. Fragment Assay

The fragment assay (Fig. 4) for peptide bond formation involves the reaction of α-N-blocked aminoacyl oligonucleotides with puromycin (Monro and Marcker, 1967; Maden and Monro, 1968; Monro et al., 1969; Vazquez et al., 1969). The aminoacyl oligonucleotides are derived from their respective aminoacyl-tRNA molecules by T_1 ribonuclease digestion (Monro and Marcker, 1967; Monro, 1971; Pestka, 1971b). In the presence of Mg^{2+}, K^+, ribosomes (70 or 80 S) or subunits (50 or 60 S), and ethanol, α-N-blocked aminoacyl oligonucleotides such as C-A-C-C-A(AcPhe) react with puromycin. Ethanol may be replaced by other organic solvents such as methanol, isopropanol, or acetone. Some differences between pro- and eukaryotic systems were demonstrated: K^+ was replaceable by NH_4^+, but not by Na^+ in the bacterial system, whereas by Na^+, but not by NH_4^+, in the mammalian system (Maden and Monro, 1968; Thompson and Moldave, 1974). These different ionic requirements were also reflected in the puromycin reaction on native polyribosomes from rat liver and E. coli (Pestka et al., 1972; Pestka, 1972a).

The fragment reaction produces rapid formation of α-N blocked aminoacyl puromycin on incubation at $0°$ or higher. The requirement for alcohol or acetone is believed to reside in the promotion of binding of the donor substrate to the P' site (Celma et al., 1970). Despite the anomalous requirement for alcohol, the formation of peptide bonds is believed to take place via the normal mechanism. The assay has some advantages since it utilizes small molecular weight substrates which presumably directly interact with the peptidyl transferase. In contrast, large peptidyl-tRNA derivatives have multiple ribosomal contacts. However, an unusual feature of this assay is its extreme acceptor substrate specificity. Of twenty-one 3'-N-aminoacyl puromycin aminonucleoside and nucleotidyl-(3'-5')-(3'-N-aminoacyl puromycin aminonucleoside) analogues of puromycin tested, only 3'-N-L-phenylalanyl puromycin aminonucleoside and its cytidylyl-(3'-5') derivative were active (Eckermann et al., 1974). In contrast, many of these analogues were active on model systems utilizing high salt washed ribosomes with AcPhe-tRNA (Eckermann et al., 1974) and on native polyribosomes, both unwashed (Vanin et al., 1974) and high salt washed (Harris et al., 1971). The extreme specificity was suggested to be due to the presence of ethanol (Eckermann et al., 1974).

4. Transesterification

Incubation of 80 S or 70 S ribosomes with fMet-tRNA, A-U-G, and ethanol resulted in variable formation of fMet-ethyl ester. However, the inclusion of tRNA or C-C-A greatly stimulated product formation (Scolnick *et al.*, 1970; Caskey *et al.*, 1971). In addition, α-hydroxypuromycin served as an acceptor under conditions of the fragment assay to form α-*O*-fMet-α-hydroxypuromycin ester (Fahnestock *et al.*, 1970). Similarly, in protein-synthesizing systems phenyllactyl-tRNA and lactyl-tRNA can participate in ester formation (Fahnestock and Rich, 1971a,b). Also, hydroxy-puromycin apparently served as acceptor with native polyribosomes (Pestka *et al.*, 1973).

5. Termination (Hydrolysis)

In addition to the catalysis of transpeptidation, peptidyl transferase is also involved in the hydrolysis of peptidyl-tRNA. A convenient assay involves the reaction of fMet-tRNA · A-U-G · ribosome complexes with water in the presence of a termination codon and release factors. The reaction results in the formation of free fMet (Caskey *et al.*, 1968). Another assay that utilizes fMet-tRNA, A-U-G, ribosomes, and C-C-A or tRNA in the presence of acetone does not require release factors (Caskey *et al.*, 1971). Perhaps the configuration of the enzyme is affected by release factors, acetone, or tRNA (or C-C-A) thus allowing water, normally excluded from the active site, to act as an acceptor, with concomitant hydrolysis of the peptidyl-tRNA.

C. Partial Reactions of Peptidyl Transferase

Transpeptidation proceeds via three steps: (1,2) binding of the 3'-ends of aminoacyl- and peptidyl-tRNA to the A' and P' sites, respectively; and (3) the transfer of the nascent peptide (Fig. 4). Assays for transpeptidation (3) have been described above. Assays for (1) and (2) are outlined below.

1. Binding of Analogues of Aminoacyl-tRNA

Aminoacyl oligonucleotides such as C-A-C-C-A(Phe) bind to 70 S and 80 S ribosomes in the presence of Mg^{2+} and K^+ (Pestka, 1969b; Pestka *et al.*, 1970; Hishizawa *et al.*, 1970; Celma *et al.*, 1971; Battaner and Vazquez, 1971). Ethanol is stimulatory but not essential for binding of C-A-C-C-A(Phe) to 70 S ribosomes, but essential for binding to 50 S subunits (Celma *et al.*, 1971; Hishizawa and Pestka, 1971). Binding to 50 S subunits was stimulated by tRNA (Hishizawa and Pestka, 1971). Several antibiotics inhibit binding at concentrations which are compatible with their known association constants whereas other antibiotics have no effect (Pestka, 1969b; Hishizawa and Pestka, 1971; Harris and Pestka, 1973). Indeed the assay has been used to determine the association constants of antibiotics which inhibit binding (Harris and Pestka, 1973).

Although puromycin is an analogue of aminoacyl-tRNA, only weak binding of puromycin to ribosomes has been reported. An important technical feature of these assays is the necessity of using puromycin-treated ribosomes. False results due to the formation of peptidyl puromycin products will result unless all peptidyl-tRNA is removed (Lessard and Pestka, 1972a). From inhibition of chloramphenicol binding to ribosomes, Lessard and Pestka (1972b) estimated the dissociation constant for puromycin binding to ribosomes as $4 \times 10^{-4} M$. Cooperman *et al.* (1975) reported photoactivated binding of puromycin to ribosomes with saturation kinetics yielding a dissociation constant of about $6 \times 10^{-4} M$ for puromycin binding to ribosomes.

2. Binding of Analogues of Peptidyl-tRNA

Peptidyl-tRNA analogues such as C-A-C-C-A(AcPhe) and C-A-C-C-A-(AcLeu) bind to 70 S and 80 S ribosomes in the presence of Mg^{2+}, K^+, and high levels of ethanol (Celma *et al.*, 1970; Hishizawa *et al.*, 1970; Battaner and Vazquez, 1971; D. J. Eckermann, personal communication). The effects of various antibiotics suggest that the binding may be specific (Celma *et al.*, 1970). These substrate binding assays can serve as probes for the determination of the site of action of various peptidyl transferase inhibitors and for examining the substrate binding site of the enzyme. They have demonstrated different characteristics and requirements for binding of C-A-C-C-A(Phe) and C-A-C-C-A(AcPhe) to ribosomes. These compounds appear to bind to distinct sites. The site to which C-A-C-C-A(Phe) ordinarily binds is defined as the A′ site; and that to which C-A-C-C-A-(AcPhe) binds is defined as the P′ site. Under certain conditions, both compounds may bind to similar or identical sites as occurs in the presence of high sparsomycin concentrations (Harris and Pestka, 1973). Evidence for the existence of distinct A′ and P′ sites is summarized in Table I.

IV. Contributions to the Understanding of Transpeptidation by Antibiotics

Various agents have been found to inhibit transpeptidation. The extensive data on the action of antibiotics and other inhibitors on transpeptidation (see Chapter 10) will not be discussed in detail here, but some of the relevant conclusions will be summarized. Most of the inhibitors have proved to be highly specific probes, elucidating structure and function of the ribosome.

Many antibiotics directly or indirectly inhibit transpeptidation (Pestka, 1971a; Vazquez, 1974). Some of these antibiotics inhibit peptide bond synthesis by prokaryotic ribosomes only (Fig. 6). These include chloramphenicol, lincomycin, macrolides (carbomycin, erythromycin, niddamycin, oleandomycin, spiramycin III, and tylosin), streptogramins A

TABLE I

Evidence for the Existence of Distinct Acceptor (A′) and Donor (P′) Sites of Peptidyl Transferase[a]

Compound or procedure	Effect on substrate binding to A′ site	Effect on substrate binding to P′ site	References[b]
Puromycin	Inhibits	P′ site substrates react with puromycin	1, 2, 3
Ethanol	Not essential, but stimulates	Essential	4, 5, 6
K⁺	$>1.2\ M$ (optimum)	$\sim 0.02\ M$ (optimum)	4, 7, 8
Chloramphenicol	Inhibits	No effect	1, 4, 5, 9
Gougerotin	Inhibits	Stimulates	2, 5, 10
Sparsomycin	Inhibits	Stimulates	2, 4, 5
Ribonuclease T₁ treatment of ribosomes	No effect	Inhibits	11

[a] The A′ and P′ sites represent the areas where the C-C-A(amino acid) and C-C-A(peptide) termini, respectively, of AA-tRNA and peptidyl-tRNA interact with the peptidyl transferase. These represent the functional and structural acceptor and donor positions of the peptidyl transferase. Substrates that bind to the A′ site include C-A-C-C-A(Phe), C-C-A(Phe), C-C-A(Leu), and similar aminoacyl oligonucleotides of the other 18 amino acids (Pestka, 1969a,b; Lessard and Pestka, 1972a). In general, aminoacyl oligonucleotides of the structure C-C-A(amino acid) bind to the A′ site. Substrates that bind to the P′ site are aminoacyl oligonucleotides with α-N-blocked amino acids such as C-A-C-C-A(AcPhe), C-A-A-C-C-A(fMet) and $\overset{U}{\underset{C}{}}$-A-C-C-A(AcLeu) (Celma et al., 1970). They serve as donors in transpeptidation (Monro and Marcker, 1967). Although sparsomycin strongly inhibits C-A-C-C-A(Phe) binding to the A′ site, at high concentrations sparsomycin stimulates C-A-C-C-A(Phe) binding to ribosomes in the presence of ethanol, presumably to the P′ site (Yukioka and Morisawa, 1971; Harris and Pestka, 1973). At high concentrations, aminoacyl oligonucleotides can serve as donor substrates in the presence of ethanol (Yukioka and Morisawa, 1971; Lessard and Pestka, 1972a).

[b] References: (1) Lessard and Pestka, 1972a; (2) Harris and Pestka, 1973; (3) Monro and Marcker, 1967; (4) Pestka et al., 1970; Pestka, 1969a,b; (5) Monro and Vazquez, 1967; (6) Yukioka and Morisawa, 1971; (7) D. J. Eckermann, personal communication; (8) Černa, 1971; (9) Lessard and Pestka, 1972b; (10) Celma et al., 1970; (11) Černa et al., 1973.

(streptogramin A, PA114A, and vernamycin A), and althiomycin. Others such as anisomycin, tenuazonic acid, trichodermin, and cycloheximide inhibit peptide bond synthesis by eukaryotic ribosomes only. Sparsomycin and the aminonucleosides (amicetin, gougerotin, and blasticidin S) inhibit transpeptidation by both pro- and eukaryotic ribosomes. The fact that these latter agents specifically inhibit transpeptidation in both pro- and eukaryotes suggests the presence of some structural similarities in their peptidyl transferases. On the other hand, the fact that many specific

ANTIBIOTIC	SYNTHETIC MODELS FOR PEPTIDE BOND SYNTHESIS	NATIVE PEPTIDE BOND SYNTHESIS	SPECIES INHIBITED
CHLORAMPHENICOL	●	●	PROKARYOTIC
LINCOMYCIN	●	○	
CARBOMYCIN	●	○	
ERYTHROMYCIN	●	○	
NIDDAMYCIN	●	○	
OLEANDOMYCIN	●	○	
SPIRAMYCIN III	●	○	
TYLOSIN	●	○	
STREPTOGRAMIN A	●	○	
PA114A	●	○	
VERNAMYCIN A	●	○	
ALTHIOMYCIN	◐	●	
SPARSOMYCIN	●	●	BOTH
AMICETIN	●	●	
GOUGEROTIN	●	●	
BLASTICIDIN S	●	●	
ANISOMYCIN	●	●	EUKARYOTIC
TENUAZONIC ACID	●	●	
TRICHODERMIN	●	●	
CYCLOHEXIMIDE		●	

Fig. 6 Summary of effects of antibiotics on transpeptidation. "Synthetic models for peptide bond synthesis" refers to assays involving washed ribosomes and synthetic donors. "Native peptide bond synthesis" refers to peptide bond synthesis in intact cells or to the puromycin reaction with native peptidyl-tRNA with isolated polyribosomes. The results with althiomycin in the synthetic system are highly dependent on the K^+ concentration. ●, inhibition; ○, no effect; ◐, partial inhibition.

inhibitors of transpeptidation inhibit either pro- or eukaryotes, but not both, indicates the existence of structural differences in the peptidyl transferases of pro- and eukaryotes. In addition, although most of the pro-karyotic inhibitors block protein synthesis in bacteria, blue-green algae, mitochondria, and chloroplasts, some show specificity within this group. For example, erythromycin and lincomycin inhibit protein synthesis by bacteria, but not by rat liver mitochondria (Towers *et al.*, 1972). Likewise, tenuazonic acid inhibits transpeptidation by mammalian, but not by yeast ribosomes (Carrasco and Vazquez, 1973). Thus, these inhibitors provide fine probes into the structural similarities and differences of ribosomes from diverse organisms.

Vogel *et al.* (1969) and Caskey *et al.* (1971) have concluded that peptidyl transferase catalyzes termination as well as transpeptidation from the similar sensitivity of the two reactions to antibiotics. In addition, peptidyl transferase activity of 50 S subunits can be abolished and restored by depletion and addition, respectively, of NH_4^+ or K^+ ions (Miskin *et al.*, 1968; Zamir *et al.*, 1969). Inactivation of the 50 S subunit eliminated both activities; reactivation restored both, and partial reactivation restored both activities partially and to the same degree (Vogel *et al.*, 1969).

Chloramphenicol inhibits the binding of C-A-C-C-A(Phe) to ribosomes (Pestka, 1969a,b). This has, therefore, led to the concept that chloramphenicol inhibits the binding of the C-C-A(amino acid) terminus of aminoacyl-tRNA to ribosomes. Thus, it was anticipated that derivatives of chloramphenicol with affinity groups would link the antibiotic to structures associated with the binding of the aminoacyl terminus of aminoacyl-tRNA. Appropriate studies with chloramphenicol affinity labels have identified proteins L2, L6, L16, and L27 in the chloramphenicol binding site (Sonenberg *et al.*, 1973; Pongs *et al.*, 1973a; see Chapters 3 and 4 for further details). It is, thus, expected that these same proteins are involved in the aminoacyl substrate binding site of the peptidyl transferase (Fig. 4). Protein L6 is altered in lincomycin and chloramphenicol resistant mutants (Wittmann and Apirion, 1974).

Studies with antibiotics have suggested that the peptidyl transferase exists in at least two different states or allosteric configurations as the ribosome passes through the various stages of the elongation epicycle. For example, it appears that peptidyl-tRNA exists in two puromycin-reactive states on native polyribosomes. One state is sensitive to inhibition by chloramphenicol, the other relatively insensitive (Cannon, 1968; Weber and DeMoss, 1969; Pestka, 1972b). Similarly, two states of peptidyl-tRNA with different sensitivities in inhibition by blasticidin S, gougerotin, and amicetin were demonstrated (Pestka, 1972b). In addition, some antibiotics (erythromycin, vernamycin A, vernamycin Bα, lincomycin, and chloramphenicol) and fMet-tRNA can stimulate the transition of peptidyl transferase from an inactive to an active form (Miskin *et al.*, 1970; Miskin and Zamir, 1974). Furthermore, two classes of binding of gougerotin and anisomycin to eukaryotic ribosomes (Barbacid and Vazquez, 1974a,b) and of chloramphenicol to bacterial ribosomes (Lessard and Pestka, 1972b) were reported. These results demonstrated directly possible different ribosomal states with different affinities for the antibiotics.

Some antibiotics which inhibit C-A-C-C-A(Phe) binding to ribosomes (amicetin, gougerotin, and sparsomycin) appear to stimulate binding of donor substrates as C-A-C-C-A(AcPhe) to ribosomes (Pestka, 1971a; Yukioka and Morisawa, 1971; Celma *et al.*, 1970; Harris and Symons, 1973; Vazquez, 1974). Similarly an affinity labeling derivative of puromycin may stimulate binding of donor substrates (Greenwell *et al.*, 1974). It is therefore possible that substrate binding at the C-C-A(amino acid) site (A' site) may stimulate donor substrate binding to the C-C-A(AcPhe) or C-C-A(peptide) site, the P' site. Studies of diphenylalanine synthesis led to similar conclusions. Since diphenylalanine synthesis occurred well at high ribosome to Phe-tRNA ratios, it was likely that binding of Phe-tRNA to ribosomes was not random and that binding of the second Phe-tRNA to a ribosome was facilitated by binding of the first (Pestka, 1968a). Kinetic studies of the fragment reaction, however, suggested that binding of puromycin (acceptor substrate) and C-A-C-C-A(AcLeu) (donor substrate) appear to be independent (Fernandez-Muñoz and Vazquez, 1973).

Although there are significant differences between the substrate specificity of the A' and P' sites, there are substantial similarities. For example, possibly the A' and P' sites of peptidyl transferase may be composed of similar proteins and/or segments of RNA while a third component catalyzes transpeptidation. Interestingly several 50 S subunit proteins and segments of 23 S RNA are known to be duplicated (Weber, 1972; Fellner and Sanger, 1968). Indeed two such repeat proteins (L18, L24) have been implicated as components or neighbors of peptidyl transferase by affinity labeling (Table II). In addition, affinity labeling studies using a puromycin analogue have suggested that the A' site (and possibly the P' site) may be composed of a segment of 23 S RNA. Although no specific sequence was established, a U-G-G sequence, base pairing to the C-C-A termini, was suggested (Harris *et al.*, 1973). Similarly, Budker *et al.* (1972) and Bispink and Matthaei (1973) have labeled 23 S RNA at the peptidyl transferase center (P' site) with affinity labeling donor substances.

V. Mechanism of Catalysis

The peptidyl transferase may operate via a determined order of addition of substrates. Results indicating that A' site affinity labeling appeared to promote P' site affinity labeling, as noted above, supports this idea. In addition several A' site specific antibiotics strongly stimulated substrate binding to the P' site and binding of the second Phe-tRNA to a ribosome was favored if one Phe-tRNA was already bound. However, substrates can bind independently to the A' and P' sites. Nevertheless, during the

TABLE II

Properties of 50 S Proteins Implicated in Peptidyl Transferase Activity[a]

50 S protein	Binding to 5 S or 23 S RNA	Alteration in antibiotic resistant mutants	At 30 S–50 S interface	Copies/50 S	Affinity labeling reagents
L2	3'-end of 23 S RNA			0.8–1.2	BrAcPhe-tRNA, BrAcMet-tRNA$_f^{Met}$, PNPC-Phe-tRNA, Bromamphenicol, BrAc(Gly)$_n$Phe-tRNA (n = 3), IodoAcPhe-tRNA
L4		Erythromycin resistant mutants with altered L4 have low peptidyl transferase activity		0.8–1.2	
L6	3'-end of 23 S RNA	Lincomycin, chloramphenicol	+	0.8–1.2	N-iodoacetyl puromycin
L11				0.8–1.2	NAG-Phe-tRNA, NAP-Phe-tRNA
L15			+	0.8–1.2	PNPC-Phe-tRNA
L16	5'-end of 23 S RNA			0.8–1.2	PNPC-Phe-tRNA, BrAcPhe-tRNA, iodoamphenicol
L18	3'-end of 23 S RNA and 5S RNA			1.8–2.2	NAG-Phe-tRNA, NAP-Phe-tRNA
L20	23 S RNA 5 S RNA		+	0.1–0.6	IodoAcPhe-tRNA
L24	5'-end of 23 S RNA			1.4–1.7	BrAc(Gly)$_n$Phe-tRNA (n = 12 to 16)
L26, L27			+	0.6–0.8	BrAcPhe-tRNA, BrAcMet-tRNA$_f^{Met}$, PNPC-Phe-tRNA, Bromamphenicol, BrAc(Gly)$_n$Phe-tRNA (n = 3 to 16)
L32, L33				0.1–0.6	BrAc(Gly)$_n$Phe-tRNA (n = 6 to 12)

[a] This table summarizes the properties of those 50 S proteins most implicated in peptidyl transferase activity or in its vicinity. The estimated number of copies of each of these proteins per 50 S subunit is taken from the data of Weber (1972). Specific studies describing the affinity labeling of these proteins by specific reagents are described in the text. Nonstandard abbreviations used are as follows: PNPC-Phe-tRNA, p-nitrophenylcarbamyl; NAP-Phe-tRNA, p-(2-nitro-4-azidophenoxy)phenylacetyl-Phe-tRNA; NAG-Phe-tRNA, N-(2-nitro-4-azidophenyl)glycyl-Phe-tRNA. Protein L4 is altered in erythromycin resistant mutants; and these mutants exhibit low peptidyl transferase activity (Wittmann et al., 1973). Protein L6 appears to be altered in some mutants resistant to lincomycin and chloramphenicol (Wittmann and Apirion, 1974).

normal elongation cycle there is a determined sequence to the events (Fig. 1) with a donor bound prior to entry of acceptor substrate.

An examination of the pH dependence of transpeptidation demonstrated that a group with a pK_a of 7.5–8.0 (imidazole residue or an N-terminal α-amino group) is essential in the unprotonated form for catalysis in both pro- and eukaryotes (Maden and Monro, 1968; Pestka et al., 1972; Pestka, 1972a). Possibly an additional group with a pK_a of 9.4 (ϵ-amino group or phenolic hydroxyl) is essential in the protonated state for transpeptidation in eukaryotes (Pestka et al., 1972). The group with pK_a of 7.5–8.0 appears to be involved in catalysis of peptidyl transfer and not substrate protonation or binding as supported by the following observations. The α-amino group of puromycin is not involved since the reaction of the α-hydroxypuromycin analogue with fMet-tRNA has a similar pH dependence (Fahnestock et al., 1970). In addition, termination (hydrolysis), where water acts as an acceptor, has a pH dependence similar if not identical to that of transesterification and transpeptidation (Caskey et al., 1971). Furthermore, there was no significant pH effect in the range 5.0 to 8.5 on the binding of U-A-C-C-A(AcLeu) (D. J. Eckermann, personal communication) nor in the range 6.0 to 8.4 on the binding of C-A-C-C-A(Phe) to ribosomes (S. Pestka, unpublished observations).

Evidence indicates that there are functionally and structurally distinct A' and P' sites (Table I). Consistent with the idea of a multicomponent enzyme is the implication of RNA and several proteins as components or neighbors of this enzyme by affinity labeling (Table II). In addition, 0.8c cores obtained by 0.8 M LiCl treatment of 50 S subunits have lost peptidyl transferase activity, yet retained 95% of their original P' site binding activity, but could not bind A' site inhibitors such as chloramphenicol (Nierhaus and Nierhaus, 1973). It was therefore concluded that the A' and P' sites are structurally distinct. Nierhaus and Montejo (1973) have reported that peptidyl transferase activity of the 0.8c cores could be restored by addition of protein L11. Thus, the peptidyl transferase may be composed of two or more proteins, one of which is L11; or L11 is required for active conformation of the peptidyl transferase which is a part of the 0.8c core particle; or L11 is the enzymatic center of the peptidyl transferase, which must be integrated into the 0.8c core particle for activity. In any case, since P' site binding occurred with 0.8c cores in the absence of protein L11, it is likely substrate binding sites and enzymatic active centers consist of multiple components. In contrast, Howard and Gordon (1974) and Ballesta and Vazquez (1974) have reported that ribosomal particles lacking L11 exhibit peptidyl transferase activity. Thus, additional studies will be required to resolve these discrepancies and to determine if, indeed, L11 is required for activity of the peptidyl transferase.

A. Stereochemical Specificity of Acceptor Substrates in Transpeptidation

Studies have demonstrated that small molecule active acceptor substrates contain the aminoacyl group on the 3'-OH, and 2'-OH stereoisomers have little or no activity as acceptor substrates. The 2'-analogue of puromycin (Nathans and Neidle, 1963), C-A(2'-O-Phe)-3'H and C-A(2'-O-Phe,3'-O-Me) (Ringer and Chládek, 1974), and the open-chain analogues of 2'-O-(L-phenylalanyl)adenosine (Chládek et al., 1973) were inactive as acceptor substrates. The following 3'-O-acyl esters were active acceptor substrates: puromycin, 2'-O-Me,3'-O-Phe-A (Pozdynakov et al., 1972), C-A(2'H,3'-O-Phe), C-A(2'-O-Me,3'-O-Phe) (Ringer and Chládek, 1974), and the open-chain analogue 3'-O-Phe-A (Chládek et al., 1973). These observations indicate that the peptidyl transferase utilizes 3'-O-aminoacyl-tRNA and appropriate 3'-O-acyl analogues as acceptor substrates. Nevertheless, 2'-O-aminoacyl analogues of tRNA inhibit both the puromycin reaction and binding of C-A-C-C-A(Phe) to ribosomes (Ringer et al., 1975). Thus, these 2'-O-aminoacyl-tRNA analogues can interact with the A' site and be recognized by peptidyl transferase, but are apparently in a stereochemically unfavorable position to serve as appropriate acceptors. However, both Phe-tRNA^{Phe}-C-C-3'dA and Phe-tRNA^{Phe}-C-C-A$_{ox-red}$ serve as acceptors with AcPhe-tRNA as donor, but the rate of dipeptide bond formation is much slower than that observed with Phe-tRNA^{Phe} (Chinali et al., 1974). The 2'-isomer of aminoacyl-tRNA can therefore accept a peptidyl residue despite the fact that the small molecular 2'-analogues of C-C-A(amino acid), the terminus of aminoacyl-tRNA, were inactive as acceptor substrates.

B. Stereochemical Specificity of Donor Substrates in Transpeptidation

The required stereochemistry of donor substrates indicates that, although 2'-aminoacyl analogues of tRNA appear to be bound to the ribosomal P' site, they do not serve as donor substrates. Thus, 2'-O-AcPhe-tRNA$_{ox-red}$ (Hussain and Ofengand, 1973; Chinali et al., 1974) and 2'-O-Phe-tRNA-C-C-dA (Sprinzl and Cramer, 1973; Chinali et al., 1974) were inactive as peptidyl donors. These results suggest that 3'-O-peptidyl-tRNA is the active donor substrate; however, in the absence of nonisomerizable active 3'-isomers this conclusion must be considered tentative.

C. Concerted or Acyl Enzyme Intermediate Mechanism?

Little is known about the actual mechanism of peptidyl transfer. Some proteases can catalyze both peptide and ester bond formation as well as hydrolysis under certain conditions (Waley and Watson, 1954; Bender and Kézdy, 1964). Similarly peptidyl transferase can catalyze the same three processes (see Section III). The three processes are displacement reactions of the type:

$$\text{Protease:} \quad R-\overset{\overset{\displaystyle O}{\|}}{C}-O-R' + R''-NH_2 \text{ (or } R''-OH) \rightleftharpoons R-\overset{\overset{\displaystyle O}{\|}}{C}-NH-R'' \text{ (or } R-\overset{\overset{\displaystyle O}{\|}}{C}-O-R'') + R'-OH$$

$$\begin{array}{l}\text{Peptidyl} \\ \text{Transferase:}\end{array} \quad R-\overset{\overset{\displaystyle O}{\|}}{C}-O-tRNA + R''-NH_2 \text{ (or } R''-OH) \longrightarrow R-\overset{\overset{\displaystyle O}{\|}}{C}-NH-R'' \text{ (or } R-\overset{\overset{\displaystyle O}{\|}}{C}-O-R'') + tRNA-OH$$

where R represents a peptide; $R-\overset{\overset{\displaystyle O}{\|}}{C}-O-R'$, a peptidyl or α-N-blocked aminoacyl ester, $R-\overset{\overset{\displaystyle O}{\|}}{C}-O-tRNA$, peptidyl-tRNA; and $R'-NH_2$ or $R''-OH$ are nucleophiles, namely, aminoacids, alcohols, or water for protease, and aminoacyl-tRNA, α-hydroxypuromycin, or water for peptidyl transferase. Hence, there is a logical basis for making mechanistic proposals for transpeptidation based on known protease mechanisms (Brattsten *et al.*, 1968). Carboxypeptidase is believed to achieve catalysis via a concerted mechanism while chymotrypsin (a suitable enzyme for comparison due to its very similar pH dependence) catalyzes via an acyl enzyme intermediate (Bender and Kézdy, 1964).

Various attempts have been made to determine the mechanism of ribosomal transpeptidation. Several irreversible inhibitors of trypsin and chymotrypsin including L-1-tosylamido-2-aminobutyl chloromethyl ketone and L-1-tosylamido-2-phenylethyl chloromethyl ketone have been applied to peptidyl transferase. Although the compounds were potent inhibitors, the effects were reversible (Rychlík, 1968). The synthesis of transition state analogues was attempted. The transition state is that chemical reaction intermediate most tightly bound to the active sites of an enzyme. Consequently, a molecule that closely mimics the transition state should be tightly bound with an affinity orders of magnitude greater than that of the substrate. Analogues are designed on the basis of a proposed mechanism; their tight binding is then used as evidence for the existence of the mechanism. Several potential analogues, designed to test a concerted mechanism, have been unsuccessfully applied to peptidyl transferase (P. Greenwell, personal communication).

An approach to detect the presence of an acyl enzyme intermediate by examining an exchange reaction between various P′ site compounds and the peptidyl transferase (E in equations below) was attempted:

U-A-C-C-A(Ac[^{3}H]Leu) + E $\rightleftharpoons$ U-A-C-C-A + Ac[^{3}H]Leu-E

Ac[^{3}H]Leu-E + C-C-A or tRNA $\rightleftharpoons$ C-C-A(Ac[^{3}H]Leu) or Ac[^{3}H]Leu-tRNA + E

No detectable acyl-enzyme intermediate or exchange product was detected under conditions in which pH, tRNA, Mg^{2+}, K^+, ethanol, and phenylpropionyl puromycin aminonucleoside were varied over large concentration ranges (D. J. Eckermann, personal communication). Thus, the mechanism of catalysis by peptidyl transferase remains an open question.

Raacke (1971) proposed a model for transpeptidation in which peptidyl-5 S RNA is formed as an intermediate during peptide bond formation. Fahnestock and Nomura (1972) have ruled out the possibility of such a peptidyl-5 S RNA intermediate involving the 3'-terminus of 5 S RNA. Ribosomes, containing 5 S RNA in which the 3'-terminus was modified chemically by oxidation or by conversion of the terminal ribose to a morpholine derivative, are as fully active in peptide bond formation and polypeptide synthesis as those ribosomes with unmodified 5 S RNA. Thus, these results exclude models for protein synthesis in which peptidyl-5 S RNA is formed as an intermediate.

VI. Structure of the Peptidyl Transferase

Various affinity labeling analogues of substrates and antibiotics have been used to identify the components of the peptidyl transferase. Ribosomal components labeled include proteins L2, L6, L11, L15, L16, L18, L20, L24, L26–27, and L32–33 (Table II) and 23 S RNA. The large number of proteins implicated suggests that many may be neighbors rather than strict components of peptidyl transferase. As suggested (Pestka, 1969a; Harris and Symons, 1973) the enzyme is probably composed of three components: the A' site, P' site, and catalytic center. However, neighboring proteins may influence peptidyl transferase activity, and ribosomal proteins in general interact with each other (Pestka, 1971a).

A schematic illustration of the relative positions of proteins in relation to the peptidyl transferase center is presented in Fig. 7. The A' site is

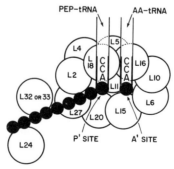

Fig. 7 Schematic illustration of the 50 S subunit proteins of the peptidyl transferase. Proteins likely to comprise or to be in the vicinity of the catalytic center, A' site, P' site, and nascent peptide groove are shown.

shown in the vicinity of proteins L6, L15, L16, and L18 and the P' site in the vicinity of L2, L11, L18, L20, and L27. The nascent peptide groove, the channel through which the elongating peptide travels, consists of proteins L2, L20, L24, L27, and L32–33 (Eilat *et al.*, 1974b).

A. The Catalytic Center of Peptidyl Transferase

L11 was proposed as the catalytic entity of peptidyl transferase (Nierhaus and Montejo, 1973), although this has been placed in doubt by later reports (Ballesta and Vazquez, 1974; Howard and Gordon, 1974) as discussed in Section V. Cores (0.8c) obtained by incubation of 50 S subunits with 0.8 M LiCl were inactive in the fragment reaction (Nierhaus and Montejo, 1973) although P' site binding activity was retained equivalent to 95% of that of native 50 S subunits (Nierhaus and Nierhaus, 1973). Restoration of peptidyl transferase activity was correlated with the amount of L11 added to reconstitution mixtures with 0.8c cores (Nierhaus and Montejo, 1973). Alternatively, L11 may constitute the A' site or it may activate peptidyl transferase already present on the cores.

Identification of the components comprising the A' and P' sites of peptidyl transferase was accomplished by affinity labeling (Table II) as well as reconstitution studies (Nierhaus and Nierhaus, 1973). Particles consisting of 0.8c cores reconstituted with protein L11 largely or totally lacked proteins L6, L7, L10, L12, L16, L27, L28, L31, and L33 which are therefore not intimately involved in peptidyl transferase activity. Some of the remaining proteins are present in reduced amounts, but since peptidyl transferase activity was restored only 65%, these proteins cannot be eliminated from involvement as parts of the peptidyl transferase. Of the twenty-five remaining proteins, comprising the active particle, it would appear that those already implicated by affinity labeling studies would be most likely to be related to the peptidyl transferase center. An additional discussion of affinity labeling of the peptidyl transferase is presented in Chapter 4.

B. Components of the P' Site

The P' site of peptidyl transferase was labeled by α-N-blocked aminoacyl-tRNA derivatives as well as antibiotics. Proteins L2 and L27 were the major products of reaction with BrAcPhe-tRNA; proteins L5, L6, L10, L11, and L14–17 were labeled to a smaller extent (Oen *et al.*, 1973; Pellegrini *et al.*, 1974). The covalently bound affinity label which reacted with L2 and L27 also served as a donor substrate for transpeptidation. With a somewhat different affinity label, *p*-nitrophenylcarbamyl-Phe-tRNA, protein L27 was chiefly labeled with L2, L14, L15, and L16 labeled to a smaller degree (Czernilofsky *et al.*, 1974a,b; Hauptmann *et al.*, 1974). Iodoacetyl-Phe-tRNA labeled proteins L2 and L20 (Bispink and Matthaei, 1973).

A similar process of elimination leads to a small group of possible P′ site candidates. As stated above 0.8c cores retain 95% of their P′ site substrate binding activity and have lost, either largely or totally, proteins L6, L7, L10, L11, L12, L16, L27, L28, L31, and L33. In addition, L1, L2, L5, L8 and/or L9, and L25 are present in reduced amounts (Nierhaus and Montejo, 1973). Since almost full P′ site binding activity was retained, it appears that the above proteins are not involved; however, one or more of the nineteen remaining proteins must comprise the P′ site. Of these candidates, those implicated in peptidyl transferase activity by affinity labeling (Table II) would be likely candidates for the P′ site. These are L15, L18, L20, L24 and 23 S RNA. Possibly other proteins not yet implicated by affinity labeling may also be involved. Photoactivated analogues of peptidyl-tRNA (NAP- and NAG-Phe-tRNA) produced somewhat different reaction patterns (Table II): proteins L11 and L18 were the major products, but L1, L2, L5, L22, L27, and L33 were also labeled (Hsiung and Cantor, 1974; Hsiung et al., 1974). The peptidyl-tRNA covalently linked to proteins L11 and L18 can participate in transpeptidation. Another photoactivated peptidyl-tRNA analogue was found to react with 23 S RNA (Bispink and Matthaei, 1973).

C. Components of the A′ Site

The A′ site was labeled also by BrAcPhe-tRNA under conditions where it was forced into the A site by high Mg^{2+} (0.03 M) and excess deacylated tRNA. Under these conditions protein L16 was the major product, but proteins L2 and L27 were also labeled (Eilat et al., 1974a). Puromycin and chloramphenicol compete with C-A-C-C-A(Phe) binding to ribosomes (Pestka, 1969a,b; Lessard and Pestka, 1972b). Thus, these antibiotics were used to label the A′ site. Bromamphenicol labeled proteins L2 and L27 of 50 S subunits (Sonenberg et al., 1973) whereas iodoamphenicol reacted with L16 of 70 S ribosomes (Pongs et al., 1973a). The puromycin analogue, 5′-O-(N-bromoacetyl-p-aminophenylphosphoryl)-3′-N-L-phenylalanylpuromycin, was covalently linked to 23 S RNA specifically (Harris et al., 1973). The bound compound was a suitable acceptor substrate indicating that it was in a functional position. N-Iodoacetylpuromycin, however, reacted with both 30 S and 50 S proteins, but not with RNA (Pongs et al., 1973b).

Although proteins L6 and L16 have been labeled at the A′ site by affinity labeling analogues of puromycin (Pongs and Erdmann, 1974), chloramphenicol (Pongs et al., 1973a), and Phe-tRNA (Eilat et al., 1974a), they are not essential components of this site since they were not necessary for the restoration of peptidyl transferase activity to 0.8c cores lacking these proteins (Nierhaus and Montejo, 1973).

D. The Nascent Peptide Groove

BrAc(Gly)$_n$Phe-tRNA ($n = 1$ to 16) reacted with several different ribosomal proteins; the specific pattern of proteins which reacted was dependent on the length of the peptidyl-tRNA reagent (Eilat et al., 1974b). Such results suggest that proteins L2, L26–27, L24, and L32–33 appear to be involved in forming a groove for the nascent peptide (Eilat et al., 1974b). Both eukaryotic and prokaryotic ribosomes are known to protect extensive stretches of nascent peptides (30–35 amino acid residues for the former) from proteases (Malkin and Rich, 1967). This is consistent with a discrete protected site for the nascent peptides.

E. The Role of RNA

A structural or functional role for RNA in peptidyl transferase is supported by the affinity labeling of 23 S RNA at both the A′ and P′ sites (Bispink and Matthaei, 1973; Harris et al., 1973; Greenwell et al., 1974). The sensitivity of the P′ site to T₁ ribonuclease (Černa et al., 1973) supports this concept. Specifically the 3′- and 5′-ends of 23 S RNA may be involved since L2 binds to the 3′-end while L16 and L24 bind toward the 5′-end (Branlant et al., 1973; Crichton and Wittmann, 1973; Zimmermann, 1974; Stöffler et al., 1971), L6 binds to 23 S RNA (Gray et al., 1972), and L18 to both 5 S and 23 S RNA (Gray et al., 1972, 1973). In addition, together L2, L6, L18, and 5 S and 23 S RNA (or the 3′-half of 23 S RNA) can form a protein · RNA complex (Gray and Monier, 1972; Gray et al., 1972). L20 also binds to 23 S RNA (Garrett et al., 1972). Furthermore, proteins L18 and L25 (and to a lesser extent L5, L20, and L30) bind to 5 S RNA (Gray et al., 1972; Horne and Erdmann, 1972). Thus, these proteins are likely to be neighbors of L18 and thus in the vicinity of the peptidyl transferase (Fig. 7). Harris et al. (1973) and Greenwell et al. (1974) have suggested that the C-C-A termini of aminoacyl- and peptidyl-tRNA bind to complementary U-G-G sequences of 23 S RNA at the A′ and P′ sites of peptidyl transferase. Additional support for the functional presence of 23 S RNA in the peptide groove comes from studies of antibiotic binding. Erythromycin as noted above interacts with peptidyl-tRNA. N^6-Dimethyladenine is normally not found in 23 S RNA of Staphlococcus aureus; however, some S. aureus strains which are resistant to erythromycin are found to have N^6-dimethyladenine in 23 S RNA (Lai et al., 1973). The presence of this methylation prevents erythromycin from binding to these ribosomes (Weisblum et al., 1971).

F. General Considerations and Interrelationships

The relative positions of the ribosomal proteins with respect to each other and the ribosomal RNA's are being defined by accumulation of data by many different techniques from numerous laboratories. Although the picture evolving does not have the precision of X-ray crystallographic measurements, it is serving to define protein neighborhoods in a gross way. Thus, many additional proteins have been placed in the area or adjacent to some of the proteins of the peptidyl transferase. For example, monovalent antibody fragments to proteins L6, L10, and L18 prevent the binding of proteins L7 and L12 to 50 S particles lacking L7 and L12 (Stöffler *et al.*, 1974). Proteins L2, L6, L18, and L25 form a complex with 5 S and 23 S RNA (Gray *et al.*, 1972). UV irradiation of the fusidic acid-stabilized complex of ribosomes · EF-G · azidophenyl-GDP resulted in selective attachment of the photoaffinity label to proteins L5, L11, L18, and L30 (Maassen and Möller, 1974). Chloramphenicol binding to 0.8c core particles devoid of L11 was stimulated by L11 as was peptidyl transferase activity (Nierhaus and Nierhaus, 1973). Protein L6 was necessary together with protein L16 for chloramphenicol binding and stimulates peptidyl transferase activity (Diedrich *et al.*, 1974). Erythromycin binds to the 50 S subunit in the vicinity of peptidyl-tRNA (Oleinick and Corcoran, 1970; Mao and Robishaw, 1972; Pestka, 1974b). Erythromycin resistant mutants show alterations in protein L4 or L22. Since peptidyl transferase activity is normal in L22 mutants, but reduced in L4 mutants, it is possible L4 is located in the vicinity of the nascent peptide chain. Studies by Wienen, Stöffler, and Pestka (unpublished observations) indicate that monovalent antibody fragments to L4 inhibit erythromycin binding to ribosomes. Collectively these results are incorporated into the rough schematic model illustrated in Fig. 7. Additional data on cross-linking of ribosomal proteins, effects of antibody binding, and numerous other ribosomal studies have been reported, and can be extended to present a more inclusive model of the 50 S subunit (see also Garrett and Wittmann, 1973). Additional discussions relating to structural and functional interrelationships on the ribosome are presented in Chapters 2, 3, 4, and 10 (for reviews, see Pongs *et al.*, 1974; Nomura *et al.*, 1974).

Several interesting features of the model (Fig. 7) are noteworthy. Peptidyl transferase is very close to or at the 50 S-30 S subunit interface. Proteins L6, L15, L20, and L27 which are near or form part of peptidyl transferase are interface proteins (Table II). This feature was not unexpected since the length of an aminoacyl-tRNA from anticodon to the aminoacyl terminus is about 80 Å (Kim *et al.*, 1974a,b; Robertus, 1974a,b), and aminoacyl-tRNA must span from the codon-anticodon site on the 30 S subunit to peptidyl transferase on the 50 S subunit. The approximate diameter of 35 Å for an average globular ribosomal protein of molecular weight 20,000 is consistent with the presence of peptidyl transferase near the interface.

Protein L11, a major component or neighbor of peptidyl transferase, may be part of an energy requiring contractile assembly involved in translocation of peptidyl-tRNA and mRNA movement. Proteins L11, L1, L3, L5, L7, L8, L9, L12, L18, and L33 are methylated (Terhorst *et al.*, 1972; Alix and Hayes, 1974; Chang and Chang, 1975), and by analogy with the methylated muscle proteins myosin and actin, they might be involved in a contractile assembly. Protein L11 is the most heavily methylated of the 50 S proteins. Proteins L1, L3, and L5 have intermediate levels of methylation containing about 0.4–0.6 methyl groups per molecule. Proteins L7, L8, L9, L12, and L18 are methylated to the extent of about 0.1 methyl group per molecule. In addition, as mentioned above, a photoaffinity labeling derivative of GDP labels proteins L5, L11, L18, and L30 suggesting they are near a GDP binding site (Maassen and Möller, 1974). It is particularly noteworthy then that L18 forms part of a 5 S protein complex which has GTPase and ATPase activity (Horne and Erdmann, 1973). Proteins L7 and L12 were among the minor proteins labeled by a photoaffinity analogue of peptidyl-tRNA (Hsiung *et al.*, 1974) demonstrating that these proteins are close to peptidyl transferase. Sequence analysis of L7 and L12 suggests these proteins have actin-like properties (Terhorst *et al.*, 1973). Proteins L7 and L12 are necessary for EF-G, EF-Tu, IF-2, and release factor dependent functions (Hamel *et al.*, 1972; Sander *et al.*, 1972; Weissbach *et al.*, 1972; Kay *et al.*, 1973; Brot *et al.*, 1974; Tate *et al.*, 1975). The hydrolysis of GTP is involved in the function of the first three factors. Perhaps these factors in combination with L7 and L12 and possibly other proteins such as L11 and L5 transduce the energy derived from GTP hydrolysis to promote conformational changes required for translocation and aminoacyl-tRNA binding (Highland *et al.*, 1973). Protein L11 may play a central role by triggering translocation when the acceptor site is occupied by peptidyl-tRNA following transpeptidation.

VII. Conclusion

The normal catalytic reaction of the peptidyl transferase is the formation of peptide bonds during elongation and hydrolysis of peptidyl-tRNA at the termination step. Its activity can be measured in a number of assays. The peptidyl transferase consists of two substrate binding sites (A', acceptor site, and P', donor site) and a catalytic center. The major gross structural components of the ribosomal peptidyl transferase sites have been outlined. The stereochemical requirements for acceptor and donor substrates have largely been defined. Yet, like most enzymes, the precise mechanisms of peptide bond synthesis and the precise molecular structures of the multicomponent particle are still to be delineated. The use of a large number of different techniques by many laboratories to examine the properties of ribosomal peptide bond synthesis has generated our integrated understanding of this part of protein biosynthesis.

REFERENCES

Alix, J. H., and Hayes, D. (1974). *J. Mol. Biol.* **86**, 139–159.
Allen, D. W., and Zamecnik, P. C. (1962). *Biochim. Biophys. Acta* **55**, 865–874.
Ballesta, J. P. G., and Vazquez, D. (1974). *FEBS Lett.* **48**, 266–270.
Barbacid, M., and Vazquez, D. (1974a). *Eur. J. Biochem.* **44**, 445–453.
Barbacid, M., and Vazquez, D. (1974b). *J. Mol. Biol.* **84**, 603–623.
Battaner, E., and Vazquez, D. (1971). *Biochim. Biophys, Acta* **254**, 316–330.
Bender, M. L., and Kézdy, F. J. (1964). *J. Am. Chem. Soc.* **86**, 3704–3714.
Benne, R., and Voorma, H. O. (1972). *FEBS Lett.* **20**, 347–351.
Bispink, L., and Matthaei, H. (1973). *FEBS Lett.* **37**, 291–294.
Branlant, C., Krol, A., Sriwidada, J., Fellner, P., and Crichton, R. (1973). *FEBS Lett.* **35**, 265–272.
Brattsten, I., Synge, R. L. M., and Watt, W. B. (1968). *Biochem. J.* **97**, 678–688.
Bretscher, M. S., and Marcker, K. A. (1966). *Nature (London)* **211**, 380–384.
Brot, N., Tate, W. P., Caskey, C. T., and Weissbach, H. (1974). *Proc. Natl. Acad. Sci. U.S.A.* **71**, 89–92.
Budker, V. G., Girshovich, A. S., and Skobel'tsyna, L. M. (1972). *Dokl. Akad. Nauk SSSR* **207**, 215–217.
Cannon, M. (1968). *Eur. J. Biochem.* **7**, 137–145.
Carrasco, L., and Vazquez, D. (1973). *Biochim. Biophys. Acta* **319**, 209–215.
Caskey, C. T., Tompkins, R., Scolnick, E., Caryk, T., and Nirenberg, M. (1968). *Science* **162**, 135–138.
Caskey, C. T., Beaudet, A. L., Scolnick, E. M., and Rosman, M. (1971). *Proc. Natl. Acad. Sci. U.S.A.* **68**, 3163–3167.
Celma, M. L., Monro, R. E., and Vazquez, D. (1970). *FEBS Lett.* **6**, 273–277.
Celma, M. L., Monro, R. E., and Vazquez, D. (1971). *FEBS Lett.* **13**, 247–251.
Černa, J. (1971). *FEBS Lett.* **15**, 101–104.
Černa, J., Rychlík, I., and Pulkrábek, P. (1969). *Eur. J. Biochem.* **9**, 27–35.
Černa, J., Rychlík, I., and Jonak, J. (1973). *Eur. J. Biochem.* **34**, 551–556.
Chang, C. N., and Chang, F. N. (1975). *Biochemistry* **14**, 468–477.
Chinali, G., Sprinzl, M., Parmeggiani, A., and Cramer, F. (1974). *Biochemistry* **13**, 3001–3010.
Chládek, S., Ringer, D., and Žemlička, J. (1973). *Biochemistry* **12**, 5135–5138.
Cooperman, B. S., Jaynes, E. N., Brunswick, D. J., and Luddy, M. A. (1975). *Proc. Natl. Acad. Sci. U.S.A.* **72**, 2974–2978.
Crichton, R. R., and Wittmann, H. G. (1973). *Proc. Natl. Acad. Sci. U.S.A.* **70**, 665–668.
Cundliffe, E. (1968). *Biochemistry* **8**, 2063–2066.
Cundliffe, E. (1972). *In* "Molecular Mechanisms of Antibiotic Action on Protein Biosynthesis and Membranes" (E. Muñoz, F. Garcia-Ferrandiz, and D. Vazquez, eds.), pp. 242–261. Elsevier, Amsterdam.
Cundliffe, E., and McQuillen, K. (1967). *J. Mol. Biol.* **30**, 137–146.
Czernilofsky, A. P., Collatz, E. E., Stöffler, G., and Kuechler, E. (1974a). *Proc. Natl. Acad. Sci. U.S.A.* **71**, 230–234.
Czernilofsky, A. P., Stöffler, G., and Kuechler, E. (1974b). *Hoppe-Seyler's Z. Physiol. Chem.* **355**, 89–92.

Diedrich, S., Schrandt, J., and Nierhaus, K. H. (1974). As communicated by Pongs *et al.* (1974).

Dubnoff, J. S., Lockwood, A. H., and Maitra, U. (1972). *J. Biol. Chem.* **247**, 2884–2894.

Eckermann, D. J., Greenwell, P., and Symons, R. H. (1974). *Eur. J. Biochem.* **41**, 547–554.

Eilat, D., Pelligrini, M., Oen, H., de Groot, N., Lapidot, Y., and Cantor, C. R. (1947a). *Nature (London)* **250**, 514–516.

Eilat, D., Pelligrini, M., Oen, H., Lapidot, Y., and Cantor, C. R. (1974b). *J. Mol Biol.* **88**, 831–840.

Ennis, H. (1970). *Proc. Int. Congr. Chemother., 6th, 1969* Vol. 2, pp. 489–498.

Fahnestock, S., and Rich. A. (1971a). *Nature (London), New Biol.* **229**, 8–10.

Fahnestock, S., and Rich, A. (1971b). *Science* **173**, 340–343.

Fahnestock, S., Neumann, H., Shashoua, V., and Rich, A. (1970). *Biochemistry* **9**, 2477–2483.

Fahnestock, S. R., and Nomura, M. (1972). *Proc. Natl. Acad. Sci. U.S.A.* **69**, 363–365.

Fellner, P., and Sanger, F. (1968). *Nature (London)* **219**, 236–238.

Fernandez-Muñoz, R., and Vazquez, D. (1973). *Mol. Biol. Rep.* **1**, 75–79.

Fico, R., and Coutsogeorgopoulos, C. (1972). *Biochem. Biophys. Res. Commun.* **47**, 645–651.

Garrett, R. A., and Wittmann, H. G. (1973). *Adv. Protein Chem.* **27**, 277–347.

Garrett, R. A., Morrison, C., Tischendorf, G., Zeichardt, H., and Stöffler, G. (1972). Communicated in Pongs *et al.* (1974).

Gottesman, M. E. (1967). *J. Biol. Chem.* **242**, 5564–5571.

Gray, P. N. and Monier, R. (1972). *Biochimie* **54**, 41–45.

Gray, P. N., Garrett, R. A., Stöffler, G., and Monier, R. (1972). *Eur. J. Biochem.* **28**, 412–421.

Gray, P. N., Bellemare, G., Monier, R., Garrett, R. A. and Stöffler, G. (1973). *J. Mol Biol.* **77**, 133–152.

Greenwell, P., Harris, R. J., and Symons, R. H. (1974). *Eur. J. Biochem.* **49**, 539–554.

Hamel, E., Koka, M., and Nakamoto, T. (1972). *J. Biol. Chem.* **247**, 805–814.

Harris, R., and Pestka, S. (1973). *J. Biol. Chem.* **248**, 1168–1174.

Harris, R. J., and Symons, R. H. (1973). *Bioorg. Chem.* **2**, 266–285.

Harris, R. J., Hanlon, J. E., and Symons, R. H. (1971). *Biochim. Biophys. Acta* **240**, 244–262.

Harris, R. J., Greenwell, P., and Symons, R. H. (1973). *Biochem. Biophys. Res. Commun.* **55**, 117–124.

Hauptmann, R., Czernilofsky, A. P., Voorma, H. O., Stöffler, G., and Kuechler, E. (1974). *Biochem. Biophys. Res. Commun.* **56**, 331–337.

Highland, J. H., Bodley, J. W., Gordon, J., Hasenbank, R., and Stöffler, G. (1973). *Proc. Natl. Acad. Sci. U.S.A.* **70**, 147–150.

Hishizawa, T., and Pestka, S. (1971). *Arch. Biochem. Biophys.* **147**, 624–631.

Hishizawa, T., Lessard, J. L., and Pestka, S. (1970). *Proc. Natl. Acad. Sci. U.S.A.* **66**, 523–530.

Horne, J. R., and Erdmann, J. A. (1972). *Mol. Gen. Genet.* **119**, 337–344.

Horne, J. R., and Erdmann, V. A. (1973). *Proc. Natl. Acad. Sci. U.S.A.* **70**, 2870–2873.

Howard, G. A., and Gordon, J. (1974). *FEBS. Lett.* **48**, 271–274.

Hussain, Z., and Ofengand, J. (1973). *Biochem. Biophys. Res. Commun.* **50**, 1143–1151.

Hsiung, N., and Cantor, C. R. (1974). *Nucleic Acids Res.* **1**, 1753–1762.

Hsiung, N., Reines, S., and Cantor, C. R. (1974). *J. Mol. Biol.* **88**, 841–855.

Kay, A., Sander, K., and Grunberg-Manago, M. (1973). *Biochem. Biophys. Res. Commun.* **51**, 979–986.

Kim, S. H., Suddath, F. L., Quigley, G. J., McPherson, A., Sussman, J. L., Wang, A. H. J., Seeman, N. C., and Rich, A. (1974a). *Science* **185**, 435–440.

Kim, S. H., Sussman, J. L., Suddath, F. L., Quigley, G. J., McPherson, A., Wang, A. H. J., Seeman, N. C., and Rich, A. (1974b). *Proc. Natl. Acad. Sci. U.S.A.* **71**, 4970–4974.

Lai, C. J., Weisblum, B., Fahnestock, S. R., and Nomura, M. (1973). *J. Mol. Biol.* **74**, 67–72.

Lessard, J. L., and Pestka, S. (1972a). *J. Biol. Chem.* **247**, 6901–6908.

Lessard, J. L., and Pestka, S. (1972b). *J. Biol. Chem.* **247**, 6909–6912.

Maassen, J. A., and Möller, W. (1974). *Proc. Natl. Acad. Sci. U.S.A.* **71**, 1277–1280.

Maden, B. E. H., and Monro, R. E. (1968). *Eur. J. Biochem.* **6**, 309–316.

Maden, B. E. H., Traut, R. R., and Monro, R. E. (1968). *J. Mol. Biol.* **35**, 333–345.

Malkin, L. I., and Rich, A. (1967). *J. Mol. Biol.* **26**, 329–345.

Mao, J. C.-H. (1973). *Biochem. Biophys. Res. Cummun.* **52**, 595–600.

Mao, J. C.-H., and Robishaw, E. E. (1971). *Biochemistry* **10**, 2054–2061.

Mao, J. C.-H., and Robishaw, E. E. (1972). *Biochemistry* **11**, 4864–4872.

Miskin, R., and Zamir, A. (1974). *J. Mol. Biol.* **87**, 121–134.

Miskin, R., Zamir, A., and Elson, D. (1968). *Biochem. Biophys. Res. Commun.* **33**, 551–557.

Miskin, R., Zamir, A., and Elson, D. (1970). *J. Mol. Biol.* **54**, 355–378.

Monro, R. E. (1967). *J. Mol. Biol.* **26**, 147–151.

Monro, R. E. (1971). *In* "Methods in Enzymology" (K. Moldave and L. Grossman, eds.), Vol. 20, pp. 472–481. Academic Press, New York.

Monro, R. E., and Marcker, K. A. (1967). *J. Mol. Biol.* **25**, 347–350.

Monro, R. E., and Vazquez, D. (1967). *J. Mol. Biol.* **28**, 161–164.

Monro, R. E., Maden, B. E. H., and Traut, R. R. (1967). *In* "Genetic Elements: Properties and Function" (D. Shugar, ed.), pp. 179–203. Academic Press, New York.

Monro, R. E., Černa, J., and Marcker, K. A. (1968). *Proc. Natl. Acad. Sci. U.S.A.* **61**, 1042–1049.

Monro, R. E., Staehelin, T., Celma, M. L., and Vazquez, D. (1969). *Cold Spring Harbor Symp. Quant. Biol.* **34**, 357–366.

Morris, A. J., and Schweet, R. S. (1961). *Biochim. Biophys. Acta* **47**, 415–416.

Nathans, D. (1964). *Proc. Natl. Acad. Sci. U.S.A.* **51**, 585–592.

Nathans, D., and Neidle, A. (1963). *Nature (London)* **197**, 1076–1077.

Nierhaus, K. H., and Montejo, V. (1973). *Proc. Natl. Acad. Sci. U.S.A.* **70**, 1931–1935.

Nierhaus, D., and Nierhaus, K. H. (1973). *Proc. Natl. Acad. Sci. U.S.A.* **70**, 2224–2228.

Nomura, M., Tissières, A., and Lengyel, P., eds. (1974). "Ribosomes," pp. 930. Cold Spring Harbor Lab., Cold Spring Harbor, New York.

Oen, H., Pelligrini, M., Eilat, D., and Cantor, C. R. (1973). *Proc. Natl. Acad. Sci. U.S.A.* **70**, 2799–2803.

Ohta, T., Sarkar, S., and Thach, R. E. (1967). *Proc. Natl. Acad. Sci. U.S.A.* **58**, 1638–1644.

Oleinick, N. L., and Corcoran, J. W. (1970). *Proc. Int. Congr. Chemother., 6th, 1969*, Vol. 1, pp. 202–208.

Pelligrini, M., Oen, H., Eilat, D., and Cantor, C. R. (1974). *J. Mol. Biol.* **88**, 809–829,

Pestka, S. (1968a). *J. Biol. Chem.* **243**, 2810–2820.

Pestka, S. (1968b). *Proc. Natl. Acad. Sci. U.S.A.* **61**, 726–733.

Pestka, S. (1969a). *Proc. Natl. Acad. Sci. U.S.A.* **64**, 709–714.

Pestka, S. (1969b). *Biochem. Biophys. Res. Commun.* **36**, 589–595.

Pestka, S. (1969c). *Cold Spring Harbor Symp. Quant. Biol.* **34**, 395–410.

Pestka, S. (1970a). *Arch. Biochem. Biophys.* **136**, 80–88.

Pestka, S. (1970b). *Arch. Biochem. Biophys.* **136**, 89–96.

Pestka, S. (1971a). *Annu. Rev. Microbiol.* **25**, 487–562.

Pestka, S. (1971b). *In* "Methods in Enzymology" (K. Moldave and L. Grossman, eds.), Vol. 20, pp. 502–507. Academic Press, New York.

Pestka, S. (1972a). *Proc. Natl. Acad. Sci. U.S.A.* **69**, 624–628.

Pestka, S. (1972b). *J. Biol. Chem.* **247**, 4669–4678.

Pestka, S. (1974a). *In* "Methods in Enzymology" (L. Grossman and K. Moldave, eds.), Vol. 30, pp. 261–289. Academic Press, New York.

Pestka, S. (1974b). *Antimicrob. Agents & Chemother.* **5**, 255–267.

Pestka, S., and Brot, N. (1971). *J. Biol. Chem.* **246**, 7715–7722.

Pestka, S., and Hintikka, H. (1971). *J. Biol. Chem.* **246**, 7723–7730.

Pestka, S., Hishizawa, T., and Lessard, J. L. (1970). *J. Biol. Chem.* **245**, 6208–6219.

Pestka, S., Goorha, R., Rosenfeld, H., Neurath, C., and Hintikka, H. (1972). *J. Biol. Chem.* **247**, 4258–4263.

Pestka, S., Vince, R., Daluge, S., and Harris, R. (1973). *Antimicrob. Agents & Chemother.* **4**, 37–43.

Pongs, O., and Erdmann, V. A. (1974). Communicated in Pongs *et al.* (1974).

Pongs, O., Bald, R., and Erdmann, V. A. (1973a). *Proc. Natl. Acad. Sci. U.S.A.* **70**, 2229–2233.

Pongs, O., Bald, R., Wagner, T., and Erdmann, V. A. (1973b). *FEBS Lett.* **35**, 137–140.

Pongs, O., Nierhaus, K. H., Erdmann, V. A., and Wittmann, H. G. (1974). *FEBS Lett.* **40**, S28–S37.

Pozdnyakov, V. A., Mitin, Yu. V., Kukhanova, M. K., Nikolaeva, L. V., Krayevsky, A. A., and Gottikh, B. P. (1972). *FEBS Lett.* **24**, 177–180.

Pulkrábek, P., and Rychlík, I. (1968). *Biochim. Biophys. Acta* **155**, 219–227.

Raacke, I. D. (1971). *Proc. Natl. Acad. Sci. U.S.A.* **68**, 2357–2360.

Rabinovitz, M., and Fisher, J. M. (1962). *J. Biol. Chem.* **237**, 477–481.

Ringer, D., and Chládek, S. (1974). *Biochem. Biophys. Res. Commun.* **56**, 760–766.

Ringer, D., Quiggle, K., and Chládek, S. (1975). *Biochemistry* **14**, 514–520.

Robertus, J. D., Ladner, J. E., Finch, J. T., Rhodes, D., Brown, R. S., Clark, B. F. C., and Klug, A. (1974a). *Nature (London)* **250**, 546–551.

Robertus, J. D., Ladner, L. E., Finch, J. T., Rhodes, D., Brown, R. S., Clark, B. F. C., and Klug, A. (1974b). *Nucleic Acids Res.* **1**, 927–932.

Rychlík, I. (1966a). *Biochim. Biophys. Acta* **114**, 425–427.

Rychlík, I. (1966b). *Collect. Czech. Chem. Commun.* **31**, 2583–2595.

Rychlík, I. (1968). *In* "Molecular Genetics" (H. G. Wittmann and H. Schuster, eds.), pp. 69–80. Springer-Verlag, Berlin and New York.

Sander, G., Marsh, R. C., and Parmeggiani, A. (1972). *Biochem. Biophys. Res. Commun.* **47**, 866–873.

Scolnick, E., Milman, G., Rosman, M., and Caskey, T. (1970). *Nature (London)* **225**, 152–154.

Sonenberg, N., Wilchek, M., and Zamir, A. (1973). *Proc. Natl. Acad. Sci. U.S.A.* **70**, 1423–1426.

Sprinzl, M., and Cramer, F. (1973). *Nature (London), New Biol.* **245**, 3–5.

Stöffler, G., Daya, L., Rak, K. H., and Garrett, R. A. (1971). *J. Mol. Biol.* **62**, 411–414.

Stöffler, G., Hasenbank, R., Bodley, J. W., and Highland, J. H. (1974). *J. Mol. Biol.* **86**, 171–174.

Tanaka, K., Teraoka, H., and Tamaki, M. (1971). *FEBS Lett.* **13**, 65–67.

Tate, W. P., Caskey, C. T., and Stöffler, G. (1975). *J. Mol. Biol.* **93**, 375–389.

Teraoka, H., Tanaka, K., and Tamaki, M. (1969). *Biochim. Biophys. Acta* **174**, 776–778.

Terhorst, C., Möller, W., Laursen, R. A., and Wittmann-Liebold, B. (1972). *FEBS Lett.* **28**, 325–328.

Thompson, H. A., and Moldave, K. (1974). *Biochemistry* **13**, 1348–1353.
Towers, N. R., Dixon, H., Kellerman, G. M., and Linnane, A. W. (1972). *Arch. Biochem. Biophys.* **151**, 361–369.
Traut, R. R., and Monro, R. E. (1964). *J. Mol. Biol.* **10**, 63–72.
Vanin, E. F., Greenwell, P., and Symons, R. H. (1974). *FEBS Lett.* **40**, 124–126.
Vazquez, D. (1974). *FEBS Lett.* **40**, S63–S84.
Vazquez, D., Battaner, E., Neth, R., Heller, G., and Monro, R. E. (1969). *Cold Spring Harbor Symp. Quant. Biol.* **34**, 369–375.
Vogel, A., Zamir, A., and Elson, D. (1969). *Biochemistry* **8**, 5161–5168.
Waley, S. G., and Watson, J. (1954). *Biochem. J.* **57**, 529–538.
Watson, J. D. (1964). *Bull. Soc. Chim. Biol.* **46**, 1399–1425.
Weber, H. G. (1972). *Mol. Gen. Genet.* **119**, 233–248.
Weber, M. J., and DeMoss, J. A. (1969). *J. Bacteriol.* **97**, 1099–1105.
Weisblum, B., Siddhikol, C., Lai, C. J., and Demohn, V. (1971). *J. Bacteriol.* **106**, 835–847.
Weissbach, H., Redfield, B., and Brot, N. (1968). *Arch. Biochem. Biophys.* **127**, 705–710.
Weissbach, H., Redfield, B., Yamasaki, E., Davis, R. C., Jr., Pestka, S., and Brot, N. (1972). *Arch. Biochem. Biophys.* **149**, 110–117.
Wittmann, H. G., and Apirion, D. (1974). Communicated in Pongs *et al.* (1974).
Wittmann, H. G., Stöffler, G., Apirion, D., Rosen, L., Tanaka, K., Tamaki, M., Takata, R., Dekio, S., Otaka, E., and Osawa, S. (1973). *Mol. Gen. Genet.* **127**, 175–189.
Yarmolinsky, M., and de la Haba, G. L. (1959). *Proc. Natl. Acad. Sci. U.S.A.* **45**, 1721–1729.
Yukioka, M., and Morisawa, S. (1971). *Biochim. Biophys. Acta* **254**, 304–315.
Zamir, A., Miskin, R., and Elson, D. (1969). *FEBS Lett.* **3**, 85–88.
Zimmermann, R. A. (1974). Communicated in Pongs *et al.* (1974).

9

Peptide Chain Termination

C. THOMAS CASKEY

I. Introduction

Peptide chain termination results in the release of the completed peptide from its ultimate ribosomal bound tRNA. Peptide chain termination occurs on the ribosome and requires soluble protein factor(s) in both pro- and eukaryotic extracts. Two proteins (release factor, RF) identified in *Escherichia coli* cells differ in their codon specificity; RF-1, UAA or UAG; RF-2, UAA or UGA. A single rabbit reticulocyte RF is active with UAA, UAG, and UGA. Binding and release of bacterial RF-1 and RF-2 from ribosomes involves a third protein factor, RF-3, which interacts with GDP and GTP. Reticulocyte RF activity is stimulated by GTP and expresses a ribosomal dependent GTPase activity. The hydrolysis of peptidyl-tRNA at chain termination also requires the ribosomal enzyme peptidyl transferase.

II. Terminator Codons

The occurrence of peptide chain termination codons was predicted from genetic studies of nonsense mutations as reviewed by Garen (1968).

443

A nonsense mutation is now equated with the conversion of an amino acid codon to a peptide chain termination codon. Nonsense mutations are best characterized in bacteria, but also now recognized in eukaryotes (Gorini, 1970). Nonsense mutations have a number of effects and characteristics including: reversion by base change mutagens; extreme negative phenotype (Benzer and Champe, 1962; Garen and Siddiqui, 1962); phenotypic correction by extragenic suppression; reduced translation of prokaryotic distal cistrons in a contiguous mRNA (polarity); and with premature peptide chain termination (Sarabhai *et al.*, 1964) appearance of amino terminal fragments of the affected gene products. These characteristics of premature peptide chain termination mutations are discussed below with reference to three genetic systems: (1) alkaline phosphatase mutants of *E. coli*, (2) T4 phage mutants, particularly in the *rII* gene, and (3) iso-1-cytochrome *c* of yeast.

The biochemical character of termination codons was initially determined by study of nonsense mutations occurring in the *Escherichia coli* alkaline phosphatase gene (Weigert *et al.*, 1966). They initially developed a large number of alkaline phosphatase-deficient strains which had no detectable immunological cross-reactivity to anti-alkaline phosphatase (*CRM⁻*). Many of these mutants regained enzyme activity and *CRM⁺* if the mutant gene was crossed into another nonisogenic cell line (Garen and Siddiqi, 1962). This genetic correction was subsequently identified to result from extragenic suppressor for nonsense mutations harbored by the recipient cell.

Garen used the specificity of suppressors to extend this work by subdividing the nonsense mutants and suppressors into two groups, the *amber* (*N1*) class (Garen *et al.*, 1965) and the *ochre* (*N2*) class (Gallucci and Garen, 1966). The *amber* suppressor strains corrected one group of nonsense mutations while the *ochre* suppressor strains suppressed, in addition, another group of mutants. Thus, mutants suppressible exclusively by *ochre* suppressors were identified as *ochre* mutants, and mutants suppressed by both *ochre* and *amber* suppressors were identified as *amber* mutants. These findings suggested nonsense mutations probably corresponded to at least two different terminator codons.

By a combination of genetic and biochemical techniques, the codon corresponding to the *amber* mutation was determined. The amino acid sequence of the wild-type alkaline phosphatase protein, and of a number of different alkaline phosphatase (+) revertants from one *amber* nonsense mutant were compared. The wild-type protein contained a tryptophan at the site of the nonsense mutation, while revertants of the *amber* mutant contained lysine, glutamine, glutamic acid, serine, tyrosine, leucine, and tryptophan residues (Fig. 1) (Weigert and Garen, 1965). Assuming that

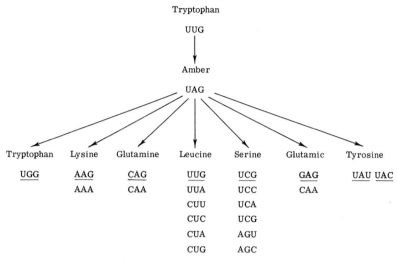

Fig. 1   Relation of *amber* codon to amino acid codons.

one of the synonym codons for these amino acids differed from the *amber* codon by only a single base change, the codon UAG was the predicted trinucleotide corresponding to the *amber* codon. The trinucleotide corresponding to the *ochre* codon (*N2*) was identified as UAA by a similar approach (Weigert *et al.*, 1967b).

The investigation of nonsense mutations in the bacteriophase T4 *rII* gene provided a second system for the genetic analysis of premature peptide chain termination and its suppression. The *rII* gene consists of two cistrons, *A* and *B*, both required for phage growth in *E. coli* strain KB, but not in *E. coli* strain B. Although the proteins coded by the *rII* cistrons are not yet identified, the *rII* mutant (*r1589*) studied by Benzer and Champe (1962) proved particularly useful for the examination of nonsense mutations and nonsense suppression. This *rII* mutant has a deletion extending from the terminal portion of the *A* cistron into the initial portion of the *B* cistron, resulting in the loss of *A* but not *B* cistron activity. By the coinfection of *E. coli* KB with *r1589* (*B* function) and a second T4 phage carrying a normal *A*, but a deleted *B* cistron, T4 phage production follows as a result of complementation. Benzer and Champe (1962) observed that the introduction of nonsense mutations into the remnant of *A* cistron of *r1589* eliminated *B* cistron activity. The loss of *B* activity could be restored by a separate mutation in the KB host. Garen's original alkaline phosphatase suppressor also suppressed the *rII* nonsense mutants. While the molecular basis of these events was not known, Benzer conceptualized it in the following manner. He suggested that the *r1589* deletion, by linking *A* and *B* cistrons, resulted in the transcription of a hybrid protein containing the amino acid sequences of a portion of *A* and a portion of *B* protein, but possessing only *B* activity. He suggested that the occurrence of a nonsense mutation in the A region prevented translation of the mRNA beyond the mutation site. More recent studies of nonsense or premature chain termination mutations are compatible with this original model.

Brenner *et al.* (1965) combined studies of mutants of T4 phage with established information on the genetic code and mutagen specificity to elegantly derive UAG and UAA as the *amber* and *ochre* codons, respectively. Their complex arguments were lucidly described in the original paper. Thus, Brenner and Garen's studies in different genetic systems agreed that UAA and UAG were peptide chain termination codons.

A few years later Brenner *et al.* (1967), again using *rII* mutants, identified an additional class of nonsense mutants which were not corrected by *amber* or *ochre* suppressors, and consequently did not correspond to UAA or UAG. Again it was possible to deduce the third nonsense codon base sequence, UGA. Sambrook *et al.* (1967) reported the isolation of a UGA specific suppressor. Thus, genetic studies indicated that UAA, UAG, and UGA are nonsense codons, and if they occur in mutant positions in mRNA cause premature peptide chain termination.

More recently, the nucleotide sequence for the naturally occurring chain termination signal has been determined by Nichols (1970) and Nichols and Robertson (1971) for the RNA bacteriophages R17 and f2, respectively (Fig. 2). In each case a T1 RNase fragment was sequenced using techniques developed by Sanger *et al.* (1965). The nucleotide sequence corresponding to the intercistronic region including the carboxyl terminus and terminator for coat protein, and the initiator region for the synthetase gene is known. As shown in Fig. 2 a tandem arrangement of terminator codons (UAAUAG) follows the codon for the C-terminal amino acid. A third terminator codon (UGA) occurs in translational frame eight codons beyond UAA. R17 and f2 are closely related, and it is uncertain whether tandem terminator codons will be a common occurrence in other organisms. Such an arrangement might ensure natural peptide chain termination even in cells with efficient nonsense suppressors.

The naturally occurring chain termination sequence of eukaryotic

Coat protein   Replicase

Ala—Asn—Ser—Gly—Ile—Tyr   fMet—Ser

f₂   (G)CA AAC UCC GGC AUC UAC UAA UAG ACG CCG GCC AUU CAA ACA UG

R17 (G)CA AAC UCC GGU AUC UAC UAA UAG AUG CCG GCC AUU CAA ACA UGA GGA UUA CCC AUG UCG

MS2   GCA AAC UCC GGC AUG UAC UAA UAG ACG CCG GCC AUU CAA ACA UGA CGA UUA CCC AUG UCG

Fig. 2   Intercistronic nucleotide sequence for bacteriophages R17, f2, and MS2.

mRNA has been indirectly estimated from study of naturally occurring hemoglobin mutants of man. Clegg *et al.* (1971) have found the α-chain of hemoglobin constant spring (CS) to have a chain length of 172 rather than the expected 141. Amino acid sequence analysis indicates the terminal residue corresponding to the predicted chain termination position is glutamine. These results suggest CS hemoglobin is the result of a mutation of UAG or UAA to CAG or CAA, and thus producing chain elongation. More recently, Clegg *et al.* (1974) have reported the amino acid sequence of hemoglobin ICARIA, another elongated α-hemoglobinopathy. Since lysine (AAA or AAG) occurs in the mutant 142 position, Clegg is able to propose with strong evidence that UAA is the naturally occurring chain termination signal. This conclusion also rests upon sequence data arrived at by study of the frame shift mutant α *Wayne*. The next termination codon in sequence is at position 173, and is unknown.

III. Nonsense Suppression

A. tRNA

Benzer and Champe (1962) and Garen and Siddiqui (1962), in their original articles on nonsense mutants, reported certain bacterial cells carried genes (suppressors) with the capacity to phenotypically correct nonsense mutations. These suppressor genes mapped separately from the mutant positions. Furthermore, the suppressor gene product would phenotypically correct nonsense mutants at several loci (pleotropic). It was suggested quite early that such a phenotypic corrective event could occur upon translation of the mutant mRNA if the cell contained an AA-tRNA species which recognized peptide chain termination codons. Such AA-tRNA species could arise either by mutational events affecting existing AA-tRNA genes or by mutation of an AA-tRNA synthetase resulting in acylation of a previously cryptic tRNA.

Capecchi and Gussin (1965) resolved these alternatives and provided initial evidence that the product of the suppressor gene was a species of tRNA. These investigators studied the *in vitro* synthesis of coat protein from mRNA extracted from the bacteriophage R17. The R17 strain used contained an *amber* (UAG) mutation in the sixth codon position of coat protein (a glutamine codon in the wild-type). The RNA extracted from this mutant phage directs *in vitro* synthesis of coat protein fragments in bacterial su⁻ extracts. Some *in vitro* synthesis of complete coat protein occurs, however, when the mutant R17 mRNA is translated in bacterial extracts from su^+ cells. This event is equivalent to *in vivo* nonsense suppression and established the first *in vitro* assay for suppression. The specific requirement was identified by examining the ability of individual components, isolated from the su$^+$ extract, to effect *in vitro* suppression when added to the crude su$^-$ extract. The addition of a tRNA species isolated from the *su*$^+$ cells converted the su$^-$ extract to su$^+$, indicating that nonsense suppression was mediated by a modified tRNA. The data supported the concept that the suppressor gene corresponded to a mutant tRNA gene or an enzyme involved in tRNA modification.

The exact molecular basis for suppression remained to be elucidated for at least one instance by Goodman *et al.* (1968). It was shown that the *su3*$^+$ mutation corresponded to single critical base change, G → C change in the anticodon, in the structure of a minor tRNA^{Tyr} species. The anticodon of su3$^+$ tRNA^{Tyr} is CUA recognizing UAG, while the su3$^-$ anticodon is G*UA recognizing UAU and UAC. Table I summarizes information on nonsense suppressors in *E. coli*. The mechanism for *ochre* suppressors translating both the UAG and UAA codons is presumably related to the wobble mechanism of base pairing (Crick, 1966). Person and Osborn (1968) converted *amber* to *ochre* suppressors suggesting that anticodon alteration was the mechanism in both cases. Altman *et al.* (1971) have shown that the anticodon for one ochre suppressor is UUA, which would pair with UAA and UAG.

The potential for suppressor tRNA's to occur could be restricted if a cell contained only one gene for a tRNA which was a potential suppressor, but also essential in the su$^-$ state. Such mutations would be lethal

TABLE I

E. coli **Nonsense Suppressors**

Suppressor[a]	Suppressed codon	Anticodon	Amino acid substituted	Comments
su1$^+$	UAG	C-U-A	Ser	
su2$^+$	UAG	C-U-A	Gln	
su3$^+$	UAG	C-U-A	Tyr	tRNA sequenced
su4$^+$	UAA, UAG	U-U-A	Tyr	Anticodon sequenced
su5$^+$	UAA, UAG	U-U-A	Lys	
su6$^+$	UAG	C-U-A	Leu	
su7$^+$	UAG	C-U-A	Gln	Recessive lethal
su8$^+$	UAA, UAG	U-U-A		Recessive lethal
su9$^+$	UAG	C-C-A	Trp	tRNA sequenced, change apart from anticodon

[a] Anticodon is presumptive except for su3$^+$, su4$^+$, and su9$^+$, su1$^+$ (Garen *et al.*, 1965); su2$^+$ (Garen *et al.*, 1965); su3$^+$ (Goodman *et al.*, 1968); su4$^+$ (Weigert *et al.*, 1967a); su5$^+$ (Weigert and Garen, 1965); su6$^+$ (K. Gopinathan, 1970); su7$^+$ (Soll and Berg, 1969); su8$^+$ (Soll and Berg, 1969); and su9$^+$ (Teh-Sheng, 1970).

and difficult to identify. Soll and Berg (1969) used partial diploid strains of *E. coli* to isolate such recessive lethal *amber* and *ochre* suppressors. Similarly, Miller and Roth (1971) have isolated recessive lethal *amber* and UGA suppressors in partial diploids of *Salmonella typhimurium*. Similar observations have been made in yeast. Thus, although many different suppressors are theoretically possible, cellular genetic restrictions of the possibilities can occur.

At least one UGA suppressor inserts tryptophan (Chan and Garen, 1970; Chan *et al.*, 1971; Chou, 1970) by a slightly different mechanism as shown by Hirsch (1970, 1971) and Hirsch and Gold (1971). The tRNA[Trp] structure in this instance has a change from G → A in a region apart from the anticodon, causing suppression of the UGA codon. Since structure function relationships are only partially understood for tRNA, the effect on coding of a change outside the anticodon is not explained. The su9[+](su UGA[+]) tRNA retains the ability to translate the UGG (tryptophan) codon. The su9[-] tRNA may, in fact, translate the UGA codon at a low frequency accounting for the leakiness of UGA mutations.

The mechanism of suppression in yeast is not as well investigated. Gilmore *et al.* (1968) have reported that eight different suppressor mutants (called supersuppressors in yeast) insert tyrosine at the site of nonsense mutations. It has been suggested that these eight separate sites represent identical structural genes for tRNA[Tyr]. Sherman *et al.* (1970) have studied revertants of nonsense mutants in the iso-1-cytochome *c* gene of yeast suggesting on the basis of amino acid substitution in revertants that the nonsense codon involved is UAA. These suppressors may be UAA specific, and thus differ from *E. coli ochre* suppressors which correct both UAA and UAG mutations. More recently, Steward and Sherman (1972) have obtained evidence which supports the codon UAG as a chain terminating codon, and furthermore, have isolated a suppressor specific for UAG. Gesteland *et al.* (1976) has recently obtained *in vitro* evidence that this *amber* suppression is also mediated via a modified tRNA. Evidence for UGA as a suppressible terminator codon is presently lacking.

The isolation of animal cell lines with nonsense mutants and corresponding suppressors has been the subject of intense study. Using the X-linked locus, hypoxanthine-guanine phosphoribosyltransferase, Gillin *et al.* (1972) and Beaudet *et al.* (1973) have identified mutants with promising characters. Chinese hamster clones which lack HGPRT and CRM reactivity are established. Many of these mutants revert with great facility. Thus, these mutants have several of the characteristics which are commonly attributed to nonsense mutation in prokaryotes. Further study is needed to delineate the precise molecular basis of these mutants.

B. Polarity

Nonsense mutations not only affect the synthesis of the protein coded by the mutant gene, but in prokaryotes also affect the synthesis of proteins whose genes are normal. The proteins of several operons, believed to be transcribed as a polycistronic mRNA, are known to be affected in this manner (Gorini, 1970; Newton *et al.*, 1965; Martin *et al.*, 1966; Yanofsky and Ito, 1966). The introduction of nonsense mutations into certain gene positions reduces the level of synthesis of proteins whose genes are normal, but occur distal in the operon, and between sites of the mutation and the end of the operon. As a result of these special positional requirements, the effect is polarized; hence, the term "polarity." For example, the *C* gene of the histidine operon in *Salmonella typhimurium* which codes for transaminase, is operator distal to the *D* gene (dehydrogenase), and operator proximal to the *H* gene (amidotransferase). The introduction of nonsense, but not missense mutations into the *C* gene, was found to lower the synthesis of amidotransferase phosphatase relative to dehydrogenase. Fine structure mapping on nonsense mutations within the *C* gene indicated that mutations occurring closest to the *D* gene were more polar than those closest to the *B* gene. Thus, a gradient of polarity for nonsense mutations existed within the *C* gene (Whitfield *et al.*, 1966). The polar effect of nonsense mutations is corrected by nonsense suppressors, and the degree of alleviation of polarity correlates with the level of phenotypic suppression.

The mechanism of polarity has stimulated great interest and controversy. Some facts appear clear. First, an intragenic gradient of polarity exists such that mutants near the amino terminal (initiator end) are more polar than mutants at loci nearer to the carboxyl terminus (Martin *et al.*, 1966; Yanofsky and Ito, 1966; Newton *et al.*, 1965; Whitfield *et al.*, 1966). It has been suggested that ribosomes leave the mRNA at the site of nonsense mutations, and Webster and Zinder (1969) have demonstrated *in vitro* that the ribosome · mRNA complex dissociates with translation of the terminator codon.

An extension of this concept currently under investigation is that polar effects results from ribosomal dissociation from mRNA at nonsense mutations, thus exposing distal mRNA to premature degradation by an endonuclease as outlined in Fig. 3. Recent studies have shown that an unusual suppressor isolated by Beckwith (1963) relieves the polarity of nonsense mutations without phenotypic correction of the mutated gene. This suppressor (suA) relieves the polarity of UAA and UAG mutation in the *lac* operon of *E. coli*, but does not suppress the nonsense mutation (neither active enzyme nor *CRM*+ material is produced) (Scaife and Beckwith, 1966). Similar observations have been made for the effect of this

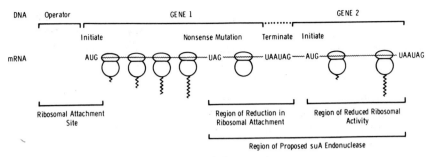

Fig. 3   Polarity model.

suppressor on nonsense mutations and polar effects occurring in the tryptophan operon. Using the latter system, Morse and Yanofsky (1969) and Morse and Primakoff (1970) showed that *trp* operon mRNA distal to a nonsense mutation is prematurely degraded in wild type *E. coli,* but not in suA cells. Similarly, Summers (1971) reports that T4 mRNA is more stable *in vivo* in suA than in wild-type cells. Morse argues that suA, known to be a recessive suppressor of polarity, represents an endonuclease mutant. Thus, suA would be nuclease deficient and mRNA when exposed would not be as rapidly degraded. Kuwano *et al.* (1971) have reported *amber* mutants of suA indicating that the *suA* gene product is a protein. In support of the endonuclease hypothesis, Kuwano *et al.* (1971) report that T4 mRNA is degraded less rapidly *in vitro* by suA extracts than by wild-type extracts. Kuwano *et al.* (1971) also report that suA is not defective in RNAse's I–V and tentatively designate its gene product as endonuclease A.

However, evidence is still being presented that polarity results from a different mechanism. Imamoto and Kano (1971), also studying the *trp* operon, interpret their mRNA kinetic data as indicating that mRNA synthesis is linked to mRNA translation in an obligatory fashion. They suggest that polarity results from a lack of mRNA transcription beyond the nonsense mutation. Thus, the mechanism of polarity remains controversial.

IV. Soluble Release Factors

A. Prokaryotes

The recognition of peptide chain termination codons requires the participation of special protein factors whose existence was initially suggested by Ganoza (1966). Using randomly ordered poly (A_3,U) to direct the synthesis and release of peptides from ribosomes, Ganoza observed dissociation of protein synthesis and protein release from tRNA when reaction mixtures contained purified sources of transfer enzymes. The substitution of purified tRNA or ribosomes for crude preparations was without effect on peptide release. These studies suggested that the supernatant, not the tRNA or ribosomal fraction of *E. coli* extracts, contained a factor(s) essential for peptide chain termination which differed from transfer factors. A protein factor was later identified in the supernatant fraction of bacterial extracts by Capecchi (1967). Capecchi used mRNA from an *amber* mutant of the bacteriophage R17 to direct the *in vitro* protein synthesis. The mutant contains a premature chain termination codon, UAG, at the sixth amino acid of coat protein. The occurrence of UAG results in synthesis and release of the hexapeptide (fMet-Ala-Ser-Asn-Phe-Thr) rather than intact coat protein. By selectively omitting one or several amino acids necessary for synthesis of the hexapeptide, the synthesis of the oligopeptidyl-tRNA can be stopped prior to terminator codon translation. For example, when threonine is depleted from crude extracts, they can be used for synthesis of ribosomal bound pentapeptidyl-tRNA. By the addition of Thr-tRNA, GTP and purified transfer enzymes to this pentapeptidyl-tRNA · mRNA · ribosomal complex, the UAG codon is brought into position, but not translated. The release of the hexapeptide from ribosomes was found to require a supernatant protein factor, RF.

The requirement for a protein factor(s) participation in peptide chain termination has been confirmed by release of fMet from an (fMet-tRNA · AUG · ribosome) intermediate using terminator trinucleotides. The expression of all terminator codons was found to be dependent upon RF, and the rate of fMET release was proportional to its concentration (Caskey *et al.*, 1968). This crude *E. coli* RF preparation was fractionated by DEAE-Sephadex column chromatography (Scolnick *et al.*, 1968) and was found to contain two codon-specific RF fractions. The first eluting fraction, RF-1, is active with UAA or UAG, but not UGA. The RF-2 is active with UAA or UGA, but not UAG. Each RF has been purified to homogeneity without any change in this codon specificity (Klein and Capecchi, 1971; Milman *et al.*, 1969). Studies which utilize phage mRNA as *in vitro* templates to direct protein synthesis have confirmed the codon specificity of RF-1 and RF-2 (Capecchi and Klein, 1970; Beaudet and Caskey, 1970). Klein and Capecchi (1971) report a molecular weight of 44,000 for RF-1 and 47,000 for RF-2, and calculate that one *E. coli* cell contains 500 molecules of RF-1 and 700 molecules of RF-2. Both RF's are insensitive to treatment with ribonuclease A or T1. Analysis of RF-1 has revealed < 1.0 atom of phosphorus/R molecule (Capecchi and Klein, 1969). Analysis of RF-2 for nucleic acid by Smrt *et al.* (1970) did not detect <0.1 residue Up or Ap/RF molecule.

The relationship between structure of the terminator codons and activity as mRNA templates for R factors has been explored in a limited way by Smrt *et al.* (1970). Using modified trinucleotides (3me-UAG, 3me-UAA, 5 me-UAG, 5me-UAA, Br-UAG, h-UAG, and UAI) as template in the formylmethionine release assay, they showed that the R factor specificity for codon recognition closely resembles the specificity of Watson-Crick and wobble base pairing.

Release factors can be bound to *E. coli* ribosomes with codon specificity in the absence of peptidyl-tRNA hydrolysis, as shown by Scolnick and Caskey (1969). The binding of RF to ribosomes is determined by Millipore filter retention of radioactive terminator trinucleotides. Ribosomes, R factor, and trinucleotide codon form a stable intermediate in the presence of ethanol. Factor RF-1 binds to ribosomes with [³H]UAA or [³H]UAG and radioactivity is subject to competition by nonradioactive UAA or UAG, but not UGA. Capecchi and Klein (1969) in experiments which do not use ribosomes have reported through equilibrium dialysis study that RF molecules can recognize oligonucleotides containing terminator codons. The specificity of this recognition is not as precise in the absence of ribosomes.

Since suppressor AA-tRNA species are known to recognize terminator codons, RF and su⁺ tRNA might be expected to compete for the translation of terminator codons in mRNA. Studies which examine such postulated competition have been performed, earlier, by addition of su⁺ AA-tRNA to crude bacterial extracts; and later, by varying the concentration of RF (Beaudet and Caskey, 1970; Ganoza and Tompkins, 1970). By varying the concentration of RF or su⁺ AA-tRNA, the level of chain termination (RF mediated) or suppression (su⁺ AA-tRNA mediated) is varied. The results of these studies do indicate that R and suppressor AA-tRNA compete, and therefore support the concept that RF-1 and RF-2 are codon recognition molecules.

These conclusions have been challenged by two recent reports by Dalgarno and Shine (1973, 1974). These investigators have demonstrated a structural homology of the 3'-terminal sequence of ribosomal RNA isolated from the smaller subunit of *Saccharomyces cerevisiae, Drosophilia melanogaster,* rabbit reticulocytes, avian myeloblasts, and *E. coli.* All terminate in the sequence GAUCAUUA A-OH. The Watson-Crick base pair of UAA, therefore, is found at a conserved 3'-termini of these molecules. On the basis of this observation, it is suggested that the terminator codons actually interact with the sequence of RNA in the ribosome and not RF. There does not appear to be accommodation for the codons UAG or UGA in this model to date.

During the development of procedures for purification of *E. coli* R factors, an additional peptide chain termination protein factor was discovered (Milman *et al.,* 1969). Factor RF-3 has no release activity, but stimulates fMet release mediated by RF-1 or RF-2 and terminator codons in the formylmethionine release assay. Since peptide chain termination involves at least two events (binding of RF to ribosomes upon recognition of terminator codon, and hydrolysis of ribosomal bound peptidyl-tRNA), RF-3 could stimulate release by acting at either or both events. The available data indicate that RF-3 affects terminator codon recognition. In the absence of RF-3, the K_m for UAA and UGA for R2 is 8.3×10^{-5} M and 5.6×10^{-5} M, respectively. The K_m for both codons is lowered to 1.3×10^{-5} M when RF-3 is added without effecting the maximal rate of hydrolysis (V_{max}), which suggests RF-3 acts at codon recognition, and not during peptidyl-tRNA hydrolysis (Goldstein *et al.,* 1970a).

Protein RF-3 has been shown to stimulate the binding and release RF-1 or RF-2 from ribosomes (Goldstein and Caskey, 1970) using radioactive terminator codons to quantitate the formation of intermediates. Factor RF-3 stimulates the formation of (RF-1 or -2 · [^{3}H]UAA · ribosome) intermediates which are actively dissociated by the addition of GTP, GDP, and less well by GDPCP, but not by GMP. Factor RF-3 becomes incorporated into the (RF-1 or RF-2 · UAA · ribosome) complex during the binding reaction and dissociates upon the addition of GTP or GDP. These studies suggest that RF-3 acts in two ways in peptide chain termination: (1) binding of RF-1 or -2 to ribosomes and/or (2) dissociation of (RF-1 or RF-2 · UAA · ribosome) intermediates.

The exact *in vivo* role of RF-3 is uncertain because of apparent *in vitro* conditional effects. At low trinucleotide concentration RF-3 stimulates fMet release, and this stimulation is eliminated by GTP or GDP. At 20-fold higher trinucleotide concentration, RF-3 alone reduces the rate of release while RF-3 plus GTP or GDP stimulates release. This stimulation probably is due to the more rapid dissociation of RF-1 or RF-2 from ribosomes. The lack of guanine nucleotide specificity may suggest that other factors (e.g., additional proteins, intact mRNA) will be required for clarification of the *in vivo* role of RF-3. Our precise understanding of RF-3 is further complicated by the report by Capecchi and Klein (1969) of purification of a protein factor which stimulates fMet release at low trinucleotide concentration and is inhibited by GTP. They suggest that RF-3 and the transfer factor EF-Tu are equivalent since this protein preparation contains both EF-Tu and stimulatory activity, and appears homogeneous by analytical disc gel analysis. Milman *et al.* (1969) have contrasting observations since their most purified RF-3 preparations contain no detectable EF-Tu activity, and their purified EF-Tu preparations contain no RF-3 activity. They therefore have concluded that RF-3 is a protein factor involved specifically in peptide chain termination, not chain elongation.

B. Eukaryotes

Goldstein *et al.* (1970b) first reported evidence for a release factor from rabbit reticulocyte extracts using a modification of the formylmethionine release assay. Formylmethionine release required a protein release factor; was stimulated by GTP, but not GDP; was inhibited by GDPCP; and required tetranucleotides UAAA, UAGA, UGAA, or UAGG (Beaudet and Caskey, 1971). Release factor purified from guinea pig liver, Chinese hamster liver, and rabbit reticulocytes functioned similarly. Using similar techniques Ilan (1973) has observed an insect soluble factor with characteristics compatible with RF. At least two forms, partially resolved by DEAE-Sephadex chromatography (Tate and Caskey, 1975), have been noted. The molecular weight of reticulocyte RF, as determined by Sephadex G-200 chromatography and equilibrium sedimentation suggest it is large (115,000). Molecular weight determinations by sucrose gradient analysis and SDS polyacrylamide gel analysis indicate RF subunit molecular weight is smaller (56,500). Thus, the available information suggest rabbit reticulocyte RF can exist as a dimer. It is unclear presently if the dimer or monomer interact at the ribosomal surface.

Reticulocyte RF ribosomal binding can be assessed with reticulocyte RF, ribosomes, and [^{3}H]UAAA in a manner analagous to that discussed for prokaryotic RF-1 or RF-2 (Tate *et al.*, 1973). This method has been used to give functional confirmation that a single reticulocyte RF recognizes all terminator codons. Reticulocyte RF · ribosome · [^{3}H]UAAA complex is measured alone and in the presence of increasing amounts of nonradioactive UAAA, UAGA, and UGAA. The complete competition by all templates suggests that any RF molecule which binds with [^{3}H]UAAA also recognizes UAGA and UGAA. The results differ if a similar experiment is performed with a mixture of RF-1 (UAAA and UAG) and RF-2 (UAA and UGA). UAGA competes out only that portion of complex formation due to RF-1, and UGAA that portion due to RF-2. The tetranucleotide UAAA, which recognizes RF-1 and RF-2, competes out all radioactive complex formation. These functional studies support the results of purification of pro- and eukaryotic RF which have indicated prokaryotes have two RF molecules (RF-1 and RF-2), while eukaryotic reticulocytes possess a single RF molecule capable of recognizing all terminator codons.

Mammalian peptide chain termination is stimulated by GTP and inhibited by GDPCP. These results suggested γ-phosphate hydrolysis by GTP is requisite for the event. A ribosomal dependent GTPase activity has been demonstrated to be associated with highly purified reticulocyte RF. This GTP hydrolysis does not occur with RF or ribosomes alone and is stimulated by the terminator oligonucleotide UAAA but not AAAA,

TABLE II
Soluble Chain Termination Factors

Designation	Molecular weight	Codon specificity	GTP recognition
Prokaryotic			
RF-1	44,000	UAA and UAG	None
RF-2	47,000	UAA and UGA	None
RF-3	46,000	None	Yes
Eukaryotic			
RF	56,500 SDS acrylamide gel electrophoresis; 115,000, G200 Sephadex analysis and equilibrium sedimentation	UAAA, UAGA, and UGAA	Yes

suggesting that GTP hydrolysis is somehow related to terminator codon recognition. Since GTP hydrolysis occurs on ribosomes not carrying nascent peptidyl-tRNA, and furthermore, is not inhibited by antibiotics which inhibit peptidyl-tRNA hydrolysis, it appears that GTP hydrolysis is not directly linked to peptidyl-tRNA hydrolysis. The binding of reticulocyte R to ribosomes is, however, stimulated by the guanine nucleotides GTP and GDPCP, but not GDP. This nucleotide specificity for R factor ribosomal binding is analagous to the protein factor-dependent ribosomal binding of initiator tRNA and aminoacyl-tRNA. In all three cases, initiation, elongation, and termination, protein factor-dependent binding utilizes GTP or GDPCP, but completion of all three protein synthetic events requires phosphate hydrolysis of the specific nucleotide, GTP.

A summary of the eukaryotic and prokaryotic RFs is given in Table II.

V. The Ribosomal Role

The necessity of ribosomal participation in peptide chain termination can be shown from a variety of studies. Ribosomal inactivation by ionic strength variation, antibiotic inhibition, pH variation, and disaggregation indicate that a ribosome must be "active" (i.e., able to form peptide bonds) to function in peptide chain termination. Vogel et al. (1969) showed a requirement for "active" 50 S ribosomal particles. They found that ribosomes prepared in buffers devoid of K^+ or NH_4^+ could not form peptide bonds or participate in peptide chain termination. The peptide bond forming and peptide chain termination activity of these "inactive" ribosomes or ribosomal subunits could be restored by addition of K^+- or NH_4^+-containing buffers. Initially, fMet-tRNA was bound to "active" 30 S ribosomal subunits, "active" or "inactive" 50 S ribosomal subparticles were added, and such intermediates examined for the capacity to form peptide bonds and participate in peptide chain termination. These studies indicated requirement of the 50 S ribosomal subparticle for peptidyl-tRNA hydrolysis, and suggested peptidyl transferase may catalyze this hydrolysis. Erdmann *et al.* (1971) have reconstituted 5 S RNA-deficient 50 S ribosomal subunits. These deficient particles will not bind R factor, (measured by [³H]UAA binding) nor will they form peptide bonds.

A. Antibiotic Inhibitors

The antibiotics tetracycline, streptomycin, sparsomycin, chloramphenicol, gougerotin, amicetin, and lincocin are all inhibitors of codon-directed peptide release in *E. coli* (Scolnick *et al.*, 1968; Vogel *et al.*, 1969, Caskey *et al.*, 1971). With the development of methods for evaluating terminator codon recognition and peptidyl-tRNA hydrolysis independently, the site of action of these inhibitors was determined. Antibiotics usually affect the codon recognition event or the peptidyl-tRNA hydrolysis event, but not both. Tetracycline and streptomycin inhibit terminator codon recognition. Amicetin, lincocin, chloramphenicol, sparsomycin, and gougerotin inhibit release of fMet without significant effect on terminator codon recognition. These latter antibiotics have been shown to be inhibitors (Monro *et al.*, 1969) of the peptide bond forming enzyme, peptidyl transferase. The relative inhibition of the codon-directed formylmethionine release and peptide bond formation has been compared at a variety of concentrations for these four antibiotics. Amicetin, lincocin, chloramphenicol, and sparsomycin inhibit each reaction in parallel (Vogel *et al.*, 1969). Menninger (1971) has obtained similar antibiotic effects with a different assay for peptide chain termination. He finds, however, that poly (U,A) stimulated release of oligolysine from oligolysyl-tRNA is more sensitive to sparsomycin, gougerotin, and erythromycin than is release of oligolysine by puromycin. The significance of this difference is uncertain. Goldstein *et al.* (1970b) examined release of formylmethionine from mammalian ribosomes by puromycin and reticulyoctye RF for antibiotic sensitivity. Sparsomycin, gougerotin, anisomycin, and, less well, amicetin, inhibited both mammalian peptidyl transferase and R factor-mediated fMet release. Thus, peptidyl transferase and RF-mediated peptidyl-tRNA hydrolysis activity are inhibited by the same antibiotics for two types of ribosomes which differ in their antibiotic sensitivity.

The putative role of peptidyl transferase in peptide chain termination is enhanced by the reports of Scolnick *et al.* (1970) and Fahnestock *et al.* (1970) that ribosomes (peptidyl transferase) can form ester bonds with tRNA or puromycin analogues. In addition, Caskey *et al.* (1971) have found that ribosomes can catalyse peptidyl-tRNA hydrolysis. A useful substrate is f[³H]Met-tRNA bound to *E. coli* or reticulocyte ribosomes.

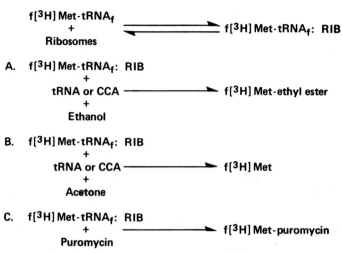

Fig. 4   Peptidyl transferase activities.

The addition of puromycin yields f[^{3}H]Met-puromycin, i.e., peptide bond formation. In the presence of ethanol, the peptidyl transferase can catalyze ester formation yielding f[^{3}H]Met-ethyl ester dependent on the presence of deacylated tRNA or its 3′-terminal sequence CCA. If acetone is substituted for ethanol, peptidyl-tRNA hydrolysis results, yielding f[^{3}H]Met again dependent on deacylated tRNA or CCA. Results are similar with *E. coli* and reticulocyte ribosomes. With *E. coli* components each reaction will occur with 50 S ribosomal subunits, is inhibited by the same antibiotics, has identical cation requirements, and has identical pH optima suggesting mediation by a common enzyme, peptidyl transferase. Thus, as shown in Fig. 4, the peptidyl transferase appears capable of peptide bond formation with an amino group as the nucleophilic agent, ester formation with an alcohol as the nucleophilic agent, and hydrolysis with water as the nucleophilic agent. The ability of peptidyl transferase to hydrolyze peptidyl-tRNA strengthens the case for its involvement in peptide chain termination (see also Chapters 8 and 10).

In considering possible mechanisms for the peptidyl-tRNA hydrolysis of peptide chain termination some observations by Caskey *et al.* (1971) may be pertinent. The antibiotics lincocin with *E. coli* ribosomes and anisomycin with reticulocyte ribosomes completely inhibit peptide bond formation and ester formation (reactions A and B, Fig. 4) while stimulating the peptidyl-tRNA hydrolysis with acetone (reaction C, Fig. 4). There is an 11-fold stimulation by anisomycin of peptidyl-tRNA hydrolysis with acetone using reticulocyte substrate and a 40% stimulation by lincocin using *E. coli* substrate. Recent reports by Innanen and Nichols (1974) have confirmed these antibiotic effects on peptidyl transferase. Such modification of peptidyl transferase specificity could be operative in chain termination where RF, rather than lincocin or anisomycin, is the modifying component.

B. Antibody Inhibitors

Using antibodies directed against specific *E. coli* ribosomal proteins, Tate *et al.* (1975) has developed additional evidence for peptidyl transferase participation in peptide chain termination. Antibodies against the *E. coli* 50 S ribosomal protens L11 and L16 inhibit *in vitro* termination as a result of affecting RF-dependent peptidyl-tRNA hydrolysis. The antibodies do not affect the interaction of RF with the ribosome. Recently, peptidyl transferase activity was restored to 50 S core particles (0.8c cores) by addition of L11 and it was suggested that L11 was the ribosomal protein possessing peptidyl transferase enzyme activity (Nierhaus and Montejo, 1973). It seems likely, therefore, from these studies that the peptidyl transferase enzyme, L11, does have some role in the cleavage of the ester bond of peptidyl-tRNA at termination. The studies do not preclude that two different sites on L11 might be involved in peptide bond formation and in hydrolysis of peptidyl-tRNA or that the hydrolysis event involves a synergism between RF and the peptidyl transferase enzyme. L16 has been identified as the chloramphenicol binding protein and deduced to be part of the A site moiety of the peptidyl transferase center (Nierhaus and Montejo, 1973). The inhibition of the peptidyl-tRNA hydrolysis partial reaction of *in vitro* termination by antibody to L16 suggests that the RF molecule may occupy a region of the A site during this event since the sites occupied by the other known components of the reaction (peptidyl-tRNA P site; peptidyl transferase L11) are not affected by chloramphenicol. If L16 and L11 were neighbors as is suggested from their roles in the peptidyl transferase center then this could support the concept that RF and the peptidyl transferase may functionally interact at chain termination.

On the basis of evidence already discussed, it is likely that the peptidyl transferase participates in peptide chain termination. Given this involvement, there are still at least two possible mechanisms for RF participation in peptidyl-tRNA hydrolysis. First, a nucleophilic group of the protein R factor could participate in the attack and transiently accept the nascent peptide in covalent linkage. Alternately, the role of R factor may be only to modify the peptidyl transferase or restrict the choice of nucleophilic agent to promote hydrolysis without the direct participation suggested above. This would be similar to the anisomycin and lincocin stimulatory effects described. There is little evidence at the moment to choose between these or other models, except that preliminary attempts to demonstrate an R factor-nascent peptide intermediate have been unsuccessful.

VI. RF Ribosomal Interaction

There is evidence using a number of approaches (Capecchi and Klein, 1969; Tompkins *et al.,* 1970) that, not only are ribosomes required, but the peptidyl-tRNA substrate must be in a configuration that permits reactivity with puromycin (P site). This implies that translocation must occur subsequent to formation of the last peptide bond and prior to termination. The P site requirement for release may also be viewed as additional evidence for participation of peptidyl transferase in hydrolysis. RF-1 and RF-2 participate in peptide chain termination by its interaction at the ribosomal A site (Tompkins *et al.,* 1970). These conclusions are further supported by the RF's ability to bind to ribosomes with occupied P sites. Both 30 S and 50 S ribosomal subunits are required for stable RF ribosomal binding. The antibiotics streptomycin and tetracycline are effective inhibitors of *E. coli* RF binding.

The guanine nucleotide GTP is clearly involved in the binding of reticulocyte RF to reticulocyte ribosomes (Tate *et al.,* 1973). Furthermore, GDP in interaction with *E. coli* RF-3 facilitates the displacement of RF-1 and RF-2 from ribosomes. These results are similar to those now reported for binding and release of IF-2 and EF-Tu from *E. coli* ribosomes (Lockwood *et al.,* 1972; Yakosawa *et al.,* 1973). It appears that the soluble factors indicated above cannot occupy the ribosome simultaneously.

Tate *et al.* (1973) have shown that *E. coli* ribosomes containing EF-G bound as a stable fusidic acid · EF-G · GDP · ribosome complex will not interact with either RF-1 or RF-2. A number of other studies have shown that stable EF-G binding to the *E. coli* ribosome can exclude the ribosomal binding of aminoacyl-tRNA with and without EF-Tu (Miller, 1972; Richman and Bodley, 1972; Cabrer *et al.,* 1972; Richter, 1972). Conversely ribosomes containing aminoacyl-tRNA bound enzymatically with EF-Tu or nonenzymatically will not associate with EF-G. Collectively, the studies indicate that these soluble factors cannot be accommodated simultaneously on a single ribosome and suggest that the cycle of events or protein biosynthesis proceed in a stepwise fashion requiring the displacement of one factor from the ribosome before the protein factor involved in the next intermediate event can interact.

The ribosomal proteins required for peptide chain termination have been investigated in three ways: (1) antibiotics, (2) antibodies directed against ribosomal proteins, and (3) ribosomes depleted of proteins.

A summary of antibiotics which inhibit partial reactions of the peptide chain termination has been described earlier in this manuscript and elsewhere (Caskey and Beaudet, 1971). At the present time, we are able to make only limited use of this information in assignment of specific ribosomal proteins to antibiotic binding sites. Vasquez (1974) has summarized the functional affects of many of these antibiotics and described functional maps of the eukaryotic and prokaryotic ribosome on the basis of a variety of studies using antibiotics. Correlations of altered ribosomal components have been made for antibiotic resistance. For example, resistant bacteria to streptomycin and erythromycin have been attributed to mutations affecting the ribosomal proteins S4 (Traub and Nomura, 1968) and L4 (Otaka et al., 1971), respectively. We may therefore assume that S4 and L4 are involved in peptide chain termination since these antibodies are inhibitors of the codon recognition and peptide hydrolysis, respectively.

Antibodies have been developed to specific prokaryotic ribosomal proteins (Stoffler and Wittman, 1971) and used extensively to identify functional activities for these proteins. These methods have identified ribosomal proteins occurring at the ribosomal subunit interface (Morrison et al., 1973), neighboring ribosomal proteins (Lutter et al., 1972), and ribosomal proteins involved in the binding of EF-Tu (Highland et al., 1974b) and EF-G (Highland et al., 1974a). Tate et al. (1975) have employed these specific ribosomal antibodies to determine those ribosomal proteins involved in peptide chain termination. The L7/L12 and S9 proteins were found to be required for RF binding. As discussed earlier L11 and L16 antibodies inhibited peptidyl-tRNA hydrolysis but not RF ribosomal binding. These latter studies support the concept of peptidyl transferase's involvement in peptidyl-tRNA hydrolysis of peptide chain termination. Using predissociated ribosomes, the antibodies against L2, L4, L6, L14, L15, L17, L18, L20, L23, L26, and L27 inhibited all termination partial reactions. Since these antibodies had no effect on 70 S ribosomes, it was concluded that they interfered with association of ribosomal subunits, requisite for chain termination.

In different studies (Brot et al., 1974) has demonstrated the requirement of L7/L12 in peptide chain termination. The removal of L7/L12 eliminated the ability of RF to bind to the ribosome. This activity was fully restored by addition of purified L7/L12 to the depleted ribosomes. Thus, by two methods L7/L12 were found critical for RF-1 and RF-2 ribosomal binding. These results are similar to those found for fMet-tRNA and AA-tRNA ribosomal binding.

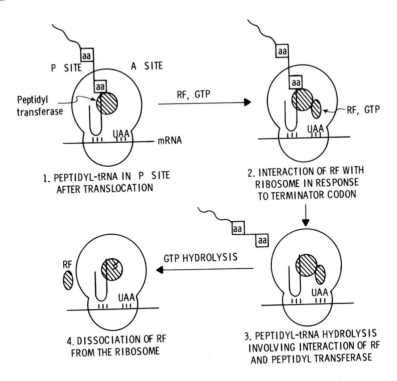

Fig. 5 A model for the intermediate steps of peptide chain termination.

VII. Conclusion

A model for the intermediate events of peptide chain termination is presented in Fig. 5. The model is based primarily upon data obtained from *in vitro* studies with both mammalian and bacterial cells and assumes a common mechanism for the two cells. All intermediate events and requirements have not been demonstrated for each cell type, and the sequence of intermediate events should be regarded as tentative.

Both mammalian and bacterial cells utilize the same terminator codons (UAA, UAG, and UGA). In bacterial cells these codons are recognized by protein release factors which are codon specific (RF-1, UAA or UAG; RF-2, UAA or UGA). In mammalian cells the RF apparently recognizes all three terminator codons. A separate protein factor RF-3, identified in bacterial extracts, has the capacity to facilitate the binding of RF to ribosomes and interacts with GDP and GTP. Since mammalian RF alone is stimulated by GTP, a functional RF-3 equivalent may not be involved. The data from both cell types favor the involvement of GTP and its γ-phosphate hydrolysis in RF binding and dissociation rather than peptidyl-tRNA hydrolysis. The release factors from both bacterial and mammalian sources cannot interact with ribosomes containing the translocation factors which are required for the event immediately preceding termination. In *E. coli* the ribosomal proteins L7, L12, S9, and the thiostrepton-binding proteins are involved in the interaction of RF with the ribosome. These and other data suggest peptidyl-tRNA hydrolysis is inhibited by antibodies to L11 and L16. Peptidyl-tRNA hydrolysis requires both an active ribosomal peptidyltransferase enzyme and the RF. The exact mechanism of this complementation is uncertain. Following peptidyl-tRNA hydrolysis RF dissociates from the ribosome, possibly dependent on the hydrolysis of GTP. The requirements for deacylated tRNA · mRNA · ribosome complex dissociation are not clear at present.

REFERENCES

Altman, S., Brenner, S., and Smith, J. (1971). *J. Mol. Biol.* **56**, 195–197.
Beaudet, A. L., and Caskey, C. T. (1970). *Nature (London)* **227**, 38–40.
Beaudet, A. L., and Caskey, C. T. (1971). *Proc. Natl. Acad. Sci. U.S.A.* **68**, 619–624.
Beaudet, A. L., Roufa, D., and Caskey, C. T. (1973). *Proc. Natl. Acad. Sci. U.S.A.* **70**, 320–324.
Beckwith, J. (1963). *Biochim. Biophys. Acta* **76**, 162–164.
Benzer, S., and Champe, S. P. (1962). *Proc. Natl. Acad. Sci. U.S.A.* **48**, 1114–1121.
Brenner, S., Stretton, A. O. W., and Kaplan, S. (1965). *Nature (London)* **206**, 994–998.
Brenner, S., Barnett, L., Katz, E. R., and Crick, F. H. C. (1967). *Nature (London)* **213**, 449–450.
Brot, N., Tate, W. P., Caskey, C. T., and Weissbach, H. (1974). *Proc. Natl. Acad. Sci. U.S.A.* **71**, 89–92.
Cabrer, B., Vazquez, D., and Modolell, J. (1972). *Proc. Natl. Acad. Sci. U.S.A.* **69**, 733–736.
Capecchi, M. R. (1967). *Proc. Natl. Acad. Sci. U.S.A.* **58**, 1144–1151.
Capecchi, M. R., and Gussin, G. N. (1965). *Science* **149**, 417–422.
Capecchi, M. R., and Klein, H. A. (1969). *Cold Spring Harbor Symp. Quant. Biol.* **34**, 469–477.
Capecchi, M. R., and Klein, H. A. (1970). *Nature (London)* **226**, 1029–1033.
Caskey, C. T., and Beaudet, A. L. (1971). *In* "Molecular Mechanisms of Antibiotic Action on Protein Biosynthesis and Membranes" (E. Muñoz, F. Garcia-Ferrandez, and D. Vazquez, eds.), *Proc. Symp., Granada, Spain,* pp. 326–336. Elsevier, Amsterdam.
Caskey, C. T., Scolnick, E., Caryk, T., and Nirenberg, M. (1968). *Science* **162**, 135–138.
Caskey, C. T., Scolnick, E., Tompkins, R., Goldstein, J., and Milman, G. (1969). *Cold Spring Harbor Symp. Quant. Biol.* **34**, 479–488.
Caskey, C. T., Beaudet, A., Scolnick, E. and Rosman, M. (1971). *Proc. Natl. Acad. Sci. U.S.A.* **68**, 3163.
Chan, T., Webster, R., and Zinder, N. (1971). *J. Mol. Biol.* **56**, 101–116.
Chan, T. S., and Garen, A. (1970). *J. Mol. Biol.* **49**, 231–234.
Chou, J. (1970). *Biochem. Biophys. Res. Commun.* **41**, 981–986.
Clegg, J. B., Weatherall, D. J., and Milner, P. F. (1971). *Nature (London)* **234**, 337–340.
Clegg, J. B., Weatherall, D. J., Contopolon-Guiva, I., Caroutsos, K., Poungouras, P., and Tsevenis, H. (1974). *Nature (London)* **251**, 245–247.
Crick, F. H. C. (1966). *J. Mol. Biol.* **19**, 548–555.

Dalgarno, L., and Shine, J. (1973). *Nature (London)* **245**, 261–262.

Dalgarno, L., and Shine, J. (1974). *Proc. Natl. Acad. Sci. U.S.A.* **71**, 1342–1346.

Erdmann, V., Fahnestock, S., Higo, K., and Nomura, M. (1971). *Proc. Natl. Acad. Sci. U.S.A.* **68**, 2932–2936.

Fahnestock, S., Neumann, H., Shashoua, V., and Rich, A. (1970). *Biochemistry* **9**, 2477–2483.

Gallucci, E., and Garen, A. (1966). *J. Mol. Biol.* **15**, 193–200.

Ganoza, M. C. (1966). *Cold Spring Harbor Symp. Quant. Biol.* **31**, 273–278.

Ganoza, M. C., and Nakamoto, T. (1966). *Proc. Natl. Acad. Sci. U.S.A.* **55**, 162–169.

Ganoza, M. C., and Tompkins, J. K. N. (1970). *Biochem. Biophys. Res. Commun.* **40**, 1455–1467.

Garen, A. (1968). *Science* **160**, 149–159.

Garen, A., and Siddiqi, O. (1962). *Proc. Natl. Acad. Sci. U.S.A.* **48**, 1121–1127.

Garen, A., Garen, S., and Wilhelm, R. C. (1965). *J. Mol. Biol.* **14**, 167–178.

Gesteland, R. F., Wolfner, M., Grisali, P., Fink, G., Botstein, D., and Roth, J. R. (1976). *Cell* **7**, 381–390.

Gillin, F., Roufa, D., Beaudet, A. L., and Caskey, C. T. (1972). *Genetics* **72**, 239–252.

Gilmore, R., Stewart, J., and Sherman, F. (1968). *Biochim. Biophys. Acta* **161**, 270–272.

Goldstein, J. L., and Caskey, C. T. (1970). *Proc. Natl. Acad. Sci. U.S.A.* **67**, 537–543.

Goldstein, J. L., Milman, G., Scolnick, E., and Caskey, C. T. (1970a). *Proc. Natl. Acad. Sci. U.S.A.* **65**, 430–437.

Goldstein, J. L., Beaudet, A. L., and Caskey, C. T. (1970b). *Proc. Natl. Acad. Sci. U.S.A.* **67**, 99–106.

Goodman, H. M., Abelson, J., Landy, A., Brenner, S., and Smith, J. D. (1968). *Nature (London)* **217**, 1019–1024.

Gopinathan, K. P., and Gaven, A. (1970). *J. Mol. Biol.* **47**, 393–395.

Gorini, L. (1970). *Annu. Rev. Genet.* **4**, 107–134.

Highland, J. H., Oschsner, E., Gordon, J., Bodley, J. W., Hasenbank, R., and Stöffler, G. (1974a). *Proc. Natl. Acad. Sci. U.S.A.* **71**, 627–630.

Highland, J. H., Oschsner, E., Jordan, J., Hasenbank, R., and Stöffler, G. (1974b). *J. Mol. Biol.* **86**, 175–178.

Hirsch, D. (1970). *Nature (London)* **228**, 57.

Hirsch, D. (1971). *J. Mol. Biol.* **58**, 459–468.

Hirsch, D., and Gold, H. (1971). *J. Mol. Biol.* **58**, 439–458.

Ilan, J. (1973). *J. Mol. Biol.* **77**, 437–448.

Imamoto, F., and Kano, Y. (1971). *Nature (London), New Biol.* **232**, 169–173.

Innanen, V. T., and Nichols, D. M. (1974). *Biochim. Biophys. Acta* **361**, 221–230.

Klein, H. A., and Capecchi, M. R. (1971). *J. Biol. Chem.* **246**, 1055–1061.

Kuwano, M., Schlessinger, D., and Morse, D. (1971). *Nature (London), New Biol.* **231**, 214–217.

Lockwood, A. H., Sarkar, P., and Maitra, U. (1972). *Proc. Natl. Acad. Sci. U.S.A.* **69**, 3602–3605.

Lutter, L. C., Zeichardt, H., and Kurland, C. C. (1972). *Mol. Gen. Genet.* **119**, 357–366.

Martin, R. G., Silbert, D. F., Smith, D. W. E., and Whitfield, H. J., Jr. (1966). *J. Mol. Biol.* **21**, 357–369.

Menninger, J. R. (1971). *Biochim. Biophys. Acta* **240**, 237–243.

Miller, C. G., and Roth, J. R. (1971). *J. Mol. Biol.* **59**, 63–75.

Miller, D. L. (1972). *Proc. Natl. Acad. Sci. U.S.A.* **69**, 752–755.

Milman, G., Goldstein, J., Scolnick, E., and Caskey, C. T. (1969). *Proc. Natl. Acad. Sci. U.S.A.* **63**, 183–190.

Monro, R. E., Staehelin, T., Celmá, M. L., and Vazquez, D. (1969). *Cold Spring Harbor Symp. Quant. Biol.* **34**, 357–366.

Morrison, C. A., Garrett, R. A., Zeichkardt, H., and Stöffler, G. (1973). *Mol. Gen. Genet.* **127**, 359–368.

Morse, D. E., and Primakoff, P. (1970). *Nature (London)* **266**, 28–31.

Morse, D. E., and Yanofsky, C. (1969). *Nature (London)* **224**, 329–331.

Newton, W. A., Beckwith, J. R., Zipser, D., and Brenner, S. (1965). *J. Mol. Biol.* **14**, 290–296.

Nierhaus, K. H., and Montejo, V. (1973). *Proc. Natl. Acad. Sci. U.S.A.* **70**, 1931–1935.

Nichols, J., and Robertson, H. (1971). *Biochim. Biophys. Acta* **228**, 676–681.

Nichols, J. L. (1970). *Nature (London)* **225**, 147–151.

Otaka, E., Itoh, T., Osawa, S., Tanaka, K., and Tamahi, M. (1971). *Mol. Gen. Genet.* **114**, 14–22.

Person, S., and Osborn, M. (1968). *Proc. Natl. Acad. Sci. U.S.A.* **60**, 1030–1037.

Rechler, M., and Martin, R. G. (1970). *Nature (London)* **226**, 908–911.

Richman, N., and Bodley, J. W. (1972). *Proc. Natl. Acad. Sci. U.S.A.* **69**, 686–689.

Richter, D. (1972). *Biochem. Biophys. Res. Commun.* **46**, 1850–1856.

Sambrook, J. F., Fan, D. P., and Brenner, S. (1967). *Nature (London)* **214**, 452–453.

Sanger, F., Brownlee, G. G., and Barrell, B. G. (1965). *J. Mol. Biol.* **13**, 373–398.

Sarabhai, A. S., Stretton, A. O. W., Brenner, S., and Bolle, A. (1964). *Nature (London)* **201**, 13–17.

Scaife, J., and Beckwith, J. R. (1966). *Cold Spring Harbor Symp. Quant. Biol.* **31**, 403–408.

Scolnick, E. M., and Caskey, C. T. (1969). *Proc. Natl. Acad. Sci. U.S.A.* **64**, 1235–1241.

Scolnick, E. M., Tompkins, R., Caskey, C. T., and Nirenberg, M. (1968). *Proc. Natl. Acad. Sci. U.S.A.* **61**, 768–774.

Scolnick, E. M., Milman, G., Rosman, M., and Caskey, C. T. (1970). *Nature (London)* **225**, 152–154.

Sherman, F., Stewart, J., Parker, J., Putterman, G., Agrawal, B., and Margoliash, E. (1970). *Symp. Soc. Exp. Biol.* **24**, 85–107.

Smrt, J., Kemper, W., Caskey, C. T., and Nirenberg, M. (1970). *J. Biol. Chem.* **245**, 2753–2757.

Soll, L., and Berg, P. (1969). *Proc. Natl. Acad. Sci. U.S.A.* **63**, 392–399.

Steward, J. W., and Sherman, F. (1972). *J. Mol. Biol.* **68**, 429–443.

Stöffler, G., and Wittman, H. G. (1971). *J. Mol. Biol.* **62**, 407–409.

Summers, W. (1971). *Nature (London), New Biol.* **230**, 208.

Tate, W. P., and Caskey, C. T. (1975). In preparation.

Tate, W. P., Beaudet, A. L., and Caskey, C. T. (1973). *Proc. Natl. Acad. Sci. U.S.A.* **70**, 2350–2355.

Tate, W. P., Caskey, C. T., and Stöffler, G. (1975). *J. Mol. Biol.* (in press).

Teh-Sheng, C., and Gaven, A. (1970). *J. Mol. Biol.* **49**, 231–234.

Tompkins, R., Scolnick, E., and Caskey, C. T. (1970). *Proc. Natl. Acad. Sci. U.S.A.* **65**, 702–708.

Traub, P., and Nomura, M. (1968). *Science* **160**, 198–199.

Vazquez, D. (1974). *FEBS Lett.* **40**, 563–564.

Vogel, Z., Zamir, A., and Elson, D. (1969). *Biochemistry* **8**, 5161–5168.

Webster, R. E., and Zinder, N. D. (1969). *J. Mol. Biol.* **42**, 425–439.

Weigert, M. G., and Garen, A. (1965). *Nature (London)* **206**, 992–994.

Weigert, M. G., Gallucci, E., Lanka, E., and Garen, A. (1966). *Cold Spring Harbor Symp. Quant. Biol.* **31**, 145–150.

Weigert, M. G., Lanka, E., and Garen, A. (1967a). *J. Mol. Biol.* **23**, 391–400.

Weigert, M. G., Lanka, E., and Garen, A. (1967b). *J. Mol. Biol.* **23**, 401–404.

Whitfield, H. J., Martin, R. G., and Ames, B. N. (1966). *J. Mol. Biol.* **21**, 335–355.

Yakosawa, H., Inone-Yokosawa, N., Arai, K., Kasahita, M., and Kaziro, V. (1973). *J. Mol. Biol.* **248**, 375–377.

Yanofsky, C., and Ito, J. (1966). *J. Mol. Biol.* **21**, 313–334.

10

Inhibitors of Protein Synthesis

SIDNEY PESTKA

I. Introduction

Antibiotics and other inhibitors have been found which block protein synthesis in various steps. Although the mechanism and precise mode of action of most agents have not been definitively determined, their use can often help elucidate not only their own mode of action, but the steps in protein biosynthesis itself. It is wise to note that antibiotics and other inhibitors cannot be used indiscriminately by blindly following procedures of other laboratories, for their use often requires specific reaction conditions. It is therefore advisable when using an inhibitor to use appropriate controls to ascertain that the inhibitions produced are indeed those desired. Often the concentration of an inhibitor is critical, producing different effects at different concentrations. Variation in components of cell-free extracts also requires adjustment of inhibitor concentrations.

In this chapter, I shall summarize the current knowledge about the mode of action of various inhibitors on pro- and eukaryotic protein synthesis. No doubt, these views shall be modified as our understanding of their mode of action increases and our knowledge of protein synthesis expands and changes.

For the purposes of this chapter, protein synthesis will be considered in terms of the individual reactions which can be measured. These are schematically summarized in Fig. 1. The effects of inhibitors on protein synthesis will be considered in terms of these steps. Inhibition of amino acid activation and aminoacyl-tRNA synthesis is not included in this chapter. It should be noted that in general most experiments in protein synthesis are interpreted in terms of the acceptor-donor site model for protein synthesis (Fig. 2). The illustrations of Fig. 1 have been drawn also in terms of

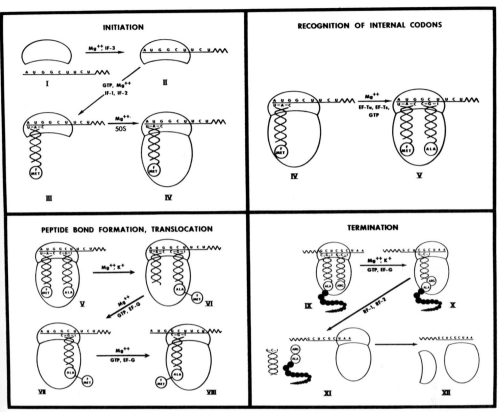

Fig. 1　Schematic summary of protein synthesis. The semilunar cap (I) represents the free 30 S subunit. Initiation of protein synthesis involves attachment of mRNA to a 30 S subunit (I) to form complex II; this process requires Mg^{2+} as well as initiation factor IF-3. Subsequent attachment of fMet-tRNA in response to initiation codon AUG to form complex III requires GTP and initiation factors IF-1 and IF-2. Junction of the 50 S subunit to complex III produces complex IV. Enzymatic recognition of internal codons involves factors EF-Tu, EF-Ts, and GTP. The EF-Tu · GTP · Ala-tRNA complex binds to the ribosomes in response to the GCU codon to form complex V. Peptide bond formation occurs by transfer of the fMet (peptidyl) group to form fMet-Ala (VI); peptidyl transfer requires only ribosomes and K^+. Translocation involves several coordinate processes: release of deacylated tRNA^{fMet} to form state VII; one codon movement of mRNA and ribosome with respect to each other, precisely positioning the next codon, UCU, into position for translation; and coordinate movement of peptidyl-tRNA (fMet-Ala) from the A to P site resulting in state VIII. By repetition of the codon recognition step Ser-tRNA would enter the A site in response to the codon UCU. Complex IX represents a peptidyl-tRNA with a polypeptide almost completed. Transpeptidation and translocation produces complex X with a completed protein still attached to tRNA and a termination codon, UAA, in the next recognition site. In response to a release factor the completed protein is released and perhaps also tRNA after translocation (XI). After termination, the ribosome may be dissociated into 30 S and 50 S subunits with release of mRNA (XII).

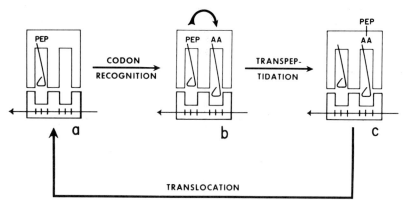

Fig. 2 Schematic illustration of the donor (P) and acceptor (A) site model of ribosome function.

the acceptor-donor model for ribosomal function. Recent experiments on protein synthesis with antibiotics have thrown some doubt as to the suitability of the simple donor-acceptor model for interpreting all experiments on protein synthesis. Another model for ribosomal function which appears to be consistent with most of the facts is shown in Fig. 3. This is presented so that the reader can readily follow the discussion concerning several antibiotics. In addition, the presentation of both models in this place presents an opportunity for the reader to evaluate their suitability. It is hoped as well, however, that their presentation will stimulate some individuals to test the predictions of and to discriminate between the two models.

II. Classification of Inhibitors

Inhibitors of protein synthesis can be classified in many ways. In Table I, inhibitors are classified according to their site of action in protein synthesis: supernatant, small subunit (30 S, 40 S), or large subunit (50 S, 60 S). In addition, in the table the inhibitors are classified as to whether they inhibit protein synthesis in prokaryotes, eukaryotes, or both. The classification is simplified in that inhibition of mitochrondrial or chloroplast ribosomes has not been included as a distinct entity. In general, prokaryotic inhibitors block protein synthesis in bacteria, blue-green algae, mitochondria, and chloroplasts. However, there are significant exceptions where inhibitors show selectivity within this group. For example, erythromycin, lincomycin, and paromomycin inhibit protein biosynthesis by bacteria, but not by rat liver mitochondria (Towers *et al.*, 1972). Also, generally inhibitors of eukaryotic protein synthesis are active in most cells having

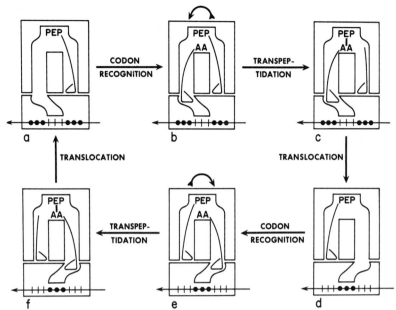

Fig. 3 The ribosome epicycle; a model of ribosome sites and ribosome function for codon recognition, transpeptidation, and translocation. Major features of the model are as follows: (a) There are two or more sites for tRNA binding on the 50 S subunit. Although the model would be essentially similar for any number of sites greater than two, the minimum number of sites on the 50 S subunit is two. (b) The two sites on the 50 S subunit are functionally similar but not identical. Transpeptidation can occur in both directions as illustrated (b, e). As a consequence of this, each site can contain aminoacyl-tRNA, peptidyl-tRNA, or deacylated tRNA. The direction of transpeptidation depends on the nature of the constituents of the sites rather than on the sites themselves. (c) There is only one site for tRNA binding on the 30 S subunit; this is the decoding site. By a conformational change (shown in the figure) or by a rotation of the subunit the site can align with either 50 S site. Because of this separate relative movement over many angstroms, the subunits should be separate particles. (d) Translocation involves movement of mRNA along the 30 S subunit, realignment of the 30 S subunit decoding site with the second 50 S site, and removal of deacylated tRNA. It is probable that removal of tRNA is the step dependent on factor EF-G and GTP; realignment of the 30 S site with the second 50 S site may or may not be dependent on EF-G. It is possible that factor EF-Tu may have a role here. (e) In order for the active center of the peptidyl transferase to interact in identical ways with peptidyl-tRNA and aminoacyl-tRNA in the two sites, it may be necessary that the active site of the peptidyltransferase rotate with respect to the sites. However, since the C-C-A(peptidyl) end of peptidyl-tRNA does not bind firmly to ribosomes, it may be possible for the 3'-terminal portion of peptidyl-tRNA to align identically with the catalytic center of the peptidyl transferase without rotation of the catalytic center. This may be possible because of free rotation around each of the bonds of the C-C-A(peptidyl) end. Analogously, although the C-C-A(amino acid) of aminoacyl-tRNA is the fixed moiety, the free rotation around each bond of this end may place the C-C-A(amino acid) from either 50 S site in the same stereochemical position with respect to the peptidyl transferase.

TABLE I

Inhibitors of Protein Synthesis[a]

Supernatant	30 S (40 S)	50 S (60 S)
Prokaryotes		
Folic acid antagonists	Aminoglycosides	Chloramphenicol
L-1-Tosylamido-2-phenylethyl	Streptomycin	
chloromethyl ketone	Dihydrostreptomycin	Macrolides
(TPCK)	Paromomycin	Niddamycin
		Carbomycin
Kirromycin	Neomycin	Spiramycin III
Mocimycin	Kanamycin	Tylosin
X-5108	Gentamycin	Leucomycin
	Bluensomycin	Erythromycin
		Chalcomycin
	Spectinomycin	Oleandomycin
		Lankamycin
	Kasugamycin	Methymycin
	Negamycin	Lincosamides
	Edeine	Celestosaminides
	Edeine A	
	Edeine B	Streptogramin A group
		Ostreogrycin A
	Colicin E3	PA114A
		Streptogramin A
	Phenanthrene alkaloids	Vernamycin A
	Cryptopleurine	Mikamycin A
	Tylocrebrine	
	Tylophorine	Streptogramin B group
		Ostreogrycin B
		PA114B, etc
		Viridogrisein (etamycin)
		Thiostrepton group
		Thiostrepton
		(bryamycin, thiactin)
		Siomycin A
		Thiopeptin B
		Multhiomycin
		Sporangiomycin
		A-59
		Althiomycin
		Micrococcin
		Bottromycin A$_2$
		Pleuromutilin
Prokaryotes and Eukaryotes		
Aminoalkyl adenylates	Pactamycin	Puromycin
Guanylyl-5'-methylene	Aurintricarboxylic acid	Aminoacyl oligonucleotides
diphosphonate		

472

TABLE I (*Continued*)

Supernatant	30 S (40 S)	50 S (60 S)

Prokaryotes and Eukaryotes (*continued*)
  Fusidic acid
		4-Aminohexose pyrimidine
		  nucleosides
		  Gougerotin
		  Amicetin
		  Blasticidin S
		  Plicacetin
		  Bamicetin
		  Hikizimycin

		Sparsomycin
	Tetracyclines Chlortetracycline	

Eukaryotes
  Diphtheria toxin
		Glutarimides
		  Cycloheximide (actidione)
		  Acetoxycycloheximide
		  Streptovitacin A

		Ipecac alkaloids Emetine
		Anisomycin

		Trichothecenes
		  Trichodermin
		  Trichothecin
		  Trichodermol
		  Crotocin
		  Nivalenol
		  T-2 Toxin
		  Verrucarin A
		  Diacetoxyscirpenol

		Tenuazonic acid
		Abrin and ricin
		Pederine

		Cephalotaxus alkaloids
		  Harringtonine
		  Homoharringtonine
		  Isoharringtonine

		MDMP
		NaF, KF

[a] Inhibitors of protein synthesis have been classified according to their site of inhibition as noted in the table. MDMP is the abbreviation for 2-(4-methyl-2,6-dinitroanilino)-*N*-methylpropionamide. In addition, although the major sites of action of all these inhibitors have not been definitively established, they have been placed according to best present estimates.

80 S ribosomes. Nevertheless, there are exceptions such as tenuazonic acid, which inhibits protein synthesis by mammalian but not by yeast ribosomes (Carrasco and Vazquez, 1973). In a few cases which will be clarified in the text, the location of the inhibitors with respect to a given sector of the table is arbitrary and represents a best estimate from the present data. For example, tetracycline probably inhibits both the small and large subunit functions, for evidence suggests that tetracycline may not be a specific inhibitor of the 30 S subunit as has been previously considered. Functions associated with the large subunit are inhibited at high tetracycline concentrations.

Inhibitors can also be classified according to their chemical structure. In Table II inhibitors are grouped according to their chemical structure. In this table molecular weights and general information regarding various inhibitors such as the site of inhibition and the specific reactions inhibited are included. Some antibiotics are active only in cell-free systems, others only on intact cells, and most on both (Table II). For example, aurintricarboxylic acid is active in cell-free extracts but inactive on intact bacteria or eukaryotic cells. Fusidic acid and guanylyl-5'-methylene diphosphonate inhibit protein synthesis in cell-free extracts but not on intact cells. Although edeine inhibits intact bacteria, it is DNA synthesis rather than protein synthesis which is inhibited. In addition, sparsomycin and tetracycline inhibit protein synthesis in eukaryotic and prokaryotic extracts as well as in intact prokaryotic cells; but they do not inhibit protein synthesis in intact mammalian cells. The inhibitor, L-1-tosylamido-2-phenylethyl chloromethyl ketone (TPCK) has an even different pattern of inhibition. It inhibits protein synthesis in cell-free extracts from pro- but not eukaryotes; in contrast, however, it inhibits intact eukaryotic cells but not intact bacteria. As can be seen by the steps and the reactions inhibited (Table II) almost every step of translation can be blocked specifically by choosing an appropriate inhibitor.

III. Ribosome Structure and Function

Throughout protein biosynthesis, ribosomes and subunits continually interact with one another; and subunits recycle (Fig. 4) through the pool of free monomers and subunits (Kaempfer and Meselson, 1969; Subramanian et al., 1969; Schlessinger et al., 1969; Phillips and Franklin, 1969). Of all the enzymatic reactions in protein synthesis (Figs. 1–4), only the formation of the peptide bond can be clearly designated as a ribosomal function not requiring other nonribosomal protein factors (Monro, 1967). The structure and function of ribosomes are discussed in Chapters 2, 3, 4, and 8. For the majority of translational operations, the ribosome provides

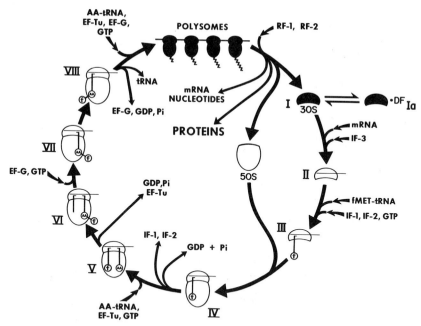

Fig. 4 Ribosomal states and the ribosome cycle. The individual complexes I to VIII are described in the legend to Fig. 1. The initiation sequence comprises stages I to IV. For every addition of a single amino acid, elongation, the ribosome cycles from state V to VIII. State IV is formally equivalent to VIII. Thus, the elongation epicycle (V to VIII) is superimposed on the overall ribosome cycle and repeats itself on each amino acid addition (Figs. 2 and 3). The termination sequence is not given in detail for it has not been precisely delineated. Dissociation factor (DF) bound to 30 S subunits (Ia) may be identical to one of the initiation factors and may be involved in dissociation of 70 S ribosomes following termination. The symbol "f" represents fMet.

a scaffold upon which appropriate apposition of mRNA codons and tRNA can occur while coordinately peptidyl- and aminoacyl-tRNA are oriented into position for peptide bond formation. To accomplish these tasks ribosomes provide precise coordination for relatively distant molecular events such as codon and peptidyl-RNA translocation (Figs. 1 and 4, steps VI–VIII). The ribosome, therefore is a multimacromolecular complex containing numerous sites for binding mRNA, tRNA, and various factors and cofactors which interact with ribosomes in a precise temporal and spacial sequence as shown in Figs. 1–4. However, ribosomes are far from being inert scaffolds upon which protein synthesis occurs, for they appear to participate intimately and to modulate the reactions while they themselves undergo conformational modifications as protein synthesis progresses. During the various stages, ribosomes and subunits probably

TABLE II

Inhibitors of Protein Synthesis[a]

Inhibitor	MW	Site of inhibition	Step inhibited	Reaction inhibited	Inhibition of pro- or eukaryotes	Inhibition in	
						Intact cells	Cell-free extracts
Althiomycin	439.47	50 S	E	Transpeptidation	Pro	+	+
Aminoglycosides							
Dihydrostreptomycin	583.62	30 S	E (I,T)	AA-tRNA binding	Pro	+	+
Streptomycin	581.58	30 S	E (I,T)	AA-tRNA binding	Pro	+	+
Bluensomycin	585.58	30 S	E	AA-tRNA coding	Pro	+	+
Gentamycin C_1	477.61	30 S	E	AA-tRNA coding	Pro	+	+
Kanamycin	484.50	30 S (50 S)	E	AA-tRNA coding; transpeptidation	Pro (Eu)	+	+
Neomycin B	614.67	30 S (50 S)	E	AA-tRNA coding; transpeptidation	Pro (Eu)	+	+
Neomycin C	614.67	30 S (50 S)	E	AA-tRNA coding; transpeptidation	Pro (Eu)	+	+
Paromomycin	615.65	30 S	E (I,T)	AA-tRNA binding	Pro (Eu)	+	+
Spectinomycin	332.36	30 S	E	Translocation	Pro	+	+
Kasugamycin	379.38	30 S	I	fMet-tRNA$_f$ binding	Pro	+	+
4-Aminohexose pyrimidine nucleosides							
Amicetin	618.69	50 S	E	Transpeptidation	Pro, Eu	+	+
Bamicetin	604.66	50 S	E	Transpeptidation	Pro, Eu	+	+
Plicacetin	517.59	50 S	E	Transpeptidation	Pro, Eu	+	+
Blasticidin S	422.45	50 S	E	Transpeptidation	Pro, Eu	+	+
Gougerotin	443.42	50 S	E	Transpeptidation	Pro, Eu	+	+
Hikizimycin	583.55	50 S	E	Transpeptidation	Pro, Eu	+	+
Anisomycin	265.31	60 S	E	Transpeptidation	Eu	+	+
Aurintricarboxylic acid	422.35	30 S, 40 S	I (E)	mRNA binding to Ribosomes	Pro, Eu	−	+
Bottromycin A_2	823.08	50 S	E	Translocation	Pro	+	+
Chloramphenicol	323.13	50 S	E	Transpeptidation	Pro	+	+
Cryptopleurine	377.47	40 S, 60 S	E	Transpeptidation	Eu	+	+

Inhibitor	Mol. wt.	Factor / Subunit	System	Reaction affected	Organism		
Diphtheria toxin (intact)	63,000	EF-T2	E	Translocation	Eu	+	–
Diphtheria toxin (fragment A)	24,000	EF-T2	E	Translocation	Eu	–	+
Edeine A	738.89	30 S, 40 S	I, E	fMet-tRNA$_f$, AA-tRNA Binding	Eu / Pro	+ / –	+ / +
Edeine B	780.93	30 S, 40 S	I, E	fMet-tRNA$_f$, AA-tRNA Binding	Eu / Pro	+ / –	+ / +
Emetine	480.63	60 S	E	(Translocation)	Eu	+	+
Fusidic acid	515.72	EF-G, EF-T2	E	Translocation	Eu / Pro	– / +	+ / +
Glutarimides							
Acetoxycycloheximide	339.39	60 S	E	Transpeptidation, (translocation)	Eu	+	+
Cycloheximide	281.35	60 S	E (I,T)	Transpeptidation, (translocation)	Eu	+	+
Streptovitacin A	297.35	60 S	E	Transpeptidation, (translocation)	Eu	+	+
Streptimidone	293.37	60 S	E	Transpeptidation, (translocation)	Eu	+	+
Guanylyl-5'-methylene diphosphonate	521.22	EF-Tu, EF-G, IF-2, EF-T1, EF-T2	I, E	GTP hydrolysis	Pro, Eu	–	+
Harringtonine	531.61	40 S	I	Met-tRNA$_f$ binding	Eu	+	+
Kirromycin (mocimycin)	796.98	EF-Tu	E	EF-Tu·GDP release from ribosomes. Other EF-Tu-dependent reactions affected	Pro	+	+
X5108	811.01	EF-Tu	E		Pro	+	+
Lincosamides							
Lincomycin	406.55	50 S	E (I,T)	Transpeptidation, (termination)	Pro	+	+
Clindamycin	424.99	50 S	E (I,T)	Transpeptidation, (termination)	Pro	+	+

(continued)

TABLE II (*Continued*)

Inhibitor	MW	Site of inhibition	Step inhibited	Reaction inhibited	Inhibition of pro- or eukaryotes	Inhibition in	
						Intact cells	Cell-free extracts
Celestosaminides							
Celesticetin	528.63	50 S	E	Transpeptidation	Pro	+	+
Macrolides							
Carbomycin	842.00	50 S	E	Transpeptidation, (translocation)	Pro	+	+
Erythromycin A	733.95	50 S	E	Transpeptidation, (translocation)	Pro	+	+
Leucomycin A$_1$	785.98	50 S	E	Transpeptidation, (translocation)	Pro	+	+
Niddamycin	783.96	50 S	E	Transpeptidation, (translocation)	Pro	+	+
Oleandomycin	687.88	50 S	E	Transpeptidation, (translocation)	Pro	+	+
Spiramycin III	899.14	50 S	E (I)	Transpeptidation, (translocation)	Pro	+	+
Tylosin	904.11	50 S	E	Transpeptidation, (translocation)	Pro	+	+
2-(4-Methyl-2,6-dinitro-anilino)-N-methyl-propionamide (MDMP)	282.26	40 S	I	40 S·mRNA·Met-tRNA coupling with 60 S	Eu	+	+
Micrococcin	~1119	50 S	E	Translocation, (AA-tRNA binding)	Pro	+	+
Negamycin	248.28	30 S	I (E,T)	Initiation, AA-tRNA coding, termination	Pro (Eu)	+	+
Pactamycin	560.66	30 S, 40 S	I (E)	fMet- and Met-tRNA$_f$ binding	Pro, Eu	+	+

Name	MW	Subunit	Inhibition	Process affected	Organism		
Puromycin	471.52	50 S, 60 S	E	Causes release of nascent peptides	Pro, Eu	+	+
Sparsomycin	361.44	50 S, 60 S	E (T)	Transpeptidation, (termination)	Eu / Pro	– / +	+ / +
Streptogramins A							
Griseoviridin	477.54	50 S	E (I,T)	Transpeptidation, (termination)	Pro	+	+
Mikamycin A	525.61	50 S	E (I,T)	Transpeptidation, (termination)	Pro	+	+
Ostreogrycin A	525.61	50 S	E (I,T)	Transpeptidation, (termination)	Pro	+	+
PA114 A	525.61	50 S	E (I,T)	Transpeptidation, (termination)	Pro	+	+
Staphylomycin M₁	525.61	50 S	E (I,T)	Transpeptidation, (termination)	Pro	+	+
Streptogramin A	525.61	50 S	E (I,T)	Transpeptidation, (termination)	Pro	+	+
Vernamycin A	525.61	50 S	E (I,T)	Transpeptidation, (termination)	Pro	+	+
Virginiamycin M	525.61	50 S	E (I,T)	Transpeptidation, (termination)	Pro	+	+
Streptogramins B							
Mikamycin B	866.98	50 S	E	(Translocation)	Pro	+	+
Ostreogrycin B	866.98	50 S	E	(Translocation)	Pro	+	+
PA 114 B-1	866.98	50 S	E	(Translocation)	Pro	+	+
Staphylomycin S	823.91	50 S	E	(Translocation)	Pro	+	+
Streptogramin B	866.98	50 S	E	(Translocation)	Pro	+	+
Vernamycin Bα	866.98	50 S	E	(Translocation)	Pro	+	+
Viridogrisein	879.03	50 S	E	(Translocation)	Pro	+	+
Virginiamycin S	823.91	50 S	E	(Translocation)	Pro	+	+
Tetracycline	444.43	30 S, 40 S, (50 S)	E (T)	AA-tRNA Binding	Eu / Pro	– / +	+ / +
Chlortetracycline	478.88	30 S, 40 S, (50 S)	E (T)	AA-tRNA Binding	Eu / Pro	– / +	+ / +
Tenuazonic acid	197.24	60 S	E	Transpeptidation	Eu	+	+
Thiostrepton group							
Multhiomycin	1064.24	50 S	E	Initiation, GTP hydrolysis, AA-tRNA binding, translocation	Pro	+	+
Siomycin A	1711.99	50 S	E	Initiation, GTP hydrolysis, AA-tRNA binding, translocation	Pro	+	+

(continued)

TABLE II (*Continued*)

Inhibitor	MW	Site of inhibition	Step inhibited	Reaction inhibited	Inhibition of pro- or eukaryotes	Inhibition in	
						Intact cells	Cell-free extracts
Sporangiomycin	~1850	50 S	E	Initiation, GTP hydrolysis, AA-tRNA binding, translocation	Pro	+	+
Thiopeptin B	1752.01	50 S	E	Initiation, GTP hydrolysis, AA-tRNA binding, translocation	Pro	+	+
Thiostrepton	1646.90	50 S	E	Initiation, GTP hydrolysis, AA-tRNA binding, translocation	Pro	+	+
L-1-Tosylamido-2-phenyl-ethyl chloromethyl ketone (TPCK)	351.85	EF-Tu	E	AA-tRNA binding	Eu	+	−
					Pro	−	+
Trichothecenes							
Crotocin	332.40	60 S	E	Transpeptidation, (termination)	Eu	+	+
Trichodermin	292.38	60 S	E	Transpeptidation, (termination)	Eu	+	+
Trichodermol	250.34	60 S	E	Transpeptidation, (termination)	Eu	+	+
Trichothecin	332.40	60 S	E	Transpeptidation, (termination)	Eu	+	+
Crotocol	264.32	60 S	I	Initiation	Eu	+	+
Diacetoxyscirpenol	366.41	60 S	I	Initiation	Eu	+	+
Nivalenol	312.32	60 S	I	Initiation	Eu	+	+
T-2 Toxin	466.53	60 S	I	Initiation	Eu	+	+
HT-2 Toxin	424.50	60 S	I	Initiation	Eu	+	+
Verrucarin A	502.57	60 S	I	Initiation	Eu	+	+

[a] Inhibitors of protein synthesis are listed alphabetically as individual compounds or as major chemical groups if appropriate. Molecular weights of the inhibitors are given. If the molecular weight was not calculated from a molecular formula, but is only approximate, this is so indicated by ~. The "Site of inhibition" refers to the site of interaction of the inhibitor producing the major inhibition. These include 30 or 50 S subunits and EF-G for prokaryotes; and correspondingly 40 and 60 S subunits and EF-T1 and EF-T2 for eukaryotes. The "Step inhibited" refers to one of the three major general stages of protein biosynthesis: initiation (I), elongation (E), and termination (T). "Reaction inhibited" refers to the specific partial reaction inhibited or otherwise affected. The partial reactions include those of initiation (mRNA binding to ribosomes, fMet-tRNA or Met-tRNA binding to ribosomes, coupling of the 30 S–50 S or 40 S–60 S subunits, initiation factor dependent GTPase activity), elongation (AA-tRNA binding, translocation, EF-Tu-dependent GTPase, EF-G-dependent GTPase, tRNA release, mRNA-ribosome movement, transpeptidation), and termination (fMet release from fMet-tRNA, terminator codon binding, release factor binding). Inhibition of whole cells or cell-free extracts from pro- and eukaryotes is more or less self-explanatory; "Pro" refers to procaryotes and "Eu" to eucaryotes; + means the antibiotic inhibits, – means that it does not. In general, inhibition of ribosomes from blue-green algae, chloroplasts, and mitochondria follow the inhibitions of prokaryotic ribosomes, but there are some exceptions as noted in the text. In some cases, as fusidic acid, the inhibitor did not inhibit whole cells and cell-free extracts from pro- and eukaryotes similarly; thus, this portion of the table had to be further subdivided. There are frequently procedures which can be used to permit entry of an agent into cells which are ordinarily impermeable. If such special procedures are required, an inhibitor was considered inactive with intact cells. In addition, some agents are active with some bacteria, but not with others. If activity was observed with some, it was considered active. Statements in the table which are surrounded with parentheses refer to effects which occur at high concentrations of the inhibitors, may not be relevant in intact cells, or are questionable for other reasons.

occupy discrete states related to their position in protein synthesis (Fig. 4). Each state, I to VIII, may represent a discrete conformation and may have a characteristic affinity for antibiotic or other factors. In addition, the presence of an initiation factor, an elongation factor, or tRNA on a ribosome may sterically impede or enhance the attachment of a particular antibiotic as will be discussed later.

Examination of polysome patterns in cells treated with an antibiotic provides a convenient method for localizing the step in the ribosome cycle blocked by an inhibitor. An inhibitor that may find ribosomes accessible during some stages of the ribosome cycle may find ribosomes inaccessible during other stages. Thus, a specific inhibitor of initiation would be expected to produce polysome breakdown; an inhibitor of translocation (viz., movement of mRNA and ribosomes with respect to each other), stabilization of the polysome pattern; and an inhibitor of termination should produce polysomes of larger size than normal as ribosomes pile up along mRNA (Hori and Rabinovitz, 1968; Tscherne and Pestka, 1975). It should be kept in mind that inhibition of protein synthesis may reflect inhibition of any of the individual steps comprising initiation, elongation, or termination. Analysis of the ribosomal cycle can help localize the area of inhibition.

As will be discussed further, antibiotics that affect ribosomes produce numerous pleiotropic effects. Thus, antibiotics that predominantly affect 30 S subunits may also influence 50 S function. Conversely, antibiotics that predominantly influence 50 S subunit function can also produce effects on 30 S subunits. Some of these pleiotropic effects are demonstrable in intact cells as well as in cell-free extracts. In other cases the effects observed in cell-free extracts may simply be pleiotropic manifestations of ribosomal topology, but only remotely related to antibiotic action in intact cells. No doubt, the marked interdependence of structural ribosomal components gives rise to the observed pleiotropy. For example, studies of Apirion and Saltzman (1974) demonstrated that certain alterations of the 30 S subunit could affect erythromycin binding to the 50 S subunit of ribosomes.

Purification of ribosomal proteins and RNA and their reconstitution to active subunits permit an insight into ribosome function and structure not previously amenable to analysis (Kurland, 1970; Traut et al., 1969; Nomura et al., 1969; Traub and Nomura, 1968a,b, 1969; Nashimoto and Nomura, 1970). Ultimately, it is hoped the three-dimensional structure of the ribosome will be known. Thus, for each antibiotic, factor, and cofactor there should be a precise physical site on the ribosome for its interaction and precise temporal positions during protein synthesis when this interaction can occur. At present we can localize antibiotics to particular subunits and, in a few cases, to specific proteins (see Chapters 2, 3, 4, and 8). Yet, so far we have little idea as to the exact position in or on the intact ribosome. However, antibiotics and other molecules that bind with great specificity to ribosomes can be used to help map the ribosome with respect to function and mutual interaction. For example, a fluorescent derivative of erythromycin, which binds to ribosomes, has been synthesized (Vince *et al.*, 1976a). Fluorescent transfer measurements with this fluorescent erythromycin probe have enabled an estimate of the distance between the erythromycin binding site and protein L7 of *Escherichia coli* ribosomes of 70Å (Langlois *et al.*, 1976).

IV. Localization of Antibiotic Action

In localizing the site of action of an inhibitor, it is generally first useful to determine if the inhibitor reacts with an S-100 fraction or with the ribosome. Since many individual assays can be performed with isolated ribosomes or S-100, the action of inhibitors on these partial reactions can be examined. Of particular use is the finding that oligopeptide synthesis can occur with isolated ribosomes in the absence of additional supernatant enzymes (Pestka, 1968, 1969a,b, 1974c; Gavrilova and Spirin, 1971, 1974). Thus, by examination of polyphenylalanine synthesis in the presence and absence of S-100 or purified elongation factors one can localize the action of an antibiotic (Pestka, 1969a,b, 1970a,b, 1975c). For example, thiostrepton and micrococcin inhibit phenylalanine incorporation with ribosomes in the presence or absence of S-100 (Pestka, 1970a, 1975c; Pestka and Brot, 1971; Pestka and Bodley, 1975). Fusidic acid, however, inhibits phenylalanine incorporation in the presence but not in the absence of S-100 or EF-G (Pestka, 1969a,b). Thus, thiostrepton and micrococcin clearly inhibit and interact with ribosomes, whereas fusidic acid interacts with EF-G.

In addition, localization of antibiotic action to S-100 or ribosome fractions can be accomplished with the use of antibiotic-resistant strains. With the use of S-100 and ribosomes from antibiotic-sensitive and -resistant strains in the four possible combinations, one can localize resistance (and thus antibiotic action) to S-100 or ribosomes.

To localize antibiotic action to a particular subunit or site, numerous direct and indirect criteria may be used. Binding of the radioactive antibiotic to a particular subunit or interference with binding of a radioactive antibiotic known to interact with a particular subunit are frequently used. Interference with a function ascribed to a particular subunit is another useful criteria, particularly if the assay can be performed with the isolated subunits. Use of 70 S ribosomes reassociated in all four combinations from subunits obtained from antibiotic-resistant and -sensitive bacterial strains is also useful. However, extensive pleiotropic effects can sometimes yield misleading conclusions. The use of reconstituted subunits lacking a specific protein or containing a substitution for a specific protein, replacing that protein from an antibiotic-sensitive strain with that from a strain resistant to the antibiotic, has been employed to define the ribosomal proteins involved in sensitivity and resistance to inhibitors. Taken together these criteria generally localize the site of action of an inhibitor.

V. Molecular Basis of Specificity and Activity of Inhibitors

Antibiotics and other inhibitors are often highly specific, not only for the step inhibited but also with respect to the ribosomal subunit and species. Selectivity of an inhibitor results from a number of variables. For example, tetracycline inhibits prokaryotic but not eukaryotic cells which are impermeable to the antibiotic. However, cell-free extracts from both pro- and eukaryotes are sensitive to it. In most cases, however, antibiotics that selectively inhibit pro- and eukaryotes are selective because of specific differences in the component inhibited from the various species. For example, chloramphenicol inhibits protein synthesis in prokaryotes but not in eukaryotes because it is able to bind to prokaryotic but not to eukaryotic ribosomes. Similarly, erythromycin and other macrolides, streptogramin A and B group of antibiotics, the aminoglycosides, the lincosamides, the thiostrepton group, and micrococcin bind to prokaryotic but not to eukaryotic ribosomes; thus, they inhibit the former, not the latter. Analogously, the glutarimides (cycloheximide, acetoxycycloheximide), phenanthrene alkaloids, harringtonine alkaloids, the trichodermin group, and MDMP inhibit eukaryotic but not prokaryotic ribosomes. In addition, a number of antibiotics inhibit both prokaryotic and eukaryotic protein synthesis and interact with both pro- and eukaryotic ribosomes. These include puromycin, sparsomycin, the aminonucleosides (gougerotin, amicetin and blasticidin S), pactamycin, and aurintricarboxylic acid. The molecular basis of cellular specificity of a compound thus depends on a number of factors: permeability of cells, interaction with cellular organelles or other components, and metabolism of the compounds.

Although antibiotics and many inhibitors of protein synthesis show extraordinarily specific interactions for ribosomes or other components, not infrequently processes other than protein synthesis are inhibited. For example, although chloramphenicol and streptogramins A inhibit bacterial and mitochondrial protein synthesis, they also inhibit respiration

(Firkin and Linnane, 1968; Dixon *et al.*, 1972; Freeman, 1970a,b). Many other examples of multiple inhibitions are known. Few inhibitors exist which block only a single reaction. Thus, agents used on intact cells performing innumerable enzymatic conversions may often produce multiple inhibitions and should be employed with constant awareness of these limitations.

Concentration of inhibitors is often another significant variable. Whereas there may be relative specificity at low concentrations, higher concentrations may exhibit additional inhibitions. Numerous examples fall into this category. For example, low concentrations of pactamycin inhibit initiation in reticulocyte lysates; however, higher concentrations inhibit both initiation and elongation (Lodish *et al.*, 1971).

VI. Comparison of Cell-Free Assays with Assays in Intact Cells

In order to correlate results in cell-free assays with antibiotic action in intact cells, it is useful to correlate the concentration at which these effects are observed in cell-free assays with that necessary for inhibition of growth or of protein synthesis in intact cells. To this end, it is useful to know the intracellular antibiotic concentration. In fact, many antibiotics that are antibacterial in gram-positive but not gram-negative organisms simply are not able to penetrate gram-negative organisms readily. These include antibiotics such as macrolides, thiostrepton, micrococcin, streptogramins A and B, fusidic acid, and many others. For many antibiotics it is frequently possible to demonstrate many effects over a wide concentration range for, at high concentrations, charged antibiotics may act as polyanions or polycations. Nevertheless, it is noteworthy that specific effects of antibiotics often occur at a ratio of one bound antibiotic molecule per ribosome.

Classification of an inhibitor as active or inactive on intact cells or cell-free extracts is often arbitrary. For example, cryptopleurine is considered inactive on cell-free extracts of prokaryotes (Table II). However, at high concentrations it was found that cryptopleurine could inhibit protein synthesis in cell-free extracts from *E. coli* (Haslam *et al.*, 1968). In some cases, as fusidic acid, the macrolides, the thiostrepton group, the streptogramin group, micrococcin, and the linocosamides, the inhibitors do not inhibit intact gram-negative cells, but readily inhibit gram-positive bacteria. Nevertheless, cell-free extracts from gram-negative cells are as sensitive as gram-positive extracts to inhibition by these agents. If activity was observed with gram-positive or gram-negative cells, the inhibitor has been classified in the table as being active, but obviously there is a specificity even within the prokaryotes and eukaryotes themselves. Frequently, gram-negative cells can be treated with Tris and EDTA (Leive, 1965) to make them permeable to antibiotics to which they ordinarily are not.

VII. Inhibitors of the Small Ribosome Subunit

A. Aurintricarboxylic Acid

Aurintricarboxylic acid (ATA) inhibits an early step of initiation, namely, the binding of mRNA to ribosomes from both pro- and eukaryotes. When ATA is added after protein synthesis has already begun, peptides are released on completion and ribosomal subunits accumulate at the expense of polysomes (Stewart et al., 1971). Grollman and Stewart (1968) showed that aurintricarboxylic acid prevents the attachment of f2 or Qβ mRNA to E. coli ribosomes. If phage Qβ RNA is incubated with ribosomes, fMet-tRNA, GTP, and initiation factors to form an initiation complex, neither Qβ RNA nor fMet-tRNA is displaced by later addition of ATA (Grollman and Stewart, 1968). Two other triphenylmethane dyes, gallein and pyrocathechol violet, were approximately equal to ATA in inhibitory capacity (Grollman, 1969). ATA interferred with initiation in extracts from wheat embryos (Marcus et al., 1970), and inhibited binding of poly(U) (Huang and Grollman, 1972a; Grollman and Huang, 1973) and globin mRNA-ribonucleoprotein complex (Lebleu et al., 1970) to reticulocyte ribosomes. Also, binding of f2 RNA to E. coli 30 S subunits (Grollman and Stewart, 1968) and globin mRNA to rabbit reticulocyte 40 S subunits was inhibited (Lebleu et al., 1970). Huang and Grollman (1972a) have shown that [³H]ATA binds to 80 S and 40 S, but not to 60 S particles.

ATA inhibited complex formation between 30 S protein S1 and poly(U) (Tal et al., 1972). Protein S1 is essential for binding mRNA to ribosomes by binding to the 3'-end of 16 S RNA; ATA disrupts this S1-RNA complex, thus preventing mRNA binding to the 30 S subunit (Dahlberg, 1974; Dahlberg and Dahlberg, 1975).

Reports indicate that ATA inhibits charging of tRNA (Siegelman and Apirion, 1971), GTP hydrolysis (Siegelman and Apirion, 1971), and elongation factor EF-Ts (Weissbach and Brot, 1970). However, these effects are apparently minimal at low ATA concentrations during translation of natural mRNA since chain elongation is unaffected by ATA concentrations inhibitory to initiation (Grollman and Stewart, 1968; Lebleu et al., 1970; Stewart et al., 1971; Webster and Zinder, 1969). The ability of ATA to bind to various proteins (Lindenbaum and Schubert, 1956) may account for the nonspecific inhibitory effects produced at high concentrations. Results with various sources of ATA suggest that some of the reported variability may be due to differences in the components present in the commercial preparations (Huang and Grollman, 1972a).

Ordinarily, intact cells are impermeable to ATA so that experiments on whole cells are precluded. In particular, ATA does not appear to affect intact HeLa cells (Tscherne and Pestka, 1975). However, at very high concentrations ATA inhibits protein synthesis in intact cells but probably by some nonspecific mechanisms (Tscherne and Pestka, 1975).

B. Aminoglycoside Antibiotics

The aminoglycoside antibiotics consist of a large group of structurally and functionally diverse compounds (Fig. 5). Because of their structural and functional diversity, several of the compounds will be discussed separately. Although there is a vast literature concerning the isolation and antimicrobial properties of many aminoglycosides, detailed studies of their effects on protein synthesis have been confined to only relatively few of them.

1. Streptomycin

It was established by Fitzgerald et al. (1948) that streptomycin (Str) inhibited protein biosynthesis in bacteria sensitive to the antibiotic. In Str-dependent E. coli strains, Spotts (1962) demonstrated that the synthesis of various enzymes was inhibited from 0–95% when these strains were deprived of antibiotic. Inhibition of protein synthesis was not seen in cells resistant to the antibiotic.

Using all possible combinations of supernatant, ribosomes, and ribosomal subunits from Str-sensitive and -resistant E. coli strains, several laboratories localized the streptomycin-sensitive site to the ribosome (Speyer et al., 1962; Flaks et al., 1962; Mager et al., 1962) and subsequently to the 30 S subunit (Cox et al., 1964; Davies, 1964). Str did not interfere with the binding of labeled poly(U) to Str-sensitive ribosomes (Davies, 1964). In other studies, Kaji et al. (1966) and Pestka (1967) showed directly that recognition of aminoacyl-tRNA (AA-tRNA) by isolated 30 S subunits prepared from Str-sensitive E. coli was inhibited by Str. EF-Tu dependent binding of AA-tRNA to ribosomes is inhibited by streptomycin (Okuyama et al., 1972).

a. *Mechanism of Inhibition of Protein Synthesis.* Modolell and Davis (1968, 1969a,b, 1970) have examined the effects of Str on various aspects of f2 RNA-directed polypeptide synthesis in cell-free extracts. With f2 RNA as a template, Str (3.4×10^{-5} M) rapidly inhibits chain extension before two amino acids are incorporated per active ribosome. Str also rapidly inhibits endogenous amino acid incorporation with natural E. coli mRNA (Modolell and Davis, 1968). After addition of Str to extracts ac-

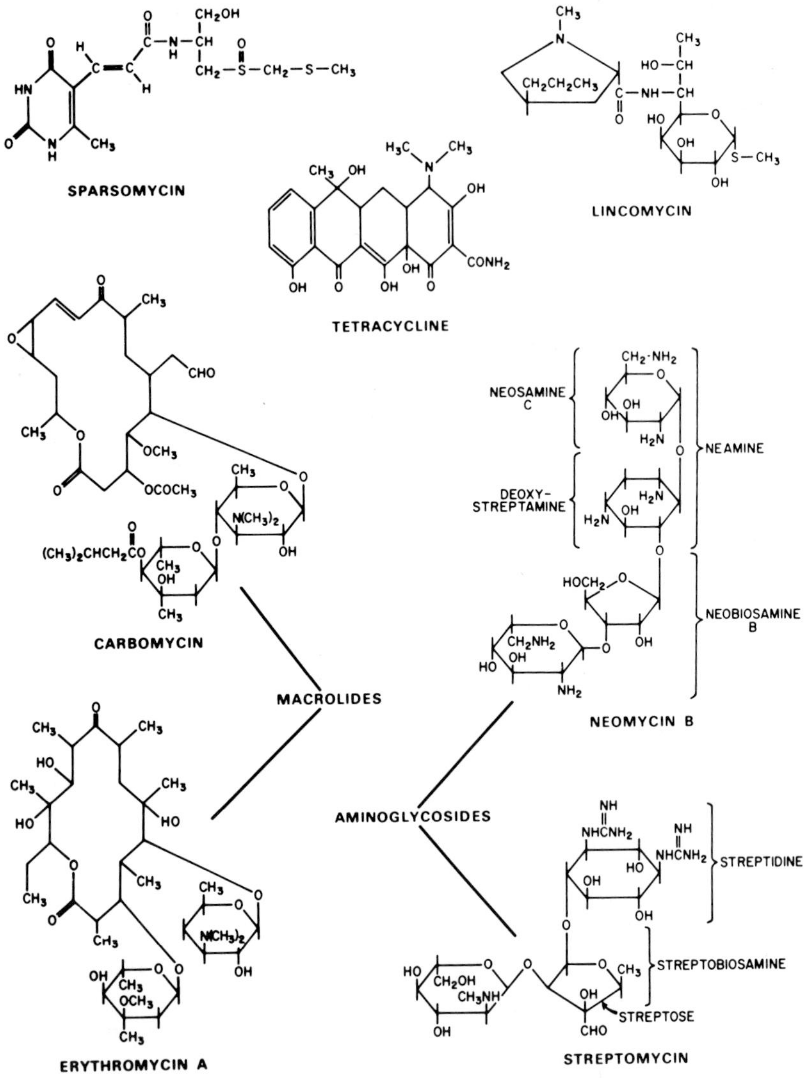

Fig. 5 Chemical structures of streptomycin, neomycin B, tetracycline, sparsomycin, erythromycin A, carbomycin, and lincomycin. References are given in the text.

tively incorporating amino acids, polysomes appear stabilized at first, but then break down after incubation for 4 min and longer. Str probably halts peptidyl-tRNA in the P site because puromycin releases all bound peptide from peptidyl-tRNA within 15 sec at the same rate in the presence or absence of Str. This indicates that $3.4 \times 10^{-5} M$ Str does not inhibit peptidyl transfer as is also supported by direct assay of peptidyl-puromycin formation (Pestka, 1972a,b). Polysome breakdown in the presence of Str was inhibited by chlortetracycline, chloramphenicol, sparsomycin, and fusidic acid.

Formation of the fMet-tRNA · poly(AUG) · 30 S complex is not inhibited by Str; however, after these complexes form in the presence of 50 S subunits, Str stimulates dissociation of the complexes so that fMet-tRNA is released (Modolell and Davis, 1970; Lelong *et al.*, 1971). When the analogue GDPCP was used, Str could not dissociate the complexes and the fMet-tRNA does not become puromycin reactive. These studies suggest that Str rapidly inhibits chain elongation, probably through inhibition of AA-tRNA binding. The bound Str distorts ribosomes so that peptidyl-tRNA is slowly released with concomitant release of ribosomes from polyribosomes. The accumulation of monosomes in Str-treated cells has been reported frequently (Luzzatto *et al.*, 1968, 1969; Kogut and Prizant, 1972; Wallace and Davis, 1973).

These concepts have been extended by Wallace and Davis (1973), who have described these effects of Str as a cyclic blockade of protein synthesis. Their results show that the Str monosomes which are released from polysomes can reinitiate protein synthesis by binding mRNA and fMet-tRNA. However, the Str monosomes exhibit an impaired response to the dissociation factor (Wallace *et al.*, 1973). The block in dissociation is incomplete, for these ribosomes can reinitiate at a reduced rate. Nevertheless, significant chain elongation does not occur and the monosomes are soon released to recycle. Thus, the polysomes of Str-treated cells are largely these rapidly turning over abortive initiation complexes. Luzzatto *et al.* (1968, 1969) previously suggested the presence of abortive initiation complexes in Str-treated cells.

If the modified complexes cannot properly initiate and translocate along the messenger, addition of further ribosomes is inhibited and the normal cycle of ribosome function is interrupted. Str monosomes which accumulate are incapable of protein synthesis either *in vivo* or *in vitro;* they do not arise during cell lysis in Str; they do not form in Str- resistant cells; nor are they nonspecific consequences of growth arrest, for glucose starvation does not produce them (Schlessinger *et al.*, 1969; Luzzatto *et al.*, 1968, 1969; Apirion *et al.*, 1969). Rate of loss of viability is well correlated with increase in Str monosomes. Str monosomes are very likely identical to the unusually stable 70 S ribosomes previously observed by Herzog (1964); 70 S ribosomes of Str- sensitive *E. coli* treated with Str did not dissociate readily at low Mg^{2+} concentration, unlike normal 70 S ribosomes or those from Str- resistant bacteria treated with Str.

b. Miscoding during Protein Synthesis. During studies of conditional

Str dependence, Gorini and Kataja (1964) suggested that Str could influence the accuracy of translation of the genetic code. Using cell-free extracts from Str-sensitive *E. coli* strains, Davies *et al.* (1964) and Van Knippenberg *et al.* (1965) showed that Str and other aminoglycoside antibiotics can stimulate miscoding. Specifically, in the presence of poly(U), Str stimulates leucine, isoleucine, serine, and tryosine incorporation. Miscoding was small or negligible with extracts prepared from Str-resistant strains. Several groups (Schwartz, 1965; Old and Gorini, 1965; Bissell, 1965, Bodley and Davie, 1966) presented evidence that in the presence of some aminoglycoside antibiotics miscoded amino acids were incorporated into protein as substitutions for the proper amino acids. Thus, Bissell (1965) predictably found that in the presence of neomycin or Str, cells produced some altered β-galactosidase.

Davies *et al.* (1965b) surveyed the effect of seven aminoglycoside antibiotics on the coding specificity of the homopolymers poly(U), poly(A), poly(C), and poly(I). The spectrum of misreading varied with the antibiotic as well as with the polynucleotide template. Using synthetic polynucleotide templates of alternating nucleotide sequence, Davies *et al.* (1966) found that the coding changes produced by Str are more limited than those produced by neomycin. They concluded that Str stimulates the misreading of the 5′-terminal and internal positions of a trinucleotide codon, of only one base at a time, and of pyrimidines more frequently than purines. Neomycin, however, produces the Str coding errors as well as misreading of the 3′-terminal nucleoside and sometimes of two bases of a codon at one time. Moreover, it appears that reading frame shifts need not be invoked to explain any of the antibiotic-induced misreading phenomena. In the case of alternating copolymers, poly(AC)$_n$ and poly(AG)$_n$, Str produced no miscoding but only inhibition of amino acid incorporation stimulated by these templates.

Pestka *et al.* (1965), Kaji and Kaji (1965), and Pestka (1966) showed that Ile-, Leu-, and Ser-tRNA binding to ribosomes was stimulated by Str when ribosomes from Str-sensitive *E. coli* were used, but was affected only slightly when ribosomes were prepared from Str-resistant *E. coli*. The effects on AA-tRNA recognition occurred at low levels of Str where the Str:ribosome ratio was approximately 1.

It appears that only in certain states during protein synthesis are ribosomes sensitive to Str action. White and White (1964) and Yamaki and Tanaka (1963) noted that puromycin, which produces polysome breakdown, enhances killing by Str. Pestka (1966) and Kaji (1967) reported that, in the presence of Str, ribosomes that were never dissociated miscoded more than ribosomes that had been dissociated. In contrast, Str hardly influenced the Phe-tRNA binding to nondissociated ribosomes, but inhibited the binding of Phe-tRNA in response to poly(U) to reassociated ribosomes from Str-sensitive *E. coli* approximately 50% (Pestka, 1966). Str-resistant bacteria were resistant to both these effects. Studying intact cells of *E. coli*, Kogut and Harris (1969) concluded that the interaction of intracellular Str with ribosomes in intact cells also must be restricted to a certain state of the cycle of ribosome function, for growth inhibitory effects were dependent on growth rate. In addition, Hatfield (1966) and Hatfield and Nirenberg (1971) showed that Str not only inhibited the binding of Phe-tRNA to ribosomes stimulated by [³H]UUC and [³H]UUUC, but also inhibited the binding of these templates.

Ribosomes prepared from Str-resistant *E. coli* strains, although relatively insensitive to inhibitions and stimulations of protein synthesis and codon recognition produced by Str, show reproducible slight effects in the presence of Str (Davies *et al.*, 1964; Van Knippenberg *et al.*, 1965; Pestka *et al.*, 1965; Pestka, 1966). In fact, Str stimulated miscoding with ribosomes prepared from Str-resistant *E. coli* strains (Anderson *et al.*, 1965); and the ability of the ribosomes to miscode in the presence of Str in cell-free extracts correlated with the ability of Str to correct or substitute for an arginine requirement (ornithine transcarbamylase) in intact cells.

   c. Physical Effects of Bound Streptomycin on Ribosomes. Some direct effects of Str on physical properties of ribosomes have been demonstrated. Thus, Leon and Brock (1967) showed that Str and neomycin protect the ribosomes of a Str-sensitive, neomycin-sensitive *E. coli* strain against thermal denaturation, whereas ribosomes from a Str-resistant, neomycin-sensitive strain are only protected by neomycin. Wolfe and Hahn (1968) also showed that the ribosomes from Str-sensitive, but not from Str-resistant, *E. coli* were stabilized to heat in the presence of Str. Sherman and Simpson (1969) reported that Str produces changes in the hydrogen exchange pattern of ribosomes: at low Str to ribosome ratios there is apparent loosening of the ribosomal structure; at high Str to ribosome ratios, there appears to be tightening of the ribosome structure. Dahlberg *et al.* (1973) showed that Str alters the electrophoretic mobility of polyribosomes from Str-sensitive strains. None of these effects is observed with Str-resistant ribosomes.

   d. Binding of Radioactive Streptomycin to Ribosomes. Str binds to both 70 S and 30 S particles from Str-sensitive bacteria (Wolfgang and Lawrence, 1967; Kaji and Tanaka, 1968; Vogel *et al.*, 1970; Chang and Flaks, 1972a,b). Binding to 30 S subunits was inhibited by the initiation factors IF-1, IF-2, and IF-3 (Lelong *et al.*, 1972). Studies by Chang and Flaks (1972a,b) examined the kinetics and equilibria of dihydrostreptomycin binding to *E. coli* ribosomes. The only apparent requirement for binding of dihydrostreptomycin to ribosomes and subunits from susceptible strains was $0.01\ M\ Mg^{2+}$. Binding of the antibiotic was rapid, reversible, and relatively temperature independent. Polynucleotides did not stimulate the binding. With concentrations of dihydrostreptomycin up to $10^{-5}\ M$, greater than 95% of the native 70 S ribosomes bound exactly one molecule of the antibiotic tightly with a K_{diss} for the bound complex at 25° of $9.4 \times 10^{-8}\ M$. There was insignificant binding to 70 S ribosomes or subunits from Str-resistant or dependent strains, or to 50 S subunits from susceptible strains. The binding to 30 S subunits from sensitive strains was weaker than that to 70 S particles with a K_{diss} of $10^{-6}\ M$ at 25°C. At antibiotic concentrations above $10^{-5}\ M$, streptomycin-sensitive 70 S and 30 S particles bound additional molecules of the antibiotic, and binding also occurred to ribosomes from Str-resistant and -dependent strains as well as to 50 S subunits from all strains. The K_{diss} for all of these binding equilibria were greater than or equal to $10^{-4}\ M$. This weaker nonspecific binding coincided with the beginning of aggregation phenomena involving the particles and occurred at sites distinct from the single site which binds the antibiotic tightly. These kinetic studies seem to indicate a reversible transition between active and inactive forms of ribosomal particles with respect to Str binding. The active form of the 30 S subunit could be stabilized by poly(U) or by association with 50 S subunits. From the thermodynamic parameters, it appeared likely that binding of the antibiotic induces a conformational change in ribosomal structure to one that is less ordered than the native particle.

 e. Other Effects on Protein Synthesis. At high concentrations, Str $(3 \times 10^{-4}\ M)$ can inhibit termination reactions of protein synthesis (Scolnick *et al.,* 1968). It is unlikely, however, that Str inhibits transpeptidation in cells, for $10^{-4}\ M$ Str did not interfere with the rate of peptidyl-[^{3}H]puromycin synthesis in cell-free *E. coli* extracts containing active polyribosomes (Pestka, 1972b); $10^{-4}\ M$ neomycin inhibited peptidyl-[^{3}H]puromycin synthesis 50%.

 In addition, Igarashi *et al.* (1969) have suggested that Str inhibits translocation as well as binding of AA-tRNA. They found that Str does not inhibit peptide bond formation from Phe-tRNA bound in the donor site; at high magnesium concentrations when Phe-tRNA is bound both in the donor and acceptor sites, maximal formation of phenylalanyl-puromycin is dependent on GTP and EF-G translocation and this reaction is inhibited by Str; also, release of tRNA^{Phe} from ribosome · poly(U) · tRNA^{Phe} complexes is inhibited by Str.

 f. The Streptomycin Protein. Using procedures for the reconstitution of ribosomal subunits, several laboratories showed that the distinction between 30 S ribosomes from Str-sensitive and Str-resistant bacteria resides in the proteins of the 30 S subunit (Traub *et al.,* 1966; Staehelin and

Meselson, 1966; Traub and Nomura, 1968a,b; Tanaka and Kaji, 1968). Ozaki *et al.* (1969) identified protein S12 as the specific protein constituent of the 30 S particle to be controlled by the *str* locus in *E. coli.* Although protein S12 is required for streptomycin binding to ribosomes, Str binds in the vicinity of proteins S3 and S5 (Schreiner and Nierhaus, 1973).

Birge and Kurland (1969) identified a single protein that is functionally altered in ribosomes obtained from Str-dependent *E. coli.* This protein is the same protein that is altered in ribosomes resistant to Str. Reversion from Str dependence appears to involve mutation in either 30 S protein S4 or S5 (Apirion *et al.,* 1969; Stöffler *et al.,* 1971).

2. Neomycin and Kanamycin

Some of the results described in the previous section discussing streptomycin apply to neomycin and kanamycin; however, there are many significant differences. Unfortunately, studies of the effects of these antibiotics on protein synthesis are not as extensive as those with Str. In any case, some of the similarities and differences will be summarized. The ribosomal localization of sensitivity and resistance to neomycin and kanamycin is similar to that of Str and involves alteration of the S12 protein of the 30 S subunit. Mutations to dependence on neomycin and kanamycin have not been described.

Miscoding by neomycin and kanamycin is more extensive than miscoding by streptomycin. Neomycin produces the Str coding errors as well as misreading of the 3′-terminal nucleoside and sometimes of two bases of a codon at one time (Davies *et al.,* 1965b). Kanamycin also produces substantial miscoding with extracts from kanamycin-sensitive *E. coli* (Tanaka *et al.,* 1967). Aminoglycoside antibiotics vary widely in degree of misreading produced as a function of their concentration (Davies and Davis, 1968). Thus, neomycin, kanamycin, and gentamicin show a three- to five-fold increase in level of misreading as drug concentration was raised from 10^{-6} to 10^{-4} M. This suggested that each of these drugs may act on more than one site and is consistent with the absence of one-step mutants with high level resistance to these antibiotics. In contrast, little increase in misreading over the same concentration range was seen with Str and paromomycin for which one-step high level resistant mutants are known. These latter drugs probably act only on a single site.

Polynucleotides that ordinarily do not serve as templates for protein synthesis can function as templates in the presence of some aminoglycoside antibiotics, particularly neomycin and kanamycin. For example, in the presence of neomycin denatured DNA, rRNA, and tRNA could serve as direct templates for protein synthesis in extracts from *E. coli* (Mc-

Carthy and Holland, 1965; Holland *et al.,* 1966). Masukawa and Tanaka (1967) also showed that the incorporation of amino acids into protein in the presence of DNA was markedly increased by kanamycin and neomycin. Less stimulation was observed with Str; and kasugamycin did not significantly stimulate amino acid incorporation by DNA. Ribosomes prepared from aminoglycoside-resistant strains were resistant to these effects. Price and Rottman (1970) showed that neomycin and Str enhanced the template activity of oligonucleotides containing 2'-O-methyladenosine in binding Lys-tRNA to ribosomes. Dunlap and Rottman (1971), furthermore showed that polynucleotides containing 2'-O-methyl analogues could serve as templates for amino acid incorporation in cell-free extracts and that neomycin stimulated the incorporation. Under normal conditions the homopolymers containing 2'-O-methyl analogues, poly(A_m), poly(C_m), and poly(U_m), were inactive as templates; however, neomycin was able to induce template activity of poly(U_m), although poly(A_m) and poly(C_m) remained inactive. Thus, some aminoglycoside antibiotics clearly alter the structural requirements for the template in protein synthesis and cause substantial perturbation of the ribosome-template interaction, permitting as templates structures that ordinarily do not serve this function.

Also, these antibiotics inhibit transpeptidation. Suzuki *et al.* (1970) showed that kanamycin, like chloramphenicol, prevents ribosomes from moving along mRNA through inhibition of peptide bond synthesis: 4×10^{-5} M kanamycin produced 44% inhibition of the puromycin reaction with polyphenylalanyl-tRNA. Neomycin inhibited peptidyl-puromycin synthesis with polyribosomes from both *E. coli* (Pestka, 1972b) and rat liver (Schneider and Maxwell, 1973).

There is some disagreement in the literature regarding the effect of these antibiotics on initiation. Neomycin, paromomycin, or gentamicin were reported to have no effect on preformed fMet-tRNA complexes or fMet-tRNA binding to ribosomes (Lelong *et al.,* 1972). However, Okuyama *et al.* (1972) reported that kanamycin and gentamicin stimulated release of fMet-tRNA from 70 S initiation complexes. In contrast, 6×10^{-6} M neomycin but not kanamycin inhibited fMet-tRNA binding to ribosomes dependent on IF-1, IF-2 and IF-3, and poly(UG) (P. Sarkar, U. Maitra, and S. Pestka, unpublished observations). However, IF-2-dependent GTPase activity was not inhibited by neomycin or kanamycin.

The ribosomal alteration responsible for sensitivity and resistance to kanamycin has been localized to the S12 protein (Masukawa *et al.,* 1968; Masukawa, 1969). It is of note that neither neomycin nor kanamycin interferes significantly with binding of dihydrostreptomycin to ribosomes (Chang and Flaks, 1972a). Thus, these antibiotics are not binding to the Str site. In addition, kanamycin and gentamicin do not interfere with

EF-Tu dependent binding of AA-tRNA to ribosomes (Okuyama *et al.*, 1972).

Eukaryotic cells are relatively insensitive to Str. However, neomycin and paromomycin affected yeast growth and inhibited amino acid incorporation by mitochondrial and cytoplasmic ribosomes (Davey *et al.*, 1970); in contrast, Str and kanamycin had little effect on either. The mitochondrial system was more sensitive than the cytoplasmic ribosomes to neomycin and paromomycin in cell-free extracts. Protein synthesis by mitochondria prepared from rat liver, however, was insensitive to neomycin and paromomycin.

3. Spectinomycin

Unlike most of the aminoglycosides, spectinomycin (Spc) is not bacteriocidal and its inhibitory effects can be reversed by washing exposed cells (Davies *et al.*, 1965a; Anderson, 1969). One-step mutants with highly resistant 30 S subunits are readily isolated, suggesting that a single site for Spc action exists. Sensitivity and resistance to Spc are properties of the 30 S subunit (Anderson *et al.*, 1967) and protein S5 of that subunit (Bollen *et al.*, 1969; Dekio and Takata, 1969; Herzog and Bollen, 1971; Bollen and Herzog, 1970; Nashimoto *et al.*, 1971; Funatsu *et al.*, 1972).

Binding of [^{3}H]dihydrospectinomycin to ribosomes was examined (Bollen *et al.*, 1969). At low antibiotic concentrations, ribosomes and 30 S subunits from Spc-sensitive *E. coli* bind more drug than those particles from Spc-resistant strains. Neomycin, kanamycin, or Str do not affect its binding to ribosomes, indicating that the binding site of these antibiotics is different from that of spectinomycin.

Inhibition of protein synthesis by Spc was maximal at 10^{-6} *M* and depended on nucleotide composition of synthetic templates; was absent with poly(U) or poly(A); increased with C and G content of polynucleotides; and was complete with poly(I) (Anderson *et al.*, 1967). Effects were independent of Mg^{2+}. Spectinomycin does not inhibit peptide bond formation, binding of AA-tRNA to ribosomes, initiation or termination reactions, nor does it produce miscoding. Although there was a striking requirement for the presence of C or G in templates for inhibition, this was unrelated to any specific AA-tRNA for the inclusion of G in a template equally sensitized to inhibition amino acids whose codons contain G as those with codons containing no G.

Wallace *et al.* (1974) have observed that spectinomycin inhibits chain elongation immediately following chain initiation. Subsequently, these blocked initiation complexes break down and recycle. Consequently, the polysomes accumulated in spectinomycin-treated cells turn over rapidly (Gurgo *et al.*, 1969; Wallace *et al.*, 1974). The dominance of sensitivity in

spc^r/spc^s heterozygotes is explained by this cyclic blockade of initiating polysomes. Studies of effects of antibiotics on intact protoplasts have suggested that translocation is the specific step of the elongation process inhibited by spectinomycin (Burns and Cundliffe, 1973).

When spectinomycin is present during initiation, 1 μM spectinomycin inhibits protein synthesis 60% (Wallace et al., 1974). With preformed polysomes, half-maximal inhibition required 300 μM spectinomycin. From these results, it might be expected that spectinomycin binds more strongly to free ribosomes or subunits than to the ribosome monomers of polyribosomes.

4. Kasugamycin

Kasugamycin is an aminoglycoside antibiotic that inhibits protein synthesis, but produces no miscoding (Umezawa et al., 1966; N. Tanaka et al., 1966a,b). Kasugamycin, like spectinomycin, is bacteriostatic and its effects are reversible. Protein synthesis is inhibited through interaction with the 30 S subunit. Kasugamycin inhibits initiation complex formation on 30 S subunits as well as on 70 S ribosomes (Okuyama et al., 1971). Although enzymatic and nonenzymatic binding of Phe-tRNA is inhibited by kasugamycin, AA-tRNA binding is not as sensitive to the antibiotic as a fMet-tRNA binding. The antibiotic does not inhibit peptide bond synthesis (Pestka, 1972b; Tai et al., 1973).

Three genetic loci have been found for kasugamycin resistance in E. coli. Mutants of the ksgA locus exhibit resistant 30 S subunits (Sparling, 1970; Sparling et al., 1973). These strains of E. coli resistant to kasugamycin have been found to be deficient in methylation of 16 S ribosomal RNA (Helser et al., 1971; Zimmermann et al., 1973). Methylation of 16 S RNA in cell-free extracts followed by reconstitution into 30 S particles converted 16 S RNA from kasugamycin-resistant strains into 16 S RNA which behaved similar to 16 S RNA from strains sensitive to the antibiotic (Helser et al., 1972). It was concluded that the ksgA locus involved in these mutations corresponds to an RNA methylase. Mutants in the ksgB locus have sensitive ribosomes; ksgB mutants probably have an altered permeability to kasugamycin. A third kasugamycin-resistant locus, ksgC, was found; ksgC mutants have 30 S subunits resistant to kasugamycin and contain an alteration in 30 S protein S2 (Okuyama et al., 1974; Yoshikawa et al., 1975).

C. Pactamycin

Pactamycin is an antitumor antibiotic that is a potent inhibitor of protein synthesis in cells and cell-free extracts of pro- and eukaryotes (Bhuyan, 1967a,b; Young, 1966; Colombo et al., 1966; Felicetti et al.,

1966; Kersten et al., 1967, 1968; Cohen and Goldberg, 1967). The complete structure of pactamycin has been reported (Wiley et al., 1970).

Felicetti et al. (1966) demonstrated that the ribosome rather than the supernatant fraction contained the site sensitive to pactamycin. Pactamycin does not interfere with the puromycin-stimulated release of nascent peptides from reticulocyte polyribosomes (Felicetti et al., 1966), nor does it inhibit peptide bond formation in bacterial systems (Cohen and Goldberg, 1967; Cohen et al., 1969a; Cundliffe and McQuillen, 1967).

In intact eukaryotic cells and extracts, conversion of polysomes to monoribosomes occurs rapidly as proteins are completed in the presence of pactamycin (Colombo et al., 1966; Goldberg et al., 1973; Taber et al., 1971). In intact reticulocytes and reticulocyte lysates, pactamycin has been shown to be an effective inhibitor of initiation at concentrations which permit completion and release of globin chains (Macdonald and Goldberg, 1970; Stewart-Blair et al., 1971). By following the differential incorporation of initiator fMet-tRNA$_f$ and internal amino acids, it was confirmed that pactamycin can block initiation in reticulocyte lysates at concentrations at which there was little inhibition of elongation (Lodish et al., 1971). However, in HeLa cells, no concentration of pactamycin was found which would specifically inhibit initiation without some inhibition of elongation (Tscherne and Pestka, 1975).

In extracts of E. coli, although initiation reactions can be inhibited (Cohen et al., 1969a), concentrations that inhibit initiation overlap those that inhibit elongation so that it is unlikely the antibiotic can be useful as a specific inhibitor of initiation (Stewart and Goldberg, 1973; Tai et al., 1973).

Pactamycin binds to 40 S subunits and 80 S ribosomes from reticulocytes at 0° (Macdonald and Goldberg, 1970); however, under these conditions it does not bind well to polyribosomes active in protein synthesis. At 37° some binding to polysomes is observed. Analogous studies with E. coli components demonstrated that pactamycin binds to 30 S subunits and 70 S ribosomes, but not to 70 S ribosomes engaged in protein synthesis containing f2 RNA at 0°. At 35°, binding to E. coli polysomes can occur (Stewart and Goldberg, 1973). Thus, it appears that pactamycin binds most readily to the small subunit or to monosomes, but relatively poorly to polysomes participating in protein synthesis. This may account for the relative specificity of pactamycin for the initiation step.

The primary effect of pactamycin in eukaryotes is on initiation. By binding to prokaryotic 30 S and eukaryotic 40 S subunits, it alters the structure of the initiation complex. Pactamycin inhibits initiation in eukaryotes by interfering with appropriate functional attachment of initiator tRNA to the initiation complex (Cohen et al., 1969b; Goldberg et al.,

1973; Macdonald and Goldberg, 1970; Stewart-Blair *et al.*, 1971; Stewart and Goldberg, 1973; Ayuso and Goldberg, 1973; Kappen *et al.*, 1973). In general, however, pactamycin does not block the binding of Met-tRNA$_f$ to ribosomes, but interferes with the conversion of initiator Met-tRNA$_f$ into a puromycin-reactive form (Goldberg *et al.*, 1973; Seal and Marcus, 1972). Specifically, it was shown that pactamycin selectively inhibits association of the 60 S subunit with the 40 S · Met-tRNA$_f$ · mRNA complex (Kappen *et al.*, 1973; Levin *et al.*, 1973). Thus, pactamycin causes the accumulation of an initiation complex on the smaller ribosomal subunit to which the 60 S subunit cannot join or with which it forms an inactive 80 S initiation complex that dissociates under the conditions of zonal centrifugation used for isolation. However, the formation of monosomes in cells treated with pactamycin suggests that nonfunctional initiation complexes are probably formed. Apparently then, pactamycin does not block the formation of initiation complexes, but interferes with their conversion into a form which can initiate and participate in the subsequent elongation events.

In the presence of pactamycin, nascent proteins are completed on mammalian polysomes. After completion of the protein, the free monosome which binds a molecule of pactamycin cannot reinitiate protein synthesis. Thus, if labeled amino acids are added to incubations of intact cells or cell-free extracts, the peptides closer to the carboxy-terminal end will be labeled to a greater extent than those toward the amino-terminal end. In this way, pactamycin was used to determine the gene order of some RNA viruses infecting animal cells (Summers and Maizel, 1971; Taber *et al.*, 1971; Butterworth and Rueckert, 1972).

D. The Tetracyclines

The tetracyclines have been shown to inhibit protein synthesis in cell-free extracts of both mammalian and bacterial cells (Gale and Folkes, 1953; Nikolov and Ilkov, 1963; Hash *et al.*, 1964; Rendi and Ochoa, 1962; Franklin, 1963, 1964). In addition, it was shown that the inhibition by tetracycline of phenylalanine incorporation directed by poly(U) correlated very well with its activity in inhibiting *E. coli* (Laskin and Chan, 1964). Tetracycline seems to bind preferentially to 30 S subunits although binding to both 30 S and 50 S subunits was found (Day, 1966; Connamacher and Mandel, 1968; Maxwell, 1968).

1. Tetracycline Inhibits AA-tRNA Binding to the A Site

The major effect of tetracycline at the lowest concentrations effectively inhibiting protein synthesis is the interference with binding of AA-tRNA to the ribosomal acceptor site. Tetracycline inhibits nonenzymatic binding of AcPhe-tRNA (Suarez and Nathans, 1965), Phe-tRNA (Hierowski, 1965; Seeds *et al.*, 1967; Sarkar and Thach, 1968; Bodley and Zieve, 1969; Obrig *et al.*, 1971a), and Lys-tRNA (Gottesman, 1967) to the ribosomal acceptor or A site. In addition, Suzuka *et al.* (1966) and Pestka and Nirenberg (1966) reported that binding of AA-tRNA to 30 S subunits in response to templates is inhibited by tetracycline. The enzymatic EF-Tu dependent binding of AA-tRNA is also inhibited by tetracycline (Lucas-Lenard and Haenni, 1968; Gordon, 1969; Ravel *et al.*, 1969; Skoultchi *et al.*, 1970; Weissbach *et al.*, 1971a). Thus, the major inhibition of tetracycline is on AA-tRNA binding to ribosomes, although many other steps can be inhibited when sufficiently large concentrations of tetracycline are used (see below).

Tetracycline inhibits termination of protein synthesis in cell-free extracts as measured by release of fMet from fMet-tRNA bound to ribosomes or by release of the amino terminal coat protein fragment with R17 RNA (Vogel *et al.*, 1969; Capecchi and Klein, 1969; Scolnick *et al.*, 1968). At $5 \times 10^{-4} M$ tetracycline, there is greater than 90% inhibition of termination. Inhibition of termination by tetracycline is a consequence of its ability to block binding of release factor to ribosomes (Scolnick *et al.*, 1968; Tompkins *et al.*, 1970).

2. Additional Effects of Tetracycline

At high concentrations of tetracycline it is possible to demonstrate other effects on various steps of protein synthesis: inhibition of initiation and of peptide bond synthesis. Only at very high concentrations ($> 5 \times 10^{-4} M$) does tetracycline inhibit binding of initiator fMet-tRNA to 30 S subunits or 70 S ribosomes (Sarkar and Thach, 1968; Zagorska *et al.*, 1971; Benne and Voorma, 1972). This, among other observations, led to the concept that initiator fMet-tRNA binds directly to the peptidyl or P site. In addition, oligolysyl-tRNA binding to ribosomes is inhibited by high tetracycline concentrations (Jonák and Rychlík, 1970; Szwaj *et al.*, 1972). Peptide bond synthesis can also be inhibited by high tetracycline concentrations (Traut and Monro, 1964; Monro and Vazquez, 1967; Černá *et al.*, 1969; S. Pestka, unpublished observations) as well as aminoacyl-oligonucleotide binding to ribosomes (Pestka, 1969d).

Intact eukaryotic organisms and bacterial strains resistant to tetracycline have a permeability barrier to tetracycline (Izaki and Arima, 1963; Franklin, 1967). When the permeability barrier of yeast cells for tetracycline is breached by low levels of polymyxin B, tetracycline can enter and kill these cells (Schwartz *et al.*, 1972). To date, no bacterial ribosomal mutants resistant to tetracycline have been definitively documented.

E. Edeine

Bacillus brevis Vm4 produces at least two closely related basic poly-peptide antibiotics, edeine A and B. Generally, edeine A has been used in most studies (Kurylo-Borowska, 1962, 1964; Kurylo-Borowska and Hierowski, 1965). It is likely that edeine B will have similar properties as edeine A, for they are similar in structure (Hettinger and Craig, 1968; 1970; Hettinger *et al.*, 1968, 1970). Edeine inhibits the synthesis of bacterial DNA, but not RNA in intact bacteria (Kurylo-Borowska, 1962, 1964). In intact bacterial cells, the inhibition of protein synthesis did not appear to be as significant as the inhibition of DNA synthesis. Nevertheless, Hierowski and Kurylo-Borowska (1965) demonstrated that edeine had a potent inhibitory effect on polypeptide synthesis in cell-free extracts from *E. coli*. In contrast, edeine appears to inhibit protein synthesis in intact mammalian cells (Georgiades *et al.*, 1975). In mammalian cells protein synthesis seemed to be more sensitive to inhibition by edeine A than was DNA synthesis.

As noted, edeine is a potent inhibitor of protein synthesis in bacterial cell-free extracts. It inhibits aminoacyl-tRNA and fMet-tRNA binding to the donor or P site at 0.01 to 0.02 M Mg^{2+}, but to both sites at lower Mg^{2+} (Szer and Kurylo-Borowska, 1970). Thus, at high Mg^{2+} edeine and tetracycline are complementary in that they inhibit donor and acceptor sites, respectively. If AA-tRNA is first bound to ribosomes, however, edeine is ineffective in removing it. Edeine inhibits binding of AA-tRNA to 30 S subunits almost completely at low and high Mg^{2+}. EF-Tu-dependent binding of Phe-tRNA 70 S ribosomes is inhibited only if no initiator tRNA is bound to ribosomes (Weissbach *et al.*, 1971b; Szer and Kurylo-Borowska, 1972). The antibiotic stabilizes the association of subunits in 70 S ribosomes and binds to both 30 S and 50 S subunits, but somewhat preferentially to 30 S subunits (Szer and Kurylo-Borowska, 1972). Binding to ribosomes and subunits is very tight for edeine once bound cannot easily be removed by dialysis. Once the initiation complex has formed, edeine cannot inhibit AA-tRNA binding to ribosomes and polysomes. Edeine does, however, prevent initiation and reinitiation, indicating that its major effect in a dynamic system synthesizing protein is on initiation. Presumably, this is a consequence of its inability to bind to polyribosomes or to monosomes with fMet-tRNA, and its ability to bind tightly to free subunits or 70 S ribosomes.

Analogous studies with eukaryotic ribosomes have led to similar conclusions: edeine is a specific and effective inhibitor of initiation of eukaryotic protein synthesis. Edeine blocks binding of initiator Met-tRNA to reticulocyte (Obrig *et al.*, 1971b) and to rat liver 40 S subunits (Schroer and Moldave, 1972). It inhibits initiation without significant inhibition of peptide chain extension in reticulocyte lysates (Obrig *et al.*, 1971b).

F. Colicin E3

Colicin E3 is a protein with a molecular weight of 60,000 produced by an *E. coli* strain (Nomura, 1963; Herschman and Helinski, 1967). A few molecules bound to the cell surface of susceptible *E. coli* causes inhibition of protein synthesis without a significant early effect on DNA, RNA, or energy metabolism. The inhibition of protein synthesis results from specific inactivation of 30 S subunits, whereas 50 S subunits remain unaffected (Konisky and Nomura, 1967). Interaction of colicin E3 with isolated 70 S ribosomes, but not with 30 S subunits or 16 S ribosomal RNA, results in a single cleavage of the 16 S RNA (Bowman *et al.*, 1971a,b; Senior and Holland, 1971; Boon, 1971, 1972; Bowman, 1972). The smaller fragment is about 50 nucleotides long and contains the 3'-end of the molecule (Bowman *et al.*, 1971b; Senior and Holland, 1971). The ability of colicin E3-treated ribosomes to initiate translation with natural mRNA is impaired more severely than is poly(U)-dependent phenylalanine incorporation (Tai and Davis, 1974). Although colicin E3 does not inactivate bacteria other than susceptible enterobacteriaceae, isolated ribosomes from various bacteria are inactivated by purified colicin E3 (Sidikaro and Nomura, 1973). In addition, in extracts 80 S ribosomes from mouse ascites cells are inactivated by colicin E3 (Turnowsky *et al.*, 1973). Although both the 40 S and 60 S subunits are required for inactivation, only the 60 S subunit is altered. Cloacin, a protein produced by *Enterobacter cloacae,* resembles colicin E3 in its action (De Graaf *et al.*, 1970, 1971).

G. Negamycin

Negamycin is a unique hydrazide antibiotic produced by a *Streptomyces* strain (Hamada *et al.*, 1970; Kondo *et al.*, 1971) and has antibacterial activity against gram-positive and gram-negative organisms. Negamycin inhibits protein synthesis in cells as well as in cell-free extracts (Mizuno *et al.*, 1970a,b). Although negamycin inhibited protein synthesis in bacterial extracts, it also slightly inhibited phenylalanine incorporation directed by poly(U) in an extract from rat liver (Mizuno *et al.*, 1970b). It inhibited binding of fMet-tRNA to *E. coli* ribosomes dependent on MS2 RNA. It also inhibited MS2 RNA-dependent protein synthesis if added prior to initiation, but interfered little or not with elongation. Peptide bond formation (AcPhe-puromycin synthesis) was not affected. Negamycin also stimulates miscoding in the presence of synthetic polyribonucleotides (Mizuno *et al.*, 1970b; Uehara *et al.*, 1972). The pattern of misreading was similar to that produced by streptomycin. However, streptomycin-resistant ribosomes were sensitive to negamycin. Although binding of Phe-tRNA to ribosomes in response to poly(U) was only weakly inhibited, Ile-tRNA binding was stimulated by negamycin. In addition, negamycin inhibited chain termination of protein synthesis directed by f2 RNA in an *E. coli* extract (Uehara *et al.*, 1974). However, the effects of negamycin on the individual termination reactions has not yet been reported. Thus, all the specific and preferential reactions which negamycin inhibits are still to be clarified.

H. Phenanthrene Alkaloids: Tylophorine, Tylocrebrine, and Cryptopleurine

These phenanthrene alkaloids irreversibly inhibit protein biosynthesis in mammalian cells (Donaldson *et al.*, 1968; Haslam *et al.*, 1968; Huang and Grollman, 1972b). Tylocrebrine inhibited DNA and RNA synthesis in HeLa cells, but to a lesser degree than its inhibition of protein synthesis (Huang and Grollman, 1972b). In reticulocyte lysates, tylocrebrine prevented breakdown of polyribosomes. Its action resembled that of cycloheximide. These alkaloids did not inhibit transpeptidation with C-A-C-C-A(AcLeu) as a donor (Battaner and Vazquez, 1971). However, 10^{-4} M cryptopleurine inhibited transpeptidation on native polyribosomes from rat liver (Pestka *et al.*, 1972b). At high concentrations, cryptopleurine, but not tylocrebrine or tylophorine, inhibited leucine incorporation in *E. coli* extracts (Haslam *et al.*, 1968); and high concentrations of cryptopleurine inhibited peptidyl-puromycin formation with *E. coli* polyribosomes very slightly (S. Pestka, unpublished studies). In addition, Haslam *et al.* (1968) and Skogerson *et al.* (1973) reported that the target of cryptopleurine is the ribosome. Specifically, it appears that mutants resistant to cryptopleurine have alterations in the small ribosomal subunit (Skogerson *et al.*, 1973; Grant *et al.*, 1974). K. Bucher and L. Skogerson (personal communication) have observed that low concentrations of cryptopleurine (10^{-6} M) inhibit translocation by yeast ribosomes. Huang and Grollman (1972b) and Grant *et al.* (1974) also concluded that these alkaloids inhibit translocation. It is clear that the principal effect of these alkaloids is on chain elongation, although the precise step of inhibition is yet to be defined.

VIII. Inhibitors of the Large Ribosomal Subunit

A. Puromycin and Aminoacyl-Oligonucleotides

Puromycin has been shown to be an inhibitor of protein synthesis in intact cells as well as cell-free extracts from diverse organisms (Nathans, 1967). Its chemical structure (Waller *et al.*, 1953; Baker *et al.*, 1954, 1955), crystal conformation (Sundaralingam and Arora, 1969), and its conformation in solution (Jardetzky, 1963; Johnson and Bhacca, 1963) have been reported. Yarmolinsky and de la Haba (1959) pointed out the structural resemblance between puromycin and the aminoacyl-adenylyl terminus of AA-tRNA (see Chapter 8, Fig. 3). This provided a basis for understanding the mechanism of puromycin action. Puromycin, like AA-tRNA, can serve as an acceptor of the nascent peptide chain of ribosome-bound peptidyl-tRNA (Rabinovitz and Fisher, 1962; Morris and Schweet, 1961) and is incorporated into the prematurely released nascent chains (Allen and Zamecnik, 1962; Nathans, 1964). The entire puromycin molecule is linked to the released peptides through the amino group of the side chain by a peptide bond through the carboxy-terminal end of the polypeptides in a reaction similar to the transfer of peptide from peptidyl-tRNA to the next AA-tRNA (Fig. 1). On analysis of lysine peptides released by puromycin, Smith *et al.* (1965) also concluded that one puromycin molecule was bound to each lysine peptide. They also noted that although dilysyl-, trilysyl-, and tetralysyl-puromycin were found, no lysyl-puromycin was detectable. In short, puromycin inhibits protein synthesis by substituting for the incoming coded AA-tRNA, in effect competing with the aminoacyl-adenylyl terminus of AA-tRNA. Puromycin does not interfere with AA-tRNA formation (Yarmolinsky and de la Haba, 1959) or with mRNA or AA-tRNA binding to ribosomes (Spyrides, 1964). Since puromycin becomes linked to the released peptides by means of a peptide bond, the reaction has been used extensively as a model of peptide bond formation in protein synthesis (see Chapter 8 for details).

Both the aminonucleoside and the amino acid side chain are necessary; for maximal inhibition the amino acid should be in the L-configuration and the amino group must be unsubstituted although at high concentrations *N*-acetyl-puromycin was inhibitory to termination (Vogel *et al.*, 1969). Also for maximal activity the amino acid side chain should contain an aromatic residue (Nathans and Neidle, 1963; Symons *et al.*, 1969; Waller *et al.*, 1966; Rychlík *et al.*, 1969; Pestka *et al.*, 1970). It is possible therefore that the aromatic amino acids can participate in some interactions which may enhance their binding to ribosomes compared to nonaromatic aminoacyl nucleosides.

Other compounds have puromycin-like activity. Takanami (1964) reported that digests of unfractionated AA-tRNA, containing aminoacyl-oligonucleotide fragments ranging in length from 4 to 9 nucleotides, were considerably more effective than pancreatic ribonuclease digests, containing aminoacyl-adenosines, in stimulating the release of nascent peptide chains. Studies with C-A-C-C-A(Phe), C-A-C-C-A(AcPhe), and similar derivatives have indicated that stable binding of these compounds requires a free amino group (Hishizawa *et al.*, 1970). The dissociation constant for binding of C-A-C-C-A(Phe) to ribosomes was reported to be 3.6×10^{-8} M in the presence of 20% ethanol (Harris and Pestka, 1973).

Nascent peptides can be transferred to C-A-C-C-A(Phe) and similar aminoacyl-oligonucleotides as to puromycin (Lessard and Pestka, 1972a,b). Also Rychlík *et al.* (1969) have shown that aminoacyl-nucleosides are most inhibitory to protein synthesis when adenosine comprises the terminal base. Phenylalanyl-adenosine, A(Phe), was almost as active as puromycin; A(Phe), I(Phe), and C(Phe) had about 70%, 15%, and 3% of the acceptor activity of puromycin in transferring the peptide residue from polylysyl-tRNA to the acceptor substrates. The aminoacyl-nucleosides, U(Phe) and G(Phe), were inactive as acceptors. The addition of a cytidine 3'-phosphoryl group to the inactive A(Gly) (Rychlík *et al.*, 1967) or 3'-N-glycyl-puromycin aminonucleoside (Symons *et al.*, 1969) converted these inactive compounds to molecules with 80–90% of the puromycin activity. The 2'-O-aminoacyl analogues of puromycin (Nathans and Neidle, 1963) and of dinucleoside phosphates (Chládek *et al.*, 1974) were inactive as acceptor substrates. Although only the 3'-O-aminoacyl dinucleoside phosphates and analogues of tRNA were active as acceptor substrates for the peptidyl transferase, the 2'-derivatives were recognized by the ribosomal A site and competed with C-A-C-C-A(Phe) for ribosomal binding (Chládek *et al.*, 1973; Ringer and Chládek, 1974; Ringer *et al.*, 1975) (see Chapter 8 for additional details).

Puromycin can directly inhibit the binding of the aminoacyl end of AA-tRNA to isolated ribosomes (Pestka, 1969a,b; Lessard and Pestka, 1972a). Binding of C-A-C-C-A(Phe) to ribosomes is inhibited by puromycin. This localized inhibition of binding of the AA-tRNA thus may be another action of puromycin distinct from, but related to, its stimulation of release of nascent protein chains from ribosomes.

The K_m for puromycin depends on the particular cell-free system. When polyribosomes are used to study the incorporation of puromycin, the K_m for puromycin on bacterial and rat liver polyribosomes are 4 μM and 8 μM, respectively (Pestka, 1972a,b; Pestka *et al.*, 1972). When fMet- or acetyl-Phe-puromycin is the product being studied, the K_m for puromycin is approximately 2 mM (Fahnestock *et al.*, 1970; Pestka, 1970a).

There are additional analogues of puromycin which function well in cell-free systems or in intact cells. Substitution of the amino group of puromycin with a hydroxyl group permits puromycin to function, and the products on the ribosomes are peptidyl esters (Fahnestock *et al.*, 1970). The K_m for hydroxy-puromycin is about two orders of magnitude higher than that of puromycin in comparable systems. Recently, the synthesis of an interesting carbocyclic analogue of puromycin has been described (Daluge and Vince, 1972). The compound contains a cyclopentane carbon skeleton replacing the ribose ring. It competes with puromycin and peptidyl-puromycin synthesis on polyribosomes almost as well as puromycin itself. The K_i for its inhibition of peptidyl-puromycin synthesis is $1 \times 10^{-5} M$ (Pestka et al., 1973).

Although the mode of action of puromycin is one of the best understood of the antibiotics, there are some uncertainties centered around it. Is release of the nascent chain from the ribosome after reaction with puromycin dependent on translocation? Can ribosomes continue translation after puromycin releases the nascent chain? Are ribosomes released from polyribosomes at the position where the nascent chain is released or must they travel to a terminator codon before release? Is a supernatant factor required for release of ribosomes from mRNA after the puromycin reaction? Thus, although the action of puromycin is understood, many questions still remain.

B. 4-Aminohexose Pyrimidine Nucleoside Antibiotics

The 4-aminohexose pyrimidine nucleoside antibiotics discussed here consist of gougerotin, amicetin, blasticidin S, bamicetin, plicacetin, and hikizimycin. A review of these and other nucleoside antibiotics has been compiled by Suhadolnik (1970). Some of these structures are given in Fig. 6.

Clark and Gunther (1963) showed that gougerotin was an inhibitor of poly(U)-dependent polyphenylalanine synthesis in cell-free extracts from *E. coli*. Subsequently, it was shown to be an inhibitor of protein synthesis in cell-free extracts from mouse liver (Sinohara and Sky-Peck, 1965), rabbit reticulocytes (Clark and Chang, 1965; Casjens and Morris, 1965), and lymphatic tissue (Bermek and Matthaei, 1970). Block and Coutsogeorgopoulos (1966) showed that amicetin could inhibit poly(U)-directed phenylalanine incorporation in cell-free extracts from *E. coli*; also, blasticidin S inhibited polyphenylalanine synthesis (Coutsogeorgopoulos, 1967a, 1971). Yamaguchi et al. (1965) and Yamaguchi and Tanaka (1966) showed that blasticidin S inhibited amino acid incorporation directed by native as well as synthetic mRNA's. By using various derivatives of and compounds similar to blasticidin S, Yamaguchi and Tanaka (1966) were able to show that removal of the amino group from the cytosine moiety or removal of the cytosine moiety reduced or eliminated the activity of the antibiotic. Additional structure-activity relationships of these nucleoside antibiotics have been discussed by Lichtenthaler and Trummlitz (1974) and Coutsogeorgopoulos et al. (1975). Yukioka and Morisawa (1969a,b,c) showed that the effect of gougerotin on polyphenylalanine synthesis varied depending on K^+ concentration: at 50 mM K^+, gougerotin inhibited polyphenylalanine synthesis; at 200 mM K^+, it was stimulatory. The stimulatory action was due to its ability to activate ribosomes.

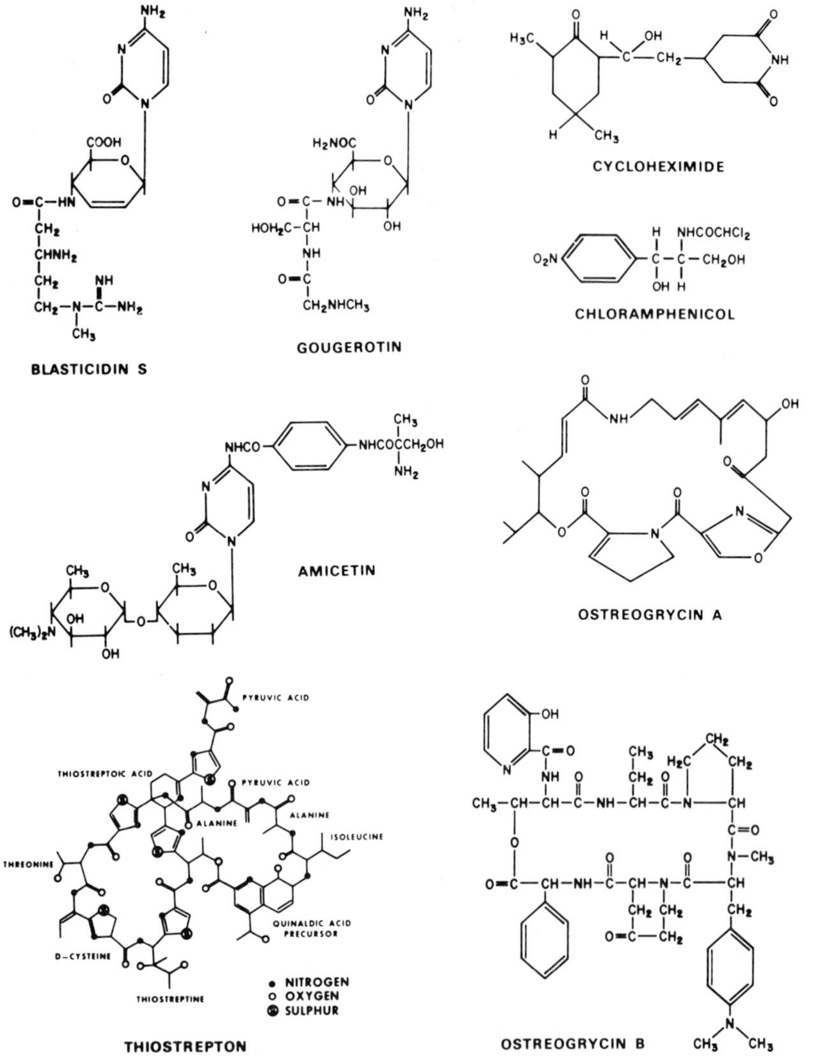

Fig. 6 Chemical structures of amicetin, gougerotin, blasticidin S, chloramphenicol, cyclo-heximide, ostreogrycin A, ostreogrycin B, and thiostrepton. References are given in the text.

Amicetin, bamicetin, gougerotin, and blasticidin S inhibit trans-peptidation reactions (Monro and Vazquez, 1967; Goldberg and Mitsugi, 1967b; Pestka, 1970a,b; Neth *et al.*, 1970; Černá *et al.*, 1971a, 1973). These antibiotics can inhibit binding of C-A-C-C-A(Phe) (Pestka, 1969a; Harris and Pestka, 1973; Černá *et al.*, 1973) and stabilize binding of C-A-

C-C-A(AcLeu) (Celma *et al.*, 1970a; Černá and Rychlík, 1972) to ribosomes from *E. coli*. Gougerotin is probably a mixed type inhibitor of the puromycin reaction (Goldberg and Mitsugi, 1967b; Pestka, 1970a). Hikizimycin, another nucleoside antibiotic, is also an inhibitor of transpeptidation with a similar inhibitory pattern (Uchida and Wolf, 1974).

Although in general these antibiotics are similar in action, they differ specifically in various systems. Thus, amicetin does not inhibit transpeptidation on rat liver polysomes as well as do gougerotin and blasticidin S (Pestka *et al.*, 1972b). Amicetin, however, seems to stabilize polyribosomes of *E. coli* very effectively (Casjens and Morris, 1965; Ennis, 1972a,b). The inhibition of peptidyl-puromycin synthesis by these antibiotics seems to show two phases, a noncompetitive and a competitive phase (Pestka, 1972b; Pestka *et al.*, 1972b). The significance of the two phases is discussed in more detail in Section VIII,C.

Capecchi and Klein (1969) reported that 5×10^{-4} M gougerotin can inhibit peptide chain termination; and Vogel *et al.* (1969) found inhibition of peptidyl transfer and peptide chain termination by amicetin correlated very well.

Kinoshita *et al.* (1970) examined the binding of blasticidin S to *E. coli* ribosomes using equilibrium dialysis. They found a single binding site for each ribosome with an association constant for binding to *E. coli* B ribosomes of 5×10^5 M^{-1}. The binding site was localized to the 50 S ribosomal subunit, and binding was reversible so that the complex with ribosomes could not be isolated by gel filtration. Binding of blasticidin S to ribosomes was inhibited by gougerotin, but was not affected by chloramphenicol, lincomycin, erythromycin, or puromycin. Mikamycin A stimulated the binding of blasticidin S to ribosomes. Similarly, binding of gougerotin was reported to the larger ribosomal subunit from *E. coli* and *Saccharomyces cerevisiae* (Barbacid and Vazquez, 1974b). Binding to *E. coli* ribosomes yielded an association constant of 6.7×10^5 M^{-1}. Its binding to ribosomes was inhibited by other peptidyl transferase inhibitors (actinobolin, blasticidin S, sparsomycin, amicetin, and puromycin). Griseoviridin stimulated its binding to ribosomes.

In general it can be stated that these 4-aminohexose pyrimidine nucleoside antibiotics inhibit transpeptidation in cell-free extracts and intact cells of both pro- and eukaryotes. They inhibit binding of C-A-C-C-A(Phe), the acceptor terminus of AA-tRNA; and stimulate binding of donor peptidyl substrates such as C-A-C-C-A(AcLeu).

C. Chloramphenicol

Chloramphenicol is chiefly a bacteriostatic agent. It is a broad spectrum antibiotic inhibiting the growth of gram-positive and gram-negative bacteria. Eukaryotic cells are generally resistant to the antibiotic although some inhibitions have been reported. The chemical structure of chloramphenicol (Fig. 6) is one of the simplest of the known antibiotics (Rebstock et al., 1949; Controulis et al., 1949; Dunitz, 1952). There are four stereoisomers, only one of which is active as an antibacterial agent, D(−)-threo-chloramphenicol (Maxwell and Nickel, 1954). A variety of electronegative groups can be substituted for the aromatic nitro group without major loss of antimicrobial activity (Hahn, 1967; Shemyakin, 1961). The dichloromethyl group can be replaced by vinyl, allenyl, ethynyl, cyclopropyl, and isopropyl groups with retention of substantial antibacterial activity (Ringrose and Lambert, 1973).

1. Inhibition of Protein Synthesis

Chloramphenicol inhibits protein synthesis (Gale and Folkes, 1953; Hahn and Wisseman, 1951; Wisseman et al., 1954; Marmur and Saz, 1953; Sorm and Grunberger, 1954; Nirenberg and Matthaei, 1961; Nathans and Lipmann, 1961; Rendi and Ochoa, 1962). Chloramphenicol was found to inhibit incorporation of amino acids into protein in cell-free extracts from E. coli (Rendi and Ochoa, 1962; Nirenberg and Matthaei, 1961; Nathans and Lipmann, 1961; Tissières et al., 1960; Lamborg and Zamecnik, 1960), but not in those from eukaryotic cells (Rendi, 1959; von Ehrenstein and Lipmann, 1961; Allen and Schweet, 1962; Borsook et al., 1957; So and Davie, 1963).

The effect of chloramphenicol on polynucleotide-directed protein synthesis was a function of the template used (Speyer et al., 1963; Kúcan and Lipmann, 1964; Vazquez, 1966c). Polypeptide synthesis directed by poly(U) and poly(UA) templates was substantially more resistant to the action of chloramphenicol than polypeptide synthesis directed by poly(A) and poly(UC) templates. It should be noted that protein synthesis directed by natural messengers such as phage f2 RNA is markedly inhibited by chloramphenicol. The apparent paradox may partially be a reflection of the differential solubility and precipitability of the various polypeptides formed under the direction of the different templates. Small phenylalanine peptides of chain length four are substantially precipitable by cold trichloroacetic acid (Pestka et al., 1969). Lysine-containing peptides tend to be soluble in trichloroacetic acid used in the usual methods of assaying protein synthesis in cell-free extracts (Gardner et al., 1962). In the presence of chloramphenicol the distribution of lysine peptides synthesized in the presence of a poly(A) template shifted to smaller nonprecipitable peptides (Julian, 1965, 1966; Teraoka et al., 1969). Similar observations with phenylalanine peptides synthesized with a poly(U) template were reported (Pestka, 1968, 1969a,b, 1970b; Coutsogeorg-opoulos, 1971a,b).

2. Binding of Chloramphenicol to Ribosomes

Radioactive chloramphenicol is bound to ribosomes (Vazquez, 1963, 1964, 1966a; Chang et al., 1969; Wolfe and Hahn, 1965; Das et al., 1966; Lessard and Pestka, 1972b). Binding is relatively unstable so that chloramphenicol can be removed by washing (Vazquez, 1967a; Cannon, 1968). The ability to remove chloramphenicol from ribosomes and intact cells by washing accounts for the reversibility of the inhibitory effects on protein synthesis.

Binding of chloramphenicol to ribosomes requires K^+ or NH_4^+ and was localized to the 50 S subunit (Vazquez, 1966a; Lessard and Pestka, 1972b). In addition, binding of [^{14}C]chloramphenicol to E. coli ribosomes, studied by equilibrium dialysis, revealed two sites for chloramphenicol binding: a high affinity and a low affinity site (Lessard and Pestka, 1972b). The dissociation constant for the major binding of chloramphenicol to ribosomes is approximately $2 \times 10^{-6}\ M$.

Many antibiotics interfere with chloramphenicol binding to ribosomes (Wolfe and Hahn, 1965; Vazquez, 1966a,b; Lessard and Pestka, 1972b; Pestka and LeMahieu, 1974a). Those that do are 50 S inhibitors. These include 4-aminohexose pyrimidine nucleosides (amicetin, gougerotin, blasticidin S), macrolides (angolamycin, carbomycin, erythromycin, lancamycin, spiramycin), streptogramins A, streptogramins B, lincomycin, sparsomycin, and puromycin. Sparsomycin and althiomycin inhibit chloramphenicol binding to polyribosomes, but do not inhibit chloramphenicol binding to monosomes or washed ribosomes (Pestka, 1974a).

Gougerotin in large excess inhibits chloramphenicol binding; blasticidin S and amicetin probably do not inhibit chloramphenicol binding to E. coli ribosomes, but do inhibit the binding to ribosomes from Bacillus stearothermophilus (Chang et al., 1969). Carbomycin and vernamycin A are strong inhibitors of chloramphenicol binding. The inhibition of chloramphenicol binding by puromycin is consistent with the ability of chloramphenicol to bind to the ribosomal acceptor site (Fernandez-Muñoz et al., 1971; Lessard and Pestka, 1972b). The inhibition of aminoacyl-oligonucleotide binding by chloramphenicol is also consistent with inhibition of binding of the acceptor end of aminoacyl-tRNA by this antibiotic. Aminoglycosides, tetracyclines, and other 30 S inhibitors do not inhibit chloramphenicol binding to ribosomes although most 50 S inhibitors, which influence the peptidyl transferase center, do. Thiostrepton is a 50 S inhibitor which does not inhibit chloramphenicol binding to ribosomes

(Chang *et al.*, 1969). Studies of this sort have led to a relatively consistent picture of the subunit localization of various inhibitors. It should be noted that although erythromycin is an excellent inhibitor of chloramphenicol binding, chloramphenicol does not inhibit erythromycin binding (Taubman *et al.*, 1966; K. Tanaka *et al.*, 1966b); and erythromycin bound to washed ribosomes can abolish the inhibitory effect of chloramphenicol on transpeptidation (Teraoka, 1970).

On irradiation of 50 S ribosomal subunits containing bound [^{14}C]-chloramphenicol, radioactivity became covalently attached to the ribosomes and irreversible loss of peptidyl transferase activity occurred (Sonenberg *et al.*, 1974). The label, however, was nonspecifically distributed among most of the subunit proteins. Analogues of chloramphenicol with reactive groups labeled 50 S proteins. Thus, bromamphenicol labeled proteins L2 and L27 (Sonenberg *et al.*, 1973) and iodoamphenicol labeled protein L16 (Bald *et al.*, 1972; Pongs *et al.*, 1973).

3. Inhibition of Peptide Bond Formation

Chloramphenicol inhibits peptide bond formation as measured by various assays. In intact bacteria, chloramphenicol inhibits peptidyl-puromycin formation (Nathans *et al.*, 1962; Nathans, 1964). Traut and Monro (1964) showed that chloramphenicol inhibited polyphenylalanyl-puromycin formation. Synthesis of polylysyl-puromycin (Goldberg and Mitsugi, 1967b; Rychlík, 1966; Coutsogeorgopoulos, 1967b; Gottesman, 1967; Černá *et al.*, 1969), fMet-puromycin (Monro, 1967; Monro and Marcker, 1967), and acetyl-phenylalanyl-puromycin (Weissbach *et al.*, 1968; Pestka, 1970a; Fico and Coutsogeorgopoulos, 1972) was also inhibited by chloramphenicol. Chloramphenicol inhibited the reaction of C-A-A-C-C-A(fMet) with puromycin (Monro and Marcker, 1967; Monro and Vazquez, 1967); 50 S subunits catalyze a similar reaction (Monro, 1967) which was also inhibited by chloramphenicol (Monro, 1969). In studying the effect of chloramphenicol on acetyl-phenylalanyl-puromycin synthesis, Pestka (1970a) showed that chloramphenicol acted as a competitive inhibitor of puromycin. Since puromycin is an analogue of the aminoacyl end of AA-tRNA, then chloramphenicol also may be. A similar suggestion was made by Coutsogeorgopoulos (1966). Additionally, after examination of the puromycin reaction in protoplasts of *Bacillus megaterium*, Cundliffe and McQuillen (1967) concluded that chloramphenicol inhibited peptidyl transfer.

The effects of chloramphenicol on the transfer of nascent chains to puromycin in cell-free extracts containing native polyribosomes was examined to evaluate the effect of the antibiotic in a more nearly physiological cell-free system than ordinarily used with washed ribosomes and synthetic donors (Pestka, 1972b). When the kinetics of chloramphenicol inhibition of peptidyl-puromycin synthesis was evaluated by the double reciprocal plot, a mixed type of competition was apparent. Analysis of the inhibition by a Dixon plot suggested that two modes of inhibition were possible, competitive inhibition at low chloramphenicol concentrations ($K_{i1} = 7 \times 10^{-5}\ M$) and a noncompetitive type of inhibition at high concentrations of the antibiotic ($K_{i2} = 2.2 \times 10^{-3}\ M$). Under different conditions, both types have been reported (Goldberg and Mitsugi, 1967b; Pestka, 1970a; Fernandez-Muñoz and Vazquez, 1973a). Thus, peptidyl-puromycin synthesis with polyribosomes seemed to show two phases of inhibition by chloramphenicol, suggesting that peptidyl-tRNA existed in two states on the ribosome with respect to sensitivity to chloramphenicol and its ability to react with puromycin (Pestka, 1972b). Similar observations were made with the 4-aminohexose nucleoside antibiotics with the use of bacterial and mammalian polyribosomes (Pestka, 1972b; Pestka *et al.,* 1972b). The model of ribosome function shown in Fig. 3 was presented to explain these observations.

The two modes of inhibition by chloramphenicol may mean there are two sites for chloramphenicol action or perhaps two classes of ribosomal states amenable to chloramphenicol inhibition. Equilibrium dialysis, in fact, suggested that there are two ribosomal sites for binding of chloramphenicol to washed ribosomes (Lessard and Pestka, 1972b). In any case, the inhibition of peptidyl-puromycin synthesis on polyribosomes by chloramphenicol appears to be asymmetric with respect to peptidyl-tRNA and puromycin. About half of the peptidyl-tRNA which is available for reaction with puromycin is in a state where chloramphenicol inhibits the reaction competitively with respect to puromycin; the other half of peptidyl-tRNA reacts with puromycin in such a way that inhibition by chloramphenicol is noncompetitive.

In the context of the current donor-acceptor model (Fig. 2), no clearly discernible mechanism is evident to explain asymmetry of inhibition of peptidyl-puromycin synthesis by antibiotics. If peptidyl-tRNA is in the donor site, it must by definition react with puromycin. If it be on the acceptor site, it cannot react with puromycin. The present evidence points to the existence of peptidyl-tRNA in two distinct states, each of which can react with puromycin to form peptidyl-puromycin. This is inconsistent with the current model (Fig. 2), but is consistent with the model illustrated in Fig. 3, where these antibiotics may interfere with one 50 S site better than the other. Other alternative explanations, however, may also be possible. Since the two models (as illustrated in Figs. 2 and 3) make different predictions in many cases, it should be possible to distinguish them experimentally.

4. Effect on Aminoacyl-Oligonucleotide Binding to Ribosomes

Binding of aminoacyl-oligonucleotides such as C-C-A(Phe), C-C-A(Lys), C-C-A(Ser), and C-C-A(Leu) to ribosomes reflects the binding of the aminoacyl termini of the respective aminoacyl-tRNA species to ribosomes. Binding of these aminoacyl-oligonucleotides to ribosomes and 50 S subunits was inhibited by chloramphenicol (Pestka, 1969b,c,d; Pestka *et al.*, 1970; Hishizawa and Pestka, 1971; Lessard and Pestka, 1972b). The biologically active isomer, D(−)-*threo*-chloramphenicol, is most inhibitory; 50% inhibition is obtained at about 3×10^{-5} *M* (Fig. 5). The remaining three enantiomers exhibit relatively little effect on the binding of C-A-C-C-A(Phe) at concentrations as high as 10^{-3} *M*. At 10^{-3} *M*, L(+)-*threo*-chloramphenicol slightly inhibits C-A-C-C-A(Phe) binding. It is interesting to note that higher concentrations of chloramphenicol are required for 50% inhibition of C-A-C-C-A(Phe) binding than are required for 50% inhibition of the lysine, leucine, and serine fragments. This probably reflects the greater affinity of the Phe-oligonucleotides than the other fragments for ribosomes.

These results demonstrated directly that chloramphenicol inhibits the binding of the aminoacyl end of aminoacyl-tRNA to ribosomes. Inhibition of functional attachment of the aminoacyl end of AA-tRNA could account for inhibition of peptide bond formation. Conversely, C-A-C-C-A(Phe) can inhibit binding of chloramphenicol to ribosomes (Yukioka and Morisawa, 1971). Also, Celma *et al.* (1970a) showed that chloramphenicol slightly stimulated binding of C-A-C-C-A(AcLeu) to ribosomes.

5. Effect on the Termination Reactions of Protein Synthesis

Chloramphenicol inhibits termination (Scolnick *et al.*, 1968; Capecchi and Klein, 1969). Vogel *et al.* (1969) correlated inhibition of peptidyl transferase activity with inhibition of termination for chloramphenicol as well as other antibiotics. It should be noted, however, that inhibition of peptidyl transferase model systems does not always correlate with inhibition of peptide bond synthesis in intact cells or on polyribosomes (Pestka, 1972a,b). It is thus probable that model systems for termination will not always reflect the mode of action of an agent in intact cells. For example, although lincomycin is a strong inhibitor of formyl-methionyl-puromycin synthesis and termination reactions in model systems (Vogel *et al.*, 1969), it is unlikely that it inhibits transpeptidation or termination in intact cells (Cundliffe, 1969) or on native polyribosomes (Pestka, 1972b).

6. Protein Synthesis Resistant to Chloramphenicol

Some phage proteins are synthesized in the presence of 30 μg chloramphenicol/ml, a concentration that virtually blocks most bacterial protein synthesis. Gene products of phage S13 and ΦX174 were synthesized in the presence of 30 μg of chloramphenicol/ml (Tessman, 1966; Sinsheimer et al., 1967). These phage proteins, however, are not made in 100–150 μg chloramphenicol/ml (Tessman, 1966; Levine and Sinsheimer, 1968). Differential inhibition of coliphage cistronic products by chloramphenicol was observed for ΦX174 and MS2 phages (Kozak and Nathans, 1972; Van der Mei et al., 1972). Analogously, an E. coli protein required for the initiation of bacterial DNA replication also can be synthesized in the presence of lower concentrations of chloramphenicol (Lark, 1966; Lark and Lark, 1966). The chloramphenicol resistant protein of ΦX174 appears to be in the membrane fraction (Levine and Sinsheimer, 1969). Other bacterial and phage proteins have also been reported to be synthesized in the presence of chloramphenicol (Aronson and Spiegelman, 1961; Sinsheimer et al., 1962). The protein which confers resistance to lysis from without by T4 bacteriophage can be synthesized in the presence of 100 μg chloramphenicol/ml (Peterson et al., 1972). Additionally, an E. coli protein involved in lysogenization was reported to be formed in the presence of 100 μg chloramphenicol/ml (Naha, 1969).

It appears that the proteins synthesized in the presence of chloramphenicol are related to functions associated with the cell membrane. Thus, it is possible that bacterial protein synthesis on polyribosomes in or near the cell membrane may be relatively resistant to chloramphenicol. Alternatively, however, chloramphenicol resistance of synthesis of particular proteins may conceivably be related to the composition of the messenger RNA template, for, as noted above, inhibition of polypeptide synthesis with synthetic templates is dependent on the base composition of those templates. Possibly both considerations may be applicable.

7. Effect of Chloramphenicol on RNA Synthesis

The synthesis of stable RNA species in many strains of E. coli is sharply reduced by deprivation of a required amino acid (Stent and Brenner, 1961; Lazzarini and Winslow, 1970; Primakoff and Berg, 1970). Amino acid starvation causes a rapid accumulation of two unusual guanosine nucleotides in stringent but not in relaxed strains (Cashel and Gallant, 1969; Cashel, 1969). The high intracellular concentrations of these compounds lead to the cessation of RNA accumulation and to other characteristics of the stringent response. Cashel and Kalbacher (1970) have identified these compounds as guanosine tetraphosphate (ppGpp) and guanosine pentaphosphate (pppGpp). Chloramphenicol inhibits the synthesis of these compounds in stringent cells and abolishes the stringent response. Thus, when bacterial cultures are inhibited by chloramphenicol, the cells accumulate considerable quantities of RNA (Pardee et al., 1957; Nomura and Watson, 1959; Kurland and Maaloe, 1962).

8. Bacterial Resistance to Chloramphenicol

There appear to be two major areas for the biochemical basis of chloramphenicol resistance. One is an acquired impermeability to the antibiotic. This has been shown to occur as a result of multistep mutations as well as episomal transfer (Okamoto and Mizuno, 1962, 1964; Hahn, 1967). A second mode of acquired resistance results from the episomal transfer of chloramphenicol acetyltransferase, the enzyme capable of acetylating chloramphenicol to form the inactive diacetyl derivative (Okamoto and Suzuki, 1965; Shaw et al., 1970; Iyobe et al., 1969, 1970).

Ribosomal mutations to chloramphenicol resistance have been reported. Černá and Rychlík (1968) reported the isolation of an E. coli mutant resistant to erythromycin by cultivation of the sensitive E. coli B in enriched medium with increasing concentrations of erythromycin. Ribosomes from this erythromycin-resistant E. coli B, exhibited cross-resistance to chloramphenicol. Neomycin- and kanamycin-resistant mutants of E. coli were reported to have decreased resistance to chloramphenicol (Apirion and Schlessinger, 1968). Also, phenotypic suppression in E. coli by chloramphenicol has been reported (Kirschmann and Davis, 1969). A mutant resistant to chloramphenicol with ribosomal protein alterations was described in B. subtilis by Osawa et al. (1973).

9. Effect of Chloramphenicol on Eukaryotes, Mitochondria, and Chloroplasts

Protein synthesis in eukaryotic organelles is inhibited by chloramphenicol. Chloroplast ribosomes from tobacco leaves (Ellis, 1969) and mitochondrial ribosomes from rat liver (Freeman, 1970b) and yeast (Gordon et al., 1972) show the same stereospecific inhibition of protein synthesis by chloramphenicol as do bacterial ribosomes. Under the same conditions cytoplasmic ribosomes are unaffected by chloramphenicol. Protein synthesis by isolated mitochondria from mammalian liver (Kroon, 1963; Ashwell and Work, 1968; Clark-Walker and Linnane, 1966, 1967), HeLa cells (Galper and Darnell, 1971), and yeast (Wintersberger, 1965; Linnane et al., 1968) is inhibited by chloramphenicol.

It should be noted that chloramphenicol and its isomers and analogues can inhibit mitochondrial respiration (Freeman, 1970a,b; Firkin and Linnane, 1968; Freeman and Haldar, 1968). In addition, a large variety of other effects of chloramphenicol and its isomers have been reported (Malik, 1972; Pestka, 1975b).

D. Sparsomycin

Sparsomycin (Fig. 5) has a broad spectrum of activity and is toxic to both prokaryotic and eukaryotic cells (Slechta, 1966, 1967; Goldberg and Mitsugi, 1966). Sparsomycin is one of the most effective inhibitors of polypeptide synthesis with both 70 S and 80 S ribosomes (Colombo et al., 1966; Goldberg and Mitsugi, 1966, 1967a). It prevents breakdown of polyribosomes. Sparsomycin specifically blocks peptide bond synthesis (Goldberg and Mitsugi, 1967b; Jayaraman and Goldberg, 1968). At low concentrations, sparsomycin inhibits C-A-C-C-A(Phe) binding to ribosomes (Pestka, 1969b,d; Yukioka and Morisawa, 1971; Harris and Pestka, 1973), but at high concentrations sparsomycin stimulates this binding similar to its stimulation of C-A-C-C-A(AcPhe) or U-A-C-C-A(AcLeu) binding to ribosomes (Monro et al., 1969; Hishizawa et al., 1970; Yukioka and Morisawa, 1971; Harris and Pestka, 1973). Sparsomycin also stimulates the formation of inert complexes between the peptidyl end of peptidyl-tRNA and ribosomes (Herner et al., 1969; Jimenez et al., 1970). BrAcPhe-tRNA covalently reacts specifically with proteins L2 and L27 of the 50 S subunit; sparsomycin perturbs the ribosome · BrAcPhe-tRNA complex in such a way so that L2 is less available and L27 more available for the covalent attachment (Oen et al., 1974).

In both pro- and eukaryotes, sparsomycin is a competitive inhibitor of the puromycin reaction whether synthetic model systems or native polyribosomes are used to study transpeptidation (Goldberg and Mitsugi, 1967a; Pestka, 1972a,b). The K_i for its inhibition of peptidyl-puromycin synthesis is about 2×10^{-7} M with bacterial polyribosomes (Pestka, 1972b) and about 1×10^{-6} M with polyribosomes from rat liver (Pestka et al., 1972b). Other partial reactions of protein synthesis which have been studied do not appear to be sensitive to sparsomycin except for release or termination which is generally inhibited by agents interfering with transpeptidation in model systems (Scolnick et al., 1968; Capecchi and Klein, 1969; Vogel et al., 1969). In addition, it appears that sparsomycin interacts best with polyribosomes and relatively poorly with ribosomes, for chloramphenicol binding to polyribosomes was inhibited by sparsomycin whereas chloramphenicol binding to monosomes was not inhibited even at concentrations of sparsomycin 1000-fold higher than that inhibiting chloramphenicol binding to polyribosomes (Pestka, 1974c). The presence of peptidyl-tRNA on ribosomes is probably necessary for optimal binding of sparsomycin to ribosomes. Accordingly, Tada and Trakatellis (1971) found that sparsomycin did not bind to washed ribosomes, but did bind to ribosomes active in protein synthesis.

E. Anisomycin

Anisomycin inhibits protein synthesis in eukaryotes only (Grollman, 1967a; Monro *et al.,* 1969; Bermek and Matthaei, 1970). The effects of anisomycin are reversible and the antibiotic stabilizes the polysome pattern in HeLa cells (Grollman, 1967a). Anisomycin inhibits transpeptidation when synthetic and natural assays are used (Vazquez *et al.,* 1969; Pestka *et al.,* 1972b; Neth *et al.,* 1970; Edens *et al.,* 1975). It binds to the 60 S subunit (Barbacid and Vazquez, 1974). Anisomycin is a competitive inhibitor of puromycin with an inhibition constant, K_i, equal to $1.2 \times 10^{-5} M$ as determined with polyribosomes from rat liver (Pestka *et al.,* 1972b). It also inhibits release of methionine from Met-tRNA^{fMet} in a model system for termination with reticulocyte ribosomes (Beaudet and Caskey, 1971). Furthermore, anisomycin was shown to inhibit the elongation process and stabilize polyribosomes in intact HeLa cells (Tscherne and Pestka, 1975).

F. Macrolides

All the macrolide antibiotics contain a large lactone ring of 12–22 atoms which contains few or no double bonds and no nitrogen atoms (Celmer, 1965; Vazquez, 1967c); in addition, they have one or more sugars attached to the lactone ring (Fig. 5). Because of their diverse structures, it is not suprising they exhibit some diverse actions on protein synthesis. As a group, macrolide antibiotics are active against prokaryotes and often show cross-resistance within the macrolide group. Yeasts and protozoa as well as animal cells are generally resistant. Despite their differences in specific mode of action, the macrolides inhibit protein synthesis (Wolfe and Hahn, 1964; Tanaka and Teraoka, 1968; Ahmed, 1968a; Mao and Wiegand, 1968).

1. Site of Action

One common feature of the macrolides is that they bind to a common region on 50 S subunits (Taubman *et al.,* 1966; K. Tanaka *et al.,* 1966a; Wilhelm *et al.,* 1968; Wilhelm and Corcoran, 1967; Mao, 1967; Mao and Putterman, 1969; Vazquez, 1967b; Fernandez-Muñoz and Vazquez, 1973b; Pestka, 1974b; Pestka and LeMahieu, 1974b; Pestka *et al.,* 1974), but not to 80 S eukaryotic ribosomes (Mao *et al.,* 1970; Teraoka, 1971). This region of the ribosome is in the vicinity of the peptidyl end of the peptidyl-tRNA (Oleinick and Corcoran, 1970; Pestka, 1974a; Tai *et al.,* 1974). By virtue of their interaction with this site, the macrolides interfere with protein synthesis.

Some bacterial mutants resistant to macrolides were shown to have 50 S subunits with an altered affinity for the macrolides, and an altered 50 S protein was identified in the erythromycin-resistant strains (Oleinick and Corcoran, 1969; Wilhelm and Corcoran, 1967; K. Tanaka et al., 1970c, 1971a,b, 1972; Ahmed, 1968b; Teraoka et al., 1970; Takata et al., 1970; Otaka et al., 1970, 1971; Černá et al., 1971b; Krembel and Apirion, 1968; Wittmann et al., 1973). Ribosomes from some erythromycin-resistant mutants of E. coli have an altered 50 S protein L4 and have a low affinity for erythromycin (Otaka et al., 1970, 1971; Dekio et al., 1970). However, ribosomes from other erythromycin-resistant mutants have an altered L22 and seem to bind erythromycin normally (Wittmann et al., 1973) (also see Chapter 3).

Because of the tight binding of erythromycin to ribosomes, a number of derivatives have been synthesized as probes of ribosome structure (Vince et al., 1976a). One such fluorescent derivative of erythromycin was utilized to determine an intraribosomal distance by fluorescent transfer measurements (Langlois et al., 1976). It is hoped that this and similar derivatives can be utilized to triangulate the ribosome through fluorescent transfer measurements to different proteins. In addition, derivatives with chemically reactive as well as photoreactive reagents are being utilized to establish the position of the erythromycin binding site with respect to ribosomal structure.

2. Inhibition of Steps of Protein Synthesis

Some macrolides can block peptide bond formation. Their influence on peptidyl transfer is highly dependent on the substrates, assays, and specific macrolides used (Mao and Robishaw, 1971, 1972; Teraoka and Tanaka, 1971; K. Tanaka et al., 1971c; Černá et al., 1969; Pestka, 1970a,b; Kubota et al., 1972). Very often they stimulate, rather than inhibit, peptidyl transfer in model systems with washed ribosomes and diverse synthetic donors. Although erythromycin does not inhibit peptide bond formation with most simple substrates such as AcPhe-tRNA, fMet-tRNA, C-A-C-C-A(fMet), and U-A-C-C-A(AcLeu), it inhibits polylysyl-puromycin synthesis (Černá et al., 1971b); and it abolishes the inhibition of transpeptidation and acceptor substrate binding by chloramphenicol, other macrolides, and lincomycin (Chang and Weisblum, 1967; Černá and Rychlík, 1972; Vogel et al., 1971). It also abolishes inhibition of termination by these antibiotics (Vogel et al., 1971).

In whole cells, however, it is unlikely that many of these macrolide antibiotics can inhibit transpeptidation on functioning polyribosomes (Cundliffe, 1969; Pestka, 1972a). In fact, studies with native polyribosomes indicate that none of the macrolides can in fact inhibit peptidyl-puromycin synthesis on polyribosomes even though model reactions in synthetic systems with washed ribosomes and synthetic donors are very strongly inhibited by some of these antibiotics. It appears that erythromycin can bind to ribosomes devoid of peptidyl-tRNA, but not to polyribosomes or ribosomes containing peptidyl-tRNA (Oleinick and Corcoran, 1970; Pestka, 1974a).

It is very likely that the specific actions of the various macrolides differ, for their effects differ from each other on intact protoplasts and in model systems. For example, erythromycin and oleandomycin apparently stabilize polyribosomes in protoplasts, in intact cells, or in extracts, whereas spiramycin, carbomycin, niddamycin, leucomycin, and tylosin produce their rapid breakdown (Cundliffe and McQuillen, 1967; Cundliffe, 1969; Ennis, 1972a; S. Pestka, unpublished observations). Erythromycin stimulates AcPhe-puromycin (Pestka, and Brot, 1971; Mao and Robishaw, 1972) and fMet-puromycin synthesis (Kubota et al., 1972) whereas spiramycin, carbomycin, and tylosin inhibited AcPhe-puromycin synthesis (Pestka, 1970a; Kubota et al., 1972; Mao and Robishaw, 1972; Pestka and Brot, 1971). Erythromycin inhibits polylysine synthesis and causes the accumulation of di- and trilysine with ribosomes translating poly(A) (Tanaka and Teraoka, 1966; Mao and Robishaw, 1971). Analogously erythromycin caused the accumulation of di-, tri-, and tetraphenylalanine peptides in ribosomal systems translating poly(U); whereas spiramycin and tylosin produced accumulation of only diphenylalanine, but blocked tri- and tetraphenylalanine synthesis; and niddamycin inhibited the synthesis of all phenylalanine peptides (Pestka, 1970c; Mao and Robishaw, 1971). Thus, the macrolide antibiotics exhibit a spectrum of inhibitions of transpeptidation depending apparently on the type and size of the donor substrate. The fact that erythromycin and other macrolides do not bind well to polyribosomes active in protein synthesis, but do bind to them after peptidyl-tRNA is removed, lends further support to the idea that erythromycin and other macrolides interact with that site of the 50 S subunit near the donor peptidyl-tRNA.

The macrolides do not inhibit initiation complex formation as noted by R17-directed fMet-tRNA binding (Tai et al., 1974), AUG-directed fMet-tRNA binding (Mao and Robishaw, 1971, 1972), or poly(U)-directed AcPhe-tRNA binding (Oleinick and Corcoran, 1970). In addition, erythromycin fails to inhibit AA-tRNA binding to ribosomes (Mao and Robishaw, 1971; Tai et al., 1974). Erythromycin did, however, inhibit EF-G dependent release of tRNA, a step of the translocation process (Igarashi et al., 1969; S. Tanaka et al., 1973). Some additional unusual effects with tylosin, spiramycin, and erythromycin have been noted. Their inhibition of reactions appears to reach a plateau at partial inhibition (Pestka, 1970a;

Celma *et al.*, 1970b; Mao and Robishaw, 1972; Harris and Pestka, 1973). Perhaps this effect is an indication of ribosomal heterogeneity, although other explanations may be invoked.

When cells are exposed to erythromycin, there is a gradual, but small, reduction in polyribosomes and the rate of ribosome movement along mRNA is reduced as judged by the stabilization of polysomes by erythromycin in the presence of rifamycin (Tai *et al.*, 1974). Peptidyl-tRNA, probably unstably attached to ribosomes in the presence of erythromycin, falls off (Menninger, 1974). In any case, although erythromycin permits reinitiation, it does not permit further elongation after reinitiation as noted by continual methionine, but no valine incorporation into polysomes remaining after 20 min exposure of cells to erythromycin (Tai *et al.*, 1974).

With all these observations, it is possible to delimit the mechanism and step inhibited by the macrolide antibiotics. It seems very likely that the macrolide antibiotics interact with a site on the 50 S subunit adjacent to the area occupied by the peptidyl moiety of peptidyl-tRNA. Depending on the size and configuration of the specific macrolide, transpeptidation with a given substrate may or may not be inhibited. Once the size of the peptidyl moiety is large, then all the macrolides inhibit transpeptidation by blocking appropriate apposition of donor and acceptor substrates for covalent linkage. This block is a steric one. The macrolides clearly do not block the catalytic center of the peptidyl transferase. Since they block appropriate positioning of peptidyl-tRNA, consequently they can inhibit some translocation associated events and they destabilize and cause release of peptidyl-tRNA. Since they cannot efficiently interfere with elongation already started, apparently, then, ribosomes complete their nascent chains in the presence of the macrolides. The macrolides then bind to ribosomes or recycling 50 S subunits devoid of peptidyl-tRNA. Initiation complexes form efficiently with these ribosomes carrying a macrolide molecule, but during the early amino acid additions, transpeptidation is blocked and peptidyl-tRNA eventually released. With macrolides (spiramycin, niddamycin, carbomycin) which block transpeptidation of small donors, monosomes form. With those that block only when the donor is large (erythromycin, oleandomycin) polysomes may appear somewhat stabilized. Some of these general ideas have been expressed previously (Pestka, 1971; Mao and Robishaw, 1972; Tai *et al.*, 1974; S. Tanaka *et al.*, 1973).

3. Inducible Resistance to Macrolides

When sensitive cells of *Staphylococcus aureus* 1206 are exposed to a subinhibitory concentration of erythromycin (10^{-8} to 10^{-7} M) in an otherwise complete medium for growth, the cells become resistant to macrolides, lincosamides, and to the streptogramin-type B antibiotics (Weisblum *et al.*, 1971; Shimizu *et al.*, 1971). In strain *S. aureus* 1206, N^6,N^6-dimethyladenine, which is not normally present in 23 S RNA, was found after resistance had been induced by erythromycin (Lai *et al.*, 1973a). Ribosome reconstitution experiments have shown that the antibiotic sensitivities of the reconstituted ribosomes depended on the source of the 23 S rRNA; 23 S rRNA containing N^6,N^6-dimethyladenine conferred upon the reconstituted ribosomes resistance to the above antibiotics (Lai *et al.*, 1973b).

G. Lincosamides and Celestosaminides

The structure of lincomycin is given in Fig. 5 (Hoeksema *et al.*, 1964). The lincosamides include lincomycin and clindamycin. Celesticetin is the only celestosaminide discussed here; it is closely related to lincomycin in structure and action. Most of the discussion here applies to lincomycin, which has been studied more extensively than celesticetin. These antibiotics inhibit protein synthesis in bacterial cells and extracts (Josten and Allen, 1964; Chang *et al.*, 1966; Vazquez, 1966a), but not in mammalian cells or extracts (Baglioni, 1966; Teraoka *et al.*, 1969; Bermek and Matthaei, 1970). Lincomycin binds to 50 S subunits (Chang and Weisblum, 1967). *Escherichia coli* mutants with increased sensitivity to erythromycin and lincomycin contain 50 S subunits with altered proteins (Krembel and Apirion, 1968).

In model systems such as fMet-puromycin formation, lincomycin inhibits peptide bond synthesis (Monro and Vazquez, 1967; Chang, 1968; Vogel *et al.*, 1969). As other inhibitors of fMet-puromycin synthesis, lincomycin inhibits termination as measured by release of formylmethionine from fMet-tRNA (Vogel *et al.*, 1969). It does not inhibit AA-tRNA binding to ribosomes (Igarashi *et al.*, 1969). With washed ribosomes the bound lincomycin interferes with proper positioning of the acceptor (and possibly donor) substrates, namely peptidyl- and AA-tRNA, so that peptide bond synthesis is interdicted and binding of C-A-C-C-A(Phe) to ribosomes is inhibited (Pestka, 1969a,b; Celma *et al.*, 1970a; Černá and Rychlík, 1972; Harris and Pestka, 1973). In the intact cell, however, lincomycin probably does not interfere with transpeptidation or termination on polyribosomes, for it is clear that transpeptidation on native polyribosomes is not inhibited by lincomycin (Pestka, 1972b) and rapid breakdown of polyribosomes occurs in the presence of lincomycin (Cundliffe, 1969). These observations indicate that lincomycin inhibits either initiation or early peptide bond formation, when the peptide chains are small. Lincomycin, however, did not inhibit enzymatic initiation complex formation (Reusser, 1975; P. Sarkar, U. Maitra, and S. Pestka, unpublished observations). Thus, it is most likely that, like the macrolides discussed in the previous section, lincomycin cannot bind functionally to polyribosomes, but only to free ribosomes or 50 S subunits; and it inhibits either the synthesis of the first or early peptide bonds. Although lincomycin and celesticetin can interact with polysomes sufficiently well to inhibit chloramphenicol binding (Pestka, 1974a), apparently this interaction is not sufficient to inhibit transpeptidation (Pestka, 1972b). An additional effect of this group of antibiotics on destabilizing peptidyl-tRNA has not been excluded.

H. Streptogramin Group

Most of the streptogramin antibiotics are produced as a complex of at least two antibiotics which are highly synergistic and fall into two structural classes. Each component alone can inhibit bacterial growth, but together they are markedly synergistic (Vazquez, 1967c; Tanaka, 1975b). Some of the most commonly used antibiotics in this group include the synergistins, mikamycins, vernamycins, PA114's, ostreogrycins, and virginiamycins. The structure of ostreogrycin A (Delpierre *et al.*, 1966; Kingston *et al.*, 1966) and ostreogrycin B (Eastwood *et al.*, 1960) are shown in Fig. 6. In general, these antibiotics inhibit bacterial protein synthesis, but not cytoplasmic protein synthesis in eukaryotes.

1. Streptogramin A Group

Although the streptogramin A antibiotics differ in structure from the lincosamides and the macrolides, these three groups have overlapping ribosomal binding sites and have similar, although not identical, modes of action. The streptogramin A antibiotics bind tightly to 50 S subunits (Ennis, 1971; Cocito and Kaji, 1971). The dissociation constant for the binding of vernamycin A to ribosomes is $2 \times 10^{-8} M$ (Ennis, 1971). The binding of the streptogramin A group is enhanced by the streptogramin B antibiotics.

With washed ribosomes and synthetic donors, streptogramin A compounds inhibit transpeptidation such as fMet- or AcPhe-puromycin synthesis (N. Tanaka *et al.*, 1970b; Pestka, 1970a; Monro and Vazquez, 1967). The streptogramin A antibiotics inhibit termination as measured by fMet release from fMet-tRNA (Vogel *et al.*, 1969). They also inhibit C-A-C-C-A(Phe) (Pestka, 1969a,b; Harris and Pestka, 1973) and C-A-C-C-A(AcLeu) binding to ribosomes (Monro *et al.*, 1970; Celma *et al.*, 1970a) as well as EF-Tu dependent binding of AA-tRNA to ribosomes (Hill, 1969; Modolell *et al.*, 1971; Cocito and Kaji, 1971). However, they do not inhibit protein synthesis readily after protein synthesis has begun (Ennis, 1966a; Yamaguchi and Tanaka, 1967) nor do they inhibit peptidyl-puromycin synthesis on polyribosomes (Pestka, 1972b).

In cells and cell-free extracts, these antibiotics produce a rapid decrease in the number of polyribosomes with concomitant increase in number of free ribosomes (Cundliffe, 1969; Ennis, 1972a; Pestka and Hintikka, 1971). In contrast to this, Cocito (1971) reported that virginiamycin M stabilized polysomes; there may thus be some variety here analogous to the observations with the macrolide antibiotics. In addition, the streptogramin A antibiotics can inhibit chloramphenicol binding to ribosomes, but not to polyribosomes (Pestka, 1974a); and Ennis (1974) has shown that vernamycin A binds to ribosomes, but not to polyribosomes. Presumably, like erythromycin they bind to polyribosomes after peptidyl-tRNA is released (Pestka, 1974a). Thus, since the streptogramin A antibiotics do not seem to interact with polyribosomes, but only with free subunits or monosomes devoid of peptidyl-tRNA, it is likely they inhibit some early events in protein synthesis.

Inhibition of nonenzymatic fMet-tRNA binding to ribosomes by vernamycin A was reported (Ennis and Duffy, 1972). However, neither mikamycin, vernamycin A, PA114A, nor virginiamycin M inhibited initiation factor dependent binding of fMet-tRNA to ribosomes nor concomitant GTP hydrolysis (Cocito *et al.*, 1974; Tanaka, 1975b; P. Sarkar, U. Maitra, and S. Pestka, unpublished studies). It is thus most likely that these streptogramin A antibiotics inhibit protein synthesis by binding to free 50 S subunits or ribosomes devoid of peptidyl-tRNA and subsequently prevent AA-tRNA binding and/or peptide bond formation.

2. Streptogramin B Group

Like the streptogramin A antibiotics, the streptogramin B antibiotics interact with the 50 S subunit (Ennis, 1970, 1974) but not with polyribosomes (Pestka, 1974a). Group B antibiotics enhance the binding of group A antibiotics to ribosomes (Ennis, 1971); and this explains in part the mechanism of synergism.

The streptogramin B antibiotics inhibit protein synthesis (Ennis, 1971; Cocito and Kaji, 1971). They stimulate fMet-puromycin synthesis, but inhibit fMet-Ala-puromycin formation (Kubota *et al.*, 1972). They do not inhibit transpeptidation with isolated 50 S subunits (the fragment reaction), indicating they do not affect the catalytic center of the peptidyl transferase (Monro and Vazquez, 1967). However, PA114B stimulates the binding of C-A-C-C-A(Phe) to ribosomes (Pestka, 1969a,b) and to 50 S subunits (Hishizawa and Pestka, 1971). Viridogrisein (another group B antibiotic) stimulates the binding of C-A-C-C-A(AcLeu) to ribosomes (Celma *et al.*, 1970a). Nevertheless, they do not inhibit transpeptidation on polyribosomes (Pestka, 1972b).

Ennis (1970) reported that synergistin B produced loss of polysomes in protoplasts of *B. subtilis* whereas Cocito (1971) reported their stabilization with virginiamycin S. Polysome breakdown may be due to the inability of the antibiotics to interact with polyribosomes, but only with ribosomes or free subunits devoid of peptidyl-tRNA (Pestka, 1974a). The streptogramin B antibiotics do not inhibit initiation factor dependent binding of fMet-tRNA to ribosomes nor concomitant GTP hydrolysis (P. Sarkar, U. Maitra, and S. Pestka, unpublished studies). Streptogramin B antibiotics may inhibit protein synthesis by fixing peptidyl- or AA-tRNA ends onto 50 S subunits of ribosomes, thus interfering with the necessary coordinated movements of these moieties on the ribosomes. Further studies with this group of antibiotics is necessary to elucidate their mode of inhibition of protein synthesis.

I. Thiostrepton Group

Thiostrepton, siomycin A, thiopeptin, sporangiomycin, and multhiomycin are similar antibiotics (Pestka, 1971; Pestka and Bodley, 1975). The structure of thiostrepton (Anderson *et al.,* 1970) is shown in Fig. 6. These antibiotics inhibit gram-positive but not gram-negative bacteria, although cell-free extracts from both gram-positive and -negative bacteria are sensitive to thiostrepton. Protein synthesis in eukarotic cells and extracts is resistant to inhibition by thiostrepton (K. Tanaka *et al.,* 1970a; Pirali *et al.,* 1972).

1. Inhibition of Protein Synthesis and Site of Action

These antibiotics inhibit protein synthesis (Weisblum and Demohn, 1970a; K. Tanaka *et al.,* 1970a,b; Pirali *et al.,* 1972). They block protein synthesis through specific interaction with the 50 S subunits of ribosomes (Weisblum and Demohn, 1970a; K. Tanaka *et al.,* 1970a,b; Pirali *et al.,* 1972; Pestka, 1970c). The ribosomal site of action was confirmed by studies of protein synthesis in extracts from thiostrepton-sensitive and -resistant strains. The ribosomes from thiostrepton-resistant *B. subtilis* strains are resistant to inhibition by thiostrepton (Goldthwaite and Smith, 1972; Pestka *et al.,* 1976), and the resistance has been localized to the 50 S subunit (Pestka *et al.,* 1976). Furthermore, thiostrepton-resistant ribosomes have a lower affinity for thiostrepton than ribosomes from thiostrepton-sensitive strains (Vince *et al.,* 1976b). A single protein (BS L11) is altered in 50 S subunits from thiostrepton-resistant strains of *B. subtilis* (Wienen *et al.,* 1976).

With the use of radioactive thiostrepton, binding of thiostrepton to ribosomes and 50 S, but not 30 S, subunits was demonstrated (Sopori and Lengyel, 1972; Highland *et al.,* 1975a). Binding of thiostrepton to ribosomes is essentially irreversible, occurs with a ribosome/antibiotic ratio of one, and is inhibited by prior binding of EF-G. Binding of EF-G also prevents ribosomal inactivation by thiostrepton (Highland *et al.*, 1971). Involvement of protein L11 in this binding was noted (Highland *et al.*, 1975b). Selective removal of protein L11 destroys the ability of the ribosome to bind thiostrepton; however, this activity can be restored by rebinding protein L11 to a ribosomal core containing only eight proteins. In addition, after rebinding L11 to this ribosomal core, antibody to protein L11 and no other ribosomal protein inhibits thiostrepton binding. Hence, it was concluded that thiostrepton binds to or in the vicinity of protein L11 on the ribosome.

2. Inhibition of Specific Steps of Protein Synthesis

These antibiotics are strong inhibitors of translocation and AA-tRNA binding through inhibition of ribosomal functions (K. Tanaka *et al.*, 1970a,b; T. Tanaka *et al.*, 1970a,b; Pestka, 1970c; Weisblum and Demohn, 1970b; Pestka and Brot, 1971; Bodley *et al.*, 1970b; Modolell *et al.*, 1971; Pestka and Bodley, 1975). They inhibit ribosomal EF-G-dependent GTP hydrolysis as well as formation of EF-G · GTP complexes with ribosomes. In addition, thiostrepton inhibits GTP hydrolysis by the RNA-protein complex consisting of 5 S RNA and two ribosomal proteins, B-L5 and B-L22 (Horne and Erdmann, 1973). These antibiotics stabilize polysomes (Pestka and Hintikka, 1971). Although they interact tightly and practically irreversibly with ribosomal 50 S subunits, these antibiotics do not inhibit transpeptidation (Pestka, 1970c, 1972b). They do, however, also inhibit EF-Tu-dependent binding of AA-tRNA to ribosomes and GTP hydrolysis (Modolell *et al.*, 1971; Kinoshita *et al.*, 1971; Weissbach *et al.*, 1972; T. Tanaka *et al.*, 1971a,b; Watanabe, 1972; Cundliffe, 1971; Cannon and Burns, 1971).

The recycling of initiation factor IF-2 is blocked by thiostrepton (Mazumder, 1973; Lockwood *et al.*, 1974; Lockwood and Maitra, 1974). Thus, IF-2 and template dependent binding of fMet-tRNA is inhibited by thiostrepton when IF-2 is used catalytically, but is unaffected by thiostrepton when IF-2 is present in stoichiometric amounts. Thiostrepton also inhibits GTP hydrolysis coupled to fMet-tRNA binding. Inhibition of initiation complex formation by thiostrepton (IF-2-dependent fMet-tRNA binding and GTP hydrolysis) occurs only when IF-1 is present to permit catalytic recycling of IF-2 (Sarkar *et al.*, 1974). In the absence of IF-1, binding of fMet-tRNA and GTP hydrolysis occur stoichiometrically with respect to IF-2, and thiostrepton has no effect on either reaction. Thus, thiostrepton prevents the catalytic recycling of IF-2 by blocking the action of IF-1, which is required for the dissociation of IF-2 from the 70 S initiation complex. Nevertheless, thiostrepton does not directly inhibit GTP hydrolysis in initiation. In addition, thiostrepton inhibits the termination reactions catalyzed by the soluble release factors (Brot *et al.*, 1974).

Studies with this group of antibiotics are a good example of the contribution inhibitory agents have made in understanding protein synthesis. The observation that these antibiotics inhibited both EF-G and EF-Tu functions led to the concept that these factors shared ribosomal sites. This suggestion was confirmed directly by the demonstration of competition between EF-Tu and EF-G (Richman and Bodley, 1972; Cabrer *et al.*, 1972; Miller, 1972; Richter, 1972). In addition, the involvement of 50 S proteins L7 and L12 in GTP hydrolysis associated with initiation (IF-2), AA-tRNA binding (EF-Tu), translocation (EF-G), and termination reactions was demonstrated (Lockwood *et al.*, 1974; Weissbach *et al.*, 1972; Brot *et al.*, 1974).

This group of antibiotics has a potent inhibitory effect on several steps of protein synthesis: initiation, enzymatic AA-tRNA binding, translocation, and termination. The fact that in intact cells and appropriate extracts thiostrepton selectively inhibits the ribosome prior to AA-tRNA binding rather than prior to translocation presumably reflects the relative rate and extent of inhibition of these individual reactions or the predominant state of the ribosome during the ribosome epicycle (Pestka, 1971, 1972a). If, as is likely, AA-tRNA binding is the limiting (slow) reaction of the elongation process, then inhibition of this reaction by these antibiotics simply reflects this fact.

J. Micrococcin

Although the complete structure of micrococcin is not known, the structures of several of the component fragments have been determined with the suggestion that the antibiotic has some structural resemblances to the thiostrepton group (Pestka, 1975c). Micrococcin is inhibitory to some prokaryotes (gram-positive bacteria, mycobacteria, actinomycetes), but not to eukaryotes.

Micrococcin inhibits protein synthesis in cell-free extracts from *E. coli* (Pestka and Hintikka, 1971). Although *E. coli* is not sensitive to micrococcin, extracts from this organism are. The antibiotic inhibits translocation through its interaction with the ribosome rather than through interaction with EF-G (Pestka and Brot, 1971). It also inhibits formation of EF-G · ribosome complexes, but not GTP hydrolysis (Cundliffe and Dixon, 1975; Smith *et al.*, 1976; Pestka and Brot, 1971). Analogous to the thiostrepton group of antibiotics, micrococcin inhibits EF-Tu-dependent AA-tRNA binding to ribosomes (Otaka and Kaji, 1974; Cundliffe and Dixon, 1975). It does not inhibit other steps in protein synthesis such as initiation complex formation (Otaka and Kaji, 1974; P. Sarkar, U. Maitra, and S. Pestka, unpublished observations), or transpeptidation (Pestka, 1972b; Pestka and Brot, 1971). Studies with polyribosomes show that micrococcin stabilizes polyribosomes even after treatment with puromycin (Pestka and Hintikka, 1971). In protoplasts AA-tRNA binding seems to be preferentially blocked, although translocation is also inhibited (Cundliffe and Dixon, 1975).

Micrococcin resistant mutants of *B. subtilis* have altered 50 S subunits which are resistant to micrococcin (Smith *et al.*, 1976; Wienen *et al.*, 1976). Thiostrepton-resistant mutants of *B. subtilis* are resistant to micrococcin; but micrococcin-resistant mutants are sensitive to thiostrepton (Goldthwaite and Smith, 1972); and ribosomes from micrococcin-resistant cells bind thiostrepton almost as well as ribosomes from thiostrepton- and micrococcin-sensitive cells (Vince *et al.*, 1976b). In addition, micrococcin slightly inhibits the binding of thiostrepton to ribosomes (Cundliffe and Dixon, 1975). These observations suggest that the thiostrepton and micrococcin binding sites on the 50 S subunit, although not identical, overlap.

Thus, micrococcin appears to interact with the 50 S ribosomal subunit. Through this interaction it inhibits both EF-G- and EF-Tu-dependent translocation and AA-tRNA binding to ribosomes, respectively.

K. Althiomycin

The structure of althiomycin has been determined (Nakamura *et al.*, 1974; Sakakibara *et al.*, 1974; Kirst *et al.*, 1975). It inhibits the growth of gram-positive and -negative bacteria, but does not inhibit eukaryotes (Pestka, 1975a).

Althiomycin inhibits protein synthesis in intact cells and cell-free extracts of *E. coli* (Fujimoto *et al.*, 1970). Protein synthesis in intact reticulocytes and reticulocyte lysates was resistant to inhibition by the antibiotic. The major step inhibited is transpeptidation. AcPhe-puromycin synthesis is partially inhibited by althiomycin (Fujimoto *et al.*, 1970); and this inhibition is dependent on the K^+ and Mg^{2+} concentration (Pestka and Brot, 1971; Pestka, 1975a). However, peptidyl-puromycin or peptidyl-Phe synthesis is strongly inhibited by althiomycin (Pestka, 1972b; Pestka and Brot, 1971). The ability to inhibit transpeptidation better with polyribosomes than with ribosomes is probably due to its ability to bind to polyribosomes, but not to ribosomes devoid of peptidyl-tRNA (Pestka, 1974a). Inhibition of transpeptidation occurs as well in bacterial protoplasts (Burns and Cundliffe, 1973). Other intermediate reactions of protein synthesis which were tested were not inhibited. Thus, AA-tRNA synthesis, Phe-tRNA binding and C-A-C-C-A(Phe) binding to ribosomes, and initiation complex formation were not inhibited by the antibiotic (Fujimoto *et al.,* 1970; Pestka and Brot, 1971; P. Sarkar, U. Maitra, and S. Pestka, unpublished observations).

The target of inhibition of protein synthesis is the ribosome, for althiomycin inhibited oligophenylalanine synthesis in the absence of elongation factors (Pestka and Brot, 1971). Furthermore, since althiomycin inhibited AcPhe-puromycin synthesis with isolated 50 S subunits, it appears the antibiotic may be classified as a 50 S inhibitor (Pestka, 1975a).

L. Bottromycin A$_2$

The bottromycins are peptide antibiotics which inhibit bacterial protein synthesis in intact cells and cell-free extracts (N. Tanaka *et al.,* 1966b). Bottromycin A$_2$ binds to 50 S ribosomal subunits, but does not inhibit chloramphenicol binding to ribosomes (Kinoshita and Tanaka, 1970; Lin and Tanaka, 1968; Chang *et al.,* 1969). It did not significantly inhibit peptide bond formation (Cundliffe and McQuillen, 1967; Lin and Tanaka, 1968; Lin *et al.,* 1968; N. Tanaka *et al.,* 1971b). Only at high concentrations could fMet-puromycin synthesis (N. Tanaka *et al.,* 1971b) and AcPhe-puromycin synthesis (Pestka, 1970a) be inhibited. In all cases, polypeptide synthesis was inhibited by concentrations of drug much less than those required to inhibit peptide bond formation so that it was concluded that bottromycin A$_2$ did not primarily inhibit peptidyl transferase activity. N. Tanaka *et al.* (1971b) suggested that bottromycin A$_2$ inhibits the translocation of peptidyl-tRNA from the A to the P site without preventing release of tRNA from the P site.

M. Glutarimide Antibiotics

The glutarimide antibiotics include cycloheximide, acetoxycycloheximide, streptimidone, the streptovitacins, and other less well-known compounds. Since all the members of this group have not been studied extensively, the conclusions derived from studies with cycloheximide (Fig. 6) may not apply to other members of this group. Cycloheximide has been reported by many groups to inhibit protein synthesis in eukaryotic cells (Kerridge, 1958; Shepherd, 1958; Gorski and Axman, 1964; Traketellis *et al.,* 1965; Young *et al.,* 1963; Bennett *et al.,* 1965) and their cell-free extracts (Colombo *et al.,* 1965, 1966; Ennis, 1966b, 1968; Ennis and Lubin, 1964; Bennett *et al.,* 1964; Siegel and Sisler, 1963; Wettstein *et al.,* 1964). Mitochondrial protein synthesis, however, is insensitive to cycloheximide (Beattie *et al.,* 1967; Borst *et al.,* 1967; Loeb and Hubby, 1968; Sebald *et al.,* 1969). Sensitivity and resistance to cycloheximide are determined by the ribosome (Siegel and Sisler, 1965; Cooper *et al.,* 1967; Haugli *et al.,* 1972). Furthermore, the glutarimide antibiotics interact with 60 S ribosomal subunits (Rao and Grollman, 1967).

It has been suggested that cycloheximide inhibits translocation (Baliga *et al.*, 1969, 1970; McKeehan and Hardesty, 1969; Obrig *et al.*, 1971a). Additionally, in cell-free extracts, it has been reported that the glutaramide antibiotics interfere with chain initiation (Obrig *et al.*, 1971a). However, in reticulocyte lysates, cycloheximide interferes with chain elongation at concentrations where initiation is not blocked (Lodish *et al.*, 1971; Lodish, 1971). Studies with native polyribosomes have indicated that cycloheximide can inhibit transpeptidation (Pestka *et al.*, 1972a,b; Schneider and Maxwell, 1973). However, acetylleucyl-puromycin synthesis with yeast ribosomes and C-A-C-C-A(AcLeu) was only slightly inhibited by cycloheximide (Vazquez *et al.*, 1969). Studies of the polyribosome patterns of intact cells have indicated that cycloheximide inhibits elongation preferentially (Wettstein *et al.*, 1964; Trakatellis *et al.*, 1965; Colombo *et al.*, 1966; Godchaux *et al.*, 1967; Grollman, 1967b; Tscherne and Pestka, 1975). Although elongation is inhibited by cycloheximide, its precise mode of action is still uncertain.

Cycloheximide is commonly used to inhibit protein synthesis in eukaryotic cells particularly since its effect is reversible. However, cycloheximide has been reported to inhibit a number of other metabolic steps such as DNA synthesis (Kerridge, 1958; Shepherd, 1958; Bennett *et al.*, 1964), DNA-dependent RNA polymerase I (Timberlake *et al.*, 1972a,b), arbovirus assembly (Friedman and Grimley, 1969), and tyrosine hydroxylase activity (Flexner *et al.*, 1973). These additional effects should be kept in mind when cycloheximide and other glutarimides are used.

N. Trichothecenes

The 12,13-epoxytrichothecenes, a group of fungal toxins, inhibit protein synthesis in eukaryotic cells (Ueno *et al.*, 1968; Ueno and Fukushima, 1968), but exhibit diverse modes of action. Trichodermin stabilizes polyribosomes and blocks the release of nascent peptides from ribosomes by puromycin (Cundliffe *et al.*, 1974; Tscherne and Pestka, 1975). Trichodermin inhibits transpeptidation in cell-free extracts as measured by the fragment reaction (Carrasco *et al.*, 1973). Trichodermin, trichothecin, trichodermol, and crotocin inhibit peptidyl- puromycin synthesis on polyribosomes (Schindler, 1974). In addition, trichodermin binds to 60 S eukaryotic subunits (Barbacid and Vazquez, 1974a). Trichodermin binding to ribosomes is inhibited by anisomycin, tenuazonic acid, and other trichothecenes (verrucarin A). Trichodermin also inhibits termination reactions in cell-free systems as do virtually all antibiotics which inhibit transpeptidation (Tate and Caskey, 1973). In intact cells, however, elongation is the chief step inhibited (Cundliffe *et al.*, 1974; Wei *et al.*, 1974; Tscherne and Pestka, 1975) although it was claimed to be a specific inhibitor of termination (Stafford and McLaughlin, 1973).

In contrast to the above, nivalenol, T-2 toxin, verrucarin A, and diacetoxyscirpenol produce breakdown of polyribosomes in intact HeLa cells as well as in yeast spheroplasts consistent with inhibition of initiation (Cundliffe *et al.*, 1974; Tscherne and Pestka, 1975). These members of the trichothecene group seem to inhibit initiation. Thus, although the trichothecenes seem to share a similar ribosomal binding site and although all inhibit transpeptidation, some inhibit elongation and others appear to inhibit initiation.

O. Tenuazonic Acid

Tenuazonic acid inhibits 60 S ribosomal subunits and prevents transpeptidation (Neth *et al.*, 1970). Tenuazonic acid, although acting on eukaryotic cells but not prokaryotic systems, preferentially inhibits mammalian systems, for it has a lesser effect on plant ribosomes and little or no significant effect on yeast ribosomes (Carrasco and Vazquez, 1973).

P. Pleuromutilin and Derivatives

The pleuromutilin derivative 14-deoxy-14-[(2-diethylaminoethyl)-mercaptoacetoxy]-mutilin interferes strongly with cell-free protein synthesis in extracts from *E. coli*. It inhibits transpeptidation assayed by fMet-puromycin synthesis from fMet-tRNA (Hodgin and Högenauer, 1974). It also inhibited EF-T-dependent binding of Ala-tRNA^{Ala} to ribosomes containing fMet-tRNA$_f$ and R17 RNA; but did not inhibit EF-T-dependent binding of Phe-tRNA to ribosomes in response to poly(U), or fMet-tRNA$_f$ binding. This derivative competes with chloramphenicol and analogues of the 3'-terminus of tRNA (puromycin, CpA, and CpCpA) for binding to ribosomes (Högenauer, 1975). However, erythromycin and lincomycin do not inhibit its binding to ribosomes. Its binding constant to ribosomes is $1.3 \times 10^{-7} M^{-1}$ and it binds to one site per ribosome.

Q. Abrin and Ricin

Abrin and ricin, toxic plant proteins, inhibit peptide chain elongation in eukaryotic cells and extracts (Olsnes and Pihl, 1972a,b). Prokaryotic extracts remain unaffected. The toxins consist of two peptide chains held together by disulfide bonds. Only one of the chains is capable of inhibiting protein synthesis in cell-free extracts, whereas the entire molecule is necessary for toxic action in intact cells (Olsnes *et al.*, 1973). Their target of action is the eukaryotic ribosome (Olsnes *et al.*, 1973). They appear to exert their effect by enzymatic action. Polysomes and ribosomes isolated from extracts treated with abrin or ricin were less active than the same components from untreated extracts (Olsnes *et al.*, 1973; Onozaki *et al.*, 1975). By hybridizing ribosomal subunits from treated and untreated cells, Onozaki *et al.* (1975) demonstrated that the 60 S subunit was inactivated by ricin. Ribosomes from cells treated with ricin could bind Phe-tRNA and exhibited no loss of peptidyl transferase activity. Definition of the specific alteration should be of interest in elucidating further 60 S ribosomal functions.

IX. Inhibitors of the Nonribosomal Factors

A. Fusidic Acid and Other Steroid Antibiotics

Fusidic acid, helvolic acid, and cephalosporin P_1 are steroid antibiotics active against gram-positive, but not against gram-negative organisms, fungi, or mammalian cells (Tanaka, 1975a). However, cell-free extracts from all these sources are sensitive. A large number of structural derivatives of fusidic acid found active in cell-free extracts have yielded some understanding of the structural requirements for optimal activity (Tanaka *et al.,* 1968, 1969; Bodley and Godtfredsen, 1972). Fusidic acid inhibits translocation in both pro- and eukaryotes through inhibition of EF-G or EF-T2 activity, respectively (N. Tanaka *et al.,* 1968, 1969, 1970a,b; Pestka, 1968, 1969a,b; Brot *et al.,* 1968; Erbe and Leder, 1968; Bodley *et al.,* 1969, 1970a; Malkin and Lipmann, 1969; Waterson *et al.,* 1972). In the presence of fusidic acid, the ribosome · EF-G · GDP or the ribosome · EF-T2 · GDP complex is fixed on the ribosome after hydrolysis of GTP to GDP. Thus, fusidic acid inhibits catalytic GTP hydrolysis by preventing the dissociation of the GDP once the complex has formed so that EF-G and GDP cannot be reutilized. However, one cycle of GTP hydrolysis stoichiometric with fusidic acid and EF-G takes place. Accordingly, radioactive fusidic acid binds to ribosomes in the presence of EF-G (Okura *et al.,* 1970, 1971). Because of the common ribosomal sites of interaction of EF-G and EF-Tu, fusidic acid secondarily inhibits AA-tRNA binding to ribosomes and this appears to be its major effect in intact cells (Cundliffe, 1972) (see also Sections VIII,I and J). Fusidic acid-resistant EF-G has been isolated from mutants resistant to the antibiotic (N. Tanaka *et al.,* 1971a; Bernardi and Leder, 1970).

B. Diphtheria Toxin

Intact diphtheria toxin has a molecular weight of 63,000 and is enzymatically inactive (Collier, 1975). Upon attachment to intact cells, the toxin may be transported into the cytosol. During this process, a small fraction of the toxin is cleaved into fragment A and fragment B of molecular weights 24,000 and 39,000, respectively. The fragment B portion of the molecule is necessary for cellular attachment of the toxin (Ittelson and Gill, 1973; Uchida *et al.,* 1972, 1973a,b,c). Fragment A inactivates the free form of EF-T2 by catalyzing the attachment of the ADPR moiety of intracellular NAD$^+$ to EF-T2. The modified factor can still attach to ribosomes and bind GTP, but is inactive in promoting translocation (Collier and Pappenheimer, 1964a,b; Collier, 1967, 1975; Honjo *et al.*, 1968; Goor and Maxwell, 1969, 1970; Everse *et al.*, 1970). The synthesis of biologically active diphtheria toxin in an *E. coli* lysate has been reported (Lightfoot and Iglewski, 1974).

C. Guanylyl-5'-methylene Diphosphonate (GDPCP)

Guanylyl-5'-methylene diphosphonate (GDPCP) with a methylene bridge connecting the β,γ phosphates is an effective analogue of GTP which does not hydrolyze to GDP (Hershey and Monro, 1966). It serves as a competitive inhibitor of GTP in all intermediate steps of protein synthesis requiring GTP hydrolysis to GDP in pro- and eukaryotes. Thus, GDPCP serves as a competitive inhibitor of GTP in initiation (Ohta *et al.*, 1967), AA-tRNA binding (Haenni and Lucas-Lenard, 1968; Shorey *et al.*, 1969), translocation (Pestka, 1968, 1969a,b), and termination (Beaudet and Caskey, 1971) steps, and accordingly is not a preferential inhibitor of any one step. Although some of the initiation, elongation, and termination factors are involved in GTP hydrolysis, so also is the ribosome so that strictly speaking GDPCP also inhibits ribosomal function as does fusidic acid. Since GDPCP does not ordinarily enter cells, intact cells are generally resistant.

D. L-1-Tosylamido-2-phenylethyl Chloromethyl Ketone (TPCK)

L-1-tosylamido-2-phenylethyl chloromethyl ketone (TPCK) irreversibly inhibits bacterial EF-Tu (Jonák *et al.*, 1971; Sedláček *et al.*, 1971; Richman and Bodley, 1973). TPCK irreversibly destroys the ability of EF-Tu · GTP to form the ternary complex with Phe-tRNA (Richmond and Bodley, 1973). Prior formation of the ternary complex protects EF-Tu from inactivation by TPCK. Neither ribosomes nor EF-G are inhibited at concentrations of TPCK which interfere with EF-Tu. In addition, TPCK has no effect on protein synthesis in mammalian cell-free extracts (Taber *et al.*, 1973; Highland *et al.*, 1974).

Since TPCK is a protease inhibitor, it has been used and reported to inhibit intracellular proteolytic enzymes in intact cells (Summers *et al.*, 1972). It has been shown, however, that TPCK inhibits initiation when intact mammalian cells are exposed to it (Pong *et al.*, 1975). This observation places in question whether TPCK can inhibit intracellular proteolytic cleavage, for the decreased labeling of shorter peptides in the presence of TPCK (Summers *et al.*, 1972) might have been due to inhibition of initiation.

E. Kirromycin (Mocimycin) and X-5108

Kirromycin is identical to the independently isolated mocimycin; and X-5108 was identified as N-methyl mocimycin (Vos and Verwiel, 1973; Maehr *et al.*, 1973). *Bacillus* species are most sensitive to these antibiotics, whereas other bacteria (including *E. coli*) are generally resistant (Wolf and Zähner, 1972; Berger *et al.*, 1973). However, cell-free extracts from *Bacillus brevis* and *E. coli* are sensitive to kirromycin (Wolf *et al.*, 1972). Kirromycin interacts with EF-Tu producing a number of effects (Wolf *et al.*, 1974). Formation of the EF-Tu · GTP complex is stimulated. Peptide bond formation is blocked because EF-Tu · GDP is not released from the ribosome in the presence of kirromycin. The antibiotic enables EF-Tu to catalyze binding of Phe-tRNA^{Phe} to ribosomes containing poly(U) in the absence of GTP. Rather dramatically the interaction of kirromycin with EF-Tu enables EF-Tu to catalyze GTP hydrolysis, a process which ordinarily requires AA-tRNA and ribosomes. These results with kirromycin suggest that the GTPase is a part of EF-Tu rather than the ribosome itself. A previous report (Wolf *et al.*, 1972) that kirromycin is a 30 S inhibitor has not been adequately explained in the subsequent publication (Wolf *et al.*, 1974).

X. Miscellaneous Inhibitors

A. Emetine

Emetine and other ipecac alkaloids are effective inhibitors of protein synthesis in eukaryotes but not in prokaryotes (Grollman, 1966). In HeLa cells the effects of emetine are irreversible, and DNA synthesis is almost as sensitive to emetine as is protein synthesis. Emetine does not prevent transpeptidation (Pestka *et al.*, 1972b) or translocation of peptidyl-tRNA from acceptor to donor sites (Baliga *et al.*, 1970); however, it does stabilize polyribosomes (Grollman, 1968). Emetine inhibits elongation in extracts (Lodish *et al.*, 1971; Lodish, 1971) and in cells (Tscherne and Pestka, 1975); however, the precise mode of action remains speculative. Binding of [³H]emetine to ribosomes or polyribosomes was not detected (Grollman, 1968). Since no definitive evidence defines the site of action of emetine, it cannot be classified at present as an inhibitor of ribosomes or supernatant function. Thus, the mode of action of emetine is enigmatic. It inhibits some elongation step irreversibly, but does not appear to bind to ribosomes or polyribosomes. It inhibits neither transpeptidation nor translocation. Although it was hypothesized that its active structure resembled that of cycloheximide (Grollman, 1966), it differs significantly from the glutarimide antibiotics in its actions. Tubulosine, an indole alkaloid, seems to resemble emetine in its actions (Grollman, 1967b). Virtually all agents which interfere with eukaryotic, but not with prokaryotic, protein synthesis do not block amino acid incorporation in mitochondria. Emetine is an exception; it inhibits protein synthesis in isolated mitochondria (Leitman, 1971; Chakrabarti *et al.*, 1972). However, mitochondrial protein synthesis is resistant to concentrations of emetine which completely block cytoplasmic protein synthesis in intact HeLa cells (Perlman and Penman, 1970). All these paradoxical observations with emetine would serve to elucidate much if they were only understood.

B. Cephalotaxus Alkaloids: Harringtonine

These alkaloids include harringtonine, isoharringtonine, and homoharringtonine (Powell *et al.*, 1970). They inhibit initiation, but not elongation, in cell-free extracts and intact cells of eukaryotes (Grollman and Huang, 1973; Tscherne and Pestka, 1975; Huang, 1975). These compounds rapidly disaggregate polyribosomes into monosomes in intact cells and prevent reformation of polyribosomes. However, they do not inhibit binding of messenger RNA to ribosomes. The specific initiation step inhibited by these alkaloids has not yet been elucidated.

C. Pederine

Pederine inhibits protein synthesis in eukaryotic cells and extracts (Brega *et al.*, 1968; Perani *et al.*, 1968; Jacobs-Lorena *et al.*, 1971). It does not inhibit protein synthesis in prokaryotes (Tiboni *et al.*, 1968) or HeLa cell mitochondria (Brega and Vesco, 1971). Low concentrations of pederine rapidly disaggregate polysomes into monosomes, whereas higher concentrations stabilize polysomes (Jacobs-Lorena *et al.*, 1971). Thus, it was suggested that low concentrations of pederine inhibit initiation. Polysomes incubated in the presence of pederine are irreversibly inactivated, indicating that its site of action is the ribosome. At a concentration which inhibits amino acid incorporation almost completely in a reticulocyte lysate, only partial inhibition of release of nascent peptides from polysomes by puromycin was observed (Jacobs-Lorena *et al.*, 1971). Pederine had little or no effect on AcPhe-puromycin synthesis with human tonsil ribosomes; but it did inhibit translocation as measured by the augmentation of AcPhe-pyromycin synthesis from AcPhe-tRNA in the presence of EF-T2 and GTP (Carrasco and Vazquez, 1972). Similar studies were performed with yeast polysomes (Barbacid *et al.*, 1975). These observations have led to the suggestion that pederine inhibits translocation as well as initiation. However, additional studies are required to define its action.

D. 2-(4-Methyl-2,6-dinitroanilino)-N-methylpropionamide

The herbicide 2-(4-methyl-2,6-dinitroanilino)-N-methylpropionamide (MDMP) inhibits protein synthesis in eukaryotes (Weeks and Baxter, 1972). Although it does not interfere with poly(U)-directed polyphenylalanine synthesis, it effectively blocks amino acid incorporation directed by TMV RNA in wheat extracts. It does not prevent formation of the 40 S initiation complex, TMV RNA · 40 S subunit · Met-tRNA. MDMP does, however, block attachment of the 60 S subunit to the 40S initiation complex. Its effect on intact HeLa cells is consistent with inhibition of initiation (Tscherne and Pestka, 1975).

E. Sodium Fluoride, Potassium Fluoride

Sodium fluoride has been used as a reversible inhibitor of initiation of protein synthesis in rabbit reticulocytes and in reticulocyte lysates (Colombo et al., 1965; Marks et al., 1965; Ravel et al., 1966; Lin et al., 1966). NaF stimulates the conversion of polyribosomes to monomeric ribosomes. In the presence of NaF, intermediate 40 S messenger RNA complexes form on polyribosomes (Hoertz and McCarty, 1971). Hoertz and McCarty (1971) suggested that NaF interferes with attachment of 60 S subunits to the mRNA · 40 S · Met-tRNA complex. Their conclusions were supported by the results of Weeks and Baxter (1972), who demonstrated that KF, like MDMP, inhibited attachment of 60 S subunits to the 40 S initiation complex.

F. Medium Hypertonicity

Hypertonicity has been shown to inhibit reversibly initiation of protein synthesis in intact HeLa cells (Robbins et al., 1970; Wengler and Wengler, 1972; Saborio et al., 1974). Appropriate medium hypertonicity has been shown to produce runoff of ribosomes from polysomes with a block at initiation. After restoration of medium isotonicity, polyribosomes reform. In poliovirus-infected HeLa cells, it has been shown that the reinitiation is synchronous and occurs at the proper site on mRNA (Saborio et al., 1974; Nuss et al., 1975). The runoff of ribosomes from polyribosomes under conditions of hypertonicity and the reformation of polyribosomes after restoration of isotonicity have been used by Tscherne and Pestka (1975) to localize the action of inhibitors of protein synthesis in intact cells.

G. Dimethyl Sulfoxide

Dimethyl sulfoxide (DMSO) inhibits protein synthesis in intact HeLa cells (Saborio and Koch, 1973). At a concentration of 12% DMSO, a complete inhibition of protein synthesis occurs in 3 to 4 min after addition of the drug. The inhibition is accompanied by rapid and complete disassembly of polyribosomes, which can be prevented by incubation of cells with cycloheximide or anisomycin. The inhibition can be reversed by washing the cells free of DMSO.

H. Dextran Sulfate, DEAE-Dextran

Dextran sulfate inhibits mRNA binding to eukaryotic ribosomes (Korner, 1969; Ascione et al., 1972) and produces rapid polyribosome breakdown in tissue culture cells (Ascione et al., 1972). DEAE-dextran has been shown to inhibit initiation in HeLa cells similar to medium hypertonicity and DMSO (Saborio et al., 1975).

I. Double-Stranded RNA

Double-stranded RNA inhibits initiation of globin synthesis in reticulocyte lysates (Hunt and Ehrenfeld, 1971; Ehrenfeld and Hunt, 1971; Hunter et al., 1971; Darnbrough et al., 1972, 1973; Hunt, 1974; Lodish and Desalu, 1973). In the presence of double-stranded RNA, polysomes are converted into monosomes after a few minutes of incubation, and protein synthesis gradually ceases. In addition, Robertson and Mathews (1973) reported inhibition of protein synthesis by double-stranded RNA in cell-free extracts of Krebs II ascites cells. Substantially higher concentrations of double-stranded RNA were required for inhibition of cell-free extracts of ascites cells compared to that required for inhibition of reticulocyte lysates. Double-stranded RNA causes the Met-tRNA$_f$ · 40 S subunit initiation complex to disappear (Darnbrough et al., 1972, 1973). Precisely how this disappearance is brought about is uncertain. In any case, double-stranded RNA is a powerful and specific inhibitor of initiation in these lysates.

J. Polydeoxythymidylate, Polyinosinate, and Other Polymers

Polydeoxythymidylate [poly(dT)] strongly inhibits the translation of reticulocyte globin mRNA in a system derived from Krebs II ascites cells (Miller et al., 1972, 1973). Poly(dT) had little or no effect on chain elongation and appeared to inhibit chain initiation specifically. Poly(I) inhibited chain initiation by preventing attachment of mRNA to Met-tRNA$_f$ · 40 S subunit complexes (Hunt, 1974). Poly(9-vinyladenine) and poly(1-vinyluracil) inhibited Phe-tRNA and Lys-tRNA binding to ribosomes, respectively, apparently by interacting with the complementary polynucleotides, poly(U) and poly(A), used as templates (Reynolds et al., 1972).

K. Proteolytic Enzymes

Protein synthesis is inhibited by more than 50% immediately after addition of 100 μg pronase/ml or 500 μg trypsin/ml to HeLa cells in spinner cultures (Koch, 1974). Polyribosome profiles are not altered, which suggests that elongation not initiation is inhibited. Protein synthesis resumes after removal of proteases by washing cells.

L. Other Organic Solvents and Reagents

N-Butanol has been shown to inhibit globin synthesis in reticulocytes at the level of initiation (Freedman *et al.,* 1967). In addition, studies have shown that polyribosomes are disaggregated rapidly in bacteria exposed to *n*-butanol consistent with inhibition of initiation (Hori *et al.,* 1969). Koch and Koch (1974) have shown that protein synthesis in HeLa cells exposed to 4% ethanol stops immediately. The inhibition by ethanol appears to be consistent with inhibition of elongation rather than initiation. Upon washing HeLa cells free of ethanol, protein synthesis resumes, indicating that the effects at these low ethanol concentrations are reversible.

Bipyridine, an iron-chelating agent, reversibly inhibits initiation of hemoglobin synthesis (Rabinovitz *et al.,* 1969). Chelating of the iron produces an effective heme deficiency that allows formation of an inhibitor of initiation, which prevents formation of the Met-tRNA$_f$ · 40 S subunit complex (Gross and Rabinovitz, 1973; Legon *et al.,* 1973).

REFERENCES

Ahmed, A. (1968a). *Biochim. Biophys. Acta* **166,** 205–217.
Ahmed, A. (1968b). *Biochim. Biophys. Acta* **166,** 218–228.
Allen, D. W., and Zamecnik, P. C. (1962). *Biochim. Biophys. Acta* **55,** 865–874.
Allen, E. H., and Schweet, R. S. (1962). *J. Biol. Chem.* **237,** 760–767.
Anderson, B., Hodgkin, D. C., and Viswamitra, M. A. (1970). *Nature (London)* **225,** 233–235.
Anderson, P., Davies, J., and Davis, B. D. (1967). *J. Mol. Biol.* **29,** 203–208.
Anderson, P., Jr. (1969). *J. Bacteriol.* **100,** 939–947.
Anderson, W. F., Gorini, L., and Breckenridge, L. (1965). *Proc. Natl. Acad. Sci. U.S.A.* **54,** 1076–1083.
Apirion, D., and Saltzman, L. (1974). *Mol. Gen. Genet.* **135,** 11–18.
Apirion, D., and Schlessinger, D. (1968). *J. Bacteriol.* **96,** 768–776.
Apirion, D., Phillips, S. L., and Schlessinger, D. (1969). *Cold Spring Harbor Symp. Quant. Biol.* **34,** 117–128.
Aronson, A. I., and Spiegelman, S. (1961). *Biochim. Biophys. Acta* **53,** 70–84.
Ascione, R., Arlinghaus, R. B., and Van de Woude, G. F. (1972). *In* "Protein Biosynthesis in Nonbacterial Systems" (J. A. Last and A. I. Laskin, eds.), pp. 59–116. Dekker, New York.
Ashwell, M. A., and Work, T. S. (1968). *Biochem. Biophys. Res. Commun.* **32,** 1006–1012.
Ayuso, M., and Goldberg, I. H. (1973). *Biochim. Biophys. Acta* **294,** 118–122.
Baglioni, C. (1966). *Biochim. Biophys. Acta* **129,** 642–645.
Baker, B. R., Schaub, R. E. Joseph, J. P., and Williams, J. H. (1954). *J. Am. Chem. Soc.* **76,** 4044–4045.
Baker, B. R., Schaub, R. E., and Williams, J. H. (1955). *J. Am. Chem. Soc.* **77,** 7–12.
Bald, R., Erdmann, V. A., and Pongs, O. (1972). *FEBS Lett.* **28,** 149–152.
Baliga, B. S., Pronczuk, A. W., and Munro, H. N. (1969). *J. Biol. Chem.* **244,** 4480–4489.
Baliga, B. S. Cohen, S. A., and Munro, H. N. (1970). *FEBS Lett.* **8,** 249–252.
Barbacid, M., and Vazquez, D. (1974a). *Eur. J. Biochem.* **44,** 437–444.
Barbacid, M., and Vazquez, D. (1974b). *Eur. J. Biochem.* **44,** 445–453.

Barbacid, M., and Vazquez, D. (1974c). *J. Mol. Biol.* **84**, 603–623.
Barbacid, M., Fresno, M., and Vazquez, D. (1975). *J. Antibiot.* **28**, 453–462.
Battaner, E., and Vazquez, D. (1971). *Biochim. Biophys. Acta* **254**, 316–330.
Beattie, D. S., Basford, R. E., and Koritz, S. B. (1967). *Biochemistry* **6**, 3099–3106.
Beaudet, A. L., and Caskey, C. T. (1971). *Proc. Natl. Acad. Sci. U.S.A.* **68**, 619–624.
Benne, R., and Voorma, H. O. (1972). *FEBS Lett.* **20**, 347–351.
Bennett, L. L., Jr., Smithers, D., and Ward, C. T. (1964). *Biochim. Biophys. Acta* **87**, 60–69.
Bennett, L. L., Jr., Ward, V. L., and Brockman, R. W. (1965). *Biochim. Biophys. Acta* **103**, 478–485.
Berger, J., Lehr, H. H., Teitel, S., Maehr, H., and Grunberg, E. (1973). *J. Antibiot.* **26**, 15–22.
Bermek, E., and Matthaei, H. (1970). *Hoppe-Seyler's Z. Physiol. Chem.* **351**, 1377–1383.
Bernardi, A., and Leder, P. (1970). *J. Biol. Chem.* **245**, 4263–4268.
Bhuyan, B. K. (1967a). *In* "Antibiotics" (D. Gottlieb and P. D. Shaw, eds.), Vol. 1, pp. 169–172. Springer-Verlag, Berlin and New York.
Bhuyan, B. K. (1967b). *Biochem. Pharmacol.* **16**, 1411–1420.
Birge, E. A., and Kurland, C. G. (1969). *Science* **166**, 1282–1284.
Bissell, D. M. (1965). *J. Mol. Biol.* **14**, 619–622.
Bloch, A., and Coutsogeorgopoulos, C. (1966). *Biochemistry* **5**, 3345–3351.
Bodley, J. W., and Davie, E. W. (1966). *J. Mol. Biol.* **18**, 344–355.
Bodley, J. W., and Godtfredsen, W. O. (1972). *Biochem. Biophys. Res. Commun.* **46**, 871–877.
Bodley, J. W., and Zieve, F. J. (1969). *Biochem. Biophys. Res. Commun.* **36**, 463–468.
Bodley, J. W., Zieve, F. J., Lin, L., and Zieve, S. T. (1969). *Biochem. Biophys. Res. Commun.* **37**, 437–443.
Bodley, J. W., Zieve, F. J., and Lin, L. (1970a). *J. Biol. Chem.* **245**, 5662–5667.
Bodley, J. W., Lin, L., and Highland, J. H. (1970b). *Biochem. Biophys. Res. Commun.* **41**, 1406–1411.
Bollen, A., and Herzog, A. (1970). *FEBS Lett.* **6**, 69–72.
Bollen, A., Davies, J., Ozaki, M., and Mizushima, S. (1969). *Science* **165**, 85–86.
Boon, T. (1971). *Proc. Natl. Acad. Sci. U.S.A.* **68**, 2421–2425.
Boon, T. (1972). *Proc. Natl. Acad. Sci. U.S.A.* **69**, 549–552.
Borsook, H., Fischer, E. H., and Keighley, G. (1957). *J. Biol. Chem.* **229**, 1059–1070.
Borst, P., Kroon, A. M., and Ruttenberg, G. J. C. N. (1967). *In* "Genetic Elements: Properties and Function" (D. Shugar, ed.), pp. 81–116. Academic Press, New York.
Bowman, C. M. (1972). *FEBS Lett.* **22**, 73–75.
Bowman, C. M., Sidikaro, J., and Nomura, M. (1971a). *Nature (London), New Biol.* **234**, 133–137.
Bowman, C. M., Dahlberg, J. E., Ikemura, T., Konisky, J., and Nomura, M. (1971b). *Proc. Natl. Acad. Sci. U.S.A.* **68**, 964–968.
Brega, A., and Vesco, C. (1971). *Nature (London), New Biol.* **229**, 136–139.
Brega, A., Falaschi, A., De Carli, L., and Pavan, M. (1968). *J. Cell Biol.* **36**, 485–496.
Brot, N., Ertel, R., and Weissbach, H. (1968). *Biochem. Biophys. Res. Commun.* **31**, 563–570.
Brot, N., Tate, W. P., Caskey, C. T., and Weissbach, H. (1974). *Proc. Natl. Acad. Sci. U.S.A.* **71**, 89–92.
Burns, D. J. W., and Cundliffe, E. (1973). *Eur. J. Biochem.* **37**, 570–574.
Butterworth, B. E., and Rueckert, R. R. (1972). *J. Virol.* **9**, 823–828.
Cabrer, B., Vazquez, D., and Modolell, J. (1972). *Proc. Natl. Acad. Sci. U.S.A.* **69**, 733–736.

Cannon, M. (1968). *Eur. J. Biochem.* **7**, 137–145.

Cannon, M., and Burns, K. (1971). *FEBS Lett.* **18**, 1–5.

Capecchi, M. R., and Klein, H. A. (1969). *Cold Spring Harbor Symp. Quant. Biol.* **34**, 469–477.

Carrasco, L., and Vazquez, D. (1972). *J. Antibiot.* **25**, 732–737.

Carrasco, L., and Vazquez, D. (1973). *Biochim. Biophys. Acta* **319**, 209–215.

Carrasco. L., Barbacid, M., and Vazquez, D. (1973). *Biochim. Biophys. Acta* **312**, 368–376.

Cashel, M. (1969). *J. Biol. Chem.* **244**, 3133–3141.

Cashel, M., and Gallant, J. (1969). *Nature (London)* **221**, 838–841.

Cashel, M., and Kalbacher, B. (1970). *J. Biol. Chem.* **245**, 2309–2318.

Casjens, S. R., and Morris, A. J. (1965). *Biochim. Biophys. Acta* **108**, 677–686.

Celma, M. L., Monro, R. E., and Vazquez, D. (1970a). *FEBS Lett.* **6**, 273–277.

Celma, M. L., Monro, R. E., and Vazquez, D. (1970b). *FEBS Lett.* **13**, 247–251.

Celmer, W. D. (1965). *Antimicrob. Agents & Chemother.* pp. 144–145.

Černá, J., and Rychlík, I. (1968). *Biochim. Biophys. Acta* **157**, 436–438.

Černá, J., and Rychlík, I. (1972). *Biochim. Biophys. Acta* **287**, 292–300.

Černá, J., Rychlík, I., and Pukrábek, P. (1969). *Eur. J. Biochem.* **9**, 27–35.

Černá, J., Lichtenthaler, F. W., and Rychlík, I. (1971a). *FEBS Lett.* **14**, 45–48.

Černá, J., Jonák, J., and Rychlík, I. (1971b). *Biochim. Biophys. Acta* **240**, 109–121.

Černá, J., Rychlík, I., and Lichtenthaler, F. W. (1973). *FEBS Lett.* **30**, 147–150.

Chakrabarti, S., Dube, D. K., and Roy, S. C. (1972). *Biochem. Pharmacol.* **21**, 2539–2540.

Chang, F. N. (1968). Ph.D. Thesis, University of Wisconsin, Madison.

Chang, F. N., and Flaks, J. G. (1972a). *Antimicrob. Agents & Chemother.* **2**, 294–307.

Chang, F. N., and Flaks, J. G. (1972b). *Antimicrob. Agents & Chemother.* **2**, 308–319.

Chang, F. N., and Weisblum, B. (1967). *Biochemistry* **6**, 836–843.

Chang, F. N., Sih, C. J., and Weisblum, B. (1966). *Proc. Natl. Acad. Sci. U.S.A.* **55**, 431–438.

Chang, F. N., Siddhikol, C., and Weisblum, B. (1969). *Biochim. Biophys. Acta* **186**, 396–398.

Chládek, S., Ringer, D., and Žemlička, J. (1973). *Biochemistry* **12**, 5135–5138.

Chládek, S., Ringer, D., and Quiggle, K. (1974). *Biochemistry* **13**, 2727–2735.

Clark, J. M., Jr., and Chang, A. Y. (1965). *J. Biol. Chem.* **240**, 4734–4739.

Clark, J. M., Jr., and Gunther, J. K. (1963). *Biochim. Biophys. Acta* **76**, 636–638.

Clark-Walker, G. D., and Linnane, A. W. (1966). *Biochem. Biophys. Res. Commun.* **25**, 8–13.

Clark-Walker, G. D., and Linnane, A. W. (1967). *J. Cell Biol.* **34**, 1–14.

Cocito, C. (1971). *Biochimie* **53**, 987–1000.

Cocito, C., and Kaji, A. (1971). *Biochimie* **53**, 763–770.

Cocito, C., Voorma, H. O., and Bosch, L. (1974). *Biochim. Biophys. Acta* **340**, 285–298.

Cohen, L. B., and Goldberg, I. H. (1967). *Biochem. Biophys. Res. Commun.* **29**, 617–622.

Cohen, L. B., Herner, A. E., and Goldberg, I. H. (1969a). *Biochemistry* **8**, 1312–1326.

Cohen, L. B., Goldberg, I. H., and Herner, A. E. (1969b). *Biochemistry* **8**, 1327–1335.

Collier, R. J. (1967). *J. Mol. Biol.* **25**, 83–98.

Collier, R. J. (1975). *Bacteriol. Rev.* **39**, 54–85.

Collier, R. J., and Pappenheimer, A. M., Jr. (1964a). *J. Exp. Med.* **120**, 1007–1018.

Collier, R. J., and Pappenheimer, A. M., Jr. (1964b). *J. Exp. Med.* **120**, 1019–1039.

Colombo, B., Felicetti, L., and Baglioni, C. (1965). *Biochem. Biophys. Res. Commun.* **18**, 389–395.

Colombo, B., Felicetti, L., and Baglioni, C. (1966). *Biochim. Biophys. Acta* **119**, 109–119.

Connamacher, R. H., and Mandel, H. G. (1968). *Biochim. Biophys. Acta* **166**, 475–486.

Controulis, J., Rebstock, M. C., and Crooks, H. M. (1949). *J. Am. Chem. Soc.* **71**, 2463–2468.

Cooper, D., Banthorpe, D. U., and Wilkie, D. (1967). *J. Mol. Biol.* **26**, 347–350.

Coutsogeorgopoulos, C. (1966). *Biochim. Biophys. Acta* **129**, 214–217.

Coutsogeorgopoulos, C. (1967a). *Biochemistry* **6**, 1704–1711.

Coutsogeorgopoulos, C. (1967b). *Biochem. Biophys. Res. Commun.* **27**, 46–52.

Coutsogeorgopoulos, C. (1971). *Biochim. Biophys. Acta* **240**, 137–150; **247**, 632.

Coutsogeorgopoulos, C., Bloch, A., Watanabe, K. A., and Fox, J. J. (1975). *J. Med. Chem.* **18**, 773–776.

Cox, E. C., White, J. R., and Flaks, J. G. (1964). *Proc. Natl. Acad. Sci. U.S.A.* **51**, 703–709.

Cundliffe, E. (1969). *Biochemistry* **8**, 2063–2066.

Cundliffe, E. (1971). *Biochem. Biophys. Res. Commun.* **44**, 912–917.

Cundliffe, E. (1972). *Biochem. Biophys. Res. Commun.* **46**, 1794–1801.

Cundliffe, E., and Dixon, P. D. (1975). *Antimicrob. Agents & Chemother.* **8**, 1–4.

Cundliffe, E., and McQuillen, K. (1967). *J. Mol. Biol.* **30**, 137–146.

Cundliffe, E., Cannon, M., and Davies, J. (1974). *Proc. Natl. Acad. Sci. U.S.A.* **71**, 30–34.

Dahlberg, A. E. (1974). *J. Biol. Chem.* **249**, 7673–7678.

Dahlberg, A. E., and Dahlberg, J. E. (1975). *Proc. Natl. Acad. Sci. U.S.A.* **72**, 2940–2944.

Dahlberg, A. E., Lund, E., and Kjeldgaard, N. O. (1973). *J. Mol. Biol.* **78**, 627–636.

Daluge, S., and Vince, R. (1972). *J. Med. Chem.* **15**, 171–177.

Darnbrough, C., Hunt, T., and Jackson, R. J. (1972). *Biochem. Biophys. Res. Commun.* **48**, 1556–1564.

Darnbrough, C., Legon, S., Hunt, T., and Jackson, R. J. (1973). *J. Mol. Biol.* **76**, 379–403.

Das, H. K., Goldstein, A., and Kanner, L. C. (1966). *Mol. Pharmacol.* **2**, 158–170.

Davey, P. J., Haslam, J. M., and Linnane, A. W. (1970). *Arch. Biochem. Biophys.* **136**, 54–64.

Davies, J. E. (1964). *Proc. Natl. Acad. Sci. U.S.A.* **51**, 659–664.

Davies, J., and Davis, B. D. (1968). *J. Biol. Chem.* **243**, 3312–3316.

Davies, J., Gilbert, W., and Gorini, L. (1964). *Proc. Natl. Acad. Sci. U.S.A.* **51**, 883–890.

Davies, J., Anderson, P., and Davis, B. D. (1965a). *Science* **149**, 1096–1098.

Davies, J., Gorini, L., and Davis, B. D. (1965b). *Mol. Pharmacol.* **1**, 93–106.

Davies, J., Jones, D. S., and Khorana, H. G. (1966). *J. Mol. Biol.* **18**, 48–57.

Day, L. E. (1966). *J. Bacteriol.* **91**, 1917–1923.

De Graaf, F. K., Goedvolk-De Groot, L. E., and Stouthamer, A. H. (1970). *Biochim. Biophys. Acta* **221**, 566–575.

De Graaf, F. K., Planta, R. J., and Stouthamer, A. H. (1971). *Biochim. Biophys. Acta* **240**, 122–136.

Dekio, S., and Takata, R. (1969). *Mol. Gen. Genet.* **105**, 219–224.

Dekio, S., Takata, R., and Osawa, S. (1970). *Mol. Gen. Genet.* **107**, 39–49.

Delpierre, G. R., Eastwood, F. W., Grean, G. E., Kingston, D. G. I., Sarin, P. S., Todd, L., and Williams, D. H. (1966). *Tetrahedron Lett.* **4**, 369–372.

Dixon, H., Kellerman, G. M., and Linnane, A. W. (1972). *Arch. Biochem. Biophys.* **152**, 869–875.

Donaldson, G. R., Atkinson, M. R., and Murray, A. W. (1968). *Biochem. Biophys. Res. Commun.* **31**, 104–109.

Dunitz, J. D. (1952). *J. Am. Chem. Soc.* **74**, 995–999.

Dunlap, B. E., Friderici, K. H., and Rottman, F. (1971). *Biochemistry* **10**, 2581–2587.

Eastwood, F. W., Snell, B. K., and Todd, A. (1960). *J. Chem. Soc.* pp. 2286–2299.

Edens, B., Thompson, H. A., and Moldave, K. (1975). *Biochemistry* **14**, 54–61.

Ehrenfeld, E., and Hunt, T. (1971). *Proc. Natl. Acad. Sci. U.S.A.* **68**, 1075–1078.

Ellis, R. J. (1969). *Science* **163**, 477–478.

Ennis, H. L. (1966a). *Mol. Pharmacol.* **2**, 444–453.

Ennis, H. L. (1966b). *Mol. Pharmacol.* **2**, 543–557.

Ennis, H. L. (1968). *Biochem. Pharmacol.* **17**, 1197–1206.

Ennis, H. L. (1970). *Proc. Int. Congr. Chemother., 6th, 1969* pp. 489–498.

Ennis, H. L. (1971). *Biochemistry* **10**, 1265–1270.

Ennis, H. L. (1972a). *Antimicrob. Agents & Chemother.* **1**, 197–203.

Ennis, H. L. (1972b). *Antimicrob. Agents & Chemother.* **1**, 204–209.

Ennis, H. L. (1974). *Arch. Biochem. Biophys.* **160**, 394–401.

Ennis, H. L., and Duffy, K. E. (1972). *Biochim. Biophys. Acta* **281**, 93–102.

Ennis, H. L., and Lubin, M. (1964). *Science* **146**, 1474–1476.

Erbe, R. W., and Leder, P. (1968). *Biochem. Biophys. Res. Commun.* **31**, 798–803.

Everse, J. D., Gardner, A., Kaplan, N. O., Galisinski, W., and Moldave, K. (1970). *J. Biol. Chem.* **245**, 899–901.

Fahnestock, S., Neumann, H., Shashoua, V., and Rich, A. (1970). *Biochemistry* **9**, 2477–2483.

Felicetti, L., Colombo, B., and Baglioni, C. (1966). *Biochim. Biophys. Acta* **119**, 120–129.

Fernandez-Muñoz, R., and Vazquez, D. (1973a). *Mol. Biol. Rep.* **1**, 75–79.

Fernandez-Muñoz, R., and Vazquez, D. (1973b). *J. Antibiot.* **26**, 107–108.

Fernandez-Muñoz, R., Monro, R. E., Torres-Pinedo, R., and Vazquez, D. (1971). *Eur. J. Biochem.* **23**, 185–193.

Fico, R., and Coutsogeorgopoulos, C. (1972). *Biochem. Biophys. Res. Commun.* **47**, 645–651.

Firkin, F. C., and Linnane, A. W. (1968). *Biochem. Biophys. Res. Commun.* **32**, 398–402.

Fitzgerald, F. J., Bernheim, F., and Fitzgerald, D. B. (1948). *J. Biol. Chem.* **175**, 195–200.

Flaks, J. G., Cox, E. C., Witting, M. L., and White, J. R. (1962). *Biochem. Biophys. Res. Commun.* **7**, 390–393.

Flexner, L. B., Serota, R. G., and Goodman, R. H. (1973). *Proc. Natl. Acad. Sci. U.S.A.* **70**, 354–356.

Franklin, T. J. (1963). *Biochem. J.* **87**, 449–453.

Franklin, T. J. (1964). *Biochem. J.* **90**, 624–628.

Franklin, T. J. (1967). *Biochem. J.* **105**, 371–378.

Freedman, M. L., Hori, M., and Rabinovitz, M. (1967), *Science* **157**, 323–325.

Freeman, K. B. (1970a). *Can. J. Biochem.* **48**, 469–478.

Freeman, K. B. (1970b). *Can. J. Biochem.* **48**, 479–485.

Freeman, K. B., and Haldar, D. (1968). *Can. J. Biochem.* **46**, 1003–1008.

Friedman, R. M., and Grimley, P. M. (1969). *J. Virol.* **4**, 292–299.

Fujimoto, H., Kinoshita, T., Suzuki, H., and Umezawa, H. (1970). *J. Antibiot.* **23**, 271–275.

Funatsu, G., Nierhaus, K., and Wittmann-Liebold, B. (1972). *J. Mol. Biol.* **64**, 201–209.

Gale, E. F., and Folkes, J. P. (1953). *Biochem. J.* **53**, 493–498.

Galper, J. B., and Darnell, J. E. (1971). *J. Mol. Biol.* **57**, 363–367.

Gardner, R. S., Wahba, A. J., Basilio, C., Miller, R. S., Lengyel, P., and Speyer, J. F. (1962). *Proc. Natl. Acad. Sci. U.S.A.* **48**, 2087–2094.

Gavrilova, L. P., and Spirin, A. S. (1971). *FEBS Lett.* **22**, 324–326.

Gavrilova, L. P., and Spirin, A. S. (1974). In " Methods in Enzymology" (L. Grossman and K. Moldave, eds.), Vol. 30, 452–462. Academic Press, New York.

Georgiades, J., Kurylo-Borowska, Z., and Dmochowski, L. (1975). *Abstr., Annu. Meet., Am. Soc. Microbiol.*, 164.

Godchaux, W., III, Adamson, S. D., and Herbert, E. (1967). *J. Mol. Biol.* **27**, 57–72.

Goldberg, I. H., and Mitsugi, K. (1966). *Biochem. Biophys. Res. Commun.* **23**, 453–459.

Goldberg, I. H., and Mitsugi, K. (1967a). *Biochemistry* **6**, 372–382.

Goldberg, I. H., and Mitsugi, K. (1967b). *Biochemistry* **6**, 383–391.
Goldberg, I. H., Stewart, M. L., Ayuso, M., and Kappen, L. (1973). *Fed. Proc., Fed. Am. Soc. Exp. Biol.* **32**, 1688–1697.
Goldthwaite, C., and Smith, I. (1972). *Mol. Gen. Genet.* **114**, 190–204.
Goor, R. S., and Maxwell, E. S. (1969). *Cold Spring Harbor Symp. Quant. Biol.* **34**, 609–610.
Goor, R. S., and Maxwell, E. S. (1970). *J. Biol. Chem.* **245**, 616–623.
Gordon, J. (1969). *J. Biol. Chem.* **244**, 5680–5686.
Gordon, P. A., Lowdon, M. J., and Stewart, P. R. (1972). *J. Bacteriol.* **110**, 504–510.
Gorini, L., and Kataja, E. (1964). *Proc. Natl. Acad. Sci. U.S.A.* **51**, 487–493.
Gorski, J., and Axman, M. C. (1964). *Arch. Biochem. Biophys.* **105**, 517–520.
Gottesman, M. E. (1967). *J. Biol. Chem.* **242**, 5564–5571.
Grant, P., Sánchez, L., and Jiménez, A. (1974). *J. Bacteriol.* **120**, 1308–1314.
Grollman, A. P. (1966). *Proc. Natl. Acad. Sci. U.S.A.* **56**, 1867–1874.
Grollman, A. P. (1967a). *J. Biol. Chem.* **242**, 3226–3233.
Grollman, A. P. (1967b). *Science* **157**, 84–85.
Grollman, A. P. (1968). *J. Biol. Chem.* **243**, 4089–4094.
Grollman, A. P. (1969). *Antimicrob. Agents & Chemother. -1968,* pp. 36–40.
Grollman, A. P., and Huang, M. T. (1973). *Fed. Proc., Fed. Am. Soc. Exp. Biol.* **32**, 1673–1678.
Grollman, A. P., and Stewart, M. L. (1968). *Proc. Natl. Acad. Sci. U.S.A.* **61**, 719–725.
Gross, M., and Rabinovitz, M. (1973). *Biochem. Biophys. Res. Commun.* **50**, 832–838.
Gurgo, C., Apirion, D., and Schlessinger, D. (1969). *J. Mol. Biol.* **45**, 205–220.
Haenni, A.-L., and Lucas-Lenard, J. (1968). *Proc. Natl. Acad. Sci. U.S.A.* **61**, 1363–1369.
Hahn, F. E. (1967). *In* "Antibiotics" (D. Gottlieb and P. D. Shaw, eds.), Vol. 1, pp. 308–330. Springer-Verlag, Berlin and New York.
Hahn, F. E., and Wisseman, C. L. (1951). *Proc. Soc. Exp. Biol. Med.* **76**, 533–535.
Hamada, M., Takeuchi, T., Kondo, S., Ikeda, Y., Naganawa, H., Maeda, K., Okami, Y., and Umezawa, H. (1970). *J. Antibiot.* **23**, 170–171.
Harris, R., and Pestka, S. (1973). *J. Biol. Chem.* **248**, 1168–1174.
Hash, J. H., Wishnick, M., and Miller, P. A. (1964). *J. Biol. Chem.* **239**, 2070–2078.
Haslam, J. M., Davey, P. J., and Linnane, A. W. (1968). *Biochem. Biophys. Res. Commun.* **33**, 368–373.
Hatfield, D. (1966). *Cold Spring Harbor Symp. Quant. Biol.* **34**, 619–622.
Hatfield, D., and Nirenberg, M. (1971). *Biochemistry* **10**, 4318–4323.
Haugli, F. B., Dove, W. F., and Jimenez, A. (1972). *Mol. Gen. Genet.* **118**, 97–107.
Helser, T. L., Davies, J. E., and Dahlberg, J. E. (1971). *Nature (London), New Biol.* **233**, 12–14.
Helser, T. L., Davies, J. E., and Dahlberg, J. E. (1972). *Nature (London), New Biol.* **235**, 6–9.
Herner, A. E., Goldberg, I. H., and Cohen, L. B. (1969). *Biochemistry* **8**, 1335–1344.
Herschman, H. R., and Helinski, D. R. (1967). *J. Biol. Chem.* **242**, 5360–5368.
Hershey, J. W. B., and Monro. R. E. (1966). *J. Mol. Biol.* **18**, 68–76.
Herzog, A. (1964). *Biochem. Biophys. Res. Commun.* **15**, 172–176.
Herzog, A., and Bollen, A. (1971). *FEBS Lett.* **17**, 21–22.
Hettinger, T. P., and Craig, L. C. (1968). *Biochemistry* **7**, 4147–4153.
Hettinger, T. P., and Craig, L. C. (1970). *Biochemistry* **9**, 1224–1232.
Hettinger, T. P., Kurylo-Borowska, Z., and Craig, L. C. (1968). *Biochemistry* **7**, 4153–4160.
Hettinger, T. P., Kurylo-Borowska, Z., and Craig, L. C. (1970). *Ann. N.Y. Acad. Sci.* **171**, 1002–1009.
Hierowski, M. (1965). *Proc. Natl. Acad. Sci. U.S.A.* **53**, 594–599.

Hierowski, M., and Kurylo-Borowska, Z. (1965). *Biochim. Biophys. Acta* **95**, 578–589.

Highland, J. H., Lin, L., and Bodley, J. W. (1971). *Biochemistry* **10**, 4404–4409.

Highland, J. H., Smith, R. L., Burka, E., and Gordon, J. (1974). *FEBS Lett.* **39**, 96–98.

Highland, J. H., Howard, G. A., and Gordon, J. (1975a). *Eur. J. Biochem.* **53**, 313–318.

Highland, J. H., Howard, G. A., Ochsner, E., Stöffler, G., Hasenbank, R., and Gordon, J. (1975b). *J. Biol. Chem* **250**, 1141–1145.

Hill, R. N. (1969). *J. Gen. Microbiol.* **58**, viii.

Hishizawa, T., and Pestka, S. (1971). *Arch. Biochem. Biophys.* **147**, 624–631.

Hishizawa, T., Lessard, J. L., and Pestka, S. (1970). *Proc. Natl. Acad. Sci. U.S.A.* **66**, 523–530.

Hodgin, L. A., and Högenauer, G. (1974). *Eur. J. Biochem.* **47**, 527–533.

Hoeksema, H., Bannister, B., Birkenmeyer, R. D., Kagan, F., Magerlein, B. J., MacKellar, F. A., Schroeder, W., Slomp, G., and Herr, R. R. (1964). *J. Am. Chem. Soc.* **86**, 4223–4224.

Hoerz, W., and McCarty, K. S. (1971). *Biochim. Biophys. Acta* **228**, 526–535.

Högenauer, G. (1975). *Eur. J. Biochem.* **52**, 93–98.

Holland, J. J., Buck, C. A., and McCarthy, B. J. (1966). *Biochemistry* **5**, 358–365.

Honjo, T., Nishizuka, Y., Hayaishi, O., and Kato, I. (1968). *J. Biol. Chem.* **243**, 3553–3555.

Hori, M., and Rabinovitz, M. (1968). *Proc. Natl. Acad. Sci. U.S.A.* **59**, 1349–1355.

Hori, M., Kunimoto, T., and Suzuki, J. (1969). *J. Biochem. (Tokyo)* **66**, 705–709.

Horne, J. R., and Erdmann, V. A. (1973). *Proc. Natl. Acad. Sci. U.S.A.* **70**, 2870–2873.

Huang, M.-T. (1975). *Mol. Pharmacol.* **11**, 511–519.

Huang, M.-T., and Grollman, A. P. (1972a). *Mol. Pharmacol.* **8**, 111–127.

Huang, M.-T., and Grollman, A. P. (1972b). *Mol. Pharmacol.* **8**, 538–550.

Hunt, T. (1974). *Ann. N.Y. Acad. Sci.* **241**, 223–231.

Hunt, T., and Ehrenfeld, E. (1971). *Nature (London), New Biol.* **230**, 91–94.

Hunter, A. R., Hunt, R. T., Jackson, R. J., and Robertson, H. D. (1971). *Haematol. Bluttransfus.* **10**, 133–145.

Igarashi, K., Ishitsuka, H., and Kaji, A. (1969). *Biochem. Biophys. Res. Commun.* **37**, 499–504.

Ittelson, T. R., and Gill, D. M. (1973). *Nature (London)* **242**, 330–332.

Iyobe, S., Hashimoto, H., and Mitsuhashi, S. (1969). *Jpn. J. Microbiol.* **13**, 225–232.

Iyobe, S., Hashimoto, H., and Mitsuhashi, S. (1970). *Jpn. J. Microbiol.* **14**, 463–471.

Izaki, K., and Arima, K. (1963). *Nature (London)* **200**, 384–385.

Jacobs-Lorena, M., Brega, A., and Baglioni, C. (1971). *Biochim. Biophys. Acta* **240**, 263–272.

Jardetzky, O. (1963). *J. Am. Chem. Soc.* **85**, 1823–1825.

Jayaraman, J., and Goldberg, I. H. (1968). *Biochemistry* **7**, 418–421.

Jimenez, A., Monro, R. E., and Vazquez, D. (1970). *FEBS Lett.* **7**, 103–108.

Johnson, L. F., and Bhacca, N. S. (1963). *J. Am. Chem. Soc.* **85**, 3700–3701.

Jonák, J., and Rychlík, I. (1970). *Biochem. Biophys. Acta* **199**, 421–434.

Jonák, J., Sedláček, J., and Rychlík, I. (1971). *FEBS Lett.* **18**, 6–8.

Josten, J. J., and Allen, P. M. (1964). *Biochem. Biophys. Res. Comm.* **14**, 241–244.

Julian, G. R. (1965). *J. Mol. Biol.* **12**, 9–16.

Julian, G. R. (1966). *Antimicrob. Agents & Chemother. -1965,* pp. 992–1000.

Kaempfer, R., and Meselson, M. (1969). *Cold Spring Harbor Symp. Quant. Biol.* **34**, 209–220.

Kaji, H. (1967). *Biochim. Biophys. Acta* **134**, 134–142.

Kaji, H., and Kaji, A. (1965). *Proc. Natl. Acad. Sci. U.S.A.* **54**, 213–219.

Kaji, H., and Tanaka, Y. (1968). *J. Mol. Biol.* **32**, 221–230.

Kaji, H., Suzuka, I., and Kaji, A. (1966). *J. Biol. Chem.* **241**, 1251–1256.

Kappen, L. S., Suzuki, H., and Goldberg, I. H. (1973). *Proc. Natl. Acad. Sci. U.S.A.* **70**, 22–26.

Kerridge, D. (1958). *J. Gen. Microbiol.* **19**, 497–506.

Kersten, H., Kersten, W., Emmerich, B., and Chandra, P. (1967). *Hoppe-Seyler's Z. Physiol. Chem.* **348**, 1424–1430.

Kersten, H., Chandra, P., Tanck, W., Wiedemhöver, W., and Kersten, W. (1968). *Hoppe-Seyler's Z. Physiol. Chem.* **349**, 659–663.

Kingston, D. G. I., Todd, A., and Williams, D. H. (1966). *J. Chem. Soc. C* pp. 1669–1676.

Kinoshita, T., and Tanaka, N. (1970). *J. Antibiot.* **23**, 311–312.

Kinoshita, T., Tanaka, N., and Umezawa, H. (1970). *J. Antibiot.* **23**, 288–290.

Kinoshita, T., Liou, Y.-F., and Tanaka, N. (1971). *Biochem. Biophys. Res. Commun.* **44**, 859–863.

Kirschmann, C., and Davis, B. D. (1969). *J. Bacteriol.* **98**, 152–159.

Kirst, H. A., Szymanski, E. F., Dorman, D. E., Occolowitz, J. L., Jones, N. D., Chaney, M. O., Hamill, R. L., and Hoehn, M. M. (1975). *J. Antibiot.* **28**, 286–291.

Koch, F., and Koch, G. (1974). *Res. Commun. Chem. Pathol. Pharmacol.* **9**, 291–298.

Koch, G. (1974). *Biochem. Biophys. Res. Commun.* **61**, 817–824.

Kogut, M., and Harris, M. (1969). *Eur. J. Biochem.* **9**, 42–49.

Kogut, M., and Prizant, E. (1972). *J. Gen. Microbiol.* **70**, 567–579.

Kondo, S., Shibahara, S., Takahashi, S., Maeda, K., and Umezawa, H. (1971). *J. Am. Chem. Soc.* **93**, 6305–6306.

Konisky, J., and Nomura, M. (1967). *J. Mol. Biol.* **26**, 181–195.

Korner, A. (1969). *Biochim. Biophys. Acta* **174**, 351–358.

Kozak, M., and Nathans, D. (1972). *J. Mol. Biol.* **70**, 41–55.

Krembel, J., and Apirion, D. (1968). *J. Mol. Biol.* **33**, 363–368.

Kroon, A. M. (1963). *Biochim. Biophys. Acta* **72**, 391–402.

Kubota, K., Okuyama, A., and Tanaka, N. (1972). *Biochem. Biophys. Res. Commun.* **47**, 1196–1202.

Kućan, Z., and Lipmann, F. (1964). *J. Biol. Chem.* **239**, 516–520.

Kurland, C. (1970). *In* "Protein Synthesis: A Series of Advances" (E. McConkey, ed.), Vol. I, pp. 179–228. Marcel Dekker, New York.

Kurland, C. G., and Maaloe, O. (1962). *J. Mol. Biol.* **4**, 193–210.

Kurylo-Borowska, Z. (1962). *Biochim. Biophys. Acta* **61**, 897–902.

Kurylo-Borowska, Z. (1964). *Biochim. Biophys. Acta* **87**, 305–313.

Kurylo-Borowska, Z., and Hierowski, M. (1965). *Biochim. Biophys. Acta* **95**, 590–597.

Lai, C.-J., Dahlberg, J. E., and Weisblum, B. (1973a). *Biochemistry* **12**, 457–460.

Lai, C.-J., Weisblum, B., Fahnestock, S. R., and Nomura, M. (1973b). *J. Mol. Biol.* **74**, 67–72.

Lamborg, M. R., and Zamecnik, P. C. (1960). *Biochim. Biophys. Acta* **42**, 206–211.

Langlois, R., Lee, C. C., Cantor, C. R., Vince, R., and Pestka, S. (1976). *J. Mol. Biol.* **106**, 297–313.

Lark, K. G. (1966). *Bacteriol. Rev.* **30**, 3–32.

Lark, K. G., and Lark C. (1966). *J. Mol. Biol.* **20**, 9–19.

Laskin, A. I., and Chan, W. M. (1964). *Biochem. Biophys. Res. Commun.* **14**, 137–142.

Lazzarini, R. A., and Winslow, R. M. (1970). *Cold Spring Harbor Symp. Quant. Biol.* **35**, 383–390.

Lebleu, B., Marbaix, G., Werenne, J., Burny, A., and Huez, G. (1970). *Biochem. Biophys. Res. Commun.* **40**, 731–739.

Legon, S., Jackson, R. J., and Hunt, T. (1973). *Nature (London), New Biol.* **241**, 150–152.

Leitman, P. S. (1971). *Mol. Pharmacol.* **7**, 122–128.

Leive, L. (1965). *Biochem. Biophys. Res. Commun.* **20**, 321–327.

Lelong, J. C., Cousin, M. A., Gros, D., Grunberg-Manago, M., and Gros, F. (1971). *Biochem. Biophys. Res. Commun.* **42**, 530–537.

Lelong, J. C., Cousin, M. A., and Gros, F. (1972). *Eur. J. Biochem.* **27**, 174–180.

Leon, S. A., and Brock, T. D. (1967). *J. Mol. Biol.* **24**, 391–404.

Lessard, J. L., and Pestka, S. (1972a). *J. Biol. Chem.* **247**, 6901–6908.

Lessard, J. L., and Pestka, S. (1972b). *J. Biol. Chem.* **247**, 6909–6912.

Levin, D. H., Kyner, D., and Acs, G. (1973). *J. Biol. Chem.* **248**, 6416–6425.

Levine, A. J., and Sinsheimer, R. L. (1968). *J. Mol. Biol.* **32**, 567–578.

Levine, A. J., and Sinsheimer, R. L. (1969). *J. Mol. Biol.* **39**, 655–668.

Lichtenthaler, F. W., and Trummlitz, G. (1974). *FEBS Lett.* **38**, 237–242.

Lightfoot, H. N., and Iglewski, B. H. (1974). *Biochem. Biophys. Res. Commun.* **56**, 351–357.

Lin, S.-Y., Mosteller, R. D., and Hardesty, B. (1966). *J. Mol. Biol.* **21**, 51–69.

Lin, Y.-C., and Tanaka, N. (1968). *J. Biochem. (Tokyo)* **63**, 1–7.

Lin, Y.-C., Kinoshita, T., and Tanaka, N. (1968). *J. Antibiot.* **21**, 471–476.

Lindenbaum, A., and Schubert, J. (1956). *J. Phys. Chem.* **60**, 1663–1665.

Linnane, A. W., Lamb, A. J., Christodoulou, C., and Lukins, H. B. (1968). *Proc. Natl. Acad. Sci. U.S.A.* **59**, 1288–1293.

Lockwood, A. H., and Maitra, U. (1974). *J. Biol. Chem.* **249**, 346–352.

Lockwood, A. H., Sarkar, P., Maitra, U., Brot, N., and Weissbach, H. (1974) *J. Biol. Chem.* **249**, 5831–5834.

Lodish, H. F. (1971). *J. Biol. Chem.* **246**, 7131–7138.

Lodish, H. F., and Desalu, O. (1973). *J. Biol. Chem.* **248**, 3520–3527.

Lodish, H. F., Housman, D., and Jacobsen, M. (1971). *Biochemistry* **10**, 2348–2356.

Loeb, J. N., and Hubby, B. G. (1968). *Biochim. Biophys. Acta* **166**, 745–748.

Lucas-Lenard, J., and Haenni, A.-L. (1968). *Proc. Natl. Acad. Sci. U.S.A.* **59**, 554–560.

Luzzatto, L., Apirion, D., and Schlessinger, D. (1968). *Proc. Natl. Acad. Sci. U.S.A.* **60**, 873–880.

Luzzatto, L., Apirion, D., and Schlessinger, D. (1969). *J. Mol. Biol.* **42**, 315–335.

McCarthy, B. J., and Holland, J. J. (1965). *Proc. Natl. Acad. Sci. U.S.A.* **54**, 880–886.

Macdonald, J. S., and Goldberg, I. H. (1970). *Biochem. Biophys. Res. Commun.* **41**, 1–8.

McKeehan, W., and Hardesty, B. (1969). *Biochem. Biophys. Res. Commun.* **36**, 625–630.

Maehr, H., Leach, M., Yarmchuk, L., and Stempel, A. (1973). *J. Am. Chem. Soc.* **95**, 8449–8450.

Mager, J., Benedict, M., and Artman, M. (1962). *Biochim. Biophys. Acta* **62**, 202–204.

Malik, V. S. (1972). *Adv. Appl. Microbiol.* **15**, 297–336.

Malkin, M., and Lipmann, F. (1969). *Science* **164**, 71–72.

Mao, J. C.-H. (1967). *Biochem. Pharmacol.* **16**, 2441–2443.

Mao, J. C.-H., and Putterman, M. (1969). *J. Mol. Biol.* **44**, 347–361.

Mao, J. C.-H., and Robishaw, E. E. (1971). *Biochemistry* **10**, 2054–2061.

Mao, J. C.-H., and Robishaw, E. E. (1972). *Biochemistry* **11**, 4864–4872.

Mao, J. C.-H., and Wiegand, R. G. (1968). *Biochim. Biophys. Acta* **157**, 404–413.

Mao, J. C.-H., Putterman, M., and Wiegand, R. G. (1970). *Biochem. Pharmacol.* **19**, 391–399.

Marcus, A., Bewley, J. D., and Weeks, D. P. (1970). *Science* **167**, 1735–1736.

Marks, P. A., Burka, E. R., Conconi, F. M., Perl, W., and Rifkind, R. A. (1965). *Proc. Natl. Acad. Sci. U.S.A.* **53**, 1437–1443.

Marmur, J., and Saz, A. K. (1953). *Antibiot. Chemother. (Washington, D.C.)* **3**, 613–617.

Masukawa, H. (1969). *J. Antibiot.* **22**, 612–622.

Masukawa, H., and Tanaka, N. (1967). *J. Biochem. (Tokyo)* **62**, 202–209.

Masukawa, H., Tanaka, N., and Umezawa, H. (1968). *J. Antibiot.* **21**, 517–518.

Maxwell, I. H. (1968). *Mol. Pharmacol.* **4**, 25–37.

Maxwell, R. E., and Nickel, V. S. (1954). *Antibiot. Chemother. (Washington, D.C.)* **4**, 289–295.

Mazumder, R. (1973). *Proc. Natl. Acad. Sci. U.S.A.* **70**, 1939–1942.

Menninger, J. R. (1974). *Fed. Proc., Fed. Am. Soc. Exp. Biol.* **33**, 1335 (Abstr. 631).

Miller, D. L. (1972). *Proc. Natl. Acad. Sci. U.S.A.* **69**, 752–755.

Miller, J. V. Jr., Thompson, E. B., Kuff, E. L., and Wilson, S. H. (1972). *Biochem. Biophys. Res. Commun.* **48**, 1280–1286.

Miller, J. V., Jr., Wilson, S. H., Kuff, E. L., and Thompson, E. B. (1973). *Biochim. Biophys. Acta* **294**, 507–516.

Mizuno, S., Nitta, K., and Umezawa, H. (1970a). *J. Antibiot.* **23**, 581–588.

Mizuno, S., Nitta, K., and Umezawa, H. (1970b). *J. Antibiot.* **23**, 589–594.

Modolell, J., and Davis, B. D. (1968). *Proc. Natl. Acad. Sci. U.S.A.* **61**, 1279–1286.

Modolell, J., and Davis, B. D. (1969a). *Cold Spring Harbor Symp. Quant. Biol.* **34**, 113–116.

Modolell, J., and Davis, B. D. (1969b). *Nature (London)* **224**, 345–348.

Modollel, J., and Davis, B. D. (1970). *Proc. Natl. Acad. Sci. U.S.A.* **67**, 1148–1155.

Modolell, J., Cabrer, B., Parmeggiani, A., and Vazquez, D. (1971). *Proc. Natl. Acad. Sci. U.S.A.* **68**, 1796–1800.

Monro, R. E. (1967). *J. Mol. Biol.* **26**, 147–151.

Monro, R. E. (1969). *Cold Spring Harbor Symp. Quant. Biol.* **34**, 357–366.

Monro, R. E., and Marcker, K. A. (1967). *J. Mol. Biol.* **25**, 347–350.

Monro, R. E., and Vazquez, D. (1967). *J. Mol. Biol.* **28**, 161–165.

Monro, R. E., Celma, M. L., and Vazquez, D. (1969). *Nature (London)* **222**, 356–358.

Monro, R. E., Fernandez-Muñoz, R., Celma, M. L., Jimenez, A., Battaner, E., and Vazquez, D. (1970). *Proc. Int. Congr. Chemother., 6th. 1969* pp. 473–481.

Morris, A. J., and Schweet, R. S. (1961). *Biochim. Biophys. Acta* **47**, 415–416.

Naha, P. M. (1969). *Biochem. Biophys. Res. Commun.* **35**, 920–925.

Nakamura, H., Iitaka, Y., Sakakibara, H., and Umezawa, H. (1974). *J. Antibiot.* **27**, 894–896.

Nashimoto, H., and Nomura, M. (1970). *Proc. Natl. Acad. Sci. U.S.A.* **67**, 1440–1447.

Nashimoto, H., Held, W., Kaltschmidt, E., and Nomura, M. (1971). *J. Mol. Biol.* **62**, 121–138.

Nathans, D. (1964). *Proc. Natl. Acad. Sci. U.S.A.* **51**, 585–592.

Nathans, D. (1967). *In* "Antibiotics" (D. Gottlieb and P. D. Shaw, eds.), Vol. I., pp. 259–277. Springer-Verlag, Berlin and New York.

Nathans, D., and Lipmann, F. (1961). *Proc. Natl. Acad. Sci. U.S.A.* **47**, 497–504.

Nathans, D., and Neidle, A. (1963). *Nature (London)* **197**, 1076–1077.

Nathans, D., von Ehrenstein, G., Monro, R., and Lipmann, F. (1972). *Fed. Proc., Fed. Am. Soc. Exp. Biol.* **21**, 127–135.

Neth, R., Monro, R. E., Heller, C., Battaner, E., and Vazquez, D. (1970). *FEBS Lett.* **6**, 198–202.

Nikolov, T. K., and Ilkov, A. I. (1961). *Abstr. Int. Congr. Biochem., 5th 1961*, p. 44, Sect. 2 (Abstr. 2,114).

Nirenberg, M. W., and Matthaei, J. H. (1961). *Proc. Natl. Acad. Sci. U.S.A.* **47**, 1588–1602.

Nomura, M. (1963). *Cold Spring Harbor Symp. Quant. Biol.* **28**, 315–324.

Nomura, M., and Watson, J. D. (1959). *J. Mol. Biol.* **1**, 204–217.

Nomura, M., Mizushima, S., Ozaki, M., Traub, P., and Lowry, C. V. (1969). *Cold Spring Harbor Symp. Quant. Biol.* **34**, 49–61.

Nuss, D. L., Oppermann, H., and Koch, G. (1975). *Proc. Natl. Acad. Sci. U.S.A.* **72,** 1258–1262.

Obrig, T. G., Culp, W. J., McKeehan, W. L., and Hardesty, B. (1971a). *J. Biol. Chem.* **246,** 174–181.

Obrig. T., Irvin, J., Culp, W. J., and Hardesty, B. (1971b). *Eur. J. Biochem.* **21,** 31–41.

Oen, H., Pellegrini, M., and Cantor, C. R. (1974). *FEBS Lett.* **45,** 218–222.

Ohta, T., Sarkar, S., and Thach, R. E. (1967). *Proc. Natl. Acad. Sci. U.S.A.* **58,** 1638–1644.

Okamoto, S., and Mizuno, D. (1962). *Nature (London)* **195,** 1022–1023.

Okamoto, S., and Mizuno, D. (1964). *J. Gen. Microbiol.* **35,** 125–133.

Okamoto, S., and Suzuki, Y. (1965). *Nature (London)* **208,** 1301–1303.

Okura, A., Kinoshita, T., and Tanaka, N. (1970). *Biochem. Biophys. Res. Commun.* **41,** 1545–1550.

Okura, A., Kinoshita, T., and Tanaka, N. (1971). *J. Antibiot.* **24,** 655–661.

Okuyama, A., Machiyama, N., Kinoshita, T., and Tanaka, N. (1971). *Biochem. Biophys. Res. Commun.* **43,** 196–199.

Okuyama, A., Watanabe, T., and Tanaka, N. (1972). *J. Antibiot.* **25,** 212–218.

Okuyama, A., Yoshikawa, M., and Tanaka, N. (1974). *Biochem. Biophys. Res. Commun.* **60,** 1163–1169.

Old, D., and Gorini, L. (1965). *Science* **150,** 1290–1292.

Oleinick, N. L., and Corcoran, J. W. (1969). *J. Biol. Chem.* **244,** 727–735.

Oleinick, N. L., and Corcoran, J. W. (1970). *Proc. Int. Congr. Chemother., 6th, 1969,* 202–208.

Olsnes, S., and Pihl, A. (1972a). *FEBS Lett.* **20,** 327–329.

Olsnes, S., and Pihl, A. (1972b). *Nature (London)* **238,** 459–461.

Olsnes, S., Heiberg, R., and Pihl, A. (1973). *Mol. Biol. Rep.* **1,** 15–20.

Onozaki, K., Hayatsu, H., and Ukita, T. (1975). *Biochim. Biophys. Acta* **407,** 99–107.

Osawa, S., Takata, R., Tanaka, K., and Tamaki, M. (1973). *Mol. Gen. Genet.* **127,** 163–173.

Otaka, E., Teraoka, H., Tamaki, M., Tanaka, K., and Osawa, S. (1970). *J. Mol. Biol.* **48,** 499–510.

Otaka, E., Itoh, T., Osawa, S., Tanaka, K., and Tamaki, M. (1971). *Mol. Gen. Genet.* **114,** 14–22.

Otaka, T., and Kaji, A. (1974). *Eur. J. Biochem.* **50,** 101–106.

Ozaki, M., Mizushima, S., and Nomura, M. (1969). *Nature (London)* **222,** 333–339.

Pardee, A. B., Paigen, K., and Prestidge, L. S. (1957). *Biochim. Biophys. Acta* **23,** 162–173.

Perani, A., Parisi, B., De Carli, L., and Ciferri, O. (1968). *Biochim. Biophys. Acta* **161,** 223–231.

Perlman, S., and Penman, S. (1970). *Biochem. Biophys. Res. Commun.* **40,** 941–948.

Pestka, S. (1966). *J. Biol. Chem.* **241,** 367–372.

Pestka, S. (1967). *Bull. N.Y. Acad. Med.* **43,** 126–148.

Pestka, S. (1968). *Proc. Natl. Acad. Sci. U.S.A.* **61,** 726–733.

Pestka, S. (1969a). *J. Biol. Chem.* **244,** 1533–1539.

Pestka, S. (1969b). *Cold Spring Harbor Symp. Quant. Biol.* **34,** 395–410.

Pestka, S. (1969c). *Biochem. Biophys. Res. Commun.* **36,** 589–595.

Pestka, S. (1969d). *Proc. Natl. Acad. Sci. U.S.A.* **64,** 709–714.

Pestka, S. (1970a). *Arch. Biochem. Biophys.* **136,** 80–88.

Pestka, S. (1970b). *Arch. Biochem. Biophys.* **136,** 89–96.

Pestka, S. (1970c). *Biochem. Biophys. Res. Commun.* **40,** 667–674.

Pestka, S. (1971). *Annu. Rev. Microbiol.* **25,** 487–562.

Pestka, S. (1972a). *Proc. Natl. Acad. Sci. U.S.A.* **69,** 624–628.

Pestka, S. (1972b). *J. Biol. Chem.* **247,** 4669–4678.

Pestka, S. (1974a). *Antimicrob. Agents & Chemother.* **5,** 255–267.
Pestka, S. (1974b). *Antimicrob. Agents & Chemother.* **6,** 474–478.
Pestka, S. (1974c). *In* "Methods in Enzymology" (L. Grossman and K. Moldave, eds.), Vol. 30, pp. 462–470. Academic Press, New York.
Pestka, S. (1975a). *In* "Antibiotics" (J. W. Corcoran and F. E. Hahn, eds.), Vol. 3, pp. 323–326. Springer-Verlag, Berlin and New York.
Pestka, S. (1975b). *In* "Antibiotics" (J. W. Corcoran and F. E. Hahn, eds.), Vol. 3. pp. 370–395. Springer-Verlag, Berlin and New York.
Pestka, S. (1975c). *In* "Antibiotics" (J. W. Corcoran and F. E. Hahn, eds.), Vol. 3, pp. 480–486. Springer-Verlag, Berlin and New York.
Pestka, S., and Bodley, J. W. (1975). *In* "Antibiotics" (J. W. Corcoran and F. E. Hahn, eds.), Vol. 3, pp. 551–573. Springer-Verlag, Berlin and New York.
Pestka, S., and Brot, N. (1971). *J. Biol. Chem.* **246,** 7715–7722.
Pestka, S., and Hintikka, H. (1971). *J. Biol. Chem.* **246,** 7723–7730.
Pestka, S., and LeMahieu, R. A. (1974a). *Antimicrob. Agents & Chemother.* **6,** 39–45.
Pestka, S., and LeMahieu, R. A. (1974b). *Antimicrob. Agents & Chemother.* **6,** 479–488.
Pestka, S., and Nirenberg, M. (1966). *Cold Spring Harbor Symp. Quant. Biol.* **31,** 641–656.
Pestka, S., Marshall, R., and Nirenberg, M. (1965). *Proc. Natl. Acad. Sci. U.S.A.* **53,** 639–646.
Pestka, S., Heck, B. H., and Scolnick, E. M. (1969). *Anal. Biochem.* **28,** 376–384.
Pestka, S., Hishizawa, T., and Lessard, J. L. (1970). *J. Biol. Chem.* **245,** 6208–6219.
Pestka, S., Goorha, R., Rosenfeld, H., Neurath, C., and Hintikka, H. (1972a). *J. Biol. Chem.* **247,** 4258–4263.
Pestka, S., Rosenfeld, H., Harris, R., and Hintikka, H. (1972b). *J. Biol. Chem.* **247,** 6895–6900.
Pestka, S., Vince, R., Daluge, S., and Harris, R. (1973). *Antimicrob. Agents & Chemother.* **4,** 37–43.
Pestka, S., Nakagawa, A., and Ōmura, S. (1974). *Antimicrob. Agents & Chemother.* **6,** 606–612.
Pestka, S., Weiss, D., Vince, R., Wienen, B., Stöffler, G., and Smith, I. (1976). *Mol. Gen. Genet.* **144,** 235–241.
Peterson, R. F., Cohen, P. S., and Ennis, H. L. (1972). *Virology* **48,** 201–206.
Phillips, L. A., and Franklin, R. M. (1969). *Cold Spring Harbor Symp. Quant. Biol.* **34,** 243–253.
Pirali, G., Lancini, G. C., Parisi, B., and Sala, F. (1972). *J. Antibiot.* **25,** 561–568.
Pong, S.-S., Nuss, D. L., and Koch, G. (1975). *J. Biol. Chem.* **250,** 240–245.
Pongs, O., Bald, R., and Erdmann, V. A. (1973). *Proc. Natl. Acad. Sci. U.S.A.* **70,** 2229–2233.
Powell, R. G., Weisleder, D., Smith, C. R., and Rohwedder, W. K. (1970). *Tetrahedron Lett.* 815–818.
Price, A. R., and Rottman, F. (1970). *Biochemistry* **9,** 4524–4529.
Primakoff, P., and Berg, P. (1970). *Cold Spring Harbor Symp. Quant. Biol.* **35,** 391–396.
Rabinovitz, J., Freedman, M. L., Fisher, J. M., and Maxwell, C. R. (1969). *Cold Spring Harbor Symp. Quant. Biol.* **34,** 567–578.
Rabinovitz, M., and Fisher, J. M. (1962). *J. Biol. Chem.* **237,** 477–481.
Rao, S. S., and Grollman, A. P. (1967). *Biochem. Biophys. Res. Commun.* **29,** 696–704.
Ravel, J. M., Mosteller, R. D., and Hardesty, B. (1966). *Proc. Natl. Acad. Sci. U.S.A.* **56,** 701–708.
Ravel, J. M., Shorey, R. L., Garner, C. W., Dawkins, R. C., and Shive, W. (1969). *Cold Spring Harbor Symp. Quant. Biol.* **34,** 321–330.

Rebstock, M. C., Crooks, H. M., Controulis, J., and Bartz, Q. R. (1949). *J. Am. Chem. Soc.* **71**, 2458–2462.

Rendi, R. (1959). *Exp. Cell Res.* **18**, 187–189.

Rendi, R., and Ochoa, S. (1962). *J. Biol. Chem.* **237**, 3711–3713.

Reusser, F. (1975). *Antimicrob, Agents & Chemother.* **7**, 32–37.

Reynolds, F., Grunberger, D., Pitha, J., and Pitha, P. M. (1972). *Biochemistry* **11**, 3261–3266.

Richman, N., and Bodley, J. W. (1972). *Proc. Natl. Acad. Sci. U.S.A.* **69**, 686–689.

Richman, N., and Bodley, J. W. (1973). *J. Biol. Chem.* **248**, 381–383.

Richter, D. (1972). *Biochem. Biophys. Res. Commun.* **46**, 1850–1856.

Ringer, D., and Chládek, S. (1974). *Biochem. Biophys. Res. Commun.* **56**, 760–766.

Ringer, D., Quiggle, K., and Chládek, S. (1975). *Biochemistry* **14**, 514–520.

Ringrose, P. S., and Lambert, R. W. (1973). *Biochim. Biophys. Acta* **299**, 374–384.

Robbins, E., Pederson, T., and Klein, P. (1970). *J. Cell Biol.* **44**, 400–416.

Robertson, H. D., and Mathews, M. B. (1973). *Proc. Natl. Acad. Sci. U.S.A.* **70**, 225–229.

Rychlík, I. (1966). *Biochim. Biophys. Acta* **114**, 425–427.

Rychlík, I., Chládek, S., and Žemlička, J. (1967). *Biochim. Biophys. Acta* **138**, 640–642.

Rychlík, I., Černá, J., Chládek, S., Žemlička, J., and Haladová, Z. (1969). *J. Mol. Biol.* **43**, 13–24.

Saborio, J. L., and Koch, G. (1973). *J. Biol. Chem.* **248**, 8343–8347.

Saborio, J. L., Pong, S.-S., and Koch, G. (1974). *J. Mol. Biol.* **85**, 195–211.

Saborio, J. L., Wiegers, K. J., and Koch, G. (1975). *Arch. Virol.* **49**, 81–87.

Sakakibara, H., Naganawa, H., Ohno, M., Maeda, K., and Umezawa, H. (1974). *J. Antibiot.* **27**, 897–899.

Sarkar, P., Stringer, E. A., and Maitra, U. (1974). *Proc. Natl. Acad. Sci. U.S.A.* **71**, 4896–4990.

Sarkar, S., and Thach, R. E. (1968). *Proc. Natl. Acad. Sci. U.S.A.* **60**, 1479–1486.

Schindler, D. (1974). *Nature (London)* **249**, 38–41.

Schlessinger, D., Gurgo, C., Luzzatto, L., and Apirion, D. (1969) *Cold Spring Harbor Symp. Quant. Biol.* **34**, 231–242.

Schneider, J. A., and Maxwell, E. S. (1973). *Biochemistry* **12**, 475–481.

Schreiner, G., and Nierhaus, K. H. (1973). *J. Mol. Biol.* **81**, 71–82.

Schroer, R. A., and Moldave, K. (1972). *Biochem. Biophys. Res. Commun.* **48**, 1572–1577.

Schwartz, J. H. (1965). *Proc. Natl. Acad. Sci. U.S.A.* **53**, 1133–1140.

Schwartz, S. N., Medoff, G., Kobayashi, G. S., Kwan, C. N., and Schlessinger, D. (1972). *Antimicrob. Agents & Chemother.* **2**, 36–40.

Scolnick, E., Tompkins, R., Caskey, T., and Nirenberg, M. (1968). *Proc. Natl. Acad. Sci. U.S.A.* **61**, 768–774.

Seal, S. N., and Marcus, A. (1972). *Biochem. Biophys. Res. Commun.* **46**, 1895–1902.

Sebald, W., Schwab, A. J., and Bucher, T. (1969). *FEBS Lett.* **4**, 243–246.

Sedláček, J., Jonák, J., and Rychlík, I. (1971). *Biochem. Biophys. Acta* **254**, 478–480.

Seeds, N. W., Retsma, J. A., and Conway, T. W. (1967). *J. Mol. Biol.* **27**, 421–430.

Senior, B. W., and Holland, J. B. (1971). *Proc. Natl. Acad. Sci. U.S.A.* **68**, 959–963.

Shaw, W. V., Bentley, D. W., and Sands, L. (1970). *J. Bacteriol.* **104**, 1095–1105.

Shemyakin, M. N. (1961). "Khimia Antibiotikov," Vol. 1. Academy of Sciences, Moscow.

Shepherd, C. J. (1958). *J. Gen. Microbiol.* **18**, iv–v.

Sherman, M. I., and Simpson, M. V. (1969). *Proc. Natl. Acad. Sci. U.S.A.* **64**, 1388–1395.

Shimizu, M., Saito, T., and Mitsuhashi, S. (1971). *Jpn. J. Microbiol.* **15**, 198–200.

Shorey, R. L., Ravel, J. M., Garner, C. W., and Shive, W. (1969). *J. Biol. Chem.* **244**, 4555–4564.

Sidikaro, J., and Nomura, M. (1973). *FEBS Lett.* **29**, 15–19.

Siegel, M. R., and Sisler, H. D. (1963). *Nature (London)* **200**, 675–676.

Siegel, M. R., and Sisler, H. D. (1965). *Biochim. Biophys. Acta* **103**, 558–567.

Siegelman, F. L., and Apirion, D. (1971). *J. Bacteriol.* **105**, 451–453.

Sinohara, H., and Sky-Peck, H. H. (1965). *Biochem. Biophys. Res. Commun.* **18**, 98–102.

Sinsheimer, R. L., Starman, B., Nagler, C., and Guthrie, S. (1962). *J. Mol. Biol.* **4**, 142–160.

Sinsheimer, R. L., Hutchinson, C. A., and Lindqvist, B. (1967). *In* "The Molecular Biology of Viruses" (S. J. Colter and W. Paranchych, eds.), pp. 175–192. Academic Press, New York.

Skogerson, L., McLaughlin, C., and Wakatama, E. (1973). *J. Bacteriol.* **116**, 818–822.

Skoultchi, A., Ono, Y., Waterson, J., and Lengyel, P. (1970). *Biochemistry* **9**, 508–514.

Slechta, L. (1966). *Antimicrob. Agents & Chemother -1965*, pp. 326–333.

Slechta, L. (1967). *In* "Antibiotics" (D. Gottlieb and P. D. Shaw, eds.), Vol. 1, p. 410–414. Springer-Verlag, Berlin and New York.

Smith, I., Weiss, D., and Pestka, S. (1976). *Mol. Gen. Genet.*, **144**, 231–233.

Smith, J. D., Traut, R. R., Blackburn, G. M., and Monro, R. E. (1965). *J. Mol. Biol.* **13**, 617–628.

So, A. G., and Davie, E. W. (1963). *Biochemistry* **2**, 132–136.

Sonnenberg, N., Wilchek, M., and Zamir, A. (1973). *Proc. Natl. Acad. Sci. U.S.A.* **70**, 1423–1426.

Sonnenberg, N., Zamir, A., and Wilchek, M. (1974). *Biochem. Biophys. Res. Commun.* **59**, 693–696.

Sopori, M. L., and Lengyel, P. (1972). *Biochem. Biophys. Res. Commun.* **46**, 238–244.

Sorm, F., and Grunberger, D. (1954). *Collect. Czech. Chem. Commun.* **19**, 167–173.

Sparling, P. F. (1970). *Science* **167**, 56–58.

Sparling, P. F., Ikeya, Y., and Elliot, D. (1973). *J. Bacteriol.* **113**, 704–710.

Speyer, J. F., Lengyel, P., and Basilio, C. (1962). *Proc. Natl. Acad. Sci. U.S.A.* **48**, 684–686.

Speyer, J. F., Lengyel, P., Basilio, C., Wahba, A. J., Gardner, R. S., and Ochoa, S. (1963). *Cold Spring Harbor Symp. Quant. Biol.* **28**, 559–567.

Spotts, C. R. (1962). *J. Gen. Microbiol.* **28**, 347–365.

Spyrides, G. J. (1964). *Proc. Natl. Acad. Sci. U.S.A.* **51**, 1220–1226.

Staehelin, T., and Meselson, M. (1966). *J. Mol. Biol.* **19**, 207–210.

Stafford, M. E., and McLaughlin, C. S. (1973). *J. Cell. Physiol.* **82**, 121–128.

Stent, G. S., and Brenner, S. (1961). *Proc. Natl. Acad. Sci. U.S.A.* **47**, 2005–2014.

Stewart, M. L., and Goldberg, I. H. (1973). *Biochim. Biophys. Acta* **294**, 123–137.

Stewart, M. L., Grollman, A. P., and Huang, M.-T. (1971). *Proc. Natl. Acad. Sci. U.S.A.* **68**, 97–101.

Stewart-Blair, M. L., Yanowitz, I. S., and Goldberg, I. H. (1971). *Biochemistry* **10**, 4198–4206.

Stöffler, G., Deusser, E., Wittmann, H. G., and Apirion, D. (1971). *Mol. Gen. Genet.* **111**, 334–341.

Suarez, G., and Nathans, D. (1965). *Biochem. Biophys. Res. Commun.* **18**, 743–750.

Subramanian, A. R., Davis, B. D., and Beller, R. J. (1969). *Cold Spring Harbor Symp. Quant. Biol.* **34**, 223–230.

Suhadolnik, R. J. (1970). *In* "Nucleoside Antibiotics," pp. 170–217. Wiley (Interscience), New York.

Summers, D. F., and Maizel, J. V., Jr., (1971). *Proc. Natl. Acad. Sci. U.S.A.* **68**, 2852–2856.

Summers, D. F., Shaw, E. N., Stewart, M. L., and Maizel, J. V., Jr., (1972). *J. Virol.* **10**, 880–884.

Sundaralingam, M., and Arora, S. K. (1969). *Proc. Natl. Acad. Sci. U.S.A.* **64**, 1021–1027.

Suzuka, I., Kaji, H., and Kaji, A. (1966). *Proc. Natl. Acad. Sci. U.S.A.* **55**, 1483–1490.

Suzuki, J., Kunimoto, T., and Hori, M. (1970). *J. Antibiot.* **23**, 99–101.

Symons, R. H., Harris, R. J., Clarke, L. P., Wheldrake, J. F., and Elliott, W. H. (1969). *Biochim. Biophys. Acta* **179**, 248–250.

Szer, W., and Kurylo-Borowska, Z. (1970). *Biochim. Biophys. Acta* **224**, 477–486.

Szer, W., and Kurylo-Borowska, Z. (1972). *Biochim. Biophys. Acta* **259**, 357–368.

Szwaj, M., Samiński, E. M., and Bresler, S. E. (1972). *Bull. Acad. Pol. Sci.* **20**, 457–463.

Taber, R., Rekosh, D., and Baltimore, D. (1971). *J. Virol.* **8**, 395–401.

Taber, R., Wertheimer, R., and Golrick, J. (1973). *J. Mol. Biol.* **80**, 367–371.

Tada, K., and Trakatellis, A. C. (1971). *Antimicrob. Agents & Chemother. -1970*, pp. 227–230.

Tai, P.-C., and Davis, B. D. (1974). *Proc. Natl. Acad. Sci. U.S.A.* **71**, 1021–1025.

Tai, P.-C., Wallace, B. J., and Davis, B. D. (1973). *Biochemistry* **12**, 616–620.

Tai, P.-C., Wallace, B. J., and Davis, B. D. (1974). *Biochemistry* **13**, 4653–4659.

Takanami, M. (1964). *Proc. Natl. Acad. Sci. U.S.A.* **52**, 1271–1276.

Takata, R., Osawa, S., Tanaka, K., Teraoka, H., and Tamaki, M. (1970). *Mol. Gen. Genet.* **109**, 123–130.

Tal, M., Aviram, M., Kanarek, A., and Weiss, A. (1972). *Biochim. Biophys. Acta* **281**, 381–392.

Tanaka, K., and Teraoka, H. (1966). *Biochim. Biophys. Acta* **114**, 204–206.

Tanaka, K., and Teraoka, H. (1968). *J. Biochem. (Tokyo)* **64**, 635–648.

Tanaka, K., Teraoka, H., Nagira, T., and Tamaki, M. (1966a). *J. Biochem. (Tokyo)* **59**, 632–634.

Tanaka, K., Teraoka, H., Nagira, T., and Tamaki, M. (1966b). *Biochim. Biophys. Acta* **123**, 435–437.

Tanaka, K., Watanabe, S., and Tamaki, M. (1970a). *J. Antibiot.* **23**, 13–19.

Tanaka, K., Watanabe, S., Teraoka, H., and Tamaki, M. (1970b). *Biochem. Biophys. Res. Commun.* **39**, 1189–1193.

Tanaka, K., Teraoka, H., Tamaki, M., Otaka, E., and Osawa, S. (1970c). *Proc. Int. Cong. Chemother. 6th, 1969* pp. 499–501.

Tanaka, K., Teraoka, H., Tamaki, M., Takata, R., and Osawa, S. (1971a). *Mol. Gen. Genet.* **114**, 9–13.

Tanaka, K., Tamaki, M., Itoh, T., Otaka, E., and Osawa, S. (1971b). *Mol. Gen. Genet.* **114**, 23–30.

Tanaka, K., Teraoka, H., and Tamaki, M. (1971c). *FEBS Lett.* **13**, 65–67.

Tanaka, K., Tamaki, M., Takata, R., and Osawa, S. (1972). *Biochem. Biophys. Res. Commun.* **46**, 1979–1983.

Tanaka, N. (1975a). *In* "Antibiotics" (J. W. Corcoran and F. E. Hahn, eds.), Vol. 3, pp. 436–447. Springer-Verlag, Berlin and New York.

Tanaka, N. (1975b). *In* "Antibiotics" (J. W. Corcoran and F. E. Hahn, eds.), Vol. 3, pp. 487–497. Springer-Verlag, Berlin and New York.

Tanaka, N., Yoshida, Y., Sashikata, K., Yamaguchi, H., and Umezawa, H. (1966a). *J. Antibiot.* **19**, 65–68.

Tanaka, N., Sashikata, K., Yamaguchi, H., and Umezawa, H. (1966b). *J. Biochem. (Tokyo)* **60**, 429–434.

Tanaka, N., Masukawa, H., and Umezawa, H. (1967). *Biochem. Biophys. Res. Commun.* **26**, 544–549.

Tanaka, N., Kinoshita, T., and Masukawa, H. (1968). *Biochem. Biophys. Res. Commun.* **30**, 278–283.

Tanaka, N., Kinoshita, T., and Masukawa, H. (1969). *J. Biochem.* *(Tokyo)* **65**, 459–464.
Tanaka, N., Nishimura, T., and Kinoshita, T. (1970a). *J. Biochem.* *(Tokyo)* **67**, 459–463.
Tanaka, N., Kinoshita, T., Lin, Y., Nishimura, T., and Umezawa, H. (1970b). *Proc. Int. Congr. Chemother. 6th, 1969* Vol. 2, pp. 502–513.
Tanaka, N., Kawano, G., and Kinoshita, T. (1971a). *Biochem. Biophys. Res. Commun.* **42**, 564–570.
Tanaka, N., Lin, Y.-C., and Okuyama, A. (1971b). *Biochem. Biophys. Res. Commun.* **44**, 477–483.
Tanaka, S., Otaka, T., and Kaji, A. (1973). *Biochim. Biophys. Acta* **331**, 128–140.
Tanaka, T., Endo, T., Shimazu, A., Yoshida, R., Suzuki, Y., Otake, N., and Yonehara, H. (1970a). *J. Antibiot.* **23**, 231–237.
Tanaka, T., Sakaguchi, K., and Yonehara, H. (1970b). *J. Antibiot.* **23**, 401–407
Tanaka, T., Sakaguchi, K., and Yonehara, H. (1971a). *J. Antibiot.* **24**, 537–542.
Tanaka, T., Sakaguchi, K., and Yonehara, H. (1971b). *J. Biochem.* *(Tokyo)* **69**, 1127–1130.
Tanaka, Y., and Kaji, H. (1968). *Biochem. Biophys. Res. Commun.* **32**, 313–319.
Tate, W. P., and Caskey, C. P. (1973). *J. Biol. Chem.* **248**, 7970–7972.
Taubman, S. B., Jones, N. R., Young, F. E., and Corcoran, J. W. (1966). *Biochem. Biophys. Acta* **123**, 438–440.
Teraoka, H. (1970). *Biochim. Biophys. Acta* **213**, 535–537.
Teraoka, H. (1971). *J. Antibiot.* **24**, 302–309.
Teraoka, H., and Tanaka, K. (1971). *Biochim. Biophys. Acta* **232**, 509–513.
Teraoka, H., Tanaka, K., and Tamaki, M. (1969). *Biochim. Biophys. Acta* **174**, 776–778.
Teraoka, H., Tamaki, M., and Tanaka, K. (1970). *Biochem. Biophys. Res. Commun.* **38**, 328–332.
Tessman, E. S. (1966). *J. Mol. Biol.* **17**, 218–236.
Tiboni, O., Parisi, B., and Ciferri, O. (1968). *G. Bot. Ital.* **102**, 337–345.
Timberlake, W. E., McDowell, L., and Griffin, D. H. (1972a). *Biochem. Biophys. Res. Commun.* **46**, 942–947.
Timberlake, W. E., Hagen, G., and Griffin, D. H. (1972b). *Biochem. Biophys. Res. Commun.* **48**, 823–827.
Tissières, A., Schlessinger, D., and Gros, F. (1960). *Proc. Natl. Acad. Sci. U.S.A.* **46**, 1450–1463.
Tompkins, R., Scolnick, E., and Caskey, C. T. (1970). *Proc. Natl. Acad. Sci. U.S.A.* **65**, 702–708.
Towers, N. R., Dixon, H., Kellerman, G. M., and Linnane, A. W. (1972). *Arch. Biochem. Biophys.* **151**, 361–369.
Trakatellis, A. C., Montjar, M., and Axelrod, A. E. (1965). *Biochemistry* **4**, 2065–2071.
Traub, P., and Nomura, M. (1968a). *Science* **160**, 198–199.
Traub, P., and Nomura, M. (1968b). *Proc. Natl. Acad. Sci. U.S.A.* **59**, 777–784.
Traub, P., and Nomura, M. (1969). *Cold Spring Harbor Symp. Quant. Biol.* **34**, 63–67.
Traub, P., Hosokawa, K., and Nomura, M. (1966). *J. Mol. Biol.* **19**, 211–214.
Traut, R. R., and Monro, R. E. (1964). *J. Mol. Biol.* **10**, 63–72.
Traut, R. R., Delius, H., Ahmad-Zadeh, C., Bickle, T. A., Pearson, P., and Tissières, A. (1969). *Cold Spring Harbor Symp. Quant. Biol.* **34**, 25–38.
Tscherne, J. S., and Pestka, S. (1975). *Antimicrob. Agents & Chemother.* **8**, 479–487.
Turnowsky, F., Drews, J., Eich, F., and Högenauer, G. (1973). *Biochem. Biophys. Res. Commun.* **52**, 327–334.
Uchida, K., and Wolf, H. (1974). *J. Antibiot.* **27**, 783–787.
Uchida, T., Pappenheimer, A. M., Jr., and Harper, A. A. (1972). *Science* **175**, 901–903.

Uchida, T., Pappenheimer, A. M., Jr., and Greany, R. (1973a). *J. Biol. Chem.* **248,** 3838–3844.

Uchida, T., Pappenheimer, A. M., Jr., and Harper, A. A. (1973b). *J. Biol. Chem.* **248,** 3845–3850.

Uchida, T., Pappenheimer, A. M., Jr., and Harper, A. A. (1973c). *J. Biol. Chem.* **248,** 3851–3854.

Uehara, Y., Kondo, S., Umezawa, H., Suzukake, K., and Hori, M. (1972). *J. Antibiot.* **25,** 685–688.

Uehara, Y., Hori, M., and Umezawa, H. (1974). *Biochim. Biophys. Acta* **374,** 82–95.

Ueno, Y., and Fukushima, K. (1968). *Experientia* **24,** 1032–1033.

Ueno, Y., Hosoya, M., Morita, Y., Ueno, I., and Tatsuno, T. (1968). *J. Biochem. (Tokyo)* **64,** 479–485.

Umezawa, N., Hamada, M., Suhara, Y., Hashimoto, T., Ikekawa, T., Tanaka, N., Maeda, K., Okami, Y., and Takeuchi, T. (1966). *Antimicrob. Agents & Chemother. -1965,* pp. 753–757.

Van der Mei, D., Zandberg, J., and Jansz, H. S. (1972). *Biochim. Biophys. Acta* **287,** 312–321.

Van Knippenberg, P. H., Grijm-Vos, M., Veldstra, H., and Bosch, L. (1965). *Biochem. Biophys. Res. Commun.* **20,** 4–9.

Vazquez, D. (1963). *Biochem. Biophys. Res. Commun.* **12,** 409–413.

Vazquez, D. (1964). *Biochem. Biophys. Res. Commun.* **15,** 464–468.

Vazquez, D. (1966a). *Biochim. Biophys. Acta* **114,** 277–288.

Vazquez, D. (1966b). *Symp. Soc. Gen. Microbiol.* **16,** 169–191.

Vazquez, D. (1966c). *Biochim. Biophys. Acta* **114,** 289–295.

Vazquez, D. (1967a). *Life. Sci.* **6,** 381–386.

Vazquez, D. (1967b). *Life Sci.* **6,** 845–853.

Vazquez, D. (1967c). *In* "Antibiotics" (D. Gottlieb and P. D. Shaw, eds.), Vol. 1, pp. 366–377. Springer-Verlag, Berlin and New York.

Vazquez, D., Battaner, E., Neth, R., Heller, G., and Monro, R. E. (1969). *Cold Spring Harbor Symp. Quant. Biol.* **34,** 369–375.

Vince, R., Weiss, D., and Pestka, S. (1976a). *Antimicrob. Agents & Chemother.* **9,** 131–136.

Vince, R., Weiss, D., Gordon, J., Howard, G., Smith, I., and Pestka, S. (1976b). *Antimicrob. Agents & Chemother.* **9,** 665–667.

Vogel, Z., Zamir, A., and Elson, D. (1969). *Biochemistry* **8,** 5161–5168.

Vogel, Z., Vogel, T., Zamir, A., and Elson, D. (1970). *J. Mol. Biol.* **54,** 379–386.

Vogel, Z., Vogel, T., and Elson, D. (1971). *FEBS Lett.* **15,** 249–253.

von Ehrenstein, G., and Lipmann, F. (1961). *Proc. Natl. Acad. Sci. U.S.A.* **47,** 941–950.

Vos, C., and Verwiel, P. E. J. (1973). *Tetrahedron Lett.* **52,** 5173–5176.

Wallace, B. J., and Davis, B. D. (1973). *J. Mol. Biol.* **75,** 377–390.

Wallace, B. J., Tai, P.-C., and Davis, B. D. (1973). *J. Mol. Biol.* **75,** 391–400.

Wallace, B. J., Tai, P.-C., and Davis, B. D. (1974). *Proc. Natl. Acad. Sci. U.S.A.* **71,** 1634–1638.

Waller, C. W., Fryth, P. W., Hutchings, B. L., and Williams, J. H. (1953). *J. Am. Chem. Soc.* **75,** 2025.

Waller, J.-P., Erdös, T., Lemoine, F., Guttmann, S., and Sandrin, E. (1966). *Biochim. Biophys. Acta* **119,** 566–580.

Watanabe, S. (1972). *J. Mol. Biol.* **67,** 443–457.

Waterson, J., Sopori, M. L., Gupta, S. L., and Lengyel, P. (1972). *Biochemistry* **11,** 1377–1382.

Webster, R. E., and Zinder, N. D. (1969). *J. Mol. Biol.* **42,** 425–439.

Weeks, D. P., and Baxter, R. (1972). *Biochemistry* **11,** 3060–3064.

Wei, C.-M., Hansen, B. S., Vaughan, M. H., Jr., and McLaughlin, C. S. (1974). *Proc. Natl. Acad. Sci. U.S.A.* **71,** 713–717.

Weisblum, B., and Demohn, V. (1970a). *J. Bacteriol.* **101**, 1073–1075.
Weisblum, B., and Demohn, V. (1970b). *FEBS Lett.* **11**, 149–152.
Weisblum, B., Siddhikol, C., Lai, C. J., and Demohn, V. (1971). *J. Bacteriol.* **106**, 835–847.
Weissbach, H., and Brot, N. (1970). *Biochem. Biophys. Res. Commun.* **39**, 1194–1198.
Weissbach, H., Redfield, B., and Brot, N. (1968). *Arch. Biochem. Biophys.* **127**, 705–710.
Weissbach, H., Redfield, B., and Brot, N. (1971a). *Arch. Biochem. Biophys.* **145**, 676–683.
Weissbach, H., Kurylo-Borowska, Z., and Szer, W. (1971b). *Arch. Biochem. Biophys.* **146**, 356–358.
Weissbach, H., Redfield, B., Yamasaki, E., Davis, R. C., Jr., Pestka, S., and Brot, N. (1972). *Arch. Biochem. Biophys.* **149**, 110–117.
Wengler, G., and Wengler, G. (1972). *Eur. J. Biochem.* **27**, 162–173.
Wettstein, F. O., Noll, H., and Penman, S. (1964). *Biochim. Biophys. Acta* **87**, 525–528.
White, J. R., and White, H. (1964). *Science* **146**, 772–774.
Wienen, B., Stöffler, G., Smith, I., Weiss, D., Vince, R., and Pestka, S. (1977). (in preparation).
Wiley, P. F., Jahnke, H. K., MacKellar, F., Kelly, R. B., and Argoudelis, A. D. (1970). *J. Org. Chem.* **35**, 1420–1425.
Wilhelm, J. M., and Corcoran, J. W. (1967). *Biochemistry* **6**, 2578–2585.
Wilhelm, J. M., Oleinick, N. L., and Corcoran, J. W. (1968). *Antimicrob. Agents & Chemother. -1967*, pp. 236–250.
Wintersberger, E. (1965). *Biochem. Z.* **341**, 409–419.
Wisseman, C. L., Smadel, J. E., Hahn, F. E., and Hopps, H. E. (1954). *J. Bacteriol.* **67**, 662–673.
Wittmann, H. G., Stöffler, G., Apirion, D., Rosen, L., Tanaka, K., Tamaki, M., Takata, R., Dekio, S., Otake, E., and Osawa, S. (1973). *Mol. Gen. Genet.* **127**, 175–189.
Wolf, H., and Zähner, H. (1972). *Arch. Mikrobiol.* **83**, 147–154.
Wolf, H., Zähner, H., and Nierhaus, K. (1972). *FEBS Lett.* **21**, 347–350.
Wolf, H., Chinali, G., and Parmeggiani, A. (1974). *Proc. Natl. Acad. Sci. U.S.A.* **71**, 4910–4914.
Wolfe, A. D., and Hahn, F. E. (1964). *Science* **143**, 1445–1446.
Wolfe, A. D., and Hahn, F. E. (1965). *Biochim. Biophys. Acta* **95**, 146–155.
Wolfe, A. D., and Hahn, F. E. (1968). *Biochem. Biophys. Res. Commun.* **31**, 945–949.
Wolfgang, R. W., and Lawrence, N. L. (1967). *J. Mol. Biol.* **29**, 531–535.
Yamaguchi, H., and Tanaka, N. (1966). *J. Biochem. (Tokyo)* **60**, 632–642.
Yamaguchi, H., and Tanaka, N. (1967). *J. Biochem. (Tokyo)* **61**, 18–25.
Yamaguchi, H., Yamamoto, C., and Tanaka, N. (1965). *J. Biochem. (Tokyo)* **57**, 667–677.
Yamaki, H., and Tanaka, N. (1963). *J. Antibiot.* **16**, 222–226.
Yarmolinsky, M. B., and de la Haba, G. I. (1959). *Proc. Natl. Acad. Sci. U.S.A.* **45**, 1721–1729.
Yoshikawa, M., Okuyama, A., and Tanaka, N. (1975). *J. Bacteriol.* **122**, 796–797.
Young, C. W. (1966). *Mol. Pharmacol.* **2**, 50–55.
Young, C. W., Robinson, P. F., and Sacktor, B. (1963). *Biochem. Pharmacol.* **12**, 855–865.
Yukioka, M., and Morisawa, S. (1969a). *J. Biochem. (Tokyo)* **66**, 225–232.
Yukioka, M., and Morisawa, S. (1969b). *J. Biochem. (Tokyo)* **66**, 233–239.
Yukioka, M., and Morisawa, S. (1969c). *J. Biochem. (Tokyo)* **66**, 241–247.
Yukioka, M., and Morisawa, S. (1971). *Biochim. Biophys. Acta* **254**, 304–315.
Zagorska, L., Dondon, J., Lelong, J. C., Gros, F., and Grunberg-Manago, M. (1971). *Biochimie* **53**, 63–70.
Zimmermann, R. A., Ikeya, Y., and Sparling, P. F. (1973). *Proc. Natl. Acad. Sci. U.S.A.* **70**, 71–75.

11

Messenger RNA and Its Translation

DAVID A. SHAFRITZ

In the past several years a great deal of progress has been made in studies on the synthesis, structure, metabolism, and translation of mRNA. This chapter will deal with the properties, isolation, and translation of mRNA primarily in eukaryotic systems, emphasizing various techniques which have been utilized in these systems. Hopefully, this information will be useful for application to new problems in mRNA translation. Sections dealing with identification of mRNA and recent studies on the synthesis, structure, and metabolism of mRNA are included. A final section is devoted to the properties and characterization of mRNA-protein complexes (mRNP's), which have been isolated from a variety of eukaryotic sources. The major intent of this section is to stimulate interest in questions regarding the possible role of the mRNP in eukaryotic post-transcriptional regulation and to indicate some approaches which might be utilized to examine various aspects of this problem.

I. Introduction

The concept that genetic information encoded in DNA is transcribed into a special RNA which serves as a "messenger" or template for protein synthesis was first introduced by Jacob and Monod (1961). Support for this hypothesis in bacterial systems was obtained shortly thereafter (Brenner *et al.*, 1961; Gros *et al.*, 1961; Nirenberg and Matthaei, 1961; Nathans *et al.*, 1962; Ofengand and Haselkorn, 1962), and this provided great impetus to studies of molecular events in gene expression. In the course of these studies, a variety of mRNA species have been isolated and partially characterized. A great deal of information has been gathered on the synthesis, properties, and metabolism of mRNA. The complete primary structure of the bacteriophage MS2 coat protein gene (Min Jou *et al.*, 1972) and much of the primary sequence of bacteriophage R17 mRNA have been elucidated (Weissmann *et al.*, 1973; Steitz, 1974). This chapter will deal with the isolation and identification of mRNA with major emphasis on translation properties. Except for certain examples, the bulk of this material will be taken from studies of eukaryotic systems. No attempt will be made to list all studies reporting translation of specific mRNA's. The specific topic of translational control or regulation of protein synthesis has been discussed separately in Chapter 5.

In prokaryotic organisms, there is limited subcellular compartmentalization and direct contact between the transcription and translation machinery (Hamkalo and Miller, 1973). In these cells, a tight coupling has been found between transcription and translation events. Association of mRNA with ribosomes, polypeptide chain synthesis, and mRNA degradation may all begin before mRNA synthesis is completed (Morse *et al.*, 1968; Mosteller *et al.*, 1970; Adesnik and Levinthal, 1970). Since the half-life of most bacterial mRNA's is very short (a matter of seconds to minutes), the spectrum of proteins synthesized at any given time may change very rapidly. Therefore, cellular regulation in prokaryotes is primarily at the level of transcription.

Eukaryotic cells are far more complex than prokaryotes in their degree of subcellular organization. These cells contain a well-defined nucleus, in which mRNA synthesis occurs in the nucleoplasm, whereas ribosomal

RNA synthesis occurs in the nucleolus (Darnell, 1968). After mRNA is synthesized, it passes through the nuclear membrane into the cytoplasm, where it is joined by ribosomes to form the polyribosome structure actively engaged in protein synthesis. Thus, in eukaryotes there is a physical separation of transcription and translation machinery, and additional mechanisms have been evolved to coordinate these functions. Events controlling the flow of information from the nucleus to the cytoplasm have not been well defined. However, certain studies have reported the association of mRNA with proteins in the nucleus and cytoplasm of eukaryotic cells. These mRNA-protein complexes, termed messenger ribonucleoprotein particles (mRNP's), have been the subject of considerable attention in recent studies and various functions for these particles have been proposed. In view of current interest in the mRNP and its possible role in eukaryotic regulation, a section will be devoted to present studies on these particles in relation to eukaryotic mRNA translation.

II. Identification of Messenger RNA

A. Early Studies

Initially, it was observed by pulse labeling studies that a certain portion of cellular RNA is rapidly labeled and relatively unstable compared to other types of RNA (Astrachman and Volkin, 1958; Gros et al., 1961; Brenner et al., 1961). This rapidly labeled RNA has a base composition similar to DNA and in eukaryotic cells is rather heterogenous in size (Penman et al., 1963; Staehelin et al., 1964; Munro et al., 1964; Girard et al., 1965; Henshaw et al., 1965; Perry and Kelley, 1968). As isolated from polysomes, this RNA is readily distinguished from ribosomal RNA (16 S and 23 S for prokaryotes and 18 S and 28 S for eukaryotes), 5 S RNA, and transfer RNA. This distinction is based not only on the heterogeneous character of this RNA, but also on its rapidity of labeling compared to these other RNA's (Penman et al., 1963; Staehelin et al., 1964; Munro and Korner, 1964; Latham and Darnell, 1965; Attardi et al., 1966; Scherrer et al., 1966; Darnell, 1968). The bulk of the sedimentation values for this RNA range from 10–30 S, although a portion of this material sediments both below and above this region (Penman et al., 1963; Trakatellis et al., 1964; Henshaw et al., 1965; Girard et al., 1965; Darnell, 1968). On the basis of a triplet genetic code, RNA of this size would code for polypeptides of molecular weight ranging from approximately 20,000 to greater than 200,000 daltons. A portion of this material can be isolated from the cytoplasm distinct from the ribosomal pool (Perry and Kelley, 1968; Infante and Nemer, 1968; Kafatos, 1968; Spohr et al., 1970; Henshaw and Loebenstein, 1970; Knöchel and Tiedermann, 1972). Synthesis of rapidly labeled RNA is not inhibited by low levels of actinomycin D, an antibiotic that under these conditions dramatically inhibits ribosomal RNA synthesis (Georgiev et al., 1963; Perry, 1963; Penman et al., 1968). Rapidly labeled RNA can also be distinguished from ribosomal RNA and transfer RNA by its high degree of hybridization to homologous DNA. This assay measures complementarity between the base sequences of RNA and DNA (Spiegelman, 1963; Marmur et al., 1963; Darnell, 1968). Since mRNA represents the widest spectrum of cytoplasmic RNA molecules, it should hybridize to a greater proportion of the DNA genome than other types of RNA. However, heterogeneous nuclear RNA (HnRNA) also hybridizes to a large portion of cellular DNA (Darnell et al., 1973) and ribosomal RNA hybridizes rapidly to highly reiterated sequences in DNA (Britten and Kohne, 1968). Therefore, much of the RNA hybridized to specific DNA in the original studies may not represent mRNA. In spite of these and other limitations, the various properties described above for rapidly labeled RNA are consistent with those anticipated for mRNA and provided indirect evidence for the existence of this molecule. Thus, criteria used in early studies for identification of mRNA include (1) instability and rapidity of labeling, (2) "DNA-like" base composition, (3) heterogenous sedimentation pattern, (4) resistence to actinomycin D in low doses, and (5) high degree of hybridization to DNA.

B. Stimulation of Specific Protein Synthesis

As recognized in many of the early studies, definitive evidence for mRNA in a given RNA extract would require translation of this material into a specific protein product (for reviews, see Singer and Leder, 1966; Chantrenne et al., 1967). In the case of f_2 and R17 bacteriophage RNA's, this was accomplished by synthesis of specific viral polypeptides in cell-free systems from Escherichia coli (Nathans et al., 1962; Capecchi, 1966). Since conditions can be found in which RNA fractions other than messenger RNA stimulate protein synthesis, even in heterologous systems (Hardesty et al., 1963; Kruh et al., 1964, 1966; Drach and Lingrel, 1964, 1966; Hunt and Wilkinson, 1967; Anderson, 1969; Gilbert and Anderson, 1970; Fuhr and Natta, 1972; Bogdanovsky et al., 1973; Goldstein and Penman, 1973), a mere stimulation of amino acid incorporation upon addition of exogenous RNA is not sufficient evidence for mRNA. The use of a heterologous translating system, one in which the specific protein under examination is not normally synthesized, is of crucial importance. The product encoded in the mRNA must also be identified by suitable biological and/or chemical criteria. These fundamental principles were most clearly established by the demonstration that a 16–18 S RNA fraction from reticulo-cytes stimulated protein synthesis in a bacterial cell-free system but the products obtained represented bacterial rather than reticulocyte polypeptides (Drach and Lingrel, 1966).

In eukaryotic systems, evidence for the isolation and translation of specific mRNA's was more difficult to obtain. Initially, an RNA fraction with properties expected for globin mRNA was described by Marbaix and Burny (1964), Burny and Marbaix (1965), and Chantrenne et al. (1967). This RNA fraction was isolated from polysomes and comprised about 2% of total polysomal RNA. It sedimented as a distinct band at 9 S in a sucrose gradient, which by initial approximation was close to the sedimentation value for a polyribonucleotide of 430 bases. This represents the minimum polyribonucleotide chain length required to code for either the α- or the β-globin polypeptide chain (141 and 146 amino acids, respectively). Reticulocyte 9 S RNA also demonstrated rapid labeling to a specific activity 3–4 times higher than ribosomal RNA and was degraded simultaneously with polysomes by incubation with very low levels of ribonuclease (far below the amount required for digestion of ribosomal RNA). Unfortunately, these investigators were unable to demonstrate translation of this RNA in their early studies, but they clearly indicated the limitations of their observations (Chantrenne et al., 1967).

The first studies demonstrating translation of eukaryotic RNA in a heterologous system with evidence for synthesis of a specific product were reported by Schapira et al. (1965, 1966, 1968). These investigators translated an RNA fraction from one species of rabbit into material with tryptic peptides specific for globin chains of this species in a cell-free system from a second species of rabbit. Comparable results were obtained when the sources of the RNA and the cell-free system were reversed. In each case globin tryptic peptides specific for the donor species of RNA were found in the cell-free product of the recipient system. Since it was most improbable that these different species of rabbits contained nonexpressed messenger RNA for heterologous globin polypeptides, this constituted solid evidence for translation of foreign messenger RNA. It was subsequently reported that an "8 S" RNA fraction from rabbit reticulocytes stimulated synthesis of globin tryptic peptides in a bacterial cell-free system (Laycock and Hunt, 1969). Except for a 14 S RNA fraction, which also appeared to contain mRNA activity, other RNA fractions showed little effect in stimulating protein synthesis. This bacterial system was dependent on NAcVal-tRNA as an initiator tRNA, but this observation has not been confirmed.

Lingrel and co-workers were the first to study translation of exogenous

eukaryotic mRNA in a comprehensive fashion. Again using the globin
system, these investigators initially demonstrated translation of mouse 9
S RNA in a crude lysate system from rabbit reticulocytes (Lockard and
Lingrel, 1969). In this system there was no increase in total amino acid
incorporation upon addition of RNA but exogenous mRNA competed
with endogenous mRNA for translation. Production of mouse globin β-
chains in the rabbit cell-free system was demonstrated by carboxymethylcellulose chromatography of the cell-free product. Similar experiments
were performed later with translation of mouse 9 S globin mRNA in cell-
free systems from ducks, guinea pigs, and rabbits (Lingrel et al., 1971;
Lockard and Lingrel, 1972, 1973). In spite of wide variations in the levels
of endogenous messenger activity and the fact that one of these systems
(the duck) was of nonmammalian origin, the efficiencies of exogenous
mRNA translation in these systems were approximately the same. The
authenticity of the cell-free products was verified by tryptic peptide mapping (Jones and Lingrel, 1972). Following the example set by these pio-
neering studies, numerous investigators have reported isolation and translation of eukaryotic mRNA's in a variety of heterologous systems (to be
discussed later).

C. Special Use of DNA-RNA Hybridization

As mentioned earlier, nucleic acid hybridization can be used to measure the degree of complementarity between the base sequences of two
polynucleotide molecules. In recent studies DNA-RNA hybridization has
been used to measure "mRNA content" in a given RNA sample, using
radioactive labeled cDNA (complementary DNA prepared from purified
mRNA). The synthesis of [^{3}H]cDNA is performed in vitro using the enzyme "reverse transcriptase," which makes complementary polydeox-
yribonucleotides from RNA templates [originally described by Temin and
Mizutani (1970) and Baltimore (1970)]. cDNA's of very high specific
activity have been made from globin mRNA (Verma et al., 1972; Kacian
et al., 1972; Ross et al., 1972) and ovalbumin mRNA (Sullivan et al.,
1973; Harris et al., 1973), and have been used in a wide range of molecular
studies. Preparations of α- and β-globin chain specific cDNA's have been
used to determine α- and β-globin mRNA content in reticulocyte polysomes from patients with α- and β-thalassemia. From previous studies it
had been determined that the level of functional β chain mRNA was reduced or absent in β-thalassemia (Nienhuis and Anderson, 1971; Benz
and Forget, 1971). Whether the messenger was functionally defective or
quantitatively reduced could not be distinguished by these studies.
cDNA-RNA hybridization experiments indicate a reduction of α-globin
mRNA in α-thalassemia and a reduction of β-globin mRNA in β-

thalassemia (Housman *et al.*, 1973; Kacian *et al.*, 1973; Forget *et al.*, 1974b). In homozygous α°-thalassemia (Kan *et al.*, 1974) and homozygous β°-thalassemia (Forget *et al.*, 1974a), there is an absence of the respective α- and β-globin mRNA's, and recent evidence suggests a deletion in α-globin genes in individuals affected with homozygous α°-thalassemia (*Hydrops fetalis*) (Taylor *et al.*, 1974).

III. Synthesis, Structure, and Metabolism of mRNA

A. Messenger RNA Synthesis

From many studies it would appear that eukaryotic mRNA is synthesized in the nucleus as a precursor molecule, heterogeneous nuclear RNA (HnRNA). The synthesis and properties of HnRNA have been studied most thoroughly in HeLa cells, in which a significant portion of this material is of very high molecular weight (Darnell, 1968; Darnell *et al.*, 1973; Weinberg, 1973). Although evidence for a precursor-product relationship between HnRNA and cytoplasmic mRNA has been difficult to establish, it is clear that these molecules share a number of common properties, including rapid labeling pattern, high percentage of hybridization to DNA, insensitivity to low doses of actinomycin D, and a 3′-poly(A) sequence. In contrast to ribosomal RNA, which is synthesized in the nucleolus, HnRNA is synthesized in the nucleoplasm. The vast majority of this RNA (frequently up to 90%) never enters the cytoplasm, and apparently only the 3′-terminal region of HnRNA, containing the poly(A) segment, becomes functional messenger RNA. Only 20–40% of total HnRNA contains a poly(A) segment and a portion of this poly(A) is degraded in the nucleus (Jelinek *et al.*, 1973; Greenberg and Perry, 1972a; Darnell *et al.*, 1973). The mechanism for processing of HnRNA into mRNA has not been fully clarified, and presently it is not known whether internal segments of the precursor molecule can also become adenylated or pass into cytoplasm as mRNA. For a review of this topic see Lewin (1975a,b).

Currently, it is thought that HnRNA processing may occur by cleavage at specific double-stranded or highly reiterated regions (Jelinek *et al.*, 1973). Most recent evidence would suggest that RNA sequences from both the 5′ and the 3′ end of HnRNA can be processed into mRNA (Derman and Darnell, 1974; Holmes and Bonner, 1974; Perry *et al.*, 1975; Salditt-Georgieff *et al.*, 1976; and Herman *et al.*, 1976). Such events might be controlled by enzymes with properties similar to bacterial RNase III, which specifically degrades double-stranded RNA (Robertson *et al.*, 1968; Dunn and Studier, 1973). One or more nucleotide sequences rich in oligo(U) has also been found toward the 5′-portion of very high molecular weight HeLa HnRNA (Molloy *et al.*, 1972, 1974). Similar oligo(U) sequences have not been found in fully processed cytoplasmic mRNA.

Although Ruiz-Carrillo *et al.* (1973) have reported translation of duck reticulocyte HnRNA into globin polypeptides in a Krebs II ascites tumor cell-free system, most investigators have been unsuccessful in translating HnRNA in cell-free systems. However, after injection into *Xenopus* oocytes, HnRNA fractions from mouse fetal liver and mouse myeloma cells have been reported to synthesize globin and immunoglobulin, respectively (Williamson *et al.*, 1973; Stevens and Williamson, 1972). This would suggest that HnRNA may require some modification in the intact cell prior to its function as mRNA (perhaps a cleavage of excess nucleotides or a modification of the primary or secondary structure in the initiation or preinitiation region). However, the possibility remains that these HnRNA preparations were contaminated with small amounts of cytoplasmic mRNA, although control experiments were performed to rule this out (Ruiz-Carrillo *et al.*, 1973; Williamson *et al.*, 1973).

A number of investigators (Ishikawa *et al.*, 1969, 1972; Schumm and Webb, 1972) have observed that RNA can be released from whole nuclei by incubation *in vitro* with buffer solutions containing ATP and an ATP generating system. It has also been reported that addition of nuclear RNA plus isolated nuclei from adenovirus-infected HeLa cells to an Ehrlich ascites tumor cell-free system results in synthesis of adenovirus-specific polypeptides (Chatterjee and Weissbach, 1974). This would imply that specific RNA processing is required for translation of adenovirus nuclear RNA and that enzymes for this specific function are present in the nucleus. In the bacterial system, using high molecular weight bacteriophage polycistronic RNA's, several laboratories have reported *in vitro* processing of T3 and T7 RNA by RNase III. The processed RNA has been coupled to a bacterial cell-free system with translation of specific viral polypeptides (Hercules *et al.*, 1974; Dunn and Studier, 1975). The cell-free synthesis of biologically active *S*-adenosylmethionine hydrolase (SAMase), a specific viral enzyme, has also been reported with processed T3 RNA (Hercules *et al.*, 1974).

B. Messenger RNA Structure

Fully processed messenger RNA molecules contain more nucleotides than required for synthesis of their specific encoded polypeptides. In bacteriophage R17 RNA, specific preinitiation regions have been identified and sequenced for all three cistrons (Steitz, 1969). These preinitiation regions in conjunction with secondary structural elements are important in regulation of specific viral polypeptide synthesis (Steitz, 1969; Lodish, 1970; Gralla *et al.*, 1974). In addition, gaps of approximately 30 nucleotides have been found between the cistrons of R17 RNA and there is duplication of the terminator codons. In monocistronic mRNA's of eukaryotic cells, extra nucleotides appear to be present at both the 5'- and the 3'-end of the molecule. Most eukaryotic mRNA's contain a terminal 3'-poly(A) sequence. Notable exceptions include histone mRNA's (Adesnik and Darnell, 1972; Schochetman and Perry, 1972; Greenberg and Perry, 1972a), a portion (30%) of HeLa cell mRNA's (Milcarek *et al.*, 1974), and certain early sea urchin embryo mRNA's (Nemer *et al.*, 1974). The 3'-poly(A) segment of eukaryotic mRNA's can range in length up to 170 nucleotides (Lin and Canellakis, 1970; Kates, 1970; Darnell *et al.*, 1971; Mendecki *et al.*, 1972; Brawerman, 1974). This sequence is similar to the 3'-poly(A) segment of HnRNA. Since a corresponding sequence of (dT) has not been found on eukaryotic DNA, the 3'-poly(A) segment must be added posttranscriptionally. Attempts to find a 3'-poly(A) segment with prokaryotic mRNA's have generally been unsuccessful. However, recent studies from Kramer *et al.* (1974) and Rosenberg *et al.* (1975) report a short sequence of oligo(A) in the 3'-region of bacteriophage T7 RNA. A short oligo(A) sequence has also been found at the 3'-end of pulse labeled RNA isolated from *E. coli* cells (Srinivasan *et al.*, 1975).

The enzyme responsible for adding poly(A) to RNA primers with a free 3'-OH group was described in the nucleus of eukaryotic cells long before the importance of the poly(A) sequence was recognized (Edmonds and Abrams, 1960; Edmonds and Caramela, 1969). This enzyme [poly(A) polymerase] is distinct from the terminal adenylate transferase, which adds AMP to the 3'-CCOH end of tRNA, and may actually represent a group of enzymes. Different divalent cation requirements have been observed with different preparations of poly(A) polymerase (Edmonds and Abrams, 1965; Winter and Edmonds, 1972; Tsiapalis *et al.*, 1973). One of these preparations appears to require Mn^{2+} (Giron and Huppert, 1972). In yeast, two nuclear polymerases with different substrate specificities have been identified (Haff and Keller, 1973). There may be different enzymes for addition of short versus long poly(A) segments. Other observations would suggest that poly(A) polymerase activity can be located in the cytoplasm as well as in the nucleus. Vesicular stomatitis virus (VSV) is a single-stranded RNA virus which replicates in the cytoplasm. Messenger RNA (but not virion RNA) for VSV contains a 3'-poly(A) segment (Mudd and Summers, 1970; Johnston and Bose, 1972). Therefore, it would appear that the posttranscriptional addition of poly(A) to VSV mRNA occurs in the cytoplasm. In the case of the slime mold *Dictyostelium,* the 3'-poly(A) region of the messenger contains two poly(A) segments, separated by a short sequence of non-A nucleotides (Firtel and Lodish, 1973; Jacobson *et al.*, 1974a,b). A similar observation has also been reported recently in HeLa cells (Nakazato *et al.*, 1974). The first poly(A) segment contains 25 nucleotides and is encoded in the genome. The second poly(A) segment contains ~ 100 nucleotides and is added posttranscriptionally. Pulse-chase experiments indicate that in the nucleus label appears more rapidly in the first poly(A) segment of *Dictyostelium* mRNA, whereas in the cytoplasm label appears more rapidly in the second poly(A) segment. Moreover, with isolated *Dictyostelium* nuclei *in vitro* the first poly(A) segment is synthesized, but the second poly(A) segment is not added. In addition, α-amanitin-sensitive and α-amanitin-resistant RNA polymerase activities have been found (Jacobson *et al.*, 1974a,b). Therefore, the two poly(A) segments are synthesized by separate mechanisms. Attempts to isolate poly(A) polymerase for addition of the second poly(A) segment from *Dictyostelium* nuclei have given negative results. At the present moment it is not entirely clear whether addition of the second poly(A) segment occurs in the nucleus, nuclear membrane, or cytoplasm. The nuclear precursor to *Dictyostelium* mRNA has an average molecular weight only 20% greater than fully processed cytoplasmic mRNA (~ 1500 nucleotides in the precursor vs. ~ 1200 nucleotides in polysomal mRNA) (Firtel and Lodish, 1973). The size of the nuclear mRNA precursor in yeast, another lower eukaryote, is also rather small and is only approximately twice the size of cytoplasmic mRNA (J. Warner, personal communication).

Other important observations on the primary structure of eukaryotic mRNA's have been made with 9 S globin mRNA. The length for the α- and β-chains of rabbit globin are 141 and 146 amino acids, respectively. To code for polypeptides of this size, a messenger RNA of approximately 430 nucleotides or a molecular weight of 145,000–150,000 daltons is required. Most recent estimates of molecular weight for 9 S globin mRNA by polyacrylamide gel electrophoresis fall into the range of 200,000–230,000 daltons (Gaskill and Kabat, 1971; Pemberton *et al.*, 1972; Gould and Hamlyn, 1973). This corresponds to an RNA chain length of 590–670 nucleotides. Except for a very small portion of early labeled mRNA with a longer poly(A) segment (Gorski *et al.*, 1974), globin mRNA has a 3′-poly(A) segment of ~ 50–60 nucleotides. This leaves 100–180 nucleotides in the unaccounted fraction. Evidence for a noncoding sequence of approximately 93 nucleotides in the region between the terminator codon and the 3′-poly(A) sequence has been obtained from studies of three human hemoglobin variants (Clegg *et al.*, 1971, 1974; Seid-Akhavan *et al.*, 1972). These hemoglobins contain extra amino acids in the carboxy terminal region of their α-chains and appear to result from mutations in or near the termination signal. Hemoglobin Constant Spring, the first variant of this type to be described, contains an additional peptide sequence of 31 amino acids beginning with glutamine. According to the genetic code, this could occur by a mutation in the terminator codon UAA or UAG in which

C is substituted for U. This would create the code word CAA or CAG, directing insertion of glutamine in place of a normal termination event. Additional nucleotides past the mutated termination signal would also be translated until a new termination signal is reached or protein synthesis stops by some other mechanism. Hemoglobin Wayne has an altered C-terminal region with 5 extra amino acids. Presently, it is thought that this variant may have occurred by a deletion affecting amino acid 139, causing a frame shift mutation (Seid-Akhavan *et al.,* 1972). Hemoglobin Icaria again contains an extra peptide segment of 31 amino acids (Clegg *et al.,* 1974). It is identical to Hb Constant Spring, except that the first extra amino acid is lysine instead of glutamine. This would appear to be the re-sult of a mutation in which the termination signal UAA has been changed to AAA (lysine). These interpretations are consistent with data obtained recently on the nucleotide sequence of human hemoglobin mRNA (Marotta *et al.,* 1974).

Globin mRNA still contains a minimum of 50 nucleotides not accounted for by the above considerations. This discrepancy may simply reflect the limitations of present methods to accurately determine nucleotide length or may suggest additional noncoding nucleotides, perhaps in the preinitia-tion 5'-region of the messenger. The best evidence for noncoding RNA se-quences in the 5'-region of mammalian mRNA comes from recent studies using DNA-RNA hybridization. By this method Dina *et al.* (1973, 1974) have observed a small portion of rapidly annealing repetitive se-quence transcripts in mRNA from *Xenopus* neuralae, and Spradling *et al.* (1974) have found appreciable amounts of repetitive sequences of the intermediate type in mRNA from HeLa cells and rat myoblasts. In *Dic-tyostelium,* approximately 10% of cytoplasmic mRNA hybridizes to re-iterated DNA sequences, and these RNA sequences are also thought to reside in the 5'-region of the mRNA (Jacobson *et al.,* 1974).

Other eukaryotic mRNA's appear to have much larger noncoding regions than globin mRNA. The sedimentation values for ovalbumin mRNA range from 16–21 S, depending on the method used (Palacios *et al.,* 1973; Rosen *et al.,* 1975). The molecular weight of such a molecule (approximately 700,000 daltons) is considerably greater than that required for synthesis of ovalbumin, a polypeptide of molecular weight 45,000 daltons. Bovine lens α-crystallin mRNA's (10 S and 16 S) are also larger than required for the coded products (Berns and Bloemendal, 1974). In this case the smaller lens mRNA codes for a larger polypeptide than the larger lens mRNA. Immunoglobulin light chain mRNA is also consider-ably larger than required for synthesis of the L chain polypeptide (Swan *et al.,* 1972; Mach *et al.,* 1973; Haines *et al.,* 1974).

Within the past two years a number of investigators have identified a unique sequence of 7-methylguanosine (m^7G) in 5'-triphosphate linkage to the 5'-terminal nucleotide of various eukaryotic and viral mRNA's (i.e., m^7 GpppNp. . . etc., where N may be any 2'-O-methyl ribonucleotide) (for general references, see Both *et al.*, 1975b,c). In addition, nuclear RNA from Novikoff hepatoma cells has been reported to contain a 5'-terminal m^7G^5 diphosphate group (Reddy *et al.*, 1974) and 5'-terminal methylase activity has been found in both the nucleus and cytoplasm of eukaryotic cells (Gröner and Hurwitz, 1975; Both *et al.*, 1975b,c), as well as in intact virions (Rhodes *et al.*, 1974; Furuichi *et al.*, 1975; Both *et al.*, 1975c). Although various studies have raised the possibility that m^7G may serve a function in mRNA processing and/or mRNA stability, current evidence indicates that this sequence is required specifically for translation of certain mRNA's (Both *et al.*, 1975b). It has been proposed that m^7Gppp Np . . . may serve as the primary recognition signal for binding of these mRNA's to the ribosome (Both *et al.*, 1975b), and recently Shafritz *et al.* (1976) have reported that $m^7G^5{}'p$ inhibits directly the interaction of eukaryotic initiation factor IF-M_3 with VSV and HeLa histone mRNA's (both containing 5'-terminal m^7G) but not the interaction of IF-M_3 with EMC virus RNA (not containing 5'-terminal m^7G).

It has also been recognized that eukaryotic mRNA's contain a considerable degree of secondary structure (as observed previously with other types of RNA). In the case of globin mRNA this has been demonstrated by classical alkaline hydrolysis and thermal denaturation experiments (Lingrel *et al.*, 1971; Williamson *et al.*, 1971), as well as by measurements of ethidium bromide fluorescense (Favre *et al.*, 1974). With ovalbumin mRNA secondary structure apparently accounts for certain variations in migration through solid support gels and sucrose gradients under denaturing vs. nondenaturing conditions (Palacios *et al.*, 1973; Rosen *et al.*, 1975). Appropriate references to such techniques are given in Section IV,B. The function(s) of mRNA secondary structure has not been elucidated, although its possible relationship to mRNA processing or metabolism, specific protein binding sites, and/or mRNA recognition regions for initiation of protein synthesis is well appreciated (Williamson *et al.*, 1971; Jelinek *et al.*, 1973; Scherrer, 1973; Favre *et al.*, 1974; Both *et al.*, 1975c).

C. Metabolism of Eukaryotic mRNA

In eukaryotic cells limited information is available regarding the mechanism for mRNA transport, the control of mRNA-ribosome interaction, and the regulation of mRNA degradation. These processes may be related to formation of specific mRNA-protein complexes, termed messenger ribonucleoprotein particles (to be discussed separately in Section VII). From many studies it is clear that the average half-life ($t_{1/2}$) for eukaryotic mRNA's is considerably longer than that observed for prokaryotic mRNA's. Most eukaryotic mRNA's have $t_{1/2}$'s on the order of several hours or more. A problem with many of the early studies on the $t_{1/2}$'s of eukaryotic mRNA's is that they were based on indirect evidence. These studies generally measured disaggregation of polysomes or cessation of protein synthesis after addition of sufficient quantities of actinomycin D to block transcription of all types of RNA. Since this agent also inhibits initiation of protein synthesis and may have other toxic effects on cells (Singer and Penman, 1973) and since a decay in structural integrity or translational activity of polyribosomes may reflect changes other than mRNA degradation, evidence of this type may be unreliable (Singer and Penman, 1972; Brawerman, 1974). A current estimate for the $t_{1/2}$ of reticulocyte globin mRNA in the anemic mouse, based on direct labeling studies, is approximately 15–17 hours (Hunt, 1974). In HeLa cells, original data suggested a $t_{1/2}$ for mRNA of approximately 3–4 hr (Penman *et al.*, 1963). In recent studies, two separate classes of HeLa mRNA with $t_{1/2}$'s of 7 and 24 hr have been reported (Singer and Penman, 1973). In another study, the $t_{1/2}$ for HeLa mRNA is given as several days (Murphy and Attardi, 1973). In exponentially growing L cells, the mean $t_{1/2}$ for poly(A)-containing mRNA has been estimated at 14.4 hr (Greenberg, 1972). One general exception to the rather long $t_{1/2}$ of many eukaryotic mRNA's is the $t_{1/2}$ for histone mRNA's. This has been estimated at approximately 1–2 hr (Borum *et al.*, 1967; Gallwitz and Mueller, 1969; Craig *et al.*, 1971) and is particularly interesting in view of the absence of a 3'-poly(A) sequence in histone mRNA's (Adesnik and Darnell, 1972; Schochetman and Perry, 1972; Greenberg and Perry, 1972a). In a very recent study (Huez *et al.*, 1974), it has been observed that deadenylated globin mRNA is degraded within 5–10 hr after injection into frog oocytes, whereas control adenylated globin mRNA continues to produce globin for up to several days. Marked differences in translational stability of adenylated vs. deadenylated globin mRNA have also been found in the Krebs II ascites cell-free system (Soreq *et al.*, 1974). From these and other considerations it has been proposed that the function of the 3'-poly(A) segment may be to stabilize eukaryotic mRNA or protect it from ribonuclease destruction.

At present it is not clear whether the degradation of mRNA occurs by a random process, by specific cleavage from either the 5'- or the 3'-end, or by endonucleolytic cleavage at specific internal sites. In bacterial cells a number of investigators have reported evidence of cleavage in the 5' to 3' direction (Adesnik and Levinthal, 1970; Mosteller *et al.*, 1970), but a specific 5'-endonuclease has not been identified. In eukaryotic cells it has been shown that the 3'-poly(A) segment of mRNA's becomes progressively shorter as the messengers age (Mendecki *et al.,* 1972; Greenberg and Perry, 1972b; Sheiness and Darnell, 1973; Gorski *et al.,* 1974). There are also several discrete size classes for the poly(A) segment of mRNA (35–45 nucleotides, 55–65 nucleotides, and 75–120 nucleotides), rather than a random distribution (Gorski *et al.,* 1974). It has also been found that the average length of the poly(A) segment of cytoplasmic mRNA in HeLa cells decreases with time during translation (Sheiness and Darnell, 1973). Certain mRNA's may be completely deadenylated after a single cell generation time (Murphy and Attardi, 1973). These studies would suggest a specific pathway for the metabolism of the poly(A) segment of mRNA.

IV. Isolation of Messenger RNA

A. General Procedures for mRNA Extraction

A number of methods have been developed for extraction of total cellular RNA. Most of these procedures utilize phenol to denature protein and extract it from RNA, which remains in the aqueous phase. Since most crude cytoplasmic extracts contain ribonuclease activity, additional measures have been taken to inactivate or remove these enzymes. Some investigators have homogenized their tissue directly in aqueous-phenol mixtures (Schutz *et al.,* 1973; Merkel *et al.,* 1974). In most cases, 0.5–1.0% sodium dodecyl sulfate (SDS) is added before phenol extraction. This detergent serves the dual function of releasing RNA from protein and inactivating ribonucleases. Other agents used to remove or inactivate RNase include EDTA, heparin, diethyl pyrocarbonate, dextran sulfate, polyvinyl sulfate, RNase inhibitor from liver supernatant, bentonite, macaloid, and 4-aminosalicylic acid. Phenol extraction is generally repeated at least once to remove as much protein as possible. An interface layer of denatured proteins collects between the organic and the aqueous phase. The phenol layer and the interface material also contain trapped RNA, frequently enriched for messenger RNA (Brawerman *et al.,* 1972). The explanation given for differential loss of mRNA into the phenol-interface layer is that at acidic pH or relatively high concentration of monovalent or divalent cations, mRNA 3′-poly(A) sequences tend to nonspecifically aggregate with denatured proteins (Brawerman *et al.,* 1972). Formation of these aggregates can be avoided in part by performing the phenol extraction under mild alkaline conditions (pH 9.0). The phenol layer should be back extracted with aqueous buffer, and the resulting aqueous phase combined with the original aqueous layer. Phenol extraction is generally repeated until no additional material is seen at the interface. Back extraction is usually omitted on the repeat phenol extracts. Phenol extraction at elevated temperatures (60°) (Georgiev and Mantieva,

1962) or addition of m-cresol to the extraction solvent (Kirby, 1965) has also been reported to improve the separation of RNA and protein.

Another problem encountered with mRNA extraction is aggregation of RNA molecules. This has led to considerable difficulties both in the isolation and characterization of individual mRNA species. Aggregation has been attributed by various investigators to the use of phenol, to extraction at acidic pH, or to the presence of Mg^{2+} during the extraction procedure. Therefore, extraction at alkaline pH, exclusion of Mg^{2+} from buffers wherever possible, and/or addition of EDTA prior to phenol extraction may be helpful in reducing mRNA aggregation. In our experience, however, none of these methods has been entirely successful in eliminating mRNA aggregates, although alkaline pH during phenol extraction is quite important for recovering reticulocyte mRNA in the aqueous phase. This has been monitored both by poly(A) content and by cell-free translation of reticulocyte mRNA (J. Mendecki and D. A. Shafritz, unpublished observations). In one modification of the phenol procedure which has been used widely, an equal volume of chloroform-isoamyl alcohol (25:1 v/v) is added to the phenol. Extraction with this mixture has been reported to preserve 3'-poly(A) segments of mRNA by reducing cleavage or shearing of these segments at the organic aqueous interface (Perry et al., 1972). A relatively new method for RNA extraction is addition of 0.5% SDS-proteinase K to isolated polysomes, followed by digestion of protein at 25° (Wiegers and Hilz, 1971). After protein digestion is completed, intact mRNA can be isolated without the need for phenol extraction. Under these circumstances mRNA aggregates are not observed (Macnaughton et al., 1974).

Once mRNA aggregates are formed, they cannot be dissociated completely by centrifugation in an SDS sucrose gradient. Aggregates of mRNA can be dissociated by centrifugation in 99% DMSO, which eliminates RNA secondary structure (Strauss et al., 1968; Sedat et al., 1969; Holmes and Bonner, 1973). However, aggregates have been noted to reform under certain ionic conditions after DMSO has been removed (Birnboim, 1972). This can be demonstrated by repeat centrifugation in the absence of DMSO. Formamide gel electrophoresis under denaturing conditions has also been used as an analytical and preparative tool for mRNA (Boedtker et al., 1973; Gould and Hamlyn, 1973; Morrison et al., 1974; Kazazian et al., 1974; Rosen et al., 1975). However, in higher eukaryotes it has been difficult to recover mRNA with good biological activity after formamide gel electrophoresis or DMSO centrifugation. Electrolytic elution of mRNA from formamide gels has recently been reported to provide good recovery of biologically active α- and β-chain globin mRNA (Kazazian et al., 1974).

In the case of certain mRNA's with sedimentation values distinct from other species of RNA, sucrose gradient centrifugation can be performed directly with SDS treated polysomes, thus avoiding problems with phenol and aggregation. This method was used in the original studies isolating 9 S globin mRNA (Chantrenne et al., 1967; Lockard and Lingrel, 1969). Since mRNA represents only a small percentage of total polysomal RNA (1–2%), it is generally necessary to follow pulse labeled RNA or RNA labeled in the presence of low doses of actinomycin D, rather than the A_{260} pattern of the RNA extract. Sedimentation of SDS treated polysomes is frequently performed at room temperature (23°), although lower temperatures can be employed if LiCl is substituted for NaCl in the gradient or sarcosyl (sodium lauryl sarcosinate) is used in place of SDS.

B. Purification of Messenger RNA

1. Affinity Columns

As mentioned earlier, most eukaryotic mRNA's, contain a segment of poly(A) at their 3'-terminus. The length of the 3'-poly(A) segment may vary considerably or be absent, depending on preparative methods or normal cellular metabolic events (see Section III,C). In spite of this limitation, the 3'-poly(A) segment has been valuable both as an analytical tool and a preparative handle for purifying messenger RNA. This results from the ability of poly(rA) to form stable hybrids with homopolymers of (dT) or (rU). The preparation of affinity columns with poly(dT), poly(rU), or other artificial polynucleotides cross-linked to a supporting matrix of cellulose was first described over a decade ago (Bautz and Hall, 1962; Gilham, 1962; Adler and Rich, 1962). These columns were used initially for examining various parameters in artificial polynucleotide hybridization and their potential usefulness for the study of natural RNA's was predicted. If a solution containing RNA with a poly(A) sequence is applied to a small column of oligo(dT)-cellulose at relatively high ionic strength (0.1 M KCl or above), the RNA will rapidly anneal to the oligo(dT) side arm by hydrogen bond base pairing. The column can be washed to remove nonannealed RNA and hybridized material eluted by reducing the ionic strength. Intermediate washing steps with buffer containing progressively lower ionic strengths can also be used to remove nonspecifically adsorbed or loosely bound material. If the poly(A) segment of mRNA is very short (less than 8–10 nucleotides), a significant portion may not bind to the oligo(dT)-cellulose. The value of this method for separation of mRNA from the bulk of cellular RNA with good recovery of biological activity was first reported by Leder and co-workers (Aviv and Leder, 1972; Swan et al., 1972). These investigators reported a 20- to 50-fold enrichment in globin mRNA and immunoglobulin light chain mRNA by a combination of oligo(dT)-cellulose chromatography and sucrose gradient centrifugation.

Messenger activity during both of these procedures was monitored by stimulation of protein synthesis in a Krebs II ascites cell-free system. As noted in these studies, small amounts of 18 S and 28 S RNA may adsorb to oligo(dT)-cellulose and 28 S RNA contamination has also been noted by other investigators. Such contaminants can be reduced by repeating the oligo(dT)-cellulose step (Singer and Penman, 1973; Gorski et al., 1974).

Numerous applications of the affinity method, utilizing oligo(dT)-cellulose (Edmonds and Caramela, 1969; Nakazato and Edmonds, 1972; Scolnick et al., 1973; Kazazian et al., 1973; Singer and Penman, 1973; Higgins et al., 1973; Prives et al., 1974; Gorski et al., 1974; Favre et al., 1974; Zelenka and Piatigorsky, 1974; Soreq et al., 1974; Cancedda and Schlesinger, 1974; Eron and Westphal, 1974; Shafritz and Drysdale, 1975; Green et al., 1975), poly(U)-cellulose (Kates, 1970; Sheldon et al., 1972; Morrison et al., 1972), and poly(U)-Sepharose (Adesnik et al., 1972; Lanyon et al., 1972; Lindberg and Persson, 1972; Firtel et al., 1972) have been reported. Oligo(dT)-cellulose columns are quite stable and may be used repeatedly by washing the column with 0.1 M KOH between experiments. When not in use the column may be stored in sodium azide or the cellulose removed and stored frozen. Poly(U)-Sepharose is somewhat less stable than oligo(dT)-cellulose, and a portion of the poly(U) may be lost with each usage of the column. Elution of poly(U) with mRNA may cause undesirable effects in subsequent study of the mRNA. It must be remembered that mRNA containing only a short segment of poly(A) (less than 8–10 nucleotides) may be lost in the affinity step. With reticulocyte mRNA, such losses are less than 10% of total messenger activity. However, with mRNA from other eukaryotic sources, as much as 30% or more of the total messenger may be found in the fraction not adsorbed to oligo(dT)-cellulose (Milcarek et al., 1974). With RNA extracts from either rat or rabbit liver polysomes, the bulk of mRNA activity is not adsorbed to oligo(dT)-cellulose (D. A. Shafritz, unpublished observations).

2. Nitrocellulose Filters

Poly(A) and eukaryotic RNA molecules containing poly(A) also have a special affinity for adsorption to nitrocellulose filters (Lee et al., 1971a; Brawerman et al., 1972; Rosenfeld et al., 1972; Palacios et al., 1973). This interaction again requires a rather high ionic strength. For effective mRNA adsorption, the length of the poly(A) segment must be at least 50–60 nucleotides (Sullivan and Roberts, 1973; Gorski et al., 1974). Although the poly(A) segment of eukaryotic mRNA's and their precursors may range up to 150–200 nucleotides, there is great variability in poly(A) length. This limits the usefulness of the nitrocellulose filter technique as a preparative method. This method, however, has been used for separating and studying the metabolism of globin mRNA molecules with long vs. intermediate poly(A) length (Gorski et al., 1974). Exposure to ultraviolet light can also be used to adsorb poly(U) to glass fiber filter discs, and these have been used as a rapid assay for poly(A)-containing RNA (Sheldon et al., 1972). Certain preparations of cellulose powder have been used for mRNA purification (Kitos et al., 1972; Schutz et al., 1972; Sullivan and Roberts, 1973; DeLarco and Guroff, 1973). These preparations appear to work through interaction of poly(A) with impurities contained within the cellulose fiber and have not found wide usage.

3. Zone Sedimentation

In some instances molecular size can be of assistance in purification of mRNA's. In the case of rabbit reticulocytes and other differentiated cell types synthesizing predominantly one protein (e.g., immunoglobins in the myeloma cell, ovalbumin in the estrogen-induced chick oviduct, and α-crystallins in the lens of the eye), specific messenger RNA's are increased in amount relative to other mRNA's. This has been of particular value in the identification and isolation of these messengers. In the rabbit reticulocyte, mRNA released from polysomes by EDTA is complexed to proteins. These mRNP complexes sediment in a sucrose gradient at approximately 15 S. This region of the gradient is well separated from other fractions containing nucleic acids, such as ribosomal subunits and transfer RNA. Therefore, zone sedimentation provides a high level of purification for this messenger as an mRNP. Messenger RNA released from the mRNP by either SDS or SDS-phenol sediments at 9 S. The 15 S mRNP and 9 S RNA isolated from this complex direct the synthesis of globin in the various cell-free systems (Lingrel et al., 1971; Housman et al., 1971; Lanyon et al., 1972; Sampson et al., 1972; Schreier et al., 1973). The advantage of first isolating globin mRNA as the mRNP particle is that this procedure will eliminate nonmessenger components of RNA which normally sediment in the 9 S region (Williamson et al., 1971; Morrison et al., 1972; Lanyon et al., 1972; Kazazian et al., 1973). The mRNP method has also been used for preparation of bovine lens α-crystallin mRNA's (Berns and Bloemendal, 1974). When an mRNA-protein complex is isolated from the supernatant fraction of rabbit reticulocytes, it also sediments in the 15–20 S region. With this material and its RNA extract, 85–90% of the cell-free product is α-globin chains (Jacobs-Lorena and Baglioni, 1972; Gianni et al., 1972; Olsen et al., 1972; Bonanou-Tzedaki et al., 1972). Therefore, the reticulocyte supernatant mRNP fraction is highly enriched in α-globin mRNA. Polysomes enriched in β-globin mRNA have been obtained by treatment of reticulocytes with O-methylthreonine. This amino acid analogue acylates $tRNA^{Ile}$ and renders it inactive as a donor of isoleucine to growing polypeptide chains. Since the α-chain in the rabbit contains three residues of isoleucine whereas the β-chain contains only one isoleucine residue, and since the isoleucine residue of the β-chain is located toward the C-terminal end of the molecule, whereas the isoleucine residues of the α-chain are distributed throughout the molecule, ribosomes tend to pile up more on β-globin messenger. This creates very large β-chain polysomes. These can be separated from smaller α-chain polysomes by differential sucrose gradient centrifugation. The RNA fraction isolated from large polysomes after O-methylthreonine treatment is enriched up to 70–75% in β-globin mRNA. The α- and β-globin mRNA fractions obtained by these procedures have been used for synthesis of α- and β-globin chain-specific cDNA's (see Section II,C).

4. Electrophoresis

Electrophoresis of RNA in gels containing polyacrylamide, agarose, or other supporting materials has been used to separate RNA species and estimate their molecular weights (for original references to this method, see Bishop et al., 1967; Peacock and Dingman, 1967, 1968; Loening, 1967). This method has been used primarily as an analytical tool but has also been applied for preparative purposes (Labrie, 1969; Williamson et al., 1971). More recently, gel electrophoresis in the presence of formamide (a denaturing agent) has been used to separate α- and β-globin mRNA's (Gould and Hamlyn, 1973; Morrison et al., 1974; Kazazian et al., 1974). When used in a very precise fashion, this method takes advantage of small differences in molecular weight of α- and β-globin mRNA. Isoelectric focusing in polyacrylamide gel has also been used to obtain partial resolution of α- and β-globin mRNA, although the basis for separation of RNA by this procedure has not been established (Shafritz and Drysdale, 1975).

5. Immunoprecipitation

Immunological properties have also been helpful in purification of mRNA. As polysomes are isolated from the cell, they contain nascent polypeptide chains of the protein encoded in their mRNA. In some cases polysomes containing a specific mRNA can be isolated by interaction of the nascent chains with purified antibodies to the completed protein. Early demonstrations of the use of immunological procedures to enrich polysomes fractions for a specific mRNA were reported by Miller et al. (1971) and Oenoyama and Ono (1972). In order to obtain mRNA that is not degraded and that demonstrates high biological activity, extreme care must be taken to rid the antibody preparation of ribonuclease contamination (Palmiter et al., 1972). Some investigators have advocated the use of antibody Fab fragment, as the Fc component may cause some nonspecific interaction (Holme *et al.*, 1971; Ikehara and Pitot, 1973). Extensive use of immunoprecipitation and its value for isolation and purification of mRNA has been demonstrated by Schimke and co-workers (for reviews, see Schimke *et al.*, 1973, 1974). These investigators have purified ovalbumin mRNA, conalbumin mRNA, and albumin mRNA by immunological precipitation coupled with other methods and have used this material for a wide range of molecular studies. Immunoglobulin light chain (L chain) mRNA has also been purified using immunoprecipitation of polysomes as one of the steps (Schechter, 1973), and rabbit reticulocyte polysomes containing α- or β-globin mRNA have been separated and isolated by solid phase immunoadsorption (Boyer *et al.*, 1974).

V. Translation of Exogenous mRNA

Many systems have been developed for translation of exogenous mRNA. These systems vary widely in their degree of fractionation of protein synthesis components, their level of background amino acid incorporation (endogenous messenger activity), and their responsiveness to individual mRNA species. As indicated previously, the use of a heterologous translating system with specific identification of product is essential for demonstrating exogenous messenger activity. For these purposes purification of the mRNA is not required. If the system is also dependent on exogenous mRNA (low background activity under the conditions used and not stimulated significantly by other types of RNA), then a stimulation of total protein synthesis (5- to 10-fold or more) can be used as a rapid and routine assay for the presence of mRNA. It should be noted that the efficiency of mRNA translation may vary with amount of mRNA or total RNA added. Excess mRNA may actually cause inhibition of protein synthesis, and other types of RNA (e.g., double-stranded RNA, tcRNA) may suppress mRNA translation under certain circumstances (Hunt and Ehrenfeld, 1971; Heywood *et al.*, 1974). The level of initiation factors or other components in the system as well as other experimental variables are also important (see Chapters 5, 6, and 7).

Differences in monovalent and divalent cation requirements for translation of various mRNA's have been found (Rourke and Heywood, 1972; Mathews *et al.*, 1972b; Roberts and Paterson, 1973), and levels giving the highest total incorporation may not be best for translation of specific mRNA. Therefore, selection of the appropriate experimental conditions as well as the system may be critical for translation of a given mRNA. For example, to simply demonstrate the presence of mRNA in a crude RNA sample, one might chose a system with a high level of translation (high efficiency), in spite of a high level of background activity. However, to examine questions of mechanism or specificity in mRNA translation, a highly fractionated system, which may have a considerably reduced efficiency, would be desirable. The more frequently used systems for translating exogenous mRNA are described in this section, indicating the advantages and disadvantages of each system. These are listed in Table I, together with examples of mRNA's which have been translated in these systems.

A. Preincubated *Escherichia coli* S30 System

One of the earliest and most widely used systems for translation of exogenous mRNA is that originally reported by Nirenberg and Matthaei (1961). This system is obtained from *E. coli* and can be utilized for translation of artificial polyribonucleotides as well as natural mRNA's. Briefly, a crude cytoplasmic extract is prepared from *E. coli;* a mutant strain deficient in one or more ribonucleases may be desirable (e.g., MRE600, Q13). Cells are lysed by grinding with alumina or by passage through a French press. Large debris is removed by centrifugation, and DNA is degraded by incubation in the cold with highly purified DNase. This treatment markedly reduces the viscosity of the solution, and the resulting material is centrifuged at 30,000 g for 30 min, collecting the supernatant layer. This fraction contains ribosomes and other components required for protein synthesis and is designated the *E. coli* "S30" fraction. The S30 fraction is then incubated at 37° for 30–60 min (80 min in the original procedure) in the presence of the complete protein synthesis mixture, including all 20 required amino acids in unlabeled form. During this period, protein synthesis proceeds to complete preexisting nascent polypeptide chains. The completed polypeptide chains and mRNA are released from the ribosome, and mRNA is degraded. The system is now dependent on exogenous mRNA for stimulation of protein synthesis and is designated the "preincubated" S30 system (Nirenberg, 1963). This fraction is dialyzed (or in more recent studies passed over a Sephadex G-25 column) to remove unused amino acids and other low molecular weight components. The preincubated S30 system can be stored frozen in liquid nitrogen with good retention of biological activity for long periods, and has been used for numerous studies on cell-free translation of viral mRNA's (Nathans *et al.,* 1962; Ofengand and Haselkorn, 1962; Clark *et al.,* 1965; Capecchi, 1966; Salser *et al.,* 1967; Lodish, 1968, 1970). Bacterial systems have also been utilized for translation of animal virus and other eukaryotic mRNA's (Laycock and Hunt, 1969; Rekosh *et al.,* 1970; Siegert *et al.,* 1972). However, under these circumstances, translation efficiencies are considerably lower than for bacteriophage RNA's, and incomplete polypeptides are generated (Rekosh *et al.,* 1970). This may be due at least in part to nucleolytic cleavage of exogenous mRNA added to the bacterial system. Deter-

TABLE I

Most Commonly Used Systems for Translation of Specific Exogenous mRNA's[a]

mRNA	Reference
A. *Escherichia coli*	
TMV	Nirenberg and Matthaei, 1961; Roberts *et al.*, 1973.
STNV	Clark *et al.*, 1965.
R17	Capecchi, 1966.
T4 (synthesizing active lysozyme)	Salsar *et al.*, 1967.
φX174	Gelfand and Hayashi, 1969.
f2	Nathans *et al.*, 1962; Lodish, 1968, 1970.
Rabbit globin	Laycock and Hunt, 1969.
Poliovirus	Rekosh *et al.*, 1970.
AMV	Siegert *et al.*, 1972.
Qβ (synthesizing active replicase)	Happe and Jockusch, 1973.
T3 (synthesizing active SAMase)	Hercules *et al.*, 1974.
B. Reticulocyte lysate	
Mouse globin	Lockard and Lingrel, 1969.
Rabbit globin	Lockard and Lingrel, 1973; Stewart *et al.*, 1973.
Duck globin	Pemberton *et al.*, 1972; Stewart *et al.*, 1973.
Ovalbumin (chicken or hen)	Rhoads *et al.*, 1971; Means *et al.*, 1972; Haines *et al.*, 1974.
Albumin	Taylor and Schimke, 1973.
Mouse myeloma L chain	Stavnezer and Huang, 1971.
Bovine lens α-crystallins	Berns *et al.*, 1972a.
Chicken lens crystallins	Williamson *et al.*, 1972.
Chicken embryo lens δ-crystallin	Zelenka and Piatigorsky, 1974.
Histone mRNA (HeLa cell)	Breindl and Gallwitz, 1973.
Sindbis virus	Simmons and Strauss, 1974.
Ferritin and albumin[b]	Shafritz *et al.*, 1973; Shafritz, 1974b.
Sindbis virus[b]	Cancedda and Schlesinger, 1974.
Human globin[b]	Nienhuis and Anderson, 1971.
Brain mRNA[b]	Gilbert, 1973.
Rabbit globin[b]	Schreier and Staehelin, 1973a.
C. Ascites tumor cell	
EMC	Smith *et al.*, 1970; Mathews and Korner, 1970; Boime *et al.*, 1971.
Mouse globin	Mathews *et al.*, 1971; Metafora *et al.*, 1972; Sampson *et al.*, 1972.
Rabbit globin	Housman *et al.*, 1971; Aviv and Leder, 1972; Metafora *et al.*, 1972; Sampson *et al.*, 1972.
Rabbit α- and β-globin	Temple and Housman, 1972.
Rabbit globin (deadenylated)	Soreq *et al.*, 1974.
Human globin	Benz and Forget, 1971; Metafora *et al.*, 1972.
Tryptophan oxygenase	Schutz *et al.*, 1973.
Mouse immunoglobulin L chain	Swan *et al.*, 1972; Schechter, 1973.
Mouse immunoglobulin H chain	Green *et al.*, 1974.
Bovine lens α-crystallin	Mathews *et al.*, 1972a.

TABLE I (*Continued*)

mRNA	Reference
Chicken lens crystallin	Williamson *et al.*, 1972.
Chick embryo lens δ-crystallin	Zelenka and Piatigorsky, 1974.
Histone (sea urchin)	Gross *et al.*, 1973.
Histone (HeLa cell)	Jacobs-Lorena *et al.*, 1973.
Qβ	Aviv *et al.*, 1972; Morrison and Lodish, 1973.
Reovirus	McDowell *et al.*, 1972.
Adenovirus type 2	Eron and Westphal, 1974; Anderson *et al.*, 1974.
Sindbis virus	Cancedda and Schlesinger, 1974.
Semliki Forest virus 42 S RNA	Smith *et al.*, 1974.
D. Wheat germ	
TMV	Marcus *et al.*, 1968; Roberts *et al.*, 1973.
Rabbit globin	Efron and Marcus, 1973; Roberts and Paterson, 1973.
SV40	Prives *et al.*, 1974.
Globin (deadenylated)	Sippel *et al.*, 1974.
L cell (adenylated, deadenylated)	Bard *et al.*, 1974.
VSV	Both *et al.*, 1975a.
E. Xenopus oocyte	
Rabbit, mouse, and duck globin	Lane *et al.*, 1971, 1973.
Rabbit globin (deadenylated)	Huez *et al.*, 1974.
Bovine lens A₂ α-crystallin	Berns *et al.*, 1972b.
Mouse myeloma L chain	Stevens and Williamson, 1972; Smith *et al.*, 1973.
Mouse myeloma H chain	Stevens and Williamson, 1973.
EMC	Laskey *et al.*, 1972
Rat brain	Lim *et al.*, 1974.
F. Miscellaneous	
Reovirus [L cell][c]	Graziadei and Lengyel, 1972.
Reovirus [L cell, CHO, HeLa]	McDowell *et al.*, 1972.
Myosin and globin [erythroblast and muscle]	Rourke and Heywood, 1972.
Brain [brain]	Zomzely-Neurath *et al.*, 1972.
Mouse and rabbit globin [mouse and rat liver]	Sampson *et al.*, 1972.
Myoglobin and myosin [chick embryo red and white muscle]	Thompson *et al.*, 1973.
Rabbit and duck globin [liver and brain]	Schreier and Staehelin, 1973a,c; Schreier *et al.*, 1973.
Rat forebrain [rat brain]	Gilbert, 1974.
Rabbit globin [chick embryo brain]	Hendrick *et al.*, 1974.

[a] This listing is not intended to be exhaustive for either translation systems or for mRNA's which have been translated in these systems.

[b] Fractionated reticulocyte cell-free systems.

[c] The system is given in brackets.

mination of the proper divalent cation concentration for optimal translation of eukaryotic mRNA's in prokaryotic systems has also been stressed (Rekosh et al., 1970).

B. Reticulocyte Systems

Similar principles have been applied in developing eukaryotic cell-free systems for translation of exogenous mRNA's. A widely used example of such a system is rabbit reticulocyte lysate, as described by Schweet et al. (1958) and Allen and Schweet (1962). This system is obtained by inducing reticulocytosis in rabbits, obtaining the immature red cells, washing these cells free from serum contaminants, and lysing the reticulocytes by suspension in H_2O or low ionic strength buffer. Membrane ghosts are removed by centrifugation at 10,000 g for 20 min. This produces "reticulocyte lysate," a eukaryotic equivalent to the E. coli S30 fraction. Once the necessary skills are developed to prepare this system, it is highly reproducible. However, unlike the E. coli system, it is difficult to obtain exogenous messenger dependence in reticulocyte lysate by the preincubation method. Protein synthesis is markedly reduced after preincubation (Miller and Schweet, 1968), but a good portion of this reduction in activity apparently results from inactivation of one or more of the initiation factors and/or activation of protein synthesis inhibitors (see Chapter 5). Rabbit reticulocyte lysate has been utilized for translation of a wide variety of eukaryotic mRNA's. Under optimal conditions, this system is quite stable and protein synthesis proceeds in a linear fashion for 30–60 min. The system has high levels of initiation factors, and reinitiation of exogenous mRNA is quite good. Hemin is also thought to be important for maintaining high levels of initiation in this system (Zucker and Schulman, 1968; Adamson et al., 1968, 1972; Gross and Rabinovitz, 1972; Legon et al., 1973; Beuzard and London, 1974). Although reticulocyte lysate does contain endogenous ribonuclease activity, this is apparently inactivated by natural ribonuclease inhibitors (Priess and Zellig, 1967; Rowley and Morris, 1967).

Reticulocyte lysate appears to have the highest translational efficiency of all eukaryotic cell-free systems so far developed. Although the level of endogenous messenger activity is high, approximately 90% of the product represents α- and β-globin chains. These polypeptides are readily identified and easily separated from other proteins. Therefore, small amounts of heterologous product under the direction of exogenous mRNA (as low as 0.1% of total protein synthesis) can be identified. Exogenous mRNA does not actually stimulate amino acid incorporation above background levels in the rabbit reticulocyte lysate system, and the presence of mRNA in a given exogenous RNA preparation cannot be accessed directly. Exogenous mRNA competes with endogenous mRNA for translation on the polysomes. As the input of exogenous mRNA is increased, total protein synthesis activity actually decreases (Lockard and Lingrel, 1969). The mechanism for this effect has not been determined. The efficiency of exogenous mouse globin mRNA translation in rabbit reticulocyte lysate is less than that obtained with endogenous messenger already present on the polysomes. If the lysate system is prepared from the duck, this material demonstrates much lower levels of endogenous messenger activity and exogenous mRNA stimulates protein synthesis (Lingrel et al., 1971; Lockard and Lingrel, 1973). In spite of these differences in endogenous messenger activity, when heterologous mouse globin mRNA is translated in either rabbit or duck lysate, the same efficiency is obtained (Lockard and Lingrel, 1972).

Since crude reticulocyte lysate has limited usefulness in examining questions of mechanism for eukaryotic mRNA translation, a number of fractionated systems have been developed. These systems are based upon removal of initiation factors from the polysomes by treatment with elevated concentrations of KCl (usually 0.5 M) (Miller and Schweet, 1968). After this procedure, ribosomes are reisolated and the system is separately dependent on ribosomes and ribosomal wash protein (crude intiation factors). However, a considerable amount of mRNA activity remains with the ribosomes. These ribosomes can be stimulated up to 3- to 5-fold by the addition of exogenous mRNA, although the final level to which these ribosomes can be stimulated is far less than the activity of crude lysate. In a modification of this system, mRNA dependence is produced by treating a separate portion of crude polysomes with low doses of pancreatic ribonuclease (0.02–0.05 ug/ml). The polysomes are treated first with puromycin to release mRNA (Crystal et al., 1972), but this step is not essential to obtain good messenger dependence (Shafritz et al., 1973). The RNase-treated ribosomes are centrifuged through a cushion of 1 M sucrose containing 0.5 M KCl to remove residual ribonuclease (as well as initiation factors). This procedure may require repetition, as some batches of ribosomes retain residual ribonuclease activity. This may inactivate larger heterologous mRNA's more readily than 9 S globin mRNA. Therefore, it is not sufficient to merely demonstrate good activity for translation of exogenous reticulocyte mRNA to be certain that all residual ribonuclease has been removed. This can be monitored by checking hydrolysis of high specific activity [^{3}H]poly(U). After ribonuclease treatment, the system is dependent on both initiation factors and mRNA (50- to 100-fold stimulation with homologous mRNA); and tRNA is added separately to ensure that it does not become rate limiting. The 105,000 g supernatant fraction is also included, although most of the components present in this material are also present in the crude ribosomal wash fraction. In this fractionated system there is a direct relationship between the level of exogenous mRNA translation and the amount of initiation factors added. A concentrated preparation of ammonium sulfate precipitated initiation factors is required for optimal activity (Shafritz, 1974a). The level of initiation factors required for optimal translation of exogenous mRNA is also much higher than that required for maximal stimulation of endogenous mRNA. Since endogenous mRNA is translated with higher efficiency than exogenous mRNA (Lockard and Lingrel, 1972), it would appear that much greater quantities of both mRNA and initiation factors are required to prime the translation system than to keep it going. Possible explanations for these effects will be discussed later (Section VII).

Other methods for obtaining initiation factor and mRNA dependence in a variety of mammalian systems have been described. Crude lysates or polysomes can be incubated with puromycin and $0.5\ M$ KCl and either ribosomes or ribosomal subunits (40 S and 60 S) collected by sucrose gradient centrifugation (Blobel and Sabatini, 1971; Mechler and Mach, 1971; Brown et al., 1974). Ribosomal wash factors are prepared from separate aliquots of polysomes. The addition of 2–3 mM Mg^{2+} during the $0.5\ M$ KCl ribosomal washing procedure has been reported to preserve better ribosomal function (Schreier and Staehelin, 1973a). Systems similar to the reticulocyte have been prepared from liver, brain, muscle, embryonal, and other tissue culture cells (Martin and Wool, 1968; Rourke and Heywood, 1972; Jorgensen and Heywood, 1974; Ascioni and Vande Woode, 1971; Sampson et al., 1972; Zomzely-Neurath et al., 1972; Levin et al., 1971, 1973; Schreier and Staehelin, 1973c; Gilbert, 1974).

C. Ascites Tumor Cell-Free Systems

Cell-free systems, dependent on exogenous mRNA, have also been obtained from both Krebs II and Ehrlich ascites tumor cells grown in the peritoneal cavity of the mouse (Mathews and Korner, 1970; Aviv et al., 1971; Housman et al., 1971; Metafora et al., 1972; Jacobs-Lorena and Baglioni, 1972). These systems are generally unfractionated and in many ways are similar to the E. coli preincubated S30 system. After ascitic fluid is removed and the cells are washed and broken by homogenization, a postmitochondrial supernatant fraction is prepared. This material is then incubated at 37° for 30–45 min in the presence of the complete protein synthesis system containing 20 unlabeled amino acids. Exogenous messenger RNA dependence is produced and the material is passed over a Sephadex G-25 column to remove unused amino acids. This material is then stored frozen at low temperatures. In one case this system is dependent on exogenous tRNA as a separate component (Aviv et al., 1971), and in a modification of this procedure Hepes buffer, pH 7.5, is substituted for Tris-HCl, pH 7.5, in some steps (McDowell *et al.*, 1972).

The ascites tumor systems demonstrate good messenger dependence (low background) and can be used to access mRNA activity by direct amino acid incorporation (Mathews and Korner, 1970; Boime *et al.*, 1971; Aviv and Leder, 1972). Very little exogenous mRNA is required to obtain stimulation of amino acid incorporation. However, these systems are quite low in translational efficiency compared to reticulocyte lysate. The ascites tumor system is also stimulated by addition of reticulocyte crude initiation factors (Metafora *et al.*, 1972; Mathews *et al.*, 1972a), but addition of these factors may increase complexity in analyzing certain types of results. The ascites tumor systems have been quite effective for demonstrating translation of globin mRNA's, lens α-crystallin mRNA's, myeloma cell mRNA, and certain viral mRNA's, but additional problems can be encountered. A major difficulty is that incomplete polypeptide fragments are frequently produced in this system either by premature termination or by nuclease cleavage of the mRNA (Mathews and Korner, 1970; McDowell *et al.*, 1972; Kerr *et al.*, 1972; Mathews and Osborn, 1974). In many cases it has been difficult to synthesize large authentic polypeptides with this system, although synthesis of such polypeptides has certainly been reported (Aviv *et al.*, 1971; McDowell *et al.*, 1972). Recently, it has been found that addition of polyamines, such as spermidine, to the ascites tumor cell-free system markedly enhances translation of viral mRNA into specific polypeptide chains (Atkins *et al.*, 1975).

D. Wheat Germ System

In the past two years wheat germ cell-free systems have become quite popular for translation of exogenous mRNA (Efron and Marcus, 1973; Roberts and Paterson, 1973; Bard *et al.*, 1974; Prives *et al.*, 1974; Anderson *et al.*, 1974; Roberts *et al.*, 1973; Packman *et al.*, 1974; Benveniste *et al.*, 1974; Last *et al.*, 1974; Sippel *et al.*, 1974; Both *et al.*, 1975a). The preparation of the wheat germ system is quite simple, although considerable variation in activity has been found from lot to lot of the germ. Details for the preparation of this system are well described (Marcus, 1972; Roberts and Paterson, 1973). As prepared directly from the germ, the system may be exogenous mRNA dependent, but a preincubation procedure is generally performed. The wheat germ system translates a wide spectrum of viral and other eukaryotic mRNA's. It has a translational efficiency at least as good as the ascites tumor cell-free systems. Translation occurs with good fidelity without large quantities of partial fragments, and high molecular weight products have been identified (Roberts *et al.*, 1973; Both *et al.*, 1975a). It has been observed with this system (and also others) that polypeptides synthesized under the direction of exogenous mRNA may be larger than the naturally occurring product. This may be related to a lack of normal posttranslational modification of polypeptides in the wheat cell-free system or other minor aberrations in translation. Another point that has also been emphasized in the wheat system (Roberts and Paterson, 1973) is the variation of monovalent and divalent cation requirements for translation of different exogenous mRNA's.

E. Intact *Xenopus* Oocytes

Translation of exogenous mRNA from a variety of sources has also been obtained in an intact whole cell system utilizing *Xenopus* oocytes. The RNA is injected by micropuncture and, after a lag time of approximately 30 min, synthesis of heterologous protein can be demonstrated. This system, originally described by Gurdon and co-workers for translation of reticulocyte globin mRNA (Lane *et al.*, 1971), is quite stable. Synthesis of heterologous protein may continue for up to two weeks (Gurdon, 1973). This system has also been useful for translation of nuclear mRNA precursors, where cell-free systems have generally failed to demonstrate activity (Williamson *et al.*, 1973; Stevens and Williamson, 1972; Ruiz-Carrillo *et al.*, 1973). Of all the systems used for translation of exogenous mRNA, the *Xenopus* oocyte is the most sensitive. In this system 1.0–10 ng of purified globin mRNA are used to obtain an mRNA saturation curve. This is close to the sensitivity of cDNA-RNA hybridization. In the *Xenopus* oocyte there is no requirement for mRNA-specific factors and there is little competition between exogenous mRNA and endogenous mRNA unless very high levels of mRNA are added. Apparently, this system has excess ribosomes, not associated with endogenous oocyte mRNA. These ribosomes are utilized for translation of exogenous mRNA. With the great stability of foreign mRNA injected into the frog oocyte, this system has great potential for studying the metabolism of mRNA (Gurdon, 1973; Huez *et al.*, 1974; Allende *et al.*, 1974).

VI. Poly(A) and Messenger RNA Translation

With the observation that eukaryotic mRNA's and their precursors contain a unique sequence of poly(A) at their 3'-OH end, the possibility was raised that the 3'-poly(A) segment may be important for mRNA function. It was soon discovered, however, that mRNA's for histones did not contain a 3'-poly(A) segment (Adesnik and Darnell, 1972; Schochetman and Perry, 1972; Greenberg and Perry, 1972a). These mRNA's were still translated in cell-free systems from Krebs ascites tumor cells (Jacobs-Lorena *et al.*, 1972; Gross *et al.*, 1973; Breindl and Gallwitz, 1973). Therefore, the 3'-poly(A) segment was not required for function of every type of eukaryotic mRNA. However, it was still possible that translation of other eukaryotic mRNA's might require a 3'-poly(A) segment or might occur with greater efficiency in the presence of a 3'-poly(A) segment. The former possibility has been virtually excluded by recent experiments demonstrating translation of eukaryotic mRNAs from which the 3'-poly(A) segment has been removed. Deadenylated L-cell mRNA (Bard *et al.*, 1974) and both rabbit reticulocyte (Sippel *et al.*, 1974) and mouse reticulocyte mRNA (Williamson *et al.*, 1974) are translated effectively in cell-free systems from wheat embryos and Krebs ascites cells, respectively. A separate class of nonadenylated (nonhistone) mRNA, not derived from adenylated mRNA, has also been found in HeLa cells, and such mRNA's are also present in other eukaryotic cell types (Milcarek *et al.*, 1974).

It has also been found that, although the 3'-poly(A) segment of reticulocyte messenger RNA shortens with aging of the cell, there is no difference in cell-free translation of mRNA from young vs. old reticulocytes (Gorski *et al.*, 1974). In addition, after L cell polysomes are disaggregated by temperature shock, they reassemble in a random manner regardless of messenger poly(A) length (Bard *et al.*, 1974). Other experiments indicate that mRNA with the 3'-poly(A) segment blocked by hybridization to poly(U) translates at the same efficiency as unblocked mRNA (Munoz and Darnell, 1974). However, somewhat better function has been reported for control as compared to deadenylated mRNA (Bard *et al.*, 1974; Williamson *et al.*, 1974; Humphries *et al.*, 1974; Soreq *et al.*, 1974). This may be related to partial degradation of the mRNA during the deadenylation procedure or to a decrease in stability of deadenylated mRNA (Huez *et al.*, 1974). In spite of these observations, experiments are still needed to show that there is no preference for polyadenylated vs. deadenylated (or nonadenylated) mRNA when both types of mRNA are presented to the same system simultaneously. It would be helpful to demonstrate this phenomenon in the intact cell, as, for example, in the *Xenopus* oocyte.

VII. Messenger Ribonucleoprotein Particles

In eukaryotic cells, the nuclear membrane serves as a boundary to separate the site of mRNA synthesis (the nucleoplasm) from its site of utilization (the cytoplasm). Initially, it was thought that mRNA simply attaches to the small (40 S) ribosomal subunit as it passes from the nucleolus to the cytoplasm (Joklik and Becker, 1965a,b; Henshaw *et al.*, 1965; McConkey and Hopkins, 1965). More recent evidence suggests that separate mechanisms may be involved in the transport of these entities (Perry and Kelley, 1968; Kafatos, 1968; Spirin, 1969; Spohr *et al.*, 1970; Sugano *et* *al.,* 1971; Gander *et al.,* 1973) and that mRNA may be associated with specific classes of proteins during different stages in its function in mammalian cells (for general reviews, see Williamson, 1973; Samarina *et al.,* 1973; Scherrer, 1973). These mRNP's have been isolated from nucleoplasmic, cytoplasmic, and polysomal fractions.* A number of functions for these proteins in messenger storage, transport, stability, and initiation have been suggested. Since nonspecific complexes between RNA and basic proteins can be formed, especially at low ionic strength (Girard and Baltimore, 1966; Baltimore and Huang, 1970; Ovchinnikov *et al.,* 1968), the physiological significance of the mRNP remains unclear. Nonetheless, many studies are being reported on the isolation and identification of mRNP's and nuclear mRNP precursors (HnRNP's) under a variety of extraction conditions, and studies on the possible relationship of these particles to mRNA translation will be reviewed.

A. Proteins of the mRNP

Attempts to identify and establish a specific function for proteins associated with mRNA have been largely unsuccessful. Many studies have shown that mRNA isolated from polysomes is associated with proteins (Weisberger and Armentrout, 1966; Huez *et al.,* 1967; Henshaw, 1968; Perry and Kelley, 1968; Temmerman and Lebleu, 1969; Kafatos, 1969; Cartouzou *et al.,* 1968; Spohr *et al.,* 1970; Lebleu *et al.,* 1971; Olsnes, 1971; Gander *et al.,* 1973; Morel *et al.,* 1973). The mRNA-protein complexes (mRNP's) can be released by a variety of techniques. These include treatment with Mg^{2+} chelating agents such as EDTA (Huez *et al.,* 1967), temperature shock (Schochetman and Perry, 1972), amino acid deprivation or other ribosomal runoff techniques (Lee *et al.,* 1971b; Christman *et al.,* 1973), and puromycin-induced premature termination (Blobel, 1972). The proteins found in these complexes are generally of higher molecular weight and are apparently distinct from ribosomal structural proteins (Morel *et al.,* 1973; Blobel, 1972, 1973). In the rabbit reticulocyte, mRNP's obtained from polysomes by EDTA release contain three major protein bands on SDS gels with molecular weights of approximately 45,000, 68,000, and 130,000 daltons, respectively (Lebleu *et al.,* 1971). The 45,000 dalton polypeptide is considered a contaminant from a 5 S RNA-protein complex. A number of minor bands are also present on the gels and these have not been further characterized. Messenger RNP's, released from duck reticulocyte polysomes by EDTA treatment, contain two major bands of molecular weights approximately 49,000 and 73,000

---

* Other terms, such as mlRNP, dRNP, dlRNP, "informosome," nRNP, HnRNP, pre-mRNP, and nuclear pre-mRNP, have been used to describe these various particles.

daltons, respectively (Morel *et al.*, 1973). There are also at least five additional minor components. Two major bands have been found on EDTA-released mRNP's of mouse reticulocyte polysomes (Lukanidin *et al.*, 1971), but the molecular weights of these polypeptides have not been reported. When mRNP's are released from polysomes by puromycin treatment in the presence of 0.5 M KCl, two polypeptides of MW's 52,000 and 78,000 daltons have been identified (Blobel, 1972, 1973). Since these complexes are resistant to dissociation by 0.5 M KCl, they form rather tight complexes and are less likely to represent nonspecific RNA-protein aggregates. Similar polypeptide components have also been identified in puromycin-released mRNP's from L cells, liver, and embryonal chicken brain cells (Blobel, 1973; Bryan and Hayashi, 1973). Duck globin mRNP's are also stable at high ionic strength (Morel *et al.*, 1973). Free reticulocyte mRNP's have been isolated from the cytoplasm of rabbit reticulocytes (Jacobs-Lorena and Baglioni, 1972; Olsen *et al.*, 1972; Bonanou-Tzedaki *et al.*, 1972; Gianni *et al.*, 1972) and duck reticulocytes (Spohr *et al.*, 1972). In the duck, mRNP proteins have been compared in fractions obtained from nuclei, cytoplasm, and polysomes, and have been found to be distinct in each case (Spohr *et al.*, 1972; Scherrer, 1973). In rat liver, a 160,000 dalton polypeptide is found in the polysomal mRNP fraction (Olsnes, 1971). The proteins of liver polysomal mRNP's and nuclear mRNP precursors are also distinct (Lukanidin *et al.*, 1972). Polypeptides associated with mRNP's and HnRNP's from other eukaryotic sources including KB cells (Lindberg and Sundquist, 1974), HeLa cells (Pederson, 1974; Kumar and Pederson, 1975), *Dictyostelium discoideum* (Firtel and Pederson, 1975), ascites tumor cells (Quinlan *et al.*, 1974; Barrieux *et al.*, 1975), and mouse kidney (Irwin *et al.*, 1975) are currently under investigation and show remarkable consistency of polypeptides in the 50,000–80,00 dalton range, as well as additional higher molecular weight polypeptides.

B. Messenger RNP's and mRNA Storage

Studies reported a decade ago suggest that preformed mRNA was present in unfertilized eggs of the sea urchin. Initially, Denny and Tyler (1964) observed that protein synthesis could be activated parthenogenically in eggs from which the nucleus had been removed. Gross and Cousineau (1964) reported that protein synthesis and early development of the sea urchin embryo could proceed in face of transcription block with actinomycin D. These observations led to the hypothesis that mRNA synthesized during oogenesis is stored in a dormant form in the cytoplasm of the unfertilized egg. This dormant or "masked" messenger RNA, as it has been called (Gross, 1967; Nemer, 1967), becomes activated during embryogenesis and directs synthesis of certain proteins during early development. This general hypothesis has been recently confirmed for microtubulin, a specific protein constituent of sea urchin membranes (Raff *et al.*, 1972).

Since the unfertilized egg demonstrates only a very low level of protein synthesis, despite the presence of ample ribosomes and other components necessary for protein synthesis (Hultin, 1961; Nemer, 1962; Wilt and Hultin, 1962; Tyler, 1963; Maggio and Catalano, 1963; Nemer and Bard, 1963), the immediate question is what regulates translation or lack of translation in these cells. Initially, Spirin *et al.* (1964) and Spirin and Nemer (1965) reported that mRNA is associated with proteins in embryos of both the loach and the sea urchin. These particles were cytoplasmic in nature, were apparently distinct from ribosomes, and were termed "informosomes" (Spirin, 1969). The unique character of sea urchin mRNP's and their distinction from ribosomal subunits were confirmed by Infante and Nemer (1968). These investigators demonstrated that sea urchin mRNP's had a characteristic buoyant density in CsCl different from ribosomal subunits, a diffuse sedimentation pattern on sucrose gradients in contrast to ribosomal subunits, and products distinct from ribosomal subunits on mild digestion with Pronase or pancreatic ribonuclease.

Evidence for preformed informational RNA or RNA containing "messengerlike" sequences in embryonal systems was based initially on DNA-RNA hybridization studies (Spirin *et al.*, 1964; Spirin and Nemer, 1965; Gross *et al.*, 1965; Whiteley *et al.*, 1966, 1970; Glisin *et al.*, 1966; Crippa *et al.*, 1967; Davidson and Hough, 1967). More recent studies using competition-hybridization (Skoultchi and Gross, 1973) and stimulation of specific protein synthesis in a heterologous cell-free system (Gross *et al.*, 1973) have confirmed that "maternal" or pretranscribed mRNA for histones is present in unfertilized sea urchin eggs. A significant portion of this mRNA exists as messenger RNP's in the sedimentation range of 20–40 S, clearly distinct from free histone mRNA (9 S) or histone mRNA-ribosome complexes (≥ 40 S). The existence of mRNA in storage form has also been reported in other developmental systems, such as wheat embryos (Marcus, 1969; Weeks and Marcus, 1971), Blastocladiella (Silverman *et al.*, 1974), cotton seedlings (Waters and Dure, 1966; Ihle and Dure, 1972), and fetal calf myoblasts (Buckingham *et al.*, 1976). Whether a similar process occurs in the normal function of mRNA in differentiated cells remains to be determined, although one study provides evidence for nontranslated mRNA in membrane-bound and free cytoplasmic polysomes of rabbit liver (Shafritz, 1974b). As indicated in this study, it is not clear whether the nontranslated mRNA is present in ribosomes or in mRNP's which have cosedimented with the liver polysome fractions.

C. Messenger RNP's and Protein Synthesis

A relationship between mRNP proteins and factors required for protein synthesis has not been established. Since deproteinized mRNA can be translated effectively in a wide variety of heterologous cell-free systems, unique factors would not appear to be required for translation of individual mRNA's. Therefore, it is unlikely that mRNP proteins are specific for translation of individual mRNA species. This in no way excludes a more generalized function for the mRNP in the translation process. As mentioned earlier, several investigators have reported polypeptides of similar size classes in mRNP's from a variety of tissues including reticulocytes, liver, L cells, and embryonal brain cells (Blobel, 1972, 1973; Bryan and Hayashi, 1973).

Evidence for a unique polypeptide in mRNP's from adenovirus-infected KB cells has also been reported (Lindberg and Sundquist, 1974). This polypeptide has a molecular weight of approximately 110,000 daltons and is not present on mRNP's from control KB polysomes. The function of this polypeptide is not known, but one exciting possibility is that it may be specific for attachment of viral mRNA to host cell ribosomes. Alternatively, this component may serve as a factor for synthesis of specific viral polypeptides in a manner similar to the specific factor reported for translation of EMC virus (Wigle and Smith, 1973; Wigle, 1973), or it may be involved in the mechanism for "shutoff" of host protein synthesis. Other investigators have found differential synthesis of glycoprotein "G" of vesicular stomatitis virus (VSV) on membrane-bound as opposed to free polysomes of infected HeLa cells (Grubman et al., 1974). mRNA for VSV glycoprotein "G" is found exclusively on membrane-bound polysomes of infected cells (Grubman et al., 1975; Morrison and Lodish, 1975). DNA-RNA hybridization studies indicate a similar preferential distribution of Rausher leukemia virus mRNA on membrane-bound polysomes of virally infected JLS-V9 bone marrow cells of Balb/c mice (Gielkens et al., 1974). Since the mechanism for viral replication may represent a specialized utilization of host cell machinery, these observations do not necessarily imply a similar process in the distribution of normal cellular mRNA's. If this were found to be the case, however, then it might be possible to identify specific factors for interaction of certain mRNA's with membranes or membrane-bound polysomes.

Most investigators have found no significant difference in translation of mRNP's vs. free mRNA (Lingrel et al., 1971; Sampson et al., 1972; Nudel et al., 1973; Freienstein and Blobel, 1974; Ernst and Arnstein, 1975), although differences have been reported (Olsen et al., 1972; Hendrick et al., 1974). This would imply that protein constituents of the mRNP are not essential for initiation or translation. As indicated by some of these authors, limitations of these studies are that they have been performed in relatively unfractionated cell-free systems. These systems already contain all the necessary components for mRNA translation, including at least five specific initiation factors (see Chapter 5). Under such circumstances, subtle differences in individual factor requirements for messenger-ribosome interaction or for other steps in initiation may be overlooked. Therefore, an individual initiation factor and mRNA dependent cell-free system will be needed to reexamine this question. In addition, present studies have been performed almost exclusively with globin mRNA vs. globin mRNP's. Since this messenger is translated readily in a wide variety of systems with greater efficiency than many other mRNA's, the mRNP may show no advantage in this particular case. The use of other messengers and their mRNP complement may be required to demonstrate a preference for the mRNP in translation.

Direct studies on the relationship between mRNP proteins and specific eukaryotic initiation factors have been quite limited. Nonetheless, a number of observations leave the impression that specific protein-mRNA interactions may be involved in the initiation process. Reticulocyte 40 S ribosomal subunits, washed with sodium deoxycholate, bind globin mRNP's but not free globin mRNA (Lebleu et al., 1971). Therefore, a component necessary for mRNA binding to 40 S subunits has been removed by deoxycholate treatment. This component is present on reticulocyte mRNP's. These findings are consistent with observations that deoxycholate treatment reduces endogenous messenger activity of eukaryotic ribosomes (Prichard et al., 1971; Picciano et al., 1973). Therefore, mRNA and mRNA binding factor(s) may be removed simultaneously from polyribosomes by deoxycholate. In contrast, reticulocyte ribosomes washed with 0.5 M KCl retain most of their mRNA, although most of the initiation factors are released into the soluble phase (Nudel et al., 1973). Addition of 30 mM EDTA to washed reticulocyte ribosomes releases this mRNA as an mRNP, qualitatively similar to the mRNP isolated from unwashed polysomes. Therefore, it would appear that the specific mRNA binding factor, which is lost by deoxycholate treatment of ribosomes, is not lost by 0.5 M KCl treatment.

In other studies, a preference for the mRNP over free mRNA has been reported in formation of an 80 S ribosome initiation complex (Cashion and Stanley, 1974; Cashion et al., 1974). The mRNP fraction is also favored in initial peptide bond formation, as measured by the puromycin model reaction (Cashion and Stanley, 1974; Cashion et al., 1974). These results suggest that the mRNP may contain a protein constituent(s) important for mRNA recognition. Although other mechanisms are equally possible, such a factor could function by modifying the secondary structure of mRNA to increase accessibility of the ribosomal binding region or initiation region of the messenger. Such a mechanism is consistent with recent observations indicating differences in factor requirements for translation of rabbit vs. duck globin mRNA in a liver cell-free system, and preferential translation of rabbit globin mRNA over duck globin mRNA when both are presented simultaneously to a liver cell-free system (Schreier et al., 1973). Association of proteins with specific regions of secondary structure in duck globin mRNA has also been observed by dark-field electron microscopy (Morel et al., 1973).

Recently, it has been found that one of the proteins of the globin mRNP demonstrates properties of a reticulocyte initiation factor, the Met-tRNA$_f$ binding protein (Hellerman and Shafritz, 1975). This factor forms a ternary complex with Met-tRNA$_f$ (the initiator tRNA) and GTP in a reaction considered the first step in eukaryotic initiation (Levin et al., 1973; Dettman and Stanley, 1973; Gupta et al., 1973; Schreier and Staehelin, 1973b). This reaction is followed by joining of the ternary complex to the 40 S ribosomal subunit in the absence of added mRNA (Cashion et al., 1974; Levin et al., 1973; Schreier and Staehelin, 1973b). Messenger RNP's obtained from rabbit reticulocyte polysomes also bind eukaryotic initiator tRNA (Hellerman and Shafritz, 1975) and contain the polypeptide components of Met-tRNA$_f$ binding factor (molecular weights approximately 35,000 and 52,000 daltons). Both the purified factor and the mRNP also bind poly(A). The binding of both Met-tRNA$_f$ and poly(A) is inhibited by aurintricarboxylic acid (ATA), and the binding of Met-tRNA$_f$, but not poly(A), is inhibited by N-ethylmaleimide (NEM). The relationship between binding of Met-tRNA$_f$ and mRNA by the same factor is not entirely clear, but this observation has certain features in common with the properties of E. coli initiation factor IF-2 and its function in prokaryotic initiation (Herzberg et al., 1969; Revel et al., 1970; Rudland et al., 1971; Gröner and Revel, 1971, 1973; Lockwood et al., 1971; Verneer et al., 1973).

An association of poly(A) with proteins in mRNP's has also been reported by other investigators. Kwan and Brawerman (1972) reported a complex between protein and the 3'-poly(A) segment of mouse sarcoma 180 mRNA. A complex with similar properties could also be formed between synthetic [^{3}H]poly(A) and crude cytoplasmic protein, but the factors involved in these complexes were not further characterized. Blobel (1973) has observed that when puromycin-released reticulocyte mRNP's are digested with pancreatic ribonuclease, the 78,000 dalton polypeptide can be isolated with the 3'-poly(A) segment of the messenger. This protein, however, is distinct from the polypeptide subunits of Met-tRNA$_f$ binding factor (35,000 and 52,000 daltons). Other investigators have also identified protein factors in various subcellular fractions which form complexes with poly(A) (Blanchard et al., 1974; Quinlan et al., 1974). The role of such factors in eukaryotic regulation or mRNA function has not yet been clarified, but it has been observed that RNA-protein complexes are degraded more slowly by ribonuclease than deproteinized RNA (Spirin, 1969; Kwan and Brawerman, 1972; Blobel, 1973).

Several reports indicate that greater quantities of initiation factors are required for maximal translation of exogenous mRNA as compared to endogenous mRNA (Prichard et al., 1971; Shafritz, 1974a). This appears to be related to initiation factor IF-M$_3$, which is required for initiation of natural messenger translation but not for poly(U) translation (Prichard et al., 1970; Shafritz and Anderson, 1970). Crude preparations of IF-M$_3$, as well as similar factors reported by other investigators for binding of mRNA to ribosomes (Heywood, 1970; Rourke and Heywood, 1972b; Kaempfer and Kaufman, 1973; Schreier and Staehelin, 1973b; Levin et al., 1973), are most probably contaminated with Met-tRNA$_f$ binding factor and other components (Anderson et al., 1974). Therefore, the exact nature of the factor specificity for eukaryotic mRNA binding to ribosomal subunits is not clear. Ilan and Ilan (1973) reported the identification of an "IF-3-like" factor on mRNP's from the insect Tenebrio. This factor, termed I$_3$, could be released from the mRNP by treatment with very high concentrations of LiCl and KCl and is required for natural messenger translation. Messenger RNP complexes of this type may contain Met-tRNA$_f$ binding factor activity as well as IF-M$_3$. The in vitro formation of such complexes from deproteinized mRNA and ribosomal wash factors, as a first step in utilization of exogenous mRNA, may require high levels of these factors. However, once the mRNP complexes have been formed and the mRNA is incorporated into polysomes, much lower levels of these factors may be required to sustain messenger function.

The above information has been presented primarily to stimulate interest in questions regarding the role of mRNP's in eukaryotic cells. It is not meant to imply that a direct relationship between mRNP's and eukaryotic translation has been established. Interference factors, which have been found in both prokaryotic and eukaryotic systems (discussed separately in Chapter 5), and a variety of other changes in the intracellular environment may also play an important role in translational control. It is also possible that specific proteins associated with eukaryotic mRNA may function in other cellular events related to mRNA processing, transport, stability, and metabolism. Therefore, a more detailed study of mRNP proteins and factors forming stable complexes with mRNA will be required, and it is entirely possible that more than one specific function will be found for mRNP's in posttranscriptional regulation.

ACKNOWLEDGMENTS

The author would like to express his appreciation to Drs. A. Skoultchi and J. Mendecki for their helpful suggestions during the preparation of this manuscript and to his wife, Sharon, for her patience during this period. Studies cited from the author's laboratory were supported in part by NIH grant AM-17609 and the Irma T. Hirschl Charitable Trust of New York.

REFERENCES

Adamson, S. D., Herbert, E., and Godchaux, W. (1968). *Arch. Biochem. Biophys.* **125,** 671.

Adamson, S. D., Yau, P. M. P., Herbert, E., and Zucker, W. V. (1972). *J. Mol. Biol.* **63,** 247.

Adesnik, M., and Darnell, J. E. (1972). *J. Mol. Biol.* **67,** 397.

Adesnik, M., Levinthal, C. (1970). *Cold Spring Harbor Symp. Quant. Biol.* **35,** 451.

Adesnik, M., Salditt, M., Thomas, W., and Darnell, J. E. (1972). *J. Mol. Biol.* **71,** 21.

Adler, A. J., and Rich, A. (1962). *J. Am. Chem. Soc,* **84,** 3977.

Allen, E., and Schweet, R. (1962). *J. Biol. Chem.* **237,** 760.

Allende, C. C., Allende, J. E., and Firtel, R. A. (1974). *Cell* **2,** 189.

Anderson, C. W., Lewis, J. B., Atkins, J. F., and Gesteland, R. F. (1974). *Proc. Natl. Acad. Sci. U.S.A.* **71,** 2756.

Anderson, W. F. (1969). *Proc. Natl. Acad. Sci. U.S.A.* **62,** 566.

Anderson, W. F., Bosch, L., Grunberg-Manago, M., Ochoa, S., Rich, A., and Staehelin, T. (1974). *FEBS Lett.* **48,** 1.

Ascione, R., and Vande Woode, G. F. (1971). *Biochem. Biophys. Res. Commun.* **45,** 14.

Astrachman, L., and Volkin, E. (1958). *Biochim. Biophys. Acta* **29,** 536.

Atkins, J. F., Lewis, J. B., Anderson, C. W., and Gesteland, R. R. (1975). *J. Biol. Chem.* **250,** 5688.

Attardi, G., Parnas, H., Hwang, M., and Attardi, B. (1966). *J. Mol. Biol.* **20,** 145.

Aviv, H., and Leder, P. (1972). *Proc. Natl. Acad. Sci. U.S.A.* **69,** 1408.

Aviv, H., Boime, I., and Leder, P. (1971). *Proc. Natl. Acad. Sci. U.S.A.* **68,** 2303.

Aviv, H., Boime, I., and Leder, P. (1972). *Science* **178,** 1293.

Baltimore, D. (1970). *Nature* **226,** 1209.

Baltimore, D., and Huang, A. S. (1970). *J. Mol. Biol.* **47,** 263.

Bard, E., Efron, D., Marcus, A., and Perry, R. P. (1974). *Cell* **1,** 101.

Barrieux, A., Ingraham, H. A., David, D. N., and Rosenfeld, M. G. (1975). *Biochemistry* **14,** 1815.

Bautz, E. K. F., and Hall, B. D. (1962). *Proc. Natl. Acad. Sci. U.S.A.* **48,** 400.

Benveniste, K., Wilczek, J., and Stern, R. (1974). *Fed. Proc., Fed. Am. Soc. Exp. Biol.* **33,** 1541.

Benz, E. J., Jr., and Forget, B. G. (1971). *J. Clin. Invest.* **50,** 2755.

Berns, A. J. M., and Bloemendal, H. (1974). *In* "Methods in Enzymology" K. Moldave and L. Grossman, eds.), Vol. 30, p. 675. Academic Press, New York.

Berns, A. J. M., van Kraaikamp, M., Bloemendal, H., and Lane, C. D. (1972a). *Proc. Natl. Acad. Sci. U.S.A.* **69,** 1606.

Berns, A. J. M., Strous, G. J. A. M., and Bloemendal, H. (1972b). *Nature (London), New Biol.* **236,** 7.

Beuzard, Y., and London, I. M. (1974). *Proc. Natl. Acad. Sci. U.S.A.* **71,** 2863.

Birnboim, H. C. (1972). *Biochemistry* **11,** 4588.

Bishop, D. H. L., Claybrook, J. R., and Spiegelman, S. (1967). *J. Mol. Biol.* **26,** 373.

Blanchard, J. M., Brissac, C., and Jeanteur, P. H. (1974). *Proc. Natl. Acad. Sci. U.S.A.* **71,** 1882.

Blobel, G. (1972). *Biochem. Biophys. Res. Commun.* **47,** 88.

Blobel, G. (1973). *Proc. Natl. Acad. Sci. U.S.A.* **70,** 924.

Blobel, G., and Sabatini, D. (1971). *Proc. Natl. Acad. Sci. U.S.A.* **71,** 68.

Boedtker, H., Crkvenjakov, R. B., Dewey, K. F., and Lanka, K. (1973). *Biochemistry* **12,** 4356.

Bogdanovsky, D., Hermann, W., and Schapira, G. (1973). *Biochem. Biophys. Res. Commun.* **54,** 25.

Boime, I., Aviv, H., and Leder, P. (1971). *Biochem. Biophys. Res. Commun.* **45,** 788.

Bonanou-Tzedaki, S. A., Pragnell, I. B., and Arnstein, H. R. V. (1972). *FEBS Lett.* **26,** 77.

Borum, T., Scharff, M., and Robbins, E. (1967). *Proc. Natl. Acad. Sci. U.S.A.* **58,** 1977.

Both, G. W., Moyer, S. A., and Banerjee, A. K. (1975a). *Proc. Natl. Acad. Sci. U.S.A.* **72,** 274.

Both, G. W., Banerjee, A. K., and Shatkin, A. J. (1975b). *Proc. Natl. Acad. Sci. U.S.A.* **72,** 1189.

Both, G. W., Furuichi, Y., Muthukrishnan, S., Shatkin, A. J. (1975c). *Cell* **6,** 185.

Boyer, S. H., Smith, K. D., Noyes, A. N., and McMullen, M. A. (1974). *J. Biol. Chem.* **249,** 7210.

Brawerman, G. (1974). *Annu. Rev. Biochem.* **43,** 621.

Brawerman, G., Mendecki, J., and Lee, S. Y. (1972). *Biochemistry* **11,** 637.

Breindl, M., and Gallwitz, D. (1973). *Eur. J. Biochem.* **32,** 381.

Brenner, S., Jacob, F., and Meselson, M. (1961). *Nature (London)* **190,** 576.

Britten, R. J., and Kohne, D. W. (1968). *Science* **161,** 529.

Brown, G. E., Kolb, A. J., and Stanley, W. M., Jr. (1974). *In* "Methods in Enzymology" (K. Moldave and L. Grossman, eds.), Vol. 30, p. 368. Academic Press, New York.

Bryan, R. N., and Hayashi, M. (1973). *Nature (London), New Biol.* **244,** 271.

Buckingham, M. E., Cohen, A., and Gros, F. (1976). *J. Mol. Biol.* **103,** 611.

Burny, A., and Marbaix, G. (1965). *Biochim. Biophys. Acta* **103,** 409.

Cancedda, R., and Schlesinger, M. J. (1974). *Proc. Natl. Acad. Sci. U.S.A.* **71,** 1843.

Capecchi, M. R. (1966). *J. Mol. Biol.* **21,** 173.

Cartouzou, G., Attali, J.-C., and Lissitzkey, S. (1968). *Eur. J. Biochem.* **4,** 41.

Cashion, L. M., and Stanley, W. M., Jr. (1974). *Proc. Natl. Acad. Sci. U.S.A.* **71,** 436.

Cashion, L. M., Dettman, G. L., and Stanley, W. M., Jr. (1974). *In* "Methods in Enzymology" (K. Moldave and L. Grossman, eds.), Vol. 30, p. 153. Academic Press, New York.

Chantrenne, H., Burny, A., and Marbaix, G. (1967). *Prog. Nucleic Acid Res. Mol. Biol.* **7,** 173.

Chatterjee, N. K., and Weissbach, H. (1974). *Proc. Natl. Acad. Sci. U.S.A.* **71,** 3129.

Christman, J. K., Reiss, B., Kyner, D., Levin, D. H., Klett, H., and Acs, G. (1973). *Biochim. Biophys. Acta* **294,** 153.

Clark, I. M., Chang, A. Y., Spiegelman, S., and Reichmann, M. E. (1965). *Biochemistry* **54,** 1193.

Clegg, J. B., Wheatherall, D. J., and Miller, P. F. (1971). *Nature (London)* **234,** 337.

Clegg, J. B., Wheatherall, D. J., Contopolon-Griva, I., Caroutsos, K., Poungouras, P., and Tsevrenis, H. (1974). *Nature (London)* **251,** 245.

Craig, N., Perry, R. P., and Kelley, D. E. (1971). *Biochim. Biophys. Acta* **246,** 493.

Crippa, M., Davidson, E. H., and Mirsky, A. E. (1967). *Proc. Natl. Acad. Sci. U.S.A.* **57,** 885.

Crystal, R. G., Nienhuis, A. W., Elson, N. A., and Anderson, W. F. (1972). *J. Biol. Chem.* **247,** 5357.

Darnell, J. E. (1968). *Bacteriol. Rev.* **32**, 262.
Darnell, J. E., Wall, R., and Tushinski, R. (1971). *Proc. Natl. Acad. Sci. U.S.A.* **68**, 1321.
Darnell, J. E., Jelinek, W. R., and Molloy, G. R. (1973). *Science* **181**, 1215.
Davidson, E. H., and Hough, B. R. (1971). *J. Mol. Biol.* **56**, 491.
DeLarco, J., and Guroff, G. (1973). *Biochem. Biophys. Res. Commun.* **50**, 486.
Denny, P. C., and Tyler, A. (1964). *Biochem. Biophys. Res. Commun.* **14**, 245.
Derman, E., and Darnell, J. E. (1974). *Cell* **3**, 255.
Dettman, G. L., and Stanley, W. M., Jr. (1973). *Biochim. Biophys. Acta* **299**, 142.
Dina, D., Crippa, M., and Beccari, E. (1973). *Nature (London) New Biol.* **242**, 101.
Dina, D., Meza, I., and Crippa, M. (1974). *Nature (London)* **248**, 486.
Drach, J. C., and Lingrel, J. B. (1964). *Biochim. Biophys. Acta* **91**, 680.
Drach, J. C., and Lingrel, J. B. (1966). *Biochim. Biophys. Acta* **129**, 178.
Dunn, J. J., and Studier, F. W. (1973). *Proc. Natl. Acad. Sci. U.S.A.* **70**, 1559.
Dunn, J. J., and Studier, F. W. B. (1975). *J. Mol. Biol.* **99**, 487.
Edmonds, M., and Abrams, R. (1960). *J. Biol. Chem.* **235**, 1142.
Edmonds, M., and Abrams, R. (1965). "Nucleic Acids, Structure, Biosynthesis, and Function" (Symposium), p. 1. Counc. Sci. Ind. Res., New Delhi, India.
Edmonds, M. P., and Caramela, M. G. (1969). *J. Biol. Chem.* **244**, 1314.
Efron, D., and Marcus, A. (1973). *FEBS Lett.* **33**, 23.
Ernst, V., and Arnstein, H. R. V. (1975). *Biochim. Biophys. Acta* **378**, 251.
Eron, L., and Westphal, H. (1974). *Proc. Natl. Acad. Sci. U.S.A.* **71**, 3385.
Favre, A., Bertazzoni, U., Berns, A. J. M., and Bloemendal, H. (1974). *Biochem. Biophys. Res. Commun.* **56**, 273.
Firtel, R. A., and Lodish, H. F. (1973). *J. Mol. Biol.* **79**, 289.
Firtel, R. A., and Pederson, T. (1975). *Proc. Natl. Acad. Sci. U.S.A.* **72**, 301.
Firtel, R. A., Jacobson, A., and Lodish, H. F. (1972). *Nature (London), New Biol.* **239**, 225.
Forget, B. G., Baltimore, D., Benz, E. F., Jr., Housman, D., Lebowitz, P., Marotta, C. A., McCaffrey, R. P., Skoultchi, A., Swerdlow, P. S., Verma, I. M., and Weissman, S. M. (1974a). *Ann. N. Y. Acad. Sci.* **232**, 76.
Forget, B. G., Benz, E. J., Jr., Skoultchi, A., Baglioni, A., and Housman, D. (1974b). *Nature (London)* **247**, 379.
Freienstein, C., and Blobel, G. (1974). *Proc. Natl. Acad. Sci. U.S.A.* **71**, 3435.
Fuhr, J. E., and Natta, C. (1972). *Nature (London), New Biol.* **240**, 274.
Furuichi, Y., Morgan, M., Muthukrishnan, S., and Shatkin, A. J. (1975). *Proc. Natl. Acad. Sci. U.S.A.* **72**, 362.
Gallwitz, D., and Mueller, G. C. (1969). *J. Biol. Chem.* **244**, 5947.
Gander, E., Stewart, A., Morel, C., and Scherrer, K. (1973). *Eur. J. Biochem.* **38**, 443.
Gaskill, P., and Kabat, D. (1971). *Proc. Natl. Acad. Sci. U.S.A.* **68**, 72.
Gelfand, D. H., and Hayashi, M. (1969). *Proc. Natl. Acad. Sci. U.S.A.* **63**, 135.
Georgiev, G. P., and Mantieva, V. L. (1962). *Biochim. Biophys. Acta* **61**, 153.
Georgiev, G. P., Samarina, O. P., Lerman, M. I., Smirnov, M. N., and Severtzov, A. N. (1963). *Nature (London)* **200**, 1291.
Gianni, A. M., Giglioni, B., Ottolenghi, S., and Guidotti, G. G. (1972). *Nature (London), New Biol.* **240**, 183.
Gielkens, A. L. J., Salden, M. H. L., and Bloemendal, H. (1974). *Proc. Natl. Acad. Sci. U.S.A.* **71**, 1093.
Gilbert, J. M. (1973). *Biochem. Biophys. Res. Commun.* **52**, 79.
Gilbert, J. M. (1974). *Biochim. Biophys. Acta* **340**, 140.
Gilbert, J. M., and Anderson, W. F. (1970). *J. Biol. Chem.* **245**, 2342.
Gilham, P. T. (1962). *J. Am. Chem. Soc.* **84**, 1311.
Girard, M., and Baltimore, D. (1966). *Proc. Natl. Acad. Sci. U.S.A.* **56**, 999.

Girard, M., Latham, H., Penman, S., and Darnell, J. E. (1965). *J. Mol. Biol.* **11**, 187.

Giron, M. L., and Huppert, J. (1972). *Biochim. Biophys. Acta* **287**, 438.

Glišin, V. R., Glišin, M. V., and Doty, P. (1966). *Proc. Natl. Acad. Sci. U.S.A.* **56**, 285.

Goldstein, E. S., and Penman, S. (1973). *J. Mol. Biol.* **80**, 243.

Gorski, J., Morrison, M. R., Merkel, C. G., and Lingrel, J. B. (1974). *J. Mol. Biol.* **86**, 363.

Gould, H. J., and Hamlyn, P. H. (1973). *FEBS Lett.* **30**, 301.

Gralla, J., Steitz, J. A., and Crothers, D. M. (1974). *Nature (London)* **248**, 204.

Graziadei, W. D., III, and Lengyel, P. (1972). *Biochem. Biophys. Res. Commun.* **46**, 1816.

Green, M., Graves, P. N., Zehiva-Willner, T., McInnes, J., and Pestka, S. (1975). *Proc. Natl. Acad. Sci. U.S.A.* **72**, 224.

Greenberg, J. R. (1972). *Nature (London) New Biol.* **240**, 102.

Greenberg, J. R., and Perry, R. P. (1972a). *J. Mol. Biol.* **72**, 91.

Greenberg, J. R., and Perry, R. P. (1972b). *Biochim. Biophys. Acta* **287**, 361.

Gröner, Y., and Hurwitz, J. (1975). *Proc. Natl. Acad. Sci. U.S.A.* **72**, 2930.

Gröner, Y., and Ravel, M. (1971). *Eur. J. Biochem.* **22**, 144.

Gröner, Y., and Revel, M. (1973). *J. Mol. Biol.* **74**, 407.

Gros, F., Hiatt, H., Gilbert, W., Kurland, C. G., Risebrough, R. W., and Watson, J. D. (1961). *Nature (London)* **190**, 581.

Gross, K. W., Jacobs-Lorena, M., Baglioni, C., and Gross, P. R. (1973). *Proc. Natl. Acad. Sci. U.S.A.* **70**, 2614.

Gross, M., and Rabinovitz, M. (1972). *Proc. Natl. Acad. Sci. U.S.A.* **69**, 1565.

Gross, P. R. (1967). *N. Engl. J. Med.* **276**, 1230.

Gross, P. R., and Cousineau, G. H. (1964). *Exp. Cell Res.* **33**, 368.

Gross, P. R., Malkin, L. I., and Hubbard, M. (1965). *J. Mol. Biol.* **13**, 463.

Grubman, M. J., Ehrenfeld, E., and Summers, D. F. (1974). *J. Virol.* **14**, 560.

Grubman, J. J., Moyer, S. A., Banerjee, A. K., and Ehrenfeld, E. (1975). *Biochem. Biophys. Res. Commun.* **62**, 531.

Gupta, N. K., Woodley, C. L., Chen, Y. C., and Bose, K. K. (1973). *J. Biol. Chem.* **248**, 4500.

Gurdon, J. B. (1973). *Acta Endocrinol. (Copenhagen), Suppl.* **180**, 255.

Haff, L. A., and Keller, E. B. (1973). *Biochem. Biophys. Res. Commun.* **51**, 704.

Haines, M. E., Carey, N. H., and Palmiter, R. D. (1974). *Eur. J. Biochem.* **43**, 549.

Hamkalo, B. A., and Miller, O. L., Jr. (1973). *Annu. Rev. Biochem.* **42**, 379.

Happe, M., and Jockusch, H. (1973). *Nature (London)* **245**, 141.

Hardesty, B., Arlinghaus, R., Schaeffer, J., and Schweet, R. (1963). *Cold Spring Harbor Symp. Quant. Biol.* **28**, 215.

Harris, S. E., Means, A. R., Mitchell, W. M., and O'Malley, B. W. (1973). *Proc. Natl. Acad. Sci. U.S.A.* **70**, 3776.

Hellerman, J. A., and Shafritz, D. A. (1975). *Proc. Natl. Acad. Sci. U.S.A.* **72**, 1021.

Hendrick, D., Schwartz, W., Pitzel, S., and Tiedermann, H. (1974). *Biochim. Biophys. Acta* **340**, 278.

Henshaw, E. C. (1968). *J. Mol. Biol.* **36**, 401.

Henshaw, E. C., and Loebenstein, J. (1970). *Biochim. Biophys. Acta* **199**, 405.

Henshaw, E. C., Revel, M., and Hiatt, H. H. (1965). *J. Mol. Biol.* **14**, 241.

Hercules, K., Schweiger, M., and Sauerbier, W. (1974). *Proc. Natl. Acad. Sci. U.S.A.* **71**, 840.

Herman, R. C., Williams, J. G., and Penman, S. (1976). *Cell* **7**, 429.

Herzberg, M., Lelong, J. C., and Revel. M. (1969). *J. Mol. Biol.* **44**, 297.

Heywood, S. M. (1970). *Proc. Natl. Acad. Sci. U.S.A.* **67**, 1782.

Heywood, S. M., Kennedy, D. S., and Bester, A. J. (1974). *Proc. Natl. Acad. Sci. U.S.A.* **71**, 2428.

Higgins, T. J. A., Mercer, J. F. B., and Goodwin, P. B. (1973). *Nature (London), New Biol.* **246**, 68.

Holme, G., Boyd, S. L., and Sehon, A. H. (1971). *Biochem. Biophys. Res. Commun.* **45**, 240.

Holmes, D. S., and Bonner, J. (1973). *Biochemistry* **12**, 2330.

Holmes, D. S., and Bonner, J. (1974). *Proc. Natl. Acad. Sci. U.S.A.* **71**, 1108.

Housman, D., Pemberton, R., and Tabor, R. (1971). *Proc. Natl. Acad. Sci. U.S.A.* **68**, 2716.

Housman, D., Forget, B. G., Skoultchi, A., and Benz, E. J. (1973). *Proc. Natl. Acad. Sci. U.S.A.* **70**, 1809.

Huez, G., Burny, A., Marbaix, G., and Lebleu, B. (1967). *Biochim. Biophys. Acta* **145**, 629.

Huez, G., Marbaix, G., Hubert, E., Leclercq, M., Nudel, U., Soreq, H., Salomon, R., Lebleu, B., Revel, M., and Littauer, U. (1974). *Proc. Natl. Acad. Sci U.S.A.* **71**, 3143.

Hultin, T. (1961). *Exp. Cell Res.* **25**, 405.

Humphries, S., Doel, M., and Williamson, R. (1974). *Biochem. Biophys. Res. Commun.* **58**, 927.

Hunt, J. (1974). *Biochem. J.* **138**, 487.

Hunt, J., and Wilkinson, B. R. (1967). *Biochemistry* **6**, 1688.

Hunt, T., and Ehrenfeld, E. (1971). *Proc, Natl. Acad. Sci. U.S.A.* **68**, 1075.

Ihle, J. N., and Dure, L. S., III. (1972). *J. Biol. Chem.* **247**, 5048.

Ikehara, Y., and Pitot, H. C. (1973). *J. Cell Biol.* **59**, 28.

Ilan, J., and Ilan, J. (1973). *Nature (London), New Biol.* **241**, 176.

Infante, A. A., and Nemer, M. (1968). *J. Mol. Biol.* **32**, 543.

Irwin, D., Kumar, A., and Malt, R. A. (1975). *Cell* **4**, 157.

Ishikawa, K., Kuroda, C., and Ogata, K. (1969). *Biochim. Biophys. Acta* **179**, 316.

Ishikawa, K., Ueki, M., Nagai, K., and Ogata, K. (1972). *Biochim. Biophys. Acta* **259**, 138.

Jacob, F., and Monod, J. (1961). *J. Mol. Biol.* **3**, 318.

Jacobs-Lorena, M., and Baglioni, C. (1972). *Proc. Natl. Acad. Sci. U.S.A.* **69**, 1425.

Jacobs-Lorena, M., Baglioni, C., and Borun, T. W. (1972). *Proc. Natl. Acad. Sci. U.S.A.* **69**, 2095.

Jacobs-Lorena, M., Gabrielli, F., Borun, T. W., and Baglioni, C. (1973). *Biochim. Biophys. Acta* **324**, 275.

Jacobson, A., Firtel, R., and Lodish, H. F. (1974a). *Brookhaven Symp. Biol.* **26**, 307.

Jacobson, A., Firtel, R. A., and Lodish, H. F. (1974b). *J. Mol. Biol.* **82**, 213.

Jelinek, W., Adesnik, M., Salditt, M., Sheiness, D., Wall, R., Molloy, G., Phillipson, L., and Darnell, J. E. (1973). *J. Mol. Biol.* **75**, 515.

Johnston, R. E., and Bose, H. R. (1972). *Proc. Natl. Acad. Sci. U.S.A.* **69**, 1514.

Joklik, W. K., and Becker, Y. (1965a). *J. Mol. Biol.* **13**, 496.

Joklik, W. K., and Becker, Y. (1965b). *J. Mol. Biol.* **13**, 511.

Jones, R. F., and Lingrel, J. B. (1972). *J. Biol. Chem.* **247**, 7951.

Jorgensen, A. Q., and Heywood, S. M. (1974). *Proc. Natl. Acad. Sci. U.S.A.* **71**, 4278.

Kacian, D. L., Spiegelman, S., and Bank, A., Terada, M., Metafora, S., Dow, L., and Marks, P. A. (1972). *Nature (London), New Biol.* **235**, 167.

Kacian, D. L., Gambino, R., Dow, L. N., Grossbard, E., Natta, C., Ramierz, F., Spiegelman, S., Marks, P. A., and Bank, A. (1973). *Proc. Natl. Acad. Sci. U.S.A.* **70**, 1886.

Kaempfer, R., and Kaufman, J. (1973). *Proc. Natl. Acad. Sci. U.S.A.* **70**, 1222.

Kafatos, F. C. (1968). *Proc. Natl. Acad. Sci. U.S.A.* **59**, 1251.

Kan, Y. W., Todd, D., Holland, J., and Dozy, A. (1974). *J. Clin. Invest.* **53**, 37a.

Kates, J. (1970). *Cold Spring Harbor Symp. Quant. Biol.* **35**, 743.

Kazazian, H. H., Jr., Snyder, P. G., and Cheng, T.-C. (1974). *Biochem. Biophys. Res. Commun.* **51**, 564.

Kazazian, H. H., Jr., Snyder, P. G., and Cheng, T. -C. (1974). *Biochem. Biophys. Res. Commun.* **59**, 1053.

Kerr, I., Brown, R. E., and Tovell, D. R. (1972). *J. Virol.* **10**, 73.

Kirby, K. S. (1965). *Biochem. J.* **96**, 266.

Kitos, P. A., Saxon, G., and Amos, H. (1972). *Biochem. Biophys. Res. Commun.* **47**, 1426.

Knöchel, W., and Tiedermann, H. (1972). *Biochim. Biophys. Acta* **269**, 104.

Kramer, R. A., Rosenberg, M., and Steitz, J. A. (1974). *J. Mol. Biol.* **89**, 767.

Kruh, J., Dreyfus, C., and Schapira, G. (1964). *Biochim. Biophys. Acta* **91**, 494.

Kruh, J., Dreyfus, C., and Schapira, G. (1966). *Biochim. Biophys. Acta* **114**, 371.

Kumar, A., and Pederson, T. (1975). *J. Mol. Biol.* **96**, 353.

Kwan, S. W., and Brawerman, G. (1972). *Proc. Natl. Acad. Sci. U.S.A.* **69**, 3247.

Labrie, F. (1969). *Nature (London)* **221**, 1217.

Lane, C. D., Marbaix, B., and Gurdon, J. B. (1971). *J. Mol. Biol.* **61**, 73.

Lane, C. D., Gregory, C., and Morel, C. M. (1973). *Eur. J. Biochem.* **34**, 219.

Lanyon, W. G., Paul, J., and Williamson, R. (1972). *Eur. J. Biochem.* **31**, 38.

Laskey, R. A., Gurdon, F. B., and Crawford, L. V. (1972). *Proc. Natl. Acad. Sci. U.S.A.* **69**, 3665.

Last, J. A., Crkvenjakov, R., and Doty, H. B. (1974). *Fed. Proc., Fed. Am. Soc. Exp. Biol.* **33**, 1541.

Latham, H., and Darnell, J. (1965). *J. Mol. Biol.* **14**, 1.

Laycock, D. G., and Hunt, J. A. (1969). *Nature (London)* **221**, 1118.

Lebleu, B., Marbaix, G., Huez, G., Temmerman, J., Burny, A., and Chantrenne, H. (1971). *Eur. J. Biochem.* **19**, 264.

Lee, S. Y., Mendecki, J., and Brawerman, G. (1971a). *Proc. Natl. Acad. Sci. U.S.A.* **68**, 1331.

Lee, S. Y., Krsmanovič, V., and Brawerman, G. (1971b). *Biochemistry* **10**, 895.

Legon, S., Jackson, R. J., and Hunt, T. (1973). *Nature (London), New Biol.* **241**, 150.

Levin, D. H., Kyner, D., Acs, G., and Silverstein, S. (1971). *Biochem. Biophys. Res. Commun.* **42**, 454.

Levin, D. H., Kyner, D., and Acs, G. (1973). *J. Biol. Chem.* **248**, 6416.

Lewin, B. (1975a). *Cell* **4**, 11.

Lewin, B. (1975b). *Cell* **4**, 77.

Lim, L., and Canellakis, E. S. (1970). *Nature (London)* **227**, 710.

Lim, L., White, J. O., Hall, C., Berthold, W., and Davison, A. N. (1974). *Biochim. Biophys. Acta* **361**, 241.

Lindberg, U., and Persson, T. (1972). *Eur. J. Biochem.* **31**, 246.

Lindberg, U., and Sundquist, B. (1974). *J. Mol. Biol.* **86**, 451.

Lingrel, J. B., Lockard, R. E., Jones, R. F., Burr, H. E., and Holder, J. W. (1971). *Ser. Haematol.* **4**, 3.

Lockard, R. E., and Lingrel, J. B. (1969). *Biochem. Biophys. Res. Commun.* **37**, 204.

Lockard, R. E., and Lingrel, J. B. (1972). *J. Biol. Chem.* **247**, 4174.

Lockard, R. E., and Lingrel, J. B. (1973). *Biochim. Biophys. Acta* **299**, 148.

Lockwood, A H., Chakrabosty, P. R., and Maitra, U. (1971). *Proc. Natl. Acad. Sci. U.S.A.* **68**, 3122.

Lodish, H. F. (1968). *Nature (London)* **220**, 345.

Lodish, H. F. (1970). *J. Mol. Biol.* **50**, 689.

Loening, U. E. (1967). *Biochem. J.* **102**, 251.

Lukanidin, E. M., Georgiev, G. P., and Williamson, R. (1971). *FEBS Lett.* **19**, 152.

Lukanidin, E. M., Olsnes, S., and Pihl, A. (1972). *Nature (London), New Biol.* **240**, 90.

McConkey, E. H., and Hopkins, J. W. (1965). *J. Mol. Biol.* **14**, 257.

McDowell, M. J., Joklik, W. K., Villa-Komaroff, L., and Lodish, H. F. (1972). *Proc. Natl. Acad. Sci. U.S.A.* **69**, 2649.

Mach, B., Faust, C., and Vassali, P. (1973). *Proc. Natl. Acad. Sci. U.S.A.* **70**, 451.

Macnaughton, M., Freeman, K. B., and Bishops, J. O. (1974). *Cell* **1**, 117.

Maggio, R., and Catalano, C. (1963). *Arch. Biochem. Biophys.* **103**, 164.

Marbaix, G., and Burny, A. (1964). *Biochem. Biophys. Res. Commun.* **16**, 522.

Marcus, A. (1969). *Symp. Soc. Exp. Biol.* **23**, 143.

Marcus, A. (1972). *Methods Mol. Biol.* **2**, 127.

Marcus, A., Luginbill, B., and Feeley, J. (1968). *Proc. Natl. Acad. Sci. U.S.A.* **58**, 1243.

Marmur, J., Rownd, R., and Schildkraut, C. L. (1963). *Prog. Nucleic Acid Res.* **1**, 232.

Marotta, C. A., Forget, B. G., Weissman, S. M., Verma, I. M., McCaffrey, R. P., and Baltimore, D. (1974). *Proc. Natl. Acad. Sci. U.S.A.* **71**, 2300.

Martin, T. E., and Wool, I. G. (1968). *Proc. Natl. Acad. Sci. U.S.A.* **60**, 569.

Mathews, M. B., and Korner, A. (1970). *Eur. J. Biochem.* **17**, 328.

Mathews, M. B., and Osborn, M. (1974). *Biochim. Biophys. Acta* **340**, 147.

Mathews, M. B., Osborn, M., and Lingrel, J. B. (1971). *Nature (London)* **233**, 206.

Mathews, M. B., Osborn, M., Berns, A. J. M., and Bloemendal, H. (1972a). *Nature (London), New Biol.* **236**, 5.

Mathews, M. B., Pragnell, I. B., Osborn, M., and Arnstein, H. R. V. (1972b). *Biochim. Biophys. Acta* **287**, 113.

Means, A. R., Comstock, J. P., Rosenfeld, G. C., and O'Malley, B. W. (1972). *Proc. Natl. Acad. Sci. U.S.A.* **69**, 1146.

Mechler, B., and Mach, B. (1971). *Eur. J. Biochem.* **21**, 552.

Mendecki, J., Lee, S. Y., and Brawerman, G. (1972). *Biochemistry* **11**, 792.

Merkel, C. G., Kwan, S.-P., and Lingrel, J. B. (1974). *Fed Proc., Fed. Am. Soc. Exp. Biol.* **33**, 1542.

Metafora, S., Terada, M., Dow, L. W., Marks, P. A., and Bank, A. (1972). *Proc. Natl. Acad. Sci. U.S.A.* **69**, 1299.

Milcarek, C., Price, R., and Penman, S. (1974). *Cell* **3**, 1.

Miller, J. V., Cuatrecasas, P., and Thompson, E. B. (1971). *Proc. Natl. Acad. Sci. U.S.A.* **68**, 1014.

Miller, R. L., and Schweet, R. (1968). *Arch. Biochem. Biophys.* **125**, 632.

Min Jou, W., Haegeman, G., Ysebaert, M., and Fiers, W. (1972). *Nature (London)* **237**, 82.

Molloy, G. R., Thomas, W. L., and Darnell, J. E. (1972). *Proc. Natl. Acad. Sci. U.S.A.* **69**, 3684.

Molloy, G. R., Jelinek, W., Salditt, M., and Darnell, J. E. (1974). *Cell* **1**, 43.

Morel, C., Gander, E. S., Herzberg, M., Dubochet, J., and Scherrer, K. (1973). *Eur. J. Biochem.* **36**, 455.

Morse, D. E., Baker, R. F., and Yanofsky, C. (1968). *Proc. Natl. Acad. Sci. U.S.A.* **60**, 1428.

Morrison, M. R., Gorski, J., and Lingrel, J. B. (1972). *Biochem. Biophys. Res. Commun.* **49**, 775.

Morrison, M. R., Brinkley, S. A., Gorski, J., and Lingrel, J. B. (1974). *J. Biol. Chem.* **249**, 5290.

Morrison, T., and Lodish, H. F. (1973). *Proc. Natl. Acad. Sci. U.S.A.* **70**, 315.

Morrison, T. G., and Lodish, H. (1975). *J. Biol. Chem.* **250**, 6955.

Mosteller, R. D., Rose, J. K., and Yanofsky, C. (1970). *Cold Spring Harbor Symp. Quant. Biol.* **35**, 461.

Mudd, J. A., and Summers, D. F. (1970). *Virology* **42**, 958.

Munoz, R. F., and Darnell, J. E. (1974). *Cell,* **2**, 247.

598

David A. Shafritz

Munro, A. J., and Korner, A. (1964). *Nature (London)* **201**, 1194.

Munro, A. J., Jackson, R. J., and Korner, A. (1964). *Biochem. J.* **92**, 289.

Murphy, W., and Attardi, G. (1973). *Proc. Natl. Acad. Sci. U.S.A.* **70**, 115.

Nakazato, H., and Edmonds, M. (1972). *J. Biol. Chem.* **247**, 3365.

Nakazato, H., Edmonds, M., and Kopp, D. W. (1974). *Proc. Natl. Acad. Sci, U.S.A.* **71**, 200.

Nathans, D., Notani, G., Schwartz, J. H., and Zinder, N. D. (1962). *Proc. Natl. Acad. Sci. U.S.A.* **48**, 1424.

Nemer, M. (1962). *Biochem. Biophys. Res. Commun.* **8**, 511.

Nemer, M. (1967). *Prog. Nucleic Acid Res. Mol. Biol.* **7**, 243.

Nemer, M., and Bard, S. G. (1963). *Science* **140**, 664.

Nemer, M., Graham, M., and Dubroff, L. M. (1974). *J. Mol. Biol.* **89**, 435.

Neinhuis, A. W., and Anderson, W. F. (1971). *J. Clin. Invest.* **50**, 2458.

Nirenberg, M. (1963). *In* "Methods in Enzymology" (S. P. Colowick and N. O. Kaplan, eds.), Vol. 6, p. 17. Academic Press, New York.

Nirenberg, M. W., and Matthaei, J. H. (1961). *Proc. Natl. Acad. Sci. U.S.A.* **47**, 1588.

Nudel, U., Lebleu, B., Zehavi-Willner, T., and Revel, M. (1973). *Eur. J. Biochem.* **33**, 314.

Oenayama, K., and Ono, T. (1972). *J. Mol. Biol.* **65**, 75.

Ofengand, J., Haselkorn, R. (1962). *Biochem. Biophys. Res. Commun.* **6**, 469.

Olsen, G. D., Gaskill, P., and Kabat, D. (1972). *Biochim. Biophys. Acta* **272**, 297.

Olsnes, S. (1971). *Eur. J. Biochem.* **23**, 557.

Ovchinnikov, L. P., Voronina, A. S., Stepanov, A. S., Belitsina, N. V., and Spirin, A. S. (1968). *Mol. Biol. (Moscow)* **2**, 448.

Packman, S., Honjo, T., Swan, D., Nau, M., and Leder, P. (1974). *Fed. Proc., Fed. Am. Soc. Exp. Biol.* **33**, 1541.

Palacios, R., Sullivan, D., Summers, N. M., Kiely, M. L., and Schminke, R. T. (1973). *J. Biol. Chem.* **248**, 540.

Palmiter, R. D., Palacios, R., and Schminke, R. T. (1972). *J. Biol. Chem.* **247**, 3296.

Peacock, A. C., and Dingman, C. W. (1967). *Biochemistry* **6**, 1818.

Peacock, A. C., and Dingman, C. W. (1968). *Biochemistry* **7**, 668.

Pederson, T. (1974). *J. Mol. Biol.* **83**, 163.

Pemberton, R. W., Housman, D., Lodish, H. F., and Gaglioni, C. (1972). *Nature (London), New Biol.* **235**, 99.

Penman, S., Scherrer, K., Becker, Y., and Darnell, J. (1963). *Proc. Natl. Acad. Sci. U.S.A.* **49**, 654.

Penman, S., Vesco, C., and Penman, M. (1968). *J. Mol. Biol.* **34**, 49.

Perry, R. P. (1963). *Exp. Cell Res.* **29**, 400.

Perry, R. P., and Kelley, D. E. (1968). *J. Mol. Biol.* **35**, 37.

Perry, R. P., LaTorre, J., Kelley, D. E., and Greenberg, J. R. (1972). *Biochim. Biophys. Acta* **262**, 220.

Perry, R. P., Kelley, D. E., Friderici, K. H., and Rottman, F. M. (1975). *Cell* **6**, 130.

Picciano, D. J., Prichard, P. M., Merrick, W. C., Shafritz, D. A., Graf, H., Crystal, R. G., and Anderson, W. F. (1973). *J. Biol. Chem.* **248**, 204.

Prichard, P. M., Gilbert, J. M., Shafritz, D. A., and Anderson, W. F. (1970). *Nature (London)* **226**, 511.

Prichard, P. M., Picciano, D. J., Laycock, D. G., and Anderson, W. F. (1971). *Proc. Natl. Acad. Sci. U.S.A.* **68**, 2752.

Priess, H., and Zellig, W. (1967). *Hoppe-Seyler's Z. Physiol. Chem.* **348**, 817.

Prives, C. L., Aviv, H., Paterson, B. M., Roberts, B. E., Rozenblatt, S., Revel, M., and Winocour, E. (1974). *Proc. Natl. Acad. Sci. U.S.A.* **71**, 302.

Quinlan, T. J., Billings, P. B., and Martin, T. E. (1974). *Proc. Natl. Acad. Sci. U.S.A.* **71**, 2632.

Raff, R. A., Colot, H. V., Selvig, S. E., and Gross, P. R. (1972). *Nature (London)* **235**, 211.

Reddy, R., Ro-Choi, T. S., Henning, D., and Busch, H. (1974). *J. Biol. Chem.* **249**, 6486.

Rekosh, D. M., Lodish, H. F., and Baltimore, D. (1970). *J. Mol. Biol.* **54**, 327.

Revel, M., Greensphan, H., and Herzberg, M. (1970). *Eur. J. Biochem.* **16**, 117.

Rhoads, R. E., McKnight, G. S., and Schminke, R. T. (1971). *J. Biol. Chem.* **246**, 7407.

Rhodes, D. P., Moyer, S. A., and Banerjee, A. K. (1974). *Cell* **3**, 327.

Roberts, B. E., and Paterson, B. M. (1973). *Proc. Natl. Acad. Sci. U.S.A.* **70**, 2330.

Roberts, B. E., Mathews, M. B., and Bruton, C. J. (1973). *J. Mol. Biol.* **80**, 733.

Robertson, H. D., Webster, R. E., and Zinder, N. D. (1968). *J. Biol. Chem.* **243**, 82.

Rosen, H. M., Woo, S. L. C., Holder, J. W., Means, A. R., and O'Malley, B. W. (1975). *Biochemistry* **14**, 69.

Rosenberg, M., Kramer, R. A., and Steitz, J. A. (1974). *J. Mol. Biol.* **89**, 777.

Rosenfeld, G. C., Comstock, J. P., Means, A. R., and O'Malley, B. W. (1972). *Biochem. Biophys. Res. Commun.* **47**, 387.

Ross, J., Aviv, H., Scolnick, E., and Leder, P. (1972). *Proc. Natl. Acad. Sci. U.S.A.* **69**, 264.

Rourke, A. W., and Heywood, S. M. (1972). *Biochemistry* **11**, 2061.

Rowley, P. T., and Morris, J. A. (1967). *J. Biol. Chem.* **242**, 1533.

Rudland, P. S., Whybrow, W. A., and Clark, B. F. C. (1971). *Nature (London), New Biol.* **231**, 76.

Ruiz-Carrillo, A., Beato, M., Schutz, G., Feigelson, P., Allfrey, V. G. (1973). *Proc. Natl. Acad. Sci. U.S.A.* **70**, 3641.

Salditt-Georgieff, M., Jelinek, W., Darnell, J. E., Furuichi, Y., Morgan, M., and Shatkin, A. (1976). *Cell* **7**, 227.

Salser, W., Gesteland, R. F., and Bolle, A. (1967). *Nature (London)* **215**, 588.

Samarina, O. P., Lukanidin, E. M., and Georgiev, G. P. (1973). *Acta Endocrinol. (Copenhagen), Suppl.* **180**, 130.

Sampson, J., Mathews, M. B., Osborn, M., and Borghetti, A. F. (1972). *Biochemistry* **11**, 3636.

Schapira, G., Padieu, P., Maleknia, N., Kruh, J., and Dreyfus, J. C. (1965). *Bull. Soc. Chim. Biol.* **47**, 1687.

Schapira, G., Padieu, P., Maleknia, N., Kruh, J., and Dreyfus, J. C. (1966). *J. Mol. Biol.* **20**, 427.

Schapira, G., Dreyfus, J. C., and Maleknia, N. (1968). *Biochem. Biophys. Res. Commun.* **32**, 558.

Schechter, I. (1973). *Proc. Natl. Acad. Sci. U.S.A.* **70**, 2256.

Scherrer, K. (1973). *Acta Endocrinol. (Copenhagen), Suppl.* **180**, 95.

Scherrer, K. Marcaud, L., Zajdela, F., Breckenridge, B., and Gros, F. (1966). *Bull. Soc. Chim. Biol.* **48**, 1037.

Schimke, R. T., Rhoads, R. E., Palacios, R., and Sullivan, D. (1973). *Acta Endocrinol. (Copenhagen), Suppl.* **180**, 357.

Schimke, R. T., Palacios, R., Sullivan, D., Kiely, M. L., Gonzales, C., and Taylor, J. M. (1974). *In* "Methods in Enzymology" (K. Moldave and L. Grossman, eds.), Vol. 30, p. 631. Academic Press, New York.

Schochetman, G., and Perry, R. P. (1972). *J. Mol. Biol.* **63**, 577.

Schreier, M. H., and Staehelin, T. (1973a). *J. Mol. Biol.* **73**, 329.

Schreier, M. H., and Staehelin, T. (1973b). *Nature (London) New Biol.* **242**, 35.

Schreier, M. H., and Staehelin, T. (1973c). *Proc. Natl. Acad. Sci. U.S.A.* **70**, 462.

Schreier, M. H., Staehelin, T., Stewart, A., Gander, E., and Scherrer, K. (1973). *Eur. J. Biochem.* **34**, 213.

Schumm, D. E., and Webb, T. E. (1972). *Biochem. Biophys, Res. Commun.* **48**, 1259.

Schutz, G., Beato, M., and Feigelson, P. (1972). *Biochem. Biophys. Res. Commun.* **49**, 680.

Schutz, G., Beato, M., and Feigelson, P. (1973). *Proc. Natl. Acad. Sci. U.S.A.* **70**, 1218.

Schweet, R., Lamform, H., and Allen, E. (1958). *Proc. Natl. Acad. Sci. U.S.A.* **44**, 1029.

Scolnick, E. M., Benveniste, R., and Parks, W. P. (1973). *J. Virol.* **11**, 600.

Sedat, J., Lyon, A., and Sinsheimer, R. L. (1969). *J. Mol. Biol.* **44**, 415.

Seid-Akhavan, M., Winter, W. P., Abramson, R. K., and Rucknagel, D. L. (1972). *Blood* **40**, 927.

Shafritz, D. A. (1974a). *J. Biol. Chem.* **249**, 81.

Shafritz, D. A. (1974b). *J. Biol. Chem.* **249**, 89.

Shafritz, D. A., and Anderson, W. F. (1970). *J. Biol. Chem.* **245**, 5553.

Shafritz, D. A., and Drysdale, J. W. (1975). *Biochemistry* **14**, 61.

Shafritz, D. A., Drysdale, J. W., and Isselbacher, K. J. (1973). *J. Biol. Chem.* **248**, 3220.

Shafritz, D. A., Weinstein, J. A., Safer, B., Merrick, W. C., Weber, L. A., Hickey, E. D., and Baglioni, C. (1976). *Nature (London)* **261**, 291.

Sheiness, D., and Darrell, J. E. (1973). *Nature (London), New Biol.* **241**, 265.

Sheldon, R., Jurale, C., and Kates, J. (1972). *Proc. Natl. Acad. Sci. U.S.A.* **69**, 417.

Siegert, W., Konings, R. N. H., Bauer, H., and Hofschneider, P. H. (1972). *Proc. Natl. Acad. Sci. U.S.A.* **69**, 888.

Silverman, P. M., Huh, M. M. O., and Sun, L. (1974). *Dev. Biol.* **40**, 59.

Simmons, D. T., and Strauss, J. H. (1974). *J. Mol Biol.* **86**, 397.

Singer, M. F., and Leder, P. (1966). *Annu. Rev. Biochem.* **35**, 195.

Singer, R. H., and Penman, S. (1972). *Nature (London)* **240**, 100.

Singer, R. H., and Penman, S. (1973). *J. Mol. Biol.* **78**, 321.

Sippel, A. E., Stavrianopoulos, J. G., Schutz, G., and Feigelson, B. (1974). *Proc. Natl. Acad. Sci. U.S.A.* **71**, 4635.

Skoultchi, A., and Gross, P. (1973). *Proc. Natl. Acad. Sci. U.S.A.* **70**, 2840.

Smith, A. E., Marcker, K. A., and Mathews, M. B. (1970). *Nature (London)* **225**, 184.

Smith, A. E., Wheeler, T., Glanville, N., and Kääriäimen, L. (1974). *Eur. J. Biochem.* **49**, 101.

Smith, M., Stavnezer, J., Huang, R. C., Gurdon, J. B., and Lane, C. D. (1973). *J. Mol. Biol.* **80**, 553.

Soreq, H., Nudel, U., Salomon, R., Revel, M., and Littauer, U. Z. (1974). *J. Mol. Biol* **88**, 233.

Spiegelman, S. (1963). *Fed. Proc.. Fed. Am. Soc. Exp. Biol.* **22**, 36.

Spirin, A. S. (1969). *Eur. J. Biochem.* **10**, 20.

Spirin, A. S., and Nemer, M. (1965). *Science* **150**, 214.

Spirin, A. S., Belitsina, N. V., and Ajtkhozhin, M. A. (1964). *Zh. Obshch. Biol.* **25**, 321.

Spohr, G., Granboulan, N., Morel, C., and Scherrer, K. (1970). *Eur. J. Biochem.* **17**, 296.

Spohr, G., Kayibanda, B., and Scherrer, K. (1972). *Eur. J. Biochem.* **31**, 194.

Spradling, A., Penman, S., Campo, M., and Bishop, J. O. (1974). *Cell* **3**, 23.

Srinivasan, P. R., Ramanarayan, M., and Rabbani, E. (1975). *Proc. Natl. Acad. Sci. U.S.A.* **72**, 2910.

Staehelin, T., Wettstein, F. O., Oura, H., and Noll, H. (1964). *Nature (London)* **201**, 264.

Stavnezer, J., and Huang, R. C. C. (1971). *Nature (London), New Biol.* **230**, 172.

Steitz, J. A. (1969). *Nature (London)* **224**, 957.

Steitz, J. A. (1974). *Fed. Proc., Fed. Am. Soc. Exp. Biol.* **33**, 1221.

Stevens, R. H., and Williamson, A. R. (1972). *Nature (London)* **239**, 143.

Stevens, R. H., and Williamson, A. R. (1973). *Proc. Natl. Acad. Sci. U.S.A.* **70**, 1127.

Stewart, A. G., Gander, E. S., Morel, C., Luppis, B., and Scherrer, K. (1973). *Eur. J. Biochem.* **34**, 205.

Strauss, J. H., Kelley, R. B., and Sinsheimer, R. L. (1968). *Biopolymers* **6**, 793.

Sugano, H., Suda, S., Kawada, T., and Sugano, I. (1971). *Biochim. Biophys. Acta* **238**, 139.

Sullivan, D., Palacios, R., Stavnezer, J., Taylor, J. M., Faras, A. J., Kiely, M. L., Summers, N. M., Bishop, J. M., and Schimke, R. T. (1973). *J. Biol. Chem.* **248**, 7530.

Sullivan, N., and Roberts, W. K. (1973). *Biochemistry* **12**, 2395.

Swan, D., Aviv, H., and Leder, P. (1972). *Proc. Natl. Acad. Sci. U.S.A.* **69**, 1967.

Taylor, J. M., and Schimke, R. T. (1973). *J. Biol. Chem.* **248**, 7661.

Taylor, J. M., Dozy, A., Kan, Y. W., Varmus, H. E., Lie-Injo, L. E., Ganesan, J., and Todd, D. (1974). *Nature (London)* **251**, 392.

Temin, H., and Mizutani, S. (1970). *Nature (London)* **226**, 1211.

Temmerman, J., and Lebleu, B. (1969). *Biochim. Biophys. Acta* **174**, 544.

Temple, G. F., and Housman, D. E. (1972). *Proc. Natl. Acad. Sci. U.S.A.* **69**, 1574.

Thompson, W. C., Buzash, E. A., and Heywood, S. M. (1973). *Biochemistry* **12**, 4559.

Trakatellis, A. C., Axelrod, A. E., and Montjar, M. (1964). *J. Biol. Chem.* **239**, 4237.

Tsiapalis, C. M., Dorson, J. W., De Sante, D. M., and Bollum, F. J. (1973). *Biochem. Biophys. Res. Commun.* **50**, 737.

Tyler, A. (1963). *Am. Zool.* **3**, 109.

Verma, I., Temple, G. F., Fan, H., and Baltimore, D. (1972). *Nature (London), New Biol.* **235**, 163.

Verneer, C., van Alphen, W., van Knippenberg, P., and Bosch, L. (1973). *Eur. J. Biochem.* **40**, 293.

Waters, L. C., and Dure, L. S., III. (1966). *J. Mol. Biol.* **19**, 1.

Weeks, D. P., and Marcus, S. (1971). *Biochim. Biophys. Acta* **232**, 671.

Weinberg, R. A. (1973). *Annu. Rev. Biochem.* **42**, 329.

Weisberger, A. S., and Armentrout, S. A. (1966). *Proc. Natl. Acad. Sci. U.S.A.* **56**, 1612.

Weissmann, C., Billeter, M. A., Goodman, H. M., Hindley, J., and Weber, H. (1973). *Annu. Rev. Biochem.* **42**, 303.

Whiteley, A. H., McCarthy, B. J., and Whiteley, H. R. (1966). *Proc. Natl. Acad. Sci. U.S.A.* **55**, 519.

Whiteley, A. H., McCarthy, B. J., and Whiteley, H. R. (1970). *Dev. Biol.* **21**, 216.

Wiegers, U., and Hilz, H. (1971). *Biochem. Biophys. Res. Commun.* **44**, 513.

Wigle, D. T. (1973). *Eur. J. Biochem.* **35**, 11.

Wigle, D. T., and Smith, A. E. (1973). *Nature (London), New Biol.* **242**, 136.

Williamson, R. (1973). *FEBS Lett.* **37**, 1.

Williamson, R., Morrison, M., Lanyon, G., Eason, R., and Paul, J. (1971). *Biochemistry* **10**, 3014.

Williamson, R., Clayton, R., and Truman, D. E. S. (1972). *Biochem. Biophys. Res. Commun.* **46**, 1936.

Williamson, R., Drewienkiewicz, C., and Paul, J. (1973). *Nature (London), New Biol.* **241**, 66.

Williamson, R., Crossley, J., and Humphries, S. (1974). *Biochemistry* **13**, 703.

Wilt, F. H., and Hultin, T. (1962). *Biochem. Biophys. Res. Commun.* **9**, 313.

Winter, M. A., Sr., and Edmonds, M. (1972). *Fed. Proc., Fed. Am. Soc. Exp. Biol.* **31**, 4031.

Zelenka, P., and Piatigorsky, J. (1974). *Proc. Natl. Acad. Sci. U.S.A.* **71**, 1896.

Zomzely-Neurath, C., York, C., and Moore, B. W. (1974). *Proc. Natl. Acad. Sci. U.S.A.* **69**, 2326.

Zucker, W. V., and Schulman, H. M. (1968). *Proc. Natl. Acad. Sci. U.S.A.* **59**, 582.

DNA-Dependent Cell-Free
Protein Synthesis

BENOIT de CROMBRUGGHE

I. Introduction*

   Some of the previous chapters are devoted to a detailed analysis of the individual reactions involved in protein synthesis. In contrast, this article will deal with the *in vitro* transfer of genetic information from DNA to RNA to protein resulting in the synthesis of active enzymes. Such complex synthesis takes place in systems consisting of crude bacterial extracts able to support both transcription of a specific DNA and translation of the synthesized RNA. The development of such systems has obviously numerous potential applications which should lead to a more comprehen-

---

   * Abbreviations used in this chapter: *lac*, lactose; *gal*, galactose; *ara*, arabinose; *trp*, tryptophan; *arg*, arginine; *CM*, chloramphenicol acetylase; cyclic AMP, cyclic adenosine 3',5'-monophosphate.

sive understanding of protein synthesis. This chapter will focus on the contributions of the DNA-dependent protein-synthesizing systems in advancing our understanding of regulation of gene expression in bacteria. A few typical examples will illustrate these achievements. We will also attempt to evaluate both the efficiency and the fidelity of these cell-free systems in reproducing the reactions which lead to the synthesis of complete enzymes in intact cells.

II. The Cell-Free "Coupled" Systems for RNA and Protein Synthesis

Two somewhat different systems have been used to demonstrate the DNA-dependent synthesis of active proteins. The first system, developed by Zubay and his co-workers (Lederman and Zubay, 1968; Zubay *et al.*, 1970a), consists of a crude unfractionated *Escherichia coli* lysate from which debris and other fast sedimenting materials are removed by low speed centrifugation (30,000 *g*, hence the name S30). Such extracts will support *de novo* synthesis of β-galactosidase and other enzymes upon addition of DNA containing the appropriate genes. Extracts are made from cells with a deletion in the gene specifying the enzyme to be synthesized. Thus, any enzyme activity detected after the incubation for protein synthesis will be due to *de novo* synthesis of this enzyme. The second system, developed by Gold and Schweiger (1969a), consists of preincubated ribosomes and a protein fraction partially purified by DEAE-cellulose chromatography. It has been used for the synthesis of enzymes specified by the DNA of bacteriophages T4 (Gold and Schweiger, 1969b), T3 (Dunn *et al.*, 1972) and T7 (Schweiger *et al.*, 1972). This system is much less efficient in supporting the synthesis of bacterial enzymes. It presumably lacks one or more factors such as the L factor, needed for optimal β-galactosidase synthesis (Kung *et al.*, 1974). Detailed descriptions of both cell-free systems have been reported (Zubay *et al.*, 1970a; Gold and Schweiger, 1971; Herrlich and Schweiger, 1974).

Both systems are programmed with DNA containing the genes or operon under study including their adjacent regulatory region. Usually this DNA is either bacteriophage DNA or DNA from a transducing phage. Although synthesis of specific *E. coli* proteins has been achieved using *E. coli* DNA as template (Kaltschmidt *et al.*, 1974), a much higher concentration of a particular bacterial gene or operon is obtained with the DNA of a transducing phage. Techniques are available for the isolation of a large number of specialized transducing phages (Shimada *et al.*, 1972; Gottesman and Beckwith, 1969; Press *et al.*, 1971) so that the DNA-dependent synthesis of a variety of *E. coli* proteins can now be studied *in vitro*.*

Several modifications have been brought to the systems originally described. The protein synthesizing reactions described by Zubay *et al.* (1970a; Lederman and Zubay, 1968) contain, in addition to the substrates for RNA and protein synthesis, a number of cofactors. These have been found to be without effect in some studies and consequently omitted (Pouwels and Van Rotterdam, 1972; Wetekam *et al.*, 1972). The divalent cation calcium can also be omitted from the reaction if replaced in equivalent concentration by magnesium ions (B. de Crombrugghe, unpublished results). Active enzyme synthesis can also be obtained with S30 extracts which are fractionated into ribosomal subunits, a salt wash of the ribosomes, and a 200,000 *g* supernatant fraction (Kung *et al.*, 1973). The supernatant fraction has been partially purified by DEAE-cellulose chromatography (Kung *et al.*, 1974). Similar modifications have been brought to the system devised by Gold and Schweiger (Herrlich and Schweiger, 1974).

III. Studies on Regulation of Protein Synthesis

A. Regulation of the Synthesis of the Enzymes of the E. coli Lactose Operon

Many of our concepts on the structure and regulation of bacterial operons derive in large part from genetic and biochemical studies on the lactose operon of *E. coli*. The *lac* operon consists of three structural genes and an adjacent regulatory region. The three structural genes specify the enzymes β-galactosidase, lactose permease, and thiogalactoside transacetylase. β-Galactosidase hydrolyzes lactose to its monosaccharide components glucose and galactose; lactose permease is active in the transport and accumulation of lactose in the cell; the function of thiogalactoside transacetylase is not known. The synthesis of the three enzymes is induced in intact cells when lactose or an analogue of lactose like isopropyl-β-D-thiogalactoside (IPTG) is added to the growth medium. In the presence of inducer the rate of synthesis of these enzymes is increased approximately 1000-fold. The protein product of a regulatory gene, the *lac i* gene, exerts a negative control on the synthesis of the *lac* enzymes. Indeed many *lac i* mutants are constitutive, i.e., they are able to synthesize the *lac* enzymes in the absence of inducer.

---

* Bacterial (and any other) genes can also be introduced in plasmid or bacteriophage DNA by *in vitro* recombination techniques.

The DNA segment that immediately precedes the *lac* structural genes includes sequences which define the *lac* operator and the *lac* promoter. Modulation of the rate of synthesis of *lac* RNA results from the interaction between various proteins and the *lac* operator and promoter. The enzyme RNA polymerase associates with and initiates transcription of *lac* DNA at the *lac* promoter. Promoter mutations result in an altered rate of initiation of the synthesis of *lac* RNA and *lac* enzymes (Scaife and Beckwith, 1966; Contesse *et al.*, 1969). The *lac* repressor, a tetrameric protein, which is the product of the *lac i* gene, binds specifically and with a very high affinity to *lac* operator DNA (Gilbert and Muller-Hill, 1967; Riggs *et al.*, 1970a). Inducers like isopropyl-β-D-thiogalactoside in turn bind to *lac* repressor and cause an increase in the rate of dissociation of the repressor-operator complex (Riggs *et al.*, 1970b). *lac* operator mutations result in a reduced ability of the operator to bind repressor (Gilbert and Muller-Hill, 1967; Riggs *et al.*, 1968b) and cause constitutive synthesis of the *lac* enzymes.

Cyclic AMP is essential for *lac* operon expression: very little β-galactosidase is synthesized in the absence of cyclic AMP. Indeed mutants deficient in the enzyme adenyl cyclase (the enzyme which makes cyclic AMP from ATP) have a greatly reduced capacity to synthesize β-galactosidase and many other inducible enzymes (Perlman and Pastan, 1969). The intracellular levels of cyclic AMP vary according to the growth conditions. When *E. coli* grow with glucose as carbon source they make less β-galactosidase and other inducible enzymes, and have lower levels of cyclic AMP than cells which grow on succinate or glycerol. Addition of cyclic AMP to cells growing on glucose overcomes this repression in enzyme synthesis (Makman and Sutherland, 1965; Perlman and Pastan, 1968; Ullmann and Monod, 1968; Pastan and Perlman, 1970).

The prior development (Lederman and Zubay, 1968) of a system for the DNA-dependent synthesis of β-galactosidase was very important in the further elucidation of the role of cyclic AMP. Zubay and his colleagues indeed found that in this system cyclic AMP stimulates the cell-free synthesis of β-galactosidase 10- to 30-fold, suggesting a direct role for cyclic AMP in *lac* operon expression (Chambers and Zubay, 1969). The cell-free system thus provided an extremely valuable tool to investigate biochemically the mechanism of action of cyclic AMP. Further progress in understanding the function of cyclic AMP in regulation of protein synthesis was stimulated by the discovery of a new class of mutants (Emmer *et al.*, 1970; Schwartz and Beckwith, 1970). Indeed, among the pleiotropic mutants which are unable to synthesize the *lac* and a variety of other inducible enzymes, some have a normal adenyl cyclase activity and are unaffected by the addition of cyclic AMP. This raised the possibility that such mutants may be deficient in a factor needed for cyclic AMP action. A protein which binds cyclic AMP specifically was indeed found in extracts from wild-type cells (Emmer *et al.*, 1970). This protein, the cyclic AMP receptor protein (CRP), was purified by assaying for its ability to bind cyclic AMP. Mutant extracts lack this protein or contain it in greatly reduced amounts. The coupled system for β-galactosidase synthesis made it possible to test the role of this protein. Extracts from mutant cells are unable to support the DNA dependent synthesis of β-galactosidase *in vitro*. This defect is corrected however if an S30 from mutant cells is complemented with an extract of wild-type cells, or with the purified cyclic AMP receptor protein (Emmer *et al.*, 1970; Zubay *et al.*, 1970b). This complementation assay for *in vitro* β-galactosidase synthesis was used by Zubay *et al.* (1970b) to monitor the purification of the protein missing in the mutant extracts. These experiments proved that the effects of cyclic AMP in regulating *lac* operon expression are mediated by CRP. CRP has been purified to homogeneity. The protein has a molecular weight of 45,000 daltons and is composed of two identical subunits (Anderson *et al.*, 1971; Riggs *et al.*, 1971). It binds to DNA and this binding is enhanced by cyclic AMP (Riggs *et al.*, 1971; Nissley *et al.*, 1972).

The DNA dependent system for cell-free β-galactosidase synthesis has thus been essential for the demonstration of the role of CRP and cyclic AMP in regulating *lac* enzyme synthesis. It also provided a direct proof for the action of the *lac* repressor. Indeed, if *lac* repressor is present in the S30 extracts and the inducer IPTG is omitted, the synthesis of β-galactosidase is reduced by 95%; IPTG prevents this repression (Zubay *et al.*, 1970a).

Studies with the cell-free system for β-galactosidase synthesis also revealed the existence of an additional regulatory element. The tetranucleotide guanosine 5'-diphosphate, 3'-diphosphate (ppGpp), which accumulates in bacterial cells under conditions of amino acid deprivation, stimulates the cell-free synthesis of β-galactosidase three- to tenfold (de Crombrugghe *et al.*, 1971a; Yang *et al.*, 1974). Another protein, the L protein, which is different from any known transcriptional or translational factor, was recently shown to be required for optimal *in vitro* β-galactosidase synthesis (Kung *et al.*, 1974).

The demonstration that *lac* repressor acts by preventing directly initiation of *lac* transcription and that CRP in conjunction with cyclic AMP is essential for initiation of *lac* transcription came from experiments using a completely defined system for *lac* transcription (de Crombrugghe *et al.*, 1971b; Eron and Block, 1971). The precise molecular interactions between these controlling proteins and the DNA of the *lac* promoter and the *lac* operator region are now under intensive study. The nucleotide sequence of the entire *lac* regulatory region has been established, and the base changes caused by several mutations in the *lac* operator and the *lac* promoter have been determined (Dickson *et al.*, 1975; Gilbert *et al.*, 1973; Gilbert and Maxam, 1973). The complete amino acid sequence of the *lac* repressor is also known (Beyreuther *et al.*, 1973), as well as the sequence of the first 25 amino acids at the N-terminal end of CRP (I. Pastan, personal communication).* *lac* RNA appears to be initiated within the *lac* operator (Maizels, 1973). The structure of the *lac* operator is highly symmetrical. This symmetry could be important for recognition by the tetrameric *lac* repressor. No obvious relationship exists, however, between alterations in this symmetry produced by mutations and degree of constitutivity. The promoter region should include a segment for RNA polymerase association and for initiation of transcription in addition to a site recognized by CRP. Interaction of CRP with this segment presumably allows a stable complex to be formed between RNA polymerase and DNA. Formation of such a stable complex appears to be required for proper initiation of *lac* transcription (de Crombrugghe *et al.*, 1971a). Three classes of *lac* promoter mutations have been isolated and genetically mapped: one class appears to affect a site in the DNA where CRP interacts (Beckwith *et al.*, 1972); another where RNA polymerase interacts (Hopkins, 1974); a third class of mutants creates a hyperactive *lac* promoter, which in some cases becomes independent of cyclic AMP (Silverstone *et al.*, 1970).

B. Regulation of the Synthesis of the Enzymes of the E. coli Arabinose Operon

The *ara* operon like the *lac* operon is an inducible operon and consists of three structural genes and an adjacent regulatory segment. The three structural genes code for enzymes active in the metabolism of L-arabinose to D-xylulose 5-phosphate. Synthesis of the three enzymes is coordinately induced by L-arabinose. The operon is controlled both by the CRP-cyclic AMP system as well as by the protein product of the *araC* gene, a regulatory gene linked to the operon. The product of *araC* has been postulated to possess the parodoxical properties of serving both as a repressor and as an activator for *ara* enzyme synthesis. The *araC* gene product is indeed required for expression of the *ara* operon (Englesberg, 1970; Irr and Englesberg, 1970). If *araC* protein is absent as a result of a deletion of the *araC* gene or a nonsense mutation in the *araC* gene the operon cannot be induced. That the *araC* protein also functions as a repressor for

* Sequences of several internal cyanogen bromide cleaved peptides of CRP have been determined and the sequence of the entire CRP protein will probably be resolved soon (I. Pastan, personal communication).

the *ara* operon was suggested by the finding that some other *araC* mutants (C^c) lead to constitutive synthesis of the *ara* enzymes (Englesberg *et al.*, 1965). These *araC^c* mutants are recessive with respect to the wild-type allele since cells containing both the C^c allele and the C^+ allele are not constitutive but normally inducible. The model proposed by Englesberg and his co-workers (1969) postulates that the product of the *araC* gene interacts with the *ara* operator region as a repressor (P) to prevent expression of the operon; addition of the inducer L-arabinose converts the repressor P1 to an activator P2. The activator (P2) interacts with *ara* DNA at a site for positive control of the operon (the *ara I* locus). The activator is required for expression of the operon.

Experiments with the cell-free system for arabinose enzyme synthesis appear to confirm the main features of this model (Greenblatt and Schleif, 1971; Zubay *et al.*, 1971; Wilcox *et al.*, 1974; Yang and Zubay, 1973). The DNA-dependent synthesis of the enzymes of the *ara* operon thus provided evidence for a mode of regulation distinctly different from the *lac* model.

The DNA-dependent *in vitro* synthesis of the enzymes of the *ara* operon, ribulokinase and L-arabinose isomerase, is dependent on *araC* protein as well as on cyclic AMP. The *araC* protein can be synthesized *in vitro* or can be prepared from whole cells. If the *araC* gene is present in the DNA template, stimulation of *ara* enzyme synthesis is due to the *araC* protein synthesized *in vitro*. If the *araC* gene is deleted from the template, the *in vitro* system has to be supplemented with *araC* protein extracted from whole cells. In kinetic experiments ribulokinase is synthesized earlier when *araC* protein is added to the system than when *araC* protein has to be made *in vitro* (Wilcox *et al.*, 1974). No *ara* enzyme synthesis occurs in the absence of L-arabinose. D-Fucose, an anti-inducer *in vivo*, completely inhibits enzyme synthesis *in vitro* (Greenblatt and Schleif, 1971; Wilcox *et al.*, 1974). The properties of the *araC* protein made *in vivo* or *in vitro* are identical with respect to their responses to increasing concentrations of L-arabinose or D-fucose (Wilcox *et al.*, 1974). The *in vitro* system for *ara* enzyme synthesis has also been used as an assay for *araC* protein during its purification (Wilcox *et al.*, 1974).

The *araC* protein also appears to function as a repressor in the absence of arabinose (Greenblatt and Schleif, 1971; Wilcox *et al.*, 1974). The most direct demonstration of this role was provided by an experiment using a DNA template with an *ara I^c* mutation. The *ara I^c* mutation causes some constitutive expression of the *ara* enzymes *in vivo* in the absence of an active *araC* gene. With this DNA template a certain level of *ara* enzyme is synthesized in the absence of *araC* protein *in vitro*. However, if *araC* protein is added this synthesis is repressed. Addition of L-arabinose together with *araC* protein restores a normal level of enzyme synthesis (Wilcox *et al.*, 1974).

Thus, experiments with the DNA-dependent system for *ara* enzymes synthesis support the previously proposed model for the regulation of this operon.

C. Regulation of the Synthesis of the Enzymes of the E. coli Tryptophan Operon

The *trp* operon of *E. coli* contains five structural genes specifying enzymes essential for the biosynthesis of tryptophan. *In vivo*, repression of this operon requires the protein product of the *trpR* gene, a regulatory gene, unlinked to the *trp* operon, and tryptophan or an analogue of tryptophan. The structural genes of the operon are preceded by a regulatory region which includes an operator and a promoter. *trp* mRNA contains at its 5'-end a so-called "leader sequence" of about 160 nucleotides preceding the translation initiation codon of the first known structural gene of the operon (*trpE*) (Cohen *et al.*, 1973). The DNA segment corresponding to this leader sequence plays an important regulatory function tion (Jackson and Yanofsky, 1973; Bertrand *et al.*, 1975). Recent experiments by Yanofsky and his co-workers (Bertrand *et al.*, 1975) indicate that the *trp* leader region contains a site for termination of transcription named the *trp* attenuator. This site controls the proportion of transcripts initiated at the *trp* promoter which transcribe the entire *trp* operon. Apparently when the cell needs to synthesize more tryptophan, a larger proportion of initiating polymerase molecules transcribe the *trp* genes beyond the attenuator. Regulation at this site is affected by the transcription termination protein *rho* and probably by tryptophan-tRNA synthetase or tryptophan-tRNA itself.

An elegant system where the synthesis of the *lac* enzymes is regulated by the *trp* regulatory elements was used to demonstrate the effect of the *trp* repressor *in vitro* (Zubay *et al.*, 1972). A trp-lac fusion, λh80d*trp-lac* DNA, was used as template in the cell-free system. In λh80d*trp-lac* DNA genetic transposition and deletion has fused the *lac* and *trp* DNA in one operon. This operon contains the *trp* promoter-operator region, part of the first *trp* structural gene (*trpE*) and the *lac*, Z, Y, and A genes in that order. *In vivo*, with this fused operon, synthesis of β-galactosidase is regulated by tryptophan in a *trpR⁺* strain. In a *trpR⁻* strain synthesis of the *lac* enzymes is constitutive. Cell-free synthesis of β-galactosidase with the *trp-lac* DNA as template is unaffected by cyclic AMP and occurs with S30 extracts made from a *trpR⁻* strain. If S30 extracts from a *trpR⁺* strain replace the *trpR⁻* extracts no β-galactosidase is synthesized in the presence of tryptophan. Extracts of *trpR⁺* and *trpR⁻* are, however, equally active in supporting β-galactosidase synthesis from a *lac* DNA template. Thus, extracts of *trpR*⁺ cells specifically inhibit β-galactosidase synthesis when the fused operon is used as template. This assay was used for the purification of the *trp* repressor (Zubay *et al.*, 1972).

Several important experiments were now possible due to the availability of a *trp* repressor preparation. Addition of the partially purified repressor together with tryptophan to an otherwise defined *in vitro* transcription system specifically blocks *trp* operon transcription (Rose *et al.*, 1973). Repression occurs with a *trp* O⁺ template (wild-type *trp* operator) not with a *trp* Oᶜ template (operator constitutive). The *trp* enzymes themselves can also be synthesized *in vitro* if λh80d*trp* DNA is used as template in the coupled system (Zalkin *et al.*, 1974). With this template the synthesis of the *trp* enzymes occurs only with extracts from *trpR*⁻ cells. Substituting *trpR*⁻ extracts with *trpR*⁺ extracts abolishes *trp* enzyme synthesis. However with φ80 *trp* DNA as template the amounts of *in vitro* synthesized *trp* enzymes are increased and this synthesis is equally effective with *trpR*⁺ and *trpR*⁻ extracts (Pouwels and Van Rotterdam, 1972; Zalkin *et al.*, 1974). With this template and in the experimental conditions of the coupled system, *trp* RNA synthesis appears to originate mainly at a phage promoter approximately 7000 to 8000 nucleotides from *trpE* (Zalkin *et al.*, 1974). Thus, when *trp* RNA synthesis results from read-through initiated at an upstream phage promoter, expression of the *trp* enzymes is not regulated by the *trpR* product.

D. Regulation of the Expression of the Bacteriophage λ Genome by the *N* Gene Product

Upon infection of its host the temperate bacteriophage λ can undergo one of two responses, a lytic response or a lysogenic response. In the lytic response the phage DNA replicates many times leading to the production and release of a large number of phage particles and lysis of the cell. In the lysogenic response the phage coexists with its host; its DNA integrates in the host chromosome by a site-specific recombination process, and the viral genes are functionally repressed. A phage DNA-coded repressor protein prevents expression of the λ genes by blocking transcription initiation at two sites (O_L and O_R) located immediately to the left and right of the repressor gene, *C1*. The action of the λ repressor has been demonstrated *in vitro* (Steinberg and Ptashne, 1971; Wu *et al.*, 1972), and the interaction between this protein and its cognate operator regions has been intensively studied (Maniatis *et al.*, 1973), as well as the structure of these operators. The two corresponding promoters (P_L and P_R) are inserted within the operator regions (Maurer *et al.*, 1974). The nucleotide sequence analysis of these segments has been determined (Maniatis *et al.*,

1974). Inactivation of the λ repressor or its insufficient synthesis leads to initiation of transcription at these two promoters and vegetative growth of the phage. However, for vegetative development another λ gene product, the *N* protein, is required in order to express the viral genes. In the absence of *N* gene product, only the two genes immediately adjacent to the two operators the *N* gene and the *tof* gene are transcribed; no distal genes are expressed (Kourilsky *et al.*, 1968). *In vitro* if λ DNA is transcribed in a purified system, the *N, tof,* and distal genes are transcribed. Addition of the *E. coli* termination factor, *rho,* to this *in vitro* system causes arrest of transcription at the distal end of the *N* and *tof* genes as *in vivo* in the absence of *N* gene product (Roberts, 1969). The *N* gene product is believed to act by allowing transcription to proceed beyond the distal ends of the *N* and *tof* genes antagonizing the function of *rho* (Roberts, 1969). Such transcription can even propagate into the neighboring bacterial chromosome when excision of the prophage is prevented (Adhya *et al.*, 1974; Franklin, 1974). This form of regulation of gene expression, namely, antitermination, differs markedly from the type of regulation which is exerted on initiation of transcription. Antitermination may be important in other aspects of λ phage regulation as well as in control of the activity of bacterial genes.

The DNA-dependent cell-free system has been used to demonstrate the role of *N* protein. In one study the synthesis of a λ-specific enzyme, the product of gene R, endolysin, is analyzed (Greenblatt, 1972, 1973). In another study the DNA of a λ*trp* transducing phage was used as template for the synthesis of anthranilate synthetase, an enzyme of the *trp* operon (Dottin and Pearson, 1973). In this phage a portion of the viral genes to the left of *N* are substituted by DNA from the *trp* operon of *E. coli*. This DNA includes the *trpE* and *trpD* genes but lacks the *trp* promoter and the *trp* operator. With this DNA, synthesis of anthranilate synthetase *in vivo* is no longer regulated by the *trp* repressor but is dependent on *N* function. The cell-free synthesis of both endolysin and anthranilate synthetase is also dependent on *N* protein (Greenblatt, 1972, 1973; Dottin and Pearson, 1973). *N* protein has to be provided by an extract of induced lysogenic cells if the template contains a mutation in *N*. It can also, however, be synthesized *in vitro* from λN⁺ DNA. The *N*-dependent *in vitro* synthesis of the two enzymes is blocked by addition of the λ repressor, thus reproducing the *in vivo* regulation. Cyclic AMP has no effect. Using the cell-free synthesis of endolysin as assay for the *N* gene product, the synthesis and fate of *N* protein in whole cells was analyzed. The results of these experiments indicate that both the products of genes *cI* and *tof* repress the synthesis of *N* and that the *N* gene product is also metabolically unstable, thus verifying previous more indirect *in vivo* evidence (Schwartz, 1970).

Any of the two *in vitro* assays could be used to monitor the purification of *N* and help elucidate the mechanism of action of this regulatory protein. The products of the genes *tof, Q, cII,* and *cIII* are other phage-specific proteins which appear to regulate expression of the λ genome. Their role in the control of λ development has been determined mainly by genetic studies. Biochemical studies with *in vitro* systems, such as the coupled DNA-dependent protein synthesis system, are obviously needed for a better understanding of the role of these proteins.*

IV. Fidelity of the Coupled System

One of the aims of the DNA-dependent protein synthesis system is to reproduce as faithfully as possible the events of transcription and translation as they take place inside the cell. The fidelity of the coupled system will be examined here by discussing some of the steps leading to the synthesis of active enzymes.

A. Efficiency of Transcription and Translation

The overall efficiency of the DNA-dependent system is obviously low compared to the enzyme-synthesizing capacity of whole cells. *In vitro* about 2 to 10 molecules of monomeric subunits of β-galactosidase and galactokinase can be synthesized per gene copy of *lac Z* and *gal K* DNA per hour (Zubay, 1973; Wetekam *et al.,* 1971), and about 50 molecules of L-arabinose isomerase per copy of *ara A* DNA per hour (Zubay *et al.,* 1971). Wilcox *et al.* (1974) estimated that the synthesis of L-arabinose isomerase was about 5% as efficient *in vitro* as in induced whole cells. The number of active polypeptides synthesized will depend on a number of factors including the rate of initiation and elongation of transcription of a specific gene as well as the rate of initiation and elongation of translation of the corresponding RNA, the half-life of this mRNA, and the ratio of complete to incomplete RNA and polypeptide chains synthesized.

Little is known about the true *in vivo* initiation rate of transcription for defined mRNA's. It is believed that about one round of transcription of *trp* mRNA occurs every 0.2 to 1 min in cells derepressed for *trp* expression (Baker and Yanofsky, 1972). This figure may however be an underestimate of the true rate of initiation. The *trp* "leader DNA" indeed modulates transcription of the *trp* operon by allowing only a fraction of the transcripts initiated at the *trp* promoter to extend into the *trp* structural genes *E, D, C, B, A* (Jackson and Yanofsky, 1973; Bertrand *et al.,*

---

* The *in vitro* synthesis of the λ*cI* gene product and its autogenous regulation have recently been reported (Dottin *et al.,* 1975).

1975). *In vitro,* in the coupled system, the rate of initiation of transcription has not been determined and many mRNA molecules could be incompletely synthesized. In a purified system for transcription of the *E. coli gal* DNA, this operon is transcribed about once per *gal* DNA copy during a 30 min incubation period at 37° (B. de Crombrugghe, unpublished observations). The rate of elongation of transcription in the coupled system is about half the *in vivo* rate. *In vitro,* the rate of elongation of transcription is about 22 nucleotides/sec at 37° for *lac* RNA (Zubay, 1973) and 19 nucleotides/sec at 37° for *trp* RNA (Squires *et al.,* 1973). The *in vivo* elongation rate was estimated to be between 37 and 45 nucleotides/sec for *trp* RNA (Rose *et al.,* 1970), and about 50 nucleotides/sec for *lac* RNA (Kepes, 1967) at 37°.

In vivo each mRNA molecule is translated by a considerable number of ribosomes. Clusters of ribosomes translating a presumptive mRNA during its own transcription have been visualized by electron microscopy (Miller *et al.,* 1970). *In vitro* in the coupled system the rate of initiation of translation is presumably much slower. The elongation rate for translation *in vitro* is about one-third of the *in vivo* elongation rate. Indeed, an *in vitro* translation elongation rate of four amino acids/sec at 34° was estimated from the time needed to complete synthesis of *trp A* enzyme after completion of *trp E* enzyme, and the length of the DNA from *trp D* through *trp A* (Squires *et al.,* 1973). The *in vivo* polypeptide elongation rate of *E. coli* β-galactosidase is approximately 15 amino acids/sec at 37° (Lacroute and Stent, 1968).

B. Half-Life of mRNA

Both the chemical half-life and the functional half-life of the RNA synthesized in the coupled system *in vitro* is much longer than *in vivo*. *In vivo* the chemical half-life of *lac* (Varmus *et al.,* 1970), *gal* (Miller *et al.,* 1971), and *trp* RNA (Morse *et al.,* 1969) at 37° is 1½ to 2 min. In the coupled cell-free system the chemical half-life of *trp* RNA was estimated to be more than two hr at 34° (Squires *et al.,* 1973). The functional half-life of *gal* RNA *in vitro,* i.e., the capacity to synthesize *gal* enzymes after arrest of transcription, is about 6 to 7 min (Schumacher and Ehring, 1973); the chemical half-life of *gal* RNA *in vitro* has not been determined. A possible cause for the greater stability of the *in vitro* synthesized mRNA is the presence of calcium ions in the reaction mixtures. Calcium ions indeed stabilize the RNA synthesized in the coupled cell-free system when *E. coli,* T4, or ϕ80 DNA are used as templates (Gremer and Schlessinger 1974). Calcium ions are known to inhibit specific ribonucleases such as ribonuclease II and ribonuclease III (Gremer and Schlessinger, 1974). If calcium is omitted the half-life of both *E. coli* and T4 mRNA is considerably shorter (Gremer and Schlessinger, 1974). The experiments measuring the half-life of *trp* and *gal* mRNA *in vitro* were performed in reactions containing calcium ions.

C. Fidelity of Regulation

1. Transcription

Considerable evidence has accumulated to indicate that regulation of gene expression in bacteria occurs mainly at the level of transcription particularly at the stage of initiation of transcription. Indeed regulatory proteins such as the *lac*, *gal*, *trp* and λ repressors prevent initiation of transcription by interacting with defined DNA segments called operators. The specificity of these interactions is determined by the nucleotide sequences of these segments. Mutations in these sites result in altered interactions between repressors and their respective operators. Some of these regulatory sites have been more completely defined by nucleotide sequence analysis. CRP, a pleiotropic regulatory protein which in conjunction with cyclic AMP stimulates initiation of transcription of a number of genes or operons, interacts with a portion of the promoter of these operons.

With the DNA-dependent coupled system, control of initiation of transcription appears to respond to the same regulatory elements which are operative *in vivo*. Indeed, the presence or absence of regulatory molecules such as repressors, inducers, anti-inducers, CRP, and cyclic AMP results in the modulation of active enzyme synthesis. These molecules were shown to control initiation of transcription in defined *in vitro* systems. In fact, the effect of these molecules was often first demonstrated *in vitro* in the more complex cell-free DNA-dependent coupled system.

In a few cases, active enzyme synthesis occurs *in vitro* but no or a reduced level of regulation is observed. This absence of *in vitro* regulation is probably due to the existence on the DNA template of a strong promoter, located upstream with respect to the promoter of the operon under study. If φ80 *trp* DNA is used as template the *trp* enzymes are synthesized, but the *trp* repressor has no effect on this synthesis. With this template *trp* mRNA initiates mainly at a promoter located 7400 to 8000 nucleotides from *trp* E, the first structural gene of the *trp* operon (Squires *et al.*, 1973). However, with the DNA of another *trp* transducing phage as template, λh80*trp* DNA transcription appears to initiate at the *trp* promoter and *trp* enzyme synthesis responds to *trp* repressor control (Squires *et al.*, 1973). If λp*lac* DNA is used as template instead of λh80d*lac* DNA, considerable synthesis of β-galactosidase is seen in the absence of cyclic AMP; addition of cyclic AMP has only a small effect (Yang and Zubay, 1974). Thus, the structure of the DNA of the transducing phage or the relative position of phage and bacterial promoters probably determines how well initiation of transcription is regulated in the coupled system.

Both the tetranucleotides, guanosine 5'-diphosphate,3'-diphosphate (ppGpp), and the pentanucleotide, guanosine 5'-triphosphate,3'-diphosphate (pppGpp), stimulate the cell-free synthesis of the *lac* (de Crombrugghe *et al.*, 1971a; Yang *et al.*, 1974), *CM* (de Crombrugghe *et al.*, 1973a; Dottin *et al.*, 1973), *ara* (Greenblatt and Schleif, 1971), *trp* (Yang *et al.*, 1974), and *his* (Artz and Broach, 1975) enzymes in the coupled system, while inhibiting the synthesis of the *gal* (Parks *et al.*, 1971) and *arg* (Yang *et al.*, 1974) enzymes. These unusual nucleotides accumulate in *E. coli* upon amino acid deprivation. At the same time, a number of metabolic changes take place in the cell among which is a severe inhibition of RNA synthesis. With mutants in the *rel* locus no ppGpp nor pppGpp accumulate, and RNA synthesis is not inhibited when the cells are starved for an amino acid (Cashel, 1969; Cashel and Gallant, 1969). Thus, these nucleotides have been postulated to cause the inhibition of RNA synthesis in starved cells. The stimulation of the synthesis of β-galactosidase and other enzymes by pppGpp is probably due to its hydrolysis to ppGpp (Yang *et al.*, 1974). ppGpp stimulates transcription of *lac* DNA but the mechanism of this stimulation is not understood (de Crombrugghe *et al.*, 1971a). Recently, a protein factor was purified from *E. coli* and shown to stimulate the synthesis of *lac* RNA in the presence of ppGpp, probably by enhancing the rate of initiation of *lac* transcription. This stimulation occurs only when CRP and cyclic AMP are also present in the reaction (Aboud and Pastan, 1975). Whether the same factor mediates the stimulatory effects of ppGpp on the synthesis of the *trp*, *ara*, *CM*, and *his* enzymes has not been determined. Also, the biological significance of the stimulation by ppGpp of the synthesis of these enzymes is not well understood.

Since bacterial operons are discrete transcriptional units, which can be activated individually, a mechanism is required for termination of transcription at the promoter distal end of operons. The size of *lac* (Contesse and Gros, 1968), *gal* (Hill and Echols, 1966; Adhya *et al.*, 1974), and *trp* (Imamoto, 1968) RNA molecules, synthesized *in vivo,* is indeed consistent with a polycistronic transcript of the operon. Defined regulatory DNA sequences presumably constitute signals for arrest of transcription. In purified *in vitro* systems termination of transcription is in some cases directly mediated by the DNA template acting on RNA polymerase (Lebowitz *et al.*, 1971; Rosenberg *et al.*, 1975). In other cases termination requires the intervention of an additional protein, such as the factor *rho* (Roberts, 1969; de Crombrugghe *et al.*, 1973b). Whether termination of transcription occurs faithfully in the coupled DNA-dependent system has not been well studied. The main size of *trp* mRNA synthesized in the coupled system from the *trp* promoter, as measured after a 10 min pulse with radioactive UTP, was about 23 S (Squires *et al.*, 1973). This size is somewhat smaller than predicted for a polycistronic RNA corresponding to the entire operon; it could, however, represent incompletely synthesized or partially degraded RNA molecules.

2. Translation

Although much evidence has accumulated showing that control of protein synthesis occurs at the level of transcription in prokaryotes, recent experiments have suggested that protein synthesis may also be regulated at the level of initiation of translation. The best evidence for this type of control comes from studies on the synthesis of proteins specified by the genome of some bacteriophages (Kozak and Nathans, 1972). At this date, no studies have been reported where the coupled cell-free system is used to indicate a role for regulation of translation in the control of bacterial gene expression.*

D. Polarity in the Coupled System

An intriguing aspect of protein synthesis apparently related both to transcription and translation is the phenomenon of polarity. Polar mutations in bacteria do not only inactivate the product of the mutated genes, but they also abolish or reduce the expression of promoter distal genes. Several types of polar mutations exist. Among these are nonsense mutations, deletions (when frame shifts result in the production of nonsense codons), and insertion mutations (Shapiro, 1967; Adhya and Shapiro, 1969; Jordan *et al.*, 1968; Malamy, 1970). Insertions are always very polar regardless of their location within a gene. (Two types of insertions, containing, respectively, about 800 and 1400 base pairs, have been studied in the coupled DNA-dependent protein system.) Both nonsense and insertion mutations in bacteria are associated with a reduced level (or the absence) of RNA homologous to the segment of the operon distal to the site of mutation. Only RNA corresponding to the mutated cistron is found, and in nonsense mutants the portion beyond the mutation is very labile (Contesse *et al.*, 1966; Imamoto and Yanofsky, 1967). No transcripts of the adjacent promoter distal genes are detected (Hiraga and Yanofsky, 1972). Polarity presumably results from the unavailability of translatable RNA distal to the mutation. With nonsense mutations, ribosomes are probably released at the site of mutation. The unavailability of translatable RNA could be caused by termination of transcription at or near the site of the mutation due to a tight coupling of transcription and translation

---

* *Note added in proof:* A recent study has shown that the synthesis of gene 32 protein of bacteriophage T4 is subject to translational autogenous regulation [Russel, M., Gold, L., Morrissett, H., and O'Farrell, P. Z. (1976). *J. Biol. Chem.* **251**, 7263].

(Imamoto and Kano, 1971), or to rapid degradation of the untranslated RNA after discharge of ribosomes at the nonsense codon (Morse and Guertin, 1971). Alternatively, polarity may be caused by *rho*-directed termination of transcription at specific sites distal to a nonsense codon (Adhya *et al.*, 1974). Such transcription termination sites exist at or near the end of the first cistron of the *gal* operon, within the *lacZ* gene, and within one type of polar insertion. Indeed, in a defined *in vitro* transcription system, the termination factor *rho* causes arrest of transcription at these sites in the *gal* and *lac* operons and within the IS2 insertion (de Crombrugghe *et al.*, 1973b; B. de Crombrugghe, unpublished results). The recent findings that the *suA* gene specifies the termination protein *rho* supports the latter model (Richardson *et al.*, 1975). *suA* mutants indeed relieve the polar effects of nonsense mutations and of some insertion mutations (Morse *et al.*, 1969; Malamy, 1970; O. Reyes and S. Adhya, personal communication). Other *in vivo* experiments also suggest a role for the transcription termination factor *rho* in polarity. *In vivo,* a transcript initiated at the sex promoter of the prophage λ has the ability to override normal phage and bacterial transcription termination signals and to release the polar properties of both nonsense and insertion polar mutations (Adhya *et al.*, 1974). Such read-through or "juggernaut" transcription occurs only if the product of the phage gene *N* is present in the cells. Since the *N* protein has been postulated to antagonize the action of the host factor *rho,* these results provide additional evidence that *rho* causes the polarity of both nonsense and insertion mutations.

With the coupled DNA-dependent protein synthesis system, the introduction of a nonsense mutation in the *gal* or *trp* operon does not cause polarity (Wetekam *et al.*, 1972; Squires *et al.*, 1973; B. de Crombrugghe, unpublished results). The reason for the absence of polarity in this system is unclear. Obviously RNA distal to the mutation must be available for translation in contrast to the *in vivo* situation. Whether ribosomes discharge at a intracistronic nonsense codon in the S30 system has not been examined. Studies with the *trp* operon suggest that ribosomes discharge from a polycistronic RNA following termination of translation at a natural chain termination codon (Squires *et al.*, 1973). In these studies, kasugamycin, an inhibitor of initiation of translation, was used. Indeed, kasugamycin prevents translation of *trp A* mRNA (specified by the promoter most distal gene of the *trp* operon) after *trp E* polypeptide synthesis was initiated (Squires *et al.*, 1973). This suggests that ribosomes are discharged before they can reinitiate translation of the promoter distal portion of a polycistronic *trp* mRNA.

With polar insertion mutations in *gal* DNA, contradictory results have been obtained. One group did not observe polarity with unsupplemented S30 extracts (Wetekam *et al.,* 1972). In the author's laboratory, however, severe polarity was demonstrated *in vitro* with the same insertions in *gal* DNA using the coupled system. Preliminary evidence indicates that the presence of calcium ions in the coupled cell-free system relieves the polar properties of an insertion mutation in *gal* DNA (D. Court, personal communication). Calcium ions greatly increase the stability of the synthesized RNA in this system. Several purified *E. coli* ribonucleases are also inhibited by calcium ions (Gremer and Schlessinger, 1974). Calcium ions, however, are not required for active enzyme synthesis in the coupled system. The presence or absence of calcium ions in the cell-free system could be the reason for the contradictory observations by the two different laboratories.

Polarity is obviously a complex phenomenon *in vivo.* The RNA distal to a nonsense codon could be a target for several competing processes, namely, intracistronic reinitiation of translation by ribosomes, recognition by the termination factor *rho,* and digestion by nucleases. Intracistronic reinitiation of translation distal to a nonsense codon has been well documented in the *rIIB* cistron of bacteriophage *T*4 and for RNA's specified by the *lac z* and *i* genes (Sarabhai and Brenner, 1967; Grodzicker and Zipser, 1968; Platt *et al.,* 1972). Recently, *rho* has been shown to possess an RNA-dependent ribonucleotidase (ATPase) activity (Lowery-Goldhammer and Richardson, 1974). This activity is required for *rho*-mediated termination of transcription *in vitro* (Howard and de Crombrugghe, 1975). It is conceivable that *rho* may utilize its ATPase activity to move along the RNA. [Some restriction endonucleases have a DNA-dependent ATPase activity; these restriction enzymes are also believed to move along the DNA from their recognition site to a site of action for their restrictive property (Shulman, 1974).]

Events that take place after ribosomes dissociate from mRNA at an intracistronic nonsense codon could be similar to those occurring after ribosomes discharge at the distal end of a cistron or at the distal end of an operon. At sites between cistrons within an operon ribosomes presumably reinitiate at a high frequency, while at the distal end of an operon no reinitiation of translation occurs. The decision as to which process takes place, reinitiation by ribosomes or arrest of transcription by *rho* (or eventually endonucleolytic digestion) will probably be determined by the nucleotide sequences distal to the nonsense codon. The inability of the S30 system to faithfully reproduce polarity indicates that some of the components closely related to protein synthesis *in vivo* are less active or limiting in the cell-free system.

V. Conclusions

The development of systems able to support both transcription of a specific DNA segment and translation of its RNA product into an active protein and the many successful applications of these coupled systems have been instrumental in our understanding of several important aspects of gene regulation in *E. coli*. Indeed the very existence of such systems for enzyme synthesis allows the investigator to attempt to modify quantitatively some of the regulatory elements believed to participate in the regulation of specific genes in intact cells. The DNA-dependent protein synthesis system has thus provided a direct demonstration for the role of a number of regulatory molecules *in vitro*. The cyclic AMP-CRP complex, for instance, was shown to activate the *lac, gal* (Wetekam *et al.*, 1971; Parks *et al.*, 1971), *ara*, and *CM* (de Crombrugghe *et al.*, 1973a) genes and to be essential for their expression. *In vitro* repression of the *lac* (Zubay *et al.*, 1970a), *gal* (Wetekam *et al.*, 1971), and *trp* (Zubay *et al.*, 1972) genes is caused by the presence of their respective repressors in the reaction mixtures. Addition of inducers or anti-inducers results in modulation of enzyme synthesis. Introduction of promoter mutations in the template causes alteration in the rate of enzyme synthesis, and templates containing operator mutations respond poorly or not at all to the addition of their cognate repressor.

Studies on the mechanisms utilized by bacteriophages to regulate the expression of their own genome and repress the activity of the host genes have been analyzed *in vitro* (Schweiger *et al.*, 1972; O'Farrell and Gold, 1973a,b). Cell-free studies on the regulation of the synthesis of such essential macromolecules as ribosomal proteins (Kaltschmidt *et al.*, 1974; Jaskunas *et al.*, 1975) and RNA polymerase subunits (Kirschbaum, 1973) are now in progress. With the availability of transducing phages for many bacterial genes, the S30 systems constitute a tool of choice for the *in vitro* study of the regulation of the synthesis of a large number of bacterial proteins.

The DNA-dependent enzyme synthesis system constitutes further a very useful assay to monitor the purification of factors active in regulation of protein synthesis in bacteria. One method consists of making an S30 extract from cells which harbor a mutation in the gene specifying the regulatory factor under study. Such mutant extracts are complemented with protein extracts from wild-type cells. Both positive regulatory proteins such as CRP (Zubay *et al.*, 1970b) and negative regulatory proteins such as the trp repressor (Zubay *et al.*, 1972) were purified with this assay method. Another method consists of complementing extracts from wild-type cells, which become deficient for the regulatory protein under study, during the preparation of the S30. Indeed some proteins are very labile and are inactivated during the course of an S30 preparation. The inactive extracts can then be complemented with protein extracts prepared otherwise. A new protein with the properties of a termination factor has recently been discovered with this second method (Yang and Zubay, 1974).

Recently studies have also been initiated which consist in fractionating the S30 extracts and purifying the individual components involved in RNA and protein synthesis. These very interesting studies should provide a more comprehensive understanding of the relative role of the individual components for RNA and protein synthesis. They may also lead to the discovery of new factors related to aspects of RNA and protein synthesis which are presently not well understood. Such a new factor, L, was recently discovered: it is required for the optimal *in vitro* synthesis of β-galactosidase and is different from known transcriptional and translational factors (Kung *et al.*, 1974).

Cell-free DNA-dependent protein synthesis has also been applied to heterologous systems. The DNA from the mammalian transforming polyomavirus has been used with an *E. coli* S30 (Crawford and Gesteland, 1973). In these experiments, the size of the *in vitro* synthesized proteins and their tryptic peptides were compared with the size of the viral proteins and the proteolytic products of the virion proteins. Many tryptic peptides obtained *in vitro* were similar to digests of virion protein. One polypeptide made *in vitro* appears to have a size equal to that of the major virion protein. However, the majority of the polypeptides synthesized *in vitro* had smaller sizes. On the other hand, extracts of yeast mitochondria are able to support the T3 DNA directed synthesis of two bacteriophage specific enzymes (Richter *et al.*, 1972). The synthesis of these enzymes occurs, however, with much less efficiency than with *E. coli* extracts.

This review has emphasized some of the achievements as well as the imperfections of the DNA-dependent protein synthesis system. Such systems will continue to be an essential instrument in demonstrating the effects of molecules that regulate the expression of specific genes and helping to understand the diversity that characterizes the regulation of individual genes or operons in bacteria.

REFERENCES

Aboud, M., and Pastan, I. (1975). *J. Biol. Chem.* **250**, 2189–2195.

Adhya, S., and Shapiro, J. A. (1969). *Genetics* **62**, 231–247.

Adhya, S., Gottesman, M., and de Crombrugghe, B. (1974). *Proc. Natl. Acad. Sci. U.S.A.* **71**, 2534–2538.

Anderson, W. B., Schneider, A. B., Emmer, M., Perlman, R. L., and Pastan, I. (1971). *J. Biol. Chem.* **246**, 5929–5937.

Artz, S. W., and Broach, J. (1975). *Proc. Natl. Acad. Sci. U.S.A.* **72**, 3453–3457.

Baker, R. F., and Yanofsky, C. (1972). *J. Mol. Biol.* **69**, 89–102.

Beckwith, J., Groszicker, T., and Arditti, R. (1972). *J. Mol. Biol.* **69**, 155–160.

Bertrand, K., Korn, L., Lee, F., Platt, T., Squires, C. L., Squires, C. and Yanofsky, C. (1975). *Science* **189**, 22–26.

Beyreuther, K., Adler, K., Geisler, N., and Klem, A. (1973). *Proc. Natl. Acad. Sci. U.S.A.* **70**, 3576–3580.

Cashel, M. (1969). *J. Biol. Chem.* **244**, 3133–3141.

Cashel, M., and Gallant, I. (1971). *Nature (London)* **221**, 838–841.

Chambers, D. A., and Zubay, G. (1969). *Proc. Natl. Acad. Sci. U.S.A.* **63**, 118–122.

Cohen, P. T., Yaniv, M., and Yanofsky, C. (1973). *J. Mol. Biol.* **74**, 163–177.

Contesse, G., and Gros, F. (1968). *C. R. Hebd. Seances Acad. Sci.* **266**, 262–265.

Contesse, G., Naono, S., and Gros, F. (1966). *C. R. Hebd. Seances Acad. Sci.* **263**, 1007–1011.

Contesse, G., Crépin, M., and Gros, F. (1969). *C. R. Hebd. Seances Acad. Sci.* **268**, 2301–2304.

Crawford, L. V., and Gesteland, R. F. (1973). *J. Mol. Biol.* **74**, 627–634.

de Crombrugghe, B., Chen, B., Gottesman, M., Pastan, I., Varmus, H. E., Emmer, M., and Perlman, R. L. (1971a). *Nature (London), New Biol.* **230**, 37–40.

de Crombrugghe, B., Chen, B., Anderson, W., Nissley, P., Gottesman, M., Pastan, I., and Perlman, R. (1971b). *Nature (London), New Biol.* **231**, 139–142.

de Crombrugghe, B., Pastan, I., Shaw, W. V., and Rosner, J. L. (1973a). *Nature (London), New Biol.* **241**, 237–239.

de Crombrugghe, B., Adhya, S., Gottesman, M., and Pastan, I. (1973b). *Nature (London), New Biol.* **241**, 260–264.

Dickson, R. C., Abelson, J., Barnes, W. M., and Reznikoff, W. S. (1975). *Science* **187**, 27–35.

Dottin, R. P., and Pearson, M. L. (1973). *Proc. Natl. Acad. Sci. U.S.A.* **70**, 1078–1082.

Dottin, R. P., Shiner, L. S., and Hoar, D. I. (1973). *Virology* **51**, 509–511.

Dottin, R. P., Cutler, L. S., and Pearson, M. L. (1975). *Proc. Natl. Acad. Sci. U.S.A.* **72**, 804–808.

Dunn, J. J., McAllister, W. T., and Bautz, E. K. F. (1972). *Eur. J. Biochem.* **29**, 500–508.

Emmer, M., de Crombrugghe, B., Pastan, I., and Perlman, R. (1970). *Proc. Natl. Acad. Sci. U.S.A.* **66**, 480–487.

Englesberg, E. (1970). *J. Bacteriol.* **81**, 996–1006.

Englesberg, E., Irr, J., Power, I., and Lee, N. (1965). *J. Bacteriol.* **90**, 946–957.

Englesberg, E., Squires, G., and Meronk, F. (1969). *Proc. Natl. Acad. Sci. U.S.A.* **62**, 1100–1107.

Eron, L., and Block, A. (1971). *Proc. Natl. Acad. Sci. U.S.A.* **68**, 1828–1832.

Franklin, N. (1974). *J. Mol. Biol.* **89**, 33–48.

Gilbert, W., and Maxam, A. (1973). *Proc. Natl. Acad. Sci. U.S.A.* **70**, 3581–3584.

Gilbert, W., and Muller-Hill, B. (1967). *Proc. Natl. Acad. Sci. U.S.A.* **58**, 2415–2421.

Gilbert, W., Maizels, N., and Maxam, A. (1973). *Cold Spring Harbor Symp. Quant. Biol.* **38**, 845–855.

Gold, L. M., and Schweiger, M. (1969a). *Proc. Natl. Acad. Sci. U.S.A.* **62**, 892–898.

Gold, L. M., and Schweiger, M. (1969b). *J. Biol. Chem.* **244**, 5100–5104.

Gold, L. M., and Schweiger, M. (1971). *In* "Methods in Enzymology" (K. Moldane and G. Grossman, eds.), Vol. 20, pp. 537–542. Academic Press, New York.

Gottesman, S., and Beckwith, J. R. (1969). *J. Mol. Biol.* **44**, 117–127.

Greenblatt, J. (1972). *Proc. Natl. Acad. Sci. U.S.A.* **69**, 3606–3610.

Greenblatt, J. (1973). *Proc. Natl. Acad. Sci. U.S.A.* **70**, 421–424.

Greenblatt, J., and Schleif, R. (1971). *Nature (London), New Biol.* **233**, 166–170.
Gremer, K., and Schlessinger, D. (1974). *J. Biol. Chem.* **249**, 4730–4736.
Grodzicker, T., and Zipser, D. (1968). *J. Mol. Biol.* **38**, 305–314.
Herrlich, P., and Schweiger, M. (1974). *In* "Methods in Enzymology" (L. Grossman and K. Moldane, eds.), Vol. 30, pp. 654–669. Academic Press, New York.
Hill, C., and Echols, H. (1966). *J. Mol. Biol.* **49**, 515–519.
Hiraga, S., and Yanofsky, C. (1972). *J. Mol. Biol.* **72**, 103–110.
Hopkins, J. D. (1974). *J. Mol. Biol.* **87**, 715–724.
Howard, B., and de Crombrugghe, B. (1975). *J. Biol. Chem.* **251**, 2520–2524.
Imamoto, F. (1968). *Proc. Natl. Acad. Sci. U.S.A.* **60**, 305–312.
Imamoto, F., and Kano, Y. (1971). *Nature (London), New Biol.* **232**, 169–173.
Imamoto, F., and Yanofsky, C. (1967). *J. Mol. Biol.* **28**, 1–35.
Irr, J., and Englesberg, E. (1970). *Genetics* **65**, 27–39.
Ito, K. (1972). *Mol. Gen. Genet.* **115**, 349–363.
Jackson, E. N., and Yanofsky, C. (1973). *J. Mol. Biol.* **76**, 89–101.
Jaskunas, S. R., Lindahl, L., and Nomura, M. (1975). *Proc. Natl. Acad. Sci. U.S.A.* **72**, 6–10.
Jordan, E., Saedler, H., and Starlinger, P. (1968). *Mol. Gen. Genet.* **102**, 353–363.
Kaltschmidt, E., Kahan, L., and Nomura, M. (1974). *Proc. Natl. Acad. Sci. U.S.A.* **74**, 446–450.
Kepes, A. (1967). *Biochim. Biophys. Acta* **138**, 107–123.
Kirschbaum, J. B. (1973). *Proc. Natl. Acad. Sci. U.S.A.* **70**, 2651–2655.
Kourilsky, P., Marcoud, L., Sheldrick, P., Luzzati, D., and Gross, F. (1968). *Proc. Natl. Acad. Sci. U.S.A.* **61**, 1013–1020.
Kozak, M., and Nathans, D. (1972). *Bacteriol. Rev.* **36**, 109–134.
Kung, H. F., Fox, J. E., Spears, C., Brot, N., and Weissbach, H. (1973). *J. Biol. Chem.* **248**, 5012–5015.
Kung, H. F., Spears, G., and Weissbach, H. (1974). *J. Biol. Chem.* **250**, 1556–1562.
Lacroute, F., and Stent, G. S. (1968). *J. Mol. Biol.* **35**, 165–173.
Lebowitz, P., Weissman, S. M., and Radding, C. M. (1971). *J. Biol. Chem.* **246**, 5120–5139.
Lederman, M., and Zubay, G. (1968). *Biochem. Biophys. Res. Commun.* **32**, 710–714.
Lowery-Goldhammer, C., and Richardson, J. P. (1974). *Proc. Natl. Acad. Sci. U.S.A.* **71**, 2003–2007.
Maizels, N. M. (1973). *Proc. Natl. Acad. Sci. U.S.A.* **70**, 3585–3589.
Makman, R. S., and Sutherland, E. W. (1965). *J. Biol. Chem.* **240**, 1309–1314.
Malamy, M. H. (1970). *In* "The Lactose Operon" (J. D. Beckwith and D. Zipser, eds.), pp. 359–373. Cold Spring Harbor Lab., Cold Spring Harbor, New York.
Maniatis, T., Ptashne, M., and Maurer, R. (1973). *Cold Spring Harbor Symp. Quant. Biol.* **38**, 857–868.
Maniatis, T., Ptashne, M., Barrell, B. G., and Donelson, J. (1974). *Nature (London)* **250**, 394–397.
Maurer, R., Maniatis, T., and Ptashne, M. (1974). *Nature (London)* **249**, 221–223.
Miller, O. L., Jr., Hamkalo, B. A., and Thomas, C. A., Jr. (1970). *Science* **169**, 392–395.
Miller, Z., Varmus, H. E., Parks, J. S., Perlman, R. L., and Pastan, I. (1971). *J. Biol. Chem.* **246**, 2898–2903.
Morse, D. E., and Guertin, M. (1971). *Nature (London), New Biol.* **232**, 165–169.
Morse, D. E., Baker, R. F., and Yanofsky, C. (1968). *Proc. Natl. Acad. Sci. U.S.A.* **60**, 1428–1435.
Morse, D. E., Mosteller, R. D., and Yanofsky, C. (1969). *Cold Spring Harbor Symp. Quant. Biol.* **34**, 725–740.

Mosteller, R. D., and Yanofsky, C. (1971). *J. Bacteriol.* **105**, 268–275.
Nissley, P., Anderson, W. B., Gallo, M., and Pastan, I. (1972). *J. Biol. Chem.* **247**, 4264–4269.
O'Farrell, P. Z., and Gold, L. M. (1973a). *J. Biol. Chem.* **248**, 5502–5511.
O'Farrell, P. Z., and Gold, L. M. (1973b). *J. Biol. Chem.* **248**, 5512–5519.
Parks, J. S., Gottesman, M., Perlman, R. L., and Pastan, I. (1971). *J. Biol. Chem.* **246**, 2419–2424.
Pastan, I., and Perlman, R. L. (1970). *Science* **169**, 339–344.
Perlman, R. L., and Pastan, I. (1968). *Biochem. Biophys. Res. Commun.* **30**, 656–664.
Perlman, R. L., and Pastan, I. (1969). *Biochem. Biophys. Res. Commun.* **37**, 151–157.
Platt, T., Weber, K., Genem, D., and Miller, J. H. (1972). *Proc. Natl. Acad. Sci. U.S.A.* **69**, 897–901.
Pouwels, P. J., and Van Rotterdam, J. (1972). *Proc. Natl. Acad. Sci. U.S.A.* **69**, 1786–1790.
Press, R., Glansdorff, N., Miner, P., de Vries, J., Kadner, R., and Maas, W. K. (1971). *Proc. Natl. Acad. Sci. U.S.A.* **68**, 795–798.
Richardson, J. P., Grimley, C., and Lowery, G. (1975). *Proc. Natl. Acad. Sci. U.S.A.* **72**, 1725–1728.
Richter, D., Herrlich, P., and Schweiger, M. (1972). *Nature (London), New Biol.* **238**, 74–76.
Riggs, A. D., Bourgeois, S., Newby, R. F., and Cohn, M. (1968). *J. Mol. Biol.* **34**, 365–368.
Riggs, A. D., Reiness, G., and Zubay, G. (1971). *Proc. Natl. Acad. Sci. U.S.A.* **68**, 1222–1225.
Riggs, A. D., Suzuki, H., and Bourgeois, S. (1970a). *J. Mol. Biol.* **34**, 365–368.
Riggs, A. D., Newby, R. F., and Bourgeois, S. (1970b). *J. Mol. Biol.* **51**, 303–314.
Roberts, J. W. (1969). *Nature (London)* **224**, 1168–1174.
Rose, J. K., Mosteller, R. D., and Yanofsky, C. (1970). *J. Mol. Biol.* **51**, 541–550.
Rose, J. K., Squires, C. L., Yanofsky, C., Yang, H.-L., and Zubay, G. (1973). *Nature (London), New Biol.* **245**, 133–137.
Rosenberg, M., Weissman, S., and de Crombrugghe, B. (1975). *J. Biol. Chem.* **250**, 4755–4764.
Sarabhai, A., and Brenner, S. (1967). *J. Mol. Biol.* **27**, 145–162.
Scaife, J. G., and Beckwith, J. R. (1966). *Cold Spring Harbor Symp. Quant. Biol.* **31**, 403–408.
Schumacher, G., and Ehring, R. (1973). *Mol. Gen. Genet.* **124**, 329–344.
Schwartz, D., and Beckwith, J. R. (1970). *In* "The Lactose Operon" (J. D. Beckwith and D. Zipser, eds.), pp. 417–422. Cold Spring Harbor Lab., Cold Spring Harbor, New York.
Schwartz, M. (1970). *Virology* **40**, 23–33.
Schweiger, M., Herrlich, P., Scherzinger, E., and Rahmsdorf, H. J. (1972). *Proc. Natl. Acad. Sci. U.S.A.* **69**, 2203–2207.
Shapiro, J. A. (1967). *J. Mol. Biol.* **40**, 93–105.
Shimada, K., Weisberg, R. A., and Gottesman, M. E. (1972). *J. Mol. Biol.* **63**, 483–503.
Shulman, M. (1974). *Nature (London)* **252**, 74–75.
Silverstom, A. E., Arditti, R., and Magusanik, B. (1970). *Proc. Natl. Acad. Sci. U.S.A.* **66**, 773.
Squires, C. L., Rose, J. K., Yanofsky, C., Yang, H.-L., and Zubay, G. (1973). *Nature (London), New Biol.* **245**, 131–133.
Steinberg, R. A., and Ptashne, M. (1971). *Nature (London), New Biol.* **230**, 76–80.
Ullmann, A., and Monod, J. (1968). *FEBS* **2**, 57–63.
Varmus, H. E., Perlman, R. L., and Pastan, I. (1970). *J. Biol. Chem.* **245**, 2259–2267.

Wetekam, W., Staack, K., and Ehring, R. (1971). *Mol. Gen. Genet.* **112**, 14–17.
Wetekam, W., Staack, K., and Ehring, R. (1972). *Mol. Gen. Genet.* **116**, 258–276.
Wilcox, G., Meuris, P., Bass, R., and Englesberg, E. (1974). *J. Biol. Chem.* **249**, 2946–2952.
Wu, A. M., Ghosh, S., Echols, H., and Spiegelman, W. G. (1972). *J. Mol. Biol.* **67**, 407–421.
Yang, H.-L., and Zubay, G. (1973). *Mol. Gen. Genet.* **122**, 131–136.
Yang, H.-L., and Zubay, G. (1974). *Biochem. Biophys. Res. Commun.* **56**, 725–731.
Yang, H.-L., Zubay, G., Urm, E., Reiness, G., and Cashel, M. (1974). *Proc. Natl. Acad. Sci. U.S.A.* **71**, 63–67.
Zalkin, H., Yanofsky, C., and Squires, C. L. (1974). *J. Biol. Chem.* **249**, 465–475.
Zubay, G. (1973). *Annu. Rev. Genet.* **7**, 267–287.
Zubay, G., Chambers, D. A., and Cheong, L. C. (1970a). *In* "The Lactose Operon" (J. D. Beckwith and D. Zipser, eds.), pp. 375–391. Cold Spring Harbor Lab., Cold Spring Harbor, New York.
Zubay, G., Schwartz, D., and Beckwith, J. (1970b). *Proc. Natl. Acad. Sci. U.S.A.* **66**, 104–110.
Zubay, G., Gielow, L., and Englesberg, E. (1971). *Nature (London), New Biol.* **233**, 164–165.
Zubay, G., Morse, D. E., Schrenk, W. J., and Miller, J. H. M. (1972). *Proc. Natl. Acad. Sci. U.S.A.* **69**, 1100–1103.

13

Genetics of the Translational Apparatus

ISSAR SMITH

Over two hundred cell components are involved in protein synthesis, and the purpose of this Chapter is to inquire: How is the synthesis of so many components regulated? Is there any significance to the organization of the genes for these components? How have the genetic determinants for the translational apparatus changed during evolution?

There is usually one aminoacyl synthetase for each amino acid, even though there are multiple species of isoaccepting tRNA's. Several genes determining primary structure of synthetases have been mapped in pro- and eukaryotes and they are not clustered. Some aspects of synthetase structure suggest that enzymes of different specificity or subunits of polymeric enzymes could have arisen by gene duplication and subsequent divergence. Mitochondria and chloroplasts possess synthetases that are different from the eukaryote cytoplasmic species, but they are synthesized on cytoplasmic ribosomes and are probably nuclear encoded.

There are roughly fifty to sixty different species of tRNA in pro- and eukaryotes, but the redundancy for each isocoding tRNA gene increases as the evolutionary ladder is ascended. There is some clustering of tRNA cistrons in prokaryotes, and in several cases, precursor RNA's with more than one tRNA sequence have been demonstrated, including the tRNA's of coliphage T4. However, tRNA genes in bacteria are widely scattered throughout the genome. In eukaryotes, specifically *Xenopus,* there is clustering of redundant genes for isocoding tRNA families, but no evidence for precursors with multiple tRNA sequences has been found. Organelles have genes for 15 to 20 "4 S" or tRNA species. Prokaryote tRNA synthesis, *in vivo,* seems to be under stringent control, i.e., when bacteria are deprived of amino acids, wild-type cells synthesize ppGpp and pppGpp and consequently tRNA, rRNA, and ribosomal protein synthesis ceases. Uncharged tRNA is involved in the stringent response and in some cases charged tRNA's are also involved in the repression of amino acid biosynthesis. The overall similarities in tRNA structure, even in T4 tRNA's, suggest that gene duplication also played a role in the appearance of tRNA molecules with different specificities for amino acids and synthetases. Several tRNA duplications are found in *E. coli,* and the existence of high levels of isocoding tRNA gene redundancy in eukaryotes can also be explained by gene duplication, in this case, with conservation of sequence and specificity.

rRNA genes are redundant in all organisms except *Mycoplasma* (and also chloroplasts and mitochondria), and the level of redundancy increases during evolution, as for tRNA cistrons. In pro- and eukaryotes, as well as mitochondria, rRNA is transcribed as a polycistronic molecule with bacterial 16 S, 23 S and 5 S rRNA synthesized in that order, while in eukaryotes, 28 S, 5.8 S, and 18 S rRNA are part of the same transcript with 28 S rRNA at the 3'-end. Though 5 S rRNA is in the larger ribosomal subunit of eukaryotes and presumably has a similar function as in prokaryotes, i.e., the binding of tRNA, its genetic determinants are not linked to those of the larger rRNA species, as in prokaryotes. On the other hand, in eukaryotes the larger ribosomal subunits also contain 5.8 S rRNA which is part of the rRNA precursor molecule, but its function is unknown. There is some clustering of rRNA genes in prokaryotes, and in eukaryotes the redundant cistrons are usually clustered in one region of the chromosome, the nucleolar organizer. rRNA synthesis is under stringent control and is probably regulated by positive and negative control elements which interact with RNA polymerase and/or promoter sites in DNA. The homologies in rRNA, both intra- and interspecifically, as shown by sequence analysis and reconstitution studies, demonstrate conservation of rRNA sequence and conformation. This also indicates that the origin of redundancy, both within species and during evolution, can be explained by gene duplication.

Ribosomal protein genes, also under stringent control, show a high degree of clustering in bacteria, and some of these genes are expressed by means of a polycistronic mRNA. Several *Drosophila* ribosomal protein genes are clustered near the nucleolar organizer region, but no evidence for clustering of similar cistrons exists in other eukaryotes. Organelle ribosomal proteins are synthesized on cytoplasmic ribosomes, but some of the genes are in the organelle genome. There is extensive homology between ribosomal proteins in prokaryotes as determined by comparative amino acid sequence analysis and heterologous ribosome reconstitution studies. There are interactions between ribosomal proteins, *in situ,* and studies on S12, the *strA* gene product, have indicated it interacts with several other 30 S ribosomal proteins and 16 S rRNA in the recognition of tRNA's and initiation sites on mRNA.

I. Introduction

The synthesis of proteins requires the interaction of a complex of cell factors, necessary to decode the biological information in the primary sequence of DNA. In bacteria a multicomponent RNA polymerase with appropriate initiation and termination factors functions to make an RNA complementary to the coding strand of DNA and, in some cases, other regulatory elements like the *lac* or λ repressors serve to prevent initiation of messenger RNA synthesis, or positive regulatory agents, like cyclic AMP binding protein, facilitate initiation of transcription at promoter sites.

The messenger RNA is then complexed to the smaller subunit of the ribosome in the presence of the initiation factor, IF-3, which also may be involved in the dissociation of 70 S ribosomes after termination of protein synthesis. fMet-tRNA$_f^{Met}$ (the initiator aminoacyl tRNA) forms a complex with IF-2, another initiation factor, and binds to the 30 S ribosome in the presence of GTP, the anticodon of the tRNA binding to AUG or GUG (initiator triplets), near, but not necessarily at, the 5'-end of the mRNA. The third initiation factor, IF-1, binds to the 30 S ribosome complex, and facilitates the association of 50 S subunit to form a 70 S particle; during this process the initiation factors are eliminated and GTP is cleaved. At this point, the fMet-tRNA is probably bound to the peptidyl (or puromycin reactive) site on the 50 S ribosomes. The second codon of the message, following the AUG initiation codon, directs the binding of the next aminoacyl transfer RNA, which in the presence of the elongation factor EF-Tu and GTP forms a ternary complex (aminoacyl-tRNA · GTP · EF-Tu). The ternary complex binds to the A or aminoacyl site on the 50 S ribosome, cleaving another molecule of GTP in the process, and an EF-Tu · GDP complex is released from the ribosomes. EF-Tu is regenerated by complexing with EF-Ts, eliminating GDP.

At this point, the protein synthesis apparatus consists of the 50 S–30 S ribosomes with the fMet-tRNA positioned in the P site, also bound via its anticodon to the AUG triplet, while the next aminoacyl-tRNA is bound to the second codon and is positioned at the A site. The peptidyl transferase, an integral part of the 50 S ribosomes, then catalyzes the formation of a peptide bond between the COOH group of the fMet and the α-amino group of the second amino acid. This results in deacylated tRNA$_f^{Met}$ at the P site and the new peptidyl-tRNA at the A site. Elongation factor EF-G and GTP are then involved in the coordinated complex of reactions involved in translocation, during which another molecule of GTP is cleaved. The deacylated tRNA at the P site is ejected; there is movement of one codon of the messenger RNA in relation to the ribosome, placing the second codon in juxtaposition to the P site, and leaving the third codon of the message in position with regard to the A site. Thus, the peptidyl-tRNA, still bound to the second codon, is in the P site. The A site is free for the positioning of the third aminoacyl-tRNA, and the process is continued until the polypeptide chain is terminated. When a proper termination signal in the message is reached, either a UAA, UAG, or UGA triplet, possibly in context with another triplet, and specific protein release factors R1, specific for UAA and UAG, or R2, specific for UAA and UGA, bind to the messenger, and presumably peptidyl transferase catalyzes the hydrolysis of the peptide from the ultimate tRNA at the P site, freeing the protein, and in the process tRNA is released. The ribosome is then released from the message, either as 50 S and 30 S subunits or as a 70 S particle which is dissociated by IF-3, as a new round of protein synthesis begins (see previous chapters for details).

In oversimplified terms, this is our current state of knowledge of the mechanism of protein synthesis in bacteria. Essentially the same picture is observed in eukaryotes with some minor differences. But many other factors are involved. There are over twenty aminoacyl-tRNA synthetases which activate amino acids and transfer them to over fifty cognate tRNA's, which, in turn, are able to participate in protein synthesis only after they have been extensively modified by several enzymes subsequent to their transcription. The most complex structure in the translational apparatus is the ribosome, consisting of two subunits, the large and small component, which in bacteria contain 55 proteins, 34 and 21, respectively, in 50 S and 30 S ribosomes, with three types of RNA, 5 S and 23 S in the large subunit and 16 S in the smaller subunit, each coded for by multiple cistrons. rRNA is also modified by specific enzymes. Thus, it is apparent that products of at least 200 genes are involved in translation in prokaryotes. How is the synthesis of so many components regulated? Is there any significance to the organization of the genes determining these components? How have the genetic determinants for protein synthesis changed during evolution? These questions and many others will be addressed in the course of this chapter.

II. Methodology

To study the genetics of any system usually requires the isolation of mutants so that alternate states of alleles can be used in the mapping of the lesions. In the case of nonessential functions, utilization of lactose where cells ordinarily grow on glucose, or supplementable functions like amino acid biosynthesis, this is a simple matter since cells can be grown in the mutated state. With mutations in essential functions, like translation, however, mutated cells are frequently nonviable. This necessitates different approaches to the isolation of such mutants. The mapping of rRNA genes is further complicated by the absence of a selection method for rRNA mutants, and by the multiplicity of rRNA cistrons.

A. Conditional Lethal Mutations

Conditional lethal mutations were first isolated in T4 (Epstein *et al.*, 1963) using temperature as the selective agent, and this has been a valuable tool in the production of mutations in the translational apparatus. Cells are mutagenized at a nonrestricting temperature, grown at this same temperature so that the mutations can be expressed, and are then tested, by replica plating, for cells which do not grow at the restrictive temperature. This enables one to isolate mutations which cause irreversible killing at the selective temperature. The use of penicillin, to kill off growing cells at the restricted temperature, can be used to enrich for mutants which are not irreversibly killed (Davis, 1948; Lederberg, 1959). Suicide treatments, in which tritiated uridine or an amino acid of very high specific activity are added to mutagenized cells at the restrictive temperature and allowed to decay (Person, 1963), have been used to enrich for reversible mutations in RNA and protein biosynthesis, as cells not actively incorporating the tritiated precursors into their macromolecules will not be damaged by the tritium disintegration (Tocchini-Valentini and Mattoccia, 1968). There could be mutations which render the synthesis of components of the translational apparatus temperature-sensitive, but which allow these components to function at the restrictive temperature. By starving cells at elevated temperatures, the translational apparatus becomes depleted and cannot recover at these temperatures when returned to enriched medium. Using this more stringent selection, mutations in genes for translational components have been isolated (Phillips et al., 1969). Usually, elevated temperatures have been used to isolate conditional mutants (Eidlic and Neidhart, 1965; Yaniv et al., 1965), but low temperatures have also been utilized successfully to isolate ribosomal assembly mutants (Tai et al., 1969; Guthrie et al., 1969). While conditional temperature-sensitive mutations are usually in genes determining proteins, temperature-sensitive suppressor mutations have been found which are due to altered tRNA genes (Galluci et al., 1970; Smith et al., 1970).

In general, the isolation of suppressor mutations has permitted the mapping of genes for minor tRNA species which are dispensable for normal cell function (Smith, 1972). Suppression at the level of the ribosome has led to the isolation of ribosomal mutations (Rosset and Gorini, 1969).

Partial diploid strains in E. coli using F' merogenotes have been used to isolate and map mutations in essential genes which would otherwise be lethal. Lethal suppressor mutations (e.g., su_7^+, su_8^+) have been characterized (Soll and Berg, 1969), and rRNA genes have been isolated and mapped, using F' (Deonier et al., 1974). amber mutations in the rif gene of coli, coding for the β subunit of RNA polymerase, have been isolated (Austin et al., 1971); and the genetics of ribosomal protein expression has been studied with partial diploids in E. coli, heterozygous for the ribosomal protein gene region (Nomura and Engbaek, 1972).

B. Specialized Transducing Bacteriophages

Specialized transducing coliphages, $\phi80$ and λ, which contain specific segments of the bacterial chromosome have been used to obtain large quantities of genes for tRNA (Squires et al., 1973a), rRNA (Ohtsubo et al., 1974), and ribosomal proteins (Jaskunas et al., 1975a) and to map these genetic determinants in E. coli.

C. Resistance to Amino Acid Analogues and Antibiotics

Amino acid analogues have been used to isolate resistant mutants with altered aminoacyl-tRNA synthetases (Fangman and Neidhardt, 1964). These are generally due to synthetases which have a much higher K_m for the analogue, and can thus grow in the presence of the inhibitor. Numerous antibiotics inhibit the growth of prokaryotes and eukaryotes by binding to ribosomes or other components of the translational apparatus (Pestka, 1971). Mutations to drug resistance or dependence frequently result from changes in the primary structure of the binding protein or in other components of the ribosomes.

Ability of cells to grow in the presence of the antibiotic or analogue can be used as a selective agent for genetic studies in bacteria (Davies and Nomura, 1972), as well as higher organisms (Skogerson et al., 1973; Mortimer and Hawthorne, 1973).

D. Cytogenetic Methods

A combination of classical genetic techniques, cytogenetics, and hybridization has related the bobbed locus in *Drosophila* to the nucleolar organizer where rRNA genes are located (Ritossa and Spiegelman, 1965). In similar fashion, mutants lacking the nucleolar organizer and consequently not possessing rRNA genes were isolated in *Xenopus* (Birnstiel et al., 1966).

E. Strain and Species Differences

The above techniques generally employ direct selection for the isolation of mutations in the translational apparatus. Sets of genetically compatible species or strains frequently possess phenotypic differences when their cell components are compared. Corresponding ribosomal proteins, because of amino acid substitutions, insertions, or deletions, may possess different migration rates on columns or acrylamide gels. RNA molecules similarly may give different nucleotide fingerprints after nuclease digestion and two-dimensional electrophoresis or homochromatography. These phenotypic differences can be used as unselected markers in genetic crosses and can be mapped in bacteria (Leboy et al., 1964; Smith et al., 1969; Sypherd et al., 1969; Jarry and Rosset, 1970; Matsubara et al., 1972) as well as in higher organisms (Steffensen, 1973).

F. Nongenetic Methods

No other genetic methods exist, at present, for the isolation of mutants or the mapping of ribosomal RNA genes. DNA:RNA hybridization in conjunction with genetic transformation has been used to map rRNA and tRNA genes in *Bacillus subtilis* (Oishi and Sueoka, 1965; Dubnau et al., 1965b). In eukaryotes, the isolation of nuclear components in conjunction with molecular hybridization has been used to discover the chromosomal localization of rRNA and tRNA genes (Birnstiel et al., 1966; Huberman and Attardi, 1967; Finkelstein et al., 1972; Goldberg et al., 1972).

Cytological methods in conjunction with autoradiography ("*in situ*" hybridization) have been used to study the chromosomal localization of rRNA and tRNA genes in eukaryotes (Gall and Pardue, 1971; Pardue *et al.*, 1973; Pardue, 1973; Grigliatti *et al.*, 1973).

Electron microscope heteroduplex mapping of rRNA and tRNA genes in prokaryotes and organelle genomes has also proved to be of great value (Wu *et al.*, 1972; Chow and Davidson, 1973; Wu and Davidson, 1973).

III. Aminoacyl-tRNA Synthetases

General reviews on aminoacyl synthetases have appeared (Neidhardt, 1966; Novelli, 1967; Söll and Schimmel, 1974), and the mechanisms of aminoacylation of tRNA have been reviewed (Loftfield, 1972).

The earliest step in the synthesis of proteins, i.e., the "activation" of amino acids to form an aminoacyl-tRNA, is crucial to the fidelity of the transfer of genetic information. A rare mischarging event would ordinarily be lethal as the specificity resides in the tRNA, not the amino acid (Chapeville *et al.*, 1962), e.g., *Neurospora* phenylalanyl-tRNA synthetase will transfer phenylalanine to *E. coli* tRNA^{Val}, and this phenylalanyl-tRNA^{Val} will insert phenylalanine into hemoglobin peptides where valine ordinarily is found (Jacobson, 1966). In general, heterologous systems, in which aminoacyl-tRNA synthetases from one species are used to charge tRNA's from different organisms, show the same rigorous amino acid specificity found in the homologous reactions (Jacobson, 1971; Loftfield, 1972). In the few examples of mischarging in the literature, very unique conditions must be employed for charging, e.g., changes in pH or temperature, presence of organic solvents, such as ethanol or dimethyl sulfoxide, extremely high levels of synthetase, etc.—conditions which clearly would not be present *in vivo* (Jacobson, 1971; Giegé *et al.*, 1974).

The aminoacyl synthetases are generally in the molecular weight range of 100,000 and are found in the form of single polypeptides as well as dimers or tetramers of identical or different subunits (Loftfield, 1972). Individual polypeptide chains are usually approximately 50,000 or 100,000 daltons, and the gene coding for at least one of the longer polypeptides, the leucyl-tRNA synthetase of *E. coli*, may have arisen from a duplication (Waterson and Koningsberg, 1974).

A. Mapping of Aminoacyl-tRNA Synthetase Genes

Several genes coding for aminoacyl-tRNA synthetases have been mapped in *E. coli* and *Salmonella typhimurium* (Table I), and, in general, they are not clustered and are not necessarily near the operons for the synthesis of the specific amino acid. For example, in *E. coli*, *serS* and *gltE*

TABLE I

Mapping of Aminoacyl-tRNA Synthetases

Gene	Aminoacyl-tRNA synthetase	Map position (minutes)
E. coli[a]		
alaS	Alanyl	61
argS	Arginyl	35
glnS[b]	Glutaminyl	15
gltE	Glutamyl	72
gltM	Glutamyl	38
gltX[c]	Glutamyl	45
glyS	Glycyl	70
ileS	Isoleucyl	1
leuS	Leucyl	14
pheS	Phenylalanyl	33
serS	Seryl	20
trpS	Tryptophanyl	65
tyrS	Tyrosyl	32
valS	Valyl	85
S. typhimurium[d]		
hisS	Histidyl	78
ilvS	Isoleucyl	3
leuS	Leucyl	26
metG	Methioninyl	67

[a] The data for *E. coli* was taken from Taylor and Trotter (1972) except where otherwise indicated.

[b] Korner *et al.* (1974).

[c] Russell and Pittard (1971).

[d] *Salmonella typhimurium* information was from Sanderson (1972).

are close to the operator loci for serine (*serO*) and glutamate (*gltC*) (Taylor and Trotter, 1972), whereas the other mapped genes for aminoacyl-tRNA synthetases are not so linked.

Genes for aminoacyl-tRNA synthetases have been mapped in other organisms. In *B. subtilis* the gene for a temperature-sensitive, fluoro-tryptophan-resistant tryptophanyl-tRNA synthetase has been mapped between *argC* and *metA,* far from the other genes of tryptophan biosynthesis (Steinberg and Anagnostopoulus, 1971) and unlinked to a lysyl-tRNA synthetase gene, found between $purA_{16}$ and *sul* (Racine and Steinberg, 1974). Temperature-sensitive mutants have been isolated in yeast that show impaired leucyl- and methioninyl-tRNA synthetase activities *in vitro* (Hartwell and McLaughlin, 1968a; McLaughlin and Hartwell, 1969). Mapping studies have shown that the two loci are unlinked and that the *ileS* gene is on chromosome II. A mutant (*trp-5*) with a defective tryptophanyl synthetase has been described in *Neurospora* (Nazario *et al.*, 1971), and another gene (*leu-5*) seems to affect the leucyl-tRNA synthetases of both cytoplasm and mitochondria (Printz and Gross, 1967).

Single amino acid tRNA synthetases are found for charging tRNA's even when multiple isoacceptor tRNA's are found (Novelli, 1967; Kan and Sueoka, 1971). Evidence for multiple synthetases has appeared in eukaryotes (Kisselev, 1972); however, mitochondria possess aminoacyl-tRNA synthetase activities with properties different from the cytoplasmic enzymes (Barnett and Brown, 1967; Epler *et al.*, 1970). Three loci for glutamyl-tRNA synthetase have been mapped in *E. coli* (Murgola and Adelberg, 1970; Russell and Pittard, 1971), but the relationship of these loci to the two dissimilar subunits, of which the enzyme is composed (Lapointe and Söll, 1972), is not known at present.

A duplication in a mutated glycyl-tRNA synthetase has been reported which changes the gly⁻ phenotype to gly⁺ (Folk and Berg, 1971). This could be a model for the evolution of duplicate regions in synthetases (Waterson and Koningsberg, 1974) or multiple subunits in these enzymes.

B. Phenotypic Effects of Mutated Aminoacyl-tRNA Synthetases

In bacteria, the phenotypic effects of loss of aminoacyl-tRNA synthetase activity mimic starvation for amino acids. For example, when *E. coli*, with a *ts* valyl-tRNA synthetase is grown at 42°, there is a decrease in the level of valyl-tRNA, a cessation in rRNA and tRNA synthesis, and a derepression of the isoleucine and valine biosynthetic pathways (Neidhardt, 1966). Similar effects have been noted with the *ts* isoleucyl-tRNA synthetase of yeast (McLaughlin *et al.*, 1969) and the mutated tryptophanyl-tRNA synthetases of *B. subtilis* (Steinberg, 1974) and *Neurospora* (Nazario *et al.*, 1971). These effects are due to lack of charged tRNA, not alteration of the aminoacyl synthetases per se, and this is demonstrated most clearly in the regulation of histidine biosynthesis. Though *hisS* mutants show derepression of histidine synthesis, other histidine regulatory mutants have been found and, in general, all result in lowered levels of histidyl-tRNA (Lewis and Ames, 1972). This point will be discussed more fully below.

C. Synthesis of Aminoacyl-tRNA Synthetases

Little is known about the control of aminoacyl-tRNA synthetase production. The levels of valyl- and arginyl-tRNA synthetases in *E. coli* and *Salmonella typhimurium* reflect the rate of cell growth in different media, increasing two- to fivefold in specific activity over a sevenfold increase in growth rate. The enzymes changed activity in the same manner as bulk RNA, and did not reflect levels of free amino acids, seemingly ruling out amino acid mediated repression of aminoacyl-tRNA synthetase synthesis (Parker et al., 1974). However, low levels of charged tRNA cause derepression of cognate aminoacyl-tRNA synthetase production in at least two cases, methioninyl-tRNA synthetase (Cassio, 1975) and histidyl-tRNA synthetase (McGinnis and Williams, 1972).

The aminoacyl-tRNA synthetases of eukaryote organelles are of interest in this regard. As mentioned above, mitochondria and chloroplasts possess synthetases which are distinct from cytoplasmic enzymes in terms of physical properties and tRNA specificity. Euglena possesses two phenylalanine and two isoleucine synthetases, one of each type found in the chloroplasts. The chloroplast synthetases will charge chloroplast, but not cytoplasmic tRNA's, and show different chromatographic mobility (Reger et al., 1970). Neurospora possesses two leucyl-tRNA synthetases which have different chromatographic mobilities, leucine affinity, and tRNA specificity; and one of the enzymes is mitochondrial (Weeks and Gross, 1971).

Though these organelle synthetases seem to have properties more closely related to prokaryote enzymes than the eukaryote cytoplasmic counterparts, their synthesis seems to be under nuclear control. Light induction of Euglena causes marked increases in the levels of the chloroplast aminoacyl-tRNA synthetases. Inhibitors of cytoplasmic protein synthesis, like cycloheximide, prevent this increase, while streptomycin, which inhibits chloroplast protein synthesis, has no effect. In addition, aplastic mutants of Euglena possess low levels of the chloroplastic synthetases (Hecker et al., 1974).

The leucyl synthetases of Neurospora, cytoplasmic and mitochondrial, are affected by mutations in the same nuclear gene, leu-5. When grown at elevated temperatures, cells with a ts mutation in the gene synthesize a cytoplasmic leucyl synthetase with an elevated K_m for leucine and do not make the mitochondrial enzyme. A pseudorevertant of the ts lesion loses the ts characteristic of the mutant, and shows renewed synthesis of the mitochondrial enzyme, which, however, has different properties than the wild-type mitochondrial enzyme (Weeks and Gross, 1971). It is not clear whether the leu-5 gene is the structural gene for the mitochondrial leucyl synthetase or whether it is a regulator or modifier gene. Conversely, it may code for one component of a polymeric enzyme, with the mitochondrial genome coding for some of the other units, as has been shown for the mitochondrial ATPase (Tzagoloff and Meagher, 1972).

An unusual type of aminoacyl-tRNA synthetase control is found after T4 infection of E. coli. The valyl-tRNA synthetase becomes modified by a

T4 gene product which alters several of the enzyme's physical properties (Neidhardt and Earheart, 1966; Marchin *et al.*, 1974). It is not known whether this modification has any relation to the T4-specific tRNA's synthesized during phage development. As yet, no T4 valyl-tRNA has been reported (Scherberg and Weiss, 1972).

IV. tRNA

The genetics of tRNA has been reviewed (Smith, 1972), as well as mechanisms of suppression (Gorini, 1970; Hartman and Roth, 1973). tRNA plays a key role in the transfer of genetic information. This molecule must be recognized by the correct aminoacyl synthetase, and in turn, transfers the amino acid to the growing peptidyl chain after the correct codon-anticodon interaction between the messenger RNA and the tRNA. Table II depicts the 61 amino acid-specific codons and three termination codons of the genetic code. This means at least thirty specific tRNA's should be present in all cells, allowing for the wobble variation in the third base of the codon (Crick, 1966). Thus far, over 50 different types of tRNA have been found in bacterial and eukaryotic cells, as evidenced by chromatographic separation (Gallo and Pestka, 1970), while hybridization

TABLE II

Genetic Code

First letter	Second letter				Third letter
	U	C	A	G	
U	Phe	Ser	Tyr	Cys	U
	Phe	Ser	Tyr	Cys	C
	Leu	Ser	*ochre* nonsense	Nonsense	A
	Leu	Ser	*amber* nonsense	Trp	G
C	Leu	Pro	His	Arg	U
	Leu	Pro	His	Arg	C
	Leu	Pro	Gln	Arg	A
	Leu	Pro	Gln	Arg	G
A	Ile	Thr	Asn	Ser	U
	Ile	Thr	Asn	Ser	C
	Ile	Thr	Lys	Arg	A
	Met	Thr	Lys	Arg	G
G	Val	Ala	Asp	Gly	U
	Val	Ala	Asp	Gly	C
	Val	Ala	Glu	Gly	A
	Val	Ala	Glu	Gly	G

TABLE III

Genes Coding for Stable RNA Species

Organism	tRNA cistrons	5 S rRNA cistrons	Large rRNA cistrons
Mycoplasma kid	44[a]	—	1[a]
E. coli	50[b]	6[c]	6[c]
B. subtilis	42[d]	4–5[d]	9–10[d]
S. cerevisiae	300–400[e]	150[f]	140[e]
D. melanogaster	750[g]	195–230[g]	180[g]
X. laevis	7800[h]	24,000[i]	610[h]
Human (HeLa)	1300[j]	2,000[j]	280[j]
Yeast mitochondria	14[s]	—	1[k]
Neurospora mitochondria	15[l]	—	1[m]
Xenopus mitochondria	15[n]	—	1[n]
HeLa mitochondria	12[o]	—	1[o]
Euglena chloroplasts	20[p]	—	1,[q] 3[r]

[a] Ryan and Morowitz (1969).
[b] Morell *et al.* (1967).
[c] Pace and Pace (1971).
[d] Smith *et al.* (1968).
[e] Schweitzer *et al.* (1969).
[f] Rubin and Sulston (1973).
[g] Tartof and Perry (1970).
[h] Clarkson *et al.* (1973a).
[i] Brown and Weber (1968).
[j] Hatlen and Attardi (1971).

[k] Reijnders *et al.* (1972).
[l] Epler (1969).
[m] Shaefer and Küntzel (1972).
[n] Dawid (1972).
[o] Wu *et al.* (1972).
[p] Tewari and Wildman (1970).
[q] Rawson and Haselkorn (1973).
[r] Scott (1973).
[s] Casey *et al.* (1974b).

experiments have shown that bacterial and eukaryotic cells have over 40 different kinds of tRNA (Birnstiel *et al.*, 1972). As organisms increase in genetic complexity, however, there is an increase in the redundancy for tRNA genes, so that bacteria have 40 to 60 genes for tRNA, while human cells have over 1000. Mitochondria and chloroplasts have genetic information sufficient for only 15 to 20 tRNA genes. This, in addition to the low numbers of cistrons for 16 S and 23 S rRNA in these organelles, may reflect the loss of genetic complexity as free living prokaryotes became symbionts (Table III).

A. Mapping of tRNA Genes

1. Prokaryotes

Genes for tRNA have been mapped in *E. coli* and *S. typhimurium* largely by conventional genetic techniques, usually by the mutation of the genes for minor tRNA's, so that these altered tRNA's can act as informational suppressors for nonsense, missense, and frame shift mutations. This type of analysis is usually limited to these tRNA's which are dispensable to normal cell function.

Nonsense mutations, which cause premature polypeptide chain termination, occur when a mutation results in an amino acid codon in a messenger RNA changing to UAG (*amber*), UAA (*ochre*), or UGA codons (Sarabhai *et al.*, 1964). Several nonsense suppressors have been mapped in bacteria (Smith, 1972), and yeast (Hawthorne, 1969), and in certain cases the mutationally altered tRNA's responsible for the nonsense suppression have been sequenced (Table IV).

Missense mutations, in which the mutation changes the sense of the codon so that one amino acid is substituted for another, result in a defective or phenotypically altered protein if that amino acid is in a critical position in the polypeptide chain (Guest and Yanofsky, 1965). Mutations in *E. coli* glycine-tRNA genes can cause suppression of missense mutations. There are three glycine-tRNA's of known sequence, and their structural genes have been mapped into four loci (Carbon and Fleck, 1974; Table III).

Frame shift mutations, in which the reading frame becomes altered by addition or deletion of nucleotides in DNA can be suppressed by compensating frame shift mutations within the gene itself (Crick *et al.*, 1961) or by external suppression in which altered tRNA's can function, in an informational sense, to restore the correct frame of reading (Riddle and Roth, 1970). Six frame shift suppressor loci have been found and mapped in *Salmonella* (Riddle and Roth, 1972a), and four of them are dominant. Alterations in two species of proline-tRNA (*sufA* and *sufB*) (responding to codons CCC) and one species of glycine-tRNA (responding to GGG) are found (Riddle and Roth, 1972b). Base sequence analysis has recently shown that glycine-tRNA$^1_{sufD}$ has a nucleotide quadruplet, C-C-C-C, at the anticodon position, instead of C-C-C, enabling this tRNA to suppress frame shift mutations (Riddle and Carbon, 1973). The *suf* genes are not linked to each other. *sufA, B, C, D, E, F* are found at 116, 73, 45, 95, 128, and 19 minutes on the *S. typhimurium* chromosome.

Table IV indicates the *E. coli* map position of tRNA genes. The data were obtained by suppressor analysis, base sequence analysis, and electron microscopic heteroduplex mapping. Though the genes for tRNA are scattered throughout the chromosome, there is evidence for clustering, and this will be discussed below.

Genes for tRNA have been mapped in *B. subtilis*, by a combination of hybridization and DNA-mediated transformation (Dubnau *et al.*, 1965b; Oishi *et al.*, 1966). Evidence for two major clusters has been obtained, but the possibility of tRNA genes scattered elsewhere in the chromosome has

TABLE IV

Mapping of tRNA Genes in E. coli

Locus[a]	Map position (minutes)	tRNA altered	Modification
su_1^+	38	Seryl	
su_2^+	15	Glutaminyl	
su_3^+	26	$Tyr^1_{UAU/G \to UAG}$	Anticodon: G-U-A $\to$ C-U-A
su_4^+		$Tyr^1_{UAU/G \to UAA/G}$	Anticodon: G-U-A $\to$ U-U-A
su_6^+	F' 14 region	$Leu_{UUG \to UAG}$	
$TrpT$	75		
su_7^+		$Trp^1_{UGG \to UAG}$ (inserts gln)	Anticodon: C-C-A $\to$ C-U-A[b]
su_{7ochre}^+		$Trp^1_{UGG \to UAA/G}$	Anticodon: C-C-A $\to$ U-U-A[b]
su_{UGA}^+		$Trp^1_{UGG \to UGA}$	G24 $\to$ A
$tyrT$	77		
su_{15B}^+		$Tyr^2_{UAU/C \to UAA/G}$	
$thrT$	77	Thr^3 (heteroduplex mapping linked to $glyT,tyrT$)	[c]
$glyT^d$	77		
suA_{A36}^+		$Gly^2_{GGA/G \to AGA/G}$	Anticodon: U-C-C $\to$ U-C-U
su_{UGA}^+		$Gly^2_{GGA/G \to UGA/G}$	
$glyU$	55		
suB_{A36}^+		$Gly^1_{GGG \to AGA/G}$	
$supT$		$Gly^1_{GGG \to GAG}$	Anticodon: C-C-C $\to$ C-U-C
$sufD$		$Gly^1_{GGG \to GGGG}$	Anticodon: C-C-C $\to$ C-C-C-C
$glyV$	86		
suA_{A58}^+		$Gly^3_{GGU/C \to GAU/C}$	
suA_{A78}^+		$Gly^3_{GGU/C \to UGU/C}$	Anticodon: G-C-C $\to$ G-C-A
ins		$Gly^3_{GGU/C \to GGA/G}$	Anticodon: G-C-C $\to$ U-C-C
$glyW$	37		
suB_{A58}^+		$Gly^3_{GGU/C \to GAU/C}$	
suB_{A78}^+		$Gly^3_{GGU/C \to UGU/C}$	Anticodon: G-C-C $\to$ G-C-A

[a] The nonsense suppressor data ($su_1^+ \to su_{15B}^+$) unless otherwise noted are taken from Smith (1972) and references therein.

[b] Yaniv et al. (1974).

[c] Wu et al. (1973).

[d] Data on gly tRNA's are from Carbon and Fleck (1974).

been suggested (Smith et al., 1968). One group seems to be in the same region of the chromosome as the major cluster of rRNA genes (Oishi et al., 1966; Smith et al., 1968), which is close to the ribosomal protein region of the B. subtilis chromosome (see below). Since the methods used for mapping these tRNA genes are indirect, these map positions must be considered approximations. For example, the reports of two clusters of tRNA genes in *E. coli* (Cutler and Evans, 1967), using techniques similar to those employed in *B. subtilis*, have not been borne out by direct genetic mapping of tRNA genes (Smith, 1972), though there is a small cluster of three tRNA genes adjacent to a major cluster of rRNA genes, near 79 minutes, and at least one near 75 minutes (su_7^+).

Escherichia coli bacteriophages T4 and T5 possess genes coding for tRNA. Specific aminoacyl-tRNA:DNA hybridization has shown that 14 genes are found in T5 and at least 5 were found in T4 (Scherberg and Weiss, 1972). More recently, eight different species of T4 specific RNA have been found after T4 infection, and the genes for the tRNA have been mapped. One of the tRNA's specified by the T4 genome is a serine accepting species and two suppressor mutations which insert serine have been isolated in this coliphage (McClain, 1970; Wilson and Kells, 1972). These *amber* suppressors have been mapped near ϵ and deletions of this region not only cause loss of the suppressor activity but also result in loss of the genes for all known T4 tRNA's (Wilson *et al.*, 1972; McClain *et al.*, 1972). This indicates the genes are clustered and precursors containing multiple T4 tRNA species have been isolated and sequenced (Guthrie *et al.*, 1973; Barrell *et al.*, 1974). The function of the phage specific tRNA's is not known, as the deletion mutants are viable in laboratory strains of *E. coli*, but as shown by the existence of su^+ mutants, they can function in protein synthesis. The T4 tRNA's recognize codons that are usually not efficiently recognized by host tRNA's (Scherberg and Weiss, 1972; Pinkerton *et al.*, 1972). Possibly, production of an additional tRNA recognizing the unusual codons would have given a selective advantage to the T4. *E. coli* strains isolated from hospital patients show reduced burst sizes when infected with T4 tRNA negative mutants (Wilson, 1973), so these tRNA's may have a role in the growth of T4 in wild-type *E. coli*.

2. Eukaryotes

Even though nonsense suppressors (supersuppressors) have been mapped in yeast (Hawthorne and Mortimer, 1968) and possibly *Neurospora* (Seale, 1972) and it is known which amino acids are substituted (Sherman *et al.*, 1973), it has not yet been possible to genetically map the loci for specific tRNA's in eukaryotes.

In this context, it has been postulated that the minute loci in *Drosophila* of which there are 50 to 60, randomly scattered through the sex chromosomes and autosomes, are the structural genes for tRNA (Ritossa *et al.*, 1966a). The evidence for this theory was that the number of minute loci, approximately 55, was similar to the estimated number of tRNA species in *Drosophila*, and that the phenotype of these mutants (i.e., delayed development, small bristles, homozygous lethality) indicated general translational inhibition. The minutes are frequently caused by deletions, not point mutations, which would suggest a redundant region must be eliminated and it is known that the tRNA genes are on the average re-iterated 15-fold (Table III). Clearly, the tRNA genes were unlinked to the nucleolar organizer (Ritossa *et al.*, 1966a).

In situ hybridization, in which labeled, unfractionated tRNA is used to hybridize with salivary gland chromosomes, *in vitro,* followed by autora-diography, has confirmed that DNA complementary to tRNA is not found in the nucleolar organizer area. The genes for tRNA, as defined by hybrid-ization, are scattered throughout the *Drosophila* genome (Steffenson and Wimber, 1971). The presumptive localization of tRNA genes was not conclusive enough to establish a relationship between the "minute" loci and the tRNA binding sites. Results obtained by this technique must be treated with caution, however, as highly purified tRNA^{Lys} has been shown to bind to the 5 S RNA region (56EF) as well as to the 48F-49A region in salivary gland preparations (Grigliatti *et al.*, 1973).

The arrangement of DNA sequences complementary to tRNA has been studied in *Xenopus* by means of isopycnic centrifugation of nuclear DNA and tRNA-DNA hybrids. There is evidence that the redundant genes for each isocoding species of tRNA, reiterated an average of 200 times (Clarkson *et al.*, 1973a), are clustered together, but are separated from each other by spacers. Genes for isoaccepting tRNA species, e.g., tRNA$_1^{Met}$ and tRNA$_2^{Met}$, were found to band at different densities (1.708 and 1.711 gm cm^{-3}, respectively), indicating that they are in separate clus-ters (Clarkson *et al.*, 1973b).

Heteroduplex mapping of ferritin-labeled 4 S RNA hybridized to HeLa mitochondrial DNA has been used to cytologically map the genes for this species (Wu *et al.*, 1972). Nine binding sites for 4 S RNA are seen on the H strand and three are found on the L strand. Three of the H strand sites for 4 S RNA binding are very close to the rRNA genes, and it is postulated that one of these sites might code for the as yet undetected 5 S RNA of mi-tochondria. *Xenopus* mitochondrial H DNA strands also code for rRNA and the majority of the mitochondrial 4 S molecules (Dawid, 1972). Since the 4 S RNA species described above have not been characterized as to function, it is not known whether all the sites code for tRNA or whether some might code for other structural ("5 S" RNA) or regulatory gene products. It has been shown, however, that rat liver and yeast mitochon-dria DNA do possess sites that will hybridize with specific mitochondrial aminoacyl-tRNA's (Nass and Buck, 1970; Casey *et al.*, 1972). The H strand of rat liver mitochondrial DNA specifically hybridizes with mito-chondrial leucine and phenylalanine-tRNA, while the L strand annealed with tyrosine- and serine-tRNA (Nass and Buck, 1970). Hybridization was also used to order several mitochondrial tRNA genes with respect to the chloramphenicol and erythromycin resistance markers in *Saccharomyces*. These experiments were performed with a series of petite mutants which had overlapping deletions in their mitochondrial DNA (Casey *et al.*, 1974a).

B. tRNA Modification

All tRNA's possess modified bases and the content of these nucleosides increases as one goes from mycoplasma to mammals (Söll, 1971). The functions of these alterations have been studied extensively (Söll, 1971; Nishimura, 1972).

All modifications occur after transcription of the tRNA genes and some, but not all, occur before precursor tRNA's are trimmed to final size (Bernhardt and Darnell, 1969; Moshowitz, 1970; Schaefer *et al.*, 1973). The *in vitro* and *in vivo* modification of tRNA has been studied by the use of partially unmodified precursor $tRNA_1^{Tyr}$ isolated from $\phi80\ su_3^+$ infected cells (Gefter and Russell, 1969; Schaefer *et al.*, 1973), and from *in vitro* transcripts of purified or enriched tRNA genes which are completely unmodified (Johnson and Söll, 1970; Zubay *et al.*, 1970). Methionineless mutants of *E. coli* possessing the *rel* mutation, when starved for methionine, show unbalanced RNA synthesis (Borek and Srinivason, 1966). tRNA's isolated from these strains are a mixture of methylated and non-methylated tRNA's which can be separated and characterized (Littauer and Inouye, 1973). Other mutants that possess defective methylases have also been used to isolate tRNA's lacking specific modifications (Björk and Isaksson, 1970). *Mycoplasma kid,* possessing a low level of modified nucleosides, has been used as a heterologous source of relatively unmodified tRNA (Johnson *et al.*, 1970).

Contradictory results have been obtained regarding the function of certain modifications. For example, mostly all tRNA's which respond to codons with U in the 5'-position possess a modified adenosine, usually 2-methyl thio-6(A^2-isopentyl) adenosine (MS^2I^6A) or isopentyl adenosine (I^6A), at the position adjacent to the 3'-end of the anticodon (Nishimura, 1972). This modification is absent from one of the forms of su_3^+ $tRNA_1^{Tyr}$ synthesized after $\phi80\ su_3^+$ infection of *E. coli* (Gefter and Russel, 1969). Form I possesses an unmodified adenosine at position 38; form II has a partial modification at position 38, the isopentyl group (I^6A); while form III is completely modified (MS^2I^6A). All three forms of the $tRNA_1^{Tyr}$ show normal rates of charging with tyrosine, but marked differences are seen when *in vitro* suppression and UAG-dependent binding to ribosomes are tested. Form III (completely modified) shows 100% *in vitro* suppression and UAG binding as compared to normal su_3^+ $tRNA_1^{Tyr}$; the incompletely modified form II shows intermediate activity; while the unmodified form I shows no *in vitro* suppressor activity and 14% of normal UAG binding activity.

su_3^+ tRNA$_1^{Tyr}$ synthesized *in vitro* is also dependent on the presence of I^6A for biological activity in *in vitro* suppression (Zubay *et al.*, 1970).

A striking example of these modifications *in vivo* comes from studies on the su_{A78}^+ and su_{A36}^+ mutations (Carbon and Fleck, 1974; Roberts and Carbon, 1974). The su_{A78}^+ mutation is a C → A change at the anticodon 3'-end of the tRNA$_{GGU/C}^{Gly3}$ (Table IV). The su_{A78}^+ tRNA$_{UGU/C}^{Gly3}$ possesses a MS^2I^6A modification adjacent to the 3'-end of the anticodon, while the unmutated tRNA$_{GGU/C}^{Gly3}$ is unmodified. The su_{A36}^+ mutation is a C → U change at the 3'-end of the anticodon of tRNA$_{GGA/G}^{Gly2}$. tRNA's responding to codons beginning with A usually have a modified adenosine adjacent to the 3'-end of the anticodon (Nishimura, 1972). This modification, a derivative of N^6-carbamylthreonyladenosine (T^6A) is found in the su_{A36}^+ tRNA$_{AGA/G}^{Gly2}$ but not in the wild-type tRNA$_{GGA/G}^{Gly2}$. Thus, the modification enzymes must recognize the anticodon region in a very specific manner, since a single base change at the 3'-position of the anticodon necessitates an immediate modification of the newly synthesized tRNA at the adjacent base.

On the other hand, *Mycoplasma kid* tRNA does not contain isopentyl derivatives in its tRNA, and it functions normally in protein synthesis (Hayashi *et al.*, 1969). *Lactobacillus acidophilus* mevalonic acid mutants (mevalonic acid is a precursor to the isopentyl group in I^6A) can be grown in limiting mevalonic acid concentrations which are adequate for growth but result in tRNA's with 50% of the normal level of I^6A. This tRNA does not show any differences in biological properties when compared with the fully modified tRNA (Litwack and Peterkofsky, 1971).

Yeast tRNA$_{UUU/C}^{Phe}$ possesses the " Y " modification on the adenosine adjacent to the 3'-end of the anticodon. If this modification is removed chemically *in vitro,* the tRNA will still function in protein synthesis but now recognizes UUC preferentially over UUU, whereas the normally modified tRNA^{Phe} shows no discrimination between the two (Ghosh and Ghosh, 1970).

It is clear that the modification at the 3'-end of the anticodon is important in certain tRNA's but is not an obligatory component.

Another modification which almost all tRNA's possess is the ribothymidine in the G-T-Ψ-C loop (Zamir *et al.*, 1965). This sequence, found in all tRNA's involved in protein synthesis except eukaryotic initiator tRNA's, (Simsek *et al.*, 1973a) seems to be involved in the binding of tRNA to ribosomes (Ofengand and Henes, 1969), presumably to the complementary C-G-A-A-C sequence of 5 S rRNA of the 50 S ribosome (Erdmann *et al.,* 1973). However, *Mycoplasma kid* tRNA^{Ile} has the sequence G-U-Ψ-C, and this tRNA functions normally in protein synthesis (Johnson *et al.,* 1970). The U in the sequence of the tRNA is a substrate for methylation by an *E. coli* enzyme, but the heterologously modified tRNA^{Ile} shows no difference in physical or biological properties. An *E. coli* mutant, lacking the uracil methylase involved in the U → rT modification in the G-U-Ψ-C sequence, has been isolated (Björk and Isaksson, 1970). No ribothymidine is found in the tRNA of strains with this mutation which maps near *argH* (79'). Normal growth is seen in rich medium, but there is a 15–20% growth retardation in minimal medium (Björk and Neidhardt, 1971). Similarly, *Streptococcus faecalis* can grow without folic acid. This results in the presence of the sequence G-U-Ψ-C in all tRNA molecules, as opposed to the normally occurring G-T-Ψ-C when folate is present (Delk and Rabinowitz, 1975).

Methyl-deficient tRNA isolated from *E. coli rel* has been analyzed, but most earlier studies were performed with mixtures of methylated and unmethylated tRNA's. When purified tRNA's are used, for example, tRNA$^{Phe}_{E.\,coli}$, methyl-deficient species bind 50% as efficiently to ribosomes in the presence of poly(U) or poly(UC) as does the methylated tRNA^{Phe}. Other functions, however, such as charging and transferring of phenylalanine to internal peptide linkages, are normal (Stern *et al.,* 1970). Other minor alterations in the biological and physical properties of methyldeprived tRNA's have been described (Littauer and Inouye, 1973).

A role for the pseudouridine modification has been demonstrated with the *hisT* mutant in *Salmonella. hisT* cells are derepressed for histidine biosynthesis in the presence of histidine, even though normal levels of *his*-tRNA^{His} are found (Lewis and Ames, 1972). tRNA^{His} from these mutants has been sequenced and has been shown to be lacking two pseudouridine modifications in the anticodon region, while the Ψ in the G-T-Ψ-C sequence is unaffected (Singer *et al.,* 1972). This indicates two enzymes are involved in the formation of Ψ residues in tRNA, in agreement with studies on the Ψ modifications of su_3^+ tRNA$^{Tyr}_1$ precursors. In this case, the su_3^+ precursor tRNA shows a Ψ residue at U64 (Fig. 1), but the U40 is not modified to Ψ (Schaefer *et al.,* 1973). The *hisT* gene codes for an enzyme that catalyzes the conversion of U → Ψ at the anticodon region of several tRNA's in *Salmonella* (Cortese *et al.,* 1974a). These tRNA's, all tRNA^{Leu} and one tRNA^{Ile} as well as tRNA^{His}, in the *hisT* mutants, are unable to function in the regulation of the leucine, isoleucine, and valine and the histidine biosynthetic pathways (Cortese *et al.,* 1974b).

The actual mechanism by which aminoacyl tRNA's are involved in repression of biosynthesis is presently unknown and is discussed below.

Mutations in tRNA modification have been described in *Drosophila.*

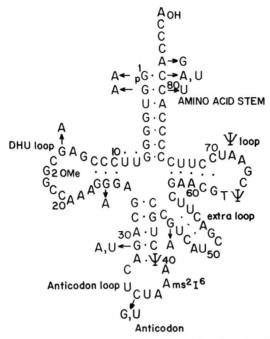

Fig. 1 The sequence of su_3^+ tRNA$_1^{Tyr}$. The sequence was taken from Goodman *et al.* (1968). Arrows refer to mutations described and references in the text. Ů at positions 8 and 9 indicates a 4-thio modification.

The vermilion mutation is caused by an altered tryptophan pyrrolase which is inactivated *in vivo* by binding to one of the three tRNA^{Tyr}. The recessive $su(s)^2$ mutation modifies tRNA$_2^{Tyr}$ so it can no longer inhibit the mutated enzyme (Twardzik *et al.*, 1971). Recently, it has been shown that the $su(s)^2$ mutation affects the enzyme responsible for the "Q" base modification of tRNA$_2^{Tyr}$ (White *et al.*, 1973). The *sufF* mutation in *Salmonella* is also recessive and is thought to be due to an altered tRNA modification enzyme (Riddle and Roth, 1972a).

A major posttranscriptional modification, at least in some tRNA species, is the addition of the C-C-A sequence at the 3'-terminus of tRNA molecules, catalyzed by tRNA-nucleotidyl transferase (Deutscher, 1973). The precursor to su_3^+ tRNA$_1^{Tyr}$ possesses the C-C-A end (Altman and Smith, 1971), but the precursor for serylprolyl T4 tRNA does not possess this 3'-terminus (Barrell *et al.*, 1974). This latter molecule serves as substrate, *in vitro,* for the *E. coli* tRNA-nucleotidyl transferase, so that the C-C-A end for the tRNA^{Ser} is formed (Schmidt, 1975). Until more tRNA precursors are sequenced, the question of primary transcription or secondary modification of the C-C-A terminus will not be clarified. At any rate, mutants with low levels of the nucleotidyltransferase have been isolated and found to map at 59.4' (Foulds *et al.*, 1974).

Cells possessing less than 15% of wild-type levels of transferase activity grow poorly, and T4 mutants, normally suppressed by T4 tRNA *amber* suppressors (see above), do not plaque in these bacterial mutants, indicating a role for the enzyme in bacterial and viral tRNA function (Deutscher *et al.*, 1974).

C. Synthesis of tRNA

tRNA precursors synthesized from DNA templates are larger than the final modified molecule (Burdon and Clason, 1969; Bernhardt and Darnell, 1969; Altman and Smith, 1971). The precursors most extensively studied, both genetically and physiologically, are su_3^+ tRNA$_1^{Tyr}$ and, more recently, the isolated precursors to T4 tRNA (McClain *et al.*, 1972). Study of the su_3^+ tRNA$_1^{Tyr}$ precursor has facilitated the isolation of mutants in the su_3^+ gene which result in the accumulation of precursor (Altman and Smith, 1971). A specific ribonuclease, ribonuclease P, is involved in the specific cleavage of the extra 41 nucleotides at the 5'-end of the precursor and it has been isolated and purified (Robertson *et al.*, 1972). Isolation of RNase P$^-$ mutants has shown that this enzyme does not remove the three additional nucleotides present at the 3'-end of the molecule (Schedl and Primakoff, 1973). While it was originally suggested that RNase II was responsible for this activity, recent experiments, using as an assay the *in vitro* synthesis of biologically active su_3^+ tRNA$_1^{Tyr}$, have demonstrated that a different RNase (PIII) is necessary for the 3' processing (Bikoff and Gefter, 1975).

The tRNA$_1^{Tyr}$ gene is ordinarily coded for by two closely linked genes (Russell *et al.*, 1970; Wu and Davidson, 1973). Mutations, like A1 and A2, which ordinarily decrease the amount of singlet precursor RNA formation in the ϕ80 su_3^+ Cambridge strain, bearing only one copy of the tRNA$_1^{Tyr}$ gene, also inhibit the formation of the closely linked tRNA$_1^{Tyr}$ gene in the doublet ϕ80 su_3^+/su^- (Ghysen and Celis, 1974a). This evidence for a common tRNA joint precursor is strengthened when precursors are synthesized in RNase P$^-$ strains and the sequences are analyzed. *In vitro* experiments have indicated that the initial transcript of su_3^+ tRNA$_1^{Tyr}$ is much larger than the RNase P substrate, and that another enzyme, possibly RNase III (or PIV), is involved in the initial cleavage (Bikoff *et al.*, 1975).

Polycistronic transcription of tRNA genes occurs elsewhere. The isolation of the defective transducing phage λh80t, bearing three tRNA genes, *glyT* su_{36}^+, *tyrT* (su_{15B}^+), and *thrT*, from the *argH* region of the coli chromosome (77'), has permitted the genetic and physical mapping of these tRNA genes on the chromosome (Squires *et al.*, 1973a; Wu *et al.*, 1973). Heteroduplex mapping in conjunction with 3'-ferritin labeling of hybridizing tRNA suggested the gene order and spacings for the tRNA genes as: $tRNA_3^{Thr}$, 200 nucleotides; $tRNA_2^{Tyr}$, 140 nucleotides; $tRNA_2^{Gly}$, suggesting spaces of approximately 60 and 180 nucleotides between the genes. This compares with the spacing of about 60 nucleotides between the su_3^+ $tRNA_1^{Tyr}$ genes (Wu and Davidson, 1973; Landy *et al.*, 1974). The three tRNA's are synthesized in equal amounts during infection with the defective phage (Squires *et al.*, 1973a). *In vitro* synthesis of tRNA, using λh80t DNA has recently shown that $tRNA_2^{Tyr}$ sequences were found in RNA molecules with an average MW of 230,000. This would be sufficient for three tRNA genes and spacers (Grimberg and Daniel, 1974). A 170 nucleotide precursor for $tRNA_2^{Gly}$-$tRNA_3^{Thr}$ has now been sequenced (Chang and Carbon, 1975), but the sequence for $tRNA_2^{Tyr}$ is not in the precursor as would be expected from the conclusions of the heteroduplex mapping discussed above.

Multiple copies of genes for $tRNA_3^{Gly}$ are found at the *glyV* locus (86') (Squires and Carbon, 1971), and a duplication of the $tRNA_2^{Gly}$ genes at the *glyT* locus can also occur (Hill *et al.*, 1969). No information on precursor forms containing multiple tRNA transcripts in these gene clusters is available, however.

Escherichia coli mutants have been described with a *ts* RNase P, which accumulate several precursors with multiple tRNA transcripts (Schedl and Primakoff, 1973; Sakano *et al.*, 1974). As shown for the *in vitro* processing of su_3^+ $tRNA_1^{Tyr}$ (see above), three enzymes are involved in the maturation of these multiple tRNA transcripts [RNase O (possibly RNase III-PIV), RNase P, and RNase Q (possibly RNase PIII)] (Sakano and Shimura, 1975). Isolation of tRNA:DNA hybrids by Cs_2SO_4 isopycnic centrifugation has provided evidence that most of the *E. coli* tRNA cistrons are in clusters of 3 to 4 with spacers of 120 to 140 nucleotides (Fournier *et al.*, 1974).

The tRNA genes found in T4 are clustered, and three precursor molecules have been isolated, each containing the sequences for two different tRNA's: serine-proline-tRNA; glutamine-leucine-tRNA; and α-δ-tRNA (Guthrie *et al.*, 1973). The serine-proline-tRNA precursor has been sequenced and is 175 nucleotides long with additional nucleotides at the 5'- and 3'-ends, but without the sequence C-C-A (Barrell *et al.*, 1974). Point mutations in the psu_1^+ serine-tRNA gene eliminate serine- and proline-tRNA's while mutations in psu_2^+ glutamine-tRNA genes cause loss of glutamine- and leucine-tRNA's (Guthrie *et al.*, 1973). Presumably, this results from inaccurate cleavage of the dimeric tRNA precursors with subsequent enzymatic destruction of both tRNA's. This is analagous to the A1 and A2 mutations in the su_3^+ tRNA gene which inhibit the synthesis of the duplicate su^- tRNA$_1^{Tyr}$ (see above). RNase P (Sakano *et al.*, 1974) and, possibly, RNase III (Paddock and Abelson, 1973) are involved in the specific cleavage of T4 tRNA multimeric precursors as well as primary bacterial tRNA transcripts.

Though tRNA genes for isocoding tRNA's cluster in *Xenopus laevis* (Clarkson *et al.*, 1973b), little is known about the nature of the precursors, i.e., whether larger precursors similar to those found in HeLa or prokaryotes exist, or if multiple tRNA transcripts are found in the same precursor. The calculated average spacer distance between isocoding tRNA genes is large, approximately 10 times the size of the mature tRNA gene molecule, in comparison to the much smaller spacers separating the adjacent su_3^+ tRNA genes and the *glyT, tyrT,* and *thrT* tRNA genes in λh80t. Though little is known about the enzymology of the processing of eukaryotic tRNA precursors, *E. coli* extracts, presumably containing ribonuclease P, specifically cleave *Bombyx mori* rapidly labeled RNA species to give 4 S products (Chen and Siddiqui, 1973). The conformation of these molecules must be similar to prokaryote tRNA precursors.

D. Control of tRNA Synthesis

tRNA synthesis is stopped when a required amino acid is withdrawn in wild-type *E. coli* (Neidhardt and Eidlic, 1963), and this can be altered with a mutation in the *relA* gene (Stent and Brenner, 1961). The *relA* gene (53′) codes for a protein, loosely associated with the 50 S ribosome (Ramagopal and Davis, 1974) which synthesizes pppGpp and ppGpp from ATP and GTP (Haseltine *et al.*, 1972) when deacylated tRNA is bound to the A site of the ribosome in the presence of its specific codon (Haseltine and Block, 1973; Pederson *et al.*, 1973). A 50 S structural protein, probably L11 (Parker *et al.*, 1976), the product of the *relC* gene, is also involved in the synthesis of ppGpp (Friesen *et al.*, 1974). The synthesis of su_3^+ tRNA$_1^{Tyr}$ is under rel^+ or stringent control *in vivo* (Primakoff and Berg, 1970), and synthesis of all tRNA's seems to be coordinately repressed during starvation for amino acids in rel^+ cells (Ikemura and Dahlberg, 1973). ppGpp and pppGpp are believed to be involved in the specific repression of stable RNA synthesis (Cashel and Gallant, 1969), but conflicting reports have appeared regarding the specific role of these nucleotides in tRNA biosynthesis. In one case, *in vitro* synthesis of su_3^+ tRNA$_1^{Tyr}$ was unaffected by the presence of ppGpp (Zubay *et al.*, 1970), while another report suggested that ppGpp does inhibit the *in vitro* synthesis of this tRNA (Travers, 1973; see below), but the effect was quite small. Recent experiments, studying the *in vitro* synthesis of rRNA, tRNA, and other gene products, have shown that ppGpp, while showing a fivefold selective inhibition of rRNA synthesis, has no effect on tRNA production (Reiness *et al.*, 1975).

Mutants controlling the synthesis of T4 tRNA which map outside of the T4 tRNA structural gene cluster have been isolated (Wilson and Abelson, 1972; Guthrie and McLain, 1973). These mutations seem to affect the synthesis of some, but not all, T4 tRNA's. The role of the genes defined by these mutations is not yet known.

Conditional mutants which do not make tRNA and/or rRNA at elevated temperatures have been isolated in *E. coli* by means of $\phi80\,su_3^+$ infection of *lac amber* cells (Schedl and Primakoff, 1973). This seems like a powerful system for elucidating the genetics of tRNA synthesis. Clearly, more work is needed in this area before a clear understanding of the mechanisms of control of tRNA synthesis is obtained. The recent synthesis of the gene coding for su_3^+ tRNA, including precursor and promoter regions, is a valuable step forward for this type of study (Loewen *et al.*, 1974; Sekiya and Khorana, 1974).

E. Role of tRNA in Regulation

As discussed above, charged histidine-tRNA with the proper modification at the anticodon region of the molecule is involved in the regulation of the histidine biosynthetic pathway, and this seems to be similar to the control of leucine, isoleucine, and valine biosynthesis (Cortese *et al.*, 1974b). Histidyl-tRNA will bind specifically to the first enzyme of the histidine pathway, the phosphoribosyl transferase (Vogel *et al.*, 1972), and this protein, *in vitro*, is a specific repressor of *his* operon transcription (Blasi *et al.*, 1973).

It has been suggested that the transferase, complexed with charged histidyl-tRNA is the repressor in this system (Goldberger, 1974). However, the *in vitro* transcription studies do not show any dependence on charged histidyl-tRNA in the transferase mediated inhibition of *his* mRNA synthesis (Blasi *et al.*, 1973), so this theory remains to be substantiated. The tryptophan operon is clearly not regulated in the above manner. Tryptophanyl-tRNA does not seem to be the corepressor of the *trp* operon (Mosteller and Yanofsky, 1971), and almost complete deletions of the first gene in the *trp* operon have no effect on *trp* regulation (Jackson and Yanofsky, 1972). Regulation of *in vitro* transcription of the *trp* operon is also not dependent on the presence of charged $tRNA^{Trp}$ (Squires *et al.*, 1973b).

F. Genetic Analysis of tRNA Structure

The $\phi80\,su_3^+$ system has proved to be valuable for the isolation of mutations in the primary sequence of $tRNA_1^{Tyr}$, thus gaining insight into the functions of the various regions of the tRNA molecule and also providing evidence for interactions between the various loops. $\phi 80\,su_3^+$ has been mutagenized *in vitro*, or cells, during $\phi 80\,su_3^+$ infection, have been mutagenized, and $\phi 80\,su^-$ have been obtained (Fig. 1). Several classes of mutants are observed: (a) *ochre* suppressors—anticodon C-U-A → U-U-A (Altman *et al.*, 1971); (b) temperature-sensitives—mutations in the amino acid acceptance stem—G_1 → A, G_2 → A, C_{81} → A (Smith *et al.*, 1970); (c) mischarging mutants, also in the amino acid stem, which show a different suppression pattern than su_3^+—G_1 → A, G_2 → A, C_{81} → U, C_{80} → U, A_{82} → G (Shimura *et al.*, 1972; Hooper *et al.*, 1972; Ghysen and Celis, 1974b), and it has been shown that glutamine is inserted, instead of tyrosine, in these mutants (Smith and Celis, 1973; Ghysen and Celis, 1974b); (d) defects in later stages in protein synthesis G_{15} → A (Abelson *et al.*, 1970). In general, the *su* negative mutations show an absence, or a markedly decreased amount, of the mature form of su_3^+ tRNA, and frequently an accumulation of precursor is observed. Conformation of the precursor seems to be an important element in correct processing of the precursor tRNA as many of the *su⁻* mutations occur in base pairing regions. For example, G_2 → A results in lack of synthesis of the mature tRNA at 42°. A mutation in the normal pairing base, C_{80} → U, restores the wild-type phenotype. The specificity of the base pair is not important, but the maintenance of helicity is (Smith *et al.*, 1970). Second site reversions of this type are observed at G_{25} where a transition to A, causing a lack of su_3^+ tRNA synthesis, is reversed by C_{11} → U mutation. Reversion by the restoral of base pairing occurs in the normal base pairs G_{46}—C_{54} → A_{46}—U_{54} and G_{31}—C_{41} → A_{31}—C_{41} or U_{31}—A_{41} (Smith, 1972; Anderson and Smith, 1972). Mutations in the precursor or promoter sequences of su_3^+ $tRNA_1^{Tyr}$ have been reported (Smith *et al.*, 1970; Ghysen and Celis, 1974b) which increase the amounts of A_1 and A_2 su_3^+ tRNA's synthesized up to 12-fold. This again stresses the importance of conformation in the maturation of the tRNA molecule.

These results confirm the essential features of the "cloverleaf" pattern proposed by Holley *et al.* (1965). Partial reversion is obtained by mutations in bases not normally paired in the two-dimensional tRNA model. G_{25} → A is partially reversed by C_{19} → U, G_{15} → A by C_{19} → DHU or C_{20} → DHU, and C_{31} → U by C_{16} → U or C_{45} → U (Anderson and Smith, 1972). This indicates interactions between the anticodon loop and the dihydrouridine loop and "d" or extra loop in su_3^+ tRNA. The 3 Å resolution electron density maps recently described for yeast tRNA^{Phe}, while differing as to the orientation of the dihydrouridine loop and stem, both propose tertiary interactions between these three regions of the tRNA molecule (Suddath *et al.*, 1974; Robertus *et al.*, 1974).

The importance of the amino acid stem in synthetase recognition is shown by the effects of mutation in this region, e.g., the mischarging caused by A_1, A_2, etc. The anticodon stem and the "d" loop are also important in synthetase interaction because A_{31} and A_{46} mutations cause elevations in K_m. These effects must be conformational since they can be reverted by second site mutations. A change in the anticodon of su_7^+ tRNA$_1^{Trp}$ so that it responds to UAG instead of UGG (C-C-A → C-U-A) causes mischarging, inserting glutamine (Yaniv *et al.*, 1974). Thus, a single nucleotide change can affect both codon binding and synthetase recognition. The dihydrouridine loop is also involved in codon recognition, e.g., mutation G_{24} → A in tRNA^{Trp} changes the coding properties so that this molecule recognizes the UGA codon, and becomes a UGA suppressor, inserting tryptophan (Hirsch, 1971).

G. Evolution of tRNA

Over forty tRNA molecules from prokaryotes and eukaryotes have been sequenced and the evolutionary relationships between them have been analyzed (Holmquist *et al.*, 1973). It is suggested that all tRNA's evolved from an ancestral molecule by means of duplication and transient inactivation of one of the duplicate genes followed by neutral mutations with a subsequent activation of this modified gene. This could result in a tRNA with altered specificity. This scheme has been proposed for protein evolution as well, to account for the appearance of isoenzymes and enzyme families (Rigby *et al.*, 1974).

There are many constraints in tRNA structure, i.e., maintenance of the cloverleaf structure and several invariant loci (e.g., the U adjacent to the 5′-end of the anticodon and the G-T-Ψ-C sequence) and evolutionary divergence has reflected this, showing statistically fewer differences between tRNA's than would be predicted on the basis of strictly stochastic events. An exception which, as the saying goes, proves the rule, is the case of eukaryotic methionyl initiator tRNA's. These molecules, isolated from yeast, rat, rabbit, sheep, mice, and wheat, all possess a similar G-A-U-C-C sequence, replacing the G-T-Ψ-C found in all other tRNA's involved in protein synthesis (RajBhandary and Ghosh, 1969; Simsek *et al.*, 1973a,b; Petrissant, 1973; Piper and Clark, 1974). Sequences of these tRNA's show a high degree of homology even though the parent organisms have diverged significantly. In fact, all mammalian initiator tRNA's thus far sequenced have identical sequences (Piper and Clark, 1974; Simsek *et al.*, 1974). Interestingly, coli tRNA$_f^{Met}$ shows marked homology, as well. The functional reason for the absence of the T-Ψ sequence is unknown, but may be related to differences in the mechanisms of initiation. For example, mammalian initiator tRNA can bind to 40 S ribosomes in the absence of template (Schreir and Staehelin, 1973; Darnbrough *et al.*,

1973). There is also an absence of formylation of Met-tRNA$_f^{Met}$ in eukaryotes, even though mammalian initiator tRNA's are substrates for prokaryote or organelle transformylases.

We have seen duplications of tRNA genes, arising in the laboratory or naturally occuring in bacteria. There is evidence for joint transcription of closely linked tRNA genes, the mature products showing specificity for different tRNA's. This is possibly a model of how evolutionary divergence began. tRNA$_3^{Gly}$ has 59 out of 76 bases in common with tRNA$_{2B}^{Val}$ (Squires and Carbon, 1971), as well as showing a high degree of homology with tRNA$_2^{Gly}$, tRNA$_3^{Gly}$, and T4-specified tRNA^{Gly} (Roberts and Carbon, 1975). A duplication of one ancestral gene with subsequent evolution could explain this. Clusters of isocoding tRNA genes (Clarkson *et al.*, 1973a) could have arisen by similar duplications with selective advantage in growth maintaining homology in these genes. For example, duplications of the su_3^+ tRNA$_1^{Tyr}$ gene possessing the A$_1$ or A$_2$ mutations have been reported (Ghysen and Celis, 1974b) which show three copies of the weak suppressor. These occur at high frequency, probably by means of unequal recombination, since *rec* mutations eliminate the appearance of the repeated genes. The existence of duplications in the mutated *glyS* gene, giving a gly$^+$ phenotype (Folk and Berg, 1971), is possibly another example of this type of mechanism. Directed evolutionary studies, beginning now with genes for proteins, could be very fruitful in this regard (Rigby *et al.*, 1974).

V. Ribosomes

The genetics and physiology of bacterial ribosomes have been reviewed recently (Davies and Nomura, 1972; Nomura *et al.*, 1974).

A. rRNA

rRNA has been reviewed previously (Attardi and Amaldi, 1970; Pace, 1973). Ribosomes from all species, in which the proper analysis has been performed, contain three types of RNA, the 23–28 S and 5 S molecules which are found in the larger ribosomal subunit, and the 16–18 S species of the smaller subunit. In addition, eukaryote larger subunits possess one 5.8 S RNA molecule. However, the rRNA present in mitochondria of higher organisms is smaller (Stewart, 1973) and no 5 S rRNA has been found yet in these organelles.

There are multiple cistrons for rRNA's in all organisms, except for *Mycoplasma kid* and organelles which seem to have one cistron for the larger and smaller rRNA species. As genetic complexity increases, there is increased redundancy in the rRNA cistrons, as is true for the tRNA genes

(Table III; Birnstiel *et al.*, 1971). The genes for rRNA are usually present in equal numbers (Pace and Pace, 1971), but, in some eukaryotes, 5 S rRNA genes are in vast excess over the numbers of genes for the larger rRNA species. This is due to the different nature of the primary rRNA transcriptional product in eukaryotes and prokaryotes and is discussed below.

1. Heterogeneity of rRNA

There is a high degree of homology in the rRNA molecules produced within a species, the presence of multiple cistrons notwithstanding. However, microheterogeneity has been found in 5 S rRNA from *E. coli* (Brownlee *et al.*, 1968; Jarry and Rosset, 1971), in 16 S rRNA (Fellner *et al.*, 1972), and also 23 S rRNA (Branlant *et al.*, 1973). In 16 S rRNA the regions where variations occur frequently show more than one change, as though these were regions of permissible change (nonessential areas) or that a mutation necessitated a corresponding compensatory change (Fellner *et al.*, 1972). This would be analagous to the second site suppression of su_3^+ tRNA negative mutations.

Heterogeneity exists in 5 S RNA of *Xenopus*. There is a major 5 S RNA sequence in kidney cells, while there are three additional sequences found in the ovary (Ford and Southern, 1973; Wegnez *et al.*, 1973). There is evidence for heterogeneity in KB (Forget and Weissman, 1969) and HeLa cell 5 S rRNA (Hatlen *et al.*, 1969).

2. Strain and Species Homologies

There is a marked conservation of homology between rRNA species from different strains, species and genera as measured by rRNA:DNA hybridization (Dubnau *et al.*, 1965a; Doi and Igarashi, 1965; Chilton and McCarthy, 1969; Brenner *et al.*, 1969; Sinclair and Brown, 1971), and by base sequence analysis of 5 S rRNA (DuBuy and Weissman, 1971) and 16 S rRNA (Fellner, 1974). rRNA's in *Xenopus laevis* and *Xenopus mulleri* seem to be identical, whereas the intercistronic spacers are very different (Brown *et al.*, 1972). There are differences, however, in 5 S rRNA sequences between strains of *E. coli* (Brownlee *et al.*, 1968; Jarry and Rosset, 1971) and between *B. subtilis* strains (C. A. Marotta, S. M. Weissman, and I. Smith, unpublished experiments, 1973). Strain differences have been found in one 16 S rRNA oligonucleotide sequence when *E. coli* B and K are compared (Muto *et al.*, 1971), and there are primary sequence differences in 23 S rRNA between *E. coli* and *Salmonella typhosa* (Sypherd *et al.*, 1969).

The absence of a known rRNA function makes it difficult to analyze the physiological significance of sequences in rRNA. Studies of specific binding of ribosomal proteins to rRNA, discussed elsewhere is this volume, are beginning to shed light on this problem. The one sequence in rRNA which seems to have a known role is the C-G-A-A-C sequence starting at position 42 or 43 in 5 S rRNA of *E. coli* (Brownlee *et al.*, 1968), *Pseudomonas* (DuBuy and Weissman, 1971), *Bacillus stearothermophilus* (Marotta *et al.*, 1973), and *B. subtilis* (Marotta *et al.*, 1976). *Chlamydomonas* cytoplasmic 5 S rRNA has a G-A-A-C at position 41 (Jordan *et al.*, 1974). Other eukaryote 5 S rRNA's do not have a similar sequence in this region of the molecules, though G-A-A-C sequences are found elsewhere in these species, but nuclease digestion studies have shown that region 40 in eukaryote as well as prokaryotic 5 S rRNA is the most exposed part of these molecules in solution (Vigne *et al.*, 1973). It is clear that this sequence is important, at least in bacteria, since the complementary sequence G-T-Ψ-C, found in all tRNA's involved in protein synthesis except mammalian initiator tRNA's (see above), plays a key role in the specific attachment of tRNA to 50 S subunits, presumably by binding to the G-A-A-C sequence of the 5 S rRNA. The tetranucleotide T-Ψ-C-G will bind to 50 S subunits or to 5 S rRNA-50 S ribosomal-protein complexes and, in the process, inhibits EF-Tu-directed aminoacyl-tRNA binding, GTP hydrolysis, and ribosome-dependent ppGpp synthesis (Richter *et al.*, 1973). Selective modification of two adenine residues in ribosome bound 5 S rRNA prevents T-Ψ-C-G binding (Erdmann *et al.*, 1973).

Another regularity in rRNA structure occurs at the 3'-end of 18 S rRNA. It has been reported that yeast, *Drosophila*, and rabbit 18 S rRNA have the terminal 3'-sequence G-A-U-C-A-U-U-A (Dalgarno and Shine, 1973), and it was postulated that the terminal 3'-U-U-A$_{OH}$ could be a recognition site for terminator triplets U-A-G and U-A-A in mRNA. U-U-A$_{OH}$ is also at the 3'-terminus of *E. coli* 16 S rRNA (Shine and Dalgarno, 1974; Ehresmann *et al.*, 1974), which would stress the importance of this sequence. However, other bacteria do not have this 3'-terminal sequence (Shine and Dalgarno, 1974), and now it seems clear that the 3'-end is involved in the initiation of translation. Ribosomal proteins S1 and S21, involved in the specificity of initiation, bind to the 3'-end of 16 S rRNA (Czernilofsky *et al.*, 1975; Dahlberg and Dahlberg, 1975) and form part of the mRNA binding site (Fiser *et al.*, 1975), while RNA:RNA hybrids containing short mRNA fragments and the 3'-end of 16 S rRNA have been isolated from *E. coli* ribosome initiation complexes (Steitz and Jakes, 1975).

3. Mapping of rRNA Genes

a. Prokaryotes. As discussed above, it is difficult to directly map rRNA genes, in the absence of a selection for rRNA function and because of the redundancy of rRNA cistrons. The strain and species differences described above have been used as unselected markers in genetic crosses, and this has allowed the mapping of some of the rRNA genes. In addition, F' merogenotes and specialized transducing bacteriophages have been used to localize other rRNA cistrons.

It is now clear that at least two regions for rRNA genes are present on the *E. coli* chromosome. One cluster, containing two or three rRNA cistrons is found between 74 and 79 minutes (Fig. 2). This has been shown by the P1 transduction of strain differences in 5 S rRNA, indicating two 5 S RNA gene loci, *cqsA* and *cqsB* (Jarry and Rosset, 1970, 1973a,b) and 16 S rRNA (Matsubara *et al.*, 1972). In addition, F'14, which encompasses this region of the chromosome, *rbs* to *argH*, contains two to three tandem sequences for 16 S and 23 S rRNA (Birnbaum and Kaplan, 1971), and it has been shown that 16 S and 23 S rRNA are synthesized in minicells containing F'14 but not KLF'41, which covers the ribosomal protein gene area, *argG-mal* (61'–66') (H. Hori *et al.*, 1974). Heteroduplex mapping of F'14 and $\phi 80su_7^+$ has shown that two 16 S–23 S tandem rRNA genes are found on this episome and that the gene order is f . . . *ilvD* . . . 16 S–23 S (*rrnA*) . . . *metB* . . . *argH* . . . 16 S–23 S(*rrnB*) . . . F (Deonier *et al.*, 1974; Ohtsubo *et al.*, 1974). *cqsA* may map with *rrnA*, but *cqsB* is not found in the same map position as *rrnB* (Fig. 2A). Possibly, another 16 S–23 S tandem gene is located between *ilv* and *rbs*, near *trkD* (Deonier *et al.*, 1974), which is where *cqsB* is located (Jarry and Rosset, 1973b). A tRNA$_2^{Glu}$ gene lies between a 16 S and 23 S rRNA gene in $\phi 80su_7^+$ (Wu and Davidson, 1975), similar to the situation in yeast mitochondria (Faye *et al.*, 1976).

In this region of the chromosome is a cluster of tRNA genes, su_7^+ (Ohtsubo *et al.*, 1974), *glyT, thrT, trypT* (Wu *et al.*, 1973), and the structural

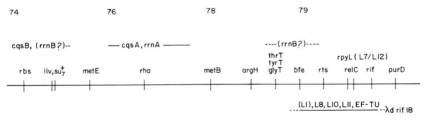

Fig. 2 Genetic map of rRNA, tRNA, and ribosomal protein gene cluster in *E. coli* (74'–79'). The map positions of the various genes were taken from references in the text and Taylor and Trotter (1972); abbreviations are as used by Taylor and Trotter (1972). The ribosomal proteins whose synthesis is stimulated *in vivo* by λd*rif18* are also indicated. L1 is indicated in parentheses since it has also been reported that the structural gene for L1 maps in the *strA-spcA* region (Fig. 3; Sypherd and Osawa, 1974). The map is not drawn to scale as the area between *argH* and *purD* has been expanded.

genes for several 50 S ribosomal proteins and ribosome-modifying enzymes (see Fig. 2).

A second cluster for rRNA genes, approximately 30% of the total, has been reported to lie between 54 and 59 minutes on the *coli* map, using *coli* F merogenotes in *Proteus vulgaris* (Unger *et al.*, 1972). However, a cistron for 5 S rRNA, *cqsD*, has been reported to lie between *argR* and *aroE*, 63'–64', and a fourth 5 S rRNA cistron, *cqsC*, is found between *aroB* and *malA*, 65'–66' (Jarry and Rosset, 1973b), quite distant from the 54'–59' cluster of 16 S–23 S rRNA genes. It had been reported that some genes for 16 S and 23 S rRNA are in the region 60'–66' (Gorelic, 1970), but other studies indicate this is not the case (Sypherd *et al.*, 1969; Yu *et al.*, 1970; Muto *et al.*, 1971; Birnbaum and Kaplan, 1971; H. Hori *et al.*, 1974). Thus, while there seems to be a rough correlation between 5 S rRNA genes and those of 16 S and 23 S rRNA, it is not an absolute one. More fine structure mapping, especially in the region around 60', is needed before a definitive answer to the question of one 5 S rRNA gene to each 16 S and 23 S rRNA gene is answered.

The presence of two major groups of rRNA genes was indicated previously in *E. coli* (Rudner *et al.*, 1965; Cutler and Evans, 1967) and seems to be true in *B. subtilis* as well. In this organism, the major cluster for rRNA genes (5 S, 16 S, and 23 S) is found in the chromosomal segment delimited by *cysA* and *lin* (Fig. 2B, Oishi and Sueoka, 1965; Dubnau *et al.*, 1965b; Smith *et al.*, 1968). In this region of the chromosome are several antibiotic resistance genes (Goldthwaite *et al.*, 1970; Harford and Sueoka, 1970; Goldthwaite and Smith, 1972a), and these are discussed more fully below. The attachment site for the lysogenic subtilis phage SP02 has been mapped between *spcA* and *lin* (Smith and Smith, 1973) and heteroduplex studies have now shown that *attSP02* is 6200 base pairs, or 4×10^6 daltons of DNA, away from a double set of tandem 16 S and 23 S rRNA genes (Chow and Davidson, 1973). Seven to nine cistrons for rRNA are postulated to occur in one linkage group in this region, depending on the interpretation of the heteroduplexes. The rRNA cistrons can lie distal or proximal to *attSP02*, but the large silent region between *attSP02* and *lin* suggests the rRNA gene cluster is to the right of the attachment site. In addition, if the cluster were proximal to *attSP02*, rRNA cistrons would intermingle with antibiotic resistance genes, a possibility which has been ruled out (Smith *et al.*, 1971).

Evidence for a second cluster of rRNA genes in *B. subtilis* was obtained, as in the case of the major cluster, by means of rRNA–DNA hybridization using DNA isolated from synchronously replicating cells after a density shift (Smith *et al.*, 1968) and also was suggested by an analysis of genetic and base sequence homologies between *B. subtilis* and a related species, *B. globigii* (Chilton and McCarthy, 1969). This cluster, containing approximately one-third of the rRNA cistrons, replicates late in the division cycle and may be near *leu*.

b. Eukaryotes. Unlike the situation in prokaryotes (Colli *et al.*, 1971), 5 S rRNA genes are unlinked to 18 S and 28 S genes in eukaryotes. Most work in the chromosomal localization of rRNA genes has been performed in *Xenopus* and *Drosophila*. Anucleate tadpoles of the former species, lacking nucleoli, do not synthesize rRNA (Brown and Gurdon, 1964) and do not contain DNA sequences complementary to 18 S and 28 S rRNA (Birnstiel *et al.*, 1966). Similarly, strains of *Drosophila*, with one, two, or four nucleolar organizer regions normally found on the X and Y chromosomes, show corresponding amounts of DNA sequences complementary to 18 S and 28 S rRNA (Ritossa and Spiegelman, 1965), and the bobbed mutation mapping in the "NO" region shows lower amounts of DNA complementary to rRNA (Ritossa *et al.*, 1966b).

Since all the cistrons for rRNA are affected by the mutations or alterations of the nucleolar organizer regions of both orgainsms, this indicated clustering for these genes. *In situ* hybridization, in which radioactive nucleic acids are annealed to cytological preparations and autoradiography is carried out, has confirmed the localization of 18 S and 28 S rRNA genes in the nucleolar organizer region of *Drosophila* (Pardue *et al.*, 1970) and *Xenopus* (Pardue *et al.*, 1973). Genes for 5 S rRNA, on the other hand, are not found in the nucleolar organizer region. The anucleolate mutant of *Xenopus* contains normal amounts of 5 S rDNA (Brown and Weber, 1968). In addition, deletions and duplications in the nucleolar organizer region of *Drosophila* have no effect on the level of DNA complementary to 5 S rRNA (Tartof and Perry, 1970).

In situ hybridization has shown that 5 S rDNA sequences are localized at the telomeres of the long arms of the majority of the 18 chromosomes of *Xenopus* (Pardue *et al.*, 1973), and on chromosome 2 at region 56 EF in *Drosophila* (Wimber and Steffensen, 1970; Procunier and Tartof, 1975). Similarly, studies on HeLa chromosomes, isolated on the basis of size, have shown that while the genes for 18 S and 28 S rRNA are found on smaller chromosomes, including those bearing nucleolar organizers (Huberman and Attardi, 1967; Spadari *et al.*, 1973), genes for 5 S rRNA are found on all size classes of chromosomes (Aloni *et al.*, 1971), with some clustering on chromosome I (Johnson *et al.*, 1974). Even though 5 S rRNA genes are not linked to 18 S–28 S rRNA genes in most eukaryotes, the DNA sequences for 5.8 S rRNA are linked and 5.8 S rRNA is part of the rRNA primary transcript (Speirs and Birnstiel, 1974; Maden and Robertson, 1974; see below).

In yeast, a majority of the 17 S–25 S rRNA cistrons are found on chromosome I (Goldberg *et al.*, 1972; Finkelstein *et al.*, 1972), but in contrast to other eukaryotes, 5 S rRNA genes are closely linked to 17 S and 25 S rRNA cistrons (Retel and Planta, 1972; Cryer *et al.*, 1973; Rubin and Sulston, 1973). However, it seems unlikely that 5 S rRNA is part of a single transcriptional unit with 17 S and 25 S rRNA in yeast, since it shows different kinetics of synthesis (Udem and Warner, 1972) and is isolated with a 5'-terminal pppG (Hindley and Page, 1972), indicating it is a primary transcription product, and not cleaved from a larger precursor.

In several eukaryotes, there is an amplification in the number of genes coding for 18 S and 28 S rRNA during various stages in the life cycle. Certain amphibians and invertebrates, during oogenesis, possess multiple nucleoli in their oocytes with corresponding increases in the amounts of rDNA, in some cases up to 2000-fold (Brown and Dawid, 1968; Gall, 1969). The macronucleus of *Tetrahymena* also contains ten fold amplified rRNA genes (Gall, 1974; Yao *et al.*, 1974).

There is good evidence that these amplified copies arise from chromosomal genes (Brown and Blackler, 1972). They are found in the form of circles and concatamers (Hourcade *et al.*, 1973; Gall and Rochaix, 1974; Gall, 1974), and seem to replicate by means of "rolling circle" intermediates (Hourcade *et al.*, 1973; Rochaix *et al.*, 1974). An RNA-dependent DNA polymerase which synthesizes new rDNA from rRNA also has been proposed to account for oocyte rDNA amplification (Tocchini-Valentini *et al.*, 1973), but other experiments have not demonstrated the existence of this enzyme and the expected rRNA–DNA hybrid intermediates of the reaction (Bird *et al.*, 1973).

c. Organelles. Organelle genomes have one gene each for the two major types of ribosomal RNA (Table III), except in chloroplasts of certain species in which two to three cistrons have been reported, i.e., *Euglena* (Stutz and Vendrey, 1971; Scott, 1973) and higher plants (Thomas and Tewari, 1974). However, other investigations indicate single rRNA cistrons in *Euglena* chloroplasts (Rawson and Haselkorn, 1973).

The rRNA genes are found on the H strand of *Xenopus* (Dawid, 1972), HeLa mitochondria (Wu *et al.*, 1972), and *Euglena* chloroplasts (Stutz and Rawson, 1970). Heteroduplex mapping of HeLa mitochondrial DNA has positioned 12 S and 16 S rRNA genes relative to tRNA genes (Wu *et al.*, 1972), and a recent report in yeast has indicated that the mitochondrial erythromycin resistance gene may code for 23 S rRNA. A deletion in the *ery*[r] gene, causing a petite phenotype, also has a partial deletion of the DNA sequence complementary to 23 S rRNA. 16 S rRNA sequences were unaffected (Faye *et al.*, 1974). More recently, hybridization studies in various petite mutants have shown that the 16 S rRNA gene is separated from the 23 S rRNA gene by several tRNA genes (Faye *et al.*, 1976).

Yeast is an ideal cell for this type of study, since the mitochondria have only one cistron for the ribosomal RNA's, which eliminates the problem of rRNA redundancy. In addition, rRNA gene mutations would not be lethal, as yeast can dispense with mitochondrial function.

4. Synthesis of rRNA

rRNA is transcribed in the form of a polycistronic precursor, in bacteria, consisting of 16 S rRNA, 23 S rRNA (Pettijohn et al., 1970; Doolittle and Pace, 1971; Dunn and Studier, 1973; Nikolaev et al., 1973) and 5 S rRNA (Ginsburg and Steitz, 1975; Hayes et al., 1975). While the location of the 5 S rRNA sequences relative to those of 16 S and 23 S rRNA are not yet known, the extremely close linkage between B. subtilis 23 S and 5 S rRNA genes suggests the order 16 S, 23 S, 5 S (Colli et al., 1971). In eukaryotes, 18 S, 28 S, and 5.8 S rRNA are in the primary transcript (Maden and Robertson, 1974) in the order 28 S, 5.8 S, 18 S (Speirs and Birnstiel, 1974). Neurospora mitochondria also synthesize a precursor containing 25 S and 19 S rRNA (Kuriyama and Luck, 1973). No high molecular weight precursor to chloroplast rRNA has been found as yet (Miller and McMahon, 1974), though the existence of such a molecule has been postulated, on the basis of kinetic evidence, with 16 S rRNA transcribed first (Surzycki and Rochaix, 1971).

The size of the primary transcription unit increases throughout evolution (Perry et al., 1970; Schibler et al., 1975). In E. coli the precursor has a molecular weight of 2×10^6 (Dunn and Studier, 1973; Nikolaev et al., 1973, 1974); in Neurospora mitochondria, 2.4×10^6 (Kuriyama and Luck, 1973); yeast, 2.5×10^6 (Udem and Warner, 1972); Xenopus, 2.6×10^6 (Loening et al., 1969); $3.7–3.9 \times 10^6$ in birds (Schibler et al., 1975); and HeLa, 4×10^6 (Weinberg et al., 1967).

The rRNA precursor contains single copies of the rRNA molecules, indicating a separation between adjacent tandem genes coding for rRNA. This has been confirmed by electron microscopy and isopycnic centrifugation of rDNA. In bacteria, the intergenic spacers are variable in length (Chow and Davidson, 1973; Deonier et al., 1974) with large distances separating some of the rRNA genes (Miller et al., 1970; Purdom et al., 1970) and nonribosomal genes interspersed among the rRNA genes (Miller et al., 1970; Deonier et al., 1974; Ohtsubo et al., 1974).

In eukaryotes, on the other hand, the rRNA genes are closely linked, in a regular array, with nontranscribed spacer regions alternating with the transcribed region, which is comprised of both sequences coding for mature RNA and nonutilized precursor segments (Miller and Beatty, 1969; Birnstiel et al., 1968). The spacer regions are similar, though not identical, in size and sequence in Xenopus laevis (Wensink and Brown, 1971; Wellauer and Reeder, 1975). However, when rDNA from two genetically compatible species, *X. laevis* and *X. mulleri,* are compared, the nontranscribed spacer regions have less than 10% homology, even though the transcribed segments, by all parameters, seem identical (Brown *et al.,* 1972; Wellauer and Reeder, 1975). This graphically illustrates the constraints against mutational change in rRNA. One would expect, however, some similarity in promoter sequences which should exist in the nontranscribed regions, unless RNA polymerase molecules move along the nontranscribed regions and reinitiate rRNA synthesis without leaving the DNA.

Though genes for eukaryote 5 S rRNA are clustered (Brown *et al.,* 1971; Wimber and Steffensen, 1970; Johnson *et al.,* 1974), the mature eukaryote molecule has a 5'-terminal ppp-purine and must be a primary transcript (Jordan *et al.,* 1974). In *Xenopus,* the 5 S rRNA gene sequences comprise one-seventh to one-eighth of the repeated sequences in the 5 S DNA cluster. Marked heterogeneity is observed in the 5 S spacers, differing from the situation in the 28 S–18 S spacer region (Brown and Sugimoto, 1973). Prokaryote pre-5 S RNA, coming from a larger transcript, has additional sequences at the 5'-end: *E. coli* with three additional nucleotides (Freunteun *et al.,* 1972) and *B. subtilis* with approximately 60 extra nucleotides at the 5'-end of the molecule (Pace *et al.,* 1973).

16 S rRNA is at the 5'-end of the primary rRNA transcript in prokaryotes (Doolittle and Pace, 1971; Pato and von Meyenburg, 1970; Kossman *et al.,* 1971). While enzymatic digestion studies and partial denaturation maps of mouse, human, and *Xenopus* rRNA precursor suggest that 28 S rRNA is at the 5'-end of the primary transcript (Perry and Kelley, 1972; Wellauer and Dawid, 1973), *in vivo* kinetic studies utilizing premature termination of rRNA precursors by cordycepin (Siev *et al.,* 1969) or ultraviolet radiation (Hackett and Sauerbier, 1975) have placed 18 S rRNA at the 5'-end of the precursor. The latter conclusion was also reached in *in vitro* studies of rRNA precursor transcription in mammals (Liau and Hurlbert, 1975) and *Xenopus* (Reeder and Brown, 1970; Hecht and Birnsteil, 1972).

5. *Maturation and Processing of rRNA*

In *E. coli,* 22% of the rRNA precursor sequences are removed during maturation of the rRNA's (Nikolaev *et al.,* 1974), while the corresponding figure is over 40% in HeLa cells (Loening *et al.,* 1969). Maturation of rRNA involves a series of nucleolytic cleavages giving rise to successively smaller molecules and modifications in both pro- and eukaryotes, in a manner similar to tRNA precursor processing (Attardi and Amaldi, 1970; Maden, 1971).

Under normal conditions, the primary rRNA transcript is not observed in *E. coli*. With the isolation of a *ts* RNase III mutant in *E. coli* (Kindler *et al.*, 1973), it was possible to demonstrate the high molecular weight precursor at the conditional temperature. After the initial cleavage of the precursor which, *in vitro* and presumably *in vivo,* is mediated by RNase III (Dunn and Studier, 1973; Nikolaev *et al.*, 1974), smaller precursors for 16 S, 23 S, and 5 S rRNA are seen (Hecht and Woese, 1968; Adesnik and Levinthal, 1969) which undergo final maturation in the form of ribonucleoprotein particles, precursors to the ribosomal subunits. RNase II has been implicated in the conversion of p16 to m16 as a *ts* RNase II mutant of *E. coli* accumulates ribosomal subunit precursors at the restricting temperature, and p16 is found in the 30 S subunit precursor. Mature 23 S rRNA is found under these conditions (Yuki, 1971; Corte *et al.*, 1971), but the involvement of this enzyme has been questioned (Weatherford *et al.*, 1972; Hayes and Vasseur, 1973). In any case, purified *coli* extract catalyzes the *in vitro* p16 → m16 cleavage (Corte *et al.*, 1971).

The enzymology of the cleavage of the eukaryotic rRNA precursor is not as clearly understood though there is some evidence for a 3'-exonuclease in L cell precursor maturation (Perry and Kelley, 1972). Intercalating agents inhibit the processing of HeLa 45 S rRNA, indicating helical regions are involved in the region of cleavage (Snyder *et al.*, 1971). RNase III acts on helical RNA regions (Robertson *et al.*, 1968), so the cleavage mechanisms in eu- and prokaryotes may be similar.

A *ts* mutant in BHK cells has been isolated which accumulates 32 S rRNA and does not form mature 28 S rRNA (Toniolo *et al.*, 1973). This mutation is complementable (Toniolo and Basilico, 1974) and could either be in a specific nuclease or modifying enzyme.

Since it has been shown that *Bombyx* pre-tRNA can be specifically cleaved by *E. coli* extracts to give "4 S RNA" (Chen and Siddiqi, 1973), it would be of interest to test the activity of purified RNase III on 45 S rRNA and RNase II on 32 S rRNA.

As with tRNA, the extent of rRNA modification increases as genetic complexity increases. In *E. coli* 16 S rRNA, 10 methylated bases are found, mostly clustered at the 3'-end of the molecule, and 24 residues. 23 S rRNA has approximately 14 methylated residues (Fellner *et al.*, 1972). Most of these modifications are at the base level. Yeast has 55 methyl groups and HeLa has 110 methylated sites, all but 5 on ribose groups (Maden and Salim, 1974). Yeast 5.8 S rRNA has one Ψ group (Rubin, 1973), while the corresponding HeLa molecule has one methyl and one Ψ residue (Maden and Salim, 1974). All 5 S rRNA molecules are unmodified. The rRNA's of chloroplasts have a high ratio of base to ribose methylation (Rijven and Zwar, 1973), as do *Neurospora* mitochondrial rRNA's

(Kuriyama and Luck, 1974), again indicating the relationship of these organelles to prokaryotes. Mitochondria from yeast, on the other hand, show only two methylated riboses in 21 S rRNA and no modifications in 15 S rRNA (Klootwijk et al., 1975).

The function of the modifications in rRNA is, at present, not known in *E. coli*. Resistance to the antibiotic kasugamycin is caused by a mutationally altered 16 S rRNA methylase which normally methylates the sequence m_2^6 Am_2^6A-C-C-U-G found near the 3'-end of the molecule (Helser et al., 1972). The methylase gene maps near *leu* (Sparling, 1970). The presence of methyl groups in the sequence renders the bacterium sensitive to the antibiotic and lack of methylation causes resistance. However, the resistant mutants show no alterations in growth or protein synthesis. There must be an important function for this sequence as similar sequences with the same methylated bases are found in yeast 17 S (Klootwijk et al., 1972), and HeLa 18 S rRNA (Maden and Salim, 1974). The methylation of this sequence occurs during ribosome assembly, and is inhibited, during *in vitro* reconstitution, by the proteins which are incorporated in the final stages of 30 S ribosome assembly (Thammana and Held, 1974).

A mutant, lacking a methylase which produces 1-methyl-G in 23 S rRNA (mapping between 77' and 82') grows relatively normally (Björk and Isaksson, 1970). Induced resistance to erythromycin in *Staphylococcus aureus* is caused by the methylation of A-A-A-G sequences in 23 S rRNA (Lai et al., 1973). This modification has no obvious effect on growth. Interestingly, all gram-positive bacteria examined do not possess N^6-methylated adenines in 23 S rRNA, while all gram-negative strains and clinically resistant (to erythromycin) *S. aureus* and *Streptococcus pyogenes* do possess N^6-methylated adenines in A-A-A-G sequences (Tanaka and Weisblum, 1975).

On the other hand, the *"poky"* mutation in *Neurospora* mitochondria, phenotypically demonstrating slow growth and a deficiency of mitochondrial cytochromes, is characterized by an undermethylation of precursor rRNA's (70% of 25 S, 55% of 17 S normal methylation). The smaller ribosomal subunit is subsequently not formed as the 17 S precursor is degraded (Kuriyama and Luck, 1974). This is similar to the situation in *E. coli* where an accumulation of rRNA precursors occurs during the amino acid starvation of a *rel* strain (Sypherd, 1968). In HeLa, deprivation of methionine, but not valine, causes an accumulation of the 32 S precursor and prevents formation of 28 S rRNA (Vaughan et al., 1967).

The precursor segments that are discarded during normal maturation of the HeLa precursor are not methylated (Maden and Salim, 1974).

In the absence of any specific function for rRNA (see below), it is difficult to assign any function for single modifications as is possible for the 3′ adjacent anticodon modifications in tRNA. The absence of methylation in excessed sequences, however, suggests that some of these modifications might be involved in the protection of the conserved RNA sequence from the action of the maturation nucleases, possibly by introducing conformational changes or base alterations. Methylation is involved in restriction and modification of prokaryote DNA's. Methylases will modify bases in DNA at or near sites where restriction enzymes normally cleave, thus rendering these sites insensitive to the action of these nucleases (Meselson *et al.*, 1972).

6. *Control of rRNA Synthesis*

The rate of cell growth and protein synthesis is determined by the amount of ribosomes and ultimately the level of rRNA in the cell (Neidhardt and Magasanik, 1960). It is therefore of crucial importance to the cell to be able to regulate the rate of rRNA synthesis. There are two general methods by which this could be effected: a change in the number of rRNA genes or regulation of initiation of rRNA transcription.

Amplification of rDNA occurs during oocyte development (see above), and other examples of variation in the rRNA gene content of eukaryotic cells during other phases of the life cycle have been described in *Drosophila* (Ritossa, 1972; Spear and Gall, 1973) and in *Tetrahymena* (Gall, 1974). During germination of *B. subtilis* spores an increase in the rate of rRNA synthesis is observed (Kobayashi *et al.*, 1965). This may be caused by a gene dosage effect as the major cluster of rRNA genes is located near the replication origin of the *B. subtilis* chromosome (Oishi and Sueoka, 1965; Dubnau *et al.*, 1965b).

Eukaryotes have different RNA polymerase activities, one of which, the α-amanitin resistant form I, is found in nucleoli and is presumed to be involved in rRNA synthesis (Roeder and Rutter, 1970; Gross and Pogo, 1974). In *E. coli* and *B. subtilis,* there is also evidence for multiple forms of rRNA polymerase, (Travers and Buckland, 1973; Diaz *et al.*, 1975) and it has been suggested that one of the forms transcribes rRNA genes.

RNA polymerase specificity is controlled both negatively and positively, e.g., *lac* and λ repressors bind to operator sites on DNA and prevent transcription of specific genes (Gilbert *et al.*, 1973; Ptashne, 1971), σ factor (Burgess *et al.*, 1969) and other positive control elements like the cyclic AMP binding protein enhance initiation of transcription at specific sites (de Crombrugghe *et al.*, 1971; Eron *et al.*, 1971).

These types of controls have been postulated for rRNA transcription as well. The phenomenon of stringent control has been extensively studied in this regard. As discussed above, starvation for an amino acid or lack of tRNA charging causes an immediate cessation of stable RNA synthesis, indicating a very tight coupling between translation and rRNA synthesis. The guanosine nucleotides ppGpp and pppGpp are produced *in vivo* upon commencement of amino acid starvation (Cashel and Gallant, 1969), or *in vitro* when an uncharged tRNA is bound to the A site of the ribosome in the presence of the appropriate mRNA codon (Haseltine and Block, 1973; Pederson *et al.,* 1973). A protein, the *relA* gene product, which is normally involved in protein synthesis and is associated with 50 S subunits, catalyzes the synthesis of the guanosine nucleotides (Hall and Gallant, 1972; Haseltine *et al.,* 1972). A functional 50 S ribosome, i.e., possessing a nonmutated *relC* protein, is also required (Friesen *et al.,* 1974). *In vitro* experiments have provided contradictory results in the effect of ppGpp on rRNA synthesis (Travers *et al.,* 1970; Haseltine, 1972). It has recently been suggested that the physical state of the DNA, affected by ionic strength and temperature, is important in these studies and that under proper conditions, in the presence of a protein factor, Ψr, ppGpp does specifically inhibit rRNA synthesis (Travers, 1973). This factor stimulates overall RNA synthesis *in vitro* (Haseltine, 1972) and seems to enhance rRNA synthesis under specific conditions (Travers, 1973). Ψr has been shown to be identical to the bacterial elongation factors EF-Tu and EF-Ts (Haseltine, 1972; Travers, 1973) which are, in turn, identical to host factors III and IV of the RNA phage Qβ replicase (Blumenthal *et al.,* 1972). EF-Tu and EF-Ts are involved in rRNA synthesis *in vivo*. A *ts* mutant of *E. coli* with a thermolabile EF-Ts does not synthesize rRNA at the restricting temperature, while mRNA synthesis is relatively unaffected (Kuwano *et al.,* 1973, 1974). The Qβ replicase synthesized in this mutant is also thermolabile and the altered enzyme does not bind GTP and cannot initiate RNA chains at the restricting temperature, while elongation is unaffected (K. Hori *et al.,* 1974). EF-Tu and EF-Ts are associated with one of the forms of RNA polymerase in *E. coli* and *B. subtilis* (Travers and Buckland, 1973; Diaz *et al.,* 1975). RNA synthesis *in vitro,* in the presence of EF-Tu-Ts, shows greater pppG initiation of RNA chains than in its absence (Travers, 1973). [A *ts* mutant of *E. coli* with a thermolabile EF-Tu, selected by resistance to RNA phage MS2, has been described (Lupker *et al.,* 1974). It would be of interest to study rRNA synthesis in this mutant.] A hypothesis that couples rRNA synthesis to protein synthesis by means of EF-Tu-Ts and the product of the *relA* and *relC* genes is extremely attractive and would explain many diverse observations. Before it is proved, however, unambiquous data from *in vitro* studies must be obtained. In this regard, recent experiments, utilizing an *in vitro* system, originally developed to study DNA directed protein synthesis, have demonstrated a very specific effect of ppGpp (Reiness *et al.,* 1975).

Under conditions where up to 25% of the *in vitro* transcripts are rRNA, ppGpp inhibits rRNA synthesis fivefold, has no effect on tRNA synthesis, and stimulates transcription of the *trp* and *lac* operons. Maximal expression of the *his* operon in *Salmonella, in vitro*, is also strongly dependent on ppGpp addition (Stephens *et al.*, 1975). Purification of similar systems should clarify the role of such factors as EF-Tu and EF-Ts and ppGpp in rRNA transcription. To again stress the relationship between transcription and translation, it has recently been shown that the third host encoded factor in the Qβ replicase is 30 S ribosomal protein, S1 (Wahba *et al.*, 1974).

Alterations of *B. subtilis* RNA polymerase during sporulation resulting in a cessation of rRNA synthesis have been described (Hussey *et al.*, 1971, 1972; Linn *et al.*, 1973). However, it also has been reported that rRNA synthesis does not stop during sporulation (Bonamy *et al.*, 1973; Testa and Rudner, 1975), and that alterations in RNA polymerases during sporulation are due to protease degradation (L. Hirschbein, personal communication, 1974).

Yeast and ascites cells show the stringent response when deprived of amino acids (Franze-Fernández and Pogo, 1971; Gross and Pogo, 1974). rRNA synthesis is more drastically affected than mRNA synthesis, and the regulation seems to occur at the level of RNA polymerase binding to DNA, which is in keeping with some of the models for stringent control in bacteria. One can postulate factors binding to DNA or to the polymerase which facilitate or inhibit polymerase interaction with promoter sites for rRNA.

ts mutants in yeast rRNA production have been described which fall into nine complementation groups (Hartwell *et al.*, 1970). These mutations seem to be at the level of processing, not synthesis, and show a very stringent control of rRNA synthesis so that no rRNA precursors accumulate (Warner and Udem, 1972).

7. Evolution of rRNA Genes

A comparison of the chromosomal organization of genes for rRNA in prokaryotes and eukaryotes suggests a similar mechanism for rDNA amplification during evolution as has been postulated for tRNA, previously. Bacteria have less than ten genes for rRNA, and while they show some clustering, they are separated by regions of the chromosome in which nonribosomal genes are found (Miller *et al.*, 1970). In most eukaryotes, the homologous genes for 28:5.8:18 S rRNA are in one site except for those cases where nucleolar organizers are found on more than one chromosome (Ferguson-Smith, 1964). In *Xenopus*, where rDNA has been most extensively studied, the rRNA genes are separated from each other by spacers of similar length and base sequence (Wensink and Brown, 1971), which however are not identical (Wellauer *et al.*, 1974).

The needs of larger eukaryotic cells for more proteins and consequently more ribosomes could have selected for duplications in rRNA genes, which would increase the amounts of rRNA synthesized. In the laboratory, overproducing enzyme mutants frequently result from gene duplication (Folk and Berg, 1971; Rigby *et al.*, 1974). There would be high selective pressure against mutation in the redundant rRNA genes because of the multiple interactions the rRNA molecules engage in with other components of the translational apparatus. This could help explain the homogeneity of rRNA genes within a species, and the nonidentical nature of the repeated spacers in *Xenopus* rDNA. These regions show marked evolutionary drift when genetically compatible organisms are compared, even though the rRNA genes themselves show complete homology (Brown *et al.*, 1972; Wellauer and Reeder, 1975). It is clear also that sequences in different rRNA molecules do not have to be identical to react with the same ribosomal proteins to give a functioning ribosome. 16 S rRNA from *Azotobacter* or *B. stearothermophilus* interacts with *E. coli* ribosomal proteins to give biologically active 30 S ribosomes (Nomura *et al.*, 1968). While it is possible that small, conserved regions of homology are responsible, conformational resemblance may also play an important role. Similar nuclease digestion patterns are obtained from diverse pro- and eukaryotic rRNA's, suggesting similar conformations (Pinder *et al.*, 1969). The maintenance of identical sequences in rDNA has been explained by theories of unequal crossing over, master-slave correction, or reduction magnification (Birnstiel and Grunstein, 1972), but the question has not been answered definitively.

No explanation at present exists for the separation of the genes for 5 S rRNA from the 18 S and 28 S rRNA genes in eukaryotes. 5 S rRNA is found in all eukaryotic larger ribosomal subunits and is of similar size and conformation as its bacterial counterpart. Deletion of 5 S rRNA genes in *Drosophila* cause "bobbed" phenotypes as do 18 S and 28 S rRNA gene deletions, indicating a ribosome function for 5 S rRNA (Procunier and Tartof, 1975). Eukaryotic 5 S rRNA interacts with specific ribosomal proteins to give complexes with ATPase and GTPase activities (Grummt and Grummt, 1974) as does bacterial 5 S rRNA (Horne and Erdmann, 1973), so it does seem to have similar physiological function. On the other hand, most eukaryotic 5 S rRNA molecules do not have a G-A-A-C sequence around position 40. In addition, eukaryote 5 S rRNA's are not active in the reconstitution of biologically active *B. stearothermophilus* 50 S ribosomes, while prokaryote 5 S's are (Wrede and Erdmann, 1973). It has been suggested that 5.8 S rRNA, which is also found in all eukaryotic ribosomes and is part of the eukaryotic primary rRNA transcript, may actually be the functional "5 S" rRNA in higher organisms (Pace, 1973). The primary sequences of yeast (Rubin, 1973) and mammalian 5.8 S rRNA (Nazar *et al.*, 1975) indicate, however, that the molecules are much longer (158 nucleotides, as opposed to approximately 120) and have no immediately apparent homology to known 5 S rRNA's, while showing 15% homology to each other. Resolution of this question must await the establishment of a ribosomal reconstitution system in yeast or other eukaryotes.

B. Ribosomal Proteins

The structure and genetics of bacterial ribosomal proteins has been reviewed previously (Davies and Nomura, 1972; Wittmann, 1972; Osawa *et al.*, 1972; Nomura *et al.*, 1974).

In recent years our knowledge of the structure and genetics of ribosomal proteins has increased greatly, in large measure due to the refinement of sensitive analytic methods such as two-dimensional gel electrophoresis (Kaltschmidt and Wittmann, 1970) and carboxymethylcellulose column chromatography (Otaka *et al.*, 1968), and immunological techniques (Stöffler, 1974), and to the development of the ribosomal reconstitution system (Traub and Nomura, 1968; Fahnestock *et al.*, 1973).

1. Mapping of Genes for Ribosomal Proteins

a. Prokaryotes. Several genes for ribosomal proteins have been mapped in *E. coli* and in *B. subtilis*, using antibiotic resistance or dependence selection or heterospecific transfer of natural variations in ribosomal proteins. The majority of ribosomal protein genes or genes for resistance to ribosomally active antibiotics is clustered, and a similar gene order is observed in both organisms (Fig. 3).

The genetic map of the major ribosomal protein gene region in *E. coli* has been compiled from many sources and deserves a few comments. The position of the *eryA* gene, coding for the 50 S ribosomal protein L4 (Takata *et al.*, 1970) was reported to be between *spc* and *aroE* (Takata *et al.*, 1970; Dekio, 1971). Experiments on the *in vivo* expression of antibiotic resistance genes also indicated the order *str, spc, ery* (Nomura and Engbaek, 1972). However, recent reports indicate that *eryA* and a closely linked erythromycin resistance gene, *eryB*, which codes for 50 S ribosomal protein L22 (Wittmann *et al.*, 1973), are between *strA* and *spcA* (Sparling and Blackman, 1973; Wittmann *et al.*, 1973; Brown and Apirion, 1974; Pardo and Rosset, 1974) in the order *strA, eryA, eryB, spcA* (Brown and Apirion, 1974). The *ery* marker used in the analysis of ribosomal protein gene expression was independently obtained by mu-

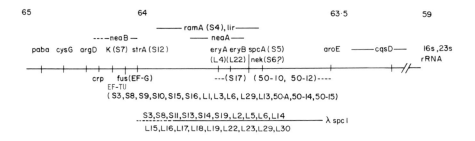

Fig. 3 Genetic map of major ribosomal protein gene regions in *E. coli* and *B. subtilis*. The map positions of ribosomal protein genes, antibiotic resistance, and nutritional markers were taken from the references cited in the text and Taylor and Trotter (1972). Other mapping data for *B. subtilis*, e.g., *aroI*, is from Harford (1975) and Lepesant-Kejzlarová *et al.* (1975). The maps have been drawn to maximize the similarities between the two organisms, and some distances have not been drawn to scale, e.g., the area between *str* and *spc* has been expanded. The upper figure represents *E. coli*, while *B. subtilis* is depicted in the lower one. The ribosomal proteins whose synthesis is stimulated *in vivo* and *in vitro* by λ*spc* are also indicated.

tagenesis and was not characterized further (Nomura and Engbaek, 1972). It could conceivably be a mutation in a different gene. This is not a trivial point, as an intermingling of 30 S and 50 S ribosomal proteins would eliminate various hypotheses of ribosome assembly (see below). The position of the *ramA* gene is still unclear as it has been placed between *strA* and *spcA* (Birge and Kurland, 1969; Brown and Apirion, 1974), or between *spcA* and *aroE* (Rosset and Gorini, 1969). In *E. coli,* the genes for at least 17 ribosomal proteins map in the *aroE-argD* region (Dekio, 1971), and λ*fus*, a specialized transducing bacteriophage carrying the bacterial segment *aroE-fus*, stimulated the *in vivo* synthesis of 30 ribosomal proteins and elongation factors EF-G and EF-Tu (Jaskunas *et al.*, 1975a,c). Some of these genes have been positioned relative to other markers (Fig. 3). Other genes conferring resistance to antibiotics which affect ribosomes have been mapped in *E. coli* in this region as well, e.g., *nek, lir* (Apirion, 1967; Apirion and Schlessinger, 1968; Brown and Apirion, 1974), and *neaA,* and *neaB,* which are different from *nek*

(Cannon *et al.*, 1974), but it is not clear which, if any, of the ribosomal components are affected by these mutations, though *nek* may code for S6 (Hitz *et al.*, 1975).

Evidence for a second cluster of ribosomal protein genes has appeared (Fig. 2). λ*drif* transducing bacteriophages, which carry *E. coli* genes in the segment around 79', including the genes for the RNA polymerase subunits β and β' (Kirschbaum and Scaife, 1974), can stimulate the *in vivo* synthesis of 50 S ribosomal proteins L1, L7/L12, L8, L10, and L11 (Lindahl *et al.*, 1975; Watson *et al.*, 1975) and elongation factor EF-Tu (Jaskunas *et al.*, 1975c) and the *in vitro* synthesis of 16 S and 23 S rRNA (Lindahl *et al.*, 1975). The latter may reflect transcription of *rrnB*, which may be between *argH* and *glyT* (Deonier *et al.*, 1974; Ohtsubo *et al.*, 1974). The gene for L7/L12 had been previously isolated as a *ts* marker affecting a 50 S ribosomal protein (Flaks *et al.*, 1966), but it has now been resolved into two mutations, *rpyL* (coding for L7/L12) and *rts,* both of which are very close to *relC* (Watson *et al.*, 1975). Two mutations, showing a cold-sensitive phenotype, which affect ribosomal maturation have also been mapped in this region. One of these markers, *rimA,* is between *pyrE* and *ilvD* (72'–75') and the other, *rimD,* if not allelic, is very close by (Bryant and Sypherd, 1974). A mutation showing defective accumulation of ribosomal precursor particles has been mapped near *xyl* (Turncock, 1969), and as previously discussed, an rRNA methylase genetic determinant is found between 77' and 82' (Björk and Isaksson, 1970). A recent report indicates that a new, unmapped kusagamycin resistance mutation, *ksgB,* alters the 30 S ribosomal protein, S2 (Okuyama *et al.*, 1974). Another mutation affecting S2 suppresses the cold sensitivity of an S5 (spectinomycin resistance) mutation. It maps at 3.5' and is linked to *polC* (Nashimoto and Uchida, 1975). The mutation is recessive in partial diploids, suggesting it codes for an S2 modifer. Its relation to *ksgB* is presently unknown. A nonsense mutation affecting the ribosomal dissociation factor (Herzog *et al.*, 1971) and a *ts* mutation which affects protein chain termination (Ganoza *et al.*, 1973) have been isolated, as well, in *E. coli* but are also unmapped. The structural gene for ribosomal protein S18 maps at 84' (Bollen *et al.*, 1973; DeWilde *et al.*, 1974) far from the two major ribosomal protein gene clusters.

A ribosomal mutation, linked to *thr* (O') which restricts the leakiness of *amber* mutations (Garvin and Gorini, 1975), is very close to *ksgA,* the gene for the 16 S rRNA methylase (Helser *et al.*, 1972), and *ramB,* which determines a modifier of ribosomal protein S4 (Zimmerman *et al.*, 1973). A mutational alteration of ribosomal protein S20, which suppresses a *ts* alanyl-tRNA synthetase mutation, is also linked to *thr* (Böck *et al.*, 1974). It is recessive in partial diploids, suggesting it codes for an S20 modifier.

Thus, the region around O' seems to have a cluster of ribosome modifying enzymes.

Recent experiments in *B. subtilis* (S. Osawa, personal communication, 1974) have confirmed the previously described map order of the genes in the *strA* region (Goldthwaite *et al.*, 1970; Harford and Sueoka, 1970; Goldthwaite and Smith, 1972a), and several ribosomal protein alterations have been related to specific antibiotic resistance mutations, i.e., *ery* (Tanaka *et al.*, 1973), *spcA* (Kimura *et al.*, 1973) and *cml*, chloramphenicol resistance (Osawa *et al.*, 1973). Several nonallelic *cml* loci are observed. This is not surprising since affinity labeling studies in *E. coli* have shown that chloramphenicol binds to 50 S ribosomal proteins L6 (Pongs *et al.*, 1973), and L2 and L27 (Sonenberg *et al.*, 1973). It has also been demonstrated that the *fus* gene, determining resistance to fusidic acid, as in *E. coli* (Kuwano *et al.*, 1971; Tanaka *et al.*, 1971), codes for elongation factor EF-G (Aharonowitz and Ron, 1972; Kimura *et al.*, 1974). The thiostrepton (bryamycin) resistance marker, *thi*, has been mapped between *cysA14* and *strA*, near *mic* (Smith and Smith, 1973; Pestka *et al.*, 1976), and it alters 50 S ribosomal protein L11 (Wienen *et al.*, 1976), which, on the basis of its two-dimensional gel electrophoresis migration, is related to *E. coli* L11. Thiostrepton resistance markers have not been obtained in *E. coli* because of natural resistance, but it has been reported that the antibiotic binds to *E. coli* ribosomal protein L11 (Highland *et al.*, 1974). However, *E. coli* mutants, resistant to thiopeptin, an antibiotic which is structurally and functionally similar to thiostrepton and siomycin, have been isolated. 50 S ribosomes are altered in these mutants (Liou *et al.*, 1973). *thi* is extremely close to *rif*, between *cysA14* and *strA* (Harford and Sueoka, 1970; Haworth and Brown, 1973), which is similar to the situation in *E. coli* where *rif* and the gene for L11 are on the same segment of bacterial DNA carried by λ*drif18* (Fig. 2) (Lindahl *et al.*, 1975; Watson *et al.*, 1975). With the previous mapping of two ribosomal protein genes, *strA* (Smith *et al.*, 1969) and the gene coding for the 6633 protein, 50-G (Smith *et al.*, 1969; Tanaka *et al.*, 1973), it is now known that at least nine ribosomal protein genes and one coding for a protein synthesis factor are clustered in this region of the *B. subtilis* chromosome. Recently, it has been shown that a genetic determinant for EF-Tu maps in this region, closely linked to *fus* (EF-G), in the order *strA, fus, EF-Tu* (Dubnau *et al.*, 1976).

There is a second region in the *B. subtilis* genome, possessing antibiotic resistance markers, which confer high level resistance to spectinomycin, *spcB* (Harford and Sueoka, 1970), and streptomycin, *strB* (Staal and Hoch, 1972). These mutations are both close to *ura*, but it is not known whether they are mutations in ribosomal components. *strB* and the unlinked marker, *strC,* show conditional resistance, becoming sensitive to streptomycin during sporulation and regaining resistance after germination (Staal and Hoch, 1972). These alleles are mapped near other sporulation genes and may have a specific role in the sporulation process. In this regard, a *strA* mutation, affecting ribosomes, renders sporulation in *B. subtilis* temperature-sensitive, while vegetative growth is normal (Leighton, 1974). *strB* and *spcB* loci, not involved with ribosomal components, also have been mapped in *Salmonella* (Yamada and Davies, 1971), while *strA* and *spcA,* which alter ribosomes, are closely linked to each other (Yamada and Davies, 1971) and to *nek* (Tyler and Ingraham, 1973) in this organism.

Antibiotic resistance markers have been described in other bacteria. In *Pneumococcus,* erythromycin and streptomycin resistance have been shown to be ribosomal alterations, and the genes are linked (Stuart and Ravin, 1968; Ravin *et al.,* 1969). *spcA* and *strA,* high level resistance mutations; map in the same part of the *Streptomyces coelicolor* genome (Hopwood *et al.,* 1973). Genetic loci for high level resistance to several antibiotics have been mapped in *Neisseria gonorrhoeae.* The genes are linked, in the order *rif* (rifampin resistance), *str, tet* (tetracycline resistance), *chl* (chloramphenicol resistance), and *spc* (Sarubbi *et al.,* 1974).

Thus, in all bacteria there seems to be a cluster of ribosomal protein genes. This has obvious implications for the regulation of ribosomal protein synthesis, as has been suggested in *E. coli* (Nomura and Engbaek, 1972).

b. Eukaryotes. Mutations which affect yeast ribosomes have been isolated by means of antibiotic resistance. A cyclohexamide resistance gene, *cyh-2,* shows altered 60 S ribosomes (Rao and Grollman, 1967) and is located in chromosome VII (Mortimer and Hawthorne, 1973). Several other cyclohexamide resistant loci have been reported, which show *in vitro* ribosomal resistance as well (Cooper *et al.,* 1967; Vomvoyanni, 1974). Altered ribosomes, possibly 40 S subunits, are also found in a mutant resistant to cryptopleurine, *cry-1,* which maps on chromosome III (Skogerson *et al.,* 1973) and in a strain resistant to trichodermin, an antibiotic which affects elongation and termination, probably by interfering with peptidyl transferase. This mutation is unlinked to the others (Schindler *et al.,* 1974). Two *ts* mutations which affect elongation (Hartwell and McLaughlin, 1968b), *prt-2* and *prt-3,* map on chromosomes VII and V, respectively (Mortimer and Hawthorne, 1973). A mutation, conferring streptomycin sensitivity to 40 S ribosomes has recently been isolated, but its map position is, as yet, unknown (Bayliss and Ingraham, 1974).

Therefore, the genes for ribosomal proteins or factors involved in protein synthesis, thus far analyzed in yeast, are not clustered and are not linked to the majority of rRNA genes on chromosome I (Finkelstein *et al.*, 1972).

On the other hand, by means of interspecific mating and analysis of the transfer of natural ribosomal protein variations between two *Drosophila* species, it has been shown that at least seven ribosomal protein genes (three 40 S and four 60 S) are in a cluster at 19E, on the X chromosome, close to the nucleolar organizer region (Steffensen, 1973).

c. Organelles. Neurospora mitochondrial ribosomal proteins, of which there are approximately 60 (Lizardi and Luck, 1972), are predominantly made on cytoplasmic ribosomes. This is shown by the insensitivity of the synthesis of the separable mitochondrial ribosomal proteins to chloramphenicol, which specifically inhibits mitochondrial protein synthesis, and the sensitivity of the process to anisomycin, which affects cytoplasmic ribosomes (Lizardi and Luck, 1972). This confirms and extends previous work in *Neurospora* (Küntzel, 1969) and yeast (Schmitt, 1971), which analyzed incorporation of amino acid into total mitochondrial ribosomal protein.

However, it is not clear where the genetic information for the mitochondrial ribosomal proteins is located. Mutations to antibiotic resistance have been isolated in *S. cerevesiae* which show uniparental, non-Mendelian inheritance and have been mapped on the mitochondrial genome (Thomas and Wilkie, 1968; Grivell *et al.*, 1973). These mutations, conferring resistance to the 50 S active antibiotics, chloramphenicol, erythromycin, and spiramycin, are genetically distinct and are expressed at the level of the ribosome, as measured by the resistance of *in vitro* protein synthesis to the antibiotics (Grivell *et al.*, 1973). The specific nature of these alterations has not been elucidated, and it is not known whether they are amino acid changes in ribosomal proteins or, as discussed above in the case of an *ery*[r] mutation, due to a base change in the mitochondrial 23 S rRNA gene (Faye *et al.*, 1974). In one case, a Mendelian mutation to erythromycin resistance causes altered mitochondrial ribosomes (Grivell *et al.*, 1971).

Erythromycin resistance and chloramphenicol resistance in paramecium is also cytoplasmic and an altered mitochondrial ribosomal protein is found associated with a specific *ery* mutation (Beale *et al.*, 1972). Resistance to mikamycin, which inhibits bacterial and not eukaryote cytoplasmic protein synthesis (Yamaguchi *et al.*, 1966), is cytoplasmically determined in yeast (Howell *et al.*, 1974) and paramecium (Beale, 1973), but no alterations in mitochondrial ribosomal proteins have been described in these mutants. Mitochondrial EF-T and EF-G seem to be coded for by nuclear genes, as mutants of *S. cerevisae*, lacking mitochondrial DNA, contain normal levels of these factors (Richter, 1971).

Chlamydomonas reinhardii has been used almost exclusively for the study of the genetics of chloroplasts. It has been clearly demonstrated that several non-Mendelian mutations to antibiotic resistance cause alterations in chloroplast ribosomal proteins (Conde *et al.*, 1975). Mutations to streptomycin resistance and dependence cause accumulation of 50 S subunits (Gilham *et al.*, 1970), and a separate *str* mutation shows an altered chloroplast 30 S ribosomal protein on carboxymethylcellulose columns (Ohta *et al.*, 1974). Two independently isolated spectinomycin resistant mutants show antibiotic resistance (1) at the 30 S ribosome level (Schlanger and Sager, 1974) and (2) the absence of a 30 S ribosomal protein (Boynton *et al.*, 1973). An erythromycin-resistant mutant, *ery-u1*, shows an altered 50 S ribosomal protein (Mets and Bogorad, 1972), and carbomycin, cleocin, and neamine resistance, *in vitro*, have been shown to be due to altered 50 S, 50 S and 30 S ribosomes, respectively (Schlanger and Sager, 1974). Some of these mutations have been mapped on the chloroplast genome, in the order *str, nea* (*ery, cleo, carb*), *spc* (Sager, 1972), which is similar to the gene sequence found in *B. subtilis* and *E. coli* (Fig. 2).

Though some of the chloroplast ribosomal proteins seem to be coded for by chloroplast genes, the nucleus plays an important role in the biogenesis of the organelles' ribosomes. Two erythromycin-resistant mutants which are nuclear in origin, *ery-m1* and *ery-m2*, and unlinked (Mets and Bogorad, 1971) cause alterations of two chloroplast 50 S ribosomal proteins (Mets and Bogorad, 1972; Davidson *et al.*, 1974) which are different from the protein altered by the cytoplasmic *ery-u1* mutation.

As in the case of mitochondria, it seems that chloroplast ribosomal proteins, even though some are coded for by mitochondrial genes, are synthesized in the cytoplasm. Cyclohexamide, but not chloramphenicol, inhibits amino acid incorporation into ribosomal structural proteins (Honeycutt and Margulies, 1973). Contradictory results on this point have been obtained in *Ochromonas* (Ellis and Hartley, 1971), but until detailed analyses of synthesis of individual r-proteins are made in these systems, as in *Neurospora*, the question will remain unresolved.

2. Synthesis of Ribosomal Proteins

In *E. coli*, all ribosomal proteins are synthesized coordinately, at a rate which is proportional to the generation time (Dennis, 1974). The level of expression of the ribosomal protein genes is controlled by the amount of mRNA available for ribosomal protein synthesis (Dennis and Nomura, 1975). The synthesis of EF-G and EF-T is also regulated so that there are equivalent amounts of these factors and ribosomes (Gordon, 1970). The *fus* to *spcA* segment contains up to 30 ribosomal protein genes and the

EF-G and EF-Tu genetic determinants. It had been suggested that this region is transcribed as a polycistronic mRNA (Nomura and Engbaek, 1972; Cabezón *et al.*, 1974). These experiments utilized the polar effect of *E. coli* bacteriophage Mu insertion on the expression of ribosomal protein genes in partial diploids. However, a more recent study has shown that the seemingly polar effects of Mu insertion are sometimes due to deletions and that *strA* and *spcA* are in separate transcription units (Cabezón *et al.*, 1975). The bacterial insertion mutation IS2 reduces the expression of S5, the *spcA* protein, while not affecting the synthesis of S4, even though their genetic determinants are closely linked. The fact that transcription seems to be in the direction S5→S4 suggests that these two genes are in separate operons (Jaskunas *et al.*, 1975b). In addition, all the ribosomal protein genes cannot be in one cistron, as there is a second cluster of ribosomal protein determinants near 79′ (see above). The genetic determinant for EF-Ts is also unlinked to the *strA* region (Gordon *et al.*, 1972). Kinetic studies of the rates of ribosomal protein synthesis after rifampicin (Molin *et al.*, 1974) or ultraviolet treatment (Hirsch-Kauffman *et al.*, 1975) indicate that there are multiple ribosomal protein operons in *E. coli* with an average length sufficient to code for approximately 10 ribosomal proteins. Obviously, there must be an overall regulatory control of these unlinked loci, to coordinate the synthesis of ribosomal proteins and protein synthesis factors. Since the pool of free ribosomal proteins is very low, varying from 1% to 3% at different growth rates (Gausig, 1974), the synthesis of these proteins must be coordinate with rRNA synthesis, also implying intercistronic regulatory elements, and it has recently been shown that ribosomal protein gene expression is under stringent control, being either directly or indirectly controlled by charged tRNA levels as is rRNA (Dennis and Nomura, 1974). Ribosomal proteins have been synthesized in a cell free system (Kaltschmidt *et al.*, 1974), and this should be a useful system for the study of regulatory factors involved in ribosomal protein synthesis. In addition, it is reported that in the same type of cell free system, using a plasmid enriched for rRNA genes, rRNA synthesis was observed, but not ribosomal protein synthesis (Kaltschmidt *et al.*, 1974). This seems to eliminate completely the hypothesis that rRNA codes for ribosomal proteins (Muto, 1968).

In eukaryotes, diverse observations have been made regarding control of ribosomal protein synthesis. In yeast, the synthesis and assembly of ribosomal components is very tightly coordinated, so that if rRNA synthesis is stopped, ribosomal proteins are not made either (Warner and Udem, 1972). On the other hand, *in vitro* synthesis of ribosomal proteins in L cells continues after inhibition of rRNA synthesis, suggesting that these processes can be separated (Craig and Perry, 1971).

As described above, the ribosomal proteins of organellar ribosomes are made, for the most part, on cytoplasmic ribosomes.

3. Genetics of Ribosomal Assembly

a. Prokaryotes. Mutations in the assembly of ribosomes have been found in *E. coli* and *Salmonella*. This type of lesion could result from either (1) an alteration in a ribosomal structural protein which would prevent proper assembly and would be dominant or partially dominant or (2) by the inactivation of a modifying enzyme, e.g., a methylase or nuclease, which would be recessive, and both types of mutations have been observed. A high proportion of cold-sensitive mutants show defects in ribosome assembly (Tai *et al.*, 1969; Guthrie *et al.*, 1969; Tai and Ingraham, 1971). Some of these mutants, showing dominance or partial dominance, have been mapped in the *strA-spcA* region and show an accumulation of ribosomal precursor particles at 20°, with a decreased synthesis of either mature 50 S or 30 S ribosomes. A coli spectinomycin resistant mutant with an altered ribosomal protein, S4, also has been isolated, showing cold sensitivity and accumulating precursors of 30 S and 50 S subunits (Nashimoto *et al.*, 1971). This phenotype could be explained by the *in vivo* attachment of 50 S and 30 S ribosomal proteins to the nascent rRNA precursor, as has been demonstrated *in vitro* (Nikolaev and Schlessinger, 1974). A mutated ribosomal protein, such as S4, which is one of the primary binding proteins to 16 S rRNA, during *in vitro* reconstitution (Held *et al.*, 1974b) may prevent processing of the p50 S–p30 S precursor rRNA complex.

Four nonallelic *cldS* mutants were also mapped in the *strA-spcA* region of *Salmonella* and may be genes for ribosomal proteins (Tyler and Ingraham, 1973). These all seem to be mutants of the first type. Other assembly mutants have been isolated in *E. coli* using cold selection (Bryant and Sypherd, 1974). These are recessive and have been mapped at 3 or possibly 4 loci: *rimA* near *ilvD* and *pyrE*; *rimB* near *aroD*; *rimC* between 20'–30'; *rimD* between *ilvD* and *malB*, close to *rimA*. At 20°, *rimA, B* and *D* accumulate 43 S particles with 5 S rRNA, and *rimC* accumulates 32 S particles. Precursor 16 S rRNA is seen in the 30 S subunits of the mutants at 20°, and 23 S pRNA is found in the 50 S subunits at the low temperature (Bryant *et al.*, 1974). This indicates that the final maturation of rRNA occurs at a very late step in ribosomal assembly. No differences in ribosomal proteins are observed in these mutants, and extracts prepared from wild-type or nonallelic *rim* mutants convert 43 S particles to 50 S components *in vitro* (Bryant *et al.*, 1974). The *rim* mutations, therefore, seem to be of the second type.

Ribosomal proteins S5 and S12, the products of the *spcA* (Bollen *et al.*,

1969) and *strA* (Ozaki *et al.,* 1969) genes, though synthesized coordinately (Dennis, 1974), are not instantaneously assembled *in vivo* via a concerted type of mechanism. F' merodiploids of *E. coli* with different combinations of *str* and *spc* alleles, i.e., $spc^r str^s/spc^s str^r$, $spc^s str^s/spc^r str^r$, etc., produce ribosomes with similar antibiotic resistance (Sparling *et al.,* 1968). This and the fact that the genes for 30 S and 50 S ribosomal proteins are intermingled (Fig. 3) indicate that the ribosomal proteins must go into a pool before ribosomal assembly. Dominance of the str^s allele over the str^r is noted in merodiploids (Sparling *et al.,* 1968; Chang *et al.,* 1974), and the effect, as judged by the results of *in vitro* reconstitution experiments, is at the level of assembly, not synthesis of these ribosomal proteins (Chang *et al.,* 1974). The presence of streptomycin during growth of str^r *E. coli* seems to affect ribosomes, possibly by changing the conformation of the ribosome during assembly (Garvin *et al.,* 1973).

In vitro reconstitution of ribosomes (discussed elsewhere in this volume) has greatly aided our knowledge of ribosomal structure, function, and assembly. It is of significance that precursor particles that accumulate at low temperatures in ribosomal assembly mutants possess a complement of proteins similar to that found in the reconstitution intermediate found during *in vitro* 30 S ribosome assembly (Nashimoto *et al.,* 1971).

b. Eukaryotes. As mentioned above, several unlinked *ts* mutations in yeast affecting ribosome synthesis have been described (Warner and Udem, 1972). Accumulation of precursor particles or rRNA was not observed in any of the mutants, indicating a stringent regulation of all the aspects of ribosome synthesis and assembly. However, a cold-sensitive mutant of *S. cerevesiae* which does not synthesize 60 S and 40 S ribosomal subunits, but does accumulate a 28 S particle at 15°, has been isolated (Bayliss and Ingraham, 1974). This strain is also streptomycin-sensitive, an alteration which is expressed at the level of the 40 S subunit, and the two phenotypic effects seem due to the same lesion. This mutation resembles the first class described above. A cold-sensitive mutation in *Neurospora crassa, crib-1,* segregates together with a defect in ribosome biogenesis at low temperature. While large and small subunits are both synthesized under the restricting condition, there is a marked reduction in the synthesis of the latter (Schlitt and Russell, 1975).

Mendelian mutations in *Chlamydomonas* have been described in which chloroplast ribosome assembly is affected (Harris *et al.,* 1974). Two classes of lesion are found, *ac-2* and *cr-4,* which cause formation of very few chloroplast ribosomes, while *cr-1, 2, 3* mutants show an accumulation of 50 S ribosomes and an inhibition of 30 S ribosome synthesis. This is different from *E. coli* cold-sensitive mutants where assembly of 50 S subunits is dependent on simultaneous assembly of 30 S components, but not the reverse (Nashimoto and Nomura, 1970).

4. Modifications of Ribosomal Proteins

Ribosomal proteins in prokaryotes show modifications, *in vivo*, such as N-terminal formylation of methionine (Hauschild-Rogat, 1968), N-terminal acetylation of L12 to give L7 (Terhorst *et al.*, 1972), methylation of L7/L12 (Terhorst *et al.*, 1972) and L11, predominantly (Alix and Hayes, 1974; Chang and Chang, 1974), and phosphorylation after T7 infection (Rahmsdorf *et al.*, 1973). Eukaryotic proteins are acetylated (Liew and Gornall, 1973) and phosphorylated (Kabat, 1970). In rat liver ribosomes, only one protein, S6, shows a phosphoserine modification Gressner and Wool, 1974). The functions of these modifications are presently unknown, but some of them may be involved in regulation (Kabat, 1970). Peptidyl transferase activity in *E. coli* is dependent on the presence of L11 (Nierhaus and Montejo, 1973), and this protein, on the basis of photoaffinity labeling, is near or is part of the peptidyl transferase center (Hsiung *et al.*, 1974). Its modification may thus be significant. *In vitro* aminoacylation of coli ribosomal proteins has been reported (Liebowitz and Soffer, 1971), and mutants lacking the enzyme, which catalyzes the transfer of leucine and phenylalanine from tRNA to basic proteins, have been isolated and the lesion has been mapped between 45' and 59' (Soffer and Savage, 1974). However, the *in vivo* significance of this enzyme is not clear.

5. Homology in Ribosomal Proteins

a. Prokaryotes. The number and average molecular weight distribution of ribosomal proteins in prokaryotes are similar (Sun *et al.*, 1972); however, the amino acid composition of these proteins varies from species to species, creating differences which can be resolved by carboxymethylcellulose column chromatography (Otaka *et al.*, 1968), gel electrophoresis (Sun *et al.*, 1972), and immunological techniques (Geisser *et al.*, 1973a,b). This implies that there are restraints on the conformation of the proteins in terms of their interactions in the ribosomes, and that variations in amino acids are permitted because they do not change overall conformation and/or certain regions of the ribosomal protein can change but some are more important: e.g., "active sites" for interaction with other ribosomal proteins, rRNA, or other components of the translational apparatus which interact transiently with the ribosome. The ribosomal proteins of most organisms are quite basic, the ratio of basic to acidic amino acids in *E. coli* being 1.6, but the extreme halophile, *Halobacterium cutirubrum*, which grows in 4 *M* salt, has a ratio of 0.5 (Strøm and Visentin, 1973).

In general, bacteria which show genetic compatibility show much greater similarity in their ribosomal proteins than more distantly related species. This has been shown in the Enterobacteriaceae (Osawa *et al.*, 1971; Geisser *et al.*, 1973a) and the Bacillaceae (Geisser *et al.*, 1973b).

Though there are differences in the primary structure of ribosomal proteins in different prokaryotes, reconstitution of functional 30 S ribosomes has been demonstrated using 16 S rRNA and 30 S ribosomal proteins from widely disparate bacterial species, but not yeast (Nomura *et al.*, 1968).

Extensive studies in the homologies between ribosomal proteins of *E. coli* and *Bacillus stearothermophilus* have been undertaken to elucidate the nature of this heterologous interaction. The N-terminal amino acid sequences of nine 30 S ribosomal proteins show 40–70% homology between the two species (Yaguchi *et al.*, 1973; Higo and Loertscher, 1974), and immunological cross-reaction is found in 14 out of 20 30 S ribosomal proteins (Isono *et al.*, 1973). Most, if not all, of the coli 30 S ribosomal proteins have functional counterparts among the stearothermophilus proteins as measured by substitution of individual proteins during *in vitro* 30 S ribosome reconstitution (Higo *et al.*, 1973). Thus, there is a conservation of primary sequence in ribosomal proteins from organisms that are very different. A comparative study of 50 S ribosomal protein L2, isolated from different species, shows an unusually high degree of structural, immunological, and functional homology, indicating that certain ribosomal proteins have evolved much less than others (Tischendorf *et al.*, 1973).

Even though coli and stearothermophilus ribosomal proteins show a high degree of homology and functional compatibility, there are differences in their specificity of initiation. Stearothermophilus ribosomes, under normal conditions, can only translate the "A" protein gene of bacteriophage f2 RNA, while coli can translate the coat protein and replicase genes as well as the "A" cistron (Lodish, 1970). This is a function of the 30 S ribosome (Lodish, 1970) and specifically the ribosomal protein fraction (Goldberg and Steitz, 1974). Reconstitution experiments, replacing *E. coli* proteins with individual counterparts from stearothermophilus indicated that stearothermophilus S12 is the only protein which has an effect on the translation of the coat protein gene *in vitro*, reducing it by 50% (Held *et al.*, 1974a). Substitution of coli 16 S rRNA by stearothermophilus RNA also lowers translation of the coat protein by about 40% and a combination of both stearothermophilus components gives an 85% reduction. This indicates an interaction between the 16 S rRNA and S12, a point discussed below. Other experiments have shown, though, that the addition of purified *E. coli* 30 S ribosomal protein, S1, which is involved in synthetic and natural mRNA binding to ribosomes (Van Duin and Van Knippenberg, 1974), causes the coat protein and replicase genes of f2 RNA to be efficiently translated *in vitro* by stearothermophilus ribosomes (Isono and Isono, 1975).

b. Eukaryotes. Eukaryote ribosomes usually have more proteins than prokaryotes, the smaller subunit of mammalian cells possessing approximately 30 proteins and the larger, 40 (Wool and Stöffler, 1974), but much less is known concerning the homologies in eukaryote ribosomal proteins. Although rat and *Tetrahymena* ribosomal proteins are very different, their subunits will interact to form functional ribosomes (Martin and Wool, 1969). Elongation factors EF-T and EF-G from mitochondria are synthesized on the cytoplasmic ribosomes and are encoded by nuclear genes (Richter, 1971), but they will only interact with mitochondrial or prokaryotic ribosomes (Richter and Lipmann, 1970; Perani *et al.*, 1971). The same relationships are seen for initiation of protein synthesis where *Neurospora* mitochondrial ribosomes will utilize *E. coli* or *B. subtilis* initiation factors for the binding of fMet-tRNA and the formation of fMet-puromycin, while cytoplasmic 80 S ribosomes are inactive (Sala and Kuntzel, 1970). Chloroplast ribosomes also will utilize *E. coli* initiation factors (Sala *et al.*, 1970).

Ribosomal subunits and factors from bacterial species are interchangeable (Takeda and Lipmann, 1966; Lodish, 1970) as are eukaryote subunits and factors (Martin and Wool, 1969; Zasloff and Ochoa, 1972). Usually prokaryote-eukaryote hybrids are not active in protein synthesis, except in one case where *Artemia-E. coli* hybrids will form fMet-puromycin (Klein and Ochoa, 1972). *Escherichia coli* ribosome subunits can be used interchangeably with spinach chloroplast ribosome subunits, but not yeast mitochondrial subunits (Grivell and Walg, 1972).

Thus, a great deal of similarity is seen between organelle ribosomes and factors and their prokaryote counterparts, in agreement with other data presented here.

6. Interactions between Ribosomal Components

The ribosome is a complex structure and many of the components must interact during ribosome assembly and protein synthesis. Interactions between antibiotic resistance genes have been noted. For example, in *E. coli,* when neomycin resistance is introduced into a *spc*[r] strain, the cell becomes phenotypically spectinomycin-sensitive, even though the genotype is still *spc,* and the reverse is also observed (Apirion and Schlessinger, 1969). Expression of erythromycin resistance, mediated by 50 S ribosomal protein L4, can be masked by mutational alteration of 30 S ribosomal proteins S4 or S12, indicating interactions between proteins of 30 S and 50 S subunits (Saltzman *et al.*, 1974; Apirion and Saltzman, 1974). Mutations in S12 can also affect nonenzymatic translocation, a 50 S function (Asatryan and Spirin, 1975). Mutations to erythromycin dependence in *E. coli* involve the interaction of the unlinked *eryA* and *macD* alleles (Sparling and Blackman, 1973).

The presence of the thiostrepton resistance marker renders *B. subtilis* twenty times more sensitive to erythromycin and oleandomycin than the wild-type (Goldthwaite and Smith, 1972b). As described earlier, *E. coli* ribosomal protein L11 is believed to be part of the peptidyl transferase center and it binds thiostrepton *in vitro*. Ribosomes from *thi^r B. subtilis* do not bind thiostrepton (Vince *et al.*, 1976) and show an altered L11. Erythromycin binds to the 50 S subunit in the vicinity of peptidyl-tRNA (Mao and Robishaw, 1972), and *E. coli ery^r* mutants with an altered L4 protein show reduced peptidyl transferase activity (Wittmann *et al.*, 1973). Mutational alteration of L11 could affect the configuration of the peptidyl transferase center, rendering the wild-type L4 or a similar protein more accessible to erythromycin, which would in turn cause increased sensitivity to the antibiotic.

Cold sensitivity, caused by mutational alteration of ribosomal protein S5, can be suppressed by secondary alterations in proteins S2 or S3 (Nashimoto and Uchida, 1975), and a *ts* alanyl-tRNA synthetase mutation can be phenotypically reverted by secondary mutations affecting proteins S5 or S20 (Wittmann *et al.*, 1974).

The clearest example of ribosomal protein interaction comes from studies of the genetics and biochemistry of the *strA* gene and its product S12. Mutations in this gene can cause resistance to or dependence on streptomycin. The resistance mutations in this gene can be classified according to their differential ability to restrict suppression *in vivo* or *in vitro*. The wild-type allele, *strA$^+$*, allows phenotypic suppression or tRNA mediated "informational" suppression, and the processes are increasingly restricted by mutations in the order, *strA60, strA40, strA2,* and *strA1* (Strigini and Gorini, 1970). *strA1, A2,* and *A60* result from the substitution of a lysine at position 42 by an asparagine, threonine, and arginine, respectively, while the *A40* mutation is an arginine substitution for a lysine at position 87 (Funatsu and Wittmann, 1972). Mutations to streptomycin dependency, also in S12 (Birge and Kurland, 1969), can be caused by a change of lysine 42 to glutamine (Funatsu and Wittmann, 1972) or by alterations in other regions of the protein (Itoh and Wittmann, 1973).

Second site mutations can change the streptomycin dependent phenotype to streptomycin independent, and in a study of 100 such revertants, 40 show detectably different ribosomal proteins, 24 with an altered S4 and 16 with a changed S5 (Hasenbank *et al.*, 1973). The S5 changes are point mutations, but in regions different from that involved in the mutations to spectinomycin resistance (Funatsu *et al.*, 1971; DeWilde and Wittmann-Liebold, 1973), and genetic recombination is observed between the *spc^r* and *str^{ind}* mutations. The alterations in S4 are more varied, giving rise, in some cases, to polypeptides with significantly greater or fewer amino acid residues than the wild-type protein at the C-terminal end. *ramA* mutations, also in the gene for S4 (Zimmerman *et al.*, 1971) which counteract the restrictive effect of *strA* mutations (Rosset and Gorini, 1969), show very similar polypeptide changes as the S4 $str^d \rightarrow str^i$ mutations (Van Acken, 1975), and introduction of *ramA* mutations into a str^d strain also produces str^i phenotypes (Bjare and Gorini, 1971). It has recently been shown that certain mutations in S5 can act like *ramA* mutations (Piepersburg *et al.*, 1975). Mutants resistant to or dependent on streptomycin have been isolated in *B. stearothermophilus* and *B. subtilis*, and they show alterations in S12. Streptomycin-independent revertants from str^d cells show altered S4 or S5 ribosomal proteins, as in *E. coli*, indicating the general nature of these ribosomal protein interactions (Isono, 1974; Itoh *et al.*, 1975).

Thus, mutational alterations in S12, causing restriction of suppression, can be counteracted by the *ramA* mutation in S4 or similar mutations in S5. Streptomycin dependence is modified by mutations similar to *ramA* in S4 or alterations in S5. Obviously, S12 must interact with S5 and S4 in the ribosome, and it has recently been shown S4 and S12 can be cross-linked *in situ*, indicating they are neighbors in the 30 S ribosome (Sommer and Traut, 1975). Streptomycin, if added in sublethal doses *in vivo* to *strA⁺* cells can phenotypically suppress various mutations (Gorini *et al.*, 1961), and causes miscoding *in vitro* (Davies *et al.*, 1964), presumably by "loosening" the structure of the ribosome as it binds to sensitive 30 S ribosomes (Chang and Flaks, 1972; Brakier-Gingras *et al.*, 1974), and it has been reported to bind to free 16 S rRNA (Biswas and Gorini, 1972) or to subribosomal particles in the presence of S3 and S5, but not S12 (Schreiner and Nierhaus, 1973). Mutations to str^r which alter S12 must change the conformation of 16 S rRNA and/or the ribosomal proteins (including S3 and S5) which form part of the streptomycin binding site. This region of the ribosome plays an important role in some of the steps of protein synthesis, as S12 and 16 S rRNA are involved in the recognition of initiation sites on mRNA's (Held *et al.*, 1974a). tRNA's covalently bind to 16 S rRNA after photoaffinity labeling of ribosomes (Schwartz and Ofengand, 1974), and it has been postulated that this site regulates the interaction of tRNA's with the ribosome during the elongation step of protein synthesis (Gorini, 1971). S3, which is part of the streptomycin binding site, is also involved in EF-T-dependent binding of tRNA to the 50 S A site (Randall-Hazelbauer and Kurland, 1971). This again stresses the interactions between S12 and other ribosomal proteins.

VI. Conclusion

With the study of the *strA* gene and its product S12 as an example of the molecular interactions between specific ribosomal proteins, rRNA, tRNA, and mRNA, we have seen how genetic analysis has aided in an understanding of how these components act together during translation. Perhaps this is a good place to stop and evaluate the masses of data presented here, in terms of the questions posed at the beginning of this chapter.

Although we do not have a clear idea of how the synthesis of all the components in the translational apparatus is regulated, it is clear that some overall control mechanisms are involved because of the coordinate synthesis of ribosomal components, protein synthesis factors, tRNA, and probably aminoacyl-tRNA synthetases, all of which, save the synthetases for which there is no clear evidence as yet, are under stringent control. A unitary hypothesis for the control of structural RNA synthesis is developing, involving positive and negative control at the level of transcription. Some of the putative control elements, e.g., EF-T and the *rel* gene product, function in protein synthesis as well, providing a firm basis for the previously postulated link between transcription and translation. In the near future, the exact nature of the interaction between the control elements and RNA polymerase and/or RNA promoter sites for rRNA and tRNA hopefully will be elucidated.

To a limited extent, the genetic organization of some genes for components of translation is important for regulation. For example, in bacteria, there are two major clusters of the genes for ribosomal proteins. This could provide a mechanism for the coordinate synthesis of these components, as there is some evidence for polycistronic mRNA's for ribosomal proteins. Likewise, in pro- and eukaryotes, genes for rRNA are linked and large and small rRNA's are also transcribed in one precursor molecule. In addition, the clustering of the eukaryote rRNA genes in the nucleolus, with a specific RNA polymerase, provides an efficient means of controlling rRNA synthesis. In bacteria, there is some clustering of tRNA genes so that two to three tRNA's are transcribed together. On the other hand, there is a high degree of scatter of tRNA and amino acid synthetase genes, and ribosomal protein and rRNA genes to a lesser extent as well, so that intercistronic control elements must act to regulate different regions of the bacterial chromosome. In eukaryotes, while clustering of genes for rRNA and tRNA families is observed, genes for 5 S rRNA are not linked, yet 5 S rRNA is present in amounts stoichiometric to 18 S and 28 S rRNA. Moreover, except for *Drosophila,* there seems to be no apparent clustering for eukaryote ribosomal protein genes.

The evolution of the protein synthetic apparatus is not characterized by an increase in the complexity of the reactions involved in translation, as a very similar mechanism is observed in all living things. The appearance of 5.8 S rRNA in eukaryote ribosomes may be a significant difference, but its function is, as yet, unknown. What has changed is the number of genes for some of the components. An increase in the redundancy for rRNA and tRNA genes is observed in the evolution of eukaryotes from prokaryotes. Presumably, the demands of a larger cell selected for this multiplication of identical genes. The mechanism of this amplification is, of course, not known with certainty, but as numerous examples in the laboratory and in wild-type bacteria have shown, gene duplication is not a rare phenomenon. A cell with redundancy in rRNA genes, for example, would have a selective advantage, and there would also be constraints against mutations in the duplicated genes, as rRNA must interact with many other components, guaranteeing similarity of these genes. As discussed previously, there is less intraspecies conservation of sequence in the non-transcribed rRNA gene spacer regions which, also, have evolved much more between species than the transcribed regions.

In addition to increased redundancy for structural RNA genes, eukaryotes have evolved an alternate system of protein synthesis in mitochondria and chloroplasts. The many similarities between the prokaryote and organelle translational apparatus have given rise to the "symbiont capture" theory of organelle origin, which suggests that proto-eukaryotic cells evolved a symbiotic relationship with certain prokaryotes which ultimately became one of complete dependence in the process, giving rise to plastids and mitochondria. The symbiotic relationships between some algae and invertebrate cells and the bacteria-like κ particles of paramecium are cited as models of possible intermediate stages in such relationships (Raven, 1970). On the other hand, it has been suggested that organelles are evolved from internal membrane systems, such as mesosomes, frequently found in prokaryotes (Raff and Mahler, 1972). This theory stresses the differences between organellar and prokaryotic translation. While part of the organellar translation apparatus is coded for by its genome, i.e., rRNA, tRNA, and possibly some ribosomal proteins, the remainder of the protein synthesis components, including aminoacyl synthetases and protein synthesis factors, are nucleus encoded, and all the protein components, whatever the location of their genes, are synthesized on cytoplasmic ribosomes. It is hoped that further studies on organelle ribosomes will clarify the relations of these structures to their prokaryotic and eukaryotic cytoplasmic counterparts. For example, it would be extremely interesting to know whether organellar $tRNA_f^{Met}$ has a G-T-Ψ-C sequence as in prokaryotes, or G-A-U-C-C, found in all eukaryotic initiator tRNA's.

ACKNOWLEDGMENTS

Work of the author described in this review was supported by grants GM-19185, GM-19693, and a career development award, GM-07871, all from the National Institutes of

Health. I am indebted to my many colleagues who sent advance copies of manuscripts and reprints. I am thankful for the helpful discussions I have had with Rosa Davidoff-Abelson, Loren Day, David Dubnau, Eugenie Dubnau, Leonard Mindich, and Barbara Scher. My greatest appreciation is reserved for Annabel Howard, without whose help this review would never have been completed.

REFERENCES

Abelson, J. N., Gefter, M. L., Barnett, L., Landy, A., Russell, R. L., and Smith, J. D., (1970). *J. Mol. Biol.* **47,** 15.
Adesnik, M., and Levinthal, C. (1969). *J. Mol. Biol.* **46,** 281.
Aharonowitz, Y., and Ron, E. Z. (1972). *Mol. Gen. Genet.* **119,** 131.
Alix, J.-H., and Hayes, D. (1974). *J. Mol. Biol.* **86,** 139.
Aloni, Y., Hatlen, L. E., and Attardi, G. (1971). *J. Mol. Biol.* **56,** 555.
Altman, S., and Smith, J. D. (1971). *Nature (London), New Biol.* **233,** 35.
Altman, S., Brenner, S. and Smith, J. D. (1971). *J. Mol. Biol.* **56,** 195.
Anderson, K. W., and Smith, J. D. (1972). *J. Mol. Biol.* **69,** 349.
Apirion, D. (1967). *J. Mol. Biol.* **30,** 255.
Apirion, D. and Saltzman, L. A. (1974). *Mol. Gen. Genet.* **135,** 11.
Apirion, D., and Schlessinger, D. (1968). *J. Bacteriol.* **96,** 768.
Apirion, D., and Schlessinger, D. (1969). *Proc. Natl. Acad. Sci. U.S.A.* **63,** 794.
Asatryan, L. S., and Spirin, A. S. (1975). *Mol. Gen. Genet.* **138,** 315.
Attardi, G., and Amaldi, F. (1970). *Annu. Rev. Biochem.* **39,** 183.
Austin, S. J., Tittawella, P. B., Hayward, R. S., and Scaife, J. G. (1971). *Nature (London), New Biol.* **232,** 133.
Barnett, W. E., and Brown, D. H. (1967). *Proc. Natl. Acad. Sci. U.S.A.* **57,** 452.
Barrell, B. G., Seidman, J. G., Guthrie, C., and McClain, W. H. (1974). *Proc. Natl. Acad. Sci. U.S.A.* **71,** 413.
Bayliss, F. T., and Ingraham, J. L. (1974). *J. Bacteriol.* **118,** 319.
Beale, G. H. (1973). *Mol. Gen. Genet.* **127,** 241.
Beale, G. H., Knowles, J. K. C., and Tait, A. (1972). *Nature (London)* **235,** 396.
Bernhardt, D., and Darnell, J. E. (1969). *J. Mol. Biol.* **42,** 43.
Bikoff, E. K., and Gefter, M. L. (1975). *J. Biol. Chem.* **250,** 6240.
Bikoff, E. K., LaRue, B. F., and Gefter, M. L. (1975). *J. Biol. Chem.* **250,** 6248.
Bird, A., Rogers, E., and Birnstiel, M. (1973). *Nature (London), New Biol.* **242,** 226.
Birge, E. A., and Kurland, C. G. (1969). *Science* **166,** 1282.
Birnbaum, L. S., and Kaplan, S. (1971). *Proc. Natl. Acad. Sci. U.S.A.* **68,** 925.
Birnstiel, M. L., and Grunstein, M. (1972). *FEBS Symp.* **23,** 349.
Birnstiel, M. L., Wallace, H., Sirlin, J. L., and Fischberg, M. (1966). *Natl. Cancer Inst., Monogr.* **23,** 431.
Birnstiel, M. L., Speirs, J., Purdom, I. F., Jones, K., and Loening, U. E. (1968). *Nature (London)* **219,** 454.
Birnstiel, M. L., Chipchase, M., and Speirs, J. (1971). *Prog. Nucleic Acid Res. Mol. Biol.* **11,** 351.
Birnstiel, M. L., Sells, B. H., and Purdom, I. F. (1972). *J. Mol. Biol.* **63,** 21.
Biswas, D. K., and Gorini, L. (1972). *Proc. Natl. Acad. Sci. U.S.A.* **69,** 2141.
Bjare, U., and Gorini, L. (1971). *J. Mol. Biol.* **57,** 423.
Björk, G. R., and Isaksson, L. A. (1970). *J. Mol. Biol.* **51,** 83.
Björk, G. R., and Neidhardt, F. C. (1971). *Cancer Res.* **31,** 706.
Blasi, F., Bruni, C. B., Avitabile, A., Deeley, R. G., Goldberger, R. F. , and Meyers, M. F. (1973). *Proc. Natl. Acad. Sci. U.S.A.* **70,** 2692.

Blumenthal, T., Landers, T. A., and Weber, K. (1972). *Proc. Natl. Acad. Sci. U.S.A.* **69**, 1313.

Böck, A., Ruffler, D., Piepersburg, W., and Wittmann, H. G. (1974). *Mol. Gen. Genet.* **134**, 325.

Bollen, A., Davies, J., Ozaki, M., and Mizushima, S. (1969). *Science* **165**, 85.

Bollen, A., Faelen, M., Lecocq, J. P., Herzog, A., Zengel, J., Kahan, L., and Nomura, M. (1973). *J. Mol. Biol.* **76**, 463.

Bonamy, C., Hirschbein, L., and Szulmajster, J. (1973). *J. Bacteriol.* **113**, 1296.

Borek, E., and Srinivasan, P. R. (1966). *Annu. Rev. Biochem.* **35**, 275.

Boynton, J. E., Burton, W. G., Gillham, N. W., and Harris, E. H. (1973). *Proc. Natl. Acad. Sci. U.S.A.* **70**, 3463.

Brakier-Gingras, L., Provost, L., and Dugas, H. (1974). *Biochem. Biophys. Res. Commun.* **60**, 1238.

Branlant, C., Krol, A., Sriwidada, J., Fellner, P., and Chrichton, R. (1973). *FEBS Lett.* **35**, 265.

Brenner, D. J., Fanning, G. R., Johnson, K. E., Citarella, R. V., and Falkow, S. (1969). *J. Bacteriol.* **98**, 637.

Brown, D. D., and Blackler, A. W. (1972). *J. Mol. Biol.* **63**, 75.

Brown, D. D., and Dawid, I. B. (1968). *Science* **160**, 272.

Brown, D. D., and Gurdon, J. B. (1964). *Proc. Natl. Acad. Sci. U.S.A.* **51**, 139.

Brown, D. D., and Sugimoto, K. (1973). *J. Mol. Biol.* **78**, 397.

Brown, D. D., and Weber, C. S. (1968). *J. Mol. Biol.* **34**, 661.

Brown, D. D., Wensink, P. C., and Jordan, E. (1971). *Proc. Natl. Acad. Sci. U.S.A.* **68**, 3175.

Brown, D. D., Wensink, P. C., and Jordan, E. (1972). *J. Mol. Biol.* **63**, 57.

Brown, M. E., and Apirion, D. (1974). *Mol. Gen. Genet.* **133**, 317.

Brownlee, C. G., Sanger, F., and Barrell, B. G. (1968). *J. Mol. Biol.* **34**, 329.

Bryant, R. E., and Sypherd, P. S. (1974). *J. Bacteriol.* **117**, 1082.

Bryant, R. E., Fujisawa, T., and Sypherd, P. S. (1974). *Biochemistry* **13**, 2110.

Burdon, R. H., and Clason, A. E. (1969). *J. Mol. Biol.* **39**, 113.

Burgess, R. R., Travers, A. A., Dunn, J. J., and Bautz, E. F. K. (1969). *Nature (London)* **221**, 43.

Cabezón, T., Bollen, A., and Faelen, M. (1974). *Arch. Int. Physiol. Biochim.* **82**, 172.

Cabezón, T., Faelen, M., DeWilde, M., Bollen, A., and Thomas, R., (1975). *Mol. Gen. Genet.* **137**, 125.

Cannon, M., Cabezón, T., and Bollen, A. (1974). *Mol. Gen. Genet.* **130**, 321.

Carbon, J., and Fleck, E. W. (1974). *J. Mol. Biol.* **85**, 371.

Casey, J. W., Cohen, M., Rabinowitz, M., Fukuhara, H., and Getz, G. S. (1972). *J. Mol. Biol.* **63**, 431.

Casey, J. W. Hsu, H.-J., Rabinowitz, M., Getz, G. S., and Fukuhara, H. (1974a). *J. Mol. Biol.* **88**, 717.

Casey, J. W., Hsu, H.-J., Getz, G. S., Rabinowitz, M., and Fukuhara, H. (1974b). *J. Mol. Biol.* **88**, 735.

Cashel, M., and Gallant, J. (1969). *Nature (London)* **221**, 839.

Cassio, D. (1975). *J. Bacteriol.* **123**, 589.

Chang, C. N., and Chang, F. N. (1974). *Nature (London)* **251**, 731.

Chang, F. N., and Flaks, J. G. (1972). *Antimicrob. Agents & Chemother.* **2**, 308.

Chang, F. N., Wang, Y. J., Fetterolf, C. J., and Flaks, J. G. (1974). *J. Mol. Biol.* **82**, 273.

Chang, S., and Carbon, J., (1975). *J. Biol. Chem.* **250**, 5542.

Chapeville, F., Lipmann, F., von Ehrenstein, G., Ray, W. J., Jr., and Benzer, S. (1962). *Proc. Natl. Acad. Sci. U.S.A.* **48**, 1086.

Chen, G. S., and Siddiqui, M. A. Q. (1973). *Proc. Natl. Acad. Sci. U.S.A.* **70**, 2610.

Chilton, M. D., and McCarthy, B. J. (1969). *Genetics* **62**, 697.

Chow, L. T., and Davidson, N. (1973). *J. Mol. Biol.* **75**, 265.

Clarkson, S. G., Birnstiel, M. L., and Serra, V. (1973a). *J. Mol. Biol.* **79**, 391.

Clarkson, S. G., Birnstiel, M. L., and Purdom, I. F. (1973b). *J. Mol. Biol.* **79**, 411.

Colli, W., Smith, I., and Oishi, M. (1971). *J. Mol. Biol.* **56**, 117.

Conde, M. F., Boynton, J. E., Gillham, N. W., Harris, E. H., Tingle, C. L., and Wang, W. L. (1975). *Mol. Gen Genet.* **140**, 183.

Cooper, D., Banthorpe, D. U., and Wilkie, D. (1967). *J. Mol. Biol.* **26**, 347.

Corte, G., Schlessinger, D., Longo, D., and Venkov, P. (1971). *J. Mol. Biol.* **60**, 325.

Cortese, R., Kammen, H. O., Spengler, S. J., and Ames, B. N. (1974a). *J. Biol. Chem.* **249**, 1103.

Cortese, R., Landsberg, R., von der Haar, R. A., Umbarger, H. E., and Ames, B. N. (1974b). *Proc. Natl. Acad. Sci. U.S.A.* **71**, 1857.

Craig, N., and Perry, R. P. (1971). *Nature (London), New Biol.* **229**, 75.

Crick, F. C. (1966). *J. Mol. Biol.* **19**, 548.

Crick, F. C., Barnett, L., Brenner, S., and Watts-Tobin, R. J. (1961). *Nature (London)* **192**, 1227.

Cryer, D. R., Goldthwaite, C. D., Zinker, S., Lam, K.-B., Storm, E., Hirschberg, R., Blamire, J., Finkelstein, D. B., and Marmur, J. (1973). *Cold Spring Harbor Symp. Quant. Biol.* **38**, 17.

Cutler, R. G., and Evans, J. E. (1967). *J. Mol. Biol.* **26**, 91.

Czernilofsky, A. P., Kurland, C. G., and Stöffler, G. (1975). *FEBS Lett.* **58**, 281.

Dahlberg, A. E., and Dahlberg, J. E. (1975). *Proc Natl. Acad. Sci. U.S.A.* **72**, 2940.

Dalgarno, L., and Shine, J. (1973). *Nature (London), New Biol.* **245**, 261.

Darnbrough, C., Legon, S., Hunt, T., and Jackson, R. J. (1973). *J. Mol. Biol.* **76**, 379.

Davidson, J. N., Hanson, M. R., and Bogorad, L. (1974). *Mol. Gen. Genet.* **132**, 119.

Davies, J., and Nomura, M. (1972). *Annu. Rev. Genet.* **7**, 203.

Davies, J., Gilbert, W., and Gorini, L. (1964). *Proc. Natl. Acad. Sci. U.S.A.* **51**, 883.

Davis, B. D. (1948). *J. Am. Chem. Soc.* **70**, 4267.

Dawid, I. B. (1972). *J. Mol. Biol.* **63**, 201.

de Crombrugghe, B., Chen, B., Anderson, W., Nissley, P., Gottesman, M., Pasten, I., and Perlman, R. (1971). *Nature (London), New Biol.* **231**, 139.

Dekio, S. (1971). *Mol. Gen. Genet.* **113**, 20.

Delk, A. S., and Rabinowitz, J. C. (1975). *Proc. Natl. Acad. Sci. U.S.A.* **72**, 528.

Dennis, P. P. (1974). *J. Mol. Biol.* **88**, 25.

Dennis, P. P., and Nomura, M. (1974). *Proc. Natl. Acad. Sci. U.S.A.* **71**, 3819.

Dennis, P. P., and Nomura, M. (1975). *J. Mol. Biol.* **97**, 61.

Deonier, R. C., Ohtsubo, E., Lee, H. J., and Davidson, N. (1974). *J. Mol. Biol.* **89**, 619.

Deutscher, M. (1973). *Prog. Nucleic Acid Res. Mol. Biol.* **13**, 51.

Deutscher, M. P., Foulds, J., and McClain, W. H. (1974). *J. Biol. Chem.* **249**, 6696.

DeWilde, M., and Wittmann-Liebold, B. (1973). *Mol. Gen. Genet.* **127**, 273.

DeWilde, M., Michel, F., and Broman, K. (1974). *Mol. Gen. Genet.* **133**, 329.

Diaz, A., Guillen, N., LeGault-Demare, L., and Hirschbein, L. (1975). *In* "Spores VI" (P. Gerhardt, R. N. Costilow, and H. L. Sadoff, eds.), p. 241, American Society for Microbiology, Washington, D.C.

Doi, R. H., and Igarashi, R. T., (1965). *J. Bacteriol.* **90**, 384.

Doolittle, W. F., and Pace, N. R. (1971). *Proc. Natl. Acad. Sci. U.S.A.* **68**, 1786.

Dubnau, D., Smith, I., Morell, P., and Marmur, J. (1965a). *Proc. Natl. Acad. Sci. U.S.A.* **54**, 491.

Dubnau, D., Smith, I., and Marmur, J. (1965b). *Proc. Natl. Acad. Sci. U.S.A.* **54,** 724.
Dubnau, E., Pifko, S., Sloma, A., Cabane, K., and Smith, I. (1976). *Mol. Gen. Genet.* **147,** 1.
DuBuy, B., and Weissman, B. (1971). *J. Biol. Chem.* **246,** 747.
Dunn, J. J., and Studier, F. W. (1973). *Proc. Natl. Acad. Sci. U.S.A.* **70,** 3296.
Ehresmann, C., Stiegler, P., and Ebel, J. P. (1974). *FEBS Lett.* **49,** 47.
Eidlic, L., and Neidhardt, F. C. (1965). *J. Bacteriol.* **89,** 706.
Ellis, R. J., and Hartley, M. R. (1971). *Nature (London), New Biol.* **233,** 193.
Epler, J. L. (1969). *Biochemistry* **8,** 2285.
Epler, J. L., Shugart, L. R., and Barnett, W. E. (1970). *Biochemistry* **9,** 3575.
Epstein, R. H., Bolle, A., Steinberg, L. M., Kellenberger, E., Boy de la Tour, E., Chevalley, R., Edgar, R. S., Susman, M., Denhardt, G., and Lielausis, A. (1963). *Cold Spring Harbor Symp. Quant. Biol.* **28,** 375.
Erdmann, V. A., Sprinzl, M., and Pongs, O. (1973). *Biochem. Biophys. Res. Commun.* **54,** 942.
Eron, L., Arditti, R. R., Zubay, G., Connaway, S., and Beckwith, J. A. (1971). *Proc. Natl. Acad. Sci. U.S.A.* **68,** 215.
Fahnestock, S., Erdmann, V., and Nomura, M. (1973). *Biochemistry* **12,** 220.
Fangman, W. L., and Neidhardt, F. C. (1964). *J. Biol. Chem.* **239,** 1839.
Faye, G., Kujawa, C., and Fukuhara, H. (1974). *J. Mol. Biol.* **88,** 185.
Faye, G., Kujawa, C., Fukuhara, H., and Rabinowitz, M. (1976). *Biochem. Biophys. Res. Commun.* **68,** 476.
Fellner, P. (1974). *In* "Ribosomes" (M. Nomura, A. Tissières and P. Lengyel, eds.) p. 169, Cold Spring Harbor Laboratory, Cold Spring Harbor, New York.
Fellner, P., Ehresmann, C., Stiegler, P., and Ebel, J.-P. (1972). *Nature (London), New Biol.* **239,** 1.
Ferguson-Smith, M. A. (1964). *Cytogenetics* **3,** 124.
Feunteun, J., Jordan, B. R., and Monier, R. (1972). *J. Mol. Biol.* **70,** 465.
Finkelstein, D. B., Blamire, J., and Marmur, J. (1972). *Nature (London), New Biol.* **240,** 279.
Fiser, I., Scheit, K. H., Stöffler, G., and Kuechler, E. (1975). *FEBS Lett.* **56,** 226.
Flaks, J. G., Leboy, P. S., Birge, E. A., and Kurland, C. G. (1966). *Cold Spring Harbor Symp. Quant. Biol.* **31,** 623.
Folk, W. R., and Berg. P. (1971). *J. Mol. Biol.* **58,** 595.
Ford, P. J., and Southern, E. M. (1973). *Nature (London), New Biol.* **241,** 7.
Forget, B., and Weissman, S. (1969). *J. Biol. Chem.* **244,** 3148.
Foulds, J., Hilderman, R. H., and Deutscher, M. P. (1974). *J. Bacteriol.* **118,** 628.
Fournier, M. J., Miller, W. L., and Doctor, B. P. (1974). *Biochem. Biophys. Res. Commun.* **60,** 1148.
Franze-Fernández, M. T., and Pogo, A. O. (1971). *Proc. Natl. Acad. Sci. U.S.A.* **68,** 3040.
Friesen, J. D., Fiil, N. P., Parker, J. M., and Haseltine, W. A. (1974). *Proc. Natl. Acad. Sci. U.S.A.* **71,** 3465.
Funatsu, G., and Wittmann, H. G. (1972). *J. Mol. Biol.* **68,** 547.
Funatsu, G., Schiltz, E., and Wittmann, H. G. (1971). *Mol. Gen. Genet.* **114,** 106.
Gall, J. G. (1969). *Genetics* **61,** Suppl., 121.
Gall, J. G. (1974). *Proc. Natl. Acad. Sci. U.S.A.* **71,** 3078.
Gall, J. G., and Pardue, M. L. (1971). *In* "Methods in Enzymology" (L. Grossman and K. Moldave, eds.), Vol. 21, p. 40. Academic Press, New York.
Gall, J. G., and Rochaix, J.-D. (1974). *Proc. Natl. Acad. Sci. U.S.A.* **71,** 1819.
Gallo, R. C., and Pestka, S. (1970). *J. Mol. Biol.* **52,** 195.

Galluci, E., Pachetti, G., and Zangrossi, S. (1970). *Mol. Gen. Genet.* **106**, 362.

Ganoza, M. C., Vandermeer, J., Debreceni, N., and Phillips, S. L. (1973). *Proc. Natl. Acad. Sci. U.S.A.* **70**, 31.

Garvin, R. T., and Gorini, L. (1975). *Mol. Gen. Genet.* **137**, 73.

Garvin, R. T., Rosset, R., and Gorini, L. (1973). *Proc. Natl. Acad. Sci. U.S.A.* **70**, 2762.

Gausing, K. (1974). *Mol. Gen. Genet.* **129**, 61.

Gefter, M. L., and Russell, R. L. (1969). *J. Mol. Biol.* **39**, 145.

Geisser, M., Tischendorf, G. W., Stöffler, G. and Wittmann, H. G. (1973a). *Mol. Gen. Genet.* **127**, 111.

Geisser, M., Tischendorf, G. W., and Stöffler, G. (1973b). *Mol. Gen. Genet.* **127**, 129.

Ghosh, C., and Ghosh, H. P. (1970). *Biochem. Biophys. Res. Commun.* **40**, 135.

Ghysen, A., and Celis, J. E. (1974a). *Nature (London)* **249**, 418.

Ghysen, A., and Celis, J. E. (1974b). *J. Mol. Biol.* **83**, 333.

Giegé, R., Kern, D., Ebel, J.-P., Grosjean, H., De Henau, S., and Chantrenne, H. (1974). *Eur. J. Biochem.* **45**, 351.

Gilbert, W., Maizels, N., and Maxam, A. (1973). *Cold Spring Harbor Symp. Quant. Biol.* **38**, 845.

Gilham, N. W., Boynton, J. E., and Burkholder, B. (1970). *Proc. Natl. Acad. Sci. U.S.A.* **67**, 1026.

Ginsburg, D., and Steitz, J. A. (1975). *J. Biol. Chem.* **250**, 5647.

Goldberg, M. L., and Steitz, J. A. (1974). *Biochemistry* **13**, 2123.

Goldberg, S., Øyen, T., Idriss, J. M., and Halvorson, H. O. (1972). *Mol. Gen. Genet.* **116**, 139.

Goldberger, R. F. (1974). *Science* **183**, 810.

Goldthwaite, C. D., and Smith, I. (1972a). *Mol. Gen. Genet.* **114**, 181.

Goldthwaite, C. D., and Smith, I. (1972b). *Mol. Gen. Genet.* **114**, 190.

Goldthwaite, C., Dubnau, D., and Smith, I. (1970). *Proc. Natl. Acad. Sci. U.S.A.* **65**, 96.

Goodman, H. M., Abelson, J. N., Landy, A., Brenner, S., and Smith, J. D. (1968). *Nature (London)* **217**, 1019.

Gordon, J. (1970). *Biochemistry* **9**, 912.

Gordon, J., Baron, L. S., and Schweiger, M. (1972). *J. Bacteriol.* **110**, 306.

Gorelic, L. (1970). *Mol. Gen. Genet.* **106**, 323.

Gorini, L. (1970). *Annu. Rev. Genet.* **4**, 107.

Gorini, L. (1971). *Nature (London), New Biol.* **234**, 261.

Gorini, L., Gundersen, W., and Burger, M. (1961). *Cold Spring Harbor Symp. Quant. Biol.* **26**, 173.

Gressner, A. M., and Wool, I. G. (1974). *Biochem. Biophys. Res. Commun.* **60**, 1482.

Grigliatti, T. A., White, B. N., Tener, G. M., Kaufman, T. C., Holden, J. J., and Suzuki, D. T. (1973). *Cold Spring Harbor Symp. Quant. Biol.* **39**, 461.

Grimberg, J. I., and Daniel, V. (1974). *Nature (London)* **250**, 320.

Grivell, L. A., and Walg, H. L. (1972). *Biochem. Biophys. Res. Commun.* **49**, 1452.

Grivell, L. A., Reijnders, L., and DeVries, H. (1971). *FEBS Lett.* **16**, 159.

Grivell, L. A., Netter, P., Borst, P., and Slonimski, P. P. (1973). *Biochim. Biophys. Acta* **312**, 358.

Gross, K. J., and Pogo, A. O. (1974). *J. Biol. Chem.* **249**, 568.

Grummt, F., and Grummt, I. (1974). *FEBS Lett.* **42**, 343.

Guest, J. R., and Yanofsky, C. (1965). *J. Mol. Biol.* **12**, 793.

Guthrie, C., and McClain, W. (1973). *J. Mol. Biol.* **81**, 137.

Guthrie, C., Nashimoto, H., and Nomura, M. (1969). *Proc. Natl. Acad. Sci. U.S.A.* **63**, 384.

Guthrie, C., Seidman, J. G., Altman, S., Barrell, B. G., Smith, J. D., and McClain, W. H. (1973). *Nature (London), New Biol.* **246**, 6.

Hackett, P. B., and Sauerbier, W. (1975). *J. Mol. Biol.* **91**, 235.
Hall, B., and Gallant, J. (1972). *Nature (London), New Biol.* **237**, 131.
Harford, N. (1975). *J. Bacteriol.* **121**, 835.
Harford, N., and Sueoka, N. (1970). *J. Mol. Biol.* **51**, 267.
Harris, E. H., Boynton, J. E., and Gillham, N. W. (1974). *J. Cell Biol.* **63**, 160.
Hartman, P. G., and Roth, J. R. (1973). *Adv. Genet.* **17**, 1.
Hartwell, L. H., and McLaughlin, C. S. (1968a). *Proc. Natl. Acad. Sci. U.S.A.* **59**, 422.
Hartwell, L. H., and McLaughlin, C. S. (1968b). *J. Bacteriol.* **96**, 1664.
Hartwell, L. H., McLaughlin, C. S., and Warner, J. R. (1970). *Mol. Gen. Genet.* **109**, 42.
Haseltine, W. A. (1972). *Nature (London)* **235**, 329.
Haseltine, W. A., and Block, R. (1973). *Proc. Natl. Acad. Sci. U.S.A.* **70**, 1564.
Haseltine, W. A., Block, R., Gilbert, W., and Weber, K. (1972). *Nature (London)* **238**, 381.
Hasenbank, R., Guthrie, C., Stöffler, G., Wittmann, H. G., Rosen, L., and Apirion, D. (1973). *Mol. Gen. Genet.* **127**, 1.
Hatlen, L. E., Amaldi, F., and Attardi, G. (1969). *Biochemistry* **8**, 4989.
Hatlen, L. E., and Attardi, G. (1971). *J. Mol. Biol.* **56**, 535.
Hauschild-Rogat, P. (1968). *Mol. Gen. Genet.* **102**, 95.
Haworth, S. R., and Brown, L. R. (1973). *J. Bacteriol.* **114**, 103.
Hawthorne, D. C. (1969). *J. Mol. Biol.* **43**, 71.
Hawthorne, D. C., and Mortimer, R. K. (1968). *Genetics* **60**, 735.
Hayashi, H., Fisher, L., and Söll, D. (1969). *Biochemistry* **8**, 3680.
Hayes, F., and Vasseur, M. (1973). *C. R. Acad. Sci. Paris* **227**, 881.
Hayes, F., Vasseur, M., Nikolaev, N., Schlessinger, D., Widada, J. S., Krol, A., and Branlant, L. (1975). *FEBS Lett.* **56**, 85.
Hecht, N., and Woese, C. (1968). *J. Bacteriol.* **95**, 541.
Hecht, R. M., and Birnstiel, M. L. (1972). *Eur. J. Biochem.* **29**, 489.
Hecker, L. I., Egan, J., Reynolds, R. J., Nix, C. E., Schiff, J. A., and Barnett, W. E. (1974). *Proc. Natl. Acad. Sci. U.S.A.* **71**, 1910.
Held, W. A., Gette, W. R., and Nomura, M. (1974a). *Biochemistry* **13**, 2115.
Held, W. A., Ballou, B., Mizushima, S., and Nomura, M. (1974b). *J. Biol. Chem.* **249**, 3103.
Helser, T. L., Davies, J. E., and Dahlberg, J. E. (1972). *Nature (London), New Biol.* **235**, 6.
Herzog, A., Ghysen, A., and Bollen, A. (1971). *Mol. Gen. Genet.* **110**, 211.
Highland, J. H., Howard, G. A., Ochsner, E., Stöffler, G., Hasenbank, R., and Gordon, J. (1974). *J. Biol. Chem.* **250**, 1141.
Higo, K., Held, W., Kahan, L., and Nomura, M. (1973). *Proc. Natl. Acad. Sci. U.S.A.* **70**, 944.
Higo, K.-I., and Loertscher, K. (1974). *J. Bacteriol.* **118**, 180.
Hill, C. W., Foulds, J., Soll, L., and Berg, P. (1969). *J. Mol. Biol.* **39**, 563.
Hindley, J., and Page, S. M. (1972). *FEBS Lett.* **26**, 157.
Hirsch, D. (1971). *J. Mol. Biol.* **58**, 439.
Hirsch-Kauffmann, M., Schweiger, M., Herrlich, P., Ponta, H., Rahmsdorf, H.-J., Pai, S.-H., and Wittmann, H.-G. (1975). *Eur. J. Biochem.* **52**, 469.
Hitz, H., Schäfer, D., and Wittmann-Liebold, B. (1975). *FEBS Lett.* **56**, 259.
Holley, R. W., Apgar, J., Everett, G. A., Madison, J. T., Marquisee, M., Merrill, S. H., Penswick, J. R., and Zamir, A. (1965). *Science* **147**, 1462.
Holmquist, R., Jukes, T. H., and Pangburn, S. (1973). *J. Mol. Biol.* **78**, 91.
Honeycutt, R. C., and Margulies, M. M. (1973). *J. Biol. Chem.* **248**, 6145.
Hooper, M. L., Russell, R. L., and Smith, J. D. (1972). *FEBS Lett.* **22**, 149.
Hopwood, D. A., Chater, K. F., Dowding, J. E., and Vivian, A. (1973). *Bacteriol. Rev.* **37**, 371.
Hori, H., Takata, R., Muto, A., and Osawa, S. (1974). *Mol. Gen. Genet.* **128**, 341.

Hori, K., Harada, K., and Kuwano, M. (1974). *J. Mol. Biol.* **86,** 699.

Horne, J. R., and Erdmann, V. A. (1973). *Proc. Natl. Acad. Sci. U.S.A.* **70,** 2870.

Hourcade, D., Dressler, D., and Wolfson, J. (1973). *Cold Spring Harbor Symp. Quant. Biol.* **38,** 537.

Howell, N., Molloy, P. L., Linnane, A. W., and Lukins, H. B. (1974). *Mol. Gen. Genet.* **128,** 43.

Hsiung, N., Reines, S. A., and Cantor, C. R. (1974). *J. Mol. Biol.* **88,** 841.

Huberman, J. A., and Attardi, G. (1967). *J. Mol. Biol.* **29,** 487.

Hussey, C., Losick, R., and Sonenshine, A. L. (1971). *J. Mol. Biol.* **57,** 59.

Hussey, C., Pero, J., Shorenstein, R. G., and Losick, R. (1972). *Proc. Natl. Acad. Sci. U.S.A.* **69,** 407.

Ikemura, T., and Dahlberg, J. E. (1973). *J. Biol. Chem.* **248,** 5033.

Isono, K. (1974). *Mol. Gen. Genet.* **133,** 77.

Isono, K., Isono, S., Stöffler, G., Visentin, L. P., Yaguchi, M., and Matheson, A. T. (1973). *Mol. Gen. Genet.* **127,** 191.

Isono, S., and Isono, K. (1975). *Eur. J. Biochem.* **56,** 15.

Itoh, T., Kosugi, H., Higo, K., and Osawa, S. (1975). *Mol. Gen. Genet.* **139,** 293.

Itoh, T., and Wittmann, H. G. (1973). *Mol. Gen. Genet.* **127,** 19.

Jackson, E., and Yanofsky, C. (1972). *J. Mol. Biol.* **71,** 149.

Jacobson, K. B. (1966). *Cold Spring Harbor Symp. Quant. Biol.* **31,** 719.

Jacobson, K. B. (1971). *Prog. Nucleic Acid Res. Mol. Biol.* **11,** 461.

Jarry, B., and Rosset, R. (1970). *Biochem. Biophys. Res. Commun.* **41,** 789.

Jarry, B., and Rosset, R. (1971). *Mol. Gen. Genet.* **113,** 43.

Jarry, B., and Rosset, R. (1973a). *Mol. Gen. Genet.* **121,** 151.

Jarry, B., and Rosset, R. (1973b). *Mol. Gen. Genet.* **126,** 29.

Jaskunas, S. R., Nomura, M., and Davies, J. (1974). *In* "Ribosomes" (M. Nomura, A. Tissières, and P. Lengyel, eds.), p. 333. Cold Spring Harbor Lab., Cold Spring Harbor, New York.

Jaskunas, S. R., Lindahl, L., and Nomura, M. (1975a). *Proc. Natl. Acad. Sci. U.S.A.* **72,** 6.

Jaskunas, S. R., Lindahl, L., and Nomura, M. (1975b). *Nature (London)* **256,** 183.

Jaskunas, S. R., Lindahl, L., Nomura, M., and Burgess, R. R. (1975c). *Nature (London)* **257,** 458.

Johnson, L., and Söll, D. (1970). *Proc. Natl. Acad. Sci. U.S.A.* **67,** 943.

Johnson, L., Hayashi, H., and Söll, D. (1970). *Biochemistry* **9,** 2823.

Johnson, L. D., Henderson, A. S., and Atwood, K. C. (1974). *Cytogenet. Cell Genet.* **13,** 103.

Jordan, B. R., Galling, G., and Jourdan, R. (1974). *J. Mol. Biol.* **87,** 205.

Kabat, D. (1970). *Biochemistry* **9,** 4160.

Kaltschmidt, E., and Wittmann, H. G. (1970). *Proc. Natl. Acad. Sci. U.S.A.* **67,** 1276.

Kaltschmidt, E., Kahan, L., and Nomura, M. (1974). *Proc. Natl. Acad. Sci. U.S.A.* **71,** 446.

Kan, J., and Sueoka, N. (1971). *J. Biol. Chem.* **246,** 2207.

Kimura, A., Kobata, K., Takata, R., and Osawa, S. (1973). *Mol. Gen. Genet.* **124,** 107.

Kimura, A., Muto, A., and Osawa, S. (1974). *Mol. Gen. Genet.* **130,** 203.

Kindler, P., Keil, T. U., and Hofschneider, P. H. (1973). *Mol. Gen. Genet.* **126,** 53.

Kirschbaum, J. B., and Scaife, J. (1974). *Mol. Gen. Genet.* **132,** 193.

Kisselev, L. L. (1972). *FEBS Symp.* **23,** 115.

Klein, H. A., and Ochoa, S. (1972). *J. Biol. Chem.* **247,** 8122.

Klootwijk, J., van den Bos, R. C., and Planta, R. J. (1972). *FEBS Lett.* **27,** 102.

Klootwijk, J., Klein, I., and Grivell, L. A. (1975). *J. Mol. Biol.* **97,** 337.

Kobayashi, Y., Steinberg, W., Higa, A., Halvorson, H. O., and Levinthal, C. (1965). *In* "Spores III" (L. L. Campbell and H. O. Halvorson, eds.), Vol. 3, p. 200. Am. Soc. Microbiol., Ann Arbor, Michigan.

Korner, A., Magee, B. B., Liska, B., Low, K. B., and Söll, D. (1974). *J. Bacteriol.* **120,** 154.

Kossman, C. R., Stamato, T. D., and Pettijohn, D. E. (1971). *Nature (London), New Biol.* **234,** 112.

Küntzel, H. (1969). *Nature (London)* **222,** 142.

Kuriyama, Y., and Luck, D. J. L. (1973). *J. Mol. Biol.* **73,** 425.

Kuriyama, Y., and Luck, D. J. L. (1974). *J. Mol. Biol.* **83,** 253.

Kuwano, M., Schlessinger, D., Rinaldi, G., Felicetti, L., and Tocchini-Valentini, G. P. (1971). *Biochem. Biophys. Res. Commun.* **42,** 441.

Kuwano, M., Ono, M., Yamamoto, M., Endo, H., Kamiya, T., and Hori, K. (1973). *Nature (London), New Biol.* **244,** 107.

Kuwano, M., Endo, H., Kamiya, T., and Hori, K. (1974). *J. Mol. Biol.* **86,** 689.

Lai, C.-J., Dahlberg, J. E., and Weisblum, B. (1973). *Biochemistry* **12,** 457.

Landy, A., Foeller, C., and Ross, W. (1974). *Nature (London)* **249,** 738.

Lapointe, J., and Söll, D. (1972). *J. Biol. Chem.* **247,** 4966.

Leboy, P. S., Cox, E. C., and Flaks, J. G. (1964). *Proc. Natl. Acad. Sci. U.S.A.* **52,** 1387.

Lederberg, J. (1959). *Methods Med. Res.* **3,** 5.

Leighton, T. (1974). *J. Mol. Biol.* **86,** 855.

Lepesant-Kejzlarová, J., Lepesant, J.-A., Walle, J., Billault, A., and Dedonder, R. (1975). *J. Bacteriol.* **121,** 832.

Lewis, J. A., and Ames, B. (1972). *J. Mol. Biol.* **66,** 131.

Liau, M. C., and Hurlbert, R. B. (1975). *J. Mol. Biol.* **98,** 321.

Liebowitz, M. J., and Soffer, R. L. (1971). *Proc. Natl. Acad. Sci. U.S.A.* **68,** 1866.

Liew, C.-C., and Gornall, A. G. (1973). *J. Biol. Chem.* **248,** 977.

Lindahl, L., Jaskunas, S. R., Dennis, P. P., and Nomura, M. (1975). *Proc. Natl. Acad. Sci. U.S.A.* **72,** 2743.

Linn, T. G., Greenleaf, A. L., Shorenstein, R. G., and Losick, R. (1973). *Proc. Natl. Acad. Sci. U.S.A.* **70,** 1865.

Liou, Y.-F., Kinoshita, T., Tanaka, N., and Yoshikawa, M. (1973). *J. Antibiot.* **26,** 711.

Littauer, U. L., and Inouye, H. (1973). *Annu. Rev. Biochem.* **42,** 439.

Litwack, M. D., and Peterkofsky, A. (1971). *Biochemistry* **10,** 994.

Lizardi, P. M., and Luck, D. J. L. (1972). *J. Cell Biol.* **54,** 56.

Lodish, H. (1970). *Nature (London)* **226,** 705.

Loening, U. E., Jones, K. W., and Birnstiel, M. L. (1969). *J. Mol. Biol.* **45,** 353.

Loewen, P. C., Sekiya, T., and Khorana, H. G. (1974). *J. Biol. Chem.* **249,** 217.

Loftfield, R. (1972). *Prog. Nucleic Acid Res. Mol. Biol.* **12,** 87.

Lupker, J. H., Verschoor, G. J., Glickman, B. W., Rörsch, A., and Bosch, L. (1974). *Eur. J. Biochem.* **43,** 583.

McClain, W. H. (1970). *FEBS Lett.* **6,** 99.

McClain, W. H., Guthrie, C., and Barrell, B. G. (1972). *Proc. Natl. Acad. Sci. U.S.A.* **69,** 3703.

McGinnis, E., and Williams, L. S. (1972). *J. Bacteriol.* **109,** 505.

McLaughlin, C. S., and Hartwell, L. H. (1969). *Genetics* **61,** 557.

McLaughlin, C. S., Magee, P. T., and Hartwell, L. H. (1969). *J. Bacteriol.* **100,** 579.

Maden, B. E. H. (1971). *Prog. Biophys. Mol. Biol.* **22,** 127.

Maden, B. E. H., and Robertson, J. S. (1974). *J. Mol. Biol.* **87,** 227.

Maden, B. E. H., and Salim, M. (1974). *J. Mol. Biol.* **88,** 133.

Mao, J. C. H., and Robishaw, E. E. (1972). *Biochemistry* **11,** 4864.

Marchin, G. L., Müller, U. R., and Al-Khateeb, G. H. (1974). *J. Biol. Chem.* **249,** 4705.

Marotta, C. A., Levy, C. C., Weissman, S. M., and Varricchio, F. (1973). *Biochemistry* **12,** 2901.

Marotta, C. A., Varricchio, F., Smith, I., Weissman, S. M., Sogin, M. L., and Pace, N. R. (1976). *J. Biol. Chem.* **251,** 3122.

Martin, T. E., and Wool, I. G. (1969). *J. Mol. Biol.* **43,** 151.

Matsubara, M., Takata, R., and Osawa, S. (1972). *Mol. Gen. Genet.* **117,** 311.

Meselson, M., Yuan, R., and Heywood, J. (1972). *Annu. Rev. Biochem.* **41,** 447.

Mets, L., and Bogorad, L. (1971). *Science* **174,** 707.

Mets, L., and Bogorad, L. (1972). *Proc. Natl. Acad. Sci. U.S.A.* **69,** 3779.

Miller, M. J., and McMahon, D. (1974). *Biochim. Biophys. Acta* **366,** 35.

Miller, O. L., Jr., and Beatty, B. R. (1969). *Science* **164,** 955.

Miller, O. L., Jr., Hamkalo, B. A., and Thomas, C. A., Jr. (1970). *Science* **169,** 392.

Molin, S., von Meyenburg, K., Gulløv, K., and Maaløe, O. (1974). *Mol. Gen. Genet.* **129,** 11.

Morell, P., Smith, I., Dubnau, D., and Marmur, J. (1967). *Biochemistry* **6,** 258.

Mortimer, R. K., and Hawthorne, D. C. (1973). *Genetics* **74,** 33.

Moshowitz, D. B. (1970). *J. Mol. Biol.* **50,** 143.

Mosteller, R. D., and Yanofsky, C. (1971). *J. Bacteriol.* **105,** 268.

Murgola, E. J., and Adelberg, E. A. (1970). *J. Bacteriol.* **103,** 178.

Muto, A. (1968). *J. Mol. Biol.* **36,** 1.

Muto, A., Takata, R., and Osawa, S. (1971). *Mol. Gen. Genet.* **111,** 15.

Nashimoto, H., and Nomura, M. (1970). *Proc. Natl. Acad. Sci. U.S.A.* **67,** 1440.

Nashimoto, H., and Uchida, H. (1975). *J. Mol. Biol.* **96,** 453.

Nashimoto, H., Held, W., Kaltschmidt, E., and Nomura, M. (1971). *J. Mol. Biol.* **62,** 121.

Nass, M. M. K., and Buck, C. A. (1970). *J. Mol. Biol.* **54,** 187.

Nazar, R. N., Sitz, T. O., and Busch, H. (1975). *J. Biol. Chem.* **250,** 8591.

Nazario, M., Kinsey, J. A., and Ahmed, M. (1971). *J. Bacteriol.* **105,** 121.

Neidhardt, F. C. (1966). *Bacteriol. Rev.* **30,** 701.

Neidhardt, F. C., and Earheart, C. F. (1966). *Cold Spring Harbor Symp. Quant. Biol.* **31,** 557.

Neidhardt, F. C., and Eidlic, L. (1963). *Biochim. Biophys. Acta* **68,** 380.

Neidhardt, F. C., and Magasanik, B. (1960). *Biochim. Biophys. Acta* **42,** 90.

Nierhaus, K. H., and Montejo, V. (1973). *Proc. Natl. Acad. Sci. U.S.A.* **70,** 1931.

Nikolaev, N., and Schlessinger, D. (1974). *Biochemistry* **13,** 4272.

Nikolaev, N., Silengo, L., and Schlessinger, P. (1973). *Proc. Natl. Acad. Sci. U.S.A.* **70,** 3361.

Nikolaev, N., Schlessinger, D., and Wellauer, D. (1974). *J. Mol. Biol.* **86,** 741.

Nishimura, S. (1972). *Prog. Nucleic Acid Res. Mol. Biol.* **12,** 49.

Nomura, M., and Engbaek, F. (1972). *Proc. Natl. Acad. Sci. U.S.A.* **69,** 1526.

Nomura, M., Traub, P., and Bechmann, H. (1968). *Nature (London)* **219,** 793.

Nomura, M., Tissières, A., and Lengyel, P., eds. (1974). "Ribosomes." Cold Spring Harbor Lab., Cold Spring Harbor, New York.

Novelli, G. D. (1967). *Annu. Rev. Biochem.* **36,** 449.

Ofengand, J., and Henes, C. (1969). *J. Biol. Chem.* **244,** 6241.

Ohta, N., Sager, R., Inouye, M. (1965). *J. Biol. Chem.* **250,** 3655.

Ohtsubo, E., Soll, L., Deonier, R. C., Lee, H. J., and Davidson, N. (1974). *J. Mol. Biol.* **89,** 631.

Oishi, M., and Sueoka, N. (1965). *Proc. Natl. Acad. Sci. U.S.A.* **54,** 483.

Oishi, M., Oishi, A., and Sueoka, N. (1966). *Proc. Natl. Acad. Sci. U.S.A.* **55,** 1095.

Okuyama, A., Yoshikawa, M., and Tanaka, N. (1974). *Biochem. Biophys. Res. Commun.* **60,** 1262.

Osawa, S., Itoh, T., and Otaka, E. (1971). *J. Bacteriol.* **107,** 168.

Osawa, S., Otaka, E., Takata, R., Dekio, S., Matsubara, M., Itoh, T., and Muto, A. (1972). *FEBS Symp.* **23,** 313.

Osawa, S., Takata, R., Tanaka, K., and Tamaki, M. (1973). *Mol. Gen. Genet.* **127,** 163.

Otaka, E., Itoh, T., and Osawa, S. (1968). *J. Mol. Biol.* **33**, 93.

Ozaki, M., Mizushima, S., and Nomura, M. (1969). *Nature (London)* **222**, 333.

Pace, B., and Pace, N. R. (1971). *J. Bacteriol.* **105**, 142.

Pace, N. R. (1973). *Bacteriol. Rev.* **37**, 562.

Pace, N. R., Pato, M. L., McKibbin, J., and Radcliffe, C. W. (1973). *J. Mol. Biol.* **75**, 619.

Paddock, G., and Abelson, J. (1973). *Nature (London), New Biol.* **246**, 2.

Pardo, D., and Rosset, R. (1974). *Mol. Gen. Genet.* **135**, 257.

Pardue, M. L. (1973). *Cold Spring Harbor Symp. Quant. Biol.* **38**, 475.

Pardue, M. L., Gerbi, S. A., Eckardt, R. A., and Gall, J. G. (1970). *Chromosoma* **29**, 268.

Pardue, M. L., Brown, D. D., and Birnstiel, M. L. (1973). *Chromosoma* **42**, 191.

Parker, J., Flashner, M., McKeever, W. G., and Neidhardt, F. C. (1974). *J. Biol. Chem.* **249**, 1044.

Parker, J., Watson, R. J., Friesen, J. D. (1976). *Mol. Gen. Genet.* **144**, 117.

Pato, M. L., and von Meyenburg, K. (1970). *Cold Spring Harbor Symp. Quant. Biol.* **35**, 497.

Pedersen, F. S., Lund, G., and Kjeldgaard, N. O. (1973). *Nature (London), New Biol.* **243**, 13.

Perani, A., Tiboni, O., and Ciferri, O. (1971). *J. Mol. Biol.* **55**, 107.

Perry, R. P., and Kelley, D. E. (1972). *J. Mol. Biol.* **70**, 265.

Perry, R. P., Cheng, T.-Y., Freed, J. J., Greenberg, J. R., Kelley, D. E., and Tartof, K. D. (1970). *Proc. Natl. Acad. Sci. U.S.A.* **65**, 609.

Person, S. (1963). *Biophys. J.* **3**, 183.

Pestka, S. (1971). *Annu. Rev. Microbiol.* **25**, 487.

Pestka, S., Weiss, D., Vince, R., Wienen, B., Stöffler, G., and Smith, I. (1976). *Mol. Gen. Genet.* **144**, 235.

Petrissant, G. (1973). *Proc. Natl. Acad. Sci. U.S.A.* **70**, 1046.

Pettijohn, D. E., Stonington, G. O., and Kossman, C. R. (1970). *Nature (London)* **228**, 235.

Phillips, S. L., Schlessinger, D., and Apirion, D. (1969). *Proc. Natl. Acad. Sci. U.S.A.* **62**, 772.

Piepersburg, W., Böck, A., and Wittmann, H. G. (1975). *Mol. Gen. Genet.* **140**, 91.

Pinder, J. C., Gould, H. J., and Smith, I. (1969). *J. Mol. Biol.* **40**, 289.

Pinkerton, T. C., Paddock, G., and Abelson, J. (1972). *Nature (London), New Biol.* **240**, 88.

Piper, P. W., and Clark, B. F. C. (1974). *Eur. J. Biochem.* **45**, 589.

Pongs, O., Bald, R., and Erdmann, V. A. (1973). *Proc. Natl. Acad. Sci. U.S.A.* **70**, 2229.

Primakoff, P., and Berg, P. (1970). *Cold Spring Harbor Symp. Quant. Biol.* **35**, 391.

Printz, D. B., and Gross, S. R. (1967). *Genetics* **55**, 451.

Procunier, J. D., and Tartof, K. D. (1975). *Genetics* **81**, 515.

Ptashne, M. (1971). *In* "The Bacteriophage Lambda" (A. Hershey, Ed.), p. 221. Cold Spring Harbor Lab., Cold Spring Harbor, New York.

Purdom, I., Bishop, J. O., and Birnstiel, M. L. (1970). *Nature (London)* **227**, 239.

Racine, F. M., and Steinberg, W. (1974). *J. Bacteriol.* **120**, 384.

Raff, R. A., and Mahler, H. R. (1972). *Science* **177**, 575.

Rahmsdorf, H. J., Herrlich, P., Pai, S. H., Schweiger, M., and Wittmann, H. G. (1973). *Mol. Gen. Genet.* **127**, 259.

RajBhandary, U. L., and Ghosh, H. P. (1969). *J. Biol. Chem.* **244**, 1104.

Ramagopal, S., and Davis, B. D. (1974). *Proc. Natl. Acad. Sci. U.S.A.* **71**, 820.

Randall-Hazelbauer, L. L., and Kurland, C. G. (1972). *Mol. Gen. Genet.* **115**, 234.

Rao, S. S., and Grollman, A. P. (1967). *Biochem. Biophys. Res. Commun.* **29**, 696.

Raven, P. H. (1970). *Science* **169**, 641.

Ravin, A. W., Rotheim, M. B., and Coulter, D. M. (1969). *Genetics* **61**, 23.

Rawson, J. R., and Haselkorn, R. (1973). *J. Mol. Biol.* **77**, 125.

Reeder, R. H., and Brown, D. B. (1970). *J. Mol. Biol.* **51,** 361.

Reger, B. J., Fairfield, S. A., Epler, J. L., and Barnett, W. E. (1970). *Proc. Natl. Acad. Sci. U.S.A.* **67,** 1207.

Reijnders, L. C., Keisen, L. M., Grivell, L. A., and Borst, P. (1972). *Biochim. Biophys. Acta* **272,** 396.

Reiness, G. Yang, H.-L., Zubay, G., and Cashel, M. (1975). *Proc. Natl. Acad. Sci. U.S.A.* **72,** 2881.

Retel, J., and Planta, R. J. (1972). *Biochim. Biophys. Acta* **281,** 299.

Richter, D. (1971). *Biochemistry* **10,** 4422.

Richter, D., and Lipmann, F. (1970). *Biochemistry* **9,** 5065.

Richter, D., Erdmann, V. A., and Sprinzl, M. (1973). *Nature (London), New Biol.* **246,** 132.

Riddle, D. L., and Carbon, J. (1973). *Nature (London), New Biol.* **242,** 230.

Riddle, D. L., and Roth, J. R. (1970). *J. Mol. Biol.* **54,** 131.

Riddle, D. L., and Roth, J. R. (1972a). *J. Mol. Biol.* **66,** 483.

Riddle, D. L., and Roth, J. R. (1972b). *J. Mol. Biol.* **66,** 495.

Rigby, P. W. J., Burleigh, B. D., and Hartley, B. S. (1974). *Nature (London)* **251,** 200.

Ritossa, F. M. (1972). *Nature (London), New Biol.* **240,** 109.

Ritossa, F. M., and Spiegelman, S. (1965). *Proc. Natl. Acad. Sci. U.S.A.* **53,** 737.

Ritossa, F. M., Atwood, K. C., and Spiegelman, S. (1966a). *Genetics* **54,** 663.

Ritossa, F. M., Atwood, K., and Spiegelman, S. (1966b). *Genetics* **54,** 819.

Rivjen, A. H. G. C., and Zwar, J. A. (1973). *Biochim. Biophys. Acta* **299,** 564.

Roberts, J. W., and Carbon, J. (1974). *Nature (London)* **250,** 412.

Roberts, J. W., and Carbon, J. (1975). *J. Biol. Chem.* **250,** 5530.

Robertson, H. D., Webster, R. D., and Zinder, N. D. (1968). *J. Biol. Chem.* **243,** 82.

Robertson, H. D., Altman, S., and Smith, J. D. (1972). *J. Biol. Chem.* **247,** 5243.

Robertus, J. D., Ladner, J. E., Finch, J. T., Rhodes, D., Brown, R. S., Clark, B. F. C., and Klug, A. (1974). *Nature (London)* **250,** 546.

Rochaix, J.-D., Bird, A., and Bakken, A. (1974). *J. Mol. Biol.* **87,** 473.

Roeder, R. G., and Rutter, W. J. (1970). *Proc. Natl. Acad. Sci. U.S.A.* **65,** 675.

Rosset, R., and Gorini, L. (1969). *J. Mol. Biol.* **39,** 95.

Rubin, G. M. (1973). *J. Biol. Chem.* **248,** 3860.

Rubin, G. M., and Sulston, J. E. (1973). *J. Mol. Biol.* **79,** 521.

Rudner, R., Rejiman, E., and Chargaff, G. (1965). *Proc. Natl. Acad. Sci. U.S.A.* **54,** 904.

Russell, R. L., Abelson, J. N., Landy, A., Gefter, M. L., Brenner, S., and Smith, J. D. (1970). *J. Mol. Biol.* **47,** 1.

Russell, R. R. B., and Pittard, A. J. (1971). *J. Bacteriol.* **108,** 790.

Ryan, J. L., and Morowitz, H. J. (1969). *Proc. Natl. Acad. Sci. U.S.A.* **63,** 1282.

Sager, R. (1972). "Cytoplasmic Genes and Organelles." Academic Press, New York.

Sakano, H., and Shimura, Y. (1975). *Proc. Natl. Acad. Sci. U.S.A.* **72,** 3369.

Sakano, H., Shimura, Y., and Ozeki, H. (1974). *FEBS Lett.* **40,** 312.

Sala, F., and Küntzel, H. (1970). *Eur. J. Biochem.* **15,** 280.

Sala, F., Sensi, S., and Parisi, B. (1970). *FEBS Lett.* **10,** 89.

Saltzman, L. A., Brown, M., and Apirion, D. (1974). *Mol. Gen. Genet.* **133,** 201.

Sanderson, K. E. (1972). *Bacteriol. Rev.* **36,** 558.

Sarabhai, A. S., Stretton, A. O. W., Brenner, S., and Bolle, A. (1964). *Nature (London)* **201,** 13.

Sarubbi, F. A., Blackman, F., and Sparling, P. F. (1974). *J. Bacteriol.* **120,** 1284.

Schaefer, K. P., Altman, S., and Söll, D. (1973). *Proc. Natl. Acad. Sci. U.S.A.* **70,** 3626.

Schedl, P., and Primakoff, P. (1973). *Proc. Natl. Acad. Sci. U.S.A.* **70,** 2091.

Scherberg, N. H., and Weiss, S. B. (1972). *Proc. Natl. Acad. Sci. U.S.A.* **69,** 1114.

Schibler, U., Wyler, T., and Hagenbüchle, O. (1975). *J. Mol. Biol.* **94,** 503.

Schindler, D., Grant, P., and Davies, J. (1974). *Nature (London)* **248**, 536.

Schlanger, G., and Sager, R. (1974). *Proc. Natl. Acad. Sci. U.S.A.* **71**, 1715.

Schlitt, S. C., and Russell, P. J. (1975). *J. Bacteriol.* **120**, 666.

Schmidt, F. J. (1975). *J. Biol. Chem.* **250**, 8399.

Schmitt, H. (1971). *FEBS Lett.* **15**, 186.

Schreier, M. H., and Staehelin, T. (1973). *Nature (London), New Biol.* **242**, 35.

Schreiner, G., and Nierhaus, K. H. (1973). *J. Mol. Biol.* **81**, 71.

Schwartz, I., and Ofengand, J. (1974). *Proc. Natl. Acad. Sci. U.S.A.* **71**, 3951.

Schweizer, E., MacKechnie, C., and Halvorson, H. O. (1969). *J. Mol. Biol.* **40**, 261.

Scott, N. S. (1973). *J. Mol. Biol.* **81**, 327.

Seale, T. W. (1972). *Genetics* **70**, 385.

Sekiya, T., and Khorana, H. G. (1974). *Proc. Natl. Acad. Sci. U.S.A.* **71**, 2978.

Shaefer, K. P., and Küntzel, H. (1972). *Biochem. Biophys. Res. Commun.* **46**, 1312.

Sherman, F., Liebman, S. W., Stewart, J. W., and Jackson, M. (1973). *J. Mol. Biol.* **78**, 157.

Shimura, Y., Aono, H., Ozeki, H., Sarabhai, A., Lanfrom, H., and Abelson, J. N. (1972). *FEBS Lett.* **22**, 144.

Shine, J., and Dalgarno, L. (1974). *Proc. Natl. Acad. Sci. U.S.A.* **71**, 1342.

Siev, M., Weinberg, R., and Penman, S. (1969). *J. Cell Biol.* **41**, 510.

Simsek, M., Ziegenmayer, J., Heckman, J., and RajBhandary, U. L. (1973a). *Proc. Natl. Acad. Sci. U.S.A.* **70**, 1041.

Simsek, M., Petrissant, G., and RajBhandary, U. L. (1973b). *Proc. Natl. Acad. Sci. U.S.A.* **70**, 2600.

Simsek, M., RajBhandary, U. L., Boisnard, M., and Petrissant, G. (1974). *Nature (London)* **247**, 518.

Sinclair, J. H., and Brown, D. D. (1971). *Biochemistry* **10**, 2761.

Singer, C. E., Smith, G. R., Cortese, R., and Ames, B. N. (1972). *Nature (London), New Biol.* **238**, 72.

Skogerson, L., McLaughlin, C., and Wakatama, E. (1973). *J. Bacteriol.* **116**, 818.

Smith, I., and Smith, H. (1973). *J. Bacteriol.* **114**, 1138.

Smith, I., Dubnau, D., Morell, P., and Marmur, J. (1968). *J. Mol. Biol.* **33**, 123.

Smith, I., Goldthwaite, C. D., and Dubnau, D. (1969). *Cold Spring Harbor Symp. Quant. Biol.* **34**, 85.

Smith, I., Colli, W., and Oishi, M. (1971). *J. Mol. Biol.* **62**, 111.

Smith, J. D. (1972). *Annu. Rev. Genet.* **7**, 235.

Smith, J. D., and Celis, J. E. (1973). *Nature (London), New Biol.* **243**, 66.

Smith, J. D., Barnett, L., Brenner, S., and Russell, R. L. (1970). *J. Mol. Biol.* **54**, 1.

Snyder, A. L., Kann, H. E., Jr., and Kohn, K. W. (1971). *J. Mol. Biol.* **58**, 555.

Soffer, R. L., and Savage, M. (1974). *Proc. Natl. Acad. Sci. U.S.A.* **71**, 1004.

Söll, D. (1971). *Science* **173**, 293.

Söll, D., and Schimmel, P. R. (1974). *In* "The Enzymes" (P. D. Boyer, ed.), 3rd ed., Vol. 10, p. 489. Academic Press, New York.

Soll, L., and Berg, P. (1969). *Proc. Natl. Acad. Sci. U.S.A.* **63**, 392.

Sommer, A., and Traut, R. R. (1975). *J. Mol. Biol.* **97**, 471.

Sonenberg, N., Wilchek, M., and Zamir, A. (1973). *Proc. Natl. Acad. Sci. U.S.A.* **70**, 1423.

Spadari, S., DiLernia, R., Simoni, G., Pedrali-Noy, G., and DeCarli, L. (1973). *Mol. Gen. Genet.* **127**, 57.

Sparling, P. F. (1970). *Science* **167**, 56.

Sparling, P. F., and Blackman, E. (1973). *J. Bacteriol.* **116**, 74.

Sparling, P. F., Modolell, J., Takeda, Y., and Davis, B. D. (1968). *J. Mol. Biol.* **37**, 407.

Spear, B. B., and Gall, J. G. (1973). *Proc. Natl. Acad. Sci. U.S.A.* **70**, 1359.

Speirs, J., and Birnstiel, M. (1974). *J. Mol. Biol.* **87**, 237.

Squires, C., and Carbon, J. (1971). *Nature (London), New Biol.* **233,** 274.
Squires, C., Konrad, B., Kirschbaum, J., and Carbon, J. (1973a). *Proc. Natl. Acad. Sci. U.S.A.* **70,** 438.
Squires, C. L., Rose, J. K., Yanofsky, C., Yang, H. L., and Zubay, G. (1973b). *Nature (London), New Biol.* **245,** 131.
Staal, S. P., and Hoch, J. A. (1972). *J. Bacteriol.* **110,** 202.
Steffensen, D. M. (1973). *Nature (London), New Biol.* **244,** 231.
Steffensen, D. M., and Wimber, D. E. (1971). *Genetics* **69,** 163.
Steinberg, W. (1974). *J. Bacteriol.* **117,** 1023.
Steinberg, W., and Anagnostopoulos, C. (1971). *J. Bacteriol.* **105,** 6.
Steitz, J. A., and Jakes, K. (1975). *Proc. Natl. Acad. Sci. U.S.A.* **72,** 4734.
Stent, G. S., and Brenner, S. (1961). *Proc. Natl. Acad. Sci. U.S.A.* **47,** 2005.
Stephens, J. C., Artz, S. W., and Ames, B. N. (1975). *Proc. Natl. Acad. Sci. U.S.A.* **72,** 4389.
Stern, R., Gonano, F., Fleissner, E., and Littauer, U. Z. (1970). *Biochemistry* **9,** 10.
Stewart, P. R. (1973). *In* "The Ribonucleic Acids" (P. R. Stewart and D. S. Letham, eds.), p. 163. Springer-Verlag, Berlin and New York.
Stöffler, G. (1974). *In* "Ribosomes" (M. Nomura, A. Tissières and P. Lengyel, eds.), p. 615. Cold Spring Harbor Lab., Cold Spring Harbor, New York.
Strigini, P., and Gorini, L. (1970). *J. Mol. Biol.* **47,** 517.
Strøm, A. R., and Visentin, L. P. (1973). *FEBS Lett.* **37,** 274.
Stuart, J. J., and Ravin, A. W. (1968). *J. Gen. Microbiol.* **51,** 411.
Stutz, E., and Rawson, J. R. (1970). *Biochim. Biophys. Acta* **209,** 16.
Stutz, E., and Vendrey, J. P. (1971). *FEBS Lett.* **17,** 277.
Suddath, F. L., Quigley, G. J., McPherson, A., Sneden, D., Kim, J. J., and Rich, A. (1974). *Nature (London)* **248,** 20.
Sun, T.-T., Bickle, T. A., and Traut, R. R. (1972). *J. Bacteriol.* **111,** 474.
Surzycki, S. J., and Rochaix, J. D. (1971). *J. Mol. Biol.* **62,** 89.
Sypherd, P. S. (1968). *J. Bacteriol.* **95,** 1844.
Sypherd, P. S., and Osawa, S. (1974). *In* "Ribosomes" (M. Nomura, A. Tissières, and P. Lengyel, eds.), p. 669. Cold Spring Harbor Lab., Cold Spring Harbor, New York.
Sypherd, P. S., O'Neil, D. M., and Taylor, M. M. (1969). *Cold Spring Harbor Symp. Quant. Biol.* **34,** 77.
Tai, P.-C., and Ingraham, J. L. (1971). *Biochim. Biophys. Acta* **232,** 151.
Tai. P.-C., Kessler, D. P., and Ingraham, J. (1969). *J. Bacteriol.* **97,** 1298.
Takata, R., Osawa, S., Tanaka, K., Teraoka, H., and Tamaki, M. (1970). *Mol. Gen. Genet.* **109,** 123.
Takeda, M., and Lipmann, F. (1966). *Proc. Natl. Acad. Sci. U.S.A.* **56,** 1875.
Tanaka, K., Tamaki, M., Osawa, S., Kimura, A., and Takata, R. (1973). *Mol. Gen. Genet.* **127,** 157.
Tanaka, N., Kawano, G., and Kinoshita, T. (1971). *Biochem. Biophys. Res. Commun.* **42,** 564.
Tanaka, T., and Weisblum, B. (1975). *J. Bacteriol.* **123,** 771.
Tartof, K. D., and Perry, R. P. (1970). *J. Mol. Biol.* **51,** 171.
Taylor, A. L., and Trotter, C. D. (1972). *Bacteriol. Rev.* **36,** 504.
Terhorst, C., Wittmann-Liebold, B., and Möller, W. (1972). *Eur. J. Biochem.* **25,** 13.
Testa, D., and Rudner, R. (1975). *Nature (London)* **254,** 630.
Tewari, K. K., Wildman, S. G. (1970). *Symp. Soc. Exp. Biol.* **24,** 147.
Thammana, P., and Held, W. A. (1974). *Nature (London)* **251,** 682.
Thomas, D. Y., and Wilkie, D. (1968). *Genet. Res.,* Camb. **11,** 33.
Thomas, J. R., and Tewari, K. K. (1974). *Proc. Natl. Acad. Sci. U.S.A.* **71,** 3147.

Tischendorf, G. W., Geisser, M., and Stöffler, G. (1973). *Mol. Gen. Genet.* **127**, 147.

Tocchini-Valentini, G. P., and Mattoccia, E. (1968). *Proc. Natl. Acad. Sci. U.S.A.* **61**, 146.

Tocchini-Valentini, G. P., Mahdavi, V., Brown, R., and Crippa, M. (1973). *Cold Spring Harbor Symp. Quant. Biol.* **38**, 551.

Toniolo, D., and Basilico, C. (1974). *Nature (London)* **248**, 411.

Toniolo, D., Meiss, H. K., and Basilico, C. (1973). *Proc. Natl. Acad. Sci. U.S.A.* **70**, 1273.

Traub, P., and Nomura, M. (1968). *Proc. Natl. Acad. Sci. U.S.A.* **59**, 777.

Travers, A. (1973). *Nature (London)* **244**, 15.

Travers, A., and Buckland, B. (1973). *Nature (London), New Biol.* **243**, 257.

Travers, A., Kamen, R. I., and Schleif, R. F. (1970). *Nature (London)* **228**, 748.

Turncock, G. (1969). *Mol. Gen. Genet.* **104**, 295.

Twardzik, D. R., Grell, E. H., and Jacobson, K. B. (1971). *J. Mol. Biol.* **57**, 231.

Tyler, B. and Ingraham, J. L. (1973). *Mol. Gen. Genet.* **122**, 197.

Tzagoloff, A., and Meagher, P. (1972). *J. Biol. Chem.* **247**, 594.

Udem, S. A., and Warner, J. R. (1972). *J. Mol. Biol.* **65**, 227.

Unger, M., Birnbaum, L. S., and Kaplan, S. (1972). *Mol. Gen. Genet.* **119**, 377.

Van Acken, J. (1975). *Mol. Gen. Genet.* **140**, 61.

Van Duin, J., and Van Knippenberg, P. M. (1974). *J. Mol. Biol.* **84**, 185.

Vaughan, M. H., Jr., Soeiro, R., Warner, J. R., and Darnell, J. E., Jr. (1967). *Proc. Natl. Acad. Sci. U.S.A.* **58**, 1527.

Vigne, R., Jordan, B. R., and Monier, R. (1973). *J. Mol. Biol.* **76**, 303.

Vince, R., Weiss, D., Gordon, J., Howard, G., Smith, I., and Pestka, S. (1976). *Antimicrob. Agents & Chemother.* **9**, 665.

Vogel, T., Meyers, M., Kovach, J. S., and Goldberger, R. F. (1972). *J. Bacteriol.* **112**, 126.

Vomvoyanni, V. (1974). *Nature (London)* **248**, 508.

Wahba, A. J., Miller, M. J., Niveleau, A., Landers, T. A., Carmichael, G. G., Weber, K., Harvey, D. A., and Slobin, L. I. (1974). *J. Biol. Chem.* **249**, 3314.

Warner, J. R., and Udem, S. A. (1972). *J. Mol. Biol.* **65**, 243.

Waterson, R. M., and Koningsberg, W. H. (1974). *Proc. Natl. Acad. Sci. U.S.A.* **71**, 376.

Watson, R. J., Parker, J., Fiil, N. P., Flaks, J. G., and Friesen, J. D. (1975). *Proc. Natl. Acad. Sci. U.S.A.* **72**, 2765.

Weatherford, S. C., Rosen, L., Gorelic, L., and Apirion, D. (1972). *J. Biol. Chem.* **247**, 5404.

Weeks, C. O., and Gross, S. R. (1971). *Biochem. Genet.* **5**, 505.

Wegnez, M., Monier, R., and Denis, H. (1972). *FEBS Lett.* **25**, 13.

Weinberg, R. A., Loening, U. E., Willems, M., and Penman, S. (1967). *Proc. Natl. Acad. Sci. U.S.A.* **58**, 1088.

Wellauer, P. K., and Dawid, I. B. (1973). *Cold Spring Harbor Symp. Quant. Biol.* **38**, 525.

Wellauer, P. K., and Reeder, R. H. (1975). *J. Mol. Biol.* **94**, 151.

Wellauer, P. K., Reeder, R. H., Carroll, D., Brown, D. D., Deutch, H., Higashinakagawa, T., and Dawid, I. B. (1974). *Proc. Natl. Acad. Sci. U.S.A.* **71**, 2823.

Wensink, P. C., and Brown, D. D. (1971). *J. Mol. Biol.* **60**, 238.

White, B. N., Tener, G. M., Holden, J., and Suzuki, D. T. (1973). *J. Mol. Biol.* **74**, 635.

Wienen, B., Stöffler, G., Smith, I., Weiss, D., Vince, R., and Pestka, S. (1976). *Mol. Gen. Genet.* (in press).

Wilson, J. H. (1973). *J. Mol. Biol.* **74**, 753.

Wilson, J. H., and Abelson, J. N. (1972). *J. Mol. Biol.* **69**, 57.

Wilson, J. H., and Kells, S. (1972). *J. Mol. Biol.* **69**, 39.

Wilson, J. H., Kim, J. S., and Abelson, J. N. (1972). *J. Mol. Biol.* **71**, 547.

Wimber, D. E., and Steffensen, D. M. (1970). *Science* **170**, 639.

Wittmann, H. G. (1972). *FEBS Symp.* **29**, 213.

Wittmann, H. G., Stöffler, G., Apirion, D., Rosen, L., Tanaka, K., Tamaki, M., Takata, R., Dekio, S., Otaka, E., and Osawa, S. (1973). *Mol. Gen. Genet.* **127**, 175.

Wittmann, H. G., Stöffler, G., Piepersburg, W., Buckel, P., Ruffler, D., and Böck, A. (1974). *Mol. Gen. Genet.* **134**, 225.

Wool, I. G., and Stöffler, G. (1974). *In* "Ribosomes" (M. Nomura, A. Tissières, and P. Lengyel, eds.), p. 417. Cold Spring Harbor Lab., Cold Spring Harbor, New York.

Wrede, P., and Erdmann, V. A. (1973). *FEBS Lett.* **33**, 315.

Wu, M., and Davidson, N. (1973). *J. Mol. Biol.* **78**, 1.

Wu, M., and Davidson, N. (1975). *Proc. Natl. Acad. Sci. U.S.A.* **72**, 4506.

Wu, M., Davidson, N., Attardi, G., and Aloni, Y. (1972). *J. Mol. Biol.* **71**, 81.

Wu, M., Davidson, N., and Carbon, J. (1973). *J. Mol. Biol.* **78**, 23.

Yaguchi, M., Roy, C., Matheson, A. T., and Visentin, L. P. (1973). *Can. J. Biochem.* **51**, 1215.

Yamada, T., and Davies, J. (1971). *Mol. Gen. Genet.* **110**, 197.

Yamaguchi, H., Yoshida, Y., and Tanaka, N. (1966). *J. Biochem. (Tokyo)* **60**, 246.

Yaniv, M., Jacob, F., and Gros, F. (1965). *Bull. Soc. Chim. Biol.* **47**, 1009.

Yaniv, M., Folk, W. R., Berg, P., and Soll, L. (1974). *J. Mol. Biol.* **86**, 245.

Yao, M.-C., Kimmel, A. R., and Gorovsky, M. A. (1974). *Proc. Natl. Acad. Sci. U.S.A.* **71**, 3082.

Yu, M. T., Vermeulen, C. W., and Atwood, K. C. (1970). *Proc. Natl. Acad. Sci. U.S.A.* **67**, 26.

Yuki, A. (1971). *J. Mol. Biol.* **56**, 439.

Zamir, A., Holley, R., and Marquisee, M. (1965). *J. Biol. Chem.* **240**, 1267.

Zasloff, M., and Ochoa, S. (1972). *Proc. Natl. Acad. Sci. U.S.A.* **69**, 1796.

Zimmerman, R. A., Garvin, R. T., and Gorini, L. (1971). *Proc. Natl. Acad. Sci. U.S.A.* **68**, 2263.

Zimmerman, R. A., Ikeya, Y., and Sparling, P. F. (1973). *Proc. Natl. Acad. Sci. U.S.A.* **70**, 71.

Zubay, G., Cheong, L., and Gefter, M. L. (1970). *Proc. Natl. Acad. Sci. U.S.A.* **68**, 2195.

Index

A

Abrin, 473, 529–530

Ac-Phe-tRNA, in prokaryotic initiation, 39, 418

Actidione, *see* Cycloheximide

Actinomycin D, mRNA not inhibited by, 558

S-Adenosyl methionine, in tRNA modification, 66

Affinity labeling, 203–241, 393, *see also* Ribosomes

  aminoacyl-tRNA affinity labels, 211, 213

  antibiotic affinity labels, 211, 213, 216

  application to ribosomes, 206–241

  biological function of affinity label, 227–229

  competition with nonreactive substrates or inhibitors, 222–224

  control of nonspecific reactions, 219–221

  design of experiments, 216

  GDP affinity label, 214

  general principles, 204–206

  identification of covalent product, 217–218

  inhibition of biological function, 225–227

  low yields in ribosome labeling, 216

  modulation of reaction, 224–225

  mRNA affinity labels, 215

  of mRNA binding site, 215–216, 234, 238

  of peptidyl transferase site, 207–213, 230–234, 236–239

  peptidyl-tRNA affinity labels, 204–211

  photoactivated labels, 205

  quantitation of covalent product, 217–218

  reaction products of similar site-directed reagents, 229

  reactive amino acid side chains near active ribosomal centers, 207

  reagent design, 204–205

  of ribosomal GTP hydrolysis site, 214, 234, 236–237

  of ribosomal polypeptide binding site, 214–215, 234, 239–240

  site specificity, 205

  size variation of label, 229

  of streptomycin binding site, 216, 238

  survey of results, 229–240

  of tRNA binding sites, 213, 234, 238

Althiomycin, 423–427, 472, 476, 526–527

Amicetin, 473, 476, 505–507, *see also* 4-Aminohexose pyrimidine nucleoside antibiotics

  inhibition of peptide bond synthesis, 423–427

  inhibition of termination, 457

Aminoacyl oligonucleotides, as inhibitors of protein synthesis, *see* Puromycin

Aminoacyl-tRNA, 11–16, 105–110, 211, 213, 323–325, 349, 350, 353–354, *see also* Aminoacylation reaction

  affinity for EF-Tu·GDP, 339

  affinity label analogues of, 211, 213

  aminoacyl stereospecificity at peptidyl transferase center, 64

  assay, 13

  binding to ribosomal A site, 104–105, 110, 323–325

  cation requirement for formation of, 13

  chemical hydrolysis of, 11

  conformational changes during ribosome binding, 106

  discrimination of cognate from noncognate tRNA's in codon binding, 105–106

  elongation factor recognition of, 58

  entrance channel on ribosome, 192

  enzymatic deacylation of, 15

  equilibrium analysis of specific binding to ribosomes, 107–110

  error frequency of codon recognition, 101, 111

  flow diagram in protein synthesis, 107

Molecular Biology

An International Series of Monographs and Textbooks

Editors

BERNARD HORECKER

Roche Institute of Molecular Biology
Nutley, New Jersey

NATHAN O. KAPLAN

Department of Chemistry
University of California
At San Diego
La Jolla, California

JULIUS MARMUR

Department of Biochemistry
Albert Einstein College of Medicine
Yeshiva University
Bronx, New York

HAROLD A. SCHERAGA

Department of Chemistry
Cornell University
Ithaca, New York

WALTER W. WAINIO. The Mammalian Mitochondrial Respiratory Chain. 1970

LAWRENCE I. ROTHFIELD (Editor). Structure and Function of Biological Membranes. 1971

ALAN G. WALTON AND JOHN BLACKWELL. Biopolymers. 1973

WALTER LOVENBERG (Editor). Iron-Sulfur Proteins. Volume I, Biological Properties—1973. Volume II, Molecular Properties—1973. Volume III, Structure and Metabolic Mechanisms—1977

A. J. HOPFINGER. Conformational Properties of Macromolecules. 1973

R. D. B. FRASER AND T. P. MACRAE. Conformation in Fibrous Proteins. 1973

OSAMU HAYAISHI (Editor). Molecular Mechanisms of Oxygen Activation. 1974

FUMIO OOSAWA AND SHO ASAKURA. Thermodynamics of the Polymerization of Protein. 1975

LAWRENCE J. BERLINER (Editor). Spin Labeling: Theory and Applications. 1976

T. BLUNDELL AND L. JOHNSON. Protein Crystallography. 1976

HERBERT WEISSBACH AND SIDNEY PESTKA (Editors). Molecular Mechanisms of Protein Biosynthesis. 1977

A 6
B 7
C 8
D 9
E 0
F 1
G 2
H 3
I 4
J 5